D0395116

ARTHROPOD PHYLOGENY

Jacket credits: From top to bottom: photos supplied by U.S. Department of Agriculture, Carolina Biological Supply Co., and Van Nostrand Reinhold Co., respectively.

ARTHROPOD PHYLOGENY

Edited by

A.P. GUPTA

Professor of Entomology
Rutgers University
New Brunswick, New Jersey

VNR **VAN NOSTRAND REINHOLD COMPANY**

NEW YORK   CINCINNATI   ATLANTA   DALLAS   SAN FRANCISCO
LONDON   TORONTO   MELBOURNE

Van Nostrand Reinhold Company Regional Offices:
New York   Cincinnati   Atlanta   Dallas   San Francisco

Van Nostrand Reinhold Company International Offices:
London   Toronto   Melbourne

Copyright © 1979 by Litton Educational Publishing, Inc.

Library of Congress Catalog Card Number: 78-16955
ISBN: 0-442-22973-9

All rights reserved. No part of this work covered by the copyright hereon may
be reproduced or used in any form or by any means—graphic, electronic, or
mechanical, including photocopying, recording, taping, or information storage
and retrieval systems—without permission of the publisher.

Manufactured in the United States of America

Published by Van Nostrand Reinhold Company
135 West 50th Street, New York, N.Y. 10020

Published simultaneously in Canada by Van Nostrand Reinhold Ltd.

15 14 13 12 11 10 9 8 7 6 5 4 3 2 1

Library of Congress Cataloging in Publication Data

Main entry under title:

Arthropod phylogeny.
    Bibliography: p.
    Includes index.
    1. Arthropoda—Evolution.   I. Gupta, A. P.,
1928–
QL434.35.A77   595'.2'0438   78-16955
ISBN 0-442-22973-9

QL
484.35
A7
1979

Preface

It is tantalizing to speculate and attempt to unravel the evolutionary history of a group. But this is an endeavor fraught with danger, for one can easily commit the error of misinterpreting genealogies. Exploring phylogeny involves intellectual sleuthing full of pitfalls. However, if we follow our "leads" and "clues" carefully and patiently, it is possible to trace back the evolutionary pathways a group of animals may have traveled over eons of time. The more diverse the avenues of approach are, the more reliable and dependable the search for those evolutionary trails is likely to be. This book attempts to provide diverse avenues for approaching the most controversial subject of arthropod phylogeny. This subject has continuously engaged the attention of students of arthropod phylogeny for nearly a century, but is yet far from being resolved. This book does not claim to provide any solution to the mono- or polyphyletic origin of arthropods. It merely attempts to bring to bear on this phylogenetic problem evidences from many areas of arthropod paleontology, embryology and development, sense organs, anatomy, morphology, and physiology. It seeks no consensus for one view or another. Each contributor has had complete freedom to develop, interpret, and present his/her views on the subject. Apart from its significance and bearing on arthropod phylogeny, each chapter provides an in-depth review of the topic it covers, and therefore stands as an important contribution in itself. Portions of the chapters by Drs. Baccetti, Boudreaux, Matsuda, and Weygoldt were presented in a symposium at the 15th International Congress of Entomology, 1976, in Washington, D.C. I invited others later to contribute to the book.

The book is organized in four sections: 1) Paleontology, 2) Embryology and Development, 3) Sense Organs, and 4) Anatomy, Morphology, and Physiology. Although taken together the topics covered under these sections might appear disparate, the common objective of all the articles has been to focus on the question of arthropod phylogeny. Since taxonomic ranking of major arthropod groups is highly controversial, each contributor has used his/her preferred taxonomic categories in the discussion of his/her topic. Both American and English spellings of words (e.g., hemocyte and haemocyte) have been used, mainly in deference to authors' preferences. The native tongues of several of the contributors is not English. And in such cases, editing has been largely confined to removing obvious infelicities in order to retain the original script style and the contents.

An endeavor such as this book cannot be undertaken and accomplished without the cooperation and assistance of many individuals. First of all, I am most grateful to all the authors who responded to my invitation to contribute, and thus made the book possible. The following individuals, journals, societies, and publishers gave their permission to reproduce published material: H. Akai (Japan), R. J. Baerwald (U.S.A.), J. Barlet (Belgium), A. Bauchau (Belgium), J. Bitsch (France), S. Campiglia (Brazil), J. L. Cisney (U.S.A.), N. M. Costin (U.K.), G. Devauchelle (France), J. N. Dumont (U.S.A.), W. H. Fahrenbach (U.S.A.), P. Favard (France), J. François (France), H. Fredericq (Belgium), J. C. Gall (Belgium), G. Goffinet (Belgium), P. P. Grassé (France), Ch. Grégoire (Belgium), R. R. Hessler (U.S.A.), C. Juberthie (France), O. Kraus (West Germany), R. Lavallard (Brazil), A. V. Loud (U.S.A., for late Dr. M. Hagopian), A. K. Raina (U.S.A.), A. Rousset (France), G. Salt (U.K.), S. Sato (Japan), L. Størmer (Norway); Academic Press, Allen Press, American Association for the Advancement of Science, Cambridge University Press, Chapman and Hall, Ltd., Cornell University Press, Elsevier/North-Holland Press, Entomological Society of America, Entomological Society of Canada, French Society of Electron Microscopy, Geological Society of America, Geologiska Foremingen (Sweden), Geology Institute (Praha), Holiday House, Inc., Institute of Comparative Anatomy (Denmark), Institute of Geology (Norway), Institute of Geology (Strasbourg), Institute of Zoology (Siena), Journal of Paleontology, Journal of Parasitology, Linnaen Society of London, McGraw-Hill Book Company, Naturwissenschaftlicher Verein (Hamburg), Palaeontological Institute of the Academy of Sciences (USSR), Pergamon Press Ltd., Rockefeller University Press, Royal Society of London, Society of Economic Paleontologists and Mineralogists (Oklahoma), Springer-Verlag, Taylor and Francis, Ltd. (London), University of Kansas Press, Zoological Institute of the University of Hamburg, Zoological Society of London.

In addition to the above acknowledgments, I must express my gratitude to my wife and children for their infinite patience and understanding in tolerating my total preoccupation with the book during its preparatory phases. Their generosity in not demanding my attention and time during that period made this work considerably easier. The infinite patience, cooperation and assistance of Mrs. Alberta Gordon and her colleagues at Van Nostrand during the production phase is greatly appreciated.

Finally, it would be presumptuous of me even to suggest that this book will bring the controversial question of the arthropod phylogeny any nearer to solution. However, I do hope that it would stimulate further discussion and research in many other areas of arthropod morphology, physiology, and behavior which would enable the students of arthropod phylogeny to probe this phylogenetic mystery with greater confidence.

A. P. Gupta

New Brunswick, New Jersey

List of Contributors

D. T. Anderson—School of Biological Sciences, University of Sydney,
Sydney, Australia (Ch. 2)

B. Baccetti—Institute of Zoology, University of Siena, Siena, Italy (Ch. 11)

J. Bergström—Department of Historical Geology and Paleontology,
University of Lund, Lund, Sweden (Ch. 1)

H. B. Bordreaux—Department of Entomology, Louisiana State University, Baton Rouge, Louisiana, U.S.A. (Ch. 9)

P. S. Callahan—Insect Attractant, Behavior and Basic Biology Research Laboratory, Gainesville, Florida, U.S.A. (Ch. 5)

K. U. Clarke—Department of Zoology, University of Nottingham,
Nottingham, United Kingdom (Ch. 8)

A. P. Gupta—Department of Entomology and Economic Zoology,
Rutgers University, New Brunswick, New Jersey, U.S.A. (Ch. 13)

S. M. Manton—Department of Zoology, British Museum (Natural
History), London, United Kingdom (Ch. 7)

R. Matsuda—Biosystematics Research Institute, Agriculture Canada,
Ottawa, Ontario, Canada (Ch. 4)

H. F. Paulus—Biological Institute I (Zoology), Albert-Ludwigs University, Freiburg, I. BR., West Germany (Ch. 6)

F. Schaller—Zoological Institute, University of Vienna, Vienna,
Austria (Ch. 10)

A. E. Tombes—Department of Biology, George Mason University,
Fairfax, Virginia, U.S.A. (Ch. 12)

P. Weygoldt—Biological Institute I (Zoology), Albert-Ludwigs University, Freiburg, I. BR., West Germany (Ch. 3)

Contents

Section II. Embryology and Development

4. Abnormal Metamorphosis and Arthropod Evolution, by R. Matsuda

Section IV. Anatomy, Morphology, and Physiology

7. Functional Morphology and the Evolution of the Hexapod Classes, by S. M. Manton

8. Visceral Anatomy and Arthropod Phylogeny, By K. U. Clarke

9. Significance of Intersegmental Tendon System in Arthropod Phylogeny and a Monophyletic Classification of Arthropoda, by H. B. Boudreaux

10. Significance of Sperm Transfer and Formation of Spermatophores in Arthropod Phylogeny, by F. Schaller

SECTION I

Paleontology

1

Morphology of Fossil Arthropods as a Guide to Phylogenetic Relationships

J. BERGSTRÖM

1.1. INTRODUCTION

The earliest fossil animals readily recognizable as arthropods are from the Cambrian and close to 600 million years old. The Early Cambrian arthropod fauna was already quite diversified, and the original splitting into main groups must have occurred even earlier. This means that the palaeozoologist is in the same predicament as the zoologist in trying to reconstruct the phylogeny of the Arthropoda; he has to extrapolate backwards from what is known from particular time levels. In principle, there is no difference between extrapolating backwards from the present and from the Cambrian. Many of the Cambrian arthropods belong to main groups still in existence, although they may be more or less primitive in particular respects. Mostly they are too advanced to reveal clear relationships with other groups.

There are two particular difficulties facing the palaeozoologist as compared with the zoologist. First, there are a number of arthropods or arthropod-like animals that cannot be placed in any existing main group, and which tend to complicate the pattern instead of elucidating it. Second, the fossil material is not adequate to reveal the structure of the once living animals. In both cases, more material is needed to help fill the gaps in the knowledge. Perhaps even more important is the use of adequate methods and scrutiny in the study of the available material.

The difficulties with the fossil material have led some authors to omit fossils in phylogenetic discussions. This can hardly be considered a proper approach. After all, the prehistoric animals are part of the evolutionary plexus, and the mere fact that some species happen to be contemporaries of man does not make them phylogenetically more interesting than those that were not. Some arthropod groups fossilized more easily than others, and in such cases the fossil record may yield a substantial addition to the understanding of the relationships of modern forms, as well as knowledge about the actual evolution of the group. In particular cases with distinct morphological steps involved in the evolution, it may even be impossible to understand the course of evolution without actual knowledge of the steps. The record of such steps in the fossil world is not rich, but is must be remembered that researchers have been trained to search for continuous change in phyletic lineages, not for discontinuities.

1.2. NATURE OF THE FOSSIL RECORD

Different arthropod groups are very differently represented in the fossil record. For instance, there is not a single described fossil onychophoran, and described pantopods number only two, while there are many thousands of species of ostracods and trilobites. As a general rule the nature of the habitat and the nature of the integument determine the number of fossils and the state of preservation. A short review may be instructive for those readers lacking geological training.

The aquatic habitat, particularly the sea, is where most of the sedimentation occurs, and where the chance for preservation of the sedimentary rocks with their fossils is largest. The terrestrial habitats are dominated by denudation, and the probability for sediments to be deposited only to be later eroded is comparatively large. In addition, the sedimentary habitat is not optimal for preservation of arthropod remains. These conditions imply that there is a tremendous preponderance of marine arthropods as compared with terrestrial forms in the fossil record.

Many of the best preserved arthropods occur in places where very unusual circumstances have led to their preservation. For instance, the Middle Cambrian Burgess Shale of British Columbia is the result of a submarine turbidity current that embedded much of the animal life on the bottom. Lack of free oxygen and a negative redox potential made the preservation of organic material possible. Similar conditions account for the preservation in the Lower Devonian Hunsrück Shale in West Germany, but in this case iron from the sediment joined sulfur from organic compounds to form pyrite. Thanks to pyrite, even anatomical features can be studied with X-rays. In the Lower (or Middle?) Devonian Rhynie Chert in Scotland, various animals and plants have been preserved in great detail and without deformation, thanks to silicification. The remarkable preservation in the Eocene (Lower Tertiary) lignite deposits of Geiseltal in East Germany was due to fast embedding in sediment and successive influence of water, rich in humus and lime after the animals were trapped in water holes or running water or drowned at flooding. Still another example is the embedding of small arthropods in resin, which turned into amber through diagenesis. The best-known example is the Baltic amber from the Eocene/Oligocene.

Arthropods with a comparatively soft integument possess low fossilization potential. Here belong onychophorans, pantopods, tardigrads, and linguatulids, among others. The linguatulids in addition are parasitic, which reduces their chances of fossilization still more. On the other hand, strong sclerotization increases the possibilities of fossilization. Some groups have calcium carbonate (calcite) or calcium phosphate in the exoskeleton. The mere presence of inorganic salts, however, is not a guarantee for good preservation. Thus, the calcitic exoskeletons of trilobites and ostracods tend to be quite well preserved, even in ultramicroscopic detail when conditions are favorable, whereas malacostacan crustaceans with calcite and aglaspidid chelicerates with phosphate are usually poorly preserved. Generally entire bodies tend to decompose or be eaten by scavengers, leaving disorganized skeletal remains; exuviae, therefore, make nicer fossils. The wings of insects are comparable to exuviae in the absence of fleshy parts. As flying insects easily come out over water, drown, and become embedded in aquatic sediments, they have better chances of preservation than other terrestrial arthropods.

Fossil arthropods are commonly difficult to treat for the palaeontologist without a keen insight in living arthropods. On the other hand, distortion and chemical alterations caused by diagenesis make the fossils hard to interpret for zoologists without a background in geological science. These are the reasons for many misinterpretations in the past. We may hope for better cooperation between specialists in the future.

1.3. MEANING OF MONOPHYLY

Bonde (1975) pointed out that Simpson's definition of monophyly (later adopted by Mayr), that monophyly means derivation from one or more lineages of an immediately ancestral taxon of the same or lower rank, is vague. For example, this could mean that Aves + Mammalia may be regarded as a monophyletic group. Bonde (1975) suggested an alternative in a simplified definition given by Hennig: "A monophyletic group includes (only) a species and all its descendants." This definition is clear and simple, but has one particular drawback, namely its inclusion of *all* descendants. Since the definition is not related to the present time level, all future descendants have to be included. A living group and its ancestors, therefore, do not constitute a monophyletic group, unless it will become extinct without leaving any future descendants. A group like the euentomate insects (Monura and Thysanura) would not be considered monophyletic, not because they did not descend from a single species but because they led to the evolution of pterygote insects (correct or not, the example shows the function of the definition). In this example, there would be no name for the euentomates, as names ought to be restricted to monophyletic taxa (this spelling of Greek *taxion*, *taxia* is generally accepted among zoologists and palaeontologists of today). A slightly modified definition may make it better adapted to practical use: *A monophyletic group is restricted to a species and its descendants. It may be construed to exclude particular descendant taxa.*

1.4. APPROACH TO PHYLOGENY

The traditional methods of establishing systematics and phylogenetic lineages is to compare the structural and/or chemical characteristics of different groups and to use the amount of similarity as a criterion of systematic and phylogenetic closeness. This is the comparative anatomy/morphology/chromosome/chemistry approach. In particular cases, palaeontologists have the key since they may follow the evolution of particular lineages in great detail through millions of years, but this is not the case for the bulk of the arthropods. Recently, the above approach has been complemented by such methods as numerical taxonomy (Sokal and Sneath, 1963; Sneath and Sokal, 1973) and cladistic phylogeny (Hennig, 1966 and others). Neither of these methods has in principle widened the field of comparison, although both make use of as many characters as possible for comparison. Numerical taxonomy, critically discussed by Schlee (1975a,b), is "objective" in treating all characters as of equal importance. Therefore, it does not discriminate between more and less important characters, nor does it distinguish coupled characters. In principle, it is unable to distinguish between a similarity caused by common ancestry and one due to convergent evolution. Cladistic phylogeny, on the other hand, includes an attempt to map the sequence of phyletic branching through a determination of characters that are shared-primitive (symplesiomorphic) and those that are shared-derived (synapomorphic). Although the logical background did not originate with Hennig, his strict formulation of the method marks a great step forward. Indeed, it is obvious that an understanding of the direction of change and the order of changes in branching phyletic lineages is quite important in any attempt to map phylogeny. Hennig's phylogenetic method must be kept distinct from his phylogenetic systematics. In its extreme form, the logic of the phylogenetic systematics may exclude, for instance, the possibility of distinguishing species among fossils (as the phyletic branching is practically never found), and it is therefore difficult to use it in practice. However, this does not diminish the value of the cladistic phylogenetic method.

The comparative anatomy approach has proved to be a valuable tool in works on classification and phylogeny. However, the frequent occurrence of features, which have evolved more than once through parallel or convergent evolution, presents severe difficulties, a fact that is well known to all arthropod students. The number of separate occurrences of particular features in arthropods is difficult to estimate, since the count is dependent on the phylogenetic model used. However, similar parallelisms are quite common in other animal groups as well, and it may be instructive to choose a few examples from fossil brachiopods, whose evolution is comparatively very well known (see Fig. 1.23; Rudwick, 1970). Thus, the primitive phosphatic shell evolved into a calcareous shell at least four times, a skeletal support for the lophophores was formed at least four times, cementation to the substrate occurred at least four times, and formation of caeca in the shell for storage purposes at least eight times, just to mention a few examples. The number of parallel occurrences may certainly be even higher. Another example is seen among the birds, where webbed feet in aquatic groups appear to have evolved at least seven times. Even sets of characters may evolve convergently, as shown by coupled characters such as sessile eyes-lack of carapace-long thoracic segments, developed at least three times in the Eumalacostraca. These examples may serve as distinct warning to those who want to see arthropod jaws, compound eyes, tracheae, or even insect wings as necessarily having evolved only once. There is no reason why evolution among arthropods, in this respect, should be different from that in other animal groups. The very different results achieved when different structures are used for a reconstruction of arthropod phylogeny, indeed clearly show that there is a high degree of convergence. The great difficulty is in distinguishing cases where a characteristic feature is a synapomorphy from those where it is due to convergence. Comparative anatomy has faced this dilemma for many decades.

Obviously, there is a need to supplement the plain comparison of structures and chemistry. The reason for morphological convergence may commonly, if not always, be a need for adaptation to a particular mode of life. Function is therefore an important trigger for morphological evolution, and the morphology may be seen as an aspect of function. Certainly the basic morphological construction sets rules and limits to the morphological adjustments to functional needs, and this no doubt is a reason why there is so much parallel (or convergent) evolution in related groups with a shared basic body construction. To put it in another way, there is a strong interdependence between morphology and function. This interdependence is now studied under the heading "Functional morphology." The result of the interdependence is that the morphology cannot be fully understood without an understanding of the function. It therefore seems logical to extend the phylogenetic arguments from pure morphology to include functional aspects. This is one of the most urgent needs generally today in all phylogenetic work, palaeontological as well as zoological. However, there is no reason to see the application of functional morphology on phylogenetic aspects as the key to all phylogenetic problems. (For further discussion on significance of function in comparative embryology, see Anderson, Chapter 2.)

In combination, the cladistic phylogenetic method of Hennig (without stress on obligate bifurcations) and comparative functional morphology may be used to visualize what ancestral forms may have looked like. A sound appreciation of characters that should have been present in ancestral forms may provide a very important platform for phylogenetic discussions.

A last method, which is less well known among phylogeneticists than comparative anatomy, numerical taxonomy, and cladistic phylogeny, is the analysis of the evolution of amino acid sequence. This method is quite promising, particularly as different amino acids appear to give almost identical phylogenetic results. The method cannot be used for fossil material, but objective erection of phylogenetic lineages for extant groups is a tremendous step forward. The method definitely brings biochemistry into the field of phylogenetic research, and arthropod students should take advantage of the opportunity to trigger an interest in arthropod evolution among biochemists.

Zoologists and palaeontologists are commonly trained to look at evolution as a morphologically continuous process, without any appreciable morphological steps. Still it is evident thay many living populations have a discontinous morphological variation (dithyrial population according to Jaanusson, 1973). It is reasonable to assume that there may be cases where a continuous evolutionary series between two morphs of the same species is theoretically impossible. The genetic background for such a discontinuity may be quite simple. Still, the further evolution of a "hopeful monster" (Fryer, 1976) will lead to a situation where ordinary methods of phylogenetic work are almost futile. Fryer (1976) stressed the importance of the presence of oddities, such as a morph of the brine shrimp, *Artemia salina*, with a functional median, stalked compound eye, and some bolyerine snakes with two maxillary bones instead of one. Such cases prove the possibility of morphological discontinuities in phylogenetic sequences. Actually, there are many indications of discontinuities in the fossil record, but few have been pointed out. The ecdysial sutures in trilobites, for instance, have a number of different patterns that may have developed only through discontinuous morphological evolution. A single well-defined case was illustrated by Jaanusson (1975). Another illustrative case, if correctly interpreted, is the sudden displacement of the prosomal/opisthosomal boundary in xiphosurids, by which the Limulina evolved from the Belinurina probably in Carboniferous times (Bergström, 1975). It is most important that the reality of discontinuous morphological evolution be understood to a much higher degree than it is presently. Evolution is certainly not *only* a morphologically continuous process, but contains discontinuous elements.

1.5. MONO- OR POLYPHYLY AT HIGHER TAXONOMIC LEVELS

Do the Arthropoda constitute a monophyletic group or not, and what is the natural state of all the major subgroups that have been proposed? What is the contribution of palaeomorphology to the controversy of monophyly or polyphyly?

Leaving some of the minor groups aside, some of the attempts to classify the arthropods in supposedly natural groups are the following: One set of groups is the Tracheata, including the Myriapoda, Insecta, Chelicerata, and possibly the Onychophora, and the Crustacea, including the Trilobita, Merostomata, and true Crustacea. Another set of groups is the Mandibulata, including the Tracheata (= Antennata) and Branchiata (= Diantennata or Crustacea), and the Amandibulata (= Chelicerophora or Arachnomorpha), including the Trilobitomorpha and Chelicerata. A third set of groups is the Uniramia, including the Onychophora, Myriapoda, Hexapoda, and probably the Linguatulida, and the Schizoramia, including the Trilobitomorpha and Chelicerata, and the Biantennata, including only the Crustacea, the latter group regarded by some authors to be allied to the schizoramians.

The Tracheata are characterized by the presence (in most but not all members) of tracheae, while those organs are absent in the Crustacea. The Mandibulata have true jaws, while the Amandibulata have, at the most, gnathobasic extensions from ordinary limbs. The Uniramia are regarded to have whole-limb mandibles, while the Schizoramia and the Biantennata may have gnathobasic extensions, in which case the main part of the limb may be preserved, be transformed into a palp, or be entirely reduced. It is important to note that every possible alternative involves numbers of cases where structures must have evolved convergently to a remarkably similar end result. Therefore, in palaeomorphology, as well as in the study of extant arthropods, a similarity in a particular respect generally does not prove relationships. On the contrary, a similarity even with a small functional or evolutionary difference may prove that the animals compared cannot be closely related to each other.

Following insects backwards in time, one finds that they are lost in the Middle or Early Devonian with the Rhynie Chert collembolan, *Rhyniella praecursor* (Fig. 1.8D), which was already of surprisingly modern design. Winged insects are known first from the Carboniferous, except possibly for *Eopterum devonicum*, which may be a Late Devonian cockroach-like form. Thus these finds have hardly any bearing on the question of the relationships of insects, except for indicating that the common origin should be pre-Devonian. On the other hand, the thysanuran-like Early Permian *Dasyleptus brongniarti* (Fig. 1.8A) has an evenly cylindrical body with nine pairs of dwarfed abdominal legs, behind the three pairs of thoracic legs. This morphology distinctly points towards a myriapod-like origin of at least the ectognathic insects. Even the earliest-known insects were terrestrial.

The myriapods as a group extend further back in time than do the insects. The earliest definite representative is *Archidesmus* from the Late Silurian and Early Devonian. There were a few similar genera in the Early Devonian. These finds are reminiscent of diplopods but are not well-enough preserved and studied to reveal much of the detailed morphology. Undescribed myriapod material from the Silurian of Australia may add to the information. Diplopod-like archi-polypods are also known from the Carboniferous; in these, large compound eyes appear to be present. Representatives of typical Diplopoda and Chilopoda are also known from the Carboniferous. The Arthropleurida have recently been de-scribed from the Early Devonian and are now considered as myriapods (Størmer, 1976). There is a possibility that at least some of the earliest forms were am-phibious or aquatic.

The "genus" *Anomalocaris*, described from Lower and Middle Cambrian strata, contains some forms reminiscent of legs, and one reminder of part of the body of a myriapod (Bergström, 1978). The leg-like structures have been re-garded as crustacean bodies, but there is never any head end, nor are there any

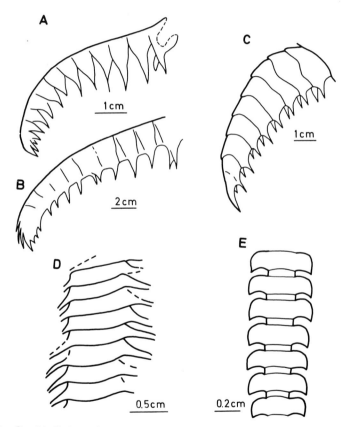

Fig. 1.1.   Possible limbs and bodies of uniramian arthropods.   A and B, Limb-like remains from Cambrian *Anomalocaris cranbrookensis* and *A. canadensis*; C, a leg of Late Carbonifer-ous myriapod, *Arthropleura*, for comparison; D, supposed part of body of Cambrian "myriapod," *Anomalocaris lineata*, compared with E, extant myriapod, *Polydesmus*. From Bergström (1978), redrawn after various sources; B, from *Treatise on Invertebrate Paleontol-ogy*, courtesy of The Geological Society of America and University of Kansas.

limbs. Instead, there are strong spines on the concave side, and, at least in some cases, two spines on each segment. In addition, the distal end has no similarity with the posterior end of any crustacean. The structures bear resemblance to the legs of arthropleurids, and it is tempting to interpret them as legs of myriapod-like arthropods, some of which must have been quite large (Fig. 1.1). The lack of associated large bodies is disturbing, but is must be remembered that these fossils are very rare and that the preservation in the Burgess Shale is due to processes that must have occurred as an accident to the fauna. The form suggestive of part of a myriapod body includes some nine segments of equal width and with long pleural extensions, similar to those found in many of the Silurian-to-Carboniferous archipolypod myriapods. If the interpretation of these forms is correct, there were already in the Early and Middle Cambrain small and large marine myriapod-like arthropods with a well-sclerotized exoskeleton and articulated uniramous legs. Furthermore, it is hardly probable that uniramians invaded land in a soft-bodied condition, an interpretation that may have a bearing on the development of jaws in the uniramians (Manton, 1973).

The odd Palaeozoic myriapods or myriapod-like arthropods indicate that there must have been an early radiation, resulting in numbers of groups, of which we know almost nothing except those that were able to survive to the present day. This interpretation favors the idea of a relationship between insects and myriapods. Although there are no known myriapod groups that fit as insect ancestors, it is quite likely that there may have been some among the unknown aquatic groups of the Palaeozoic. The case for a wide range of variation would be strongly supported if the enigmatic euthycarcinoids of the Carboniferous and Triassic (see below) were accepted as aquatic uniramians.

There are no fossil onychophorans known, except possibly for the poorly known Early Cambrian (?) *Xenusion*. The Middle Cambrian *Aysheaia pedunculata* has generally been accepted more or less as a marine ancestor of modern terrestrial onychophorans (Tiegs and Manton, 1958), but new facts indicate that the differences are much more profound than suspected (Fig. 1.2). The so-called

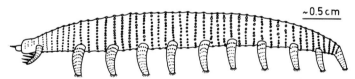

Fig. 1.2. Reconstruction of Middle Cambrian *Aysheaia pedunculata*. Note terminal position of mouth and last pair of appendages and the multiple claws, features reminiscent of Tardigrada. There are no jaws or antennae, and the only appendages of the head are a pair of laterally directed limbs with very strong spines on the anterior side, probably used for catching prey. There are four rings of nodes on the body for each appendage pair, except behind the spiniferous limbs, where there are about eight rings, suggesting the presence of a limbless segment. From Bergström (1978). Figures 1.2, 1.3, 1.4, 1.9, 1.11, 1.12, and 1.13 published with permission from Prof. Otto Kraus, Naturwissenschaftlicher Verein in Hamburg.

antennae were strong limbs, directed laterally and supplied with long and strong spines on the anterior side. Thus, they cannot be regarded as antennae but must have been strong appendages for grasping prey. There are obviously no jaws (Delle Cave and Simonetta, 1975). On each pair of legs, there are four body rings of tubercles, except between the grasping appendages and the first pair of legs, where there are about eight rings, suggesting the presence of a limbless segment. Thus, while the onychophoran head carries one pair each of antennae, jaws, and slime papillae, the arrangement in the anterior end of *Aysheaia* is one pair of grasping appendages, a probably limbless segment, and the first pair of legs. Other odd features of *Aysheaia* include the terminal positions of the mouth and the last pair of limbs, both features suggestive of the condition in the Tardigrada. Also, the multiclawed feet bear resemblance to those of the Tardigrada. Apart from the similar body shape, *Aysheaia* therefore seems to be as remote from onychophorans as the latter are from such arthropods as insects and myriapods. Thus, there is no fossil evidence that the Onychophora are allied to a myriapod-insectan group or to any other arthropod group.

Terrestrial arachnids (except for scorpions) form a parallel to insects in that they can be traced back to the Early Devonian and direct precursors are lacking. This pattern fits with the appearance of a primitive land flora in the Late Silurian and Early Devonian. The Devonian arachnids are generally considered to have been terrestrial, a conclusion that is confirmed by the discovery of well-developed book-lungs from Rhynie Chert trigonotarbids (Claridge and Lyon, 1961; Størmer, 1976). Scorpions are known back from the Middle Silurian, and all Silurian and Devonian species are regarded to have been aquatic (Kjellesvig-Waering, 1966). On the other hand, the Silurian eurypterid, *Baltoeurypterus*, appears to have had tracheal structures (Størmer, 1976). There is therefore palaeontological evidence that chelicerates acquired respiratory organs for breathing air at least three times, disregarding the possible different land invasions among the Arachnida (Kraus, 1976). The fossil evidence therefore shows that any comprehensive tracheate group must be polyphyletic.

The merostome groups Eurypterida and Xiphosura may be traced back to the Middle or Early Cambrian (Bergström, 1968). More important for the understanding of the possible relationships with nonchelicerate arthropods are the Cambrian-to-Ordovician Aglaspidida, which have chelicerae and a long series of uniform limbs in the prosoma and opisthosoma. The limbs are apparently biramous, with an inner walking branch and an outer branch with closely set lamellae (Figs. 1.3J, 1.17B; Repina and Okuneva, 1969). In the arrangement of the limbs along the body and in the morphology of the appendage, the aglaspidids are obviously very close to trilobites and other trilobitomorphs, apart, of course, from the presence of chelicerae and probable absence of antennae. The aglaspidid morphology appears to confirm Størmer's (1944) conclusion that the limb lamellae of trilobitomorphs and xiphosurids are homologous, although the lamel-

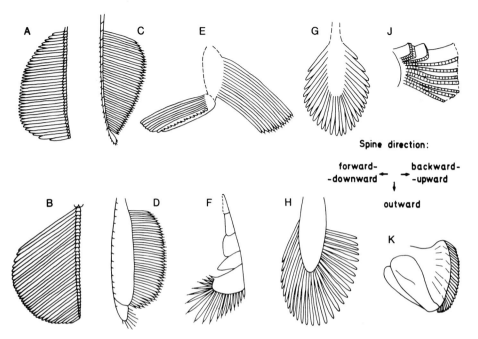

Fig. 1.3. Arrangement of lamellae in outer appendage branch of various schizoramian arthropods. In trilobitomorphs and aglaspidid chelicerates (A–J) lamellae form strong needle-shaped lamellar spines, mostly with a thickened border. In xiphosurid chelicerates (K), they are transformed into broad lamellar gills. In marrellomorphs (A and B) there is one row of ventrally directed spines, each extending from a distinct podomere. Trilobites (C–F) and merostomoids (G and H) have ventrally and/or dorsally directed spines, at least as a rule belonging to larger podomeres. Arrangement in merostomoids (G and H) is not yet well known in detail. A, *Vachonisia*, Devonian; B, *Mimetaster*, Devonian; C, *Triarthrus*, Ordovician; D, *Olenoides*, Cambrian; E, *Cryptolithus*, Ordovician; F, *Ceraurus*, Ordovician; G, *Sidneyia*, Cambrian; H, *Cheloniellon*, Devonian; J, *Khankaspis*, Cambrian; and K, *Limulus*, extant. From Bergström (1978).

lae are shaped as slender flattened spines in the trilobitomorphs (as in aglaspidids) but as plate-shaped gills in the xiphosurids (Fig. 1.3). Because of the morphological functional difference, it is wise not only to refer to those structures as lamellae, but to differentiate between lamellar spines (or spine lamellae) and lamellar gills (or gill lamellae).

The trilobitomorphs, known from the Early Cambrian to the Permian or perhaps to the Triassic, are poorly understood, as they are entirely extinct. It is usually difficult to tell if a particular fossil arthropod is a trilobitomorph or not, particularly when the appendages are not known. There is a fairly wide morphological variation. The only feature that appears to be fairly constant is the lamellar spine of the lateral appendage branch (Fig. 1.3A–H). As mentioned above, this character is also found among the chelicerates, so the combination antenna-lamellar spine is needed to distinguish trilobitomorphs. In the absence of known appendages, it may be impossible to tell if an animal is a trilobitomorph or a chelicerate, particularly since some trilobitomorphs are very merostome-like in their general appearance. The morphological gap between chelicerates and trilobitomorphs is further diminished by the Early Devonian *Cheloniellon calmani* from the Hunsrück Shale (Fig. 1.4D). This animal has two pairs of appendages in front of the "mouth," viz., one pair of uniramous antennae and one pair of grasping appendages. The latter, which are not chelate, have a position closely comparable to that of the chelicerae in chelicerates, and a loss of the antennae

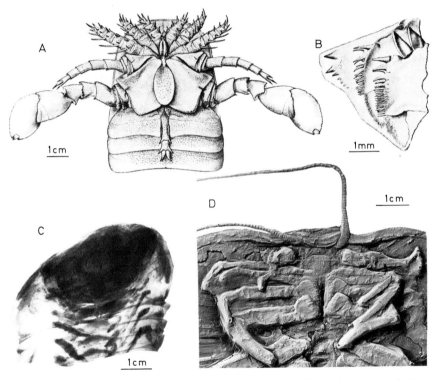

Fig. 1.4. Illustrations of similarity in arrangement of coxae with gnathobases in trilobito-morphs and chelicerates. A and B, Silurian eurypterid, *Baltoeurypterus tetragonophthalmus* (B shows gnathobases of the first to third walking legs); C, Devonian trilobite *Phacops ferdinandi*, with two pairs of gnathobases visible behind the large labrum; D, Devonian merostomoid *Cheloniellon calmani*. While trilobites have only one pair of antennae and merostomes only chelicerae as preoral appendages, *Cheloniellon* is intermediate in having both antennae and possibly grasping appendages in the preoral position. A and B, from Holm (1899), with permission from Geologiska Foereningen; C, from Stürmer and Bergström (1973); D, new.

would almost make *Cheloniellon* a chelicerate. A further chelicerate character is the small labrum with posteriorly directed spines, closely comparable morphologically to the condition in *Limulus*. In addition, the posterior gnathobases are stronger than the anterior ones, a feature typical also of eurypterids and xiphosurids, as well as of trilobitomorphs such as trilobites (Fig. 1.5). The comparatively short head with only five pairs of appendages exclusive of the antennae contrasts with the prosoma of chelicerates with six to seven pairs of appendages. *Cheloniellon* was too late in time to be a progenitor of chelicerates, but it shows how chelicerates could have evolved from trilobitomorphs through comparatively

Trilobites		Cheloniellon		Limulus	
limb	gnatho-base	limb	gnatho-base	limb	gnatho-base
antenna		antenna		——	
?	?	preoral raptorial appendage	——	preoral chelicera	——
leg	weak	leg	weak	leg	weak
leg		leg		leg	
leg	strong	leg		leg	
		leg	strong	leg	
				leg	strong
				chilarium	weak

Fig. 1.5. Comparison between limbs of head tagma in trilobites, merostomoids (*Cheloniellon*), and chelicerates (*Limulus*). A second preoral appendage has not been identified in trilobites and probably was not present. Question marks indicate the possibility of a fourth pair of legs, which may be reduced, e.g., in *Phacops* but may be indicated by dorsal segmentation in many trilobites. Suggested raptorial function of second pre-oral appendage in *Cheloniellon* needs substantiation. Note similar arrangement of gnathobases (also seen in eurypterids). Similarity between *Cheloniellon* and *Limulus* is striking and tends to substantiate the idea of a close relationship between trilobitomorphs and chelicerates. From Stürmer and Bergström (1978).

small changes. In all, the close relationship between trilobitomorphs and chelicerates seems to be well established. The gap between the two groups is not larger than between certain constituent subgroups, and is far smaller than the range of variation within the Trilobitomorpha. On the whole, the Schizoramia (= Amandibulata, Arachnomorpha, Chelicerophora) therefore must be considered to be a monophyletic group. However, there are still problems in placing a number of fossil forms systematically, and there has been a tendency to make the Trilobitomorpha a wastebasket for poorly known early arthropods.

The Crustacea may be traced back to the Early Cambrian, but the exact assignment of a number of Cambrian forms is difficult. At any rate, the Malacostraca and Branchiopoda obviously were present in the Cambrian. As pointed out by Hessler and Newman (1975), there are many basic similarities between crustaceans and arachnomorphs, and a common origin would not seem to be impossible. However, the earliest safely recognizable crustaceans are already definitely crustaceans, and are not particularly close to an arachnomorph type of morphology. Both groups have branched appendages, but the lamellar spines or gills, typical of a number of schizoramian groups, have not been observed in any crustacean. The branching of the appendage is commonly said to be more distal in crustaceans than in schizoramians. However, we do not fully understand the variation in the crustaceans, and we still do not know much of the variation in the arachnomorphs. Cisne (1975) recently reconstructed a surprisingly long coxopodite with a distally positioned outer appendage branch in the trilobite *Triarthrus.* However, this is not so in other forms. The basal podomere appears to be quite short in the trilobite *Phacops* (Stürmer and Bergström, 1973), in the marrellomorph *Mimetaster* (Stürmer and Bergström, 1976), and in *Cheloniellon* (Broili, 1933). Also, the probable outer branch in *Limulus* has a proximal position. Despite all this, I am afraid, the generalization that trilobitomorphs have a distally placed outer limb branch may be included in uncritical literature—sweeping generalizations of that kind unfortunately are much too common and tend to obscure what is really known about fossil arthropods. Finally, it may be said that the fossil material does not allow us either to unite the crustaceans with any other group, or to state the absence of any particular relationships.

Some indirect evidence of the early evolution of arthropods may be gained through an appreciation of some general aspects of morphology in relation to the basic mode of life. Two main types of body limbs may be distinguished. One is the unbranched or uniramous type, the other the branched or schizoramous type. The schizoramous type, presumably present in ancestral trilobitomorphs and crustaceans, may have been used for crawling or swimming. In either case, currents around the edges could hardly be avoided, and it is fairly easy to see how a method for collecting suspended particles behind the mouth had to develop, associated with a backwardly oriented mouth or atrium oris and a prolonged labrum. With the right kind of ancestral animals, this kind of evolution may have occurred over and over again in several lineages. A similarity in this respect, as for instance between trilobitomorphs and crustaceans, therefore is no proof of phylogenetic relationship. The uniramous appendage, probably fairly pillar-shaped in the ancestral forms, could hardly have been used to collect suspended food without strong modifications. The ancestral uniramian must have been a crawler, and there appear to have been only two ways to ingest the food. Either the food (mud or large particles) could be grasped directly on the subsurface with the mouth, or at least one pair of anterior appendages could be used to grasp large particles (including prey) and move them to the mouth. This would seem to be mere speculation, but fits the new reconstruction of the Cambrian uniramous *Aysheaia* (Fig. 1.2). In this animal, there is a pair of anterior spiny appendages that can only have been used for grasping, and the mouth is terminal. An interesting aspect is the possible evolution of masticating organs in the two main cases. In the schizoramian arthropods it is quite possible to think of how enditic processes would have evolved to treat the food particles after they left the region of the distal branches. In the uniramous model, the food would have never come close to the appendage bases, and mandibles evolving out of grasping appendages such as those in *Aysheaia* would, by necessity, have been whole-limb jaws, i.e., the tips of the appendages would have formed the edges of the jaws. It is reasonable that such whole-limb jaws would have formed at an early stage, perhaps before the segmentation of the limbs was accomplished. If this is so, the absence of an obvious segmentation in the mandibles of myriapods and insects is only to be expected, and is no proof that parts homologous with the distal parts of appendages are missing.

Although this interpretation favors the idea that the Mandibulata are diphyletic, it must be remembered that direct fossil evidence for this conclusion is missing. The conclusion is based on the same kind of extrapolations that may be made from extant arthropods, although from a level much closer to the evolutionary base. Although the Cambrian arthropods are surprisingly diversified, and may have a common root far back in the pre-Cambrian, it is fully possible, some would say almost certain, that the proper arthropodization occurred only shortly before Cambrian times. The late pre-Cambrian "Ediacarian" faunas do not contain any unquestionable arthropods, the oxygen pressure may have been too low, and trace fossils indicate that typical metazoan life began or became abundant perhaps only 100 million years before the beginning of the Cambrian. Such a late start would indicate that the Cambrian arthropods were fairly primitive in the sense that they were original. This is also indicated by the different chemical composition of the exoskeletons, showing that at least parts of the sclerotization processes evolved independently in separate groups, even within the main arthropod groups.

A postulated monophyletic group of Mandibulata would have originated either with schizoramous or uniramous appendages. The presence in the Cambrian of the primitive uniramous *Aysheaia*, lacking jaws and antennae and with a terminal mouth (Fig. 1.2), clearly shows that not all uniramous forms may be derived from schizoramous mandibulates. If *Aysheaia* was not derived from schizoramous forms, there is obviously no compelling reason to derive other uni-

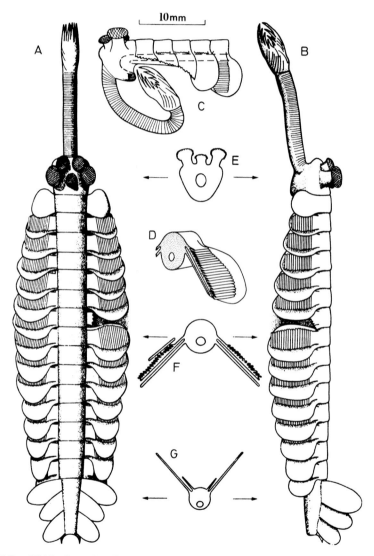

Fig. 1.6. Middle Cambrian *Opabinia regalis* from Burgess Shale. A, Dorsal view; B, lateral view; C and D, details; E to G, cross sections. Unique nature of proboscis, five compound eyes (?) and appendages indicate that this arthropod represents a discrete phylogenetic lineage with a separate origin among segmented worms. From Whittington (1975). Reprinted with permission from the Royal Society of London.

ramous arthropods from schizoramous mandibulates either, even if all the uni-ramous forms would not form a monophyletic group. There remains the possi-bility that uniramous arthropods were at the base of mandibulate radiation, which would mean that crustaceans were derived from onychophoran- or myriapod-like ancestors. I find it hard to see how this would have been managed.

Ultimately the remarkable animal *Opabinia regalis* may be mentioned (Fig. 1.6). The redescription by Whittington (1975) shows an animal with a segmented body, five large compound eyes, a long "proboscis," and two pairs of plate-like appendages on most body segments. It may be called an arthropod, but the pe-culiar structure indicates that it has nothing to do with any other known arthro-pod. I fully concur with Whittington's conclusion that it may be derived from a prearthropod and/or preannelid stock. *Opabinia* is a striking example of how arthropodization could develop along quite separate phyletic lineages. It adds substance to the conclusion hinted at above that the Arthropoda are polyphyletic.

1.6. MONO- OR POLYPHYLY AT LOWER TAXONOMIC LEVELS

In the previous section the phylogenetic homogeneity of groups above the super-class or class level was discussed. Turning to lower levels we encounter groups that are somewhat more generally accepted than the high-level groups. The lower we go on the taxonomic scale, the greater is the probability that particular groups originated in Phanerozoic times; and, therefore, there is an increasing possibility that the transition between groups may be reflected in the fossil record. For convenience, groups are discussed under separate headings, begin-ning with uniramians and proceeding with trilobitomorphs, chelicerates, crusta-ceans, and pycnogonids.

1.6.1. Uniramia

Two major groups of myriapods and hexapods are unknown as fossils. These are the Pauropoda and Protura. Of the others, the Diplopoda, Chilopoda, Collem-bola, and Insecta are known from the Palaeozoic, the Symphyla and Diplura only from the Tertiary. A number of poorly known Silurian-to-Carboniferous forms have been included in the class Archipolypoda (Hoffman, 1969). The myriapodous class Arthropleurida is known from the Devonian to the Carbonif-erous (Størmer, 1976). Cambrian arthropod remains known under the name of *Anomalocaris* are suggestive of archipolypods and arthropleurids, and may represent myriapodous uniramians (Figs. 1.1, 1.7). In addition, the Permian-to-Triassic Euthycarcinoidea (Fig. 1.9A, B) are commonly regarded to be crusta-ceans or trilobitomorphs, but may be uniramians not readily referable to myria-pods or hexapods. The supposed Carboniferous crustacean, *Tesnusocaris* (Fig. 1.9C, D) may also belong here. It is noteworthy that *Anomalocaris* was marine; euthycarcinids probably limnic; arthropleurids, and early archipolypods, prob-ably aquatic to amphibious; and *Tesnusocaris* aquatic.

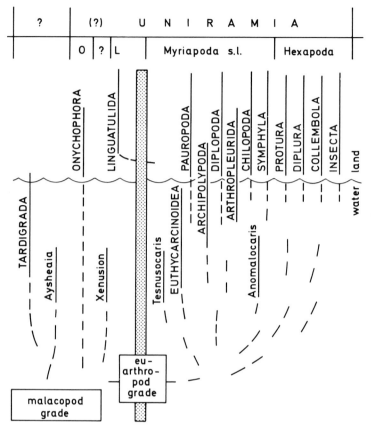

Fig. 1.7. Schematic evolutionary relationship between uniramous arthropods. Linguatulids may have reached malacopod grade secondarily through simplification. Almost nothing is known about primary aquatic radiation of uniramian euarthropods.

The position of the genital opening is unknown in the strictly fossil groups, which therefore cannot be assigned to the Opisthogoneata or Progoneata. Similarly, the mouth parts are too poorly known to give any clue to any particular position in the system. The presence of more than three pairs of walking limbs would make them myriapods, but the morphology of euthycarcinids is quite far from that of any myriapod.

What is particularly needed at present is a thorough revision of the archipolypods (in addition to that performed by Kraus, 1974), which may, for instance, include aquatic diplopod ancestors. Also, the euthycarcinoids need careful revision because of the generally important bearing they may have on discussions on uniramian evolution, if they prove to be a remnant of plesiomorphically aquatic uniramian groups.

The various classes do not appear to converge backwards in time as long as they are known from the fossil record. The Early or Middle Devonian *Rhyniella* (Fig. 1.8D) was already a typically advanced endognathous collembolan, and the same can be said about early members of other extant groups. *Rhyniella* was almost contemporary with the earliest supposed land plants, and still the morphology would indicate a long time of terrestrial evolution before such an advanced hexapod would develop. The idea of an Early Silurian or Late Ordovician terrestrial flora (Gray and Boucot, 1977) may be mentioned in this connection. Alternatively, the morphological evolution may have been very fast. However, there are also the possibilities that ancestral collembolans had already developed "land" characters in the water, or that they invaded shores with stranded seaweed long before there was any terrestrial flora.

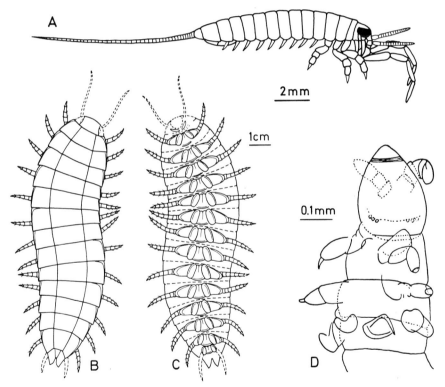

Fig. 1.8.   A, Permian monuran insect, *Dasyleptus brongniarti*, preserving large limb rudiments in abdomen and with poor distinction between thorax and abdomen; B and C, dorsal and ventral views of Devonian arthropleurid myriapod, *Eoarthroppleura*, as reconstructed by Størmer (1976); D, Devonian collembolan *Rhyniella praecursor*. A, redrawn from Sharov (1957); B and C, from Størmer (1976), with author's permission; D, simplified from Scourfield (1940).

The various aquatic-to-amphibious myriapodous groups from the Cambrian(?) and Silurian-to-Carboniferous indicate or rather prove that sclerotization was already fully established in the aquatic forebears of tracheate groups. The existence of sclerotized marine myriapods at a time before the plants invaded land seems to disqualify Manton's (1973) belief that uniramians left the sea in a soft-bodied state. A better knowledge of the earliest myriapods would hopefully make it possible to infer from fossil evidence whether invasion on land occurred only once in uniramians, or, as in the chelicerates, several times. This cannot be safely judged at present. However, the quite different positions of spiracles found, for instance, in insects, diplopods, and symphylans, and the absence of tracheae in pauropods, may indicate a multiple acquisition of the ability to breathe on land.

The absence of known primarily aquatic hexapods and the lack of other suitable aquatic ancestors in the fossil record makes the origin of the hexapods a difficult problem from the palaeontological viewpoint. The very old collembolan, *Rhyniella*, may be an indication that possible common hexapod ancestors must have been aquatic, but there are many unknown aspects. The morphology of the Carboniferous-to-Permian euentomate (monuran) *Dasyleptus* (Fig. 1.8A), with short abdominal appendages, indicates myriapodous ancestors for the Insecta (Eurentomata + Pterygota). Pterygotes likely evolved from euentomate insects. They already form a complicated morphological plexus in Carboniferous times, and it is not possible at present to tell from palaeontological evidence whether they form a monophyletic group or not. Additional Devonian material is urgently needed and should be found sooner or later. The most distinctive morphological feature, the wings, should be critically analyzed with regard to the possible mode of original development. Possibly, homologous (for instance serially homologous) features should be searched for, and analyzed in fossil and extant insects. An analysis of the morphology, anatomy, and functions might give valuable clues to the morphology and mode of life of the pre-pterygotes, and hence to the question of origin, mono- or polyphyletic.

The Euthycarcinida (Fig. 1.9A, B), known from the Late Carboniferous and Triassic, have been proposed to be almost everything except relatives of the myriapods and insects. The main reason may be that they were aquatic, while the myriapods and insects have been considered to be terrestrial tracheates from the beginning. With the acceptance of early myriapods as marine animals, there is no reason to deny relationships on the single basis that the euthycarcinoids were aquatic. My experience of this group is based primarily on one of the three known species, the Triassic *Synaustrus australis*, which I had the opportunity to study during a visit to Sydney in 1976. The supposed second antenna, regarded as a crustacean character, cannot be identified. All clearly visible appendages are distinctly uniramous and multiarticulate. There are double- and triple-segments, paralleled only in the myriapods. There are no visible divertic-

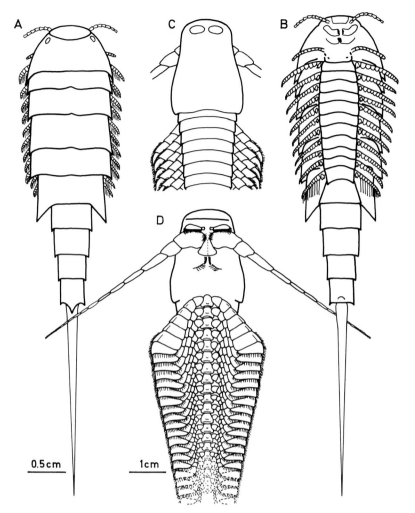

0.5cm   1cm

Fig. 1.9. Supposed primarily aquatic uniramous arthropods. A and B, Dorsal and ventral views of Triassic *Euthycarcinus kessleri*; C and D, Late Carboniferous *Tesnusocaris goldichi*. Both forms were probably adapted for swimming, *Euthycarcinus* through strong setation (shown in B only on last pair of legs), and *Tesnusocaris* through flattening of the podomeres. A and B redrawn after Gall and Grauvogel (1964), with authors' permission; C and D, copied with slight modification after Brooks (1955) with permission of Society of Economic Paleontologists and Mineralogists and *Journal of Paleontology*.

ula from the alimentary canal. The jaws are poorly preserved, if present at all. *Euthycarcinus* may show mandibles without palps. The appendage, tentatively regarded as a second antenna in *Euthycarcinus* (Gall and Grauvogel, 1964), has a markedly posterior position (Gall and Grauvogel, 1964, Pl. 6), and I agree with

Riek (1968) that it is postmandibular. Although the arguments are few, they are consistent with a myriapod-hexapod affiliation, whereas a connection with the Crustacea, Trilobitomorpha, or Chelicerata appears to be very difficult to maintain. I, therefore, suggest that the euthycarcinoids be accepted, at least provisionally, as a uniramian group, perhaps at the same level as the Myriapoda and Hexapoda. Alternatively, they may be regarded as myriapods, possibly as remnants of an early aquatic group. It is remarkable that there are only three known species, widely separated in time and space. They may have been rare. They may also have been exclusively limnic, and fully limnic habitats are rare in the fossil record. "Possibly limnic" deposits in the Ordovician (Caster and Brooks, 1956) and Cambrian (Chlupáč and Havlíček, 1965) indicate that the limnic habitats had a long biological history, which may explain the absence of particular groups from the Lower Palaeozoic fossil record.

Another systematically difficult arthropod is the Early Carboniferous(?) *Tesnusocaris goldichi*, known from a single specimen (Fig. 1.9C, D). It was originally considered to be a crustacean (Brooks, 1955), and opinions have not changed since, although there is no agreement on the exact position among the Crustacea. The most crustacean-like feature is the presence of two pairs of anterior appendages, which may be thought of as antennulae and antennae. However, both pairs are uniramous. The hardened edges of a pair of jaws (mandibles) are clearly visible. There may be additional cephalic appendages, but they are not clearly visible. The body is of uniform construction throughout the preserved portion. Each segment carries a pair of limbs with seven flattened segments. Brooks tried to find a second (inner) branch, but found no conclusive evidence. In all, the comparision with crustaceans is tenuous. The antennae are generally different in the absence of branching, and the (uniramous?) body appendages are unlike anything seen among crustaceans. The uniramous(?) condition indicates that *Tesnusocaris* may be an aquatic uniramian, with a similar evolutionary position as the euthycarcinoids. The paddle-shaped limbs are an adaptation to swimming; an alternative, namely setation of slender limbs, is found in the euthycarcinoids. The second "antenna" may be the premandibular limbs. Maybe the reason for its absence in terrestrial myriapods and hexapods is that it had a particular function in the aquatic environment and became superfluous after invasion on land.

Osche (1963) ably discussed the position of the Linguatulida, of which there is no fossil evidence. He concluded that they are close to the myriapods. Being confined as parasites to the breathing organs of amniote vertebrates, the logical suggestion was made that they were derived from the primary stock of terrestrial myriapods, perhaps from a lineage leading to the Diplopoda and Pauropoda. An original connection with amniotes was also suggested. I would like to express my agreement with the proposed ideas, and urge the non-German readers to consider the German opus. In detail, modern palaeontological knowledge would modify the discussion of the origin slightly. Since the myriapods were already in existence in the Early Palaeozoic seas, and did not originate as terrestrial tracheates, the linguatulids cannot be derived from a position close to the base. On the contrary, they must be derived from one of the separate and advanced groups on land, perhaps from the "archipolypods." This is further emphasized by the confinement to the amniotes, which would indicate that the linguatulids did not evolve earlier than very late in the Carboniferous, or in the Permian, i.e., at least about 100 million years after the first terrestrial hexapods lived at Rhynie in Scotland.

Thus, the fossil record so far provides little substance to the question of mono- or polyphyly in the myriapod-hexapod group. A modern analysis, particularly of the ancient aquatic and amphibious myriapods, and new discoveries of well-preserved material from Cambrian-to-Devonian rocks are urgently needed.

1.6.2. Trilobitomorpha

The question of monophyly or polyphyly is directly dependent on the composition of a particular taxon, and there is much less agreement on the composition of the Trilobitomorpha than on any other arthropod taxon of comparable rank. Here, the following groups are considered to belong to the Trilobitomorpha (Fig. 1.10) (Bergström, 1978): Marrellomorpha (with Marrellida, Mimetasterida, Acercostraca, and probably Halicyna), Trilobita, and Merostomoidea (with Limulavida, Cheloniellida, probably the Strabopidae, and perhaps the Emeraldellida). All these groups are extinct, a circumstance making the morphological characters particularly important as guides to relationships, but also particularly difficult to understand.

Other classifications, e.g., those of Størmer (1944, 1959), include only some of the above-mentioned groups in the Trilobitomorpha, and some others, like the Leanchoiliida, Burgessiida, Waptiida, and Opabiniida. The problem with these and other groups is that the detailed morphology is too poorly known, and what is known is partly very poorly understood, particularly in the Opabiniida. Some forms with a crustacean-like habitus have been included within a trilobitomorph class or subclass Pseudonotostraca, or Pseudocrustacea with the orders Burgessiida, Waptiida, and possibly the "Hymenocarina" (Størmer, 1944). In other cases, the crustacean-like habitus of these forms has been regarded as proof of a direct link between trilobitomorphs and crustaceans (Simonetta and Delle Cave, 1975, and Hessler and Newman, 1975, are among the latest supporters of this view).

The groups here counted as trilobitomorphs, but regarded as something else by most other authors, have been placed in various systematic positions. For instance, *Mimetaster* was treated by Rolfe (1969) among Arthropoda *incertae sedis*, but with the suggestion that it might be a crustacean larva. The

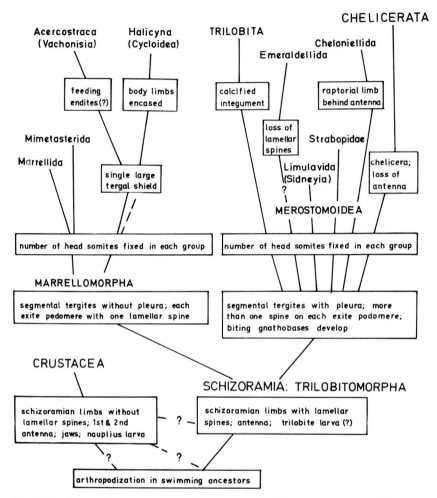

Fig. 1.10. Suggested evolutionary relationships of schizoramous groups, Schizoramia (Arachnomorpha) and Biantennata (Crustacea). Similarities between these two groups may be due to parallel evolution (the alternative to the left of the three given) or to a common origin in the pre-Cambrian. Systematic position is particularly questionable regarding the Emeraldellida, where lamellar spines appear to be missing, and Strabopidae, where appendages are unknown. From Bergström (1978).

single acercostracan *Vachonisia* is usually regarded as a phyllopod crustacean (Tasch, 1969). The Halicyna (or cycloids) have been compared to xiphosurids, eurypterids, trilobites, and various crustaceans (brachyurans, phyllocarids, branchiurans, copepods) (Glaessner, 1969; Gall and Grauvogel, 1967). *Mimetaster* and *Vachonisia* have recently been redescribed and cycloids discussed, with the result that there seems to be reason to regard them as trilobitomorphs (Stürmer and Bergström, 1976; Bergström, 1976a, 1978).

Størmer has repeatedly (1944, 1959) stressed the importance of the so-called trilobite appendage for recognizing trilobitomorphs. There is now some doubt about the significance of the supposedly different mode of branching in crustacean and trilobitomorph appendages, but the unique character of the lamellar structures of the outer limb branch is still regarded as a fact, and the evidence on this point is growing (Fig. 1.3; Stürmer and Bergström, 1976; Bergström, 1978). Very similar structures are found in the probably closely related chelicerates, but nothing even remotely similar is known from any crustacean or uniramian. The lamellar structures, known as filaments, gills, lamellae, or lamellar spines, therefore seem to be the one distinguishing character for the Trilobitomorpha, in addition to the antennae, which are lacking in chelicerates. The significance of the lamellar spines hinges on the question of whether they are shared-derived (synapomorph) structures or not. If they are, the Trilobitomorpha constitute a monophyletic group; if not, the lamellar spines, and probably the trilobitomorphs, have evolved more than once.

An understanding of the function of the lamellar spines may be important for an understanding of their origin, and for their tendency to show adaptational changes. In the past, they have commonly been regarded as gills. Although this may not be altogether wrong, the idea is based on several misconceptions. One misconception is that all crustaceans have appendage parts functioning as gills; another is that the very general similarity between the outer branches in trilobites and decapod crustaceans indicates a breathing function in the former, although in fact this is generally not the case with the compared structures in the decapods. A third is based on the gill function in branchiopod crustaceans, although morphologically there is no similarity with the lamellar spines of trilobites. A fourth is the probable homology between trilobite lamellar spines and xiphosurid lamellar gills, although the surface enlargement in xiphosurids is an obvious adaptation to the gill function not found in trilobites. A fifth misconception is that all merostomes had homologous gills on their appendages— eurypterid gills are now known to be situated on the ventral body surface, and they are not similar to or homologous with xiphosurid gills. Ultimately, the lamellar spines of trilobites have been stated to be soft structures, although in fact they tend to be preserved almost as commonly as antennae and walking legs, while the ventral body integument, apart from doublure and labral plate, is never met with. In addition, the spines tend to be straight and have a distinct outline, even in microscopic detail. The positive evidence for a gill function is therefore nonexistent. The single argument in favor of a gill function is that the spines would be large enough to house a blood vessel loop.

In recent years, some effort has been made to understand the function of the appendages of trilobitomorphs, including the lamellar spines (Bergström, 1969,

1972, 1973a,b, 1976b; Stürmer and Bergström, 1973, 1976). The lamellar spines appear to be almost invariably flattened with a thickened border, apparently providing rigidity particularly in one direction. They are invariably arranged in series, as the teeth in a comb, with the flattened sides facing each other. In some forms, the lamellar spines hang down from the outer appendage branch, and there is impressive evidence from trace fossils that appendages of this construction were used to shovel up the sediment from the bottom, with as much as 1000 mm^3 shoveled away with a single stroke of a single appendage in a large trilobite (Bergström, 1976b). The function was definitely different in other trilobites and trilobitomorphs, with lamellar spines too short to reach the bottom or, in other cases, actually directed upward-backward. In a trilobite such as *Triarthrus*, with both appendage branches almost invariably found held together and straight to the side, it is very likely that the appendages, including the lamellar spines, were used for swimming. Still, the spine morphology appears to be identical to that in burrowing forms like calymenids. In other cases there may be still other functions, such as ventilation of gills on the ventral body surface; and one function does not necessarily exclude another.

What comes out of these studies is the striking conservatism of the lamellar spine morphology (Fig. 1.3), despite considerable variation in function. Indications are that the lamellar spines probably did not evolve at different times in response to different needs, but, instead, were derived from a single source, and utilized afterwards for various new purposes. Even for the same purpose, quite different structures may be formed and used, as shown by other arthropods. I therefore agree with Størmer that the lamellar spines most probably form a good taxonomic character in being a shared-derived (synapomorphic) character for the Trilobitomorpha (and Schizoramia). This does not, however, mean that the lamellar spines are present or unmodified in all descendants of the original trilobitomorph.

It is unfortunate that the question of monophyly has to be based on a single character. There may be others as well, but the possibilities of distinguishing them are limited by the qualities of preservation. Most forms have gnathobases on the limbs of the head tagma, characteristically with a grading in size, with the weakest anteriorly and the strongest posteriorly (Fig. 1.5). However, there is no indication of gnathobases in the Marrellomorpha. Most forms have a trilobed body, with segmental tergites having well-developed pleura. Again, this character is not seen in the Marrellomorpha. These two characters are shared also with the aquatic chelicerates. They may be considered as shared-derived characters, in which case the condition in the Marrellomorpha is probably shared primitive. The shared primitive condition in the latter may be used as an indication that they represent remnants of a trilobitomorph stem group. On the other hand, the arrangement of the lamellar spines in a single line directed ventrally, each spine carried by a distinct podomere, may be a shared-derived character, as indicated in the diagram (Fig. 1.10).

Among trilobitomorph groups, only the trilobites can be seriously discussed with regard to the question of mono- or polyphyly (Fig. 1.11). A notable and almost unique character is the calcitic exoskeleton with extraordinary resistance against diagenetic changes after embedding, paralleled only in the ostracodes. In addition, only the doublure and the labral plate are sclerotized on the ventral side, and the segmental composition of the head is probably identical throughout. An additional unique feature is the presence of dorsally placed ecdysial sutures (facial sutures). Facial sutures are absent in some trilobites, notably in secondarily blind forms and in the early olenellids. The old opinion, repeated by Eldredge (1977), that the "eye lines" in olenellids represent fused facial sutures, accounts neither for the detailed topography (which is consistent with, e.g., a vascular prosopon interpretation) nor for the extension to the glabella rather than to the eye (where the facial sutures facilitated ecdysis in other trilobites) in *Biceratops* (Pack and Gayle, 1971). Primitively in all major groups there is a pair of compound eyes situated lateral to the rhachis and connected with an eye ridge, features quite different from those found, for instance, in *Cheloniellon*. Trilobites constitute a closely knit group. Among major groups, only the olenellids and agnostids have been proposed to be phylogenetically and systematically distant from other trilobites. However, olenellids are very close to early redlichiid and corynexochid trilobites in all characters, except for the absence of facial sutures, which may be a plesiomorphic feature shared with other arachnomorphs. Agnostids differ in being small and preserving larval characters, and in having only two or three articulated thoracic segments. Most of them are blind and lack facial sutures. However, in the most primitive-looking eodiscid agnostids, there are both compound eyes and facial sutures. The monophyletic character of the Trilobita, therefore, can be considered as certain.

The phylogenetic composition of various alternative orders of trilobites is more difficult to discuss, but the unusually good representation of trilobites in the fossil record makes trilobites a group of great potential interest for studies of phylogenetic processes and features such as parallel and convergent evolution. Many features commonly used as systematic characteristics, such as the morphology of the cephalic rhachis (glabella) and the course of the facial suture, have a limited number of possible shapes, and each possible alternative has tended to occur repeatedly as synapomorphies for limited trilobite groups. There is no space for any detailed discussion here. It will be enough to mention as an example that a phylogenetically delimited order Ptychopariida may be defined by one functional and twelve morphological features shared at least by virtually all Cambrian forms (Bergström, 1977), whereas "the *Treatise* diagnosis of the order Ptychopariida contains not a single character state that can be taken as a synapomorphy among even two of the five suborders" (Eldredge, 1977). Too many compromises apparently have destroyed valuable systematic ideas of individual contributors. The interested reader is referred to the discussion about different approaches given by Bergström (1973a, 1977) and Fortey

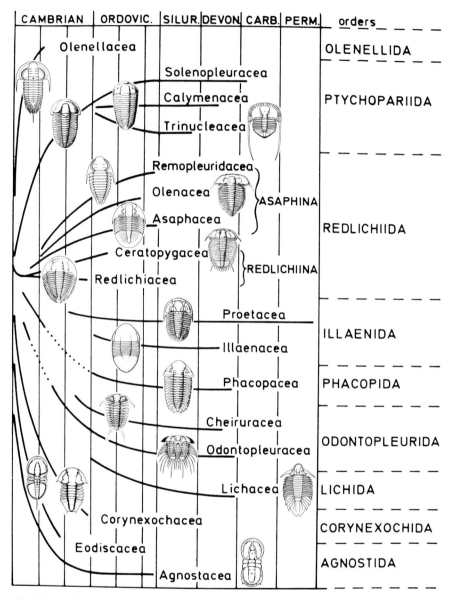

Fig. 1.11. Evolution of trilobites indicates derivation from a common ancestor, perhaps in the Early Cambrian. Trilobite drawings copied from *Treatise on Invertebrate Paleontology*, courtesy of The Geological Society of America and the University of Kansas Press.

and Owens (1975). There is a wealth of trilobite material, but new thinking is urgently needed in the systematic and phylogenetic work. Transient characters and characters that occurred repeatedly owing to convergent evolution have commonly been used. A single example may illustrate this condition. The *"Treatise"* (Moore, 1959) order Phacopida is based primarily on the proparian course of the facial suture. For stratigraphical reasons the opisthoparian type

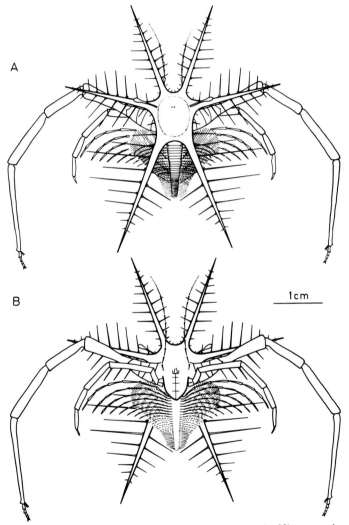

Fig. 1.12.   Dorsal and ventral views of Devonian marrellomorph, *Mimetaster hexagonalis*. From Stürmer and Bergström (1976).

of ecdysial suture is considered to be primitive in relation to the proparian type. The proparian suture, therefore, could be a synapomorphic feature for the Phacopida. However, primitive members or progenitors of all three constituent superfamilies apparently had opisthoparian sutures, and so the proparian character must have evolved independently in the three cases as well as in several other trilobite groups (Sdzuy, 1957; Thomas, 1977, and others for the Calymenina, Bergström, 1973a, for the Cheirurina, and Jaanusson, 1975, for the Phacopina). Much of the trilobite systematics is, therefore, based primarily on "technical" characteristics without much phylogenetic significance. Moreover, interest has been concentrated on the middle part of the head. It is important that characters from the entire body be considered, and that they be used as much as possible with an understanding of their nature as plesiomorphic, apomorphic, or transient characters.

Among other trilobitomorph groups, the Marrellomorpha have only three known members, *Marrella* from the Middle Cambrian, and *Mimetaster* (Fig. 1.12) and *Vachonisia* (Fig. 1.13) from the Early Devonian, possibly in addition to the enigmatic cycloids from the Early Carboniferous to the Late Triassic. As

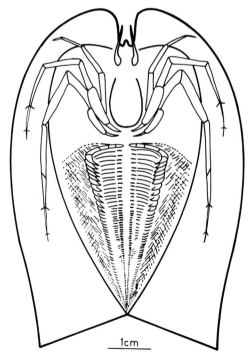

Fig. 1.13. Devonian marrellomorph, *Vachonisia rogeri,* in ventral view. From Stürmer and Bergström (1976).

Marrella has only two pairs of appendages in the head but *Mimetaster* three and *Vachonisia* four pairs, it may be surprising that they are suggested to belong to the same class. However, although the number is more or less constant in extant arthropod classes, there is a wide variation between the main groups. The differences are no doubt due to a more profound variation among early arthropods; therefore the exact number is not such an important character in the oldest arthropods as it is among living groups.

1.6.3. Chelicerata

The question of the origin of chelicerates was discussed above. Chelicerates are commonly divided into the aquatic Merostomata and the basically terrestrial Arachnida, including the scorpions. I regard this as a grade classification, terrestrial chelicerates being derived more than once from aquatic progenitors (cf. Kraus, 1976). The water-to-land transition is found within the Scorpionida, of which Silurian and Devonian members appear to have been marine. The actual transition in this group may have taken place in Devonian or Carboniferous times. As mentioned previously, tracheal structures (pseudotracheae) had already evolved independently in eurypterids in the Silurian. A separation of the closely similar and probably closely related Scorpionida and Eurypterida, as subgroups of the Merostomata and Arachnida respectively, is therefore fully unwarranted and creates polyphyly. It is worth noting that inclusion of scorpions in the Arachnida is also inconsistent with embryological evidence (Anderson, 1973). Arachnids other than scorpions invaded land in Early Devonian times or earlier, as shown by well-preserved book lungs in the Rhynie Chert (Claridge and Lyon, 1961; Størmer, 1976). It is remarkable that two quite different systems for oxygen exchange have been developed among the arachnids, the book lungs and the tracheae. This might have indicated separate aquatic origins for different arachnid groups, were it not for the co-occurrence of both types in araneids. The absence of possible direct progenitors in the known fossil record of aquatic chelicerates makes a discussion of poly- or monophyly difficult.

Various morphological characters are not equally reliable as guides to relationships. The number of segments of the chelicerae, for instance, varies between two and three in an obviously random way, as is also the case within the Eurypterida. The shape of the sternum and the position of the surrounding coxal attachments may be a good character. Thus, some orders have the coxae of the first walking legs more or less laterally disposed in comparison with the adjoining coxae. These groups are the Thelyphonida (or Uropygi, Late Carboniferous to Recent), Schizomida (Tertiary and Recent), Phrynichida (or Amblypygida, Late Carboniferous only), Kustarachnida (Late Carboniferous), Palpigradida (Late Jurassic to Recent), Ricinuleida (Late Carboniferous to

Recent), and Solpugida (or Solifugae, Late Carboniferous to Recent). The coxal disposition mentioned is probably connected with the prolongation of the first walking leg into antenna-like structures in most of these groups. It may be phylogenetically significant that the presence of a narrow posterior part of the opisthosoma (so-called pygidiosoma) and a flagellum is practically restricted to this array of orders (Thelyphonida, Schizomida, Phrynichida, Palpigradida, and Ricinuleida). Apart from the Acarida, a subdivision of the prosomal tergal shield into discrete plates is also restricted to this group (Schizomida, Palpigradida, and Solpugida). A more regular type of coxal arrangement is found in the Haptopodida (Late Carboniferous), Architarbida (or Phalangiotarbi, Late Carboniferous), Phalangiida (or Opiliones, Late Carboniferous to Recent), Trigonotarbida (Fig. 1.14C; Early Devonian to Late Carboniferous), Anthracomartida (Late Carboniferous), and Araneida (Early/Middle Devonian?, Late Carboniferous to Recent). These groups may belong together in a phylogenetic unit, possibly together with the Acarida (Fig. 1.14A, B; Early/Middle Devonian to Recent). The Pseudoscorpionida (Tertiary to Recent), as well as the Scorpionida, stand apart from the other groups in sternal shape and coxal arrangement. This grouping is based on a study of the morphology of fossil and extant forms, and not on phylogenetic lineages. Nevertheless, the separation of the true arachnids (excluding the scorpions) in at least three major groups, may indicate a corresponding very early phyletic separation.

The existence even in Early Devonian times of at least some groups may indicate either a comparatively long terrestrial prehistory, which does not seem to

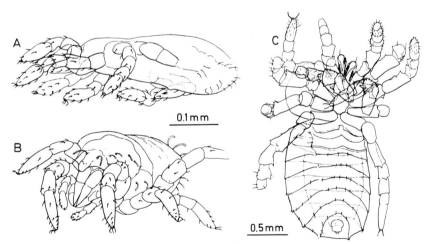

Fig. 1.14. Arachnids from the Devonian Rhynie Chert. A and B, Two views of acarid *Protacarus crani*; C, trigonotarbid *Palaeocharinoides hornei*. Both from Hirst (1923), *The Annals and Magazine of Natural History* including Zoology, Botany, and Geology, with permission from Taylor and Francis Ltd., London.

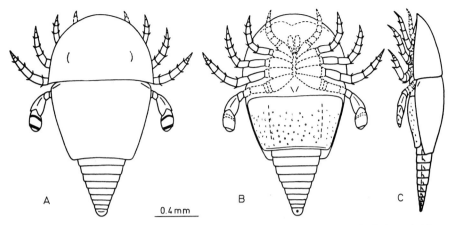

Fig. 1.15. Devonian supposed aquatic arachnid relative *Diploaspis casteri.* It is probably primitive in the absence of a tail spine. "Eyes" correspond to a faint fold line in one specimen and are added here. Presence of eyes is not proven. Ventral abdominal shield may be for protection against desiccation. From Størmer (1972), with author's permission.

fit well with our knowledge of terrestrial floras, or separate origins from distinct aquatic progenitors. However, even with a multiple invasion of land it is possible that the Arachnida (apart from the scorpions) constitute a monophyletic group, with a common aquatic ancestor. They differ from typical merostomes and scorpions in a simple but consistent way, viz., in lacking a tail spine. This is probably a primitive feature for the Chelicerata as a whole, but still useful since the typical merostomes have a keeled tail spine as a synapomorphic character (Bergström, 1975). The arachnids, therefore, must stem from the common phylogenetic root, not from the ordinary merostomes. There are very few fossils that may belong to the aquatic part of the arachnid stem. One is the Middle Cambrian *Beckwithia* (Fig. 1.17D), which has been regarded as an aglaspidid but lacks phosphatization of the integument. Another is the Early Devonian *Diploaspis* (Fig. 1.15). Both of these apparently represent phylogenetic side-lines, as their modifications would preclude them from being directly ancestral to known terrestrial arachnids.

The keeled tail spine was mentioned as a synapomorphic feature for merostome groups. The spine is probably mostly a telson, although segmental material is included in the Xiphosurida. An additional feature is possibly the inclusion in the prosoma of the pregenital segment, although the corresponding tergite is not always fused to the main prosomal tergite. Similar free prosomal tergites are found in the arachnid orders Schizomida, Palpigradida, and Solpugida. The prosomal character of the pregenital segment is most clearly seen in the Xiphosura (Fig. 1.16; Bergström, 1975). However, it may be that the pregenital segment could be counted as prosomal in arachnids as well; and if so, there is

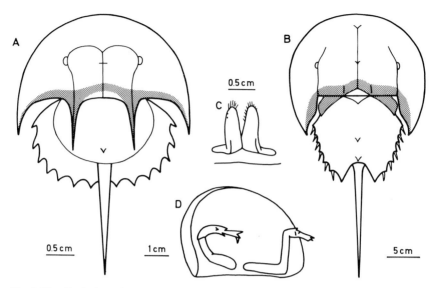

Fig. 1.16. Evolution of prosomal-opisthosomal boundary in xiphosurids. A, Carboniferous *Euproops rotundatus*; B, extant *Limulus polyphemus*. In both forms extent of segment 6, as interpreted earlier, is shown by stippling. Segment 7 (white posterior medial portion of prosoma) is reduced in size. The stippled spines on the posterior margin of the prosoma in *Euproops* appear to correspond to the anterolateral "fused spines" of the opisthosoma in *Limulus*, indicating a sudden shift in position during phylogeny. Recent study of *Limulus* embryology indicates that the shifting spines may belong to a more posterior metamere, although they would still be of prosomal origin (Scholl 1977). C, Chilaria of segment 7 in *Limulus*, probably consisting of gnathobases of reduced limb pair; D, second pair of pushing legs in Devonian *Weinbergina opitzi*, apparently belonging to segment 7 and corresponding to chilarial appendages in *Limulus*. Appendage morphology shows that segment 7 belongs to prosoma in xiphosurids. From Bergström (1975).

no difference in this respect. The pregenital tergite actually may be fused with the main prosomal tergite in the pseudoscorpions. The apparently monophyletic Merostomata include the Xiphosura with the Chasmataspidida (Ordovician), Synziphosurida (Silurian, Devonian), and Xiphosurida (Cambrian to Recent) and an unnamed unit with the Eurypterida (Cambrian to Permian), Scorpionida (Silurian to Recent), and probably the poorly known Cyrtoctenida (Devonian and Carboniferous). The aglaspidids (Fig. 1.17A, B) are commonly included in the Xiphosura. If this is done, the latter group is decidedly polyphyletic, as there are profound differences. Synapomorphic features for the Aglaspidida (Cambrian and Ordovician) include the mineralization of the exoskeleton (phosphatic), the rounded cross section of the tail spine, probably four-segmented chelicerae (against two to three segments in other chelicerates), and perhaps a postventral plate. Plesiomorphic features found in the Aglaspidida but lost in the Merostomata and Arachnida (where the corresponding features may repre-

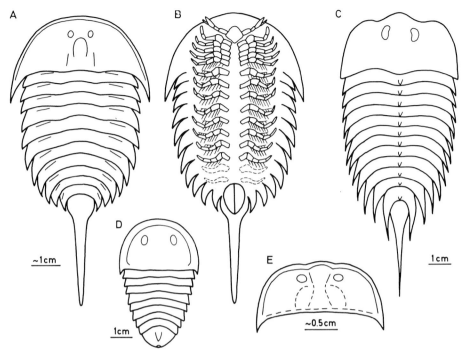

Fig. 1.17. A and B, Dorsal and ventral views of Late Cambrian aglaspidid chelicerate, *Aglaspis spinifer.* Note presence of chelicerae lateral to large labrum and presence of long series of limbs of uniform construction, with lateral branch carrying lamellae. Aglaspidids differ from all other chelicerates in having a phosphatic exoskeleton; C, Middle or Early Cambrian supposed eurypterid, *Kodymirus vagans*; D, Middle Cambrian *Beckwithia typa*; E, probably Early Cambrian xiphosurid *Eolimulus alatus*, which seems to be morphologically fairly close to *Kodymirus*, indicating common descent not much earlier. A and B, modified from *Invertebrate Fossils* by R. C. Moore, C. G. Lalicker, and A. G. Fischer (1952). Copyright 1952 by the McGraw-Hill Book Company, Inc. Used with permission of McGraw-Hill Book Company. C, from Chlupát and Havlítek (1965).

sent synapomorphies) include the biramous trilobite-like limbs with lamellar spines, the preservation of fully developed limbs in the prosoma as well as in the opisthosoma, and the preservation of a large labral plate. In addition, the number of prosomal appendage pairs is said to be six, including the chelicerae (Raasch, 1939), which apparently is one less than in primitive Merostomata. It is not known if this is a plesiomorphic or an apomorphic feature. The chelicerate prosoma may have grown through the successive addition of segments, but it may also, and perhaps more likely, have been separated as a unit at the moment tagmosis began to affect the homogeneity of the limb series.

The fossil record thus indicates that the old separation of chelicerates into the aquatic Merostomata and the terrestrial Arachnida does not fully reflect the

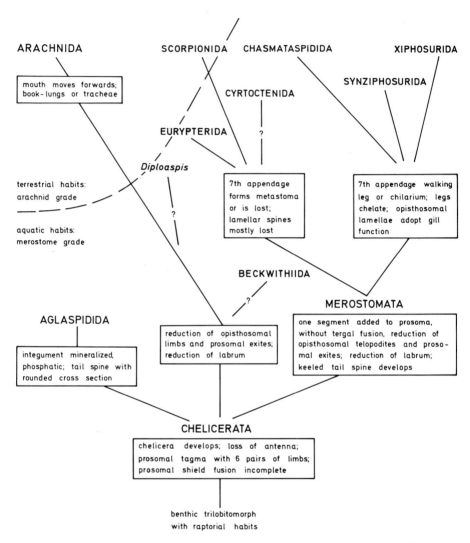

Fig. 1.18. Suggested evolutionary relationships of chelicerate groups. In addition to lineages indicated to approach or reach a terrestrial habitat, chasmataspidids may have been able to withstand desiccation, and arachnids may have reached arachnid grade along more than one line. This emphasizes the impossibility of using a terrestrial or aquatic habitat or associated adaptations as a basis for classification. Mainly after Bergström (1978).

evolutionary pattern. Instead, three main lineages may be distinguished (Fig. 1.18). One is the primitive trilobite-like Aglaspidida, which must have branched off quite early. Another is the Merostomata, including the scorpions, with separate trends towards invasion on land by scorpions and eurypterids. A third lineage is the Arachnida, of which the early aquatic history is more or less un-known, although the Cambrian *Beckwithia* (Fig. 1.17D) and the Devonian *Diploaspis* (Fig. 1.15) may belong here. New discoveries of aquatic arachnid progenitors are much needed. The available material contains many features that should be further analyzed—for instance, the presence of two kinds of gills and two kinds of structures for breathing in air, the poorly understood segmentation at the prosomal-opisthosomal boundary, particularly in eurypterids and scorpions, and so on, to mention a few.

1.6.4. Biantennata (Crustacea)

Most crustacean groups are rare as fossils. Only ostracodes are common and malacostracans moderately common. In the same way as in many other groups, various taxa may be traced far backwards in time only to be lost before the ancestry is approached. The Lower Palaeozoic is almost devoid of crustacean remains, except for ostracodes. The various crustacean-like forms found in the Middle Cambrian Burgess Shale therefore are morphologically and temporally disconnected from later crustaceans, and what is known of the appendages has not been enough to tie them safely to the crustacean lineage. Therefore, they tend to be regarded as trilobitomorphs rather than as crustaceans. However, we have to face the fact that crustaceans were already in existence in Early or Middle Cambrian times, as shown by the malacostracan *Hymenocaris*, and we must be prepared to meet an array of early "experiments" without modern descendants, just as we do in the Trilobitomorpha and Chelicerata.

Among the branchiopods, the Conchostraca and Anostraca range back to the Early Devonian, and the Lipostraca are confined to the Devonian. The Notostraca are first known from the Late Carboniferous, while the Cladocera are known with certainty only from the Tertiary. The Kazacharthra are notostracan-like forms, so far known only from the Early Jurassic, although as many as seven genera have been described. *Odaraia*, *Protocaris*, and *Branchiocaris* are three genera known from the Cambrian. Of these the last has the general morphology of an almost perfect progenitor of conchostracans and cladocerans, indicating a common origin of at least these two groups (Fig. 1.19). *Odaraia* with its stalked eyes (Simonetta and Delle Cave, 1975, Pl. 3: 6, 52: 1-7) may come closer to the Anostraca, although it has a carapace, making it in some way intermediate between anostracans and phyllopods. There is no particular reason to doubt the branchiopod character of these Cambrian forms, although we would like to know still more about the morphology. The connection between branchiopods and other crustaceans is not elucidated by these fossils, but the Devonian Lipostraca with the single known genus, *Lepidocaris*, (Fig. 1.20.) is partially intermediate in somewhat the same way as the extant Cephalocarida.

Among other crustaceans, the Copepoda are known first from the Tertiary,

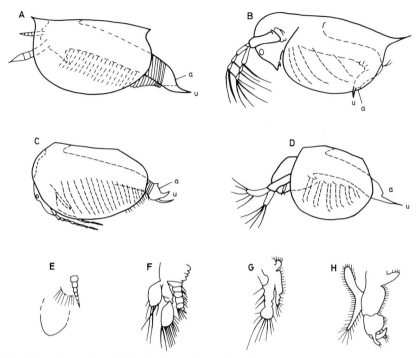

Fig. 1.19.   A and E, Middle Cambrian *Branchiocaris*, for comparison with extant branchiopods; B, *Daphnia* (Cladocera); C and G, *Cyzicus* (Conchostraca); D and H, larva of *Limnadia* (Conchostraca); F. *Hutchinsoniella* (Cephalocarida).   Note great similarities in the presence of a large terminal sclerite with terminal anal opening (*a*) and uropods (*u*). Position of anal opening indicates that terminal sclerite belongs to a true segment, while a telson is missing. Not to scale: Marine *Branchiocaris* reached at least 85 mm in length, limnic extant forms are much smaller. From Bergström (1978), redrawn after various sources. F, from Sanders (1957).

and the Mystacocarida and Branchiura are unknown as fossils. The Cirripedia range back to the Silurian. Of the Ostracoda, the Podocopida and Myodocopida range back to the Early Ordovician, where the lineages are lost without any safe trace of any ancestry. The Palaeocopida (Ordovician to Permian) and Leperditicopida (Cambrian?, Ordovician to Devonian) appear to have a similar composition of the exoskeleton and may be true ostracodes, although there are no direct morphological transitions.   The Cambrian (to Ordovician?) Bradoriida (or Archaeocopida) are morphologically different from other safe and supposed ostracodes, and the exoskeleton is only slightly (if at all) calcified. Öpik (1968) may be right in stating that the bradoriids are better regarded as malacostracans.

The situation is only slightly better in the Malacostraca.   There is a temporal succession, with the Hymenostraca known from the Cambrian and Early Ordovician, the Archaeostraca from the Early Ordovician to the Late Triassic,

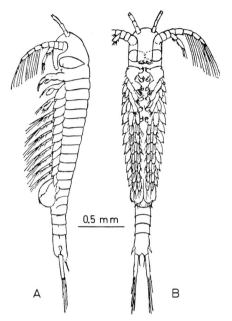

A   B

Fig. 1.20. Lipostracan *Lepidocaris rhyniensis* from Lower or possibly Middle Devonian Rhynie Chert. Like other Rhynie Chert arthropods (Figs. 1.8D, 1.14A–C) *Lepidocaris* is preserved in minute detail. From Scourfield (1926), copied with the permission of the Royal Society of London.

and the Eumalacostraca from the Middle Devonian to Recent time. The Leptostraca are not known with certainty as fossils, although there is a report from the Permian. The eumalacostracans have been thought to converge backwards to an origin perhaps in Devonian times from the Archaeostraca. However, there are two errors in this assumption. First, the idea of a convergence has to be modified after the discoveries of typical tanaidaceans and isopods in the Upper Carboniferous and of a spelaeogriphacean in the Lower Carboniferous. Second, the known archaeostracans are not acceptable as eumalacostracan ancestors, different as they are, for instance, in the number of segments, the loss of pleopods, and the highly reduced state of the uropods (the "furca"). In the two latter respects, the eumalacostracans are decidedly more primitive, despite the apparent later appearance. Palaeontology cannot at present bring together the major malacostracan groups. Instead, we have to postulate an unknown separate evolution of the eumalacostracans from perhaps Cambrian common ancestors. On the other hand, this indicates that the Eumalacostraca constitute a monophyletic group, and not only an evolutionary grade higher than the Archaeostraca (Fig. 1.21). With this knowledge at hand, we may accept a wider range of experimentation and variation among early malacostracans

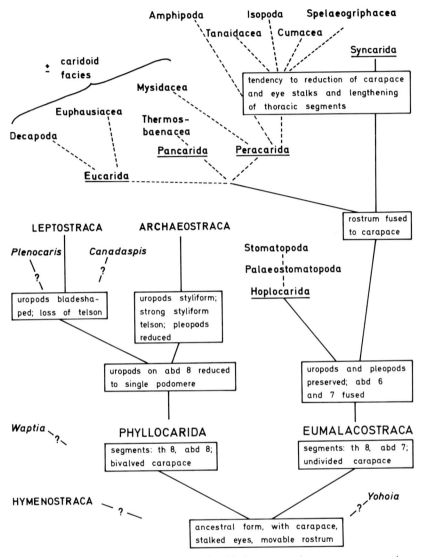

Fig. 1.21. Diagram to show possible relationship between malacostracan groups. It may be worth noting that reduction of the carapace, found as a convergent feature in various eumalacostracan groups, is also encountered in some archaeostracans. Although some Burgess Shale forms are included, it must be stressed that their position in the diagram is very difficult to evaluate.

than we thought of before, which means that forms like *Waptia* and *Canadaspis* from the Middle Cambrian Burgess Shale may be true malacostracans, although they do not fit archaeostracan or eumalacostracan concepts.

Convergent evolution is a problem in malacostracans as well as in other arthropod groups. Palaeontology may help us to understand the remarkable variation in the development of stalked eyes and the carapace, which is important for the discussion of phylogeny. It is well known that amphipods and isopods lack eye stalks as well as a carapace. It is not, however, mentioned in connection with the lack of a carapace that the thoracic segments are as long as the abdominal ones, and not considerably shorter, as in malacostracans with a well-developed carapace. Remarkably enough, the same combination of features is found in the Koonungidae within the Syncarida. In this case, there is evidence from other syncarids that the combination was secondarily derived. Thus, there is an eye stalk and a slight shortening of the thoracic segments in the Anaspididae, and the Permian *Clarkecaris* has distinctly short thoracic segments and possibly a short carapace remnant. Obviously, the presence of a carapace is a primitive feature, and the eye stalks and thoracic shortening appear to be functionally coupled with the carapace. This coupling may explain the high degree of habitual similarity between the unrelated groups, and is also a warning against the uncritical use of morphological characters, singly or in a group, in phylogenetic discussions.

While there are few definite statements that can be based on the fossil evidence regarding polyphyly versus monophyly at the level discussed, the general impression tends to confirm the conclusions based on living forms. Again we may come much further with cooperation between zoologists and palaeontologists, and with an understanding of morphological features as exponents of function.

1.6.5. Pycnogonida

The extant Pantopoda make up a well-defined group, and there can hardly be any doubt that they are monophyletic. Since there are only two described fossil species of the Pycnogonida, they cannot add much to the controversy. However, the fossil material shows some unique features and may be briefly discussed. All of it is from the Lower Devonian Hunsrück Shale. *Palaeopantopus maucheri* has been regarded as the sole member of the order Palaeopantopoda and appears to differ considerably from extant pantopods, although the details are poorly known. An undescribed species is modern in type, and obviously belongs to the Pantopoda. The third form is the large *Palaeoisopus problematicus*. After the realization that the supposed proboscis actually is a long segmented abdomen (Lehman, 1959), the morphology may be more closely compared to that of pantopods. The general construction with proboscis, chelifers, ovigers, and four pairs of legs agrees with general pantopod morphology. However, the legs are unique in having a flattened cross section, indicating that *Palaeoisopus* was a good swimmer. Another quite remarkable feature is the eye tubercle, which apparently has two median ocelli, in addition to only one pair, while modern pantopods have their four ocelli arranged in two pairs. The difference is unexplainable but cannot deny the relationship with pantopods. The segmented abdomen is interesting in showing a similarity to those in other arthropods. Unfortunately nothing in the fossil material indicates relationships with any particular arthropod group. (For a different viewpoint, see Weygoldt, Chapter 3)

1.7. PATTERN OF OCCURRENCE

There are some characteristic features in common between various groups of arthropods, that merit some consideration. One is the comparatively sudden occurrence of many groups in the Cambrian, while the pre-Cambrian is devoid of definite arthropod remains. This may be associated with the fairly sudden acquisition of a well-sclerotized exoskeleton. There was obviously much parallel evolution in this respect, as there is much variation in the chemistry from one group to another. Various groups independently introduced calcium carbonate or calcium phosphate in their exoskeletons. The latter alternative proved less successful in the long run, and is not represented today. The pre-Cambrian history may have been fairly short. Trace fossils in the sediments reveal the existence of metazoan animals only for about 100 million years before the beginning of the Cambrian.

A second characteristic feature is the apparent morphological explosion at the beginning of some groups, with a number of odd and short-lived subgroups present in addition to the still living ones. The latter may be easily recognizable from the beginning. The result is that in tracing particular taxa backwards, an increasing variation is revealed, instead of a convergence toward an ultimate origin as one might have expected. A possible explanation is that the beginning of new major lineages is the beginning of new modes of life, opening wide fields for exploitation and therefore great possibilities for expansion and experimentation.

A third feature, characteristic for particular groups, is that their trace is lost at a particular time, although there is very strong indication that they should have a long earlier history (Fig. 1.22). This is the case with the terrestrial Hexapoda and Arachnida, and to some degree with the Myriapoda. It could be argued in these cases that the aquatic progenitors had an exoskeleton that was too delicate for preservation during ordinary circumstances. However, the progenitors still should be preserved in the Devonian Hunsrück Shale and the Cambrian Burgess Shale, where even entirely soft-bodied animals are well preserved. There is a similar pattern in some aquatic groups, like the Eurypterida, which are practically lost in the Ordovician, and the Eumalacostraca, with the earliest known members in the Devonian. The latter simply cannot be derived from the Archaeostraca, and must have had a separate origin well back in time. There is no generally accepted reason for this pattern. However, there is a possibility that has not been much considered in the literature (except for early vertebrates). The only habitat that is generally represented in the Lower Palaeozoic is the marine realm. There might have been biological evolution in fresh waters, whose deposits are practically unknown, or at least unrecognized. There is a

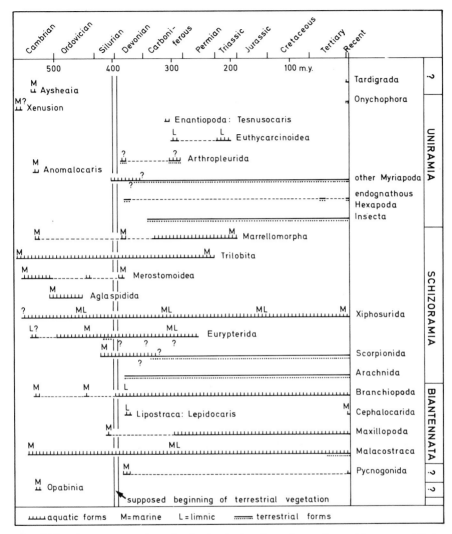

Fig. 1.22. An attempt to show distribution in time for various arthropod lineages and their habitats. Various groups, including Ostracoda, are omitted. Although Lipostraca are branchiopods, they are shown separately because of their phylogenetic importance (they are on the same line as Cephalocarida only to save space). Four hundred-million-year line is a natural boundary for terrestrial groups. Progenitors must have been primarily aquatic and should be searched for among problematic aquatic arthropods.

single occurrence in the Ordovician that may illustrate the case. Caster and Brooks (1956) reported a meager fauna and flora that they considered as marine, although the time interval concerned (Early/Middle Ordovician boundary) was characterized by a local regression, and marine sediments are found only far away from the locality in Tennessee. The reported species are *Chasmataspis laurencii, Douglasocaris collinsi, Cestites mirabilis,* and some algal threads, and a burrow. The first species is a unique merostome and the second a unique type of crustacean. The third species was considered to be a ctenophoran, but more likely consists of algal remains. There are volcanic debris, varv-like graded bedding with coaly surfaces, and mudcracks in the sediment. I am strongly inclined to regard the whole situation as continental and limnic.

A second, probably limnic, environment is represented in the Devonian Rhynie Chert, with the unique crustacean, *Lepidocaris* (Fig. 1.20), representing the aquatic fauna.   Still another, possibly limnic, environment (considered limnic or brackish also by Chlupáč and Havlíček, 1965), is represented by Lower or Middle Cambrian beds in Bohemia, yielding the primitive eurypterid, *Kodymirus* (Fig. 1.17C), as the only fossil. Ultimately, a unique and still undescribed arthropod (crustacean?) from Upper Cambrian strata at the River Lena in Siberia abounds on a surface with mudcracks in such a way that drying up of a lake or pond is suggested. All this suggests that the idea of an Early Palaeozoic evolution in fresh-water habitats is a real alternative and a possible explanation for the absence of particular groups from marine sediments. A morphological feature strongly adding to the probability is the abdominal shield in forms like *Chasmataspis* and the Devonian *Diploaspis*. Størmer (1976, p. 137) noted that a similar development of ventral shields is found among various more or less amphibious or terrestrial merostomes and scorpions as well as isopods, as a protection against damage and desiccation on land. Although it is unlikely that the Ordovician *Chasmataspis* was terrestrial, it may have been able to withstand periodic dry spells. Regarding eurypterids, Kjellesvig-Waering (1961) recognized three ecological "phases" in the Silurian, ranging from the high marine Carcinosomatidae-Pterygotidae phase to the perhaps brackish water Hughmilleridae-Stylonuridae phase.   The above-mentioned presence of tracheal structures in a Silurian eurypterid is of interest in this connection.

The trend toward fresh waters, in combination with the poor fossil representation of continental habitats, may illustrate the danger of using the order of first stratigraphical occurrence uncritically in the construction of phylogenetic diagrams.

1.8. CONCLUSION

1.8.1. Mono- or Polyphyly in the Arthropoda

The palaeontological evidence for or against monophyly is commonly of a somewhat indirect character. The fossils confirm the constant uniramous nature of the legs in myrapods and insects, and the presence of dwarfed abdominal legs in a Permian insect fits the conclusion, based mainly on extant forms, that the two groups belong together in the Tracheata (Antennata). Possibly additional groups of the Uniramia, the Onychophora and Linguatulida, are not known with certainty from fossils (*Aysheaia* appears to be closer to the Tardigrada), and the state of the entire Uniramia is therefore more difficult to discuss. Indirect conclusions indicate that the uniramous groups reached the arthropod level independently of other arthropods. Therefore, there are no palaeontological objections against an acceptance of the Uniramia as a monophyletic group. On the other hand this would mean that the Mandibulata as well as the Tracheata (including the chelicerates) are polyphyletic. The Biantennata (Crustacea, Branchiata, Diantennata) are regarded by most zoologists to be monophyletic, and there is no conflicting evidence from the fossils. They have some basic features in common with the Schizoramia, but the fossil evidence is not enough to show if those are due to a common ancestry or to parallel or convergent evolution. At any rate, the assumed separate prearthropod origin of the uniramians as well as of *Opabinia* would make the Arthropoda a polyphyletic group.

It would be reasonable to expect somewhat safer results from palaeontology at lower systematic levels. This may be the case on a comparatively low level in groups that are well represented in the fossil record. However, at a class-to-order level there are still more questions than answers. One striking feature is that the major groups were as distinct from each other at their first appearance as they are today. Instead of the intermediate forms that may have been expected, there are found, on the one hand, arthropods clearly belonging to still living groups and, on the other, a bewildering array of odd forms that cannot be placed systematically. In the case of the terrestrial Hexapoda and Arachnida, there was already a remarkable variation in Devonian and Carboniferous times. This is difficult to understand if these groups invaded land with the plants in the Early Devonian or perhaps Late Silurian. Either the evolution on land occurred very rapidly (or perhaps for a longer time among seaweed on the shores), or different terrestrial lineages were already separate before they left water. The possibilities are difficult to evaluate in the case of the Hexapoda and Arachnida, where no possible aquatic direct ancestors are known. In the case of myriapods, scorpions, and eurypterids, a sequence from aquatic to terrestrial forms can be vaguely discerned within the groups.

Palaeontological evidence may show that the Insecta form a monophyletic assemblage, but there is no evidence regarding the Hexapoda as a whole. Myriapod-like forms apparently were present in the seas back in the Cambrian, but the evidence beyond the Silurian is extremely scanty, and nothing positively can be said about the ancestry. The Linguatulids, if accepted as uniramians, must have evolved from one of the terrestrial myriapod groups, not earlier than the latest Carboniferous. The presence of the possibly uniramian euthycarcinoids in Permian and Triassic waters and of *Tesnusocaris* in the Carboniferous may indicate that there was a wide morphological variation among early aquatic uniramians.

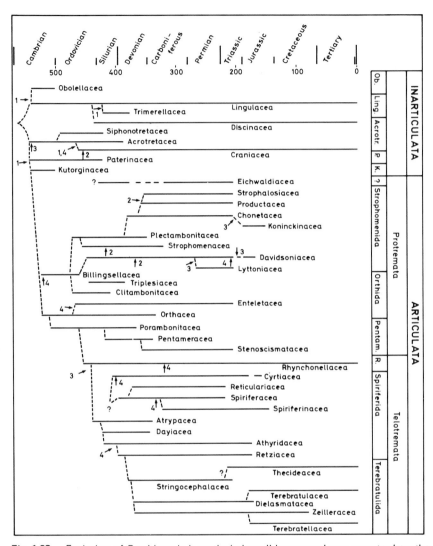

Fig. 1.23. Evolution of Brachiopoda is particularly well known and may serve to show the extent of convergent evolution that may be expected in any group. Four different kinds of innovations are recorded in the diagram, and points of introduction are indicated by arrows. 1, The phosphatic shell turned calcareous four times; 2, cementation to substrate occurred at least four times; 3, skeletal support for lophophore developed at least four times; 4, punctation (with caeca in the shell) evolved probably at least eight times. Each innovation is a radical change and may be compared to acquisition of legs, jaws, or wings in arthropods. Information mainly from Rudwick (1970).

The situation is much better as far as the Schizoramia (Trilobitomorpha and Chelicerata) are concerned. There is an almost complete morphological spectrum from the marrallomorphs, in which gnathobases were not yet developed, over trilobites and other trilobitomorphs with merostome-type gnathobases, to cheloniellids with postantennal but preoral grasping appendages in the position of chelicerate chelicerae, further to chelicerate aglaspidids, in which the antennae were lost, to the xiphosurids and eurypterids with modification and loss of the lamellar spines and strong appendage tagmosis. The almost complete lack of all signs of an antennal segment in living chelicerates is a negative proof, and cannot disqualify the overwhelming similarity between primitive chelicerates and trilobitomorphs. The greatest morphological discontinuity actually is not between trilobitomorphs and chelicerates but between the Merostomata and the Arachnida. There is, therefore, very strong palaeontological evidence that the Schizoramia form a natural monophyletic unit. The evidence of successive morphological change in the course of phylogeny is by far the best in comparison with the Uniramia and Biantennata (Crustacea).

In the Biantennata, the actual links between major groups are mostly missing; only in the branchiopods are there some morphologically intermediate forms in the Cambrian. In the same way, a connection between crustaceans and other groups, such as trilobitomorphs, is lacking or at least not fully convincing with the present amount of knowledge.

The Pycnogonida are represented only by three Devonian species, of which only one undescribed form is of modern design. One of the others has a long segmented abdomen, indicating an ancestry among more ordinary arthropods.

1.8.2. Tools in Phylogenetic Research

The common method of phylogenetic research is to compare structural and/or chemical features of different animals. This is the comparative anatomy/morphology/genetics/chemistry approach. It does not take functional background into account. Particular methods have been developed to enhance its effectiveness. One is numerical taxonomy, which is a very coarse tool, at least so far. Another method is cladistic phylogeny. This method emphasizes the very important but commonly neglected principle that only shared-derived (synapomorphic) characters can be used for distinguishing monophyletic groups. The knowledge that morphology and function are closely interdependent means that an understanding of function is necessary for a real understanding of morphology. Comparative functional morphology is, therefore, the most natural next step in phylogenetic work. It is important that the possibilities of cladistic phylogeny and comparative functional morphology be well followed. Recently, biochemists have developed the parsimony analysis of amino acid evolution. This method appears to be extremely useful for extant groups and should be applied to arthropods as soon as possible. It is to be hoped that this method will first be used to interpret the relationship between major arthropod groups, including the enigmatic tardigrades, pantopods, and linguatulids, as well as the relationship between arthropods and other prostomian groups.

1.9. SUMMARY

Knowledge of fossil arthropods extends back to the Early Cambrian. At that time the main groups were already in existence as separate entities. Also, within the main groups of arthropods were numbers of subgroups, now partly extant and partly extinct. This means that the earliest divergence at various high levels occurred before the first known arthropods were fossilized. The fossil record therefore does not yield any direct information about the possible mono- or polyphyletic state of various groups.

Among tracheates, the earliest myriapods are from the Silurian and Devonian. These are the poorly known Archipolypoda, somewhat reminiscent of the Diplopoda, and the unique Arthropleurida, which no doubt form an extinct endline. The Diplopoda and Chilopoda occur first in the fossil record in the Carboniferous, and the other groups are not known until much later. The Archipolypoda urgently need to be restudied. Among the hexapods, fully modern Collembola are known from the Early or Middle Devonian and a possible pterygote (*Eopterum*) from the Late Devonian. If the Carboniferous-to-Triassic aquatic Euthycarcinida are tracheates, they add considerably to the diversity, since they cannot be considered as myriapods or hexapods. The fossil tracheates do not provide any particular evidence regarding the ancestry of and interrelationships between the constituent groups of myriapods and hexapods. The early appearance of some groups indicates that they were already present as distinct groups in the ancestral aquatic environment. The same indication perhaps is provided by the aquatic euthycarcinoids and by possibly myriapod-like marine animals in the Middle Cambrian.

Among the chelicerates, terrestrial arachnids and semiterrestrial (?) eurypterids are known from the Early Devonian. Aquatic scorpions, eurypterids, and xiphosurids are older, and the two latter groups appear to be represented by similar forms in the Early Cambrian, indicating a separation at that time. The type of tagmosis, the tail spine, and the overall morphology in Silurian and Devonian scorpions indicate that they are close to the eurypterids. The other terrestrial arachnids cannot with certainty be traced to any aquatic ancestry. The Early Devonian merostome-like *Diploaspis* has a probably primitive tiny round telson, different from anything known in true Merostomata, and may be a late representative of an aquatic chelicerate group that gave rise to the terrestrial Arachnida, except for the scorpions. The number of groups invading land in Silurian-Devonian time is evidence of the ease of this process, and arachnids may well represent several originally aquatic groups. The Cambrian and Ordovician Aglaspidida had a comprehensive prosoma and chelicerae but apparently lacked antennae, and are considered to be chelicerates. However, the appendages behind the chelicerae appear to be similar throughout the body, and they have an outer ramus with serially arranged lamellae much as in various trilobitomorphs. This detailed similarity, in addition to an overall habitual similarity, is a strong argument for a close relationship between the Chelicerata and the Trilobitomorpha. It seems likely that at least the main stock of chelicerates arose only once from trilobitomorphs through the loss of the antenna, development of the chelicera, and addition of segments to the head tagma, and later also the distinct limb tagmosis. The aglaspidids may be thought of as a very early offshoot, or possibly as a convergently evolved group. The fossil pantopods do not approach any other group sufficiently to indicate whether they are related to the chelicerates or not.

The trilobitomorphs have been considered to include more or less all early arthropods, which cannot be placed with certainty in any extant group. Restudy of old material shows that various so-called trilobitomorphs can be placed without doubt among the crustaceans or myriapods. This being the case, the "classical" Trilobitomorpha are polyphyletic. However, there are a number of groups with the trilobite appendage as a common denominator. These are the Trilobita, Marrellida, Acercostraca, Limulavida, and Cheloniellida. The most characteristic feature is the lanceolate lamellae of the outer appendage branch. The lamellae kept their characteristic morphology with various functions, which indicates that they were not only an adaptational feature likely to evolve several times when needed. Other supposed trilobitomorphs are poorly known or lack typical lamellae (owing to reduction?) and must be placed as *incertae sedis* at present. With the present delimitation, I regard the Trilobitomorpha as a monophyletic group. Moreover, the lamellae, the morphology and arrangement of the gnathobases with the strongest elements placed posteriorly, and the development of a preoral grasping appendage closely comparable to a chelicera in the Cheloniellida are features strongly indicating chelicerate affinities.

The Crustacea have been thought to be allied to either tracheates or various arachnomorphs. There is not even a remote similarity between early crustaceans and early tracheates. The arachnomorph case is more difficult to evaluate. There are several general similarities that may be due to an original presence of branched appendages and food-collecting behind the mouth. However, it is likely that similar functional systems would evolve in basically similar animal groups, evolving under similar functional and ecological pressure; thus general similarities are no proof of monophyly. Crustaceans have been compared to various so-called trilobitomorphs. However, either the latter can now be stated to be true crustaceans, like the Cambrian *Protocaris*, *Branchiocaris*, *Canadaspis*, and *Plenocaris*, or they cannot be safely assigned to the Trilobitomorpha, like the so-called emeraldellids. A third possibility is that the comparison is generally unwarranted. No trilobite-type outer limb branch lamellae are known in any crustacean, and the differentiation of gnathobasic jaws tends to proceed backwards, while the posterior gnathobasic endites are generally better developed in trilobitomorphs when they are at all developed. A derivation of the Crustacea from the Trilobitomorpha with its possible polyphyletic significance thus cannot be safely substantiated. Among the Cambrian Crustacea, at least the Malacostraca and the Branchiopoda can be distinguished. Thus also in this case there is no direct fossil evidence of the possible mono- or polyphyletic origin and subsequent primary radiation. However, a probable pediform segmented endopod in Cambrian Branchiopoda at least indicates a convergence backwards in time with other crustaceans.

In general it may be said that there is a strong need for more discoveries of early arthropods. The fossil record is still almost devoid of known primary aquatic representatives of the tracheates and the arachnids, and this is certainly a severe drawback in any attempt to reconstruct the early evolution of arthropods. However, it is also the case that the known early fossil arthropods are in part poorly studied, and this is a point where effort is needed and should be effected. Unfortunately, many of the more interesting arthropod finds have attracted so much interest from collectors that they have described them without the aid of specialists. Thus many forms have been described by geologists or palaeontologists without a zoological background. Many of these forms have been redescribed or discussed by zoologists without an understanding of the geological and sedimentological background or of diagenetic and tectonic alteration and distortion. The result in particular cases is a number of quite disparate reconstructions, and nobody knows which of them to trust, if any. The famous Middle Cambrian fauna from the Burgess Shale provides a good example of such scattered work. The early controversies about the morphology and systematic relationship of various Burgess forms appear to have faded away, and many arthropod students of today dare not have any opinion regarding those fossils.

What is really needed is a careful redescription of already existing material, partly with the aid of new methods but, much more important, also with a simultaneous evaluation of geological and zoological aspects. Moreover, the discussion on systematics and phylogeny should not be based on a rigid comparison between morphological features (comparative morphology) alone, but also on an evaluation of the morphology as part of functional systems. In addition, it is necessary to take into account the ever-present possibility of morphologically discontinuous steps in the evolution. Ultimately, in the phylogenetic discussions it is not enough to point at similarities and dissimilarities as is commonly done, but it is also necessary to distinguish between (shared) primitive features and (shared) advanced features in order to arrive at a reasonable sequence of events and direction of the evolution.

ACKNOWLEDGMENTS

I am grateful to many persons for discussions on arthropod morphology and phylogeny. A number of authors, societies, publishers, and journals are acknowledged for their kind permissions to use their illustrations. In addition to the contributors indicated in the figure legends, I would like to thank Professor Wilhelm Stürmer in Erlangen for permission to use material in preparation. Mr. Sven Stridsberg performed the photographic work, Mrs. Christine Ebner made the drawings, Mrs. Ingrid Lineke and Mrs. Margareta Tunbjer typed the first draft of the text, and Mr. Brian Holland critically read the text. I am grateful for their efforts, without which I would not have been able to finish the manuscript in time. Finally, grants from the Swedish Natural Science Foundation made it possible for me to study some important fossil arthropod material.

REFERENCES

Anderson, D. T. 1973. *Embryology and phylogeny in annelids and arthropods.* Pergamon Press, Oxford, New York.

Bergström, J. 1968. *Eolimulus,* a Lower Cambrian xiphosurid from Sweden. *Geol. Fören. Stockh. Förh.* **90:** 489–503.

Bergström, J. 1969. Remarks on the appendages of trilobites. *Lethaia* **2:** 395–414.

Bergström, J. 1972. Appendage morphology of the trilobite *Cryptolithus* and its implication. *Lethaia* **5:** 85–94.

Bergström, J. 1973a. Organization, life, and systematics of trilobites. *Fossils Strata* **2:** 1–69.

Bergström, J. 1973b. Palaeoecologic aspects of an Ordovician *Tretaspis* fauna. *Acta Geol. Pol.* **23:** 179–206.

Bergström, J. 1975. Functional morphology and evolution of xiphosurids. *Fossils Strata* **4:** 291–305.

Bergström, J. 1976a. Early arthropod morphology and relationships. *25th Int. Geol. Congr. Abstr.:* 289. Sydney.

Bergström, J. 1976b. Lower Palaeozoic trace fossils from eastern Newfoundland. *Can. J. Earth Sci.* **13:** 1613–33.

Bergström, J. 1977. Proetida–A disorderly order of trilobites. *Lethaia* **10:** 95–105.

Bergström, J. 1978. Morphology and systematics of early arthropods. *Abh. Verh. Natur-Wiss. Ver. Hamburg* (in press).

Bonde, N. 1975. Origin of "higher groups": Viewpoints of phylogenetic systematics. *Coll. Int. CNRS* **218:** 293–324.

Broili, F. 1933. Ein zweites Exemplar von Cheloniellon. *Sitzungsber. Bayer. Akad. Wiss. Math.-Naturwiss. Abt. Jahrg.* **1933:** 11–32.

Brooks, H. K. 1955. A crustacean from the Tesnus Formation (Pennsylvanian) of Texas. *J. Paleontol.* **29:** 852–56.

Caster, K. E. and H. K. Brooks. 1956. New fossils from the Canadian-Chazyan (Ordovician) hiatus in Tennessee. *Bull. Amer. Paleontol.* **36:** 157–99.

Chlupáč, I. and V. Havlíček. 1965. *Kodymirus* n.g., a new aglaspid merostome of the Cambrian of Bohemia. *Sborník Geol. Věd. Paleontol.* **6:** 7–20.

Cisne, J. L. 1975. Anatomy of *Triarthrus* and the relationships of the Trilobita. *Fossils Strata* **4:** 45–63.

Claridge, M. F. and S. G. Lyon. 1961. Lung-books in the Devonian Palaeocharinidae (Arachnida). *Nature (Lond.)* **191:** 1190-91.

Delle Cave, L. and A. M. Simonetta. 1975. Notes on the morphology and taxonomic position of *Aysheaia* (Onychophora?) and of Skania (undetermined phylum). *Monit. Zool. Ital. (N.S.)* **9:** 67-81.

Eldredge, N. 1977. Trilobites and evolutionary patterns, pp. 305-32. *In* A. Hallam (ed.), *Patterns of evolution as illustrated by the fossil record. Developments in Palaeontology and Stratigraphy 5.* Elsevier, Amsterdam.

Fortey, R. A. and R. M. Owens. 1975. Proetida—A new order of trilobites. *Fossils Strata* **4:** 227-39.

Fryer, G. 1976. Adaptation, speciation and time. *Zool. Scripta* **5:** 171-72.

Gall, J.-C. and L. Grauvogel. 1964. Un arthropode peu connu: Le genre *Euthycarcinus* Handlirsch. *Ann. Paléontol.* **50:** 1-18.

Gall, J.-C. and L. Grauvogel. 1967. Faune du Buntsandstein. II. Les Halicynes. *Ann. Paléontol.* **53 (pt. 1):** 1-14.

Glaessner, M. F. 1969. Cycloidea. Addendum to Cycloidea, pp. R567-R570, R629. *In* R. C. Moore (ed.) *Treatise on Invertebrate Paleontology*, Part R, Arthropoda 4. Geol. Soc. Amer. and University of Kansas Press, Boulder, Colorado.

Gray, J. and A. J. Boucot. 1977. Early vascular land plants: Proof and conjecture. *Lethaia* **10:** 145-74.

Hennig, W. 1966. *Phylogenetic systematics.* University of Illinois Press, Chicago, London.

Hessler, R. R. and W. A. Newman. 1975. A trilobitomorph origin for the Crustacea. *Fossils Strata* **4:** 437-59.

Hirst, S. 1923. On some arachnid remains from the Old Red Sandstone (Rhynie Chert Bed, Aberdeenshire). *Ann. Mag. Nat. Hist.* (9) **12:** 455-74.

Hoffman, R. L. 1969. Myriapoda, exclusive of Insecta, pp. R572-R606. *In* R. C. Moore (ed.) *Treatise on Invertebrate Paleontology*, Part R, Arthropoda 4. Geol. Soc. Amer. and University of Kansas Press, Boulder, Colorado.

Holm, G. 1899. Om den yttre anatomien hos *Eurypterus Fischeri. Geol. Fören. Stockh. Förh.* **21:** 83-129.

Jaanusson, V. 1973. Morphological discontinuities in the evolution of graptolite colonies, pp. 515-21. *In* R. S. Boardman, A. H. Cheetham and W. A. Oliver (eds.) *Animal Colonies.* Dowden, Hutchinson and Ross, Inc., Stroudsburg, Pennsylvania.

Jaanusson, V. 1975. Evolutionary processes leading to the trilobite suborder Phacopina. *Fossils Strata* **4:** 209-18.

Kjellesvig-Waering, E. N. 1961. The Silurian Eurypterida of the Welsh Borderland. *J. Paleontol.* **35:** 789-835.

Kjellesvig-Waering, E. N. 1966. Silurian scorpions of New York. *J. Paleontol.* **40:** 359-75.

Kraus, O. 1974. On the morphology of Palaeozoic diplopods. *Symp. Zool. Soc. Lond.* (*1974*) **No. 32:** 13-22.

Kraus, O. 1976. Zur phylogenetischen Stellung und Evolution der Chelicerata. *Entomol. German.* **3:** 1-12.

Lehman, W. M. 1959. Neue Entdeckungen an *Palaeoisopus. Paläontol. Z.* **33:** 96-103.

Manton, S. M. 1973. Arthropod phylogeny—A modern synthesis. *J. Zool.* **171:** 111-30.

Moore, R. C. (ed.). 1959. *Treatise on Invertebrate Paleontology.* Part O, Arthropoda 1. Geol. Soc. Amer. and University of Kansas Press, Lawrence, Kansas.

Öpik, A. A. 1968. Ordian (Cambrian) Crustacea Bradoriida of Australia. *Commonw. Aust. Bur. Miner. Resour. Geol. Geophys. Bull.* **103:** 1-37.

Osche, G. 1963. Die systematische Stellung und Phylogenie der Pentastomida. *Z. Morphol. Ökol. Tiere* **52:** 487-596.

Pack, P. D. and H. B. Gayle. 1971. A new olenellid trilobite, *Biceratops nevadensis*, from the Lower Cambrian near Las Vegas, Nevada. *J. Paleontol.* **45**: 893-98.

Raasch, G. O. 1939. Cambrian Merostomata. *Geol. Soc. Amer. Spec. Pap.* No. 19.

Repina, L. N. and O. G. Okuneva. 1969. Cambrian Arthropoda of Primorye. *Paleontol. Zh.* **1969**: 106-14.

Riek, E. F. 1968. Re-examination of two arthropod species from the Triassic of Brookvale, New South Wales. *Rec. Aust. Mus.* **27**: 313-21.

Rolfe, W. D. I. 1969. Arthropoda incertae sedis, pp. R620-R625. *In* R. C. Moore (ed.) *Treatise on Invertebrate Paleontology*, Part R, Arthropoda 4. Geol. Soc. Amer. and University of Kansas Press, Lawrence, Kansas.

Rudwick, M. J. S. 1970. *Living and fossil brachiopods.* Hutchinson & Co., Ltd., London.

Sanders, H. L. 1957. The Cephalocarida and crustacean phylogeny. *Syst. Zool.* **6**: 112-29.

Schlee, D. 1975a. Review of: Numerical taxonomy. The principles and practice of numerical classification (by P. H. A. Sneath and R. R. Sokal). *Syst. Zool.* **24**: 266-68.

Schlee, D. 1975b. An analysis of numerical phenetics. *Entomol. Scan.* **6**: 1-9.

Scholl, G. 1977. Beiträge zur Embryonalentwicklung von *Limulus polyphemus* L. (Chelicerata, Xiphosura). *Zoomorphologie* **86**: 99-154.

Scourfield, D. J. 1926. On a new type of crustacean from the Old Red Sandstone (Rhynie Chert Bed, Aberdeenshire)–*Lepidocaris rhyniensis* gen. & sp. nov. *Phil. Trans.* **B. 214**: 153-87.

Scourfield, D. J. 1940. The oldest known fossil insect (*Rhyniella praecursor* Hirst & Maulik)–Further details from additional specimens. *Proc. Linn. Soc. Lond. 152nd Sess.* **1940**: 113-31.

Sdzuy, K. 1957. Bemerkungen zur Familie Homalonotidae (mit der Beschreibung einer neuen Art von *Calymenella*). *Senckenb. Leth.* **38**: 275-90.

Sharov, A. G. 1957. A unique Paleozoic wingless insect of the new order Monura (Insecta Apterygota). *Dokl. Akad. Nauk. SSSR (C.R. Acad. Sci. U.S.S.R.)* **115**: 795.

Simonetta, A. M. and L. Delle Cave. 1975. The Cambrian non trilobite arthropods from the Burgess Shale of British Columbia. A study of their comparative morphology taxonomy and evolutionary significance. *Palaeontol. Ital.* **69**: 1-37.

Sneath, P. H. A. and R. R. Sokal. 1973. *Numerical taxonomy. The principles and practice of numerical classification.* W. H. Freeman & Co., San Francisco, California.

Sokal, R. R. and P. H. A. Sneath. 1963. *Principles of numerical taxonomy.* W. H. Freeman & Co., San Francisco, California.

Størmer, L. 1944. On the relationships and phylogeny of fossil and recent Arachnomorpha. A comparative study on Arachnida, Xiphosura, Eurypterida, Trilobita, and other fossil Arthropoda. *Skrift. Vid.-Akad. Oslo, I. Math.-Nat. Kl.* No. 5: 1-158.

Størmer, L. 1959. Trilobitoidea, pp. O23-O37. *In* R. C. Moore (ed.) *Treatise on Invertebrate Paleontology*, Part O, Arthropoda 1. Geol. Soc. Amer. and University of Kansas Press, Lawrence, Kansas.

Størmer, L. 1972. Arthropods from the Lower Devonian (Lower Emsian) of Alken an der Mosel, Germany. Part 2: Xiphosura. *Senckenb. Leth.* **53**: 1-29.

Størmer, L. 1976. Arthropods from the Lower Devonian (Lower Emsian) of Alken an der Mosel, Germany. Part 5: Myriapoda and additional forms, with general remarks on fauna and problems regarding invasion of land by arthropods. *Senckenb. Leth.* **57**: 87-183.

Stürmer, W. and J. Bergström. 1973. New discoveries on trilobites by X-rays. *Paläontol. Z.* **47**: 104-41.

Stürmer, W. and J. Bergström. 1976. The arthropods *Mimetaster* and *Vachonisia* from the Devonian Hunsrück Shale. *Paläontol. Z.* **50**: 78-111.

Stürmer, W. and J. Bergström. 1978. The arthropod *Cheloniellon* from the Devonian Hunsrück Shale. *Paläontol. Z.* **52**: 57-81.

Tasch, P.   1969.   Branchiopoda, pp. R128-R191.   *In* R. C. Moore (ed.) *Treatise on Invertebrate Paleontology*, Part R, Arthropoda 4.   Geol. Soc. Amer. and University of Kansas Press, Lawrence, Kansas.

Thomas, A. T.   1977.   Classification and phylogeny of homalonotid trilobites. *Palaeontology* **20:** 159-78.

Tiegs, O. W. and S. M. Manton.   1958.   The evolution of the Arthropoda. *Biol. Rev. (Cambridge)* **33:** 255-337.

Whittington, H. B.   1975.   The enigmatic animal *Opabinia regalis*, Middle Cambrian, Burgess Shale, British Columbia. *Phil. Trans. R. Soc. Lond.* **B. 271:** 1-43.

SECTION II

Embryology and Development

2

Embryos, Fate Maps, and the Phylogeny of Arthropods

D. T. ANDERSON

2.1. PROBLEM

To a very large extent, the interpretation of arthropod phylogeny must rest on comparative functional morphology (Manton, 1972, 1977). Only in this discipline is there the wealth of functionally integrated detail that permits the interpretation of functional lineages, the elucidation of basic functional organizations within major groups, and the comparison of these basic organizations with respect to the functional feasibility of a common or disparate derivation from earlier prearthropodan forms. Clearly, the fossil evidence must also be integrated into any functionally based scheme, but the present evidence from this source adds little to the data on which we must base decisions about the phylogenetic relationships between the Uniramia, Crustacea, and Chelicerata (Manton, 1977; Bergström, 1978). In neontological terms, based on comparative morphology, these major groups are either subphyla of the Phylum Arthropoda (some authorities would exclude the Onychophora from this) or they are phyla.

Now these alternative propositions have certain implications for embryology. Since a phylum is, by definition, a monophyletic assemblage, it follows that the developmental constructional pathways through which the adult structure is formed in each species of the phylum are modifications of a common developmental pathway that produced the adult structure of the ancestral members of the phylum. In other words, for all species within a phylum, comparative embryology should reveal a basic, underlying constructional theme of morphological development shared among those species. A vast range of functional modifications of this theme is possible, associated with nutritional deviations (larval specializations, storage and release of yolk, viviparity, and so on) and with the development of differing adult end points, but the common ground of development of the basic structural organization characteristic of the phylum remains. The more complex the level of morphological organization of the phylum, the more recognizable is this common ground. There is no problem, for example, in identifying the shared features that characterize the developmental sequences of, say, the Chordata or the Echinodermata or the Mollusca as they progress from egg to adult. This is not to say that embryological evidence alone suffices to identify monophyly. Embryos, being morphologically simpler than adults, offer less evidence for interpretation, and the possibility of parallel or convergent evolution usually cannot be firmly eliminated. It does mean, however, that among the species of a phylum there is the expectation of a fundamental commonality in the functional sequence of morphological development from the egg. If the Arthropoda are a phylum, they should exhibit this fundamental commonality of development in the same way that other morphologically complex phyla do. If, on the other hand, the arthropods are polyphyletic and each phylum of arthropods (e.g., Uniramia, Crustacea, Chelicerata) had an independent ancestry, the consequent hypothesis is that each phylum should exhibit a different theme of development, functionally incompatible with those of the other phyla. The alternative hypotheses of monophyly versus polyphyly in arthopods can thus be tested embryologically, provided that the limitations of this type of evidence are kept firmly in mind.

One set of limitations concerns the way in which the evidence should be expressed and evaluated. This is dealt with in the next section of the present chapter. A more general limitation arises from the meaning that can be attached to any conclusions drawn. If the evidence reveals a common theme of development for all arthropods, it provides support for the concept of monophyly. If, in contrast, the evidence reveals three different themes, functionally irreconcilable, the case for polyphyly is strengthened, though not proved by this evidence alone. All that can be stated conclusively from the latter circumstance is that the extant arthropods fall into three distinct groups embryologically which have evolved independently from prearthropod ancestors. The relationship between these ancestors is not one that embryology can properly address.

It is perhaps appropriate to state at this point that my interpretation of the available evidence, presented in detail in a previous work (Anderson, 1973), leads me to the conclusion that the Uniramia, Crustacea, and the Chelicerata have fundamentally different patterns of embryonic development, such that the evidence of embryology refutes the concept of phylum Arthropoda and tends to support the recognition of three extant arthropod phyla, the Uniramia, the Crustacea, and the Chelicerata. Certainly, the monophyletic nature of each of these groups is strongly supported by embryological evidence. The origin of the Crustacea independently of the Uniramia is also strongly supported. The chelicerate embryological evidence is equivocal. They are certainly nonuniramian, and the modern chelicerates share nothing embryologically with the Crustacea; but the possibility of a remote phylogenetic connection between the Chelicerata and Crustacea cannot be ruled out embryologically in the same way that the Uniramia can be separated from the other arthropods. The crustacean-chelicerate question must be decided on other grounds, although whatever the decision, it must encompass the need for functionally viable development in the ancestors of both groups.

Recent embryological studies on arthropods, published in 1973-1977, have not included any information that bears directly on these conclusions established in 1973. Some important progress has been made in elucidating developmental patterns within phyla, e.g., Scholl (1977) and Weygoldt (1975) on Chelicerata, Kohler (1976) and Scheidegger (1976) on Crustacea, Knoll (1974) on Chilopoda; but these works have not furthered our understanding of the basic developmental themes characterizing the phyla, and have thus not provided material for renewed speculation. I should also point out, in answer to certain criticisms of my earlier work, that the embryology of the Pycnogonida and of the Tardigrada and Pentastomida remains so poorly known that comparison with the major groups of arthropods is not possible at the present time.

2.2. METHODS OF COMPARATIVE EMBRYOLOGY

The phylogenetic conclusions that can be drawn from comparative embryology depend heavily on the methodological approach adopted for the interpretation of the evidence. The facts of development are established in the same way in all cases, but the difficulty is that the application of these facts in phylogenetic studies is several times removed from the facts themselves. The method of derivation of meaningful generalizations from the facts is one whose validity underlies the acceptance or rejection of the resulting conclusions.

Comparative embryology, if it takes account of function, can proceed towards phylogenetic conclusions in the following way. By direct observation, the reconstruction of histological series of fixed stages, and the experimental analysis of origins and lineages, a verbal and graphic description can be constructed, which provides a conceptual model of the progressively elaborated structure built up during the development of a particular species. Each sequential structural process within the whole complex of development can also be interpreted in terms of the functions stemming from its formation. Basically, these are functions of two kinds: 1) functions in the furtherance of structural elaboration—e.g., the function of a mesodermal somite is, among other things, to provide a cellular configuration from which a subsequent configuration of segmental musculature can arise; 2) functions in the continuous maintenance of the developing organism as a living animal—e.g., the function of vitellophages is in making available the constituents of yolk as nutrients for the developing tissues.

These two types of function are elaborately interlocked. In many cases, the development of a particular structure within the embryo involves its constructional components in the performance of both functions. For example, the stomach of a planktotrophic cirripede nauplius larva is functional both in digestion in the larva and simultaneously as the basis for the further development of part of the adult midgut (Walley, 1969). In other cases, only one or the other type of function might be performed. The mesodermal somites of onychophoran embryos are functional only with respect to the further development of segmental, mesodermally derived structures (Manton, 1949). The placental stalk and plate of a viviparous onychophoran embryo are, conversely, functional only in relation to maintenance during early embryonic development, and do not enter into definitive structural development (Anderson and Manton, 1972).

The conceptual model of development stemming from the descriptive analysis of the development of a species thus becomes a complex functional model, involving the dimensions of spatial configuration, temporal change in that configuration, continuous developmental function, and continuous maintenance function. It is this model that provides the basis for embryological penetration into phylogeny. The next step in the process is a comparison of models. If two species are related, i.e., have a common ancestry, it follows that the models which can be separately constructed of their development must be variants on a common earlier model. This earlier model is entirely a mental construct. By the very nature of the evolutionary process, no direct observations can be made. Given that the differences between the species are assessed with due regard to viable changes in function, however, the common features which must have been present will inevitably be a part of the logical construction of the earlier model. Within this framework it is possible to take account of the fact that developmental functions and their associated structures can be deleted in evolution in favor of more direct pathways of further development (e.g., the somite stage of mesoderm development is absent in the cyclorrhaphan Diptera, Anderson, 1966a, 1972b), or that new temporary functions and associated structures can be interpolated into developmental sequences (e.g., the interpolation of an amnion in the development of the Thysanura and Pterygota, Anderson, 1972a,b, 1973).

As the facts of development are brought into consideration from more and more species, generalized functional models of development can be constructed for larger and larger groups of animals, e.g., the Pterygota, the Malacostraca. Problems that impede the establishment of a generalized functional model of development immediately raise questions about the degree of unity of the group under consideration. The Arachnida are a good example of this. Developmental models can be set up easily for the separate arachnid orders, but not for the Arachnida as a whole. In particular, the Scorpiones stand apart from the other orders in this respect to a degree that indicates a different origin (Anderson, 1973; Yoshikura, 1975). Taking the present knowledge of arthropod embryos into account, reasonably detailed functional models can be set up without too much difficulty for the development of the following categories: Onychophora, Chilopoda, Myriapoda other than Chilopoda, apterygote Hexapoda other than Thysanura, the Thysanura, the Pterygota, Branchiopoda, Copepoda, Cirripedia, Malacostraca, Xiphosura, Scorpiones, Araneae, Opiliones, Acarina, Pseudoscorpiones, and Amblypygi.

These models, then, provide the basis for the final step that must be taken in applying embryological tests to hypotheses concerning the phylogeny of the arthropods. By a comparison of the models for the development of the major groups of Crustacea, a basic functional model for all crustacean development can be erected. A corresponding basic functional model should be discernible for all uniramian development if the Uniramia are monophyletic, and another for the Chelicerata. At this level of construction, the features of the model, spatial temporal, and functional, approximate the essential constructional sequences that were present in the development of the early members of these groups. All variations and specializations of development arising later in evolution will now have been taken into account. If the resulting models are functionally compatible in terms of derivation from a common earlier model, monophyly will be upheld; if functionally incompatible, polyphyly. If the outcome of comparison is functionally equivocal, then the evidence of embryology cannot be applied to the problem.

Since it is clear that this functional approach reveals developmental processes which are fundamental to the construction of a major type of organization (e.g., crustacean, chelicerate, uniramian), the whole sequence of development from the egg onwards can be considered. This includes the configuration of the initial set of major presumptive rudiments established at the blastula stage or its equivalent, the functional reassortment of these rudiments into a three-dimensional spatial configuration during gastrulation, and the functional sequence through which each rudiment is then elaborated into its organogenetic products. In other words, the model includes process as well as structure, and is not a typological (thus nonfunctional) ideal.

I now propose to demonstrate the application of this approach in elucidating a basic model of development for the Crustacea, to follow this with similar analysis for the Uniramia and Chelicerata, and to consider the results in terms of a basic developmental model for all arthropods.

Throughout the ensuing discussion, much emphasis will be placed on a comparison of fate maps. The fate map, an interpretive extraction from the blastula or blastoderm, is a convenient tool for the epitomization of a whole developmental sequence. It distinguishes the basic components of an embryo. It sets out their relative juxtapositions. It foreshadows a particular set of consequent relative movements, and it implies a predictable sequence of organogeny. Thus a comparison of fate maps is a comparison of total developmental sequences. The evidence shows that if animals are related, and their developments are variants of a common ancestral pattern of development, their fate maps will be similar in configuration, even though the cellular composition of corresponding areas may be different. The obverse of this argument is that if fate maps differ fundamentally between groups, the groups are unrelated. With some reservations due to lack of evidence, arthropod relationships can be approached through the fate maps of arthropod embryos (for a different viewpoint, see Weygoldt, Chapter 3).

2.3. CRUSTACEAN EMBRYOLOGY

The crustaceans are the most useful group of arthropods with which to exemplify the principles of comparative functional embryology, because of the undisputed unitary nature of the Crustacea as a group, the wide range of types of development displayed among extant species, and the deep level of descriptive analysis to which they have been subjected. We have a detailed morphological knowledge of cleavage, gastrulation, and organogeny in many species of crustaceans. These include species with small eggs and a primary total cleavage, as well as species with larger eggs, centrolecithal cleavage, and the blastoderm formation so common among uniramians and chelicerates. Modern crustaceans are thus ideally suited for developmental model building, though their neglect in causal and molecular embryology suggests that they are less suitable as embryos for experimentation. I shall first attempt to elucidate a basic model of crutacean development, then utilize this model to illustrate the way in which a fate map of the blastula epitomizes the whole developmental theme of the group, and finally indicate how conclusions can thus be drawn about crustacean phylogenetic relationships.

2.3.1. Basic Crustacean Development

The elucidation of a basic model of crustacean development depends on the functional evaluation of class models developed mainly for the maxillopods (especially cirripedes and copepods), malacostracans, and, to a lesser extent, branchiopods. While it is clear that other classes, the Ostracoda and Cephalocarida, fit the model in general terms, much more work needs to be done on their comparative embryology before we are able to see how they might influence the construction of the current model. The key groups at the present time are, in fact, the Cirripedia and the Malacostraca. The thoracican cirripedes, although highly specialized in later development (Walley, 1969; Anderson, 1973), retain the clearest example of modified spiral cleavage and development of the basic nauplius larva of all classes of Crustacea (Anderson, 1969). The Malacostraca, with their wide range of developmental types, are much less informative about small eggs, total cleavage, and the development of a small, free-swimming nauplius (euphausiid and penaeid embryology still need detailed attention), but have provided a mass of information on cleavage, gastrulation, and organogeny in larger eggs developing through a blastoderm (summarized by Anderson, 1973; see also Dohle, 1970, 1972, 1967a,b; Fioroni and Bandaret, 1971; Stromberg, 1972; Lang, 1973; Pace et al., 1976; Scheidegger, 1976).

2.3.1.1 Cleavage

As mentioned above, the small eggs of many thoracican cirripedes undergo a modified form of spiral cleavage. The major modification (Fig. 2.1) is that the yolk of the egg is segregated, from the first cleavage division on, into a single blastomere (D, 1D, 2D, 3D, then 4D). In spite of this, as shown by Anderson (1969), every cell of the blastula can be identified and designated in accordance with the system of spiral cleavage notation routinely used for spiral cleavage in the Platyhelminthes, Nemertea, Sipuncula, Echiura, Mollusca, and Annelida. Costello and Henley (1976) have recently questioned the validity of this interpretation of cirripede cleavage, maintaining that it is best interpreted as a modification of spiral cleavage in which only the D quadrant is retained. Terminological disputes of this nature are of little import, provided that the recognition of cirripede cleavage as some form of modified spiral cleavage is not in question.

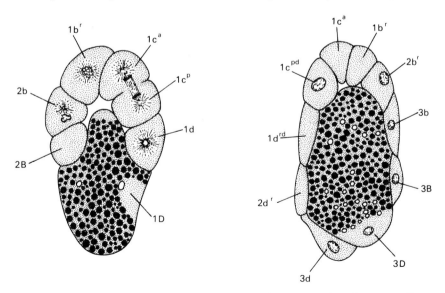

Fig. 2.1. Sagittal sections through cleavage stages of cirripede *Tetraclita rosea.* From Anderson (1973), after Anderson (1969).

The significant comparisons apply later in development, at the blastula stage. A more serious aspect of Costello and Henley's remarks, however, concerns their attempt to resurrect the interpretations of Bigelow (1902) on the development of thoracican cirripedes from the blastula stage on. Bigelow's account is demonstrably inaccurate and inadequate (Anderson, 1969), and should not be used to make guesses about the course of cirripede development when more properly validated evidence is already available.

As pointed out in previous works (Anderson, 1969, 1973), once the pattern of cirripede cleavage became clear, it was possible to look anew at cleavage in copepods (Fuchs, 1914) and cladoceran branchiopods (Kühn, 1913; Baldass, 1937) and to see in them a secondarily radial modification of an ancestral spiral cleavage of the cirripede type. The recent investigation by Kohler (1976) of the embryonic development of a parasitic copepod has again indicated a pattern of cleavage following the cirripede type, with yolk segregated into a single blastomere, though Kohler does not provide sufficient details of the process to allow a complete interpretation. The traces of modified spiral cleavage retained in certain syncarid and eucarid malacostracans (Hickman, 1937; Taube, 1909, 1915; Nair, 1949; Shiino, 1950; Fioroni, 1970; Lang and Fioroni, 1971) also became easier to interpret and understand. Finally, it became clear that intralecithal cleavage followed by direct blastoderm formation had evolved independently in several groups of crustaceans (e.g., branchiopods, Baldass, 1941; ostracods, Weygoldt, 1960; leptostracan Malacostraca, Manton, 1934; hoplocaridan, syncaridan, peracaridan, and eucaridan Malacostraca, Shiino, 1942; Manton, 1928; Nair, 1939; Scholl, 1963; Stromberg, 1965, 1967, 1971; Dohle, 1972).

Total, modified spiral cleavage leading to the formation of a blastula, and intralecithal cleavage leading to the formation of a blastoderm, are thus primitive and derived conditions, respectively, in Crustacea. In either the blastula or the blastoderm, it is possible to identify by descriptive means the developmental fates of all of the cells comprising the blastula or blastoderm. From this information, the configuration of the fate maps of this stage of development can be discerned (Fig. 2.2). Two important conclusions immediately arise (Anderson, 1973). First, the configuration of the fate map is stable and shared among all species. Second, the cellular composition of each identifiable presumptive area differs among species in the number, dimensions, and other characteristics of its component cells and in the way these cells are formed during cleavage. For example, the presumptive mesoderm of the cirripede blastula consists of three largish cells formed as a result of total, modified spiral cleavage, while that of the caridean shrimp *Caridina* consists of a patch of many small cells formed as part of a blastoderm. Associated with this difference, the activities performed by these cells during subsequent gastrulation and organogeny differ between the two types, but the developmental function of giving rise to the naupliar and postnaupliar mesoderm of the embryo is the same for both.

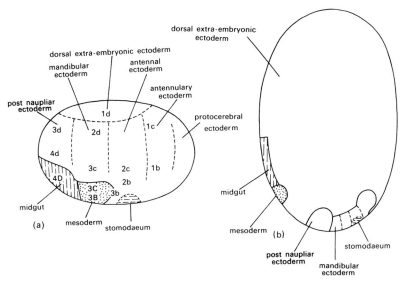

Fig. 2.2.   Fate maps.   a, Blastula of cirripedes; b, blastoderm of decapod *Caridina laevis*. From Anderson (1973).

It is this kind of analysis that makes it possible to identify a basic fate map of the blastula for the Crustacea. Adding to this the lineage information that we have about crustacean spiral cleavage, the lineage of all the cells that make up the presumptive areas of the basic crustacean blastula can also be identified with a high degree of accuracy. A pattern of cleavage, blastula formation, and presumptive area segregation, of which all extant species display variations, can thus be recognized as the first sequence of events in our model of basic crustacean development. One should not be deterred by the fact that cleavage is a highly variable process in certain ways. The important constructional aspects of the process that go to make up the beginnings of a crustacean have a high degree of stability, just as they do in other major groups of bilateral Eumetazoa. Admittedly, the pattern can break down in some highly specialized circumstances (e.g., in ectolecithal platyhelminths and viviparous placental salps), but usually it does not.

Without going into the details of the evidence (see Anderson, 1973), I will now reiterate the model of basic crustacean cleavage and the resulting blastula put forward in that work. A small but yolky egg with a diameter of 100–200 μm is envisaged. Cleavage follows a spiral pattern, but modified in such a way that the yolk of the egg is segregated into a single posteroventral cell, 4D, partially covered at the blastula stage by the remaining yolk-free blastomeres. A small blastocoel, partly occluded by the inward penetration of the yolky 4D cell, can also be envisaged. Regarding the blastula as a single layer of cells around the blastocoel, the part of the blastula wall occupied by the exposed part of the 4D cell is posteroventral in position. The yolky 4D cell constitutes the presumptive midgut. The yolk-free cells that lie ventrally in front of the exposed part of the 4D cell are, in lineage, the other three stem cells of the modified spiral cleavage, 3A, 3B, and 3C, in which the sixth cleavage division has not taken place. These cells are the presumptive mesoderm of the blastula.

In front of the presumptive mesoderm, midventral cells formed by divisions of 2b and 3b make up the constituents of a small area of presumptive stomodaeum. The remainder of the blastula wall, laterally, anteriorly, posteriorly, and dorsally, comprises presumptive ectoderm. Within this presumptive ectoderm are subzones of characteristically crustacean pattern: an anterior protocerebral area of first quartette origin; paired lateral antennulary areas also of first quartette origin; paired lateral antennal areas of second quartette origin; paired lateral mandibular areas of second and third quartette origin; a posterior area of presumptive postnaupliar ectoderm, mainly of third quartette origin but also including 4d; and a dorsal area of presumptive extraembryonic (subsequently naupliar carapace) ectoderm mainly 1d in origin. All extant crustacean embryos during their cleavage and in their blastula stage or its equivalent can be interpreted as functional variants of this model. The main point that I wish to emphasize at this stage is the degree to which a crustacean is a crustacean even at the blastula stage of its development. No other arthropod at this stage has this particular combination of features.

2.3.1.2. Gastrulation

Does the same crustacean singularity manifest itself in the way in which the presumptive areas express their further development during gastrulation? Again, as in cleavage, there are many variants in gastrulation among extant species, but all of them lead one's thinking to a consistent basic model of how gastrulation occurred in ancestral Crustacea. The components of this phase of the model (Fig. 2.3a) follow directly from the functional constraints already established in the blastula. The presumptive midgut, already partially internal, moves fully into the interior of the embryo and divides into two or more cells occupying most of the internal space. The three presumptive mesoderm cells migrate inwards to lie beneath the presumptive midgut cell, then migrate in a posterior direction along the ventral midline, dividing as they go, to come to rest as a group of mesoderm cells behind the midgut cells at the posterior end of the embryo. The presumptive stomodaeum undergoes an independent invagination in the ventral midline. The presumptive ectoderm, mainly established in its definitive position as a result of cleavage, spreads to cover the posteroventral and ventral areas vacated at the surface by the immigrating midgut and mesoderm cells, and undergoes a compensatory attenuation of its dorsal component of extraembryonic ectoderm. In relation to the subzones of presumptive ectoderm, the stomodaeum (the

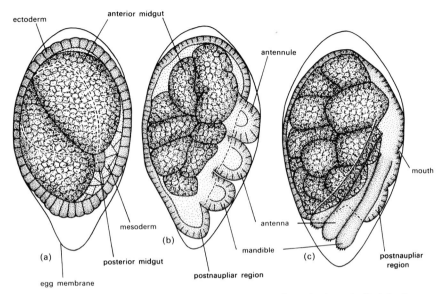

Fig. 2.3. Development of cirripede *Tetraclita rosea.* a, Gastrula; b, early limb bud stage; c, later limb bud stage. From Anderson (1973), after Anderson (1969).

invagination of which may be delayed until after other gastrulation movements are completed) lies midventrally between the two halves of the antennal ectoderm, while the mesoderm comes to rest within the posterior cap of postnaupliar ectoderm. The gastrulation movements in our crustacean embryo model thus result in a clearly defined gastrula structure. While this structure is highly modified in extant species with blastodermal development, it is never entirely lost.

2.3.1.3. Organogeny Before Hatching

The first outcome of organogeny in a crustacean is the naupliar organization (Figs. 2.3b,c), with provision for subsequent development of the postnaupliar region. Even though the naupliar organization may be fully embryonized, as in many large-egged species, its precocity and recognizability in embryonic development always endure. Basically, the nauplius organization becomes functional as the first hatched stage of development (Fig. 2.4), and the postnaupliar region develops anamorphically after hatching. This pattern of organogeny must therefore constitute the next phase of our basic crustacean developmental model. Each major component of organogeny can be considered in turn.

The single, yolky midgut cell of the gastrula undergoes several division cycles, whose products become arranged as an epithelium around a newly formed mid-

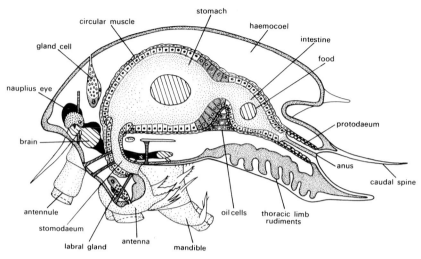

Fig. 2.4. Nauplius larva of cirripede *Balanus balanoides*. From Anderson (1973), after Walley (1969).

gut cavity. This midgut epithelium becomes functionally differentiated as the midgut of the nauplius larva by the time hatching occurs. The invaginated stomodaeum becomes connected anteroventrally with the midgut sac. A protodaeum also invaginates at the posterior midpoint of the postnaupliar ectoderm and becomes connected with the posterior end of the midgut, thus completing the functional gut of the nauplius larva. The ingrowing proctodaeum pushes through the posterior mass of mesoderm before making this connection. During the further growth in length of the animal that results from the post-hatching growth and development of the postnaupliar region, the naupliar gut elongates and gives rise to the definitive gut.

The mass of mesoderm cells at the posterior end of the gastrula proliferates a pair of lateral mesodermal bands. These grow forward on either side of the developing midgut as far as the protocerebral ectoderm. The mesodermal bands then become subdivided into preantennulary mesoderm anteriorly, three pairs of naupliar somites laterally, and a residual mass of postnaupliar mesoderm at the posterior end.

The preantennulary mesoderm gives rise to the labral and other musculature at the anterior end of the nauplius, and to the stomodaeal musculature. A pair of vestigial preantennulary somites may be components of the preantennulary mesoderm.

The naupliar somites give rise to the complex muscles of the three pairs of naupliar limbs and perhaps contribute a layer of splanchnic mesoderm to the external surfaces of the midgut. A pair of antennal glands (coelomoduct segmental organs) is formed from some of the cells of the antennal somites. These are the only signs of coelom formation within the naupliar somites.

The residual postnaupliar mesoderm develops into four main components, the maxillulary somites, the maxillary somites, a bilateral ring of mesoteloblasts behind the maxillary somites, and the mesoderm of the telson and proctodaeum behind the mesoteloblasts. The mesoteloblasts may bud off the somite rudiments of one or more postmaxillary segments before hatching occurs, but none of the postnaupliar segments is functional in the newly hatched nauplius.

The ectoderm of the gastrula continues its development in a relatively direct manner, giving rise mainly to epidermis and neural tissue. The protocerebral ectoderm proliferates the protocerebral ganglia and nauplius eye. The naupliar segmental ectoderm pouches out ventrolaterally as a pair of limb buds in each segment, the antennules being uniramous and preoral, the antennae biramous and paraoral, the mandibles biramous and postoral. These three pairs of limbs, with their associated musculature, become functionally differentiated as the locomotory and feeding apparatus of the nauplius larva. The ventral ectoderm of the naupliar segments proliferates segmental ganglia, antennulary in a preoral position, antennal paraoral, mandibular postoral and conjoined. Behind the mandibular segment, the postnaupliar ectoderm develops into the four components which correspond to those of the postnaupliar mesoderm. A ring of maxillulary ectoderm is followed by a ring of maxillary ectoderm, then a ring of ectoteloblasts, and finally a posterior cap of telson ectoderm. The ectoderm of one or more postmaxillary segments may be budded from the ectoteloblasts before hatching takes place, but none of the postnaupliar segments has functional limbs in the newly hatched nauplius larva.

2.3.1.4. Development After Hatching

From what we know of the comparative morphology and development of crustaceans hatching as a nauplius larva, there is little need to build into our model the suggestion of any fundamental changes of integrated structure of the naupliar region after hatching, other than the development of the paired compound eyes as outgrowths of the protocerebrum. Various functional modifications of existing structures take place, usually associated with a nauplius-postnauplius metamorphosis, but little is added or lost.

The development of the postnaupliar region, in contrast, has its organogenetic phase entirely after hatching. It proceeds anamorphically as a primary condition (Fig. 2.5), and the model of generalized post-hatching crustacean development must be built around this fact. Comparative considerations, based mainly on events in the Cephalocarida and Branchiopoda (Sanders, 1963; Anderson, 1967, 1973), indicate that metamorphosis basically coincides with the completion of functional development of six pairs of trunk limbs. The first pair of postnaupliar

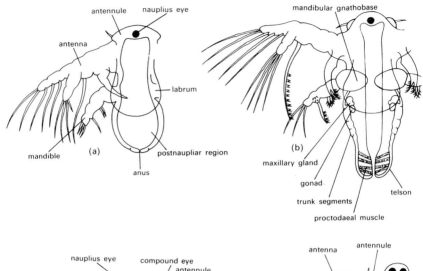

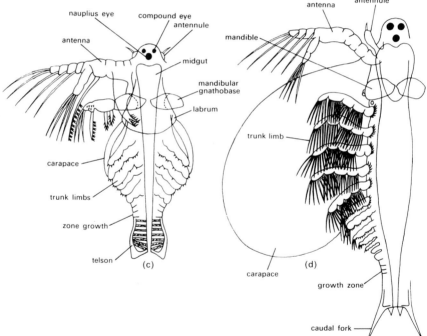

Fig. 2.5.   Larval development and metamorphosis of branchiopod *Limnadia stanleyana*. a, stage 1; b, stage 3; c, stage 5; d, stage 6. From Anderson (1973), after Anderson (1967).

limbs, the maxillules, is cephalized before metamorphosis takes place, indicating that the maxillulary segment is part of a generalized, dignathan crustacean head. The next pair of limbs, the maxillae, is basically (hence in our model) the first pair of functional trunk limbs. A sequence of further segments with similar, functional trunk limbs gradually intervenes between the maxillary segment and the telson, but there is no evidence to suggest that the model should include more than about 20 trunk segments. The genital openings develop ventrally on segments in the middle region of the trunk.

In terms of the constructional processes through which the maxillulary, maxillary, and trunk segments have their origin and attain their functional condition, the Crustacea show a characteristic basic pattern. Each segment rudiment is proliferated in succession from the teloblastic growth zone in front of the telson and is thus built around the elongating midgut (Fig. 2.6). The mesoteloblasts proliferate bilateral elongations of the mesodermal bands, which divide in anteroposterior sequence into bilateral arcs of cells, the somites, lying against the ectoderm. The ectoteloblasts proliferate a generalized cylindrical ectoderm, which

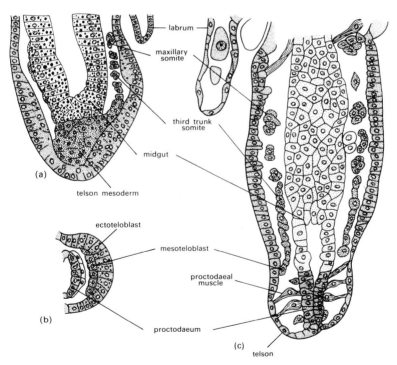

Fig. 2.6. Development of postnaupliar region of branchiopod *Limnadia stanleyana.* a and b, Stage 1; c, stage 3. From Anderson (1973), after Anderson (1967).

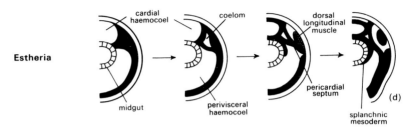

Fig. 2.7. Development of postnaupliar somite mesoderm of branchiopod *Estheria.* From Anderson (1973), after Weygoldt (1958).

becomes segmentally delineated in association with the paired somites. When first formed, the somites are separated from the midgut by a wide haemocoel.

Each postnaupliar segment then develops directly to the juvenile adult condition. The segmental ectoderm pouches out ventrolaterally as limb buds and gives rise midventrally to the paired ganglia of the segment. Invaginations of ectoderm form endophragmal skeletal bars, as well as the short exit ducts of the maxillary segmental organs and gonoducts.

The paired somites of the segment (Fig. 2.7) first produce a pair of median, dorsolaterally placed extensions, which pass medially across the haemocoel to meet the midgut. These extensions form the pericardial floor. A pair of vestigial coelomic cavities develops in each segmental portion of pericardial floor. Above the level of the pericardial floor, the median somite tissue of each side comes together above the gut to form heart wall. The dorsolateral part of each somite separates off and develops as dorsal longitudinal muscle. The ventrolateral part of each somite develops as limb musculature and ventral longitudinal muscle. The splanchnic mesoderm at each segmental level is formed by a downward growth of mesoderm over the midgut from the median edge of the pericardial floor.

In general, the information gleaned from studies of postnaupliar somite development in extant Crustacea allows a detailed and characteristic pattern of somite development to be included in the crustacean developmental model. It is even possible to add further details on certain special features. The maxillary somites give rise to a pair of segmental organs, the maxillary glands, with cavities derived from the vestigial coeloms of the somites. Perhaps in ancestral crustaceans other segments produced similar segmental organs in the same way. Gonad development is also associated with the segmental vestigial coeloms, but in a distinctive manner. One (in dioecious species) or two (in hermaphrodite species) pairs of coelomoduct gonoducts grow out from somites in the middle region of the body in the same manner as the coelomoducts of the maxillary glands. Primordial germ cells also temporarily occupy the ventral walls of the segmental coeloms of at least some of the anterior pairs of somites, but these coeloms do not develop directly as gonocoel. The germ cells proliferate and the groups of each side join up to form paired germ-cell strands. These strands, enclosed by mesoderm cells, then move down to become depended as solid strands from the ventral surface of the pericardial floor. The vestigial segmental coeloms are thus excluded from the gonads and subsequently disappear. Each solid gonad strand finally hollows out, developing a new gonocoel, and becomes secondarily connected up at its posterior end to the corresponding gonoduct. Only the gonoduct lumen is a direct extension of previous coelom.

Another special feature of basic crustacean development revealed by comparative evidence is the specialization of somite development in the three naupliar segments. These somites retain only certain aspects of the generalized pattern of crustacean somite development. They form limb musculature, perhaps splanchnic mesoderm, and, in the case of the antennal somites, coelomoduct segmental organs. They do not, however, give rise to dorsal or ventral longitudinal muscle components, heart, pericardial floor, or gonads.

2.3.2. Crustacean Embryology and Phylogeny

Comparative morphological studies thus permit a detailed model of basic crustacean development, encompassing the pattern of morphological change that leads from the egg to the juvenile adult. The model is a powerful one, derived from a large amount of detailed, well-substantiated evidence. The model is also represented in epitome by the basic fate map of the crustacean blastula (Fig. 2.2), in that the entire course of development can be described by tracing the fates of the components identified in the fate map and relating the components and their development one to another in space and time.

This being so, a comparison of fate maps between phyla, assuming that these fate maps have been established using similar criteria, becomes a comparison of functional developmental models for the phyla being compared. In the case of the Crustacea, the crucial comparison is with the Annelida, the only extant form of segmented, coelomate, spirally cleaving worms. Once this question is resolved, then the other major arthropod groups, none of which retains spiral cleavage, can be examined in relation to both the Annelida and the Crustacea, for possible evidence of phylogenetic relationships. I shall return to this comparison in later sections, carrying the conclusion that the Annelida and Crustacea are unrelated except insofar as both are components of a larger spiral cleavage assemblage embracing many phyla (Anderson, 1973). In fact, the embryological evidence indicates that the Crustacea are more remote from the annelids than the Annelida are from the Mollusca, a pair of phyla which share much of their early development in common.

This conclusion derives from a very simple fact. The Annelida, whose basic model of development is the polychaete model (Anderson, 1959, 1966b, c, 1973; Schroeder and Hermans, 1975), have a configuration of presumptive areas

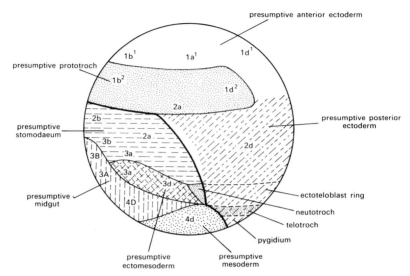

presumptive anterior ectoderm

presumptive prototroch

presumptive stomodaeum

presumptive midgut

presumptive posterior ectoderm

ectoteloblast ring

neutotroch

telotroch

pygidium

presumptive ectomesoderm

presumptive mesoderm

Fig. 2.8.   Fate map of blastula of polychaete *Scoloplos armiger.* From Anderson (1973).

which include 4d mesoderm behind a midgut component formed by 3A-3C and 4D (Fig. 2.8). This they share with molluscs (Cather, 1971), sipunculids (Rice, 1975a, b), echiuroids (Gould-Somero, 1975), and, in general terms, with platyhelminths (Henley, 1974; Anderson, 1977) and nemerteans (Riser, 1974). The origin of the mesoderm in the two latter groups is not fully understood. Although 4d is a part of the presumptive mesoderm, other cells of the A and C quadrants are also implicated, tending to raise the possibility of an ancestral spiral cleavage condition in which presumptive mesoderm was segregated as a ring around presumptive midgut. The Crustacea, while deriving from a spiral cleavage ancestry, have evolved an opposite segregation pattern, in which 3A-3C mesoderm lies in front of 4D midgut, and 4d is merely part of the presumptive ectoderm. Taking into account the fact that these segregations are associated with totally different patterns of subsequent development of the respective rudiments, it is clear that the Crustacea are only remotely related, through spiral cleavage ancestors, to all those phyla which share a 4d origin of mesoderm. The logical consequence of this conclusion is that the Crustacea were preceded by ancestors which evolved the coelom and metamerism independently of annelids and then evolved arthropodization (see also Weygoldt, Chapter 3). If these same ancestors were the progenitors of other arthropod groups, the crustacean model of embryonic development should show evidence of some retention in these groups. I shall now examine uniramian embryonic development from this point of view.

2.4 UNIRAMIAN EMBRYOLOGY

In their distinction between subphyla, the Uniramia as defined by Manton (1972) are notably more diverse than the Crustacea. Functional morphological studies, however, show that the onychophorans, myriapods, and hexapods are indeed subgroups of a monophyletic phylum (Manton, 1972, 1977). Comparative embryology also points to the same unity (Anderson, 1973). I will therefore address three questions in the present section. First, what is the basic model of development of which all uniramian modes of development are variants? Second, what does this basic model reveal about the relationship between the Uniramia and the Annelida? Third, what conclusions can then be drawn about the relationship between the Uniramia and the Crustacea?

The first of these questions requires an initial comparative survey of the embryos of onychophorans, myriapods, and hexapods. The analysis is more difficult than for the Crustacea, because the embryological differences between the subphyla of Uniramia are greater than between classes of Crustacea, and because all uniramian fate maps are blastodermal and somewhat ill-defined.

2.4.1. Onychophoran Embryonic Development

The Onychophora are well known as exemplars of yolkless viviparity (e.g., Kennel, 1884, 1885; Manton, 1949; Anderson and Manton, 1972), and the best studies on their development have been carried out on these forms. At the same time, some species retain yolky development (e.g., *Peripatoides novaezealandiae*, *Peripatus leuckarti*; Sheldon, 1888, 1889a, b; Anderson, 1966c), and enough is known of this to allow the reconstruction of a basic pattern of onychophoran development of which the various types of yolkless development are variants (Anderson, 1973). From this information, a basic model of onychophoran development can be established in the following manner.

2.4.1.1. Cleavage

The basic type of egg in Onychophora can be envisaged as relatively large, up to 2 mm in length, with uniform, dense yolk. Its cleavage is centrolecithal, with cleavage energids spreading through the yolk and giving rise to a uniform blastoderm of small cells around the yolk mass (Fig. 2.9). No trace of spiral cleavage can be discerned in any onychophoran. Knowledge of the subsequent developmental functions and fates of the blastoderm cells permits the blastoderm to be zoned as a basic configuration of presumptive areas, as shown in Fig. 2.10. A long band of presumptive midgut cells occupies the midventral surface and is functionally divisible into two parts, a long anterior midgut and a compact posterior midgut. At the anterior and posterior ends of the presumptive anterior

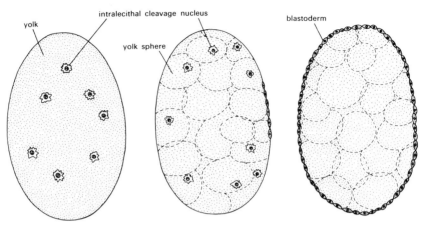

Fig. 2.9.   Cleavage and blastoderm formation in a yolky onychophoran egg. From Anderson (1973).

midgut are arcs of cells forming, respectively, the presumptive stomodaeum and presumptive proctodaeum.   Behind the presumptive posterior midgut, in a posteroventral position, is a small group of presumptive mesoderm cells. The remainder of the blastoderm constitutes presumptive ectoderm, subdivisible into three components.   Ventrolaterally, on either side of the midventral areas, is a narrow band of presumptive ventral extraembryonic ectoderm.   Above this on either side is a broader band of presumptive embryonic ectoderm.   Dorsally to laterally, the remainder of the blastoderm constitutes presumptive dorsal extra-

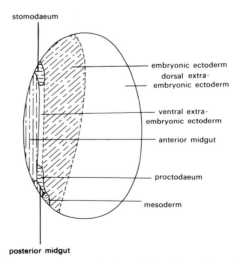

Fig. 2.10.   Fate map of a yolky onychophoran blastoderm. From Anderson (1973).

embryonic ectoderm. Whatever else might be said about the blastodermal fate map basic to Onychophora, it is certainly different from the basic fate map of the blastula of Crustacea (Fig. 2.2). In particular, it is worth noting that the presumptive mesoderm of Onychophora lies behind the presumptive midgut, in the manner of spiral cleavage fate maps with 4d mesoderm (Fig. 2.8).

2.4.1.2. Gastrulation

As is commonly the case with blastodermal arthropod development, gastrulation movements following the end of onychophoran cleavage are largely replaced by migratory invasion of the interior of the embryo by cells proliferated at the surface (Fig. 2.11). There is good evidence, however, that the basic model of yolky onychophoran development includes a unique gastrulation movement involving the presumptive anterior midgut, stomodaeum, and proctodaeum. Before any proliferative activity of these rudiments begins, the presumptive anterior midgut splits along the ventral midline, exposing the yolk, and invaginates slightly as paired cell bands. At the anterior and posterior ends of the invagination, the presumptive stomodaeal and proctodaeal arcs also turn into the interior. Later in development, the ventral split closes again, with separation of the anterior midgut wholly into the interior and coverage of the ventral surface of the embryo by ventral extraembryonic ectoderm. As this closure takes place, the

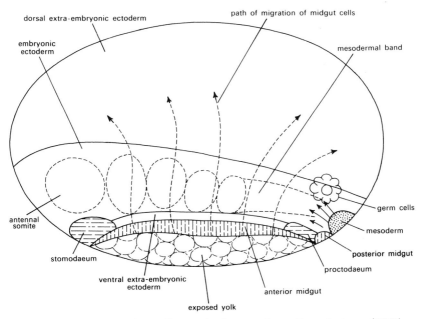

Fig. 2.11.   Gastrulation in a yolky onychophoran embryo. From Anderson (1973).

stomodaeal and proctodaeal arcs are transformed into tubes, opening by the mouth and anus respectively.

2.4.1.3. Organogeny

Development in the Onychophora is direct, with hatching or birth of a well-formed juvenile taking place. The basic model of onychophoran development therefore includes direct organogeny in the embryo, there being no evidence of vestigial retention of any form of ancestral larval rudiments.

Following the partial invagination of the paired bands of anterior midgut cells, these bands now proliferate cells that migrate over the yolk surface (Fig. 2.11), taking up part of the yolk material, and give rise to a diffuse vitellophage epithelium enclosing the yolk mass. When the ventral split closes, the vitellophage epithelium becomes continuous. After completion of its vitellophage function, this epithelium transforms directly into the definitive epithelium of the anterior part of the midgut. The posterior midgut rudiment, which remains in a superficial position behind the anus, proliferates cells that form a tubular continuation of the midgut in the growing trunk. The tubular stomodaeum and proctodaeum effect direct connections with the midgut and develop as foregut and hindgut, respectively.

The presumptive mesoderm also remains superficial. In its location at the posterior end of the embryo, it proliferates two streams of mesoderm cells (Fig. 2.11), which pass forward as paired lateral bands on either side of the midgut and eventually meet in the anterior midline. During their forward growth, the paired mesodermal bands begin to segment into an anteroposterior sequence of paired somites. Each somite is an epithelial mesodermal wall enclosing a large coelomic cavity. The first and largest pair of somites penetrates into a preoral position as antennal somites. The second pair, paraoral in position, constitutes the somites of the jaw segment; the third pair, postoral in position, those of the oral papilla segment. All somite pairs behind this are components of trunk segments. The trunk somites are all formed before hatching or birth takes place.

Further development of the somites in the Onychophora follows a basically different course from that in Crustacea. Again looking toward a basic model, each trunk somite enlarges and becomes partially subdivided into three compartments—dorsolateral, medioventral, and appendicular (Fig. 2.12). The appendicular compartment pushes out as a lobe and then becomes mesenchymatous, giving rise to the somatic musculature of the body wall and limb. The dorsolateral compartment grows up around the midgut toward the dorsal midline. Its somatic wall gives rise to heart and pericardial floor. Its splanchnic wall merges into the pericardial floor and, below the heart, spreads down around the midgut as splanchnic musculature. The medioventral compartment develops a coelomoduct

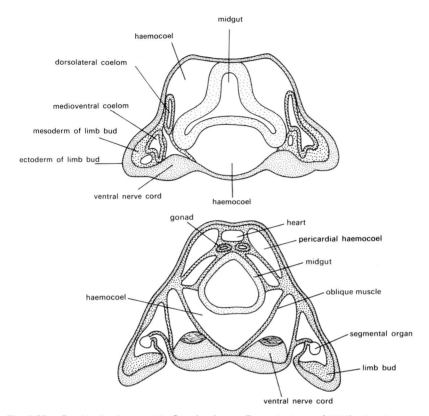

Fig. 2.12. Somite development in Onychophora. From Anderson (1973), after Sedgwick (1888).

and becomes the end sac of a segmental organ. The only other persistent coelomic spaces are those of the gonads, persistent parts of the coeloms of the dorsolateral compartments, forming the cavities of gonadal tubes depended from the pericardial floor. The gonoducts are formed by the coelomoducts of the preanal segment, and the genital opening is posterior. Thus, in addition to incorporating direct development, our model of basic onychophoran development includes opisthogoneate gonopore formation.

Unlike the crustacean model, the onychophoran model of development includes little modification of the three anterior somite pairs. In general, they develop in the same manner as those serially following behind, with only their segmental organs being deviant. The antennal segment has only transient coelomoducts, the jaw segment has none, and the slime papilla segment has its segmental organs modified as salivary glands.

Development of the ectoderm in Onychophora differs from that of the other major organ systems in showing temporary modifications in association with yolk enclosure. As I have already indicated, much of the blastoderm consists of presumptive extraembryonic ectoderm (Fig. 2.10). A major area of this ectoderm covers the large yolk mass dorsally, while paired narrow bands lie ventrolaterally. With closure of the ventral opening following gastrulation, this paired component spreads to cover the yolk mass ventrally. The embryonic ectoderm develops as two dense, lateral bands external to the mesodermal bands. As the anteroposterior succession of mesodermal somites develops from the mesodermal bands, the overlying ectodermal bands show corresponding segment delineation. Each segmental unit then gives rise to nerve cord ventrally and uniramous limb buds ventrolaterally (Fig. 2.12). Only later does the embryonic ectoderm spread towards the midline dorsally and ventrally, incorporating and replacing the extraembryonic ectoderm and completing the definitive body wall. Of the three anterior segments, the only important features of which the model must take cognizance are the enlargement of the antennal portions of the nerve cords as cerebral ganglia and the lack of good evidence of any presegmental ganglionic components in these ganglia.

2.4.2. Myriapod and Hexapod Development

If the preceding section sets up an acceptable basic model of onychophoran development, and on present evidence this seems to be the case, the recognition of the phylum Uniramia carries with it the implication that basic models of myriapod development and hexapod development must be compatible structurally and functionally with the onychophoran model. We can now consider this question, beginning with the Myriapoda. The comparative features of myriapod embryonic development were fully discussed by Anderson (1973). While some very interesting information has been adduced since then by Dohle (1974) on diplopods and Knoll (1974) on scutigeromorph chilopods, the general picture of myriapod development remains unchanged.

2.4.2.1. Chilopoda

As shown by Anderson (1973), based on the evidence of Heymons (1901), the embryonic development of scolopendromorph chilopods follows a pattern remarkably similar to that of yolky onychophoran embryos. Some further adaptations to yolk are observed, but in general the model of yolky onychophoran development outlined above is in all respects a model from which scolopendromorph development can be functionally derived. A comparison of fate maps illustrates this point (Fig. 2.13 compared to Fig. 2.10), revealing a similarity of modes of development as convincing as that which exists between different crustaceans, and is supported by considerations of the manner in which

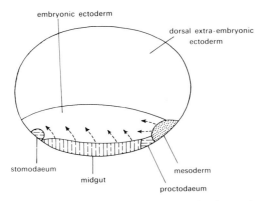

Fig. 2.13. Fate map of blastoderm of scolopendromorph Chilopoda. From Anderson (1973).

the further development of the presumptive areas of the blastoderm is expressed in the scolopendromorph embryo.

The presumptive midgut in chilopods is broadly spread along the ventral blastoderm, as in Onychophora, although it does not display the gastrulation split of the presumptive midgut of the latter group, or the distinct compact area of posterior presumptive midgut. The absence of a posterior midgut component is functionally associated with a different mode of encompassing the yolk mass during development in chilopods as compared with onychophorans. Instead of enclosure of the yolk within the anterior part of the developing embryo and addition of the more posterior part in subsequent development behind the yolk mass, the chilopod germ band flexes to embrace the yolk mass within all segments of the trunk (Fig. 2.14). The addition of further gut behind the yolk mass is redundant in this mode of development.

In the formation of the midgut epithelium in chilopods, numerous cells are proliferated from the presumptive midgut, migrate inward, and spread around the yolk mass to form the midgut epithelium. This is similar to the mode of anterior midgut development in yolky Onychophora, except that the temporary vitellophage function of this epithelium in Onychophora is not expressed in chilopods. Vitellophages arise in other ways, mainly as cleavage energids that remain in the yolk after cleavage (see Anderson, 1973) and do not subsequently participate in the formation of the midgut epithelium. This specialization has evolved independently in many groups of arthropods.

Associated with the mode of development of the chilopod midgut, the ventral surface of the embryo remains covered by a cell layer after the midgut cells have become internal. This cell layer, as in Onychophora, is ventral extraembryonic ectoderm (Fig. 2.14). In the absence of the gastrulation split, the presumptive stomodaeum and presumptive proctodaeum of chilopods invaginate independently to form the foregut and hindgut, respectively.

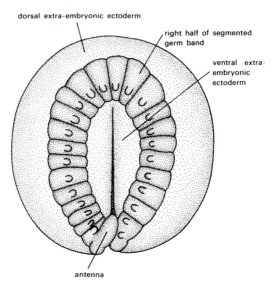

dorsal extra-embryonic ectoderm

right half of segmented germ band

ventral extra-embryonic ectoderm

antenna

Fig. 2.14. Segmental germ band of *Scolopendra* after flexure. From Anderson (1973), after Heymons (1901).

The development of the gut in chilopods, therefore, is a functionally derivable modification of that basic to Onychophora. The same conclusion can be drawn for the mesoderm. A small posteroventral area of presumptive mesoderm lies behind the presumptive proctodaeum and proliferates paired mesodermal bands, which grow forward on either side of the yolk mass as far as the stomodaeum. During their growth, the mesodermal bands subdivide in anteroposterior sequence into paired, hollow somites. Of these somites, the first two pairs, preantennal and antennal, migrate to a preoral position, while the third or premandibular pair become paraoral. The only significant modification in mesodermal band formation in chilopods as compared with Onychophora is that the bands are augmented by cells immigrating inwards from the blastoderm overlying their paths of growth (presumptive embryonic ectoderm, Fig. 2.13) and supplemented by ventral median mesoderm formed by cells immigrating inward from the ventral extraembryonic ectodermal area. Thus, presumptive mesoderm in chilopods is mainly localized posteroventrally, but is in part scattered diffusely along the length of the developing germ band. This is a minor difference, commensurate with further adaptation to yolk, when compared with Onychophora, but an important difference when it comes to interpreting presumptive mesoderm in other Uniramia (see below).

The further development of the somites in chilopods also retains much of the onychophoran pattern. The trunk somites become partially subdivided into dorsolateral, medioventral, and appendicular compartments (Fig. 2.15). The

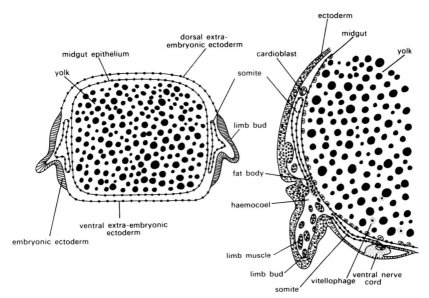

Fig. 2.15. Somite development in Chilopoda. From Anderson (1973), after Heymons (1901).

appendicular compartment gives rise to limb musculature and to fat body, a tissue not present in Onychophora. The dorsolateral compartment extends toward the dorsal midline and gives rise to dorsal longitudinal muscle, heart, pericardial floor, gonad, and splanchnic musculature. The medioventral compartment similarly extends toward the ventral midline and forms mainly ventral longitudinal muscle and splanchnic musculature. Coelomoduct segmental organs are not formed in the trunk segments of chilopods, but the gonoducts develop as coelomoducts of the last two trunk segments, in association with an opisthogoneate gonopore. The cephalic somites of chilopods are more numerous and more modified in development than those of Onychophora, but clearly derived from the same generalized condition.

For the comparison to be complete, it is necessary to compare the development of the presumptive ectoderm in chilopods with that of Onychophora. The broad area of presumptive dorsal extraembryonic ectoderm is retained, as are the ventrolateral bands of embryonic ectoderm, but the presumptive ventral extraembryonic ectoderm is not obvious in the blastoderm. The capacity to develop this ventral, temporary cover for the yolk mass is retained, however. At the blastodermal stage, the broad spread of the presumptive midgut obscures the ventral extraembryonic ectoderm, but when the midgut cells have become internal, the cells that remain at the surface develop mainly as this component. The ventral extraembryonic ectoderm then becomes broadly spread and attenuated as the two halves of the growing germ band flex over the surface of the yolk mass (Fig. 2.14), before being later reduced and incorporated into the embryonic ectoderm. The dorsal extraembryonic ectoderm suffers a similar reduction and incorporation. The major products of the embryonic ectoderm, apart from epidermis, are the nerve cord ventrally and uniramous limb buds ventrolaterally.

2.4.2.2. Other Myriapods

I have dwelt at length on the comparative functional patterns of development of scolopendromorph centipedes and onychophorans in order to demonstrate that the basic model of embryonic development which can be constructed for the Onychophora also serves as a model for the somewhat more modified development of the chilopods. The similarity is epitomized in fate maps of the two groups and provides further evidence of their identity as Uniramia. A similar argument, developed by Anderson (1973), supports the view that other patterns of development within the Myriapoda, in the Diplopoda, Pauropoda, and Symphyla, derive from a similar origin. As pointed out previously, the modifications of development in these myriapod classes are more extreme than those of chilopods and share in certain respects a different functional pattern of modification of embryonic development from that of chilopods. In spite of this, however, they can all be interpreted as modifications of the basic onychophoran model in a manner that supports their inclusion with the Chilopoda in a subphylum Myriapoda and the placement of the Myriapoda with the Onychophora in a phylum Uniramia.

2.4.2.3. Hexapoda

In keeping with the vast diversity and importance of the winged insects, the descriptive embryology of this group of hexapods, reviewed by Anderson (1972a,b, 1973), continues to be augmented space. In the last few years, new descriptions of embryonic development have been presented for species of Ephemeroptera (Tsui and Peters, 1974), Dictyoptera (Aiouaz, 1974), Cheleutoptera (Wolf-Neis, 1973), Orthoptera (DeLuca and Viscuso, 1972b; Petavy, 1975; Bate, 1976; Guita and Boufersaoui, 1976), Embioptera (Ando and Haga, 1974), Heteroptera (Choban and Gupta, 1972; Matolín, 1973; Mori, 1976, 1977), Coleoptera (Ressouches, 1973; Ando and Kobayashi, 1975; Krysan, 1976; Rempel et al., 1977), Mecoptera (Ando and Haga, 1974), Trichoptera (Miyakawa, 1973, 1974a,b,c, 1975; Anderson and Lawson-Kerr, 1977), Lepidoptera (Tanaka, 1970; Wall, 1973; Ando and Tanaka, 1976), Diptera (van der Starre-van der Molen, 1972; Fytizas and Mourikis, 1973; Bownes, 1975; Raminani and Cupp, 1975), and Hymenoptera (Amos and Salt, 1974). In general, however, with the exception of the interesting work of Mori (1976, 1977), these investigations have not been of sufficient penetration to provide the kind of information required for comparative embryological purposes, so that our understanding of pterygote fate maps has not been advanced by them. It is also notable that certain key orders, the Odonata, Plecoptera, Dermaptera, Megaloptera, and Neuroptera, still await detailed descriptive embryological analysis. A more significant event has been the recent publication of a most excellent review of experimental insect embryology by Sander (1976a; see also Sander, 1976b), with important implications for the phylogenetic interpretation of pterygote germ bands and fate maps. From a descriptive viewpoint, however, the summary of hexapod embryonic development presented by Anderson (1973) can still be upheld.

Just as the myriapods other than the Chilopoda show substantial modifications of development in relation to an onychophoran model, so too do the Hexapoda. For this group of uniramians the problem of interpretation is compounded by lack of reliable embryological description by modern methods for all classes except the Pterygota, and by the fact that no class of hexapods retains a developmental pattern as generalized as that of the scolopendromorph centipedes. All are specialized, the specializations vary between classes, and the Pterygota are the most extreme when looked at from the point of view of fate maps. At the same time, as discussed by Anderson (1973), certain firm conclusions can be drawn from comparative hexapod embryology. First, there is good embryological evidence in favor of the view that the Hexapoda are a unitary group, even though deep divergences exist between the various classes. In particular, they share a distinctive pattern of development of the labiate hexapod head. Second, the evidence of embryology supports the view that the Hexapoda are not descended from the Myriapoda. Third, the model of development presently available for the yolky Onychophora is a functional model from which the various patterns of hexapod development can be derived. The last of these conclusions is the critical one from the point of view of embryology and phylogeny, and it is this aspect that I will review briefly in the present account.

Fate maps of the blastoderm are illustrated in Fig. 2.16 for the Diplura, Collembola, Thysanura, and Pterygota, constructed mainly from the evidence of Uzel (1897a,b, 1898), Jura (1965, 1966, 1967a,b, 1972), Heymons (1896, 1897a,b), and Larink (1969, 1970) and from the many works on pterygotes reviewed by Anderson (1972a,b, 1973). The Protura cannot yet be treated in this way, though it is a pleasure to report a first brief description of a proturan embryo by Bernard (1976), indicating a mode of development similar to that of the Diplura. Continuing with our theme that fate maps can be taken to epitomize development if interpreted with regard to function, these maps provide a basis for comparing hexapod development with the onychophoran model (Anderson, 1973). Midgut, mesoderm, and ectoderm will be discussed in turn.

There is no doubt that the basic presumptive midgut in hexapods is the vitellophages which occupy the yolk mass. In the Diplura, these vitellophages enter

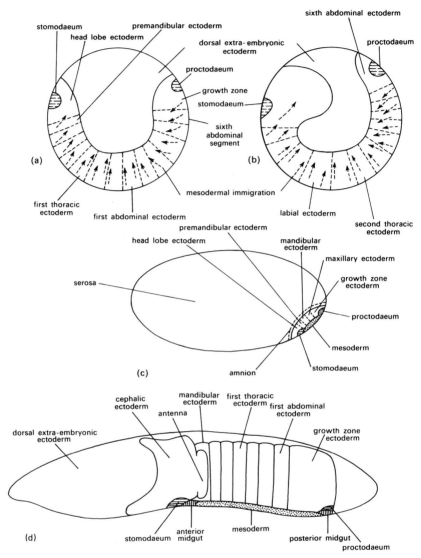

Fig. 2.16. Fate maps of blastoderm of hexapods. a, Diplura; b, Collembola; c, Thysanura; d, Odonata. From Anderson (1973).

the yolk by generalized immigration from the blastoderm, function temporarily as vitellophages, then occupy the surface of the yolk mass to form the definitive midgut epithelium. This can be envisaged as a modification of the mode of anterior midgut development of yolky Onychophora, the posterior midgut component being abandoned in association with changed relationships between the yolk mass and germ band, the anterior midgut component becoming diffuse in distribution and precocious in its entry into the yolk mass. Taken in isolation, the pattern of anterior midgut development in the Diplura could have a variety of antecedents, of which the midgut development of yolky Onychophora is only one possibility. What is important here is that the possibility exists and encompasses a functionally viable transition.

Other modes of midgut formation in hexapods show further modification of the dipluran mode. The Collembola and Thysanura form their vitellophages precociously during cleavage, in different ways, and these vitellophages subsequently develop as midgut epithelium. The Pterygota appear to be different, in that vitellophage formation and function is in most species divorced from midgut formation. Vitellophages are segregated during cleavage in the same manner as in Thysanura, and the midgut epithelium then develops from anterior and posterior midgut rudiments in the blastoderm, later associated with the internal ends of the stomodaeum and proctodaeum (Fig. 2.17). As pointed out by Anderson (1973), this pattern of midgut development seems likely to be a specialization of a prior thysanuran condition. When the thysanuran midgut develops from the vitellophages, the midgut epithelium first develops at the ends of the yolk mass in contact with the stomodaeum and proctodaeum, and later spreads towards the middle. The presumptive anterior and posterior midgut rudiments of pterygote embryos could be a further specialization of this condition. It may also be significant that some pterygote embryos appear to retain a vitellophage contribution to the middle region of the midgut epithelium (Mori, 1969, 1976, 1977).

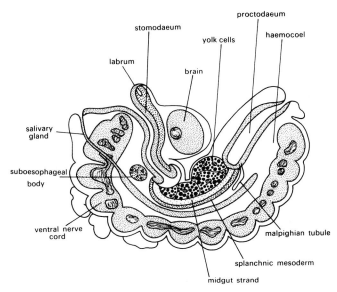

Fig. 2.17.   Development of gut in embryo of lepidopteran *Epiphyas postvittana*.   From Anderson (1973).

The presumptive mesoderm of hexapods shows its most generalized development in the Diplura and Collembola, where it is bilaterally spread along the length of the presumptive germ band (Fig. 2.16a,b) and migrates inward to give rise directly to the paired mesodermal bands. This mode of development, again, could be functionally derived from that of Onychophora, in which the presumptive mesoderm is a small posteroventral area and the mesodermal bands are proliferated forward on either side of the yolk mass. The embryos of myriapods indicate the way in which such a transition to a more direct development of mesodermal bands might occur. Chilopods retain the posteroventral presumptive mesoderm, but show supplementary immigration of mesoderm cells from the blastoderm overlying the mesodermal bands. Other myriapods, with shorter germ bands, have bilaterally distributed presumptive mesoderm and no longer retain the posteroventral component. Dipluran and collembolan hexapods appear to have evolved a parallel modification in association with short germ bands. The Thysanura and Pterygota show a further specialization of this condition, in which the presumptive mesoderm is concentrated as a midventral band (Fig. 2.16c,d) and is sharply segregated from the presumptive ectoderm on either side of it.

The development of the somites in hexapods also follows the onychophoran model, with a series of modifications paralleling those of chilopods. The somites become trilobated, with dorsolateral, medioventral, and appendicular lobes. The appendicular lobe develops as limb musculature. The dorsolateral lobe develops as dorsal longitudinal muscle, heart and pericardial floor, splanchnic musculature, and, in certain abdominal segments, gonad. The medioventral lobe develops as ventral longitudinal muscle and fat body. Coelomoduct segmental organs are not developed in hexapods, but the gonoducts develop from posterior coelomoducts and lead to an opisthogoneate gonopore.

The presumptive ectoderm and its development in hexapods is also derivable from the onychophoran model. The broad area of dorsal extraembryonic ectoderm is retained, but ventral extraembryonic ectoderm is absent. The increase in length of the hexapod germ band during development is accommodated (Figs. 2.18, 2.19) by various kinds of dorsoventral flexure or by embryonic movements (blastokinesis) that do not involve bilateral separation of the halves of the germ band. The embryonic ectoderm (Fig. 2.16) thus extends to the midline except where it is temporarily separated by a specialized, midventral presumptive mesoderm (Thysanura and Pterygota). Basically, the paired bands of presumptive embryonic ectoderm are relatively long and include the ectodermal rudiments of all except the last few abdominal segments. The major products of the embryonic ectoderm, apart from epidermis and limb buds, are the ganglia of the brain and ventral nerve cord, characteristically and uniquely developed in hexapods through the proliferative activity of neuroblasts.

In general, therefore, the embryological evidence points to a unity among the hexapods and is in accord with the concept of descent of the hexapods from the

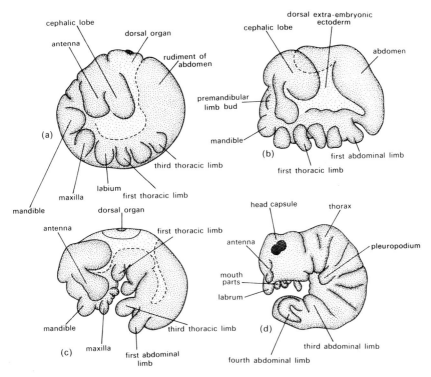

Fig. 2.18. Development of collembolan *Orchesella*. a, Segmenting germ band; b, early ventral flexure; c, later ventral flexure; d, dorsal closure. From Anderson (1973), after Bruckmoser (1965).

same ancestry as the Onychophora and Chilopoda. At the same time, the embryological evidence of itself is not adequate to confirm these relationships. As mentioned earlier, diversity in uniramian embryos far exceeds that in Crustacea, and the extant uniramians do not include species whose development exhibits any of the functional intermediate conditions postulated above. It is sufficient, however, to recognize that the onychophoran model of development is a feasible basis for the derivation of the various modes of hexapod development and thus continues to serve as a general model for the Uniramia. This, in turn, permits a comparison of crustacean and uniramian development as a test of their relationships. I will now briefly review this comparison.

2.4.3. Uniramian Embryology and Phylogeny

Crustacean development basically retains spiral cleavage but follows a fundamentally different course from that of annelids, epitomized in a fate map that has mesoderm anterior to midgut. Uniramian development shows no trace of

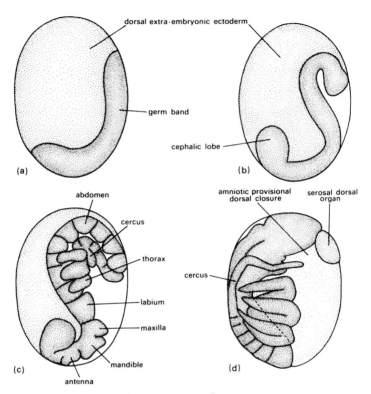

Fig. 2.19. Development of ephemeropteran *Baëtis*.   a, Embryonic primordium; b, segmenting germ band; c, flexed, segmented germ band before katatrepsis; d, embryo after katatrepsis. From Anderson (1973), after Böhle (1969).

spiral cleavage but proceeds in a manner that is functionally derivable from a spiral cleavage mode with 4d mesoderm. The annelids indicate to a remarkable degree how these developmental modifications of the Onychophora might have arisen. Among the clitellate annelids, the egg is relatively large and yolky, retains spiral cleavage and 4d mesoderm, but develops both dorsal and ventral extraembryonic ectoderm as a temporary cover for the large, yolky midgut. Viewed in terms of the formation and fates of presumptive areas, onychophoran development could have been derived from clitellate development (Anderson, 1966c, 1973). While it is improbable that the derivation was direct, the evidence is all in favor of a phylogenetic descent of onychophoran development from a pattern of spiral cleavage development with 4d mesoderm. Embryology thus groups the Uniramia with the phylum Annelida, without implying an immediate common ancestry, and separates the Uniramia widely from the Crustacea, to the point of an independent origin, in the two groups, of the coelom, metamerism, and arthropodization. It follows from this view that the comparative segmental composition of the head in Crustacea and Uniramia has no phylogenetic significance.

2.5. CHELICERATE EMBRYOLOGY

What, then, of the Chelicerata? As I have pointed out previously (Anderson, 1973), the Chelicerata are the least known embryologically of the three major arthropod groups, so that the application of embryological evidence to chelicerate relationships is still fraught with problems. Recent studies on chelicerate embryos (Kryzystofowicz and Boczek, 1970; Yoshikura, 1972; Muñoz-Cuevas, 1973; Weygoldt, 1975; Zissler and Weygoldt, 1975; Scholl, 1976, 1977; Pross, 1977), while admirably detailed and clear, have not added materially to the kind of evidence required for comparisons based on fate maps. The different modes of development found among chelicerate embryos also add to the difficulty, in that the model they suggest of basic chelicerate development is itself a specialized one. No trace of spiral cleavage occurs in any chelicerate. Development is blastodermal and direct, so we are dependent, as in Uniramia, on an interpretation of the formation and fates of presumptive areas of a blastodermal fate map. When viewed in this way, the available evidence points clearly to the view that a basic model of chelicerate development can be constructed from events in the Xiphosura (Kishinouye, 1891; Kingsley, 1892, 1893; Iwanoff, 1933; Anderson, 1973), and that the various patterns of embryonic development in the Arachnida are derivatives of this model. The pycnogonids cannot be taken into account at the present time, owing to lack of suitable evidence, but in general the unitary nature of the Chelicerata is supported embryologically. The question can be asked, therefore, what does the basic model of chelicerate development reveal about the phylogenetic relationships between the Chelicerata, the Crustacea, and the Uniramia? The answer in brief (Anderson, 1973) is that the embryological evidence argues strongly against a relationship with the Uniramia, but is equivocal about relationships with the Crustacea. I shall now outline the reasons for these conclusions.

2.5.1. Chelicerate Fate Map

The xiphosuran model of chelicerate development begins with an egg of relatively large volume, 2–3 mm in diameter, which yet retains total cleavage. The cleavage pattern shows no trace of spirality. As cleavage proceeds (Fig. 2.20), a blastoderm is segregated external to a mass of yolky cells. The blastoderm, on the basis of the evidence of subsequent development, can be zoned into the presumptive areas shown in Fig. 2.21. This fate map provides a basis for modeling the general course of chelicerate development.

The first notable feature of the xiphosuran fate map is the absence of a blastodermal area of presumptive midgut. This component of the embryo already lies internally, in the form of a mass of yolky cells established within the blastoderm as a result of cleavage (Fig. 2.20b). As development proceeds, these cells give rise directly to the epithelium of the midgut and digestive diverticula.

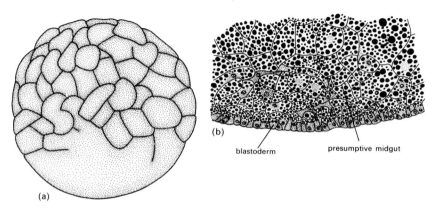

(b)

blastoderm   presumptive midgut

(a)

Fig. 2.20.   Cleavage in xiphosuran *Tachypleus*. a, Total cleavage; b, blastoderm formation. From Anderson (1973), after Iwanoff (1933).

The ventral midline of the xiphosuran blastoderm is occupied by presumptive mesoderm, as a narrow band of cells extending from the presumptive stomodaeum to the presumptive growth zone ectoderm (see below). The presumptive mesoderm retains a gastrulation movement, but of a unique kind. The mesoderm cells invaginate, forming a gastral groove (Fig. 2.22a,b), but remain superficial in the growth zone. The gastral groove soon closes over anteriorly, but is progressively formed as a groove more and more posteriorly during the ensuing growth of the germ band.

After entry in the ventral midline of each successive segment, the mesoderm spreads laterally and forms a pair of somite rudiments, which then split inter-

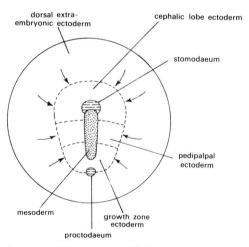

dorsal extra-embryonic ectoderm

cephalic lobe ectoderm

stomodaeum

pedipalpal ectoderm

mesoderm

growth zone ectoderm

proctodaeum

Fig. 2.21.   Fate map of blastoderm of Xiphosura. From Anderson (1973).

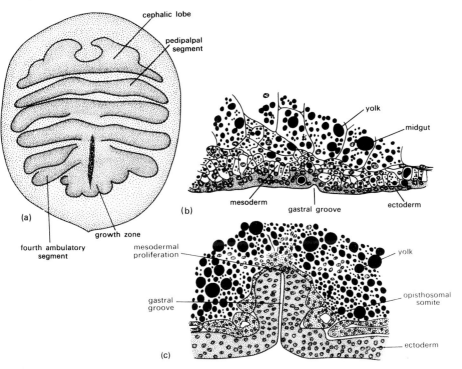

Fig. 2.22.   Gastrulation in xiphosuran *Tachypleus.* a, Segmenting germ band with posterior gastral groove; b, early gastral groove in transverse section; c, gastral groove in opisthosoma, flanked by somites. From Anderson (1973), after Iwanoff (1933).

nally, forming paired, hollow somites with thick somatic and thin splanchnic walls (Fig. 2.22c).   Anteriorly, the cheliceral pair of somites grows forward beneath the precheliceral ectoderm.   There is no evidence of a separate pair of precheliceral somites.

In the segments in which limb buds are developed, each somite extends an appendicular lobe into the corresponding limb bud.   The dorsal parts of the somites grow up toward the dorsal midline and contribute to formation of the heart and pericardial floor.   The somatic walls develop as somatic musculature, the splanchnic walls as splanchnic musculature; but the details of this aspect of somite development are not yet clear.   The ventral portions of the six pairs of prosomal somites give rise to coxal glands, modified segmental organs which open by a pair of coelomoducts on the third ambulatory segment.   Evidence from other chelicerates, still awaiting confirmation in the xiphosuran embryo, indicates that the somites of the genital (second opisthosomal) segment give rise to the gonads, with the gonoducts growing out as coelomoducts of this segment.

The presumptive stomodaeum of Xiphosura lies in the ventral midline in front of the presumptive mesoderm, and invaginates as the anterior wall of the gastral groove. When the anterior part of the gastral groove closes over, the stomodaeal invagination becomes tubular. Following further growth, the stomodeaum develops as the wall of the buccal cavity, oesophagus, and proventriculus, and secretes a cuticular lining.

The presumptive proctodaeum in Xiphosura comprises a group of cells in the midline behind the ectodermal growth zone. Later in development, an independent invagination of this area establishes the proctodaeum as a short tube that develops a short, simple hindgut.

The presumptive ectoderm, which comprises most of the blastoderm, is divisible into embryonic ectoderm ventrally and extraembryonic ectoderm elsewhere. The embryonic ectoderm surrounds the midventral areas and comprises presumptive head lobe (precheliceral and cheliceral) ectoderm anteriorly, pedipalpal ectoderm laterally, and growth zone ectoderm posteriorly. Thus the presumptive germ band in xiphosuran embryos is very short, and all segments behind the pedipalpal develop by growth zone proliferation. As the presumptive embryonic ectoderm develops and spreads dorsally, the extraembryonic ectoderm is gradually incorporated into it after a period of attenuation.

The short presumptive germ band in Chelicerata and the consequent proliferation of all except the first two segments from the growth zone, constitutes a mode of development analogous to the anamorphic development of the Crustacea, but with one fundamental difference. In the Crustacea, as we have seen, anamorphic development basically follows the formation of a specialized larval (naupliar) region anteriorly, stemming from a small egg, and the naupliar specialization persists in all larger crustacean embryos. In the Chelicerata, the large egg is basic, and there is no trace of any larval specialization of the anterior segments.

The ectodermal growth zone gives rise successively to the ectoderm of all postpedipalpal segments. The major products of the ectoderm are epidermis, limb buds, and the ganglia of the brain (precheliceral and cheliceral) and ventral nerve cord.

2.5.2 Chelicerate Embryology and Phylogeny

In view of the fact that a functional approach based on comparison of fate maps provides strong evidence of a lack of relationship between Crustacea and Uniramia, the corresponding comparisons are worth exploring for the Chelicerata using the xiphosuran model. This was first attemped by Anderson (1973). As demonstrated at that time, a chelicerate-uniramian relationship can be rejected on embryological grounds with a reasonable degree of certainty. The entire process of development has a different configuration in the two groups, beginning with a different mode of adaptation of development to increased yolk. We need, therefore, to consider whether chelicerate development reveals any evidence of derivation from spiral cleavage development, and if so, whether this is the pattern of spiral cleavage development with 4d mesoderm, the crustacean pattern, or some other pattern. In the absence of any trace of spiral cleavage in the Chelicerata, the analysis depends entirely on functional comparisons of fate maps, and here the evidence is inconclusive. Since the presumptive midgut in the chelicerate model becomes internal during cleavage, the spatial relationship between presumptive midgut and mesoderm which distinguishes the Crustacea from the Annelida is no longer a point of comparison. The chelicerate fate map, with all of its implications for subsequent development, could have been derived from a crustacean-like origin or an annelid-like origin or neither. Some features of chelicerate basic development show resemblance to corresponding features in Crustacea. They include the development of the gut with its midgut diverticulum; the coelomoduct gonoducts in the middle region of the body; and the development of excretory segmental organs, though differently placed. But these are isolated features, which could be the result of convergent evolution. Taking a broader view of the total configuration of development in the two groups, the evidence of the fate maps is that crustaceans are crustaceans and chelicerates are chelicerates, with nothing in common embryologically between them. Furthermore, while crustaceans and annelids (and also, by deduction, Uniramia) belong within the spiral cleavage assemblage, there is nothing in chelicerate development to place the chelicerates within this assemblage. It may be that further investigations on chelicerate embryos will resolve this problem, but the present evidence cannot be used to support a crustacean-chelicerate affinity, and tends by its negativity to deny such an affinity. If we take the view that a phylum is characterized, among other things, by a shared pattern of development epitomized in a common basic fate map of the blastula or blastoderm, then the embryological evidence places the modern arthropods in three phyla, Crustacea, Uniramia, and Chelicerata (Anderson 1973). The views of Manton (1972, 1973, 1977) on the polyphyletic origin of the arthropods are thus upheld when tested independently by embryological criteria based on fate maps.

2.6. NEED FOR FURTHER INVESTIGATIONS

In the further exploration of this idea and of certain attendant ideas concerning the unity of the Hexapoda and the lack of unity of the Arachnida (perhaps a grade, see Bergström 1978; Manton, 1977; Kraus, 1976), new embryological research is required in several directions. The Crustacea have already been well served by modern embryologists, although many gaps remain to be filled, but the Uniramia are in need of a number of critical studies. The development of yolky onychophoran embryos is an outstanding example of this. The interpretation of these embryos is central to our understanding of uniramian embryology, yet rests at present mainly on nineteenth century evidence. Similarly, the embryology of scolopendromorph centipedes must be reexamined in the manner so beautifully demonstrated for pauropods and symphylans by Tiegs (1940, 1947). Diplopods, too, need further attention, particularly in the early stages of development. The value of hexapod embryology, in consolidating the uniramian placement of the hexapods and in assisting the elucidation of relationships between hexapod classes, will remain tentative until detailed modern studies are carried out on the embryonic development of the Collembola, Diplura, and Protura and further information is gathered about the Thysanura, Odonata, Ephemeroptera, Dictyoptera, and Plecoptera. The Chelicerata, too, present a major challenge. In spite of the excellent recent work of Scholl (1977), the critical early stages of xiphosuran development are still not clear. Furthermore, no arachnid order can yet be said to have been explored in a satisfactory manner from the point of view of comparative embryology, though the outstanding studies of Weygoldt (1964, 1965, 1968, 1971, 1975) on pseudoscorpions and amblypygids have shown the way. The pcynogonids are also almost untouched. Comparative arthropod embryology thus present a continuing major challenge, with a direct bearing on the interpretation of arthropod phylogeny.

2.7. SUMMARY

Hypotheses concerning the phylogenetic relationships of the major arthropod groups can be tested by comparative embryology. One approach to this is through an interpretation of arthropod embryonic development based on fate maps. The embryology of the Crustacea is discussed from this point of view. Embryological evidence for the unity of the Uniramia and their origin independently of the Crustacea is then reviewed. The equivocal nature of present understanding of the embryology of the Chelicerata from a comparative standpoint is reiterated. In general, a consideration of arthropod embryology based on fate maps supports the concept of the polyphyletic origin of the major arthropod groups.

REFERENCES

Aiouaz, M. 1974. Chronologie du développement embryonnaire de *Leucophaea madesae* Fabr. (Insecta, Dictyoptère). *Arch. Zool. Exp. Gén.* **115**: 343–58.

Amos, W. B. and G. Salt. 1974. An atlas of the development of eggs of an ichneumonid wasp (Hym. Ichneumonidae). *J. Entomol.* **B45**: 11–18.

Anderson, D. T. 1959. The embryology of the polychaete *Scoloplos armiger*. *Q. J. Microsc. Sci.* **100**: 89–166.

Anderson, D. T. 1966a. The comparative embryology of the Diptera. *Annu. Rev. Entomol.* 11: 23-64.

Anderson, D. T. 1966b. The comparative embryology of the Polychaeta. *Acta Zool. (Stockh.)* 47: 1-41.

Anderson, D. T. 1966c. The comparative early embryology of the Oligochaeta, Hirudinea and Onychophora. *Proc. Linn. Soc. N.S.W.* 91: 10-43.

Anderson, D. T. 1967. Larval development and segment formation in the branchiopod crustaceans *Limnadia stanleyana* King (Conchostraca) and *Artemia salina* (L.) (Anostraca). *Aust. J. Zool.* 15: 47-91.

Anderson, D. T. 1969. On the embryology of the cirripede crustaceans *Tetraclita rosea* (Krauss), *T. purpurascens* (Wood), *Chthamalus antennatus* (Darwin) and *Chamaesipho columna* (Spengler) and some considerations of crustacean phylogenetic relationships. *Phil. Trans. R. Soc.* B256: 183-235.

Anderson, D. T. 1972a. The development of hemimetabolous insects, pp. 95-163. *In* S. J. Counce and C. H. Waddington (eds.) *Developmental Systems–Insects, Vol. I.* Academic Press, New York, London.

Anderson, D. T. 1972b. The development of holometabolous insects, pp. 165-242. *In* S. J. Counce and C. H. Waddington (eds.) *Developmental Systems–Insects, Vol. I.* Academic Press, New York, London.

Anderson, D. T. 1973. *Embryology and Phylogeny in Annelids and Arthropods.* Pergamon Press, Oxford.

Anderson, D. T. 1977. The embryonic and larval development of the turbellarian *Notoplana australis* (Schmarda, 1859) (Polycladia: Leptoplanidae). *Aust. J. Mar. Freshwater Res.* 28: 303-10.

Anderson, D. T. and C. Lawson-Kerr. 1977. The embryonic development of the marine caddis fly *Philanisus plebeius* Walker (Trichoptera, Family Chathamidae). *Biol. Bull. (Woods Hole)*: 152: 98-105.

Anderson, D. T. and S. M. Manton. 1972. Studies on the Onychophora VIII. The relationship between the embryos and the oviduct in the viviparous onychophorans *Epiperipatus trinidadensis* (Bouvier) and *Macroperipatus torquatus* (Kennel) from Trinidad. *Phil. Trans. R. Soc.* B264: 161-89.

Ando, H. and K. Haga. 1974. Studies on the pleuropodia of Embioptera, Thysanoptera and Mecoptera. *Bull. Sugadaira Lab., Tokyo Kyoiku Daig.* 6: 1-8.

Ando, H. and H. Kobayashi. 1975. Description of early and middle development stages in embryos of the firefly, *Luceola cruciata* Motshulsky (Coleoptera, Lampyridae). *Bull. Sugadaira Lab., Tokyo Kyoiku Daig.* 7: 1-11.

Ando, H. and M. Tanaka. 1976. The formation of germ rudiment and embryonic membranes in the primitive moth, *Endoclyta excrescens* Butler (Hepalidae, Monotrysia, Lepidoptera) and its phylogenetic significance. *Proc. Jap. Soc. Syst. Zool.* 12: 52-5.

Baldass, F. von. 1937. Entwicklung von *Holopedium gibberum*. *Zool. Jahrb. Anat. Ontog.* 63: 300-454.

Baldass, F. von. 1941. Die Entwicklung von *Daphia pulex*. *Zool. Jahrb. Anat. Ontog.* 67: 1-60.

Bate, C. M. 1976. Embryogenesis of an insect nervous system. I. A map of the thoracic and abdominal neuroblasts in *Locusta migratoria* (Orthoptera, Acrididae). *J. Embryol. Exp. Morphol.* 35: 107-23.

Bernard, E. C. 1976. Observations on the eggs of *Eosentomom australicum* (Protura: Eosentomidae). *Trans. Amer. Microsc. Soc.* 95: 129-30.

Bergström, J. 1978. Morphology of fossil arthropods as a guide to phylogenetic relationships, pp. 3-56. *In* A. P. Gupta (ed.) *Arthropod Phylogeny*, Van Nostrand Reinhold, New York.

Bigelow, J. A. 1902. The early development of *Lepas*. A study of cell lineage and germ layers. *Bull. Mus. Comp. Zool. Harv.* **40**: 61-144.

Böhle, H. W. 1969. Untersuchungen über die Embryonalentwicklung and die embryonalen Diapause bei *Baëtis vernus* Curtis and *Baëtis rhodani* (Picket) (Baetidae, Ephemeroptera). *Zool. Jahrb. Anat. Ontog.* **86**: 493-575.

Bownes, J. 1975. A photographic study of development in the living embryo of *Drosophila melanogaster*. *J. Embryol. Exp. Morphol.* **33**: 789-801.

Bruckmoser, P. 1965. Untersuchungen über den Kopfbau der Collembole *Orchesella villosa* L. *Zool. Jahrb. Anat. Ontog.* **82**: 299-364.

Cather, J. N. 1971. Cellular interactions in the regulation of development of annelids and molluscs. *Adv. Morphog.* **9**: 67-125.

Choban, R. G. and A. P. Gupta. 1972. Meiosis and early embryology of *Blissus leucopterus hirtus* Montandon (Heteroptera: Lygaeidae). *Int. J. Insect Morphol. Embryol.* **1**: 301-14.

Costello, D. P. and C. Henley. 1976. Spiralian development: A perspective. *Amer. Zool.* **16**: 277-92.

DeLuca, V. and R. Viscuso. 1972a. Morfologia dell'embryone di *Eyprepocnemis plorans plorans* (Chasp.) (Orth. Acrid.) nei vari stadi di sviluppo. *Redia* **53**: 123-38.

DeLuca, V. and R. Viscuso. 1972b. Prime fase dello sviluppo fecondato di *Eyprepocnemis plorans* (Chasp.) (Orth. Acrid.) *Redia* **53**: 239-49.

Dohle, W. 1970 Die Bildung und Differenzierung des post-nauplialen Keimstreifs von *Diastylis rathkei* (Crustacea, Cumacea). 1. Die Bildung der Teloblasten und ihrer Derivate. *Z. Morphol. Tiere* **67**: 367-92.

Dohle, W. 1972. Über die Bildung und Differenzierung des postnauplialen Keimstreifs von *Leptochelia* spec. (Crustacea, Tanaidacea). *Zool. Jahrb. Anat. Ontog.* **89**: 503-66.

Dohle, W. 1974. The segmentation of the germ band of Diplopoda compared with other classes of arthropods. *Symp. Zool. Soc. Lond.* **32**: 143-61.

Dohle, W. 1976a. Die Bildung und Differenzierung des postnauplialen Keimstreifs von *Diastylis rathkei* (Crustacea, Cumacea). II. Die Differenzierung und Musterbildung des Ektoderms. *Zoomorphologie* **84**: 235-77.

Dohle, W. 1976b. Zur Frage des Nachweises von Homologien durch die komplexen Zell– und Teilungsmuster in der Embryonalentwicklung höherer Krebse (Crustacea, Malacostraca, Peracarida). *Sitzb. Ges. Nat. Freunde Berlin (N.F.)* **16**: 125-44.

Fioroni, P. 1970. Die organogenetische und transitorische Rolle der Vitellophagen in der Darmentwicklung von *Galathea* (Crustacea, Decapoda, Anomura). *Z. Morphol. Tiere* **67**: 236-306.

Fioroni, P. and E. Bandaret. 1971. Mit dem Dotteraufschluss liierte Ontogenese–Abwandlungen bei einigen decapoden Krebsen. *Vie Milieu* Ser. A. *Biol. Mar.* **22**: 163-88.

Fuchs, F. 1914. Die Keimblätterentwicklung von *Cyclops viridis* Jurine. *Zool. Jahrb. Anat. Ontog.* **38**: 103-56.

Fytizas, E. and P. A. Mourikis. 1973. L'embryologie de *Dacus olea* Gmel. (Diptera: Tephritidae). *Int. J. Insect Morphol. Embryol.* **2**: 25-34.

Gould-Somero, J. 1975. Echiura, pp. 277-311. *In* A. C. Giese and J. S. Pearse (eds.) *Reproduction of Marine Invertebrates. Vol. III. Annelids and Echiurans.* Academic Press, New York, London.

Guita, M. and A. Boufersaoui. 1976. Embryonic development tables of *Pamphagus elephas* (L.) (Insecta, Orthoptera). *Arch. Zool. Exp. Gén.* **117**: 437-50.

Henley, C. 1974. Platyhelminthes (Turbellaria), pp. 267-343. *In* A. C. Giese and J. S. Pearse (eds.) *Reproduction of Marine Invertebrates. Vol. I. Acoelomate and Pseudocoelomate Metazoans.* Academic Press, New York, London.

Heymons, R. 1896. Ein Beitrag zur Entwicklungsgeschichte der Insecta apterygota. *Sitzb. K. Preuss. Akad. Wiss. Berlin* 51: 1386–89.

Heymons, R. 1897a. Ueber die Bildung und den Bau der Darmkanals bei niederen Insekten. *Sitzb. Ges. Nat. Freunde Berlin* 1897: 111–19.

Heymons, R. 1897b. Entwicklungsgeschichteliche Untersuchungen an *Lepisma saccharina* L. *Z. Wiss. Zool.* 62: 583–631.

Heymons, R. 1901. Die Entwicklungsgeschichte der Scolopender. *Zoologica (Stuttg.)* 13: 1–224.

Hickman, V. V. 1937. The embryology of the syncaridan crustacean *Anaspides tasmaniae*. *Pap. Proc. R. S. Tasmania.* 1936: 1–36.

Iwanoff, P. P. 1933. Die embryonale Entwicklung von *Limulus molluccanus. Zool. Jahrb. Anat. Ontog.* 56: 163–348.

Jura, C. 1965. Embryonic development of *Tetradontophora bielanensis* (Waga) (Collembola) from oviposition till germ band formation. *Acta Biol. Cracow Zool.* 8: 141–57.

Jura, C. 1966. Origin of the endoderm and embryogenesis of the alimentary system in *Tetradontophora bielanensis* (Waga) (Collembola). *Acta Biol. Cracow Zool.* 9: 95–102.

Jura, C. 1967a. Origin of germ cells and gonad formation in embryogenesis of *Tetradontophora bielanensis* (Waga) (Collembola). *Acta Biol. Cracow Zool.* 10: 97–103.

Jura, C. 1967b. The significance and function of the primary dorsal organ in embryonic development of *Tetradontophora bielanensis* (Waga) (Collembola). *Acta Biol. Cracow Zool.* 10: 301–11.

Jura, C. 1972. The development of apterygote insects, pp. 49–94. *In* S. J. Counce and C. H. Waddington (eds.) *Developmental Systems–Insects, Vol. I.* Academic Press, New York, London.

Kennel, J. 1884. Entwicklungsgeschichte von *Peripatus edwardsi* Blanch und *Peripatus torquatus* n.sp. *Arb. Zool.-Zootom. Inst. Würzburg* 7: 95–229.

Kennel, J. 1885. Entwicklungsgeschichte von *Peripatus edwardsi* Blanch und *Peripatus torquatus* n.sp. *Arb. Zool.-Zootom. Inst. Würzburg* 8: 1–93.

Kingsley, J. S. 1892. The embryology of *Limulus. J. Morphol.* 7: 36–66.

Kingsley, J. S. 1893. The embryology of *Limulus. J. Morphol.* 8: 195–268.

Kishinouye, K. 1891. On the development of *Limulus longispina. J. Coll. Sci. Imp. Univ. Japan* 5: 53–100.

Knoll, H. J. 1974. Untersuchungen zur Entwicklungsgeschichte von *Scutigera coleoptrata* L. (Chilopoda). *Zool. Jahrb. Anat. Ontog.* 92: 47–132.

Kohler, H. J. 1976. Embryologische Untersuchungen an Copepoden: Die Entwicklung von *Lernaeocera branchialis* L. 1767 (Crustacea, Copepoda, Lernaeoida, Lernaeidae). *Zool. Jahrb. Anat. Ontog.* 95: 448–504.

Kraus, O. 1976. On the phylogenetic position and evolution of the Chelicerata. *Entomol. German.* 3: 1–12.

Krysan, J. L. 1976. The early embryology of *Diabrotica undecimpunctata howardi* (Coleoptera: Chrysomelidae). *J. Morphol.* 149: 121–37.

Kryzystofowicz, A. and J. Boczek 1970. Embryonic development of *Tetranychus urticae* Koch (Acarina: Tetranychidae). *Zest. Prob. Postepaw Nauk Polnic* 109: 123–34.

Kühn, A. 1913. Die Sonderung der Keimsbezirke in der Entwicklung der Sommereier von *Polyphemus pediculus* de Geer. *Zool. Jahrb. Anat. Ontog.* 35: 243–340.

Land, R. 1973. Die Ontogenese von *Maja squinado* (Crustacea, Malacostraca, Decapoda, Brachyura) unter besonderer Berücksichtigung der embryonalen Ernährung und der Entwicklung der Darmtraktes. *Zool. Jahrb. Anat. Ontog.* 90: 389–449.

Lang, R. and P. Fioroni. 1971. Darmentwicklung und Dotteraufschluss bei *Macropodia* (Crustacea, Malacostraca, Decapoda, Brachyura). *Zool. Jahrb. Anat. Ontog.* 88: 84–137.

Larink, O. 1969. Zur Entwicklungsgeschichte von *Petrobius brevistylis* (Thysanura, Insecta). *Helgol. Wiss. Meeresunters.* **19:** 111–55.

Larink, O. 1970. Die Kopfentwicklung von *Lepisma saccharina* L. (Insecta, Thysanura). *Z. Morphol. Tiere* **67:** 1–15.

Manton, S. M. 1928. On the embryology of a mysid crustacean, *Hemimysis lamornae. Phil. Trans. R. Soc.* **B216:** 363–463.

Manton, S. M. 1934. On the embryology of *Nebalia bipes. Phil. Trans. R. Soc.* **B223:** 168–238.

Manton, S. M. 1949. Studies on the Onychophora VII. The early embryonic stages of *Peripatopsis* and some general considerations concerning the morphology and phylogeny of the Arthropoda. *Phil. Trans. R. Soc.* **B223:** 483–580.

Manton, S. M. 1972. The evolution of arthropod locomotory mechanisms. Part 10. *J. Linn. Soc. Zool.* **51:** 203–400.

Manton, S. M. 1973. Arthropod phylogeny – A modern synthesis. *J. Zool. Lond.* **171:** 111–30.

Manton, S. M. 1977. *The Arthropoda: Habits, Functional Morphology and Evolution.* Oxford University Press, Oxford.

Matolín, S. 1973. The embryonic development of *Pyrrhocoris apterus* (L.) (Heteroptera, Pyrrhocoridae). *Acta Entomol. Bohem.* **70:** 150–56.

Miyakawa, K. 1973. The embryology of the caddis fly *Stenopsyche griseipennis* Maclachlan (Trichoptera: Stenopsychidae). I. Early stages and changes in external form of embryo. *Kontyû* **41:** 412–25.

Miyakawa, K. 1974a. The embryology of the caddis fly *Stenopsyche griseipennis* Maclachlan (Trichoptera: Stenopsychidae). II. Formation of germ band, yolk cells and embryonic envelopes and early development of inner layer. *Kontyû* **42:** 64–73.

Miyakawa, K. 1974b. The embryology of the caddis fly *Stenopsyche griseipennis* Maclachlan (Trichoptera: Stenopsychidae). III. Organogenesis: Ectodermal derivatives. *Kontyû* **42:** 305–24.

Miyakawa, K. 1974c. The embryology of the caddis fly *Stenopsyche griseipennis* Maclachlan (Trichoptera: Stenopsychidae). IV. Organogenesis: mesodermal derivatives. *Kontyû* **42:** 451–66.

Miyakawa, K. 1975. The embryology of the caddis fly *Stenopsyche griseipennis* Machlachlan (Trichoptera: Stenopsychidae). V. Formation of the alimentary canal and other structures, general considerations and conclusions. *Kontyû* **43:** 55–74.

Mori, H. 1969. Normal embryogenesis of the waterstrider *Gerris paludum insularis* Motschulsky, with special reference to midgut formation. *Jap. J. Zool.* **5:** 53–67.

Mori, H. 1976. Formation of the visceral musculature and origin of the midgut epithelium in the embryos of *Gerris paludum insularis* Motschulsky (Hemiptera: Gerridae). *Int. J. Insect Morphol. Embryol.* **5:** 117–25.

Mori, H. 1977. Inductive role of the visceral musculature in formation of the midgut epithelium in embryos of the waterstrider *Gerris paludum insularis* Motschulsky. *Annot. Zool. Japan* **50:** 22–30.

Muñoz-Cuevas, A. 1973. Embryogenèse, organogenèse et rôle des organes ventraux et neuraux de *Pachyulus quinamavidensis* Muñoz (Arachnida, Opilions, Gonyleptidae). Comparison avec les Annelides et d'autres Arthropodes. *Bull. Mus. Natl. Hist. Nat. Paris, ser. 3 Zool.* **128:** 1517–37.

Nair, K. B. 1939. The reproduction, oogenesis and development of *Mesopodopsis orientalis* Tatt. *Proc. Indian Acad. Sci.* **B9:** 175–223.

Nair, K. B. 1949. On the embryology of *Caridina laevis. Proc. Indian Acad. Sci.* **B29:** 211–88.

Pace, F., R. R. Harris and V. Jaccarini. 1976. The embryonic development of the Mediterranean freshwater crab, *Potamon edulis* (= *P. fluviatile*) (Crustacea, Decapoda, Potamonidae.) *J. Zool. Lond.* 180: 93-106.

Petavy, G. 1975. Involution des annexes embryonnaires dans l'oeuf de *Locusta migratoria migratorioides* R. and F. (Orthoptera, Acrididae): Morphologie et histologie. *Int. J. Insect Morphol. Embryol.* 4: 1-22.

Pross, A. 1977. Diskussionsbeitrag zur Segmentierung des Cheliceraten–Kopfes. *Zoomorphologie* 86: 183-96.

Raminani, L. N. and E. W. Cupp. 1975. Early embryology of *Aedes aegypti* (L.) (Diptera, Culicidae). *Int. J. Insect Morphol. Embryol.* 4: 517-28.

Rempel, J. G., B. S. Heming and N. S. Church, 1977. The embryology of *Lytta viridana* LeConte (Coleoptera: Meloidae). IX. The central nervous system, stomatogastric nervous system and endocrine system. *Quaest. Entomol.* 13: 5-23.

Ressouches, A. P. 1973. Étude du developpement embryonnaire de *Pissodes notatus* F. (Col. Curculionidae). I. De la ponte à la bondalette embryonnaire segmentée. *Bull. Soc. Zool. Fr.* 98: 283-300.

Rice, M. E. 1975a. Sipuncula, pp. 76-127. *In* A. C. Giese and J. S. Pearse (eds.) *Reproduction of Marine Invertebrates Vol. II. Entoprocts and Lesser Coelomates.* Academic Press, New York, London.

Rice, M. E. 1975b. Observations on the development of six species of Caribbean Sipuncula with a review of development in the phylum, pp. 141-160. *In* M. E. Rice and M. Todorovic (eds.) *Proceedings of an International Symposium on the Biology of the Sipuncula and Echiura, Vol. I.* Naucno Delo, Belgrade.

Riser, N. W. 1974. Nemertinea, pp. 359-89. *In* A. C. Giese and J. S. Pearse (eds.) *Reproduction in Marine Invertebrates. Vol. I. Acoelomate and Pseudocoelomate Metazoans.* Academic Press, New York, London.

Sander, K. 1976a. Morphogenetic movements in insect embryogenesis. *Symp. R. Entomol. Soc.* 8: 35-52.

Sander, K. 1976b. Specification of the basic body pattern in insect embryogenesis. *Adv. Insect Physiol.* 12: 125-238.

Sanders, H. L. 1963. The Cephalocarida. Functional morphology, larval development, comparative external anatomy. *Mem. Conn. Acad. Arts Sci.* 150: 1-80.

Scheidegger, G. 1976. Stadien der Embryonalentwicklung von *Eupagurus prideauxi* Leach (Crustacea, Decapoda, Anomura) unter besonderer Berücksichtigung der Darmentwicklung und der am Dotterabbau beteiligten Zelltypen. *Zool. Jahrb. Anat. Ontog.* 95: 297-353.

Scholl, G. 1963. Embryologische Untersuchungen an Tanaidaceen. *Zool. Jahrb. Anat. Ontog.* 80: 500-54.

Scholl, G. 1976. Stützen embryologische Daten bei *Limulus polyphemus* die monophyletische Entstehung der Arthropoden? *Verh. Dtsch. Zool. Ges.* 1976: 228.

Scholl, G. 1977. Beiträge zur Embryonalentwicklung von *Limulus polyphemus* L. (Chelicerata, Xiphosura). *Zoomorphologie* 86: 99-154.

Schroeder, P. C. and C. O. Hermans. 1975. Annelida: Polychaeta, pp. 1-213. *In* A. C. Giese and J. S. Pearse (eds.) *Reproduction in Marine Invertebrates. Vol. III. Annelids and Echiurans.* Academic Press, New York, London.

Sedgwick, A. 1888. The development of the Cape species of *Peripatus*. Part 4. *Q. J. Microsc. Sci.* 28: 373-98.

Sheldon, L. 1888. On the development of *Peripatus novae-zealandiae*. *Q. J. Microsc. Sci.* 18: 25-37.

Sheldon, L. 1889a. On the development of *Peripatus novae-zealandiae*. *Q. J. Miscrosc. Sci.* 29: 283-94.

Sheldon, L. 1889b. The maturation of the ovum in the Cape and New Zealand species of *Peripatus*. *Q. J. Microsc. Sci.* **30**: 1-29.

Shiino, S. M. 1942. Studies on the embryology of *Squilla oratoria* de Haan. *Mem. Coll. Sci. Kyoto Univ.* **B28**: 77-174.

Shiino, A. M. 1950. Studies on the embryonic development of *Palinurus japonicus* (von Siebold). *J. Fac. Fish. Prefect. Univ. Mie* **1**: 1-168.

Stromberg, J. O. 1965. On the embryology of the iosopod *Idotea*. *Ark. Zool.* **17**: 421-73.

Stromberg, J. O. 1967. Segmentation and organogenesis in *Limnoria lignorum* (Rathke) (Isopoda). *Ark. Zool.* **20**: 91-139.

Stromberg, J. O. 1971. Contribution to the embryology of bopyrid isopods with special reference to *Bopyroides, Hemiarthrus* and *Pseudione* (Isopoda, Epicaridea). *Sarsia* **47**: 1-46.

Stromberg, J. O. 1972. *Cyathura polita* (Crustacea, Isopoda), some embryological notes. *Bull. Mar. Sci.* **22**: 463-82.

Tanaka, J. 1970. Embryonic development of the rice webworm, *Ancylolomia japonica* Zeller. I. From fertilization to germ band formation. *New Entomol.* **19**: 35-41.

Taube, E. 1909. Beiträge zur Entwicklungsgeschichte der Euphausiden. I. Die Furchung der Eies bis zur Gastrulation. *Z. Wiss. Zool.* **92**: 427-64.

Taube, E. 1915. Beiträge zur Entwicklungsgeschichte der Euphausiden. II. Von der Gastrula bis zum Furciliastadium. *Z. Wiss. Zool.* **114**: 577-656.

Tiegs, O. W. 1940. The embryology and affinities of the Symphyla, based on a study of *Hanseniella agilis*. *Q. J. Microsc. Sci.* **82**: 1-225.

Tiegs, O. W. 1947. The development and affinities of the Pauropoda, based on a study of *Pauropus sylvaticus*. *Q. J. Microsc. Sci.* **88**: 165-267, 275-336.

Tsui, P. T. and W. L. Peters. 1974. Embryonic development, early instar morphology and behaviour of *Tortopus incertus* (Ephemeroptera: Polymitacidae). *Fla. Entomol.* **57**: 349-56.

Uzel, H. 1897a. Borlaufige Mittheilungen über die Entwicklung der Thysanuren. *Zool. Anz.* **20**: 125-32.

Uzel, H. 1897b. Beiträge zue Entwicklungsgeschichte von *Campodea staphylinus*. *Zool. Anz.* **20**: 232-37.

Uzel, H. 1898. *Studies über die Entwicklung der apterygoten Insekten*. Friedlander, Berlin.

van der Starre-van der Molen, L. G. 1972. Embryogenesis of *Calliphora erythrocephala* Meigen. I. Morphology. *Neth. J. Zool.* **22**: 119-82.

Wall, C. 1973. Embryonic development of two species of *Chesias* (Lepidoptera: Geometridae). *J. Zool. Lond.* **169**: 65-84.

Walley, L. J. 1969. Studies on the larval structure and metamorphosis of *Balanus balanoides* (L.). *Phil. Trans. R. Soc.* **B256**: 237-80.

Weygoldt, P. 1958. Die Embryonalentwicklung des Amphipoden *Gammarus pulex pulex* (L.). *Zool. Jahrb. Anat. Ontog.* **77**: 15-110.

Weygoldt, P. 1960. Beitrag zur Kenntnis der Malakostrakenentwicklung. Die Keimblätterbildung bei *Asellus aquaticus* (L.). *Z. Wiss. Zool.* **163**: 340-54.

Weygoldt, P. 1964. Vergleichend-embryologische Untersuchungen an Pseudoscorpionen (Chelonethi). *Z. Morphol. Ökol. Tiere* **54**: 1-106.

Weygoldt, P. 1965. Vergleichend-embryologische Untersuchungen an Psseudoscorpionen. III. Die Entwicklung von *Neobisium muscorum* Leach (Neobisiinea, Neobisiidae). Mit dem Versuch einer Deutung der Evolution des embryonalen Pumporgans. *Z. Morphol. Ökol. Tiere* **55**: 321-82.

Weygoldt, P. 1968. Vergleichend-embryologische Untersuchungen an Pseudoscorptionen. IV. Die Entwicklung von *Chthonius tetrachelatus* Preyssl., *Chthonius ischnocheles*

Hermann (Chthoniinea, Chthoniidae) und *Verrucaditha spinosa* Banks (Chthoniinea, Tridenchthoniidae). *Z. Morphol. Ökol. Tiere* **63**: 111-54.

Weygoldt, P. 1971. Vergleichend-embryologische Untersuchungen an Pseudoscorpionen. V. Das Embryonal-stadium mit seinem Pumporgan bei verschiedenen Arten und sein Wert als taxonomisches Merkmal. *Zool. Syst. Evolutionsforsch.* **9**: 3-29.

Weygoldt, P. 1975. Untersuchungen zur Embryologie und Morphologie der Geisselspinne *Tarantula marginemaculata* C. L. Koch (Arachnida, Amblypygi, Tarantulidae). *Zoomorphologie* **82**: 137-99.

Wolf-Neis, R. 1973. Differenzierungsverlauf von *Carausius*–Embryonen in ovo und in vitro. *Zool. Jahrb. Anat. Ontog.* **91**: 1-18.

Yoshikura, J. 1975. Comparative embryology and phylogeny of Arachnida. *Kumamoto J. Sci. Biol.* **12**: 71-142.

Yoshikura, M. 1972. Notes on the development of a trapdoor spider, *Ummidia fragaria* (Doenitz). *Acta Arachnol.* **24**: 29-39.

Zissler, D. and P. Weygoldt 1975. Feinstruktur der Embryonalen Lateralorgane des Geisselspinne *Tarantula marginemaculata* C. L. Koch (Amblypygi, Arachnida). *Cytobiologie* **11**: 466-79.

3

Significance of Later Embryonic Stages and Head Development in Arthropod Phylogeny

P. WEYGOLDT

3.1. INTRODUCTION

For more than 100 years it has been an open question whether the Arthropoda are a monophyletic taxon that originated from a common ancestor or whether the subphyla have evolved independently from different unknown pre-arthropods. The contradicting opinions and their histories have been summarized by Tiegs and Manton (1958), and the question is far from being settled. Polyphyly has been strongly suggested by Tiegs and Manton (1958), Manton (1958, 1960, 1961, 1964, 1966, 1967, 1973a,b), and Anderson (1966b, 1969, 1972a, 1973). Manton has presented comprehensive morphological evidence in favor of her hypothesis, whereas Anderson has mainly used evidence from early embryonic development of articulates. Siewing (1960, 1969, 1974) and Lauterbach (1972a,b, 1973, 1974), however, have supported the idea of a monophyletic origin of arthropods. Lauterbach (1973) has convincingly reconstructed the key events of arthropod evolution. I also believe that the arthropods are a monophyletic taxon. Their evolution started in pre-Cambrian times, more than 600 million years ago. The first trilobites, crustaceans, and chelicerates are reported from the Cambrian. Myriapods appear in the Silurian and apterygote insects in the Devonian, and winged insects were established in the Carboniferous, more than 350 million years ago (see also Bergström, Chapter 1). The basic arthropod stock from which these subphyla evolved has completely disappeared, as have most pre-Cambrian organisms. Their descendants early occupied different ecological zones, and subsequently went through extensive adaptive radiations, which resulted in quite different types of arthropods, and the complete loss of any possible link between the subphyla.

During their long evolutionary history, the arthropods acquired different habits and adaptations, and have thus become increasingly different. However, convergent evolution has taken place on several occasions. Arthropods started as marine animals, and Crustacea and Chelicerata still possess marine representatives. But the Tracheata, comprising Myriapoda and Insecta, took up terrestrial life at an early stage, as did the Arachnida, and later some Crustacea followed after having evolved similar preadaptations (see also Bergström, Chapter 1 Figs. 1.7, 1.10, 1.18, 1.20). Tracheae have evolved independently more than three times: in the Onychophora, Tracheata, and different Arachnida (see Bergström, Chapter 1, Fig. 1.23, for other instances of convergence.) However, in spite of deep differences in habits and structures of the various subphyla, and even of many convergences, all arthropods have retained a number of basic structural characters, which strongly suggest their monophyletic origin.

3.2. STUDY OF RELATIONSHIPS

The question of arthropod monophyly or polyphyly is the question of taxonomic relationships of arthropods. How do we study relationships? There are two basic methods, both theoretically and fundamentally refined in recent times (see also Bergström, Chapter 1). The first method concerns the search for homologies. Homologous characters are formed because transfer of information has taken place, and, as far as inherited genetic information is concerned, are interpreted as being inherited from a common ancester (Osche, 1973). The operational criteria by which homologies can be detected have been summarized by Remane (1956, 1961). More recently, Hennig (1950, 1965, 1966, 1969) has refined the theoretical background to this approach and thereby founded a clear, logical, and rather objective method of taxonomic research. According to him, the primitive (plesiomorphic) or derived or advanced (apomorphic) state of each character has to be established for any meaningful consideration of phylogeny. Common possession of derivative characters (synapomorphies) alone can be used to establish relationships because they alone are inherited from the most recent common ancestor. This method can produce a dendogram or phylogenetic tree, showing the sequence in which the ancestors of the forms compared have split to produce new taxa. The advantage of this method is that the ways in which the results are obtained are clearly defined, and each step of the procedure can be followed. In spite of the fact that the common ancestor must have had the characters that have been found to be synapomorphic (and a number of symplesiomorphies as well), we do not know the common ancestor or its characters. Because evolution often proceeds at different speeds in different lines, or proceeds in different directions, taxa found to be closely related may be of very different appearance.

Often, only one of two divergent lines acquires new and unique characters (autapomorphy). However, organisms that share one or a number of synapomorphies are related, no matter how different they may look.

The second method involves an approach followed in evolutionary biology. It includes the study of body design, habits, functional requirements, and adaptations. Postulating that evolution is a continuous process, and that each change during evolution brings an advantage to the evolved species or at least a better, a more economic adaptation to its special habits, an attempt is made to reconstruct the pathways of evolution, and the reason why evolution proceeded along those lines. This method has been used by Clark (1964), Manton (1958, 1961, 1964, 1966, 1973a,b), and Lauterbach (1972a,b, 1973, 1974, 1975) for the articulates. It is a very interesting and fascinating approach, but it may lead to quite different results, depending on the taxonomic assumptions underlying ones interpretations, as is evident from a comparison of the works of Manton and Lauterbach. Strickly speaking, this method shows not relationships, but possible lines of evolution. Therefore it cannot replace the approach of establishing homology.

The operational criteria by which homologies are established do not include modes of development (Wilson, 1894-1896; de Beer, 1971). Indeed, early development may be very different even in related forms. This has been clearly shown by numerous embryologists. Development starts from the egg and proceeds through different modes of cleavage and germ layer formation toward a stage that shows most clearly the general body pattern of the group studied (the "Körpergrundgehalt" of Seidel, 1960). Subsequently, organogeny takes place and more specific characters become increasingly apparent.

I should think that some readers, after having read the second part of this chapter, might consider me a typologist. Therefore, two points should be stated: First, the approaches of Manton, Anderson, Hessler and Newman (1975), and Cisne (1974), who have recently published on arthropod evolution, are also typological, perhaps even more so. Second, modern phylogenetic typology has little similarity with the idealistic typology of Goethe. It is indeed an accepted scientific procedure. No generalized or idealistic typus is constructed, but an attempt is made to reconstruct the basic body pattern of the group in question, that is, the symplesiomorphies and synapomorphies of the most recent common ancestor. Since for each character the plesiomorphic or apomorphic state has to be shown, this reconstruction is open to falsification.

3.3. EARLY DEVELOPMENT OF ARTICULATA

3.3.1. Annelida

The most distinctive character of annelid development is spiral (Fig. 3.1) cleavage with a highly mosaic type determination pattern. Two cells are especially important: the 2d cell forming the adult ectoderm, and the 4d cell forming the

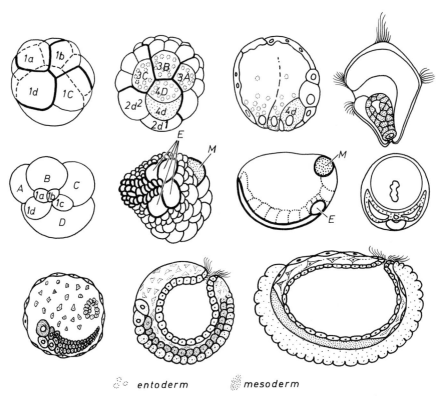

entoderm   *mesoderm*

Fig. 3.1. Early development in annelids. Upper row, from left to right: 8-cell stage from *Arenicola* seen from animal pole: late cleaveage stage of *Arenicola* from vegetative pole; gastrula of *Nereis* in longitudinal section; generalized trochophore. Middle row, from left to right; eight-cell stage of *Tubifex* from animal pole; later cleavage stages of *Tubifex* showing formation of germ bands; longitudinal and cross sections through an older embryo. Lower row, from left to right: Early embryo of *Piscicola*; later embryo of *Piscicola* with newly formed embryonic pharynx; stil later embryo absorbing nutritive fluid. Modified from Child, Wilson, Penners, and Schmidt, from Weygoldt (1963, 1966). *E* = ectoteloblasts; *M* = mesoteloblasts.

mesoderm. Gastrulation proceeds by invagination or epiboly. In Polychaeta, a trochophore is then formed with a teloblastic growth zone at its posterior end. All major organs, including stomodaeum, midgut, proctodaeum, prototroch, apical sense organ, and so on, can be traced back to individual cells, which are homologous in different species (Anderson, 1966a,b, 1973).

This basic pattern of development is modified in clitellate annelids with yolky eggs. Cleavage is highly unequal, and the spiral pattern soon gives rise to a bilateral cleavage pattern. But in most species the 2d cell and the 4d cell can still be recognized. The 2d cell divides into eight ectoteloblasts, which, together with two descendants from the 4d cell, form paired germ bands on either side of the body. There seems to be a tendency to restrict the cells of the A, B, and C quadrants to transitory structures, all definite organs being formed by the D quadrant. In the Piscicolidae and Gnathobdellidae, development is further specialized. A secondary encapsulated larva is formed by transitory larval ectoderm and mesoderm, and a precocious ciliated embryonic pharynx absorbs nutritive fluid stored in the egg capsule. This is later replaced by the definite pharynx (Anderson, 1966b, 1973; Weygoldt, 1966; Siewing, 1969).

3.3.2. Crustacea

Different patterns of development can be observed in the Crustacea. Cleavage of some Entomostraca* resembles spiral cleavage in having, from the 4-cell stage onward, alternatively oblique cell divisions (Fig. 3.2). It is questionable whether crustacean cleavage should be termed spiral, since the prospective fate of individual cells is quite different from those in annelids, molluscs, and other forms with typical spiral cleavage (see page 117). Actually, there are at least two different determination patterns. In Cirripedia, the center of the vegetative pole, shortly before gastrulation, is occupied by the 4D cell, which is the sole source of entoderm (= endoderm). Mesoderm is formed by the other three quadrants and, since the D-quadrant is situated at the future posterior body end, is formed in front of the entoderm. In Cladocera, the D-quadrant forms a primordial germ cell and a primordial entoderm cell, and the center of the vegetative pole is occupied by the germ cell. If the observations of Baldass (1937, 1941) and Kaudewitz (1950) are correct, then the D-quadrant in these eggs marks the future anterior body end, and the mesoderm, which is formed from the other three quadrants, originates posterior to the entoderm anlage. Similarly, in *Daphnia* (Baldass, 1941) and in the ostracod, *Cyprideis* (Weygoldt, 1960), mesoderm migrates inwards or invaginates at the future posterior end of the egg, whereas the entoderm is formed at the anterior end.

Development of the body pattern in *Daphnia* is not a consequence of the cleavage pattern. Although cleavage is highly mosaic, it can be changed by weak centrifugation. Yet in many cases a nearly normal body pattern ensues (Kaudewitz, 1950).

Many Crustacea have yolky eggs, and different degrees of superficial development can be found. But even in the largest eggs with direct development, the first germ band is of the short germ type, comprising only the naupliar segments (Fig. 3.3). In the Malacostraca, the first embryonic anlage is often a V-shaped germ band with a blastoporal area at its posterior end. In *Hemimysis*, Manton (1928) was able to show the spatial relationships of the future germ layers. The

---

*The Entomostraca are not a natural taxon but an assemblage of all nonmalacostracan crustaceans.

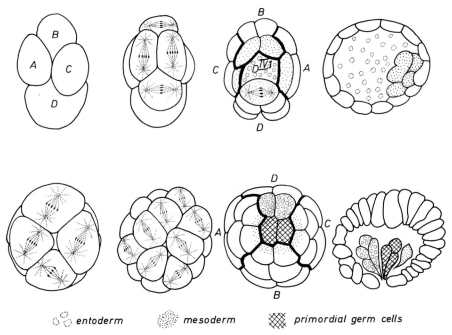

◌ entoderm   mesoderm   ▨ primordial germ cells

Fig. 3.2.   Early development in some Entomostraca. Upper row, from left to right: 4-cell stage and 8-cell stage of *Lepas*, both from animal pole; later cleavage stage immediately before gastrulation from vegetative pole; longitudinal section of gastrula. Lower row, from left to right: 8-cell stage and 16-cell stage of *Cyprideis* from animal pole; 32-cell stages of *Polyphemus* from vegetative pole; longitudinal section of *Polyphemus* gastrula. Modified from Bigelow, Kühn, and Weygoldt, from Weygoldt (1963).

entoderm and primordial germ cells originate from the center of the blastoporal area, and the naupliar mesoderm is at first situated laterally and in front of this. The postnaupliar ectoderm and mesoderm are formed by teloblasts that develop near the anterior blastoporal lip. This seems to be the basic mode of embryo formation in the Malacostraca, although there are variations in different species. In *Hemimysis*, entoderm formation is a prolonged process, new entodermal cells migrating inwards throughout the period of germ band formation. In *Gammarus* (Weygoldt, 1958), gastrulation starts well before the formation of the V-shaped germ band, and the spatial relationships of the future germ layers remain obscure. Mesoteloblasts become visible much later and posterior to the blastoporal area, and ectoteloblasts are missing (Dohle, 1976a). The teloblasts do not form segment anlagen but the material which is later divided into segments (Dohle, 1970, 1972, 1976a,b). In *Caridina*, on the other hand, germ band and blastoporal area are separated by a wide gap of extraembryonic ectoderm (Nair, 1949). Gastrulation, especially entoderm formation, may be two- or more-phasic in sev-

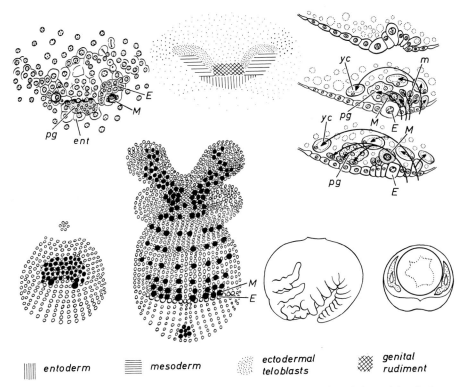

| ||||| entoderm | ≡≡≡ mesoderm | ∷∷ ectodermal teloblasts | ✖✖ genital rudiment |

Fig. 3.3.   Early development in some Malacostraca. Upper row, from left to right: Early germ band of *Hemimysis;* spatial relationships of prospective germ layers in *Hemimysis*; inward migration of germ layers during *Hemimysis* gastrulation. Lower row, from left to right: Early germ band of *Gammarus;* later germ band of *Gammarus*, showing germ band elongation through actions of teloblasts (actually, *Gammarus* is unique in not having well-differentiated ectoteloblasts); side view and cross section through late *Gammarus* embryo. Modified from Manton (1928) and Weygoldt (1963). *E* = ectoteloblast; *ent* = entoderm; *m* = mesoderm; *pg* = primordial germ cells; *M* = mesoeloblast; *yc* = yolk cell.

eral species. Vitellophages, entodermal cells that engage in yolk resorption, often originate from the early germ disc or even blastoderm, but many of them are transitory and do not participate in gut formation (Weygoldt, 1960; Fioroni, 1970a,b; Fioroni and Banderet, 1971).

3.3.3. Insecta

Development of the Insecta is superficial, except for a number of apterygotes and some secondarily specialized species with total cleavage. The germ band may be of the short- or long-germ type, comprising all future segments, or anything in between (Fig. 3.4). It is formed by two halves migrating toward and uniting on the future ventral side, and often involves extensive movements of cells and changes in shape. The short-germ type primordium may be V-shaped and elongates through the action of a posterior growth zone, which only produces a uniform superficial blastema. Germ layer formation is achieved by inward migration or infolding of the midventral cells of the germ band along nearly its entire length. In this way, a gastral grove is formed corresponding to the blastoporal area. The larger part of the infolded material gives rise to the mesoderm, only the anteriormost and posteriormost portions forming the anlagen in the midgut. In apterygote insects and some Myriapoda only the mesoderm is formed in this way, and the midgut develops from vitellophages which originated by radial cell divisions during the blastoderm stage. Insect development is further complicated by the formation of an amniotic cavity and extensive movements of the embryo anlage during blastokinesis (Seidel, 1960, 1966; Anderson, 1972a,b, 1973; Sander, 1976a, b; Siewing, 1969).

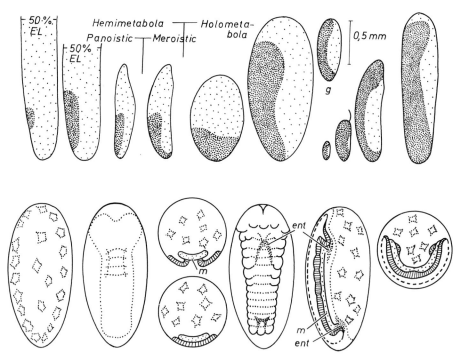

Fig. 3.4. Development in Insecta. Upper row: Eggs of different insects showing relative lengths of early germ bands, extreme short germ type on left side, long germ type on right side. Lower row, from left to right: Cleavage, germ band formation; germ layer formation; later germ band; longitudinal section and cross section through later germ band stage. Modified from Sander (1976b) and Seidel (1960). *ent* = entoderm, *m* = mesoderm.

3.3.4. Chelicerata

In Chelicerata, development sometimes starts with total cleavage, but further development is usually superficial. After the blastoderm stage, a blastoporal area (the primitive plate) is formed first, and subsequently semi- or quartercircular cell aggregations appear in front of the blastoporal area, forming the first anlagen of the prosomal segments (Fig. 3.5). The posterior segments are produced later by a caudal growth zone from which the ectoderm and mesoderm of the opisthosomal segments originate. Gastrulation involves an inward migration of a mesentodermal mass, which proliferates the mesoderm of the prosomal segments and forms the anlagen of the posterior midgut. Prior to this, the cumulus ("cumulus primitivus," "cumulus posterior") separates from the blastoporal area

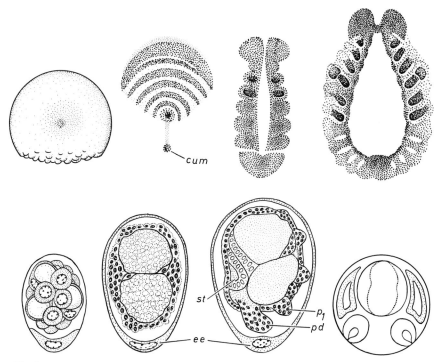

Fig. 3.5.   Development in Arachnida. Upper row, from left to right: Formation of primitive plate in *Tarantula*; appearance of prosomal segment anlagen and separation of cumulus from blastoporal area; early germ band stage; later germ band stage, of *Tarantula*. Lower row: Modified development of pseudoscorpion *Chthonius*, from left to right: Cleavage, formation of body form; later stage, showing precocious development of stomodaeum and embryonic pharynx; cross section through opisthosoma of a much later stage. Modified from Weygoldt (1968, 1975). *cum* = cumulus; *ee* = embryonic envelope; p_1 = first leg anlage; *pd* = pedipalp anlage; *st* = stomodaeum.

and migrates around the future posterior body end of the embryo toward the dorsal side, and then disintegrates. In spite of many investigations, the exact nature of the cumulus remains obscure in most arachnids; in the amblypygid *Tarantula* it proliferates vitellophages that participate in the formation of the anterior midgut (Sekiguchi, 1973; Weygoldt, 1975; Anderson, 1973).

The development of the scorpions and pseudoscorpions is different. Ovoviviparous scorpions with yolky eggs show a pattern similar to discoidal development, but unlike the discoidal cleavage of vertebrates and cephalopods; the germ band forms near the vegetative pole and future ventral side. The first germ band is of an extremely short germ type, much shorter than in other arachnids. In viviparous scorpions and pseudoscorpions, cleavage is total, and the entire development is highly modified and adapted to the requirements of early food intake (Anderson, 1973; Siewing, 1969; Weygoldt, 1964, 1965).

3.3.5. Onychophora

Cleavage in yolky eggs is superficial, but in the yolkless eggs of *Peripatopsis* is modified and somewhat resembles that of scorpions, with blastoderm formation starting from a localized area (Fig. 3.6). During gastrulation, only entoderm enters through the blastopore. The blastopore then closes, but its anterior end elongates and forms a new opening, the mouth-anus, which later divides into mouth and anus. At the posterior border of the blastopore, a growth zone proliferates ectoderm and mesoderm (Manton, 1949; Anderson, 1973).

3.4. EARLY DEVELOPMENT AND TAXONOMY

It is evident from the preceding accounts that early development is quite diverse indeed. But these different developmental patterns all result in a similar general body pattern, that is, an elongate segmented embryo or embryo anlage, with

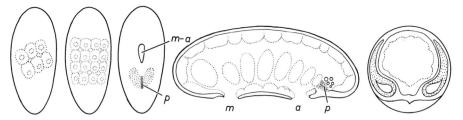

Fig. 3.6.   Development in Onychophora.   From left to right: Early cleavage stage; later cleavage stage; shortly after gastrulation, when anterior part of blastopore persists as mouth-anus, posterior part as mesoderm proliferating area; longitudinal section through later stage; cross section through later stage.   Modified from Manton (1949) and Pflugfelder (1948). *a* = anus; *m* = mouth; *m–a* = mouth-anus; *p* = mesoderm proliferating area.

paired coelomic rudiments, paired ventral nerve cord ganglia, and limb buds, thus clearly indicating the articulate nature of the groups compared.

What can be inferred from the different modes of development about the evolution and taxonomic relationships of arthropods? According to Anderson (1969, 1973), ontogenesis indicates different origins for the arthropod subphyla. Anderson reconstructs fate maps of the most important organs and structures, and from a comparison of them he finds evidence for a relationship between the Tracheata and the Onychophora, which, following Manton, he brings together as Uniramia. Because of some similarities of fate maps and entoderm formation in Onychophora and clitellate Annelida he concludes that the Uniramia have evolved from annelid-like ancestors. In all these forms, mesoderm is formed behind the entoderm, and its prospective area likewise maps behind the presumptive entoderm.

On the other hand, in the Crustacea, according to Anderson (1973), the presumptive mesoderm area is situated in front of the entoderm, or forms a semicircle along the anterior lip of the blastoporal area. Thus, the spatial relationships of entoderm and mesoderm appear reversed in crustaceans: "Since spiral cleavage occurs in Crustacea, we must contemplate the possibility of an ancestral condition of spiral cleavage in which the presumptive mesoderm was segregated as a ring around the presumptive midgut (i.e. radially symmetrical) at the posterior end of the blastula, and think of 4d mesoderm as one specialization, and crustacean mesoderm as another specialization, of the ancestral conditions" (Anderson, 1973, p. 467). Thus we are to think of an ancestral condition from which, on the one hand, all 4d Spiralia, including the so-called Uniramia, have evolved, and on the other hand the Crustacea or their ancestors, which were segmented worms with sprial cleavage, lacking the 4d determination pattern. Either we are to conclude that articulate segmentation has taken place in part of the 4d Spiralia, namely the Annelida-Uniramia, as well as in the wormlike unknown crustacean ancestors with the different mode of spiral cleavage, or we are to believe that the segmented condition is the primitive state of all forms with some sort of spiral cleavage and that segmentation has been subsequently lost in Mollusca, Turbellaria, and the other nonsegmented Spiralia. Although it may be difficult to disprove the above two possibilities, especially the latter, which is even supported by the serial arrangement of structures in the mollusc Neopilina, I find it difficult to accept Anderson's suggestion.

First of all, as long as the observations of Baldass (1937, 1941) and Kaudewitz (1950) (see page 111) are not disproved conclusively, we may have to face the fact that there are at least two very different spatial relationships of presumptive areas in the Crustacea. Furthermore, it seems to me that the fate maps used by Anderson have a rather unequal significance. In the highly mosaic development of annelids and some Entomostraca, the prospective areas can be traced back to the uncleaved egg or to the 4-cell stage. In the dragonfly *Platycnemis* a fate map of the blastoderm stage has been obtained by mapping with UV defects (Seidel, 1961); in the spider *Agelena* vital staining has been used to obtain information of cell movements during gastrulation (Holm, 1952); and in the crustacean *Hemimysis* Manton (1928) was able to elucidate the spatial relationships of the prospective germ layers in the early V-shaped germ band that had been formed by previous cell migrations. In many other cases, the fate maps that Anderson uses are projections from later stages backward to the early germ band stage. Since at least in some insects, crustaceans, and arachnids, germ band formation involves extensive cell movements along the egg surface, it remains doubtful whether these maps really show the spatial relationships of prospective areas of the blastoderm stage. At the same time, fate maps seem to me somewhat meaningless in species that, like pseudoscorpions and polyembryonic parasitic Hymenoptera, have a rather anarchic cleavage, and whose polarity and body axes are probably induced later by some internal or external stimulus.

To support the idea that fate maps are useful taxonomic characters, Anderson (1973) mentions the vertebrates. In vertebrates, fate maps are indeed rather similar, although they have never been used for taxonomic purposes. The vertebrate classes are much more closely related to one another than the subphyla of arthropods and have evolved much later; they are comparable, for example, to the different insect subclasses. But even here fate maps may be different, as becomes evident from a comparison of the classical fate map of Amphibia with the more recently established fate map of the chick embryo by Rosenquist (1966), and from comparison of fate maps of different meroblastic eggs, e.g., fishes (Witschi, 1956), with each other or with those of birds.

Fate maps have their own descriptive and heuristic values. Comparison of those of closely related forms whose relationships have already been established can help us understand evolutionary changes in developmental events. Presumptive areas are probably not as conservative as Anderson assumes. Areas can be shifted or can even completely disappear like the prospective entoderm, which is not represented on the blastoderm surface in some myriapods, apterygote insects, and chelicerates (e.g., *Limulus*). It seems to be a general rule, expressed by Sewertzoff (1931) and Remane (1956), that within undifferentiated meristematic structures, such as the cells of blastula or the segment anlagen of the body, organ-forming areas can be shifted or substituted by adjacent areas. Or, as de Beer (1971) states: "It does not seem to matter where in the egg or the embryo the living substance out of which homologous organs are formed comes from. Therefore, correspondence between homologous structures cannot be pressed back to similarity of position of the cells of the embryo or the parts of the eggs out of which these structures are ultimately differentiated." A persuasive example is the pelvic girdle of vertebrates. The ventral fins of fishes are formed by a variable number of mesodermal and nervous segments. The position of these fins varies in different families, and in the Gadidae the ventral fins are located in front of the pectoral girdle. Similarly, their prospective areas, that is, the segments forming the girdles, vary in position, and the pattern is reversed in species with ventral fins anterior to the pectoral fins. Furthermore, Anderson does not provide any hint as to which characters of fate maps are plesiomorphic or apomorphic, and thus creates a system based solely on the overall similarity of incomplete stages or of the ideas of such stages.

In summary, I do not believe that fate maps and modes of early development can be used to demonstrate a polyphyletic or monophyletic origin of arthropods. We should, however, accept as a general rule that in closely related taxa early development and the arrangement of presumptive areas are usually similar. But different arrangements and modes of development do not necessarily indicate different phyletic origins for the forms compared.

The 4d-type spiral cleavage represents a different situation. Here, homologous organs or germ layers can be traced back to homologous cells of the early blastula stage. Throughout the development, these cells and their descendants can be homologized because of their identical position.

If spiral cleavage is older than annelid segmentation and segmentation is older than arthropodization, then we have to classify the Arthropoda with the Spiralia. However, the evolution of arthropods obviously started with a severe disturbance of spiral cleavage. We can only speculate why and how the different arthropod modes of development evolved. One is usually tempted to think of an increase in the amount of yolk, causing drastic changes in cleavage and developmental pattern. Perhaps the first arthropods had large eggs rich in yolk, similar to those of *Limulus*. Since evolution is not reversible, the first crustaceans, when returning to less yolky eggs, could not resume the 4d-type spiral cleavage. However, we do not know whether the eggs of the first arthropods were poor or rich in yolk; and from the molluscs it is clear that the amount of yolk alone cannot be responsible for a change of cleavage pattern. Gastropods retain spiral cleavage even in large, yolky eggs. Cephalopods, however, have a discoidal cleavage that is retained even in eggs smaller than the largest gastropod eggs (Fioroni, 1967, 1971; Mangold-Wirz and Fioroni, 1970).

A point of special interest is the similarity—or lack of it—of the total cleavage of some Entomostraca and Annelida. In both Annelida and Cladocera, the arrangement of prospective entoderm and mesoderm is similar insofar as the mesoderm is situated behind the entoderm. However, the D cell of the 4-cell stage marks the future posterior (and dorsal) body end in Annelida, but the anterior (and ventral) end in Cladocera. In fact, the only characters by which the D cell is recognized are its large size and the fact that it reaches farthest toward the vegetative pole. Thus, this largest cell will give rise to the entoderm, which forms at the center of the vegetative pole. It is easy to envisage that a slight change of the inclination of the cleavage axis may result in another blastomere containing most of the vegetative material (Fig. 3.7). Clearly, the D cells in the

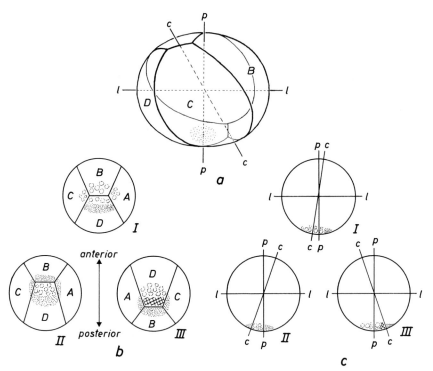

Fig. 3.7.   a, 4-cell stage of *Daphnia*, indicating the three most important axes, as seen obliquely from the left side and from the animal pole; b, prospective areas and four quadrants of the 4-cell stages of three cleavage types, I = quartet-4d-spiral cleavage, II = *Lepas*-type cleavage, III = *Polyphemus*-type cleavage; c, position of the cleavage axis in the same cleavage types in diagrammatic longitudinal sections. Modified from Baldass (1942) and Weygoldt (1962).   *c–c* = cleavage axis; *l–l* = longitudinal axis; *p–p* = polarity axis. The situation is simplified here in order to make it easier to understand. Actually, the polarity axis does not run perpendicularly to the longitudinal axis, but the polarity axis and the cleavage axis should be turned counterclockwise so that the animal pole approaches the future anterior body end. The D-quadrant then becomes the anterior and ventral quadrant in *Polyphemus* and *Daphnia* but the posterior and dorsal quadrant in spiral cleavage and *Lepas* cleavage. If cirripedian cleavage is a "one-quadrant" type (see text), then this interpretation is only valid for comparison of spiral cleavage and cladoceran cleavage.

Annelida and in different Crustacea are not homologous structures but blastomeres that were designated D cell simply because of their size and the fact that they contain most of the vegetative pole plasm. This, however, does not explain the different arrangement of prospective areas in the Cirripedia. Anderson's assumption that the crustacean mesoderm primarily originated as a ring around the entoderm is probably correct, and such a situation seems still to exist in some Crustacea, e.g., *Lepas* (Bigelow, 1902) and *Penaeus* (Zilch, personal communication). But this type of cleavage is not a typical spiral cleavage. This is also apparent from the discussion of Costello and Henley (1976) of aberrant types of spiral cleavage. According to these authors, cirripedian cleavage is a "one-quadrant" system, the first cleavage furrow separating 1D and 1d. I feel myself unable to understand this. If we assume that annelid-like ancestors and all subsequent forms up to the cirripedia had a total cleavage, then either the first four cells have to be designated A, B, C, and D, independent of what is later formed by each, or we assume with Costello and Henley that some ancestor started with dexiotrop and laeotrop cell divisions from the egg onward, instead of from the 4-cell stage onward. It is possible that this could have happened by the suppression of the first two cleavage divisions. Then, the first two cells would have to be designated 1D and 1d; but this is a formalized approach, since there is no way of homologizing these cells with those of the Cladocera, and so on.

It may seem disappointing to learn that the study of early development does not provide characters useful for taxonomic research. Indeed, relationships are best studied by morphological, ethological, and biochemical investigations, which have to be subjected to the homology criteria and the method of Hennig. How then can we use comparative embryology in taxonomic research? First, one can search for complex characters of embryonic stages that can be subjected to the morphological and taxonomic methods. This has been demonstrated by Dohle (1976a) for peracarid Crustacea. Second, one of Remane's (1961) operational criteria for homologies may involve developmental stages: two different characters are homologous when they can be linked by a continuous row of intermediates. The intermediate states may be found in living forms, fossils, or ontogenetic stages. Third, the general body pattern ("Körpergrundgestalt") and its further development provide much information relevant to phylogenetic questions.

The general body pattern indicates the articulate nature of the forms in question. It also shows specific characters of the different subphyla and classes, e.g., number of segments, number of limb buds, early formation of tagmata, and so forth. Thus this stage itself does not provide clues to the phyletic origin of the arthropods, but the study of its further development can be very helpful and sometimes is the only method by which body design and composition of certain complex structures can be investigated. Thus our understanding of the composition and evolution of the arthropod head has been greatly enhanced by study of its ontogenetic development, and I will use this example to demonstrate how comparative embryology can contribute to the problem of morphology and to the question of the arthropod interrelationships.

3.5. DEVELOPMENT OF THE ARTHROPOD HEAD

Although the papers of Manton and Tiegs and Manton are difficult to interpret because the apomorphies and plesiomorphies are not clearly stated, it seems to me that two of the most important arguments used in favor of their hypothesis of arthropod polyphyly are directly concerned with head morphology. According to their hypothesis, there is one evolutionary line leading from annelid-like ancestors to the Onychophora, Myriapoda, and Insecta, which are united as the "Uniramia." The Crustacea have a different, unknown origin, as do the Chelicerata, which may have evolved from the Trilobita. The first and most important argument is the nonhomology of the mandibles in the Tracheata and Crustacea. The masticatory process of the crustacean mandible is a coxal endite or gnathobase, but in the Tracheata, it is said to be the tip of the whole appendage. The second argument relates to the number of head segments. The Onychophora possess one pair of feeding claws. From this monognathan condition the dignathy of some Myriapoda, and finally the trignathy of most Tracheata, are said to have evolved. Similarly, a second maxilla is missing in some primitive Crustacea, and the trignathy of the Crustacea is therefore said to be the result of convergent evolution.

The crustacean head bears five pairs of appendages, suggesting the presence of the acron and five limb-bearing segments. There are two pairs of antennae, one pair of mandibles, and two pairs of maxillae. In insects, the second antenna is missing, but its corresponding segment, the premandibular segment, is clearly visible and may even develop vestigial limb buds during the late germ band stage. The homology of the premandibular and the antennal segments is emphasized by the stomatogastric nervous system arising from the ganglia of this segment (the tritocerebrum). In addition to the somites of these segments, another pair of somites, the preantennulary somites, appears during embryonic development. These somites give rise to the mesoderm and muscles of the labrum and stomodaeum in insects (Scholl, 1969; Anderson, 1972a). In the Crustacea, they also form the anterior aorta and participate in the development of the mesodermal sheath of the brain (Manton, 1934; Weygoldt, 1961). In a nearly identical manner, the precheliceral somites of the Chelicerata, which are homologous to the preantennulary somites of the Crustacea, form the anterior aorta, mesoderm of the upper lip and stomodaeum, and a mesodermal sheath of the brain (Weygoldt, 1975) (Fig. 3.8). The upper lip or labrum appears as an outgrowth in front of the developing mouth. In some Crustacea, Insecta, and Arachnida, its first anlage may be bilobed or even paired, and then resembles limb buds. The somites, ganglionic anlagen, and limb buds of the antennulary segment often first appear laterally or even behind the mouth (Fig. 3.9). This spatial relationship is soon changed by backward migration of the mouth, and forward shifting of the antennulary and even antennary (premandibular) segment halves. Thus, a backwardly directed mouth is formed and the antennulae attain a preoral position. The brain is formed by the V-shaped anterior part of the germ band. Anterior to the antennulary ganglionic masses (the anlagen of the deutocerebrum), there are usually an anterior median pair of ganglionic thickenings or lobes and lateral lobes.

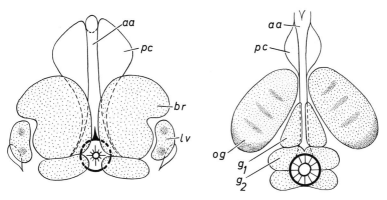

Fig. 3.8. Comparison of brain anlagen and precheliceral or preantennulary coelom in an arachnid (*Tarantula*, left) and crustacean (*Palaemonetes*, right). Modified from Weygoldt (1975). *aa* = anterior aorta; *br* = brain anlagen; g_1 = anteriormost ganglionic anlage, sometimes thought to represent ganglion of preantennulary segment; g_2 = anlage of deutocerebrum; *lv* = lateral vesicle, perhaps homologous to optic ganglia; *og* = optic ganglia; *pc* = precheliceral or preantennulary coelom, respectively.

The protecerebrum is formed by the anterior median anlagen and part of the lateral lobes, which also participate in the formation of the optic ganglia and the compound eyes. The anterior median pair of the brain anlagen is sometimes believed to belong to the segment of the preantennulary somites (Malzacher, 1968; Scholl, 1969; Siewing, 1969), but the evidence is not conclusive. In crustaceans, insects, and arachnids, the part of the brain formed by these lobes later innervates the median eyes and forms the central body. Further similarities of the insect and crustacean head are: nearly identical compound and median eyes (see Paulus, 1972, 1973, 1974, and Chapter 6) and segmental organs that were pri-

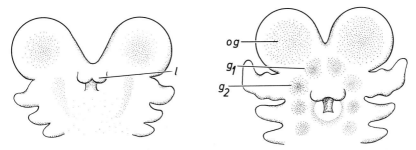

Fig. 3.9. Naupliar region of germ band of *Gammarus*. Left: Early stage; right: later stage. Note that mouth and labrum first appear in front of antennular anlagen. Modified from Weygoldt (1958). g_1 = anteriormost ganglionic anlage; g_2 = anlage of deutocerebrum; *l* = labrum; *og* = optic ganglia.

marily present in the same segments (Tiegs and Manton, 1958). Often one of the pairs of segmental organs is reduced, and in tracheates the function of the labial gland changes, so that they serve as salivary glands in most higher insects. The anlagen of the premandibular glands are present only in the embryos of some Myriapoda, and soon disintegrate and form nephrocytes.

Thus, there is ample morphological and embryological evidence to support the homology of the crustacean and tracheate head. Both the criteria of similar structure and those of identical position within a complex system apply for most structures, and it becomes highly improbable that such a complex head has evolved convergently several times. This is also true of the mandibles, whose homology is doubted by Tiegs and Manton (1958). The mandibles of the Tracheata are said to be whole-limb mandibles, not gnathobases, because there is never any trace of a palpus and because they are articulated in some myriapods. But arthropods are not always so conservative as to recapitulate structures no longer used. Isopods and amphipods, for example, lack exopods on their pereipods, and there never is a trace of them, even during embryogenesis. Further, an articulated mandible has evolved in the isopod *Rocinela* (Aegidae) (Lauterbach, 1972a), and immature stages of *Argulus* (Branchiura) also have mandibles with a movable endite. These mandibles possess a palpus, but if this palpus were missing, the mandible would really look like a whole-limb mandible. This shows that a gnathobase can become articulated. Finally, Snodgrass (1952) and Boudreaux (Chapter 9) have shown that the muscles of the articulated mandibles of Myriapoda are truly coxal muscles quite different from those of a telopodite.

The most persuasive character demonstrating the synapomorphous nature of the mandibles, however, is their position together with the fate of the second antennae. The fact that the appendages of the premandibular segment have been lost in the Tracheata, instead of being used as mouth parts, strongly suggests that in the ancestor of the Tracheata this segment had already reached a paraoral position with its appendages inserting close to the base of the labrum, perhaps similar to the condition found in the most primitive crustaceans, the Cephalocarida. Since the mandible of the common ancestor of the Mandibulata was probably quite different from that of modern insects or crustaceans—a leglike appendage with a small endite—it may be asked how a mandible could be defined, or, in other words, what is the synapomorphy of the mandible, as opposed, for example, to the first leg of *Limulus*? It now appears that the criterion of identical position is the most important one: a head appendage is a mandible if and only if a coxal endite is used to aid in food intake, if this appendage belongs to the third head segment (preantennulary segment not counted), and if it is the first or anteriormost mouth part of the adult head.

Indeed, the arthropod head is the most reliable indication of arthropod monophyly. A clear picture of its evolution can be derived from the studies of embryology, morphology, and habits.

3.6. POSITION OF THE ARTHROPOD MOUTH

One of the most conspicuous features of the primitive arthropod head is a backwardly directed mouth, as opposed to the terminal mouth of annelids. The floor of the preoral cavity was formed by a large labrum (Manton, 1964; Lauterbach, 1973; Cisne, 1974), which, according to old ideas recently revived by Lauterbach (1973) and Rempel (1975), was originally formed by the medially fused appendages of the first (preantennulary) segment. The formation of the labrum and the preoral shifting of the first antennae were part of the first key events of arthropod evolution. They took place well before the Cambrian, and are so firmly fixed now that the anlagen of the labrum never appear behind the mouth. Later, but also still in pre-Cambrian times, more segments became fused to this tagma, and the definitive head was established. Arthropods probably started as marine mud and fine particle feeders living on the surface of the substratum. Particles were transported to the preoral cavity by the movements of the thoracic and head appendages. Only the first antennae did not participate in food collection but served as sense organs. This ancestral condition was present in the trilobites, and similar situations can still be found in Cephalocarida and Branchiopoda and, in a somewhat modified form, in merostomes that became adapted to crush and feed on larger food items.

In some higher Crustacea as well as in Tracheata and Arachnida a second process convergently caused the mouth to return toward a more terminal position. This was not achieved by reversing the relative movements of the mouth and head appendages that had formed the backwardly directed mouth, but by elevation of the anterior head parts. Elevation was probably triggered by a change of habits and feeding methods, and accompanied by elaboration of mandibular structure and function as shown by Manton (1964). It was the necessary prerequisite for terrestrial life, because particle feeding of the primitive type is impossible on land. The process of elevation took place much more recently than the first steps of head formation, and is clearly recapitulated during ontogeny. In higher crustaceans, especially in the Amphipoda and Isopoda, and in the Tracheata and Arachnida, the anterior parts of the head and brain anlagen are bent upward and in arachnids even backward on the dorsal side. The originally anteriormost parts of head and brain thereby assume a dorsal position in insects and in crustaceans, such as isopods and amphipods. In Chelicerata, the anteriormost part of the developing brain is the central body. In *Limulus*, it retains this anterior position, but in arachnids, elevation causes the central body finally to occupy the posteriormost part of the supraesophageal ganglionic mass.

3.7. CONCLUSIONS

I have tried to show how and which part of comparative embryology can contribute to phylogenetic and taxonomic considerations relevant to arthropod evolution. Early development, especially if it is very diverse, does not provide useful clues that might help solve the question of monophyly or polyphyly. Adult members of a given phylum are exposed to selection. They have assumed different habits and niches and thus become very diversified. If their ontogeny is followed backward, the developmental stages become increasingly similar up to the stage that shows the basic body pattern. This was first clearly observed by Baer (1828) and has led to the so-called biogenetic law of Haeckel (1866; see also Remane, 1960). Eggs also are exposed to selection. They have to face the problems of yolk resorption and different sizes, and so on. Thus, within a given taxon, very different modes of early development converge toward a similar basic body pattern, and afterwards, development diverges again to the different adult forms (Seidel, 1960).

Using head development, I have tried to show how embryology can contribute to morphology and to an understanding of complex structures. I am well aware of the fact that the composition of the head, as outlined here, is not generally accepted. A number of investigators deny the existence of a preantennulary segment. The different theories concerning the insect head have been summarized by Weber (1952) and Rempel (1975) and aptly criticized by Manton (1949, 1960). The question of whether or not the preantennulary (or precheliceral) coelomic cavities represent a reduced segment fused with the acron and the question of whether the labrum is formed by medially fused appendages, or is simply an outgrowth of the anterior border of the mouth, are pertinent for an understanding of the evolution of the arthropod head as outlined here. But even if this theory is not accepted, the identical positions of these and the other head structures remain, and strongly support their homologous and synapomorphic nature and the monophyletic origin of the Euarthropoda.

Manton and Anderson, on the other hand, although in agreement with the proposed composition of the arthropod head, do not consider it very significant. According to Anderson (1972a, p. 148), " . . . the phylogenetic importance once attached to the comparative segmental composition of the arthropod head has now been outmoded by recent evidence (e.g., Manton, 1964, 1973a; Anderson, 1966a,b, 1969) that neither the Crustacea nor the Arachnida have played any part in the evolutionary history of the onychophoran-myriapod-hexapod assemblage of arthropods. The question of the presence or absence of a vestigial preantennal segment in the pterygote head thus becomes less significant." The most pertinent question, therefore, remains that of whether the Crustacea and Tracheata form a natural taxon, the Mandibulata (Snodgrass, 1938) (see also Bergström, Chapter 1), or whether the Onychophora are related to the Tracheata. This question cannot be answered by embryology alone. I do not see any convincing homology, or better, synapomorphy, that the Onychophora share with the Tracheata. All similarities seem to be convergent acquisitions related to similar habits of onychophorans and myriapods, and symplesiomorphies that onychophorans share with all arthropods or even articulates. The feeding claws are not at all homologous to the mandibles because of their different morphology and position. Compared with the mandibles, which are the third or even fourth (if the labrum also is counted) head appendages, the feeding claws are the second head appendage. (Some authors have claimed that there exists a preantennulary segment in the Onychophora also, and Pflugfelder, 1948, even suggests the presence of a premandibular segment, thus indicating an identical position of the manidbles of Arthropoda and feeding claws of Onychophora. However, as long as these observations are not confirmed, I consider the observations of Manton, 1949, more convincing.) The feeding claws are a clear autapomorphy of the Onychophora. This is probably also true of the gaits, which are very important in Manton's discussions. The uniqueness of the Onychophora is even appreciated by Tiegs and Manton (1958): "The development of the mesogloea-like subcutaneous connective tissue skeleton by the Onychophora or their ancestors, and the stiff scutes by other arthropods, may have been an early parting of the ways of primitive arthropods." They state further: "Moreover, an appreciation of the real specializations of the Onychophora indicates how this group could have afforded to neglect the advancement of its crawling or running, which would in any case have been difficult without the evolution of scutes." The question is not whether "Crustacea or Arachnida have played any part in the evolutionary history" of the Tracheata, but whether they have had a common ancestor. That, indeed, is strongly suggested by several synapomorphies.

3.7.1. Plesiomorphies and Apomorphies of the Arthropod Head

A primitive head with preoral antennae, four pairs of postoral biramous appendages, and a large labrum was present in the Trilobita and is still recaptitulated in many Crustacea. It was also realized in the ancestors of the Tracheata. Compared with the Annelida and Onychophora, this is a synapomorphy indicating the monophyletic nature of the Euarthropoda.* Other characters of the primitive arthropod head were: a ventral food groove, a cephalic fold surrounding the lateral and anterior margins of the head (Lauterbach, 1973), and compound eyes that might have been different from those of the Mandibulata. Within the Euarthropoda, these are all symplesiomorphic characters, which cannot be used for the study of the relationships of the different subphyla.

*Palaeozoic Arthropoda underwent intensive adaptive radiation and developed a surprising diversity. In many of these forms the number of head segments was different from that of the primitive head as outlined above. In some trilobites the last head appendage was missing, and *Chelonellion* had one additional appendage. None of these forms can be regarded as ancestors of the recent Arthropoda, and their existence does not affect the conclusions about the relationships of the arthropod subphyla. Similarly, the variable number of head (cephalothoracic) segments of recent Crustacea or Myriapoda does not affect the conclusion that in the Mandibulata there are primarily five pairs of head appendages.

The next step of evolution was the forward shifting of the first postoral appendages and the transformation of the second pair of postral appendages into the mandibles. This is the most important synapomorphy of the Mandibulata. More synapomorphies are the similarities mentioned on p. 123. Whether the ancestors of the Mandibulata already possessed two pairs of maxillae is an open question. In the most primitive Crustacea (the Cephalocarida), the second maxillae are leglike; but they are true head appendages, as is indicated by the position of their intersegmental tendon (Hessler, 1964), and a similar condition may have occurred in the first Tracheata. The second maxillae may, therefore, be homologous appendages, which convergently evolved their specialized maxillar morphology and function (that is, a homoiology). The fact that the segment of the second maxillae was primarily present in the Mandibulata is also indicated by the embryological investigations of Dohle (1964, 1965), who found that this segment is present—without appendages—in the embryos of the dignathan Diplopoda.

The Chelicerata are usually believed to have evolved from trilobites (Schulze, 1937; Störmer, 1944; Johansson, 1933; Tiegs and Manton, 1958; Bergström, Chapter 1, Fig. 1.18). They have reduced the antennae and the whole antennal segment. Any attempts to find traces of this segment (Johansson, 1933; Pross, 1966, 1977; Schoel, 1977) have failed because the evidence is not conclusive and a different interpretation seems more convincing (Weygoldt, 1975). They have further added two more segments to the head and thus formed the prosoma. But the genetic information for the formation of the old head without these two segments seems still to be present and is realized in some arachnids (schizomids, solpugids, palpigradids, and mites). However, most characters used to relate the Chelicerata to the Trilobita are symplesiomorphies, which do not indicate close affinity, and it may be difficult to find convincing synapomorphies. The cephalic fold and pleurotergites that are thought to be common characters of Trilobita and Chelicerata have been proved to be symplesiomorphies by the discovery of the Cephalocarida. Usually the leg structure is considered to be a synapomorphy. The lateral ramus of the leg of a trilobite and the flabellum of the fifth prosomal leg of *Limulus* are said to be preepipodites instead of exopods. Lauterbach, however, maintains that these lateral rami are exopods, homologous of those of the Crustacea, and that biramous appendages are the plesiomorphic character of all Arthropoda. But even if this is correct, the exopods of trilobites and of *Limulus* originate from a more basal position than those of the Crustacea; perhaps this is a synapomorphy. Both Trilobita and Chelicerata, and probably their ancestors also, acquired early larger sizes and a much stronger exoskeleton than the early Mandibulata. Olenellid-like ancestors probably gave rise to the early Chelicerata. Indeed, some olenellid species and some early species of the Limuloidea with fully segmented opisthosoma are nearly indistinguishable. A characteristic feature of these forms is the gradual replacement of the telson by a large tergal spine, which perhaps persists as the caudal spine in the Xiphosura and Gigantostraca and as a sting in scorpions (Störmer, 1944, p. 30; Fage, 1949, Fig. 33; Lauterbach, personal communication). The Pycnogonida are an early branch of the chelicerate line (see also Paulus, Chapter 6, and for a different viewpoint Bergström, Chapter 1). Their anterior complex with cheliphores and palps seems to be homologous to and synapomorphous with the chelicerae and pedipalps of the chelicerates; and their proboscis, although unique, is probably a very specialized sucking apparatus that could have evolved from less specialized preoral structures similar to those of some mites. Pycnogonids do not share any synapomorphies with any other arthropod subphylum, and they are no more different from the arachnids than are the caprellids from some entomostracan crustaceans. Cisne (1974) and Hessler and Newman (1975) found similarities of the Trilobita to the Cephalocarida, and, therefore, concluded that the Trilobita-Chelicerata-Crustacea form a natural group. But it is evident, and is admitted by Cisne, that the characters used are primitive characters or symplesiomorphies. These characters do not indicate relationship in the sense of Hennig; that is, they do not exclude the possibility of a more recent common ancestor of Crustacea and Tracheata. In fact, the urscrustacean of Hessler and Newman is not a crustacean but a creature that, perhaps except for its stalked eyes, is composed solely of plesiomorphic characters, and thus resembles the euarthropod ancestor.

3.7.2. Phyletic Relationships of the Arthropod Subphyla

In summary, all morphological and embryological data concerning head development strongly support the old view that the Arthropoda are a monophyletic taxon. They are related to the Annelida. Any attempt to link the Arthropoda or parts of them to certain groups of annelids (Sharov, 1966; Iwanoff, 1928) have failed. The arthropod branch early divided into the Onychophora and the Euarthropoda. The first splitting of the Euarthropoda occurred in pre-Cambrian times and led to the appearance of the Arachnata (=Amandibulata) and Mandibulata (Fig. 3.10). Also in the pre-Cambrian, the Arachnata branch divided into the Trilobita and Chelicerata, and the Mandibulata branch gave rise to the Crustacea and Tracheata. The data used here are not new, nor is the conclusion. But it is the only conclusion one arrives at if the data are subjected to the modern methods of taxonomy and systematics.

However, this conclusion has certain implications. The fact that synapomorphies of the Trilobita and Arachnida are difficult to state has already been mentioned. The synapomorphies of the Mandibulata are much clearer. According to the present hypothesis, the primitive Mandibulata possessed biramous appendages. Within the Mandibulata, the Tracheata are a well-defined, apomorphic group, and the synapomorphies are apparent. The Crustacea, on the other hand,

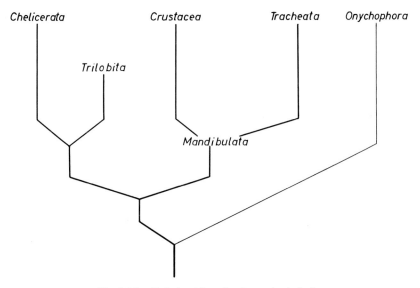

Fig. 3.10.   Relationships of arthropod subphyla.

are poorly defined, because all the characters of primitive Crustacea, e.g., those that Hessler and Newman used to reconstruct the urscrustacean, are plesiomorphies which only show that Crustacea are mandibulate arthropods. The question of the synapomorphies of the Crustacea seems indeed worth careful consideration. At first sight this implication may seem strange, because a crustacean is easy to recognize. However, careful examination and character weighting lead to similar results in many invertebrate and vertebrate animal groups. For example, within the Tracheata, the Insecta are the apomorphic group with a number of synapomorphies, but the question of whether the Myriapoda are a monophyletic, paraphyletic, or even polyphyletic assemblage is still controversial. The situation is similar in the Annelida. The Clitellata are a well-defined, apomorphic group, but what are the synapomorphies—if there are any—of the Polychaeta?

3.8. SUMMARY

Evolution of the Arthropoda started in the pre-Cambrian. The primitive stock from which the different subphyla evolved has completely disappeared, as have all possible links between the subphyla. The question of arthropod monophyly or polyphyly is the question of the taxonomic relationships of the subphyla. To establish relationships, homologies and synapomorphies that the subphyla share with each other or with different animal phyla have to be demonstrated. The operational criteria by which homologies can be established do not include modes of early development. Indeed, development of related forms can be very different. By substitution, organforming areas can be shifted from one part of an undifferentiated embryonic complex or of the blastula to another, and fate maps of related and similar forms may be of very different appearance. Study of the later stages of development is more promising. During and after segment formation, the general body patterns become apparent. Then can be studied structures and their homologies that in still later stages may become obscured. Knowledge of the composition and evolution of the arthropod head has been greatly enhanced by embryological studies. The heads of insects and crustaceans are composed of the same number of segments. Appendage rudiments of the first (preantennulary) segment form the labrum. Fusing in front of the mouth, they form a backwardly directed preoral cavity, which is the primitive condition of the arthropod head. Only later, and convergently in different arthropods, did the elevation of the anterior head parts cause the mouth to resume a more terminal position. Embryogenesis shows the formation of the head, and recapitulates some of the events that have occurred during its evolution, thus providing support for a monophyletic origin of the arthropods.

REFERENCES

Anderson, D. T. 1966a. The comparative embryology of the Polychaeta. *Acta Zool. (Stockh.)* 47: 1–42.

Anderson, D. T. 1966b. The comparative early embryology of the Oligochaeta, Hirudinea and Onychophora. *Proc. Linn. Soc. N.S.W.* 91(1): 10–43.

Anderson, D. T. 1969. On the embryology of the cirripede crustaceans *Tetraclita rosea* (Krauss), *Tetraclita purpurascens* (Wood), *Chthamalus antennatus* (Darwin) and *Chamaesipho columna* (Spengler) and some considerations of crustaceans phylogenetic relationships. *Phil. Trans. R. Soc. Lond.* B 256(806): 183–235.

Anderson, D. T. 1972a. The development of hemimetabolous insects, pp. 95–163. *In* S. J. Counce and C. H. Waddington (eds.) *Developmental Systems—Insects.* Academic Press, New York, London.

Anderson, D. T. 1972b. The development of holometabolous insects, pp. 165–242. *In* S. J. Counce and C. H. Waddington (eds.) *Developmental Systems—Insects.* Academic Press New York, London.

Anderson, D. T. 1973. *Ambryology and Phylogeny in Annelids and Arthropods.* Pergamon Press, Oxford.

Baer, K. E. von. 1928. *Über die Entwicklungsgeschichte der Tiere.* Königsberg.

Baldass, F. von. 1937. Entwicklung von *Holopedium gibberum. Zool. Jahrb. Anat. Ontog.* 63: 399–454.

Baldass, F. von. 1941. Die Entwicklung von *Daphnia pulex. Zool. Jahrb. Anat. Ontog.* 67: 1–60.

Bigelow, M. A. 1902. The early development of *Lepas.* A study of cell lineage and germ layers. *Bull. Mus. Comp. Zool. Harv.* 40: 61–144.

Cisne, J. L. 1974. Trilobites and the origin of arthropods. *Science (Wash., D.C.)* 186: 13–18.

Clark, R. B. 1964. *Dynamics in Metazoan Evolution: The Origin of the Coelom and Segments.* Oxford University Press, Oxford.

Costello, D. P. and C. Henley. 1976. Spiralian development: A perspective. *Amer. Zool.* **16**: 277-91.

de Beer, G. R. 1971. Homology, an unsolved problem, pp. 1-16. *In* J. J. Head and O. E. Lowenstein (eds.) *Oxford Biology Readers* 11. Oxford University Press, Oxford.

Dohle, W. 1964. Die Embryonalentwicklung von *Glomeris* marginata (Villers) im Vergleich zur Entwicklung anderer Diplopoden. *Zool. J. Anat. Ontog.* **81**: 241-31.

Dohle, R. 1965. Über die Stellung der Diplopoden im System. *Verh. Dtsch. Zool. Ges.* **1964**: 587-606.

Dohle, W. 1970. Die Bildung und Differenzierung des postnauplialen Keimstreifs von *Diastylis rathkei* (Crustacea, Cumacea) I. Bildung der Teloblasten und ihrer Derivate. *Z. Morphol. Tiere* **67**: 307-92.

Dohle, W. 1972. Über die Bildung und Differenzierung des postneuplialen Keimstreifs von *Leptochelia* spec. (Crustacea, Tanaidacea). *Zool. Jahrb. Anat. Ontog.* **89**: 503-66.

Dohle, W. 1976a. Zur Frage des Nachweises von Homologien durch die komplexen Zell- und Teilungsmuster in der Embryonalentwicklung höherer Krebse (Crustacea, Malacostraca, Peracarida). *Sitzb. Ges. Nat. Freunde Berlin (n.f.)* **16(2)**: 125-44.

Dohle, W. 1976b. Die Bildung und Differenzierung des postnauplialen Keimstreifs von *Diastylis rathkei* (Crustacea, Cumacea) II. Die Differenzierung und Musterbildung des Ektoderms. *Zoomorphologie* **84**: 235-77.

Fage, L. 1949. Classe des Mérostomacés (Merostomata, Woodward (1866), pp. 219-62. *In* P. P. Grassé (ed.) *Traité de Zoologie VI*. Masson et Cie, Paris.

Fioroni, P. 1967. Molluskenembryologie und allgemeine Entwicklungsgeschichte. *Verh. Naturforsch. Ges. Basel* **78(2)**: 283-307.

Fioroni, P. 1970a. Die organogenetische und transitorische Rolle der Vitellophagen in der Darmentwicklung von *Galathea* (Crustacea, Decapoda, Anomura). *Z. Morphol. Tiere* **67**: 263-306.

Fioroni, P. 1970b. Am Dotteraufschluß beteiligte Organe und Zelltypen bei höheren Krebsen; der Versuch zu einer einheitlichen Terminologie. *Zool. Jahrb. Anat. Ontog.* **87**: 481-522.

Fioroni, P. 1971. Die Entwicklungstypen der Mollusken, eine vergleichend-embryologische Studie. *Z. Wiss. Zool.* **182(3/4)**: 263-394.

Fioroni, P. and E. Banderet. 1971. Mit dem Dotteraufschluss liierte Ontogenese-Abwandlungen bei einigen decapoden Kresben. *Vie Milieu Ser. A Biol. Mar.* **22(1) A**: 163-88.

Haeckel, E. 1866. *Generelle Morphologie*. Berlin.

Hennig, W. 1950. Grundzüge einer Theorie der phylogenetischen Systematik. Deutscher Zentralverlag, Berlin.

Hennig, W. 1965. Phylogenetic systematics. *Annu. Rev. Entomol.* **10**: 97-116.

Hennig, W. 1966. *Phylogenetic Systematics*. University of Illinois Press, Urbana, Chicago, London.

Hennig, W. 1969. *Die Stammesgeschichte der Insekten*. Verl. W. Krammer, Frankfurt.

Hessler, R. R. 1964. The Cephalocarida: Comparative skeletomusculature. *Mem. Conn. Acad. Arts Sci.* **16**: 1-97.

Hessler, R. R. and W. A. Newman. 1975. A trilobitomorph origin for the Crustacea. *Fossils Strata* **4**: 437-59.

Holm, A. 1952. Experimentelle Untersuchungen über die Entwicklung und Entwicklungsphysiologie des Spinnenembryos. *Zool. Bidr. Upps.* **30**: 199-222.

Iwanoff, P. P. 1928. Die Entwicklung der Larvalsegmente bei den Anneliden. *Z. Morphol. Ökol. Tiere* **10**: 62-161.

Johansson, G. 1933. Beiträge zur Kenntnis der Morphologie und Entwicklung des Gehirns von *Limulus polyphemus*. *Acta Zool.* (Stockh.) **14**: 1-100.

Kaudewitz, F. 1950. Zur Entwicklungsgeschichte von *Daphnia pulex*. *Roux Arch. Entw.-Mech.* **144**: 410-77.

Lauterbach, K.-E. 1972a. Über die sogenannte Gansbein-Mandibel der Tracheata, insbesondere der Myriapoda. *Zool. Anz.* **188(3/4):** 145–54.

Lauterbach, K.-E. 1972b. Die morphologischen Grundlagen für die Entstehung der Entognathie bei den apterygoten Insekten in phylogenetischer Sicht. *Zool. Beitr. N.F.* **18(1):** 25–69.

Lauterbach, K.-E. 1973. Schlüsselereignisse in der Evolution der Stammgruppe der Euarthropoda. *Zool. Beitr. N.F.* **19(2):** 251–99.

Lauterbach, K.-E. 1974. Über die Herkunft des Carapax der Crustaceen. *Zool. Beitr. N.F.* **20(2):** 273–327.

Lauterbach, K.-E. 1975. Über die Herkunft der Malacostraca (Crustacea) *Zool. Anz.* **194(3/4):** 165–79.

Malzacher, P. 1968. Die Embryogenese des Gehirns paurometaboler Insekten: Untersuchungen an *Carausius morosus und Periplaneta americana. Z. Morphol. Tiere* **62:** 103–61.

Mangold-Wirz, K. and P. Fioroni. 1970. Die Sonderstellung der Cephalopoden. *Zool. Jahrb. Syst.* **97:** 522–631.

Manton, S. M. 1928. On the embryology of a mysid crustacean, *Hemimysis lamornae. Phil. Trans. R. Soc. Lond.* **B 216:** 363–463.

Manton, S. M. 1934. On the embryology of the crustacean, *Nebalia pipes. Phil. Trans. R. Soc. Lond.* **B 223:** 168–238.

Manton, S. M. 1949. Studies on the Onychophora VII. The early embryonic stages of *Peripatopsis*, and some general considerations concerning the morphology and phylogeny of the Arthropoda. *Phil. Trans. R. Soc. Lond.* **B 233:** 483–580.

Manton, S. M. 1958. Habits of life and evolution of body design in Arthropoda. *J. Linn. Soc. Lond. (Zool.)* **44(295):** 58–72.

Manton, S. M. 1960. Concerning head development in the arthropods. *Biol. Rev. (Cambridge)* **35:** 265–82.

Manton, S. M. 1961. The evolution of arthropodan locomotory mechanism. Part 7. Functional requirements and body design in Colobognatha (Diplopoda), together with a comparatice account of diplopod burrowing techniques, trunk musculature and segmentation. *J. Linn. Soc. Lond. (Zool.)* **44(299):** 383–461.

Manton, S. M. 1964. Mandibular mechanism and the evolution of arthropods. *Phil. Trans. R. Soc. Lond.* **B 247(737):** 1–183.

Manton, S. M. 1966. The evolution of arthropodan locomotory mechanism. Part 9. Functional requirements and body design in Symphyla and Pauropoda and the relationships between Myriapoda and Pterygote insects. *J. Linn. Soc. (Zool.)* **46(309):** 103–41.

Manton, S. M. 1967. The polychaete *Spinther* and the origin of the Arthropoda. *J. Nat. Hist.* **1:** 1–22.

Manton, S. M. 1973a. Arthropod phylogeny–A modern synthesis. *J. Zool. Lond.* **171:** 111–30.

Manton, S. M. 1973b. The evolution of arthropodan locomotory mechanisms Part 11. Habits, morphology and evolution of the Uniramia (Onychophora, Myriapoda, Hexapoda) and comparisons with the Arachnida, together with a functional review of uniramian musculature. *Zool. J. Linn.* **53:** 257–375.

Nair, K. B. 1949. On the embryology of *Caridina laevis. Proc. Indian Acad. Sci.* **B 29:** 211–88.

Osche, G. 1973. Das Homologisieren als eine grundlegende Methode der Phylogenetik. *Aufsätze Red. Senckenb. Naturforsch. Ges.* **24:** 155–65.

Paulus, H. F. 1972. Die Feinstruktur der Stirnaugen einiger Collembolen (Insecta, Entognatha) und ihre Bedeutung für die Stammesgeschichte der Insekten. *Z. Zool. Syst. Evolutionsforsch.* **10:** 81–122.

Paulus, H. F. 1973. Die Feinstruktur der Stirnaugen einiger Collembolen (Insecta, Entognatha) und ihre Bedeutung für die Stammesgeschichte der Mandibulaten. *Verh. Dtsch. Zool. Ges.* 1972: 56-60.

Paulas, H. F. 1974. Die phylogenetische Bedeutung der Ommatidien der apterygoten Insekten (Collembola, Archaeognatha und Zygentoma). *Pedobiologia* 14: 123-33.

Pflugfelder, O. 1948. Entwicklung von *Paraperipatus amboinensis* n. sp. *Zool. Jahrb. Anat. Ontog.* 69: 443-92.

Pross, A. 1966. Untersuchungen zur Entwicklungsgeschichte der Araneae (Pardosa hortensis (Thorell)) unter besonderer Berucksichtigung des vorderen Prosomaabschnittes. *Z. Morphol. Ökol. Tiere* 58: 38-108.

Pross, A. 1977. Diskussionsbeitrag zur Segmentierung des Cheliceraten-Kopfes. *Zoomorphologie* 86: 183-96.

Remane, A. 1956. *Die Grundlagen des natürlichen Systems der vergleichenden Anatomie und der Phylogenetik: Theoretische Morphologie und Systematik I.* Geest und Portig K.-G., Leipzig.

Remane, A. 1960. Die Beziehungen zwischen Phylogenie und Ontogenie. *Zool. Anz.* 164(7-10): 306-37.

Remane, A. 1961. Gedanken zum Problem: Homologie und Analogie. Paeadaptation und Parallelität. *Zool. Anz.* 166(9-12): 447-65.

Rempel, J. G. 1975. The evolution of the insect head: The endless dispute. *Quaest. Entomol.* 11(1): 7-25.

Rosenquist, G. C. 1966. A radioautographic study of labeled grafts in the chick blastoderm: Development from primitive-streak stages to stage 12. *Contrib. Embryol. Carnegie Inst. Wash. D.C. Publ.* 38(262): 71-110.

Sander, K. 1976a. Morphogenetic movements in insect embryogenesis, pp. 35-52. *In* P. Lawrence (ed.) *Insect Development. Symp. R. Entomol. Soc.*, London.

Sander, K. 1976b. Specification of the basic body pattern in insect embryogenesis. *Adv. Insect Physiol.* 12: 125-238.

Scholl, G. 1969. Die Embryonalentwicklung des Kopfes und Prothorax von *Carausius morosus* Br. (Insecta, Phasmida). *Z. Morphol. Tiere* 65: 1-142.

Scholl, G. 1977. Beiträge zur Embryonalentwicklung von *Limulus polyphemus* L. (Chelicerata, Xiphosura). *Zoomorphologie* 86: 99-154.

Schulze, P. 1937. Trilobita, Xiphosura, Acarina. Eine morphologische Untersuchung über Plangleichheit zwischen Trilobiten und Spinnentieren. *Z. Morphol. Ökol. Tiere* 32: 181-226.

Seidel, F. 1960. Körpergrundgestalt und Keimstruktur. Eine Erörterung über die Grundlagen der vergleichenden und experimentellen Embryologie und deren Gültigkeit bei phylogenetischen Überlegungen. *Zool. Anz.* 164(7-10): 235-305.

Seidel, F. 1961. Entwicklungsphysiologische Zentren im Eisystem der Insekten. *Verh. Dtsch. Zool. Ges.* 1960: 121-42.

Seidel, F. 1966. Das Eisystem der Insekten und die Dynamik seiner Aktivierung. *Verh. Dtsch. Zool. Ges.* 1965: 166-87.

Sekiguchi, K. 1973. A normal plate of the development of the Japanese horse-shoe crab, *Tachypleus tridentatus. Sci. Rep. Tokyo Kyoiku Daigaku* B 15 (229): 153-62.

Sewertzoff, A. N. 1931. *Morphologische Gesetzmäßigkeiten der Evolution.* Gustav Fischer, Jena.

Sharov, A. G. 1966. *Basic Arthropodan Stock, with Special Reference to Insects.* Pergamon Press, Oxford.

Siewing, R. 1960. Zum Problem der Polyphylie der Arthropoda. *Z. Wiss. Zool.* 164(3/4): 238-70.

Siewing, R. 1969. *Lehrbuch der vergleichenden Entwicklungsgeschichte der Tiere.* Paul Parey, Hemburg, Berlin.

Siewing, R. 1974. Referat über: D. T. Anderson, Embryology and phylogeny in annelids and arthropods. *Z. Zool. Syst. Evolutionsforsch.* **12**: 238.

Snodgrass, R. E. 1938. Evolution of Annelida, Onychophora and Arthropoda. *Smithson. Misc. Collect.* **138**: 1-77.

Snodgrass, R. E. 1952. *A Textbook of Arthropod Anatomy.* Comstock, Ithaca, New York.

Størmer, L. 1944. On the relationship and phylogeny of fossil and recent Arachnomorpha, a comparative study on Arachnida. *Skr. Norske Vid. Acad. Math.-Nat. Kl. Oslo* **5**: 1-158.

Tiegs, O. W. and S. M. Manton. 1958. The evolution of the Arthropoda. *Biol. Rev. (Cambridge)* **33**: 255-337.

Weber, H. 1952. Morphologie, Histologie und Entwicklungsgeschichte der Articulaten. *Fortschr. Zool.* **9**: 1-231.

Weygoldt, P. 1958. Die Embryonalentwicklung des Amphipoden *Gammarus pulex pulex* (L). *Zool. Jahrb. Anat. Ontog.* **77(1)**: 51-110.

Weygoldt, P. 1960. Embryologische Untersuchungen an Ostrakoden: Die Entwicklung von *Cyprideis litoralis* (G. S. Brady) (Ostracoda, Podocopy, Cytheridae). *Zool. Jahrb. Anat. Ontog.* **78(3)**: 369-426.

Weygoldt, P. 1961. Beitrag zur Kenntnis der Ontogenie der Decapoden: Embryologische Untersuchungen an *Palaemonetes varians* (Leach). *Zool. Jahrb. Anat. Ontog.* **79(2)**: 223-70.

Weygoldt, P. 1963. Grundorganisation und Primitiventwicklung bei Articulaten. *Zool. Anz.* **171(9/10)**: 363-76.

Weygoldt, P. 1964. Vergleichend-embryologische Untersuchungen an Pseudoscorpionen (Chelonethi). *Z. Morphol. Ökol. Tiere* **54**: 1-106.

Weygoldt, P. 1965. Vergleichend-embryologische Untersuchungen an Pseudoscorpionem III. Die Entwicklung von *Neobisium muscorum* Leach (Neobisiinea, Neobisiidae). Mit dem Versuch einer Deutung der Evolution des embryonalen Pumporgans. *Z. Morphol. Ökol. Tiere* **55**: 321-82.

Weygoldt, P. 1966. Die Ausbildung transitorischer Pharyncapparate bei Embryonen. *Zool. Anz.* **176(3)**: 147-60.

Weygoldt, P. 1968. Vergleichend-embryologische Untersuchungen an Pseudoscorpionen IV. Die Entwicklung von *Chthonius tetrachelatus* Preyssl., *Chthonius ischnocheles* Hermann (Chthoniinea, Chthoniidae) und *Verrucaditha spinosa* Banks (Chthoniinea, Tridenchthoniidae). *Z. Morphol. Tiere* **63**: 11-54.

Weygoldt, P. 1975. Untersuchungen zur Embryologie und Morphologie der Geißelspinne *Tarantula marginemaculata* C. L. Koch (Arachnida, Emblypygi, Trantulidae). *Zoomorphologie* **82**: 137-99.

Wilson, E. B. 1894--1896. The embryological criterion of homology. *Biological Lectures Delivered at the Marine Biological Laboratory in the Summer Session of 1894, Boston.*

Witschi, E. 1956. *Development of Vertebrates.* W. B. Saunders, Philadelphia, London.

Zilch, R. 1974. Die Embryonalentwicklung von *Thermosbaena mirabilis* Monod. (Crustacea, Malacostraca, Pancarida). *Zool. Jahrb. Anat. Ontog.* **92**: 462-576.

4

Abnormal Metamorphosis and Arthropod Evolution

R. MATSUDA

INTRODUCTION

Metamorphosis is a more or less conspicuous structural change during ontogeny of an animal. Various kinds of abnormal metamorphosis (defined on pp. 138-140) are more or less conspicuous deviations in the pattern of metamorphosis from more normal (more common) metamorphosis patterns in another sex or other individuals (in both sexes) of the same species, or in other species of the group (genus and above) to which the species in question belongs.

Abnormal metamorphosis always occurs, as we shall repeatedly see in the text, in changing or unusual environments. Endocrinologists, especially insect endocrinologists, have gradually become aware of the fact that abnormal metamorphosis (such as caste differentiation, some cases of neoteny) as well as some abnormal behaviors, diapause, and so on, are attributable to hormonal disturbance induced by special factors in such environments. In the latest symposium on caste and phase determination held in Washington, D.C. (1976), an increasing awareness of the importance of this aspect of morphogenesis is clear in the introductory remarks by M. Lüscher (1976a), who stated, "It has become clear that in almost all carefully investigated cases the external factors influence the endocrine system and that hormones are ultimately responsible for (morph) determination."

In this chapter an attempt is made to analyze less clearly recognized cases of abnormal metamorphosis in insects and other arthropods, and evaluate their significance on arthropod phylogeny in particular and on evolution in general. The method of analysis is as shown in Fig. 4.1, in which the genetic changes are also shown to cause changes in hormonal activity. However, this aspect is usually negligible, since in insects, and apparently also in other arthropods, the direct

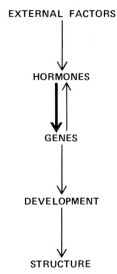

Fig. 4.1.   Diagram showing sequence of morphogenesis under influence of external factors.

influence of environmental factors upon the hormonal activity is overwhelming. As far as the insect endocrinological literature is concerned, Novák (1975) counted 6,000 references, and certainly some of the facts and established principles in this vast body of literature are useful in interpreting the cases of abnormal metamorphosis in insects.   Endocrinology of other groups of the Arthropoda (Crustacea, Myriapoda, and Arachnida), however, remains much less satisfactorily known than insect endocrinology, and the inference that can be made with regard to the hormonal mechanism of abnormal metamorphosis is much more limited.   However, a summary of relevant endocrinological knowledge is given for each group, to see how much more remains to be learned.

Despite some inconclusiveness, this study shows that integration of endocrinological knowledge into the study of evolution will clarify some principles of structural evolution, and that it may make macroevolution more plausible. This work represents an extension of my work on the neoteny in insects (Matsuda, 1976).

4.2. KINDS AND DEGREES OF ABNORMAL METAMORPHOSIS

Modes of abnormal metamorphosis can be classified into the following categories and subcategories, which are not always distinct from one another:

Neoteny.   This term refers to the condition in which the gonad completes its development prematurely before the normal differentiation of imaginal structures is completed.   In the reproductively functional adults thus produced, imaginal structures remain incompletely differentiated; some juvenile structures may be retained, and often compensatory development of structures (increased differentiation of preexisting structures and development of new structures) occurs.

Based on the number of molts these neotenous arthropods undergo, neoteny can be classified into two categories: *prothetely* and *metathetely*, the terms used in insect endocrinology. When an insect becomes a neotenous adult after undergoing normal or more than normal numbers of molts, the condition is metathetely; whereas when it becomes a neotenous adult after undergoing less than the normal numbers of molts, the condition is prothetely. The neotenous insects produced through prothetely can be essentially larval (nymphal) or pupal in somatic differentiation, and the production of such conspicuous neotenous insects is often called *paedogenesis*. Prothetely also results in reproductively nonfunctional juvenile castes in ants and termites.

Neoteny may occur only in one sex or in both sexes. In insects, neoteny occurs far more frequently in the female than in the male, for the reason discussed on p. 142. In a presumed phylogenetic sequence (e.g., Conchostraca-Cladocera complex, Blattaria-Isoptera, and so forth), the neotenous features (either in terms of arrest of development of structures alone, or in terms of both arrest of development and compensatory development of structures) have become increasingly pronounced; such a phylogenetic trend is termed here *phylogenetic neoteny*.

Acceleration. When elimination of large numbers of developmental stages occurs in the process of attaining the definitive adult stage, the condition is called *acceleration*. The consequence of acceleration may be a normal adult, a neotenous adult, or an adult with drastically different structures (halmatometamorphosis). Acceleration is, therefore, essentially the same as prothetely. The term *acceleration* is kept here as a more general term referring to the substantial elimination of developmental stages in noninsectan arthropods.

Halmatometamorphosis. This is an excessive metamorphosis that occurs in parasitic (especially endoparasitic) arthropods. When arthropods enter the parasitic stage, their metamorphic changes may be drastic, and the parasites are not comparable with any developmental stage or the adult of any free-living ancestral species. For such profound metamorphosis the term *halmatometamorphosis* (*halmatos* = leap) is proposed here. The process of halmatometamorphosis consists of degeneration of larval structures and construction of novel structures adapted for the parasitic life.

Caenogenesis. This term refers to the constructive (and adaptive) development of structures during embryonic, larval, and pupal stages. Caenogenetic structures represent either the structures that were never present in the adult ancestor (sensu Haeckel), or the embryonic, larval, and pupal modifications of the structures that were present, or could have been present, in the adult ancestor (Matsuda, 1976). The caenogenetic structures are usually not carried into the adult stage.

Hypermetamorphosis. This is multiple (and successive) modifications of structures associated with alteration of hosts during larval stages in some holometabolous insects. Hypermetamorphosis may be regarded as a special form of caenogenesis.

Production of juvenile social castes. In some groups of insects (ants, bees, and termites) two or more functionally different castes (within the same sex) coexist; caste differentiation occurs only in the female (ants and bees), or in both sexes (termites). The reproductively nonfunctional castes in ants and termites (worker and soldier) are essentially juvenile, and result from prothetely; a neotenous caste also occurs in termites and ants. In bees, the reproductively nonfunctional worker is not juvenile, except for the female gonad.

4.3. EXAMPLES OF ABNORMAL METAMORPHOSIS IN VARIOUS ARTHROPOD CLASSES

4.3.1. The Insecta

4.3.1.1. Hormonal Regulation of Normal and Abnormal Metamorphosis
To facilitate understanding of discussions of individual cases of abnormal metamorphosis, some relevant facts and principles of insect endocrinology are given briefly below.

4.3.1.1.1. A Simplified Scheme of the Hormonal Control Mechanism of Normal and Abnormal Metamorphosis. In highly simplified terms, two major hormones, juvenile hormone (JH) and ecdysone (ED), are concerned with metamorphosis in insects. JH is produced in the corpora allata (CA), and ecdysone (ED) in the ecdysial glands [prothoracic glands (PG) and comparable structures]; the activities of these glands are stimulated by the activation hormone released from the neurosecretory cells (NSC) in the brain through the corpora cardiaca (CC). Based on the generalization proposed by Williams (1961) and Slama (1975), a general scheme of hormonal balance during normal metamorphosis in the Holometabola can be stated as: ED, in the presence of high amounts of JH, apparently induces larval molt; ED, in the presence of less JH, induces pupal molt; and in the absence of JH, ED induces imaginal molt. In the Hemimetabola, the imaginal molt is triggered by ED in the absence of JH.

In the above scheme, ED is the hormone for molting and morphogenesis, and JH is a status quo hormone, which modifies the action of ED. Quite often, however, the general scheme has been stated in an oversimplified manner. Thus, many authors have formulated the scheme essentially as follows: A high titer of JH yields larval molt, low titer the pupal molt; and in the absence of JH, the imaginal molt occurs. In such a scheme the action of ED is nowhere clear, and it may give the false impression that JH performs all necessary morphogenetic functions in metamorphosis. It is this fallacious (or misleading) aspect of the generalization that was severely criticized by Slama (1975). As Slama emphasized anew, it is the ED that directly triggers morphogenesis, and the decline in JH titer at different developmental stages is a necessary condition for metamorphosis. Abnormal metamorphosis results from the disturbance in the balance between JH and ED during postembryonic development.

4.3.1.1.2. Environmental Factors Influencing the Hormonal Activity and Metamorphosis.

The action of the neuroendocrine system controlling metamorphosis is triggered by diverse environmental factors, such as photoperiod, temperature, humidity, food, population density, and so on. At least most of these factors directly stimulate the neurosecretory cells in the brain, and then the action of the endocrine organs follows. It should be remembered that usually none of these factors acts independently in nature; two or more factors usually act jointly, one factor modulating the effect or effects of the others. More direct evidence of the effect of external factors on the neurosecretory cells has often been found in the studies of diapause of certain insects (e.g., Williams, 1969; de Wilde 1969), and data concerning the factors stimulating the activity of CA have been summarized by Engelmann (1970). For more direct evidence of the influence of external factors on hormonal activity and consequent morphogenesis see below, Fig. 4.3.

As we shall see repeatedly later, abnormal metamorphosis can ultimately be attributed to alteration in environmental factors in changing seasons and different geographical areas, and to unusual factors in some extreme environments (parasitic milieu, arid area, and so forth).

4.3.1.1.3. Reprogramming of JH Action and Its Effect on Metamorphosis.

Slama (1975) showed that when a large amount of a juvenoid was applied to the fifth instar larva of *Pyrrhocoris apterus*, all degrees of incomplete (hence abnormal) metamorphosis occurred, depending on the day of the last nymphal instar on which the juvenoid was applied. When it was applied on the first day, extra nymphal instars occurred; when applied later, nymphal-adultoids of various degrees occurred; and when applied on the fourth day, normal adults resulted. Many similar experiments in other insects have yielded similar results.

The above facts indicate that when, in nature, disruption of the hormonal balance occurs slightly in favor of JH during late larval stages, a more limited action of ED and less differentiation of imaginal structures will result, and hence abnormal metamorphosis. Direct evidence for such abnormal hormonal balance inducing the production of a neotenous morph is seen in the case of locusts (Fig. 4.3). During all stages of larval development of the honey bee (*Apis mellifera*), JH titer is consistently higher in the presumptive queen larvae than in the presumptive worker larvae, and this difference apparently results in different degrees of action of ED and consequent caste differentiation (p. 182). Many similar cases are shown in dealing with individual cases of abnormal metamorphosis.

When embryos of some insects (*Hyalophora*, *Pyrrhocoris*) were exposed to JH and JH analogues (Riddiford, 1970; Riddiford and Truman, 1972), it was found that, depending on the time of application, the effect was realized late during metamorphosis, in that extra larval instars and adultoids of various degrees were produced. This phenomenon was attributed to interference with the programming of the embryonic CA; the disruption of the embryonic programming that results from such interference is that CA does not cease secretion of JH as a prelude to metamorphosis (Riddiford and Truman, 1972). The influence of environmental factors on the activity of embryonic CA was suggested for termites (p. 149). However, this hypothesis does not apply to the case of the parthenogenetic female of an aphid species (p. 170).

4.3.1.1.4. Gonadotropic Function of JH and Female Abnormal Metamorphosis. Besides being a status quo hormone (already discussed), JH has another important function, promoting egg maturation by facilitating yolk deposition (gonadotropic function). Normally, the stimulation of oogenesis occurs after an insect has emerged as an adult. Engelmann (1970) recorded the then known (or presumed) cases of this gonadotropic function of JH, which had been found in eight orders of insects (Thysanura, Orthoptera, Dermaptera, Blattaria, Hemiptera, Diptera, Coleoptera, Lepidoptera). Since the gonadotropic function occurs in Thysanura and Blattaria, and since JH(?) in the Myriapoda also has the same function (p. 197), this function was probably acquired in Carboniferous insects; Novák (1965) claimed that this function was the only function of JH in primitive insects.

As repeatedly pointed out later, neoteny occurs much more frequently in the female than in the male. As I (Matsuda, 1976) emphasized, this is most probably because of the gonadotropic function of JH, which is usually effective only in the female. When, as seen in solitary locusts, overproduction of JH occurs during late postembryonic development in certain definite environments, the hormone is likely to promote the maturation of eggs precociously. Rohdendorf and Sehnal (1972) have contended that at least in some insects differentiation of female gonads is completed prior to imaginal molt, and, therefore, the process of female gonad differentiation is affected (i.e., promoted) only when JH or its analogues are administered to larvae or pupae. This experimental fact supports the hypothesis proposed here. In such cases, the quantity of ED and the time available for insects to trigger differentiation of some imaginal structures (wings, and so on) would become more limited, hence yielding a neotenous female insect. In fact, as seen in the text, the ovaries in female insects with reduced wings (hence neotenous) are usually well developed, or even enormously developed, and have the eggs ready for fertilization upon emergence; in such cases overproduction of JH can be safely assumed (even though direct evidence is not available).

4.3.1.1.5. Male Abnormal Metamorphosis Without the Gonadotropic Effect of JH. In the male, the development of the gonads cannot be much faster than that of other structures, since usually JH has no function in promoting the formation of the male gamete. Hence, when neoteny occurs in males under the influence of excessive JH, it does not cause premature acceleration of testicular development. This may explain why reproductively viable neotenous males have less suppression of imaginal structures than females, as seen, for instance, in Strepsiptera (p. 195).

4.3.1.1.6. Loss of the Gonadotropic Function of JH and Bisexual (male and female) Abnormal Metamorphosis. In some insects, the gonadotropic function of JH has been secondarily lost, and abnormal metamorphosis can occur in both sexes equally. Thus, in termites high titers of JH do not expedite the maturation of ovaries, and instead induce the production of the reproductively nonfunctional soldier caste. In the Phasmida also the gonadotropic function of JH has been lost; the material released from the NSC in the brain took over the gonadotropic function, and apparently this has enabled phasmids to evolve by prothetely (p. 153).

4.3.1.1.7. JH and Compensatory Growth and Development of Structures. Another positive effect of JH is the compensatory development and growth of structures that is often induced by alteration in JH titer. It has been found that when the development of certain structures (such as wings) is suppressed, hind legs (and perhaps other legs) grow abnormally larger, as seen in the wing-reduced forms of Orthoptera, Siphonaptera, Coleoptera, Lepidoptera, Homoptera, and Hymenoptera. Since at least in the solitary (neotenous) locusts the enlargement of the hind legs is clearly conditioned by an increase in the JH titer during development (p. 157), this compensatory growth can perhaps be attributed to the switch of regulatory genes conditioned by the alteration in JH titer (p. 201, 202); such regulatory gene switch is of great evolutionary significance (p. 202). The development of caenogenetic structures in the soldier caste in some groups of termites may also be due to the regulatory gene switch conditioned by the significantly increased JH titer (p. 149).

4.3.1.1.8. Other Functions of JH. A more limited morphogenetic function of JH is the induction of development of the pheromone-producing structure in gregarious locusts (p. 157). The presence of JH is known to be a prerequisite for the maintainance of the ecdysial (prothoracic) glands, at least in some insects. Therefore, the persistence of the ecdysial glands in pharate adult could be interpreted as a sign of persistence of JH, and could constitute supporting evidence in regarding the insects in question as undergoing abnormal metamorphosis, such as neoteny.

A very low titer of JH, or the absence of it, leads to the death of many larval structures, as seen in degeneration of these structures during the larval-pupal molt (in Holometabola) and nymphal imaginal molt (in Hemimetabola). Pflugfelder (1938) found that removal of CA from the first and second stage nymphs of *Carausius* results in degeneration of many internal structures (which, however, is followed by encystment of dead tissues). De Kort (1969) found a clear correlation between the reduction of JH titer and degeneration of flight muscles in the diapausing Colorado beetle (*Leptinotarsa*). The degeneration of larval tissues, however, paves the way for further morphogenesis during metamorphosis. Therefore, such cell death accompanied by a decrease in JH titer is an essential aspect of morphogenesis in insects.

4.3.1.1.9. Action of Ecdysone. ED is a hormone concerned with morphogenesis and is often called molting hormone; two kinds of ED, α-ecdysone and β-ecdysone have been found in insects. ED is certainly not the only hormone involved with a complex molt cycle that consists of apolysis, molting fluid production, initiation of cuticle formation, ecdysis, tanning, and deposition of endocuticle. Bursicon concerned with postecdyseal tanning and endocuticle deposition, and an eclosion hormone governing the emergence of several saturnid moths from their cocoon (Truman, 1970), are now known to be part of the chain of the molting cycle. Yet, as Willis (1974) maintained, ED most likely sustains and coordinates the totality of actions, instead of being just a trigger of one of the several actions involved with the molt cycle.

4.3.1.1.10. Sensitivity of Tissues to JH. The sensitivity of tissues to JH is genetically determined, and different tissues in the same individuals and the rudiments of homologous structures in different individuals and species have different sensitivities to the action of JH.

A few experimentally proven cases of different tissue sensitivites are given below. In *Hyalophora cecropia* high doses of JH are necessary to suppress differentiation of the large epithelial cells, which give rise to cortices of the rectal pads; moderate doses of JH are necessary and sufficient to suppress proliferation of the hindgut epithelium, which produces the rectal wall; and the medullary cells are inhibited by low levels of JH (Judy and Gilbert, 1970). When JH is applied to postfeeding larvae of *H. cecropia*, the pupae bear tubercles, identical in sculpturing and pigmentation to those of the last larval instar; the tubercle-bearing areas of these abnormal pupae form pupal cuticle. In these pupae the internal organs are larval, and the adults thus formed, though normal externally, have the pupal viscera (Riddiford, 1972). It is clear from this experiment that different tissues and rudiments do not differentiate simultaneously under the same hormonal milieu.

With respect to the difference in tissue sensitivity to JH in different species,

Metwally and Sehnal (1973) found that beetles of different families differ one from another by their sensitivity to JH analogues. For example, *Tenebrio* is generally more sensitive than *Coryedon*, but with respect to certain compounds it is less sensitive than *Trogoderma*.

In some insects (Dermaptera, Isoptera, female aphids, female cecidomyids, and so on), development of many structures becomes arrested when high doses of JH are available during development; and this phenomenon is due at least partly to high sensitivities of various tissues in these insects. In Coleoptera (except *Micromalthus*) and Heteroptera, on the other hand, only the development of wings (and occasionally ocelli) may be arrested; in the case of Coleoptera the action of JH may not be directly involved. It is apparent, therefore, that different species and groups of insects have, during evolution, developed either increased or decreased tissue sensitivity to JH, and it must be an important reason why the frequency and degree of abnormal metamorphosis varies so much in different groups of insects.

4.3.1.2. Blattaria and Mantodea

In the neoteny of the Blattaria compensatory development of structures is not at all pronounced. In the female of this order, the stylus is known to be cast off during the molt into the adult. Therefore, the retention of the stylus in the apterous female of *Nocticola termitophila* (Chopard, 1949a) represents a neotenous feature (the male of this species is brachypterous). In many species of this order wing polymorphism occurs, and the reduced wing morphs are neotenous, as discussed below; some species are permanently apterous.

Lefeuvre (1971) found that: 1) some species with reduced wings ("paedopterous" of Lefeuvre) have fewer developmental stages than their macropterous relatives; 2) in the paedopterous species wings are more of a nymphal type than they are imaginal; 3) the tracheal system and the peripheral nervous system as well maintain juvenile features; 4) the reduction of eyes, ocelli, and cerci, and the retention of nymphal integument also accompany the wing reduction (especially in the female); and 5) the prothoracic (ventral) glands persist throughout the life of the paedopterous cockroaches.

All the above features (1–5) can be attributed to an excessive effect of JH, which suppresses molting (1), and causes retention of nymphal features (2, 3, 4) and persistence of the action of the prothoracic glands, which can be interpreted as a sign of neoteny (5). When the abbreviation of developmental stages occurs (1), the resultant neoteny is prothetely. However, as regards (5), Bodenstein (1953) and Lanzrein and Lüscher (1970) showed that in Blattaria the presence of JH is not a necessary prerequisite for the existence of the prothoracic glands.

Further, Lefeuvre (1971) was able to induce wing reduction by implantation of CA (which results in an increase in JH titer), and showed that the wing reduction can occur also when allatectomy is performed in young nymphs (premature but insufficient action of ED would occur in this case). Lefeuvre also found that the development of complete wings (macroptery) cannot be induced in the species without (normally) fully developed wings by any experimental means, and contended that the evolution of wings in cockroaches has been only one way toward reduction.

Apparently, nothing is known regarding the environmental factor or factors inducing the proper hormonal milieu (i.e., overproduction of JH) for the development of neotenous features such as reduced wings and precocious oogenesis. According to Cornwell (1968), humidity, temperature, and air movement are the major physical factors of the environment that most influence the life of cockroaches. Since the gonadotropic function of JH has been well established for this order (see Guthrie and Tindall, 1968, and Adiyodi and Adiyodi, 1974, for summaries), it is *a priori* probable that overproduction of JH would stimulate oogenesis precociously before the normal development of other structures (in macropterous individuals) has been completed.

In a mantid family, Eremiaphilidae, which occur in the desert of North Africa, wings are strongly reduced. In *Eremiaphila*, the tarsi of all legs are five-segmented as in other Mantodea. In *Heteronutarsus* of the same family, however, the tarsus is four-segmented in the front legs, and three-segmented in the other pairs of legs (Beier, 1968a). These reduced conditions of structures must result from developmental arrest; hence they are neotenous.

4.3.1.3. Isoptera

Ample anatomical and paleontological evidence points to the probable origin of this order from late Paleozoic or early Mesozoic cockroaches. However, in this order abnormal metamorphosis takes the form of production of juvenile castes, which results from premature completion of development (prothetely).

In the colony of a more primitive species, *Calotermes flavicollis*, for instance, three major castes with different degrees of somatic differentiation occur. They are soldiers, supplementary (replacement) reproductives, and primary reproductives. The soldiers are essentially nymphal, but they do not molt and they are reproductively nonfunctional; their social function is to defend the colony. The juvenile (and hence arrested) features of structures in this caste include the nondifferentiation of the copulatory vestibule in the female, the lack of ocelli and wings (although projections of posterolateral angles of meso- and metanotum appear to represent incipient wing rudiments, as seen in a figure given by Lebrun, 1970), retention of styli, juvenile nonfunctional reproductive organs, and so on. However, a constructive (and adaptive) development in this caste is the enlarged head, which is armed with a frontal gland in some families (Rhinotermitidae, Serritermitidae, and most Termitidae). These structures, occurring in the essentially nymphal caste, can be considered as caenogenetic developments that arose in association with the defense function.

The supplementary (replacement) reproductives are reproductively functional, neotenous adults in which the imaginal differentiation is incomplete. They replace or, in some species, supplement the primary reproductives. The arrested development of structures in this caste includes characteristically the wings, which remain as wing buds (or they may be without them), as well as less pigmented eyes and integument. In *C. flavicollis*, the development of male and female reproductive systems (male and female gonads, seminal vesicles in the male, the spermatheca in the female, and so forth) is slightly arrested as compared with those in the primary reproductives (Lebrun, 1967a). The worker caste occurs only in higher termites and may correspond to the pseudergate in lower termites (Miller, 1969).

The developmental pathways from larvae* and pseudergate* to the primary reproductives, to supplementary reproductives, and to soldiers in *Calotermes* are highly plastic, as seen from Fig. 4.2, although in higher termites such as *Macrotermes* caste determination requires a few decisive steps and is less plastic than in *Calotermes* (Lüscher, 1976b).

The hormonal control mechanism of caste differentiation in termites is now reasonably well known. The development of soldiers in *Calotermes* is induced by high titers of JH, as was shown earlier by Lüscher (1958), Lüscher and Springhetti (1960), and Lebrun (1967a,b) in their experiments with *Calotermes*. By implantation of CA these workers were able to induce the production of soldiers from larvae, from pseudergates, and even from the nymphs on the way to developing into adults. Likewise, application of JH analogues to *Calotermes* by Lüscher (1969) and Springhetti (1974) and similar (vapor) treatment of JH analogues with other termites by Hrdy (1976, for summary) also strongly indicated that a high titer of JH during development is indeed responsible for the soldier development. Lüscher (1976b) now strongly suspects that the JH titer in the first stage larva or in the old embryo affects the eventual developmental pathways, including that of the soldier (as discussed later).

In termites ED, and not JH, is probably the oocytes maturation hormone as in *Aedes* (Lüscher, 1976b). In termites JH has apparently lost the gonadotropic function retained in cockroaches. Consequently, when JH is abundantly present

---

*In the study of termites the earlier developmental stages are called "larvae" and the last two or more stages are called "nymphs"; the larva is without wing buds and the nymph shows wing buds. Pseudergates are individuals that have regressed from the nymphal stage by molting, and thereby they reduce or eliminate wings; they can also be derived from larvae by undergoing stationary molt (without differentiation). For more information and definitions of castes in termites refer to Miller (1969), Noirot (1969), Wilson (1971), and Lüscher (1974).

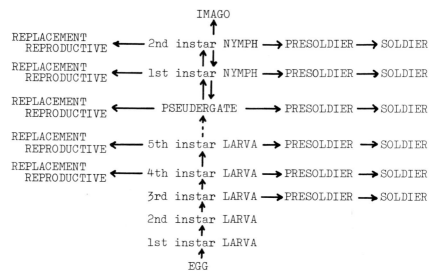

Fig. 4.2.   Possible developmental pathways in *Calotermes flavicollis* Lüscher.

during the production of the soldier, the hormone probably does not stimulate oogenesis, and this is probably the reason why the soldier never becomes neotenous.

Lüscher (1963) found that in the isolated pseudergates of *Calotermes*, CA grow very rapidly, by about four times in volume at the molt into the supplementary reproductives. The apparent involvement of JH seen in the enormously enlarged CA, however, turned out to be spurious in various later experiments, which showed that the enlarged CA are in fact inactive. Lüscher (1974), therefore, thought that the production of the supplementary reproductives results from premature release of ED at the time the CA are inactive. Similarly, Wanyonyi (1974) found that in *Zootermopsis* differentiation of the primary and supplementary reproductives occurs only in the absence of JH. The ED, which has the gonadotropic function and is abundantly available at this stage, presumably stimulates the development of oocytes to produce the female supplementary reproductives.

The alteration of the hormonal milieu in producing different castes (discussed above) is related to the pheromonal control of caste determination. In maintaining certain proportions of different castes within a colony, according to Lüscher (1976b), the pheromone of reproductives inhibits the development of the primary and supplementary reproductives, and the pheromone of the soldier, if it exists, inhibits the soldier development and stimulates the production of reproductives. The pheromone of the primary reproductives is presumably produced in the head-thorax region (in *Zootermopsis*, Light, 1942, 1943, 1944; Lüscher, 1961);

and it is given off through the anus (in *Calotermes*, Lüscher, 1974), and is circulated within a colony through the anus to mouth (Lüscher, 1974, 1975).

Springhetti (1971) and Lüscher (1975) believe that the pheromone influences caste differentiation by influencing the activity of CA. Lüscher (1975) assumes that the pheromone of the reproductives stimulates the CA of the larva to produce more JH, which (more directly) inhibits the production of reproductives. The soldier pheromone, if it exists, has an inhibitory action on the CA; therefore the absence of JH would inhibit the development of soldiers, and simultaneously stimulate the production of reproductives.

Besides the effect of pheromones, Lüscher (1976b) also emphasized the effect of seasons (which may be temperature, humidity, physiological conditions of food) on the activity of CA of the queen,* which, in turn, affects embryos and first stage larvae. Lüscher (1976b) thought that, in *Macrotermes*, the higher (or the lower) the JH titer in the queen, the higher (or the lower) the JH content in the eggs that are laid later. The varying amounts of JH would, according to Lüscher, affect the activity of CA in the first stage larva differently, and this blastogenetic determination would partly determine the fate of developing larva, either into nymphs then to imagos, or into other castes. Lüscher hypothesized further that only those eggs with low JH titers may be competent for the development into nymphs.**

The fact that, in *C. flavicollis*, any larva can develop into any one of the different developmental stages and castes (Fig. 4.2), the fact that the CA activity, which affects the developmental pathways (into different castes) is, in turn, directly influenced by pheromonal stimuli and some external factors, and the fact that caste differentiation occurs in both sexes lead us to believe that no definite genetic mechanism is involved with the production of different castes in this species. Probably, many other termite species also have essentially the same plasticity with regard to the development of castes, although the degree of plasticity may be more limited in some termites (e.g., *Macrotermes*). A conspicuous deviation appears to be that of *Schedorrhinotermes*, in which workers and soldiers are females (Lüscher, 1974); in such cases genetic determination of castes must lie at the level of the sex-determining mechanism.

The prolongation of the head (nasute head) with the frontal gland (for forcible ejection of sticky and irritating fluid onto enemies) in the essentially larval soldier caste is, as already noted, caenogenetic. This caenogenetic tendency has apparently been enhanced during the evolution of some groups of termites (Rhino-

---

*Whether or how the pheromones and other external factors first stimulate the NSC in the brain (described by Noirot, 1957, and Gillott and C.-M. Yin, 1972) before the CA become stimulated remains to be seen.

**This hypothesis is essentially the same as the concept of reprogramming of the CA activity based on some experimental facts (p. 141).

termitidae, Serritermitidae, and most Termitidae). Thus, Emerson (1961) showed that the soldier of the primitive stock of these termites had most probably the small "squirt gun" (frontal gland) and the defensive mandible. During further evolution, however, the nasute head with the squirt gun became increasingly larger and more efficient; at the same time the mandibles became increasingly smaller and inefficient, and eventually they became vestigial. Further, during the evolution of the soldier caste within the Rhinotermitidae, according to Emerson (1971), many structures, such as the eye spot, tibial spur, mandibles, and so on, became increasingly less differentiated (apparently regressed, fused, or smaller), while the frontal gland and associated structures became more elaborate and larger.

It is clear that the more certain structures (such as mandibles) become suppressed during caenogenesis, the more certain other structures (such as the squirt gun) become differentiated, and that this trend of material compensation has been enhanced during the evolution of the soldier caste in some groups of termites. Since the material compensation occurs in the presence of high titers of JH during soldier development, the increased trend of this phenomenon appears to reflect, at least in part, the increased effect of JH in evolution. A presumed physiogenetic mechanism involved with material compensation is discussed on p. 220.

Apparently, the increased effect of JH in evolution is also seen in the tendency of progressive reduction of many structures in the primary reproductives of termites. Thus, for instance, the female ovipositor has become lost, except in a primitive genus *Mastotermes*, and the penis has become completely lost in the winged Isoptera. The seven kinds of exocrine glands of the head in Blattaria have become reduced in number to two or one in Isoptera (Brossut, 1973). Within the Isoptera, as shown by Jucci and Springhetti (1952) and Springhetti (1964), four types of the seminal vesicles–accessory glands complex are recognized in the male primary reproductives of termites, and they represent four stages of evolutionary reduction. The first and most primitive type is that of *Mastotermes*, in which the accessory glands are well developed and consist of two tufts of tubules as in Blattaria. In other types, the glands are less developed or absent, and the size of the seminal vesicles has become increasingly smaller; in the least developed condition the seminal vesicles are comparable with those of the soldier caste in *C. flavicollis*.

The above sequence of reduction of adult structures thus appears to reflect, at least in part, the increased effect of JH, which increasingly limited the action of ED, and hence adult differentiation. Lebrun (1970) found that JH can suppress the development of the female accessory (colleterial) glands and spermatheca and the male seminal vesicles. However, in this phylogenetic neoteny of the primary reproductives within the Blattaria-Isoptera complex, very slight, if any, compensatory development or growth of structures has taken place.

4.3.1.4. Zoraptera and Ephemeroptera

In Zoraptera, the wings are either fully developed or absent, and this is true of all species. According to Denis (1949), a colony of Zoraptera consists of: 1) the colored adults with eyes and ocelli, either winged or wingless (apterous); 2) nymphs with wing rudiments; 3) apterous adults that are not colored and are without eyes and ocelli, although their genitalia are comparable with those in winged adults; 4) adults differing only in coloration from the adults in (1) and (3); and 5) the young nymph (larvae of Denis) without eyes and ocelli and with eight-segmented antennae (nine-segmented in 1–4). When a colony is not highly populated, the apterous adults are more numerous. Apparently, the apterous adults (3 above) are neotenous.

In some Ephemeroptera reduction of the hind wings is highly pronounced, and they may be altogether absent in some genera (*Cloeon* and *Caenis*). The female subimagos of some mayfly species are reproductively functional (W. Peters' personal communication), and they can be regarded as neotenous.

4.3.1.5. Grylloblattodea

The species of this order inhabit the alpine regions of the temperate zones (in both the old and new worlds); they are found under stones, and so on. The insects of this order are always apterous. Judging from the presence of dorsal longitudinal muscles in the pterothorax, and of the pleural suture, which reaches dorsally the wing base in other orders, the loss of wings has apparently occurred rather recently (Matsuda, 1970).

Grylloblattodea have been known to have many primitive characters. As I have pointed out (Matsuda, 1976), however, most of these "primitive features" probably represent less differentiated conditions of structures, caused by developmental retardation or arrest; hence they are neotenous. Such neotenous features include: the retention of the mesothoracic spina and the first abdominal sternum, which usually become lost during development; the subanal lobes, which are distinct from the clearly retained tenth abdominal sternum (they are usually indistinguishably fused); the absence of ocelli, which usually form very late during postembryonic development; the absence of the female subgenital plate, due probably to developmental arrest; underdeveloped eyes; paired penes, which do not undergo further division; and the arrested development of the tubular accessory glands. The array of these neotenous features makes it difficult to understand their phylogenetic relationships to other orders of insects.

4.3.1.6. Plecoptera

In Swedish stoneflies, according to Brinck (1949), two types of wing reduction (hence neoteny) occurs. The first is microptery, in which the wing length measures less than 50% of the normal wing; and the second type is brachyptery, in which the length of the reduced wing ranges from 50 to 90% of the normal wing length. In Sweden, microptery occurs only in the males of six species, and the size of the reduced wing in each of these species is rather stable. The species with micropterous males are therefore sexually dimorphic (when wing reduction does not occur in the female). Some Swedish stonefly species are polymorphic with respect to wing development; various degrees of wing reduction occur in both sexes of these species.

According to Brinck (1949), in the sexually dimorphic species the male microptery occurs in all geographical ranges of their distribution. Further, according to Brinck, micropterous males tend to emerge earlier than macropterous females. It appears probable that at the time males emerge, prevailing lower temperatures could induce the kind of hormonal balance leading to the production of microptery; such seasonally conditioned dimorphism (diphenism) is clear in the case of *Gerris odontogaster* (p. 173).

In the wing polymorphic species, according to Brinck (1949), wing reduction occurs only in certain geographical areas of Sweden (high mountains and some southern lakes); some populations of these species may consist almost entirely of brachypterous individuals. Since wing reduction occurs only in these cold-water areas, it is apparently induced (primarily) environmentally and hormonally. Aptery is known to occur in the female of *Aphanicercopsis denticulata* and in the male of *A. hawaquae* (Despax, 1949).

Hynes (1941), who observed the tendency of wing reduction at higher altitudes in Britain (ca 265 m and above), suggested that the lack of flying ability which results from reduced wings may be of some advantage to stoneflies living in open mountainous districts, in that the insects are less liable to be carried far away from the streams which are essential for their existence.

4.3.1.7. Phasmida

In great majority of phasmids, wings are absent. This winglessness (aptery), which was presumably preceded by wing polymorphism, can be regarded as neoteny. Within the apterous phasmids again, however, neoteny affecting other structures has occurred. For instance, *Timema californica* (Timeminae, Phyllidae) is a tiny phasmid measuring 16 to 17 mm in length in the male and 21 to 25 mm in the female, with a vertical range of distribution of from 2,000 ft to 6,500 ft in the Chaparral area, California (Henry, 1937). The insect has the three-segmented tarsi, instead of the five-segmented tarsi found in other phasmids, and the female ovipositor mechanism and the posterior abdominal segments exhibit juvenile features (see Matsuda, 1976). All these features most probably result from developmental arrest (and hence are neotenous), as the following experiment indicates.

Berthold (1973) found that the removal of CA from the fourth instar nymph of *Carausius morosus* results in premature molt, with the nymph developing into a dwarf adult. In this case, early exhaustion of JH is apparently followed by premature (accelerated) action of ED for imaginal differentiation; hence the prothetelous dwarf adult. It is probable, therefore, that when the dwarf adult of *Timema californica* is produced, the same (abnormal) endocrine activity (as that in the experiment of *Carausius*) is induced by some environmental factors (e.g., temperature) at the high altitudes in California. The fact that the number of molts ranges from four to eight in this order (Chopard, 1949b; Beier, 1968b) suggests the possibility that (in this order) acceleration of growth and development through prothetely has been a dominant feature of their evolution. Further, as a consequence of such evolution, JH has apparently lost its gonadotropic function, as many endocrinological studies of this order now convincingly show. The gonadotropic function has, as shown by Mouton (1969, 1970, 1971), been taken over by the endocrine factor which is synthesized by the NSC in the pars intercerebralis of the brain and stored in the corpora cardiaca.

4.3.1.8. Embioptera

In Embioptera, the female is always apterous. In the males of some genera, according to Ross (1970), brachypterous and apterous species occur in arid regions or localities with a long dry season, and in some genera (e.g., *Metoligotoma*) males are always apterous. Further according to Ross (1970), a more pronounced neoteny is seen in a new genus and species from Afghanistan, in which the apterous male is completely nymph-form, including abdominal terminalia. Rose therefore thought that such a species is difficult to place, and it seemed better to create a new family for it than to assign it without evidence to one of the established families.

4.3.1.9. Dermaptera

Ozeki (1958) found, in laboratory cultures of an apterous species, *Anisolabis maritima*, that some male nymphs (14.8%) become adults at the sixth molt with 25-35 antennal segments; some male nymphs (51.1%) become adults at the fifth molt with 23-31 antennal segments; and some other male nymphs (34.1%) become adult at the fourth molt with 21-27 antennal segments. Assuming that six molts are normal in nature, those becoming adults at the fourth and fifth stages with fewer numbers of segments represent prothetely. The highly labile morphogenesis is apparently reflected in the great variation in number of antennal segments which ranges from 10 to 50 within the order (Chopard, 1949c).

Ozeki (1958, 1959) further found that the removal of the CC-CA system from the second to last stage (which can be either the third, fourth, or fifth stage nymphs) causes precocious metamorphosis, and in the adults thus formed the efferent duct of the male reproductive system is poorly developed. For instance, in the adult males formed at the third molt the paired ajaculatory ducts were not formed. Such a poorly developed condition of the efferent duct is apparently induced by the accelerated action of ED, which probably follows the depletion of JH (which follows the removal of the CC-CA). It is expected, therefore, that the male adults formed after different numbers of molts would have the efferent ducts of different degrees of development. Whether such individual variations actually exist in *Anisolabis maritima* (and in other Dermaptera) remains to be seen. It is known, however, that the degree of development of paired ejaculatory ducts and the posterior unpaired part of the ejaculatory duct (preputial duct of Ozeki) can be very different in different species of Dermaptera (see Matsuda, 1976), and such differences may be partly due to the different developmental stages at which different species (and individuals) develop into adults.

As in other orders, wing reduction must be a good indicator of neoteny. La Greca (1954) classified the degree of wing reduction into three major categories (A, B, C), described as follows: A) typical elytra, 1) with hind wings normally developed (*Forficula auricularia* type), 2) with hind wings shortened (*Forficula decolvi* type), and 3) with hind wings vestigial (*Forficula decipiens* type); B) reduced elytra, 1) with elytra slightly shorter and hind wings vestigial and lobate, but not distinct from the metanotum (*Psalis gagatina* type), and 2) with elytra lobate and hind wings absent (*Euboriella moesta* type); C) elytra and hind wings absent (*Anisolabis* type).

In some species belonging to categories 1 and 2 of group A, dimorphism of hind wings occurs. Pantel (1917) cited a case of normal development of the flight organs in the male of *Anisolabis annulipes*. Environmental factors affecting the development of structures remain to be studied.

4.3.1.10. Orthoptera *

In this order neoteny is evident in the reduction of wings and in the concomitant modifications (reduction and compensatory development) of other structures, which have occurred in varying degrees in diverse groups within the order.

The polymorphism of some locusts (*Locusta, Schistocerca, Nomadacris*, and so on), associated with gregarious and solitary phases of their life, has long been known. The neotenous features are evident in their solitary phase, and comprise such features as the (usually) relatively short wings,** short sternal hairs, less developed pronotal lobe, green pigmentation, lack of integumentary glandular

---

*Including the Grylloptera of Kevan (1973).
**Occasionally long-winged solitary locusts occur (Kevan, personal communication).

structures (pheromone-producing glands) that are present in gregarious males (Cassier and Delorme-Joulie, 1976), retention of the prothoracic glands in high humidity (Cassier and Fain-Maurel, 1970), and so forth. A well-known progressive (compensatory) development in the neotenous solitary locusts is prolongation of the hind femur.

Postembryonic development is completed sooner in gregarious nymphs of *Locusta, Nomadacris,* and *Schistocerca,* partly because the gregarious nymphs hatch more differentiated and tend to become adults with one fewer molt (Kennedy, 1961), or sometimes with two fewer molts (Kevan, personal communication). Since the solitary locusts require at least as many molts as the gregarious locusts do, their neotenous conditions must be considered to be due to metathetely.

The most important external factor causing the polymorphism in locusts is population density. Loher (1960) first discovered that in the gregarious phase a chemical stimulation resulting from contacts between individuals causes imaginal differentiation, and that the source of the chemical stimulus is a pheromone produced from integumentary glandular structures in the male. In the absence of such glandular structures in the solitary males, the solitary locusts (both male and female) tend to be more juvenile (neotenous) than the gregarious locusts. Most probably, the pheromone influences the balance between ED and JH, which has a more direct effect on the phase differentiation.

Direct evidence of the primary role of JH titer in producing the phase differences was found by Joly and Joly (1974) in their assay of the hormone in the hemolymph of *Locusta migratoria migratorioides,*[*] which showed (Fig. 4.3) that in the solitary phase the titer of JH sharply drops during the early stage of the fifth instar, but soon sharply increases. In the gregarious individuals, on the other hand, JH becomes nearly completely depleted as they become the fifth instar nymphs, and this condition continues until the sixth day of their adult life; then JH sharply increases. The high titer of JH in the fifth instar nymphs of solitary locusts would expedite the maturation of oocytes (prematurely) at this fifth stage, while simulateneously it would limit the action of ED for imaginal differentiation; thus the nymphs would become neotenous adults. In the gregarious individuals, on the other hand, ED would be allowed to be active in the absence of JH during the fifth nymphal stage and early postimaginal development. Consequently, imaginal structures would be fully differentiated, and the ovaries would become mature after the fifth day of the adult life, as the JH titer increased sharply.

---

[*]Johnson and Hill (1973a,b, 1975) applied the *Galleria* assay to the quantification of the JH present in the gregarious migratory locust (*Locusta migratoria*) and clarified the timing of release of JH during development. An important fact revealed in these studies was that the volume of CA can be correlated with gland activity only to some extent (Johnson and Hill, 1975). For the influence of the brain hormone on phase differentiation see Girardie (1976).

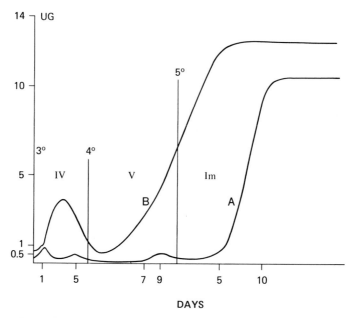

Fig. 4.3.   Titers of JH in hemolymph during late postembryonic developmental stages and the adult stage in *Locusta migratoria migratorioides*.   A, Gregarious phase; B, Solitary phase; *I* = imago.   From Joly and Joly (1974).

Some earlier experiments appear to support the above interpretation. Thus, for instance, by implantation of the CA of the young adult of *Locusta migratoria* into the fifth stage gregarious nymphs, Staal and de Wilde (1962) obtained adults with some features of the solitary locust, such as green pigmentation, crumpled wings, short sternal hairs, broad pronotal band, and some morphometric values characteristic of the solitary locusts; in this case the artificially elevated titer of JH prevented the full differentiation of imaginal structures. Further, implantation of the prothoracic glands of the fifth instar nymphs into young nymphs accelerated molting and imaginal differentiation, and the latter became normal adults at the fifth nymphal stage.

In gregarious locusts, the prothoracic glands are known to degenerate soon after the imaginal molt, as in many other insects. However, in the solitary locusts they persist in humid conditions because of the precocious and intensive action of JH (Fain-Maurel and Cassier, 1969; Cassier and Fain-Maurel, 1970). Thus, the persistence of the prothoracic glands indicates that metamorphosis has not been completed by the influence of excessive JH; hence the solitary individuals could become incompletely adult and neotenous.

In *Locusta, Nomadacris,* and *Schistocerca*, according to Kennedy (1951), gregarious females are less fecund than solitary females, because fewer ovarian cycles and fewer ovarioles per ovary occur in the gregarious females. Consequently, solitary females lay two, three, or four times as many eggs as gregarious females. This fact appears to reflect, at least in part, a strong gonadotropic effect of the excessive JH during the fifth nymphal instars of the solitary females.*

It is certainly true that differentiation of imaginal structures is triggered by the action of ED, but not that of all structures. Thus, as shown by Cassier and Delorme-Joulie (1976), in *Schistocerca gregaria* the glandular structures producing pheromone become differentiated only in the gregarious adult males, from the hypodermal glandular cells and under the influence of JH, which again becomes available at this stage. The absence of the glandular structures in the female must be determined by the genes linked to sex. In the male of solitary *Schistocerca*, the glandular cells are fewer in number and insensitive to JH.

Pigmentation of the body has been known to be green in isolation and in moist conditions. According to Joly (1972), moist environments induce green pigmentation by changing the timing of activity of the CA. Joly hypothesizes that a nongreen hopper must have the active CA during a slightly shorter period than the first half of each intermolt, and the green hopper during a slightly longer period than the first half of each intermolt.

The ratio of the hind femur to the head width (F/E ratio) has been shown to be greater in solitary nymphs and adults than in gregarious nymphs and adults, and this feature has been used as a good distinguishing character of the two phases (Uvarov, 1966). Staal (1961) was able to induce this progressive development of the hind femur by implanting extra CA into young nymphs. Apparently, JH stimulates the growth of the hind femur instead of suppressing its growth in solitary locusts. Such compensatory growth accompanying the reduction of some other structures (wings, and so on) may be due to the switch of regulatory genes in which JH can be a corepressor (see p. 217).

The influence of photoperiod and temperature on phase differentiation is also evident in an experiment on *Locusta* by Cassier (1966), which showed that when reared under long days and under crowded conditions the effect of gregarization becomes enhanced, and that higher temperatures have a similar effect. Cassier therefore contended that these factors also contribute to phase differentiation in *Locusta*. In the isolated (solitary) condition, photoperiod and temperature may play, in the absence of the pheromonal stimulus, even more important roles in development and growth; and this influence may account for the occurance of long-winged individuals in the solitary phase (see footnote on p. 154).

Photoperiod is also known to influence the sexual behavior of locusts. For instance, in *Locusta* a 12-hr day length induces intensive oviposition and vigorous sexual behavior in the male (Perez et al., 1971). Since the JH released from CA is also known to stimulate the activity of the male (reproductive) accessory glands (Hartmann, 1971; Pener et al., 1972; Gillott and Friedal, 1976), it is safe to assume that certain day lengths heighten the activity of CA by inducing more

---

*For postimaginal oocyte development in *Schistocerca* and *Locusta* (both in isolated and crowded conditions) refer to Highnam and Haskell (1964).

release of neurosecretion, which, according to Girardie and Garnier (1973) and others, stimulates the CA, and hence the vigorous male sexual behavior.

Szöllosi (1975) found that JH, when applied to the late nymphal instars of *Locusta* and *Schistocerca*, inhibits the imaginal differentiation of the accessory glands and spermiduct. This fact leads us to suspect that in the solitary locusts these male reproductive structures may be less differentiated than in gregarious locusts; the matter has apparently not been investigated so far. In *Gryllus*, as shown later, these structures are less developed in neotenous, short-winged forms than in macropterous individuals, and the inhibitory effect of the JH on the development of the male accessory glands is also known with some other insects. As Mordue et al. (1969) emphasized, feeding also affects the endocrine activity.

Polymorphism occurs also in many species of Gryllidae, the phenomenon being most obvious in the degree of wing development. However, Cousin (1938) found that in the micropterous individuals of *Gryllus campestris* the ovipositor and accessory glands of the spermatheca are poorly developed, and that the male accessory glands, which are so numerous in the fully winged males, are absent. Cousin regarded these features, quite correctly, as neotenous. A similar study certainly must be done for other species of this family. Because of the lack of accessory glands the micropterous males are, in effect, not fecund. The ovaries are normally developed.

Judging from what is known about the hormonal control mechanism in locusts, the above neotenous conditions appear to be due to the overproduction of JH in late nymphal instars, which inhibits the full differentiation of imaginal structures and expedites the maturation of ovaries at the same time.* Tanaka (1976) showed that in micropterous females of *Pteronemobius taprobanensis* rapid oocyte growth starts on the second day after emergence, while in the macropterous female oocyte growth proceeds very slowly, the peak of oviposition in the latter occurring 15 days later than in micropterous females. This difference indicates that more JH (which has the gonadotropic function) is produced in the pharate micropterous females than in the macropterous females.

Population density usually has a marked effect on wing polymorphism, as in locusts. Thus, Fuzeau-Braesch (1961) found that in two out of the six species studied (*Gryllus argentinus* and *Scapsipedus marginatus*) grouping of individuals clearly induces macroptery. Conversely, isolation induces no macroptery and results only in the production of brachypterous individuals. Saeki (1966), in *Scapsipedus aspersus*, and McFarlane (1966), in *Gryllodes sigillatus*, also found that grouping can have a marked effect on the percentage increase of macropterous individuals. It is thus clear that in crickets also, as in locusts, crowding tends to

---

*Thomas (1964) showed that, in *Gryllus domesticus*, neurosecretion from the suboesophageal ganglion apparently intervenes oogenesis. However, this does not preclude the probability that JH from CA has the gonadotropic function.

induce full wing development, presumably through activation of the pheromone-hormone system, which, however, has not been studied with Gryllidae.

Population density also affects body pigmentation in crickets as in locusts. Fuzeau-Braesch (1960) and Levita (1962) found that when the nymphs of *Gryllus bimaculatus* are reared as a group (two or three individuals in one group are necessary and sufficient), the adults acquire pale yellowish-brown pigmentation in various parts of the body, although when reared in isolation they are usually black. Levita (1962) further demonstrated that such a group effect results from the exchange of sensorial stimuli between individuals, and that antennae play the most important role in receiving the stimuli. Levita concluded that the mechanism of the group effect lies in the exchange of tactile, visual, and vibratory stimuli which supplement one another, and the resultant modification of physiology leads to either black or yellow pigmentation.

The effect of photoperiod and temperature on wing polymorphism has been studied for several species of Gryllidae, with the results tending to reveal the nature of the geographically and seasonally conditioned polymorphism. Thus, a Japanese species, *Nemobius yezoensis*, which inhabits the northern half of Japan, is always micropterous in nature. However, Masaki and Oyama (1963) found, by exposing developing nymphs to various photoperiods (11 to 16 hr), that at 12-hr and shorter photoperiods all individuals are micropterous as found in nature, while at longer photoperiods the incidence of macropterous individuals increases. Thus, they were able to expose the genetic potentials that never manifest themselves in nature.* Masaki and Oyama therefore contended that the nymphs are exposed to a short photoperiod beyond the middle of August when the day length is rapidly decreasing in the field, and that the indispensable photoperiodic condition for the development of long wings is simply nonexistent in nature. As long as the species retains its present mode of annual cycle, the nymphs necessarily encounter short days in their late instars, whereby they are destined to have degenerated hind wings and short fore wings in their adulthood. Thus, the species represents a case of photoperiodically induced neoteny, which occurs only in a certain geographical area and in a certain season.

By exposing the nymphs of *Gryllodes sigillatus* to various photoperiods (10 to 18 hr), Mathad and McFarlane (1968) found that a 14-hr photoperiod was optimal for wing polymorphism in that macropterous and micropterous individuals and intermediates occur at the higest frequency. They further showed that at a 12-hr photoperiod nearly all individuals become micropterous. Thus, their results agreed closely with the results obtained by Masaki and Oyama (1963) for *Nemobius yezoensis*.

Mathad and McFarlane (1970) further found that in the last stage nymphs of

*Macropterous individuals of *Gryllodes sigillatus* were also unknown at the time Ghouri and McFarlane (1958) produced such individuals for the first time by temperature treatment.

the macropterous form of *Gryllodes sigillatus*, the median NSC of the brain show a small amount of PF-positive material in the cell bodies and a considerable amount in axons. On the other hand, in the last stage nymphs of mircopterous form the cell bodies were densely packed with PF-positive material, but little or none was present in the axons. Mathad and McFarlane therefore inferred that the brain hormone was released in the macropterous larvae and was not released in the micropterous larvae, and they agreed with Sellier (1954), who concluded that the wing development in *Gryllus campestris* is determined at the level of the brain hormone activity. How or whether the neurosecretion (brain hormone) affects endocrine organs such as CA (as in locusts) or prothoracic glands remains unknown. Mathad and McFarlane (1970) found no significant difference in the volume of CA between macropterous and micropterous nymphs.

As already seen, at certain photoperiods the incidence of wing polymorphism increases both in *Nemobius* and *Gryllodes*. The different wing morphs produced at these photoperiods roughly reflect genetic difference with regard to the ability to develop wings at the individual level, and such a difference may lie, at least partly, in the ability of the NSC to respond to certain photoperiods. At certain lower photoperiods (12 hr and less), however, all individuals became micropterous, whereupon the genetic difference does not manifest itself and becomes meaningless.

Gryllus desertus is brachypterous in France and macropterous in Eastern Europe, and it is not immediately clear whether this geographical difference in wing development is due to genetic difference or environmental influences. Lability in wing development is also evident in some other gryllids. For instance, in *Acheta rubens*, the winter generation with diapause is brachypterous, while the summer generation without it is macropterous (Alexander, 1961); *Gryllus campestris* is usually brachypterous, but becomes macropterous when diapause is accidentally broken (Sellier, 1949). Many cricket species living in unusual environments are often completely apterous. For instance, the Myrmecophilinae living in association with ants are apterous.

In *Teleogryllus yezoemma* no clear-cut wing dimorphism occurs. Masaki (1966), however, obtained individuals with unusually long tegmina and ovipositor by exposing them to a long photoperiod (16 hr); the ratios of the tegmina and ovipositor to the head tended to be greater in them than in individuals reared under a short photoperiod (11 hr). However, the ratio of the hind femur to the head width remained the same in both groups. It is clear, therefore, that the growth of the hind femur apparently remains, as in locusts (p. 157), free from the hormonal suppression affecting other structures (such as wings). The strong development of the hind femur in apterous myrmecophiline crickets appears to parallel the compensatory growth of the hind femur in locusts.

In some other groups of Orthoptera the degree of neoteny is more pronounced. In Stenopelmatidae the wings are usually lacking, and some other structures also remain juvenile. Thus, in *Hemiandrus* (of Stenopelmatidae) the stylus is retained as in the soldier caste of termites, and the ovipositor is also incompletely developed. Furthermore, in this genus the degree of fusion of abdominal ganglia is less marked than that in other Orthoptera, indicating that the process of highly pronounced fusion of abdominal ganglia characteristic of the order Orthoptera is arrested (see Matsuda, 1976). In *Oryctopus* (Stenopelmatidae) the wings and ocelli are absent. However, the front legs are curiously modified in adaptation to their subterranean habitat, and the hind legs are also greatly developed. Apparently, in this case also the loss of wings and ocelli has been compensated for by the progressive development of the legs.

Cyphoderris (Prophalangopsidae) is brachypterous or macropterous, with the multilobed condition of the male genitalia (penis) and the absence of the male accessory glands representing juvenile (hence neotenous) conditions (see Matsuda, 1976).

Metrioptera roeselii (Tettigoniidae) is normally brachypterous and rarely macropterous in nature. Ramme (1931) found that in all macropterous females he collected the ovariolar tubes are greatly reduced and contain only small, weakly developed eggs, although the sperms in macropterous males are normally developed. This fact appears to indicate that, as in locusts (pp. 155, 156), high titers of JH induce production of the normal (in this species) brachypterous and reproductively functional females, and that reversal to the ancestral macropterous condition with lower titers of JH can occur at the cost of loss of fecundity.

Thomas and Alexander (1962) found that micropterous individuals of *Orchelimum concinnum* (Tettigoniidae) occur only in the relic area, while flourishing populations of the same speices are macropterous. Alexander (1968) thought that this variation results from one or the other of two possibilities: 1) a genetic change resulting in loss of the ability to produce macropterous individuals in the relic area, or 2) failure of the relic population to reach sufficient densities to trigger macroptery. The second alternative appears to be more probable.

La Greca (1946) pointed out that in Orthoptera, wing reduction tends to be more pronounced in the female than in the male. Thus, for instance, in *Chrysochraon dispar* and *Metrioptera roeselii* the degree of brachyptery is greater in the female than in the male, and in some Orthoptera (*Porthetis, Lamarckiana, Saussurea*) the male is winged and the female is apterous, and so on. A simple hypothesis to account for this tendency is that an excessive titer of JH would more easily suppress imaginal differentiation in the female than in the male, because JH promotes the maturation of ovaries but not of testes (see also p. 142).

Compensatory growth as exhibited by enlargement of the hind legs (and other legs) occurs also in some cavernicolous Orthoptera (*Troglophilus spinulosus, Dolichopoda Bormansi*, and so on). It is apparent that the physiogenetic mechanism producing this effect of compensatory growth is a feature shared by many Orthoptera, as well as other insects (see p. 223 for discussion).

4.3.1.11. Psocoptera

In some psocids all degrees of wing reduction occur ranging from extreme microptery to macroptery. Kalmus (1945) showed, in 19 species of the Psocoptera, a correlation between the reduction of wings (especially that of hind wings) and that of ocelli, and thought that these correlated reductions constitute a reliable criterion for neoteny. Mockford (1965) recognized wing reduction in many psocid species, and found that in these psocids some other structures are also not fully differentiated; his findings included obliteration of the trilobed condition of the mesoscutum, poorly developed ocelli and the absence or poorly developed condition of the ctenidia on the proximal tarsal segment. According to Smithers (1972), the nymphal features that may be retained in adult psocids are the two-segmented tarsus without ctenidiobothria, shorter antennae, the lack of the trichobothrial field on the paraprocts, and the presence of the marginal cone of the paraprocts. In the male of *Lachesilla*, as shown by Klier (1956), the styli are retained and the penis is completely absent; Matsuda (1976) regarded these features as neotenous.

Badonnel (1948, 1949) found that when the nymphs of an obligatorily parthenogenetic species *Psyllipsocus ramburi* are reared in isolation, they always develop into brachypterous or micropterous females and never into macropterous females; whereas when these nymphs are reared in a group of four or more, they develop into macropterous females; and when they are reared in a group of two or three nymphs, macropterous, brachypterous, and micropterous females emerge. These experiments were conducted at a temperature of 23°C. Badonnel (1949) further showed that when they are reared at 25°C or 27°C, the macropterous form never occurs, even when they are reared as a group. Clearly, at these temperatures the effect of grouping disappears. Badonnel's (1949) study also indicated that stimulation for the development of wings is chemical in nature, and that vision and contact do not play any role in the stimulation.

Badonnel (1959) further showed that for the completion of a life cycle macropterous females require a longer period of time (43–45 days) than micropterous females (36–40 days). It is apparent that in the micropterous females overproduction of JH results in retarded and insufficient action of ED to produce the reduced wing. Badonnel (1959) hypothesized that the initial stimulus in the crowded population more or less inhibits the activity of NSC in the brain or activates the inhibitory factor of the molting glands (prothoracic glands), which occur in many species of psocids (Badonnel, 1970). Actually, however, we know nothing directly about the endocrine control mechanism in this order. The effect of the photoperiod on the development of wings and other structures, which is now so well established in other major groups of Hemipteroidea and other hemimetabolous orders, also remains to be studied. It should be remembered that modifications of the structures in *Psyllipsocus ramburi* reflect the effects of extrinsic factors influencing essentially the same genome, since the insect is obligatorily parthenogenetic.

In *Archipsocus*, according to Mockford (1965), males are always very brachypterous or apterous. Females reared in isolation invariably become the shortest-winged adults, while the macropterous females appear only in large groups. Further, according to Mockford (1965), males generally pass through five numphal instars, while the brachypterous females pass through six, thus accounting for the larger size and longer wings of these females. Six larval instars in the female are presumably primary for this species, and five instars in the male secondary; and the production of the shorter wings in the male may be regarded as prothetely.

Accoring to Turner (1974), two species of Psocoptera, *Pseudocoecilius citricole* and *Hemipsocus roseus*, occur throughout the altitudinal range between 500 and 4,000 ft (150–1,200 m) in Jamaica. He found that both body size and wing length increase at higher altitudes. Apparently, lower temperatures at higher altitudes are responsible for such change, but a trend of neoteny is not evident.

4.3.1.12. Thysanoptera

In many species of Thysanoptera wing polymorphism occurs; the different wing morphs are called macropterous, hemimacropterous, brachypterous, and apterous, depending on the degrees of reduction. In the individuals with reduced wings other structures (such as ocelli, compound eyes, antennal segments, legs, and setae) also tend to become reduced. Kalmus (1945) showed convincing evidence indicating the presence of a high correlation between the absence of ocelli and reduction of wings; Kalmus believed that the reduction or loss of these structures is evidence for neoteny. In the apterous male of *Rhopalandothrips annulicornis*, the maxillary palpus is two-segmented, although it is three-segmented in the macropterous female of the same species (Pesson, 1951a).

According to Köppä (1970), the macropterous form of *Anaphothrips obscurus* increases in number when exposed to long days, whereas short days inhibit the production of macropterous individuals. These results agreed with the field observation that the proportion of brachypterous individuals increases toward autumn; here lower temperatures may also be involved. Kamm's (1972) experiment on the photoperiodic effect with the same species agreed with the results obtained by Köppä (1970). Köppä's experiment also showed that the kinds of food also affect the determination of wing morphs. Further, Kamm showed that the number of macropterous individuals increases with higher population density (as in Psocoptera, Orthoptera, and so on).

For winglessness in Phthiraptera (Mallophaga and Anoplura) see p. 226.

4.3.1.13. Auchenorrhynchous Homoptera

Müller (1954) found, by an extensive examination of specimens and repeated experiments, that *Euscelis plebejus* and *E. incisus* in Europe actually represent a seasonal dimorphism of a single species (*E. plebejus*), the former occurring in the summer and the latter in the spring. They are distinguishable by darker and lighter colorations, larger and smaller sizes of wings, and different shapes of the apical end of the aedeagus, which result from different degrees of differentiation. The short-winged condition and the other structural changes that accompany the wing reduction in the spring generation must be due to suppression of developmental process, and represent neotenous features.

Müller (1957) further found similar relations between the West European species *Euscelis lineolatus, E. stictopterus, E. bilobatus,* and four related, yet-undescribed forms, all being seasonal modifications of a single species *Euscelis lineolatus.* He showed that the *lineolatus* and *superlineolatus* forms result from exposure to a photoperiod of 8 hr per day, and that the *bilobatus, dubius, sub-stictopterus, stictopterus,* and *superstictopterus* result from a long day (16 hr of light per day). By applying these photoperiods, he produced either the *incisus* or *plebejus* form, and either the *lineolatus* or *bilobatus* form, respectively. Müller's experiments further indicated that temperature only modifies but does not cancel the effect of the photoperiod, and that the quality of food also slightly influences the photo-periodic effect on the aedeagous size in *E. plebejus* and *E. lineolatus.* Müller (1964) further studied the effect of different spectral ranges in the photoperiodic induction of the seasonal forms of the same species.

In a Japanese planthopper species, *Nilaparvata lugens,* population density is known to be the factor causing wing dimorphism. Kisimoto (1956) found that when the nymphs of this species are reared in isolation one individual per tube), they always become brachypterous in the female, although no brachypterous male is produced under the same condition (all macropterous). As the population density becomes higher (5, 10, 20 individuals per tube), the emergence of macropterous females increases, and some brachypterous males are produced. The influence of food was also found to be important in determining wing morphs in this experiment. Johno (1963) found that the production of the male wing morphs is dependent on temperature and photoperiod (and not population density) in this species; his experiment showed that the frequency of brachypterous males increases with short days and low temperatures. In *Javesella pellucida* the percentage of macropterous adults (in both sexes) increases in crowded conditions (Mochida, 1973; Ammar, 1973).

Nilaparvata lugens migrates to Japan in June, presumably from Southern

China. They are naturally macropterous as they reach Japan. However, the frequency of brachyptery is very high in the next generation; and later, as the population density becomes higher, the frequency of macropterous individuals becomes higher. This tendency in the male cannot be related to the above experiment by Johno (1963).

Kisimoto (1957) compared the macropterous form with the brachypterous forms in *Nilaparvata lugens, Sogata furcifera,* and *Delphacodes striatella.* He found that in the brachypterous form: 1) the fore wings are short and hind wings are rudimentary; 2) the scutellum is shorter than in the macropterous form; 3) the lengths of the hind femur, the hind tibia, and the ovipositor are greater than in the macropterous form, although there is no difference in the head width between the two morphs; 4) the body color is paler than in the macropterous form; and 5) the preovipositional period is shorter than in the macropterous form.

Most of these characteristics are the same as or similar to those in brachypterous forms of Orthoptera, and such characteristics can result, as in Orthoptera, from the overproduction of JH during development (the endocrine control mechanism of wing morphs in auchenorrhynchous Homoptera remains unknown). Based on what we know about the locust endocrinology, number 3 (above) appears to represent a case of compensatory growth that accompanies the reduction of wings; and this phenomenon presumably results from overproduction of JH (see p. 157 for discussion). The fact that the preovipositional period is shorter in the brachypterous form (5 above) appears to indicate that oogenesis starts earlier in the brachypterous form because of the higher titer of JH produced toward the end of nymphal development. In *Javesella pellucida* also the brachypterous form is slightly more fecund than in the macropterous form (Mochida, 1973). The fact that the wing is clearly dimorphic without intermediates (1 above) may mean that the wing morphs are genetically determined. However, such a clear wing dimorphism can also result from the difference in hormonal balance, as in *Gerris odontogaster* (see p. 173). For references to many studies on wing polymorphism (or dimorphism) by Japanese workers, see Mochida (1973).

Denno (1976) attempted to show the relationship between spatial and temporal variation in the structures of the host grasses and wing polymorphism in some fulgoroids. He showed that the proportion of brachypterous individuals increases with the seasonal increase in biomass of the host grass, while the proportion of macropterous individuals does not. His results also show that the production of macropterous individuals is correlated with high levels of crowding during nymphal stages.

4.3.1.14. Sternorrhynchous Homoptera

As is well known, aphids are polymorphic with respect to the degree of development of structures. The ancestors of aphids, were, however, most probably alate (winged) amphigonic and oviparous insects, as Hille Ris Lambers (1966) contended. From this presumed ancestral condition, more specialized modern aphids with at least one parthenogenetic viviparous morph and at least one apterous female morph must have been derived (see Hille Ris Lambers, 1966, for kinds of morphs in aphids).

In the apterous morphs (apterous fundatrix, apterous parthenogenetic vivipara, apterous ovipara, and so on) the degree of differentiation of various structures is generally much less pronounced than in alate morphs, as seen in shorter antennae with fewer rhinaria, smaller compound eyes, and the lack (nondifferentiation) of ocelli which are always present in alate forms, etc. These features, along with the absence of wings, definitely represent neotenous features that result from arrested development.

The arrest of development of imaginal structures in the apterous virginopara is clearly seen, for instance, in the ontogeny of *Megoura viciae*. The alate adult of this species, according to Lees (1961), has a complex, highly sclerotized pterothorax; a series of conspicuous black pigment spots is present on the dorsal surface of the abdomen; and there are many sensilla on the third antennal segment. In contrast, all these structures are absent in the fourth instar alata, so that the nymphs undergo a striking metamorphosis, with the various sensory-motor and other features appearing for the first time. The adult aptera, on the other hand, has only a weakly sclerotized pterothorax and a few antennal sensilla, and the abdominal pigment pattern is entirely lacking, showing that morphological changes at the final molt are slight. Since both alata and aptera undergo the same number of molts, the less-differentiated condition of structures in the latter must be considered metathetely. The occurrence of paedogenesis (prothetely) in *Schizaphis graminum* was reported by Wood and Starks (1975). This species usually undergoes five molts from birth to maturity. Under laboratory conditions, 1.8% of the alate greenbugs were found to reproduce paedogenetically. In the test, thirteen of the paedogenetic insects reproduced one day before molting into alata, four reproduced two days before molting to alata, and one reproduced three days before the final molt.

The degrees of differentiation of structures in fundatrices, when wingless, may differ in many respects from apterous viviparous females of later generations. Generally, in the fundatrix, there is more reduction in sensilla and length of antennae, often in number of antennal segments, and in length of legs, siphunculi, and cauda (Hille Ris Lambers, 1966). These differences represent different degrees of neoteny resulting from different degrees of developmental arrest. For other highly labile aspects of morphogenesis in aphids see Hille Ris Lambers (1966).

Overproduction of JH, which inhibits imaginal differentiation, is apparently related to neoteny as described above. This appears to have been evidenced in many experiments (Johnson and Birks, 1960; Lees, 1961; Dehn, 1963; White,

1968; White and Gregory, 1972; Kuhr and Cleere, 1973; Clouthier and Perron, 1975; and others, which have shown the juvenilizing effect of the JH and its analogues. When these agents are applied to, for older (third or fourth instar) nymphs of the alate form, development of wings* and some other structures becomes arrested, often with supernumerary nymphs resulting; presumably the effect of JH persists for an unusually long time, and the time for the action of ED (triggering the differentiation of imaginal structures) becomes limited.

The effect of JH in producing different morphs can also be surmised from the pattern of development of CA, which release JH.** White (1965, 1968, 1971) and Hales (1976) found that in *Brevicoryne brassicae* (Fig. 4.4) the CA are larger in apteriform nymphs than in alatiform nymphs during the third and fourth instars, and that at the imaginal molt, the CA in the former become much larger than those in the latter. The CA in the alatiform nymphs somewhat decrease in volume between the beginning of the fourth instar and the imaginal molt, but 24 hr after the imaginal molt they again increase in size. In contrast, the CA of the apteriform adult decrease sharply in volume during the adult stage. Hales (1976) thought that the enlarged condition of the CA was the sign of active JH secretion. Based on this criterion (of Hales), the above data can be interpreted as follows.

The steady increase in JH activity (accompanying the increasingly enlarged CA) until the pharate adult stage in the aptera could induce the maturation of ovaries even in the old fourth stage nymph, and in the pharate female adult; Elliott (1975) also found that in *Aphis craccivora* the higher CA activity is correlated with the precocious ovarian growth in the aptera. Further, because of the consistently higher titers of JH (as seen in consistently larger CA) throughout postembryonic developmental stages and in the pharate adult of the aptera, the action of ED would become more limited than in the alata, and consequently the differentiation of imaginal structures (including wings) would become more suppressed (than in the alata)—hence the neotenous female aptera. It should be remembered that during late nymphal stages and in the young adult of the solitary (phase) *Locusta* (Fig. 4.3) the production of higher amounts of JH occurs. Therefore, it appears probable that the enlargement of CA, as Hales contended, may well be the sign of the active secretion of JH which induces the production of the aptera.

In the alate female, on the other hand, a significant increase in the JH titer apparently occurs during the adult stage (judging from the enlarged CA) as in many winged insects, and it appears to be at this stage that ovaries become mature by a

*As Shull (1937), Kitzmiller (1950), Johnson and Birks (1960), and White (1971) have shown, wing rudiments occur in the old embryo; therefore all nymphs start out as presumptive alatae (winged forms).
**The size of CA, which is subject to change during development, may not necessarily be correlated with the secretion of JH (see Doane, 1973).

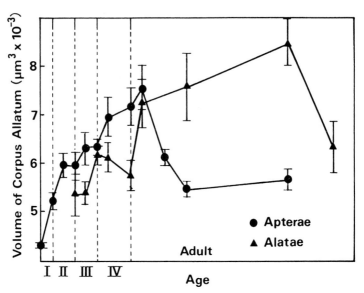

Fig. 4.4.   Change in volume of CA with age in cabbage aphid (*Brevicoryne brassicae*). Symbols I, II, III, IV refer to four nymphal stages. From Hales (1976).

sufficient gonadotropic action of the JH. Further, in the alata the lower titers of JH during postembryonic development (as seen in smaller CA), especially at the imaginal molt (no JH may be secreted at this stage), would allow the ED to trigger the normal differentiation of the imaginal structures including the wings.

Following the above interpretation, the pharate female apterae are presumably reproductively functional, while the pharate alatae are probably not functional adults; such difference occurs in many other female insects with wing dimorphism. Thus, Hale's criterion of CA size as an indicator of JH activity apparently holds. What is needed here is a bioassay of hormones (both JH and ED) during all stages of development.

White (1971) and Hales (1976) further maintained that the higher titer of JH in the adult alata (Fig. 4.4) influences the endocrine system of the embryos and young nymphs developing in the ovary of the mother, in a way that ensures higher CA activity during subsequent development of apterae. This interpretation appears to explain the strong tendency, in many aphid species, for the alate mother to produce only apterous offspring. In defense of this hypothesis Hales (1976) referred to the work by Riddiford (1971), which showed the positive feedback effect of JH applied to the embryo on the activity of the CA throughout nymphal development of *Pyrrhocoris*.

The above interpretation with regard to the prenatal determination of apterae, however, was contradicted by an experiment by Applebaum et al. (1975), which showed that application of JH and its analogues to the prenatal stage and to young nymphs of *Myzus persicae* has no apterizing effect. Furthermore, as shown below, Steel's (1976a,b) studies on the neuroendocrine control mechanism in *Megoura viciae* showed that the mother's JH has nothing to do with the endocrine activity of the developing young within the mother's body.

M. viciae is a monoecious species, and the prenatal determination of wing morphs has been well established.* Lees (1961) postulated that the maternal factor coming from the brain controls the activity of CA of the embryos developing in ovarioles, either by elevating or depressing the activity of the latter. Lees (1967) postulated further that the maternal factor from the brain may be an alata determiner which may inhibit the secretory activity of the CA during the late prenatal and early postnatal periods of larval development.

Steel (1976a,b) identified the maternal factor controlling the determination of embryos as virginoparae or oviparae in *Megoura* as a secretory product of one of the five groups (group 1 of Steel) of NSC in the brain. Steel found that ablation of this group of neurosecretory cells causes the loss of ability of the aphid to switch from production of one morph to the other (virginopara and ovipara); group 1 NSC are necessary for the production of virginoparae under long-day conditions. The product of the NSC is apparently not released into the hemolymph, but is transported intraaxonally to the abdomen, where it may be delivered directly to the reproductive system (ovarioles), which receives fine nerve branches from the abdominal nerve. Steel suggests that the secretion influences the programming of the endocrine system of the embryos. Since denervation of the CA had no effect on the determination of embryos as virginoparae or oviparae, JH appears to play no role in this system.

What Steel found was a mechanism of differentiation of the embryo into a virginopara or ovipara by the effect of the photoperiod (the first stage). Since the virginopara and ovipara are both apterous and structurally rather similar, the hormonal disturbance in the embryo caused by the secretion from the NSC could be small. At the second stage of differentiation of *Megoura*, the embryos determined as virginoparae at the first stage become determined either winged or wingless. It must be at this stage that a more significant modification of the hormonal balance occurs to produce either alate or apterous virginoparae; at this stage of differentiation crowding is known to be the major factor producing the alate form.

As seen from the foregoing discussion, the mechanism of wing morph determination is highly complex, when the mechanism involved during the prenatal stage is taken into consideration. Further complicating the matter is the phe-

---

*The prenatal determination is evidenced, for instance, by the fact that the generations of offspring of the artificially induced (by isolation) apterous virginoparous mother are always apterae, and almost all of the first generation offspring of the artifically induced (by crowding) virginoparae are winged. Hence, the prenatal determination is dependent on environment.

nomenon known as the "interval timer." It is known that virginoparae from young clones do not yield sexuales when exposed to short photoperiods, even though sexuales are readily formed when the clone is older. This phenomenon is attributed to the ability of the aphids to respond to the photoperiodic stimulus, only after the lapse of certain times unrelated to the number of generations. This type of clock or "interval timer" appears to make the action of hormones ineffective.

A review of the voluminous literature dealing with the extrinsic factors controlling polymorphism of aphids is simply beyond the scope of this study. Major factors that have been studied are photoperiod, temperature, population density, and food.

Photoperiod and temperature are usually interrelated in nature, and presumably the effect of one factor modulates the effect of the other. Their effects are more or less limited to certain stages of development, and the stage at which these factors are effective varies in different aphids. Often, short days have been found to cause production of the winged form. Crowding tends to increase the winged morph, as has been repeatedly found since the discovery of this effect by Bonnemaison (1951). Conversely, isolation causes production of the apterous morphs. This group effect therefore parallels the same phenomenon in Orthoptera, Psocoptera, auchenorrhynchous Homoptera, etc. Quality and quantity of food can have a dominant influence on the determination of morphs in some aphids. For summaries of literature on dietary influence on aphid polymorphism see Mittler and Sutherland (1969) and Mittler (1973). These extrinsic factors must influence the complex physiological processes producing the wing morphs, which, however, are poorly understood; and the physiological mechanism through which polymorphism occurs can vary significantly in various groups of aphids.

In scale insects (Coccoidea), the female has fewer nymphal instars and wings are not formed. The first female nymphal instar (crawler) is provided with three pairs of legs, which ensure its dispersal. The subsequent developmental process, however, differs considerably in different groups of female coccids. In a more primitive family, Pseudococcidae (mealybugs), two nymphal stages are known to follow the first nymphal instar, and the adult female emerges from the third stage nymph. The adult of this family usually maintains the three pairs of legs, an exception being *Antonina* spp., which are apodous and with (only) two-segmented antennae in some of them (McKenzie, 1967). Constructive developments in the female adult of mealybugs comprise various types of pores, some of which are believed to function in the production of ovisacs, and two pairs of ostioles located submarginally on the seventh abdominal segment and on the head. The submarginal ostioles secrete globules of body fluid, perhaps acting as a defense mechanism (McKenzie, 1967). The lack of wings and the presence of these pores and ostioles are neotenous features of the female mealybugs.

Since the third male nymphal instar becomes the fourth instar before emerging as an adult from the latter, the female mealybugs clearly undergo one less molt than the male mealybugs, and such a developmental process producing a neotenous adult represents prothetely (paedogenesis).

In a more derived group of Coccidae, such as Diaspinae, according to Pesson (1951b), the six-legged first nymphal instar becomes apodous, and antennae and eyes also degenerate at the next molt; then she becomes adult at the next molt. In the male, on the other hand, the second stage nymph becomes a "pupa," and one or two more molts follow before he becomes a winged male adult. The kind of metamorphosis resulting in the adult with highly degenerated conditions of structures in female Diaspinae may be called halmatometamorphosis.

The number of molts male coccids undergo during development varies considerably in different species. It is probable, therefore, that in the adult scale insects varying degrees of retention of juvenile features (neoteny) could occur as in the male Strepsiptera (p. 194).

In *Lecanium*, as shown by Pflugfelder (1936), the CA are very much larger in the female than in the male, although they are nearly equal in volume in both sexes during earlier nymphal stages. It is obvious, therefore, that the CA grow in volume much faster in the female than in the male, which means that more JH is synthesized in the female than in the male. An excessive secretion of JH from the enlarged CA presumably inhibits imaginal differentiation and expedites the maturation of ovaries at the same time, hence yielding a neotenous female.

In white flies (Aleurodidae), the first stage nymph is six-legged and mobile. In the next two instars, as shown by Weber (1931, 1934), however, antennae and legs become reduced; then in the fourth instar both legs and antennae become larger again, and the legs are now two-segmented. From the fourth instar the adult aleurodid emerges directly, but through an unusual process of transformation; the adult aleurodids are winged in both sexes. This unusual metamorphosis is sometimes called allometabolism. As in hypermetamorphosis, allometabolism does not result in polymorphic adults, although caenogenetic modification is conspicuous.

4.3.1.15. Heteroptera

In Heteroptera, neoteny is evident in the reduction of wings, which ranges from brachyptery to aptery through microptery, and in some species (e.g., *Aradus cinnamomeus*) the narrowing of wings or stenoptery occurs. These different wing morphs are sometimes fairly distinct, although in some Heteroptera all gradations of wing reduction occur. Another neotenous feature is the reduction or loss of ocelli, which sometimes accompanies the wing reduction, as in some other orders; a complete association of the wing reduction with the loss of ocelli occurs only in Microphysidae (Southwood, 1961). For modifications of the thorax in the wingless form see Matsuda (1970). Brinkhurst (1963) outlined the types of wing reduction within Heteroptera as follows: 1) species mainly wingless . . . riverine aquatic and semiaquatic species and in most families of the Cimicomorpha, especially in species inhabiting grasses, heaths, or marshes, as well as the lyctocorine and cimicine Cimicidae; 2) species with winged males, mostly or entirely short-winged females . . . in the Microphysidae, most of which inhabits lichens, and a few mirids; 3) species with polymorphic females, flightless males . . . *Aradus cinnamomeus*; 4) species polymorphic in both sexes . . . Amphibicorisae, several Miridae, Lygaeidae, as well as *Pyrrhocoris apterus* and *Chorosoma schillingi*; 5) Species usually macropterous.

The frequency of wing reduction is high in some groups. For instance, in Emesinae (Reduviidae) wing reduction occurs in more than 50% of the known genera, and often in those inhabiting semiarid and arid environments (Wygodzinsky, 1966). Reduction of wings, often involving complete loss of wings, frequently occurs in the Aradidae, which usually live under bark (Usinger and Matsuda, 1959). In Gerridae, the loss of wings is so frequent that generic descriptions must be based primarily on the apterous form (Matsuda, 1960).

Apparently, no other external structures become neotenous in the adults of Heteroptera. It is possible that in this suborder the tissues forming many adult structures have lost or decreased sensitivity to the action of JH. This interpretation is consistent with the marked trend of increasing differentiation of many thoracic and abdominal structures in this suborder (Matsuda, 1970, 1976). Arrest of development of some structures of the internal reproductive system may occur in association with wing reduction.

Many endocrinological studies on Heteroptera since Wigglesworth's pioneering works on *Rhodnius* in the 1930s now clearly show that in this suborder JH produced from the CA has the gonadotropic function, in addition to its function of suppressing imaginal differentiation, and that ED produced from the prothoracic glands promotes imaginal differentiation. As in some other orders, the NSC in the brain influence the activity of CA and perhaps other endocrine organs. In *Rhodnius prolixus*, for instance, the cerebral control of vitellogenesis occurs in less than 24 hr after the meal intake, probably by stimulating the CA to secrete JH (Baehr, 1973).

During the normal metamorphosis of this suborder, the development of imaginal structures (including the imaginal wings) occurs in the complete absence of JH (Slama, 1964a), and the prothoracic glands degenerate and disappear soon after the final molt (Wigglesworth, 1955). The reduced imaginal wings, therefore, presumably result from insufficient action of ED, which may often follow the abnormally high titer of JH during nymphal stages; the results of various experiments inducing prothetely and metathetely also indicate this probability (see, e.g., Slama et al., 1974).

The effect of photoperiod (which is more or less modulated by temperature) on the determination of wing morphs is evident in the studies of some European species of *Gerris* spp. (waterstriders). Thus, according to Vepsäläinen (1971a,b, 1974a), the first generation of *Gerris odontogaster* is dimorphic in Finland, viz., it consists of micropterous and macropterous individuals. Those individuals reaching the fourth nymphal stage at a critical photoperiod become micropterous by the effect of lengthening photoperiod, and those emerged later become macropterous; all individuals of the second generation are macropterous and diapause as adults. Thus, different photoperiods determine the production of two quite distinct types of wings, and Vepsäläinen (1971a,b, 1974a) concluded that such seasonal dimorphism cannot be the result of genetic segregation.

It appears that, in producing micropterous individuals, the lengthening photoperiod induces excessive secretion of JH in the fourth stage (or older) nymph. The effect of excess JH thus produced may persist for a long time, or the secretion of JH persists longer than it does normally. Such overproduction of JH can result in a deficiency in the action of ED, and hence the production of micropterous individuals. It is important, in this connection, to note that in the fifth instar nymphs of this species the sizes of wing rudiments are virtually the same in both presumptive macropterous nymphs and presumptive micropterous nymphs (Brinkhurst, 1963). It is obvious, therefore, that in the latter the wings simply do not grow in size, although some adult characteristics (veins, and so on) do develop in them. Further, this fact indicates that two distinct wing morphs can result simply from a disruption in hormonal balance.

The fact that the micropterous individuals reproduce soon after emergence (Vepsäläinen, 1974a) means that their ovaries are ripe and ready for fertilization at the time of emergence; and this phenomenon can be attributed to the gonadotropic effect of the JH, which must be abundantly available toward the end of postembryonic development of the micropterous individuals. In fact, according to Brinkhurst (1963), in the last stage nymphs of the summer generation of *Gerris* spp. (in which the short-winged form occurs in Britain), the gonads are well developed. Furthermore, according to S. Miyamoto (personal communication), the ovaries containing mature eggs are always enlarged in the short-winged Heteroptera; the same is certainly true of many other short-winged female insects (Orthoptera, Homoptera, and so forth). In the males of *G. odontogaster*, the same hormone presumably stimulates the function of the accessory glands, and by the time of emergence they could be ready for copulation; the fact of stimulation of the male accessory glands by JH is known for *Rhodnius* (Wigglesworth, 1936) and in some other orders, including the Orthoptera (p. 157).

In Finland seasonally restricted wing reduction occurs also in *Gerris argentatus*, *G. thoracicus*, and *G. paludum*, although in Hungary *G. thoracicus* is always macropterous (Vepsäläinen, 1974b). In Germany *G. lacustris* is seasonally dimorphic (Darnhofer-Demar, 1973), and in this species macroptery is induced, unlike in *G. odontogaster* in Finland, by lengthening days.

G. rufoscutellatus always remains macropterous. *G. najas* is also permanently apterous, and its distribution is quite localized in Southern Finland. Among 38 specimens Vepsäläinen (1974a) examined, however, one female specimen from the U.S.S.R. was brachypterous. Thus, the apterous condition of this species in Southern Finland is probably comparable to the micropterous condition of *Nemobius yezoensis*, which occurs only in the Northern half of Japan without losing the potential for developing wings fully (p. 159).

G. lacustris and *G. lateralis* are always wing-dimorphic throughout the year in Finland (Vepsäläinen, 1974a), and the effect of photoperiod and temperature on the production of wing morphs is less clear-cut than in *G. odontogaster* and others that exhibit seasonal dimorphism. Yet, the presumed segregation of alleles of a single gene resulting in wing dimorphism is, as discussed later (p. 177), not convincing.

The effect of photoperiod on wing polymorphism (macroptery, brachyptery, and microptery) is also evident in an experiment on *Pyrrhocoris apterus* by Honék (1974), which showed that under a long photoperiod (18 hr) some of the adults (derived from brachypterous parents) were macropterous (8 and 6% in males and 5 and 20% in females), whereas under a short day length (12 hr) virtually no macropterous individual emerged. Clearly, in this species, as in *G. lacustris* (Darnhofer-Demar, 1973) and some cricket species, longer photoperiods tend to induce macroptery. Honék (1976) confirmed this tendency by additional experiments, and showed further that the effect of photoperiod becomes modified significantly by temperature, and that isolation probably modifies the physiology so that brachyptery can occur.

Honék (1976) was also able to raise, by selection in the laboratory, the proportion of macropterous individuals up to 80% (whereas some of the non-selected samples comprised only about 2% of macropterous individuals at the same time). This certainly indicated, as Honék maintained, that the form of the wing is genetically influenced, and that individuals are not identical regarding the tendency toward macroptery. However, under the overriding influence of environmental factors, such genetic differences usually do not express themselves in nature (compare with a similar case in crickets, p. 160).

The possible effect of host plants on wing morphs is clearly seen in an interesting discovery of the "paper factor" by Slama and Williams (1965). When the European species, *Pyrrhocoris apterus*, was reared in the U.S., the nymphs failed to undergo metamorphosis, and all nymphs molted into sixth instars or into adultoid forms, preserving many nymphal characters. The cause of this phenomenon was soon traced to the exposure to the active principle present in a certain paper towel, which had been placed in the rearing jars. The towel is made from an American pulp tree, the balsam fir (*Abies balsamea*). It was found that this active material readily penetrates the cuticle and is distributed throughout the insect, exerting its juvenilizing effect at the time the endogenous JH becomes depleted before metamorphosis, with a consequent abnormally prolonged nymphal stage.   Slama and Williams (1965) further showed that the extracts of *Abies balsamea, Tsuga canadensis,* and *Taxus brevifolia* have high JH activity, while those from other trees have intermediate or barely detectable activity. Furthermore, the most active extracts were found to be without any detectable effect when applied to some other insects. These findings indicate that the occurrence of abnormal metamorphosis resulting in wing reduction could be dependent on host plants.

Southwood (1961) classified wing reduction into two categories: metathetely and prothetely. Metathetely is brought about, according to him, by excessive influence of JH, which leads to juvenile characters in the adult, and prothetely is brought about by depression of the influence of JH, which leads to adult characters in the nymph.

The cases of metathetely Southwood cited were the brachypterous Gerroidea, in which brachyptery is associated with high altitude (with lower temperatures), and many bivoltine species, with the first generation developing in May and June (under relatively low temperatures).   Other examples of metathetely were brachyptery in *Nabis ferus* (Nabidae), *Bryocoris pteridis* (Miridae), and *Trapezonotous* spp. (Lygaeidae) which occur in mountains where colder temperatures prevail.   An example of prothetely that Southwood cited was *Dolichonabis limbatus,* which has four nymphal instars instead of the normal five in Heteroptera; the long alate morph of this species occurs most frequently under colder mountainous conditions.   Therefore, Southwood inferred that *Dolichonabis* adults are normally an example of paedogenesis or prothetely, and only under especially cold conditions does the normal hormone condition obtain. Another example of prothetely cited by Southwood was *Microvelia* spp., which have four or five nymphal instars; the adults produced in the former condition are apterous.   Thus, Southwood (1961) related all cases of abnormal metamorphosis (prothetely and metathetely) only to low or high temperature, and this interpretation was consistent with the result of the experiment on temperature effect on the development of *Rhodnius* by Wigglesworth (1952). The latter experiment showed that when the fourth instar nymphs are reared under relatively low temperature, some structures (wing lobes, genitalia) in the fifth instar are more juvenile than is normally the case (because the hormonal balance becomes slightly in favor of JH); while when they are reared under high temperature, molting is somewhat delayed and the characters developed in the next instar are slightly "adultoid."

Although Southwood (1961) was not specific with regard to the nature of the reduced wings, the wings in question are certainly imaginal wings, not the nymphal one, as pointed out by Wigglesworth (1961) and Brinkhurst (1963). With respect to the hormonal mechanism for the production of reduced wings, Southwood's explanation was insufficient in failing to refer to the action of ED in producing the reduced wings. It is certainly true, as Southwood contended, that in bringing about metathetely excessive JH influence is necessary. However, the imaginal wings are produced only by the action of ED.* Therefore, the most probable mechanism of metathetely is a more limited action of ED as a consequence of overproduction of JH. If prothetely results, as Southwood maintained, from depression of JH influence, the limited imaginal differentiation can result from precocious action of ED, which follows. However, in such cases it is hard to conceive of the production of the reproductively functional neotenous female (in the lack of sufficient JH, which has the gonadotropic effect in the female); such early depletion of JH can result in a juvenile caste as in ants (p. 184). In addition to the effect of temperature on wing reduction (which Southwood emphasized), the effect of photoperiod must be seriously considered.

According to Brinkhurst (1963), in microphysids and some mirids the male is fully winged and the female is always or usually brachypterous. A probable underlying mechanism of this phenomenon is, as in other orders of insects, the overproduction of JH, which leads to the suppression of imaginal differentiation and to precocious oogenesis (with the female thereby becoming reproductively functional more readily than the male).

The genetic analysis of determination of wing morphs in *Gerris* spp. was first attempted by Poisson (1924), who performed the following crossing experiment: *Gerris najas*: four crossings between apterous individuals produced 78 offspring, of which only one individual was macropterous; *Gerris thoracicus*: one crossing between apterous individuals produced 6 apterous and 83 macropterous individuals; *Gerris argentatus*: two crossings between brachypterous individuals produced, in addition to brachypterous individuals, 1/3 to 1/8 macropterous individuals, and two further crossings produced only macropterous individuals. A crossing between a macropterous individual and a brachypterous individual produced 6 brachypterous and 8 macropterous individuals. From these experiments Poisson hypothesized that a single gene was involved in the production of wing morphs, and that AA was lethal, Aa short-winged, and aa long-winged. No effect of food and temperature on the wing morphs was found by Poisson. Ekblom (1927/8), who conducted similar crossing experiments, came to realize that segregation of alleles was complex and that some other genetic factors were apparently involved. Larsen (1931) and Klingstedt (1939) pointed out, however, that the pattern of segregation of alleles is so irregular in different species and so often unexpected exceptions occur in these experiments that Poisson's explanations (such as different mortalities in different morphs, atavism, pseudobrachyptery) hardly justify his hypothesis. Klingstedt (1939) attributed the wing morphs to temperature, citing the work by Mitis (1937), which showed that in three bivoltine species of *Gerris* occurring in Germany the

---

*As already seen (p. 172), imaginal wings develop only in the complete absence of JH, i.e., only under the influence of ED.

summer generation is exclusively brachypterous and the winter generation is always macropterous.

Despite the criticism by Larsen and Klingstedt and despite the hormonal theory of wing polymorphism in Heteroptera by Southwood, several authors (Ekblom, 1949, 1950; Brinkhurst, 1959, 1963; Guthrie, 1959; Anderson, 1973; Vepsäläinen, 1974c) still attempted to analyze wing polymorphism in terms of genetic segregation, and controversy sometimes centered around whether a single gene (Brinkhurst, 1959; Guthrie, 1959; Vepsäläinen, 1974c) is involved, or polygenes (Ekblom, 1949, 1950; Anderson, 1973) are involved.

In the latest crossing experiments on wing morphs of *Gerris lacustris*, Vepsäläinen (1974c) attempted to show that the genetic switch operates through one locus, with one allele for brachyptery being completely dominant. In most of his experiments, however, he failed to obtain a clear-cut Mendelian segregation of alleles, because, according to Vepsäläinen, the morph determination is strongly influenced by environmental factors, especially photoperiod. The only indication of the genetic determination was that Br X Br yielded significantly more brachypterous offspring than the Br X Ma cross, which, in turn, produced more brachypterous individuals than the Ma X Ma cross. When the determination of wing morphs is so strongly influenced by environmental factors, the presumed genetic mechanism is at least very difficult to prove. It should be pointed out further that the two distinct morphs can result purely from the disruption of hormonal balance (without regard to different genetic constitutions in different individuals), as seen in the case of *Gerris odontogaster* (p. 173). Wing dimorphism as an adaptive strategy in *Gerris* spp. was discussed by Järvinen and Vepsäläinen (1976).

4.3.1.16. Coleoptera

In some Coleoptera neoteny occurs only in the female, although less pronounced neoteny affecting mainly wing development occurs in both sexes. In some Cantharoidea the adult female is apterous and is apparently larviform. During postembryonic development of *Lampyris noctiluca*, according to Naisse (1966c), the male undergoes four or five larval molts (five or six larval stages), and then the pupal and adult stages follow. The female of the same species, however, undergoes five or six molts before the pupal molt. Clearly, the female undergoes one extra molt, which accounts for her larger size.

At the level of neuroendocrine activity, Naisse (1966a) found the following differences in producing sexual dimorphism. In the male the small granular NSC in the brain start their secretion at the time of sexual differentiation, i.e., after the third molt, and the secretion continues up to the imaginal molt. The CA remain inactive during the pupal and adult stages, and the prothoracic glands (PG) degenerate in the pupa. In the female, on the other hand, the neurosecretion from the small granular NSC starts after the fourth molt and has three periods of interruption. The CA are active in both the pupa and the adult, although inactive at the pupal and imaginal molts; the PG remain active until the end of the pupal stage and degenerate after the imaginal molt. Clearly, the activities of CA and PG in the male are abnormal, although they are normal in the female.

Naisse (1966b,c, 1969) discovered further that in the male neurosecretion from the small granules induces the formation of mesodermal apical cells at the apex of the gonad, and the androgenous hormone secreted from these cells was found to be inducing the normal male structures, including the wings and the genitalia. Clearly, during the pupal stage, the apical cells replace the morphogenetic function of the CA and PG, which become inactive or degenerate. The sexual differentiation is, therefore, ultimately attributable to different actions and numbers of the particular NSC, which must be genetically determined. Although the occurrence of androgenous hormone is common in Crustacea, its presence in *Lampyris* must be a physiological specialization. In no other insects has a sex hormone been found so far.

In the female, the activity of the CA and that of PG are comparable with those in the other insects with fully developed wings, and the absence of wings is not due to hormonal suppression of wing rudiments, which are apparently never present during postembryonic development. In fact, Davydova's (1967) CA-CC removal experiment with the same species failed to cause the development of wings. It is therefore probable that the absence of wings in the female is determined by a sex-linked gene or genes. Because of the lack of wings, the adult female of *Lampyris* appears to be larviform. However, other structures and integument are quite normal and imaginal; obviously their development is induced by a hormonal control mechanism typical of insects. This kind of genetically determined neoteny affecting only wing development is again not directly comparable to the hormonally induced neoteny in other orders of insects.

Another conspicuous case of neoteny in Coleoptera is that of *Micromalthus debilis* (Micromalthidae), in which five kinds of adults occur. They are the winged male, thelytokous (female producing) larval female, arrhenotokous (male producing) larval female, amphoterotokous (male and female producing) larval female, and the winged female (Scott, 1938). The three kinds of larviform females represent paedogenetic neoteny produced by prothetely. The male producer is known to give rise to the male by haploid pathenogenesis, and the female producer to the female by diploid parthenogenesis as in aphids and *Heteropeza*. Apparently, no sex chromosome occurs (Scott, 1936). The amphoterotokous female is essentially a male producer in which development of the male is arrested and female-producing eggs develop secondarily, and the sex may be environmentally determined (Scott, 1941).

The anomalous species, *Rhizostylops inquirendus*, described by Silvestri (1905), has certain characters and habits that seem to place it as an intermediate between Rhipiphoridae (Coleoptera) and Strepsiptera, and the larviform female bears a marked resemblance to those of *Mengenilla* and *Eoxenos* of the Strepsiptera. The female of *Ripidius* is also apterous and larviform (Clausen, 1940).

Although the reduction of the elytra occurs in numbers of families (Rhipiphoridae, Staphylinidae, Meloidae, some Cerambycidae, and so on), the reduction of hind wings is far more frequent and occurs in both sexes of many families (see Rüschkamp, 1927, and others). The metathorax (bearing hind wings) becomes reduced in many Coleoptera with reduced hind wings. Apparently, the reduction of other structures that accompanies the wing reduction is usually very limited in this order; in Carabidae, for instance, the eyes are directly affected by wing atrophy only rarely (Darlington, 1936).

In Carabidae, as shown by Oertel (1924), Darlington (1936, 1943, 1971), Lindroth (1946, 1949, 1957), and others, all degrees of reduction of hind wings occur. Goldschmidt (1940) distinguished two types of reduction of hind wings in Carabidae, viz., reduction due to insufficient growth of the imaginal discs and reduction caused by histolysis of areas of the wings.

Darlington (1936, 1943, 1971) noticed that hind wing reduction occurs frequently in the Carabidae inhabiting mountainous areas. Thus, for instance, at low altitudes in New Guinea only 4% of carabids have reduced wings. The percentage of wing reduction rises to 32% in the carabids living at 500-1,000 m, and to 95% in those living at an altitude of 3,000 m and higher (Darlington, 1971).

Since wing reduction does not occur seasonally and since its occurrence is usually not confined to certain geographical areas, Darlington thought that the wing reduction in the mountains had been induced by recurrent mutations. He denied the effect of cold temperature as a direct (physiological) cause of wing reduction, and proposed that three factors are important in favoring wing atrophy and flightlessness in the mountains: 1) an indirect effect of cold temperature, which reduces ease and usefulness of flight, 2) intensity of competition with ants correlated with altitude, and 3) the limitation of area in higher altitudes.

Darlington's hypothesis of genetic determination of wing morphs appears to have been supported by a crossing experiment of wing morphs of *Pterostichus anthracinus* (Carabidae) by Lindroth (1946). The results of crossing experiments of macropterous ♀ X macropterous ♂, brachypterous ♂ X brachypterous ♀, and macropterous ♀ X brachypterous ♂ showed that: 1) brachypterous parents may produce macropterous offspring, and 2) macropterous parents in three crossings produced macropterous offspring only (21 specimens). These results (details are not given here) justified the conclusions that: 1) the wing dimorphism in *Pterostichus anthracinus* has a hereditary basis; 2) the brachyptery is dominant, and the macropterous individuals are homozygotes; and 3) inheritance takes place in simple Mendelian fashion. Essentially, the same results were obtained by Jackson (1928) and Stein (1973) in their crossing experiments of wing-dimorphic species of Curculionidae.

Thus, at least in *Lampyris*, Carabidae and Curculionidae penetrance of genetic factors for the wing morphs is unusually good. From an endocrinological viewpoint, the reason for this tendency may be that either the CA become inactive toward the end of postembryonic development (as in the male of *Lampyris*) and imaginal differentiation proceeds according to the genetically determined program (without JH suppression), or else the sensitivity of tissues[*] to JH has decreased to the point where the genetic mechanism of wing morphs is fully expressed. In some artificial conditions, however, wing morphs do not segregate according to the presumed genetic program. Thus, Palmen (1944) reared 92 larvae derived from brachypterous parents of *Carabus melanocephalus* under the most densely populated condition possible, and found that all adults which emerged (32 individuals) were brachypterous. Exposure of larvae of *Tribolium confusum* to lower temperatures by Nagel (1934) caused the production of various intermediates between the normal larva and the normal pupa. These findings suggest the possibility that in certain extreme environments (such as the arctic area) wing reduction could possibly be induced physiologically.

In cavernicolous beetles also hind wing reduction occurs frequently, and such reduction is usually linked with reduction of eyes, as seen in Trechini (Trechidae), Bathysciinae (Catopidae), Pselaphidae, and so on. Bathyscines were, according to Vandel (1965), apparently preadapted for cavernicolous life, as is evident from the fact that their lucicolous members are depigmented and that a great majority of them are apterous. The absence of light in the cavernicolous environment must have affected the process of reduction and depigmentation. Vandel (1965) claimed that the influence of light was probably affected by the intermediary of endocrine pathways. In these cavernicolous beetles, however, appendages (legs and antennae) are usually greatly elongated, and this tendency parallels the compensatory growth of legs in other short-winged or wingless insects. Deleurance and Charpin (1970) and Deleurance (1975) have attempted to show the endocrine control mechanism involved with the contraction of the life cycle in the cavernicolous catopine beetles. In termitophilous Staphylinidae all degrees of reduction of both pairs of wings occurs (Seevers, 1957; Kistner, 1958; Campbell, 1973). Whether such reduction is due to hormonal suppression or is genetically determined remains to be seen.

Hypermetamorphosis is known to occur in Rhipiphoridae, Meloidae, some Carabidae, and some Staphylinidae. A typical hypermetamorphosis seen in

---

[*]Tissues are certainly sensitive to JH and ED and their analogues when they are applied experimentally (see Metwally and Sehnal, 1973; and Socha and Sehnal, 1973).

Sitaris humeralis, cited by Goldschmidt (1940), is as follows: From its eggs a typical coleopteran larva hatches, which will not continue development unless it succeeds in getting attached to a bee, which carries the larva into the hive. Here the larva molts and, losing eyes, legs, and so on, emerges as a primitive maggot, feeding on honey; after some time, a resting stage (pseudopupa) occurs, from which another maggot emerges. This one finally pupates regularly, and from the pupa the beetle hatches.

4.3.1.17. Hymenoptera

In Hymenoptera, morphologically and/or physiologically distinguishable castes occur only in the female. The caste differentiation is induced by environmental factors, which, in turn, influence the endocrine function. In *Apis mellifera*, structural differences between queen and worker are derived from essentially the same genome by nutritional conditions during larval development. The royal jelly, the food of the queen larvae, has been shown to determine the queen; those larvae fed with the royal jelly throughout larval life develop into the queen, and those which do not get the royal jelly develop into workers (see Rembold, 1973, 1976, for the active substance in the royal jelly). As summarized by Michener (1974), the structural differences between the two castes include: 1) 2 to 12 ovarioles per ovary in the worker bee and 150-180 ovarioles per ovary in the queen; 2) the larger spermatheca in the queen; 3) the presence of the wax gland and Nassanov's gland in the worker and their absence in the queen; 4) the proboscis, which is longer in the worker than in the queen; 5) the presence of the corbicula in the worker and its absence in the queen; 6) the pollen press, which is present in the worker and absent in the queen; and 7) some other structures (relative area of the antennal surface, chemoreceptive plates per antenna, number of facets of compound eyes, antennal lobes of the brain, hypopharyngeal lobes, and so on), which are more or better developed in the worker than in the queen. It is clear that the degree of constructive development of structures (differentiation of structures) is greater in the worker than in the queen (except for the ovary and spermatheca).

The hormonal control mechanism of caste differentiation in the honey bee was studied by Wirtz (1973), who showed that: 1) in worker larvae grafted into queen cells the volume of CA increases at the end of the third day; 2) the titer of JH in the hemolymph increases much faster in the presumptive queen larvae than in worker larvae; 3) the change in ultrastructural pattern of CA also indicates a much higher rate of JH synthesis in the queen larvae than in worker larvae; and 4) external application of JH-1 to worker larvae results in the appearance of queen characters (when it is applied shortly before the third day).

The above findings indicate that in the queen larvae a sufficient supply of JH enables the normal maturation of ovaries. In the worker larvae, on the other hand, the low titer of JH during development is apparently insufficient to stimulate the normal development of ovaries; consequently the adult worker is sterile.

The presence of a lower titer of JH in worker larvae apparently leads to more prolonged action of ED during the pupal stage, which last for nine days, instead of five days as in the queen. This presumed difference in the duration of ED action is obviously correlated with the difference in degree of structural differentiation between the two castes. It is apparent that the prolonged action of ED in the worker results in the activation of some extra genes which are not activated in the queen, and hence more differentiated conditions of many structures in this caste than in the queen; these genes could become available through a possible regulatory gene switch associated with the lowered JH titer. De Wilde (1976) was concerned with the hormonal control mechanism of queen bee development, and interpreted Wirtz's data (above) differently. He emphasized the positive morphogenetic roles JH can play during development. According to de Wilde, the active substance in the royal jelly taken before the end of the third day of larval life, heightens the activity of CA. The heightened activity of CA, in turn, leads to covert morphogenetic induction, which directs the developmental program toward the queen differentiation and becomes overt later in development, involving the shortening of development. The above concept is essentially the same as the concept of regulatory gene switch by altered (higher) JH titer, which, in turn, results in a new developmental process. The consequence of the heightened titer of JH is, however, the production of a normal, fully winged hymenopteron (which is the queen bee). It is, therefore, evident that the higher titer of JH does not result in the kind of regulatory gene switch which results in material compensation (through which typically reduction of wings and enlargement of legs occur). It is probable, therefore, that the apparently high titer of JH in the queen larvae (as compared with worker larvae) is normal for the Hymenoptera, and what is abnormal is the lowered titer of JH in the worker (discussed above): in the case of soldier development in termites, with which de Wilde compared the queen development in *Apis*, the high titer of JH is abnormal, and apparently the switch of regulatory genes occurs (p. 217).

In considering queen development, therefore, the positive morphogenetic function of JH is not important. What is more important is the balance between JH and ED during postembryonic development, and the hormonal control mechanism in the queen bee, therefore, must be compared with those in other (normal) Hymenoptera. Thus, in *Cephus* (Cephidae), according to Church (1955), postdiapause morphogenesis is initiated by the PG (i.e. by ED). Further, in several species of sawflies both pupal and adult cuticles are formed and released in the absence of JH (Slama, 1964b).

The production of castes in the stingless bee genus *Melipona* has been shown to be genetically biased (Kerr, 1950; Kerr et al., 1975, and others); unlike the case of the honey bee, no intermediates occur between the worker and the queen. According to this theory, only the doubly heterozygous (AaBb) larvae can develop into queens, provided that they are reared with a proper amount of food; with the optimal amount of food both workers and queens emerge in the expected ratio (in many of the cases 25% of them were queens). When the quantity of food given is below a certain limit, all or some larvae become workers; Kerr and Nielsen (1966) showed evidence that the larvae with the queen genome can become workers. Clearly, the expression of the genetic mechanism for caste determination is dependent on the environment, i.e., food.

Kerr et al. (1975) found that the queen development is correlated with a greater CA activity, as in *Apis*. Velthuis and Velthuis-Kluppel (1967) found that JH application is most effective in producing the queen when applied to the prepupal stage; most larvae treated this way become queens. Thus, in producing the queen, relatively high titers of JH during larval development appear to be a necessary prerequisite, as in *Apis*. It is most probable that the intake of certain amounts of food induces an optimal range of hormonal balance in which the JH titer is relatively high, and within this optimal range of hormonal balance the presumed genetic mechanism expresses itself more or less fully. As Velthuis and Velthuis-Kluppel's (1975) study indicates, if the hormonal balance is upset excessively in favor of JH at a critical stage, the genetic mechanism becomes meaningless, and most of them could (theoretically) become queens. Similarly, for JH titer below that of the optimal range for queen production, more than three-fourths of the larvae would become workers. Kerr et al. (1975) further elaborated the theory of genetic determination of castes by adopting the model proposed by Britten and Davidson (1969, 1971).

In bumble bees (*Bombus hypnorum* and *B. terrestris*), as shown by Röseler (1970) and Röseler and Röseler (1974), morphologically and/or physiologically distinguishable castes (queen and worker) occur in the female. In *B. hypnorum* castes are only physiologically differentiated. Those females with a high concentration of a specific protein are able to diapause, and they become queens; those females failing to diapause are workers (Röseler and Röseler, 1974). Röseler (1970) found that only those larvae which obtain the optimal amount of food become queens. Further, Röseler and Röseler (1974) found that when JH is applied to the fourth (last) larval instars, all of them become queens. These findings indicate that the caste differentiation in this species is physiologically induced and that a high titer of JH in the last larval instar may, as in *Apis* and *Melipona*, be the necessary prerequisite for the production of queens. Further, the high titer of JH appears, in turn, to result from sufficient intake of food during larval development.

In *Bombus terrestris* castes are both morphologically (in size) and physiologically distinct, and the caste differentiation is fixed in the first three days of their larval development. According to Röseler (1970), the queen produces a pheromone that causes female larvae to develop into workers; in the absence of the queen inhibition, larvae may, depending on nutritional conditions, develop into queens. Application of JH to the fourth instar larvae already determined to be workers does not induce them to develop into queens. As Röseler and Röseler (1974) maintained, the physiological caste differentiation in *B. hypnorum* is more primitive.

The castes of ants are highly polymorphic. Major castes are worker, soldier, and queen. Worker and soldier castes can be divided into continuously intergrading subcastes according to their sizes. The ergatogyne is a form intermediate between worker and queen; at least in some groups of ants they are reproductively functional, and therefore they can be regarded as neotenous.

In the larva of *Myrmica rubra*, as shown by Brian (1959, 1965), the imaginal discs consist of dorsal and ventral sets. The dorsal set consists of wing buds, gonad rudiments, and ocellar buds, while the ventral set comprises leg buds, mouthparts buds, and the central nervous system. In the queen development both sets of discs develop fully to form normal hymenopterous structures. In the presumptive worker larvae, however, the wings and the ovary stop growing, although the legs grow more quickly than in the queen larva. The wing rudiments never inflate and become lost into the thoracic wall, and the splitting of the ovary rudiment into ovarioles does not occur. It is clear then that the worker caste differentiation results from the arrest of development (differentiation) of the dorsal set of imaginal discs. The difference in body size between the worker and queen is discontinuous in *Myrmica rubra*, probably because of suppression of a molt in the worker (Brian, 1976); there are ways of forming intercastes artificially (Brian, 1976). As shown by Brian (1959), there is a difference between presumptive queen larvae and presumptive workers in the activity of NSC. The neurosecretion may, as in some other insects, stimulate the activity of PG. However, the glands, which are present in the second stage larva, disappear in the next instar; therefore ED may be produced elsewhere within the body (Brian, 1976). Further, according to Brian (1959), there appears to be no difference in the pattern of development of CA between the two castes. Thus, the neuroendocrine control mechanism of caste differentiation in *M. rubra* still remains unclear. Brian (1959), however, inferred that in the worker larvae the secretion of JH ceases prematurely, and that this is followed by a sharp rise in ED titer, which, however, triggers the development of the ventral set of the imaginal discs alone.[*] Thus, the incomplete differentiation of imaginal structures in the worker, which may undergo one less molt, represents a case of prothetely.

---

[*]Injection of ED into last instars of female sexual larvae resulted in the production of adults with reduced sizes; they were either small queens, or intermediates, or workers (Brian, 1974).

In the worker caste the head becomes disproportionately enlarged, as if compensating for the reduction of body size and the arrest of development of some imaginal structures. This enlarged head has a defense function, and such adaptive development in the essentially juvenile worker caste therefore can be regarded as caenogenetic. Further, being (apparently) conditioned by the abnormally lowered titer of JH, such caenogenetic development could result from a regulatory gene switch.

In *Myrmica rubra*, nutrition is the decisive extrinsic factor in determining castes. Several other factors such as temperature, perception of the queen pheromone, small eggs laid, and so on, are known to affect the process of caste differentiation (see Wilson, 1971; Brian, 1976, for summaries). The mechanism of caste differentiation has also been studied with *Formica rufa* (see Wilson, 1971) and *Leptothorax nylanderi* (Plateaux, 1971). The effects of application of JH analogues to the fire ant (*Solenopsis invicta*) have recently been studied (Troisi and Riddiford, 1974; Vinson and Robeau, 1974; Vinson et al., 1974; Robeau and Vinson, 1976).

Wing reduction and associated reduction of other sturctures frequently occur in parasitic Hymenoptera, either in only one sex or in both sexes, and food is usually the dominant factor determining morphs. According to Salt (1941), in general, hosts affect the size of parasites, and the altered size in turn affects the proportion of parts, presence or absence of wings, and some physiological aspects, such as developmental period, mating behavior, and so forth. Quantity and quality of food available for parasites are the dominant factors determining these morphological and physiological consequences. A conspicuous example cited by Salt (1941) was *Trichogramma semblidis*, which exhibits wing dimorphism; those males that parasitized the eggs of *Sialis* (Neuroptera) are incapable of expanding wing rudiments, and such males are known as "runts." In the runt the setae on antennal segments are much less developed, and leg segments are also not fully developed. These features must result from developmental arrest; hence runts are neotenous.

Another conspicuous case of neoteny is *Melittobia chalybii* (Eulophidae), studied by Schmieder (1936). This species occurs in two forms, the type form and the short-lived second form, and the production of one or the other of the two forms is determined by trophic conditions during the growth of the larvae in the host *Trypoxylon politus*. The males of the second form differ from the type form in their darker color and their more aborted condition of optic organs. The females of the second form differ from the type females in their lighter color, their failure in spreading wings upon emerging from the pupa, and their swollen abdomen, which is full of ripe eggs. The presence of ripe eggs at the time of emergence with incompletely differentiated structures appears to reflect a heightened action of JH, which has the dual functions of suppressing imaginal differentiation and promoting oogenesis.

In *Gelis corruptor* (Ichneumonidae), according to Salt (1952), the wing development is trimorphic in that females are always apterous and males are either macropterous or micropterous. Since no trace of wing buds occurs in the larval stage, the female aptery is probably a sex-limited, genetically determined character. All male larvae have wing buds and have the potential to develop the wings fully. On a host providing adequate nourishment, the male larvae develop into the macropterous form; on a host providing meager nourishment, however, the male larvae develop into micropterous adults. The two types of wings are remarkably distinct, and there are no intermediates. Therefore, the wing morphs may be genetically determined. Here again, however, the expression of the presumed genetic mechanism is clearly dependent on the quantity of food.

According to Reid (1941), reduction of eyes and ocelli (with some exceptions) accompanies wing reduction in Hymenoptera. An interesting finding by Reid was an increase in leg size in many Hymenoptera with reduced wings (Scolioidea, Bethylidae, Sclerogibbidae, Agaonidae, and some ants); excessive development of legs in the worker of *Myrmica rubra* was observed by Brian (1976). This phenomenon of compensatory growth is discussed on p. 223.

4.3.1.18. Diptera

In Diptera neoteny occurs usually in the female, the degree of arrest of development of imaginal structures varying considerably in different groups of Diptera. Paedogenetic female neoteny occurs in the Cecidomyidae, in which the bisexual generation with fully developed imaginal structures alternates with a neotenous generation. In the latter, larvae develop within the hemocoele of the mother, which may be either a larva or a pupa. In some cecidomyids the bisexual generation may not occur (Wyatt, 1961), and the stage of development at which paedogenetic neoteny occurs varies considerably in different groups of cecidomyids (Wyatt, 1961, 1963, 1964, 1967). Thus, in *Henria psalliotes* (Leptosynini) paedogenesis occurs after the third larval instar; the pupal adult thus formed has peculiarly formed spiracles, and such a pupa is called the "hemipupa" (Wyatt, 1963).* In Micromyini the evolutionary tendency has been toward larval paedogenesis. In *Mycophila nikolei*, for instance, the hemipupal stage has been eliminated, but three larval stages are retained (Nikolei, 1961); *M.*

---

*Prominent features of the hemipupa are the prothoracic and posterior spiracles. The prothoracic spiracle is short and with 4 pores, and the posterior spiracle (of the 8th abdominal segment) is not found in the normal pupa. In the hemipupa the distribution of irregularly shaped spinules of the body cuticles is more extensive than in the normal pupa. The problem of oxygen supply to the fully developed larvae within the paedogenetic mother has been solved by the caenogenetic development of these complex spiracles and associated spinules which perforate the larval cuticle.

speyeri completes paedogenesis in the second larval instar (Wyatt, 1964). For more variations in the life cycle of paedogenetic cecidomyids refer to Wyatt (1967). The alternation of generation is highly dependent on environment. Light and temperature were considered to be such environmental factors by earlier workers. More recently, Wyatt (1964) found that in *M. speyeri* the primary conditions which induce the production of (fully winged) sexual forms are, as in some hemimetabolous insects, overcrowding and food, and that *Miastor castanea* also readily enters the sexual cycle under crowded conditions (Wyatt, 1967).

Ulrich (1938, 1940) first established the fact that (in *Heteropeza pygmaea*) nutrition derived from a mushroom mycelium plays a decisive role in determining the developmental pathways, either toward the bisexual generation or to the paedogenetic generation. Kaiser (1969) found that paedogenesis can occur only under low oxygen tension, and further that the glycogen derived from the mushroom mycelium seems to be the source of energy during paedogenesis. Kaiser (1974) further found that even injury and intoxication can result in paedogenesis.

These external factors are now considered to induce paedogenesis by influencing the activity of the CA. Pohlhammer (1969a,b) and Pohlhammer and Treiblmayr (1973) were able to show that the JH from CA has the gonadotropic function and that farnesyl-methyl ether has the same capability. Kaiser (1974) found that during paedogenesis the CA become enlarged in the last instar larva, and that during imaginal differentiation (in the bisexual generation) no such enlargement occurs. Kaiser (1974) therefore concluded that during paedogenesis the activated CA in the last larval instar cause precocious maturation of ovaries and simultaneously inhibit imaginal differentiation.

In the paedogenetic female, the egg undergoes one equational meiotic division (without reduction of the chromosomal complement), although normal meiosis occurs in the male. Hence, in this family the paedogenetic females reproduce by diploid parthenogenesis. This cytological sex-determining mechanism is, however, highly labile, being dependent on environmental factors (Ulrich, 1936). Pohlhammer and Pohlhammer (1970) thought that the hormone (JH?) produced from the cerebral ganglia can suppress the reduction division in the female. It appears, then, that when external factors cause overproduction of JH(?), the latter suppresses not only imaginal differentiation but also the normal process of meiosis, and then parthenogenesis may ensue.

The marine chironomid genus *Clunio* comprises 16 species; the female has been described for 10 of these species (Hashimoto, 1971). The female is always apterous and larviform, and many other structures, such as eyes, ocelli, legs, and antennae, are reduced in varying degrees in different species of *Clunio*; the reduced conditions of the structures obviously result from the arrest of development. Further, the female cannot molt by herself, and the male is known to assist the molting of the female by tearing off the pupal skin (Hashimoto, 1957; Neumann and Dordel, 1972, and so on). All these facts show that the females of *Clunio* do not complete the normal process of metamorphosis. Yet, no elimination of developmental stages occurs; therefore, the neoteny in *Clunio* must be regarded as resulting from metathetely.

According to Hashimoto (1971), differentiation of imaginal structures is most complete in *C. marinus* and least complete in *C. takahashii*. In the female of *C. marinus*, the antenna is seven-segmented, whereas in *C. takahashii* it is represented by one segment (pedicel) only; in *C. marinus* the tibia retains the apical spur comparable with that of the male, although in all other species it is absent; in *C. marinus* each eye consists of 21–23 ommatidia, whereas in other species the number of ommatidia per eye is smaller, and *C. takahashii* is eyeless.

It appears probable that the arrest of development of these structures is due primarily to insufficient action of ED, which, in turn, results from overproduction of JH. If an unusually high titer of JH persisted longer than usual during development, the time available for the action of ED would be limited, and this would result in more or less incomplete differentiation of imaginal structures. At the same time, the hormone would accelerate the maturation of ovaries. It is relevant to refer, in this connection, to the work of Laufer and Greenwood (1969), which showed that injection of JH into the prepupa of *Chironomus thummi* (Chironomidae) causes precocious maturation of ovaries by facilitating yolk deposition. It is probable that in *Clunio* also the developing ovaries in the comparable stage are sensitive enough to the excessive JH that the female eggs become ripe and ready for fertilization before the imaginal structures become fully differentiated; hence the neotenous females in this genus.

A generalization that Hashimoto (1971) drew was that in *Clunio marinus* the degree of differentiation (hence the number) of ommatidia is greater in the individual with the longer body, and that those living at lower temperatures are larger in body size than those living at higher temperatures. According to Hashimoto, this generalization applies to the level of interspecific comparison, as seen in the fact that the northern species tend to be larger and have more ommatidia. Since, as already seen, when the reduction (due to developmental arrest) of ommatidia occurs, the reduction of many other structures occurs at the same time, it may be said that in northern areas (seas) and in certain seasons the body may become larger and more imaginal structures differentiate. This tendency appears to indicate, further, that overproduction of JH is less in the northern seas (with lower temperatures) than elsewhere. The effect of photoperiodism also must be studied in future research.

In the neotenous females of a related genus, *Pontomyia*, the degree of suppression of imaginal characters is even more pronounced than in *Clunio*. For instance, in this genus legs are represented merely by two pairs of vestigial processes, and the antennae are absent. In some species of *Tanytarsus* (Chironomidae) larviform females are known to produce eggs (Wesenberg-Lund, 1943). The larviform adult may well be, however, an incompletely differentiated adult, as in *Clunio*. Hinton (1946) found that the supposed paedogenetic (pupal) adult of *Tanypus bohemicus* was in fact an adult. He pointed out that the failure of the adult to shed the effete pupal cuticle cannot alter the fact that it is an adult at the time the eggs are laid (hence metathetely).

Metathetelous neoteny apparently occurs also in some arctic simuliids. According to Downes (1965), in eight out of nine species of the simuliids occurring in that area the female is nonfeeding. The mouthparts are weak and cannot pierce the skin. The eggs start to develop before the adult emerges from the pupal cuticle, and in at least two species they are mature at that time. Downes thought that this finding implies a considerable reorganization of the hormones controlling metamorphosis and ovarian development. It is most probable that the reorganization of the hormonal control mechanism means the overproduction of JH in late stages of development, which, as in paedogenetic cecidomyids, accelerates ovarian development and simultaneously suppresses imaginal differentiation. The lower temperatures in the area must certainly be the dominant external factor causing the overproduction of JH. The reason why wing reduction does not occur in the arctic mosquitoes may be that β-ecdysone is the gonadotropic hormone, not JH (Hagedorn et al., 1975).

When neoteny occurs in other Diptera, it is evident mainly in the reduction of fore wings and the concomitant reduction and modification of the mesothorax (including mesothoracic muscles) bearing the wings. Reduction of other structures linked with the fore wing reduction includes the haltere (Brauns, 1939), eyes and palpi in Sciaridae (Lengendorf, 1937, 1949), and sometimes also antennae. The extensive modifications of body regions that accompanied the loss of wings in *Chionea* and *Basilia* were discussed by Matsuda (1970). The problem of wing reduction in Diptera was first discussed by Bezzi (1916) and later by La Greca (1954) and Hackman (1964); numerous examples of wing reduction were cited in these works.

According to Hackman (1964), fore wing reduction of various degrees (brachyptery, stenoptery, microptery, and aptery) occurs in more than 20 families of Diptera, and is usually confined to the female sex. As Hackman's review shows, wing reduction tends to occur in the Diptera living in more or less unusual habitats, such as high altitudes, the arctic area, oceanic islands, seashore areas, marine habitats, terricolous and hypogeous habitats, parasitic habitats, the nests of Hymenoptera and termites, and so on. The unusual factors (such as low temperatures, high humidities, and so forth) available in these habitats presumably upset the hormonal milieu slightly in favor of JH. The unusually heightened titer of JH could, as in cecidomyiids, inhibit full differentiation of some imaginal structures such as the wings, and simultaneously could expedite the maturation of oocytes, and hence the metathetelous female neoteny; the gonadotropic function of JH has been proved in many Diptera (Adams, 1974; Adams et al., 1975; Kambysellis and Heed, 1974; Mjeni and Morrison, 1976; and many other earlier works). Some of the well-known cases of wing reduction are discussed below.

In Tipulidae, as Byers (1969) attempted to show, wing reduction occurs mainly at high altitudes, in winter, or in the arctic and antarctic areas, and on cold islands; and the wing reduction usually occurs in the female. Here the factors involved appear to be primarily cold temperatures and often the different photoperiods available in different seasons and areas. The tendency of wing reduction in this family may usually be a physiologically induced condition for the reason discussed above, although in certain cases such a phenotype might have been permanently fixed.

In *Limosina pullula* (Sphaeroceridae), according to Rohacek (1975), the fore wings are dimorphic (macropterous and brachypterous) regardless of sex. Therefore, Rohacek assumed that the different wing morphs are determined by a single gene with two alleles, one for the long-wing (presumably recessive) and another for the short-wing (presumably dominant). Nearly all individuals found by Rohacek in more open environments (pasture lands, moss, grass, under mowed pea plants, and so on) were macropterous; those found in more restricted and wet places (underneath fallen leaves, garden compost, detritus, and so on) were often brachypterous. It is apparent, therefore, that the differentiation of the wing morphs is heavily dependent on environmental factors. Assuming, however, that the genetic determining mechanism for wing morphs (above) actually exists, predominance of one form over another in the two different habitats is due to incomplete penetrance of the genetic mechanism (compare with *Melipona* p. 183) and/or to different selection pressures in the different habitats, under which one form survives better than another. The presumed genetic mechanism (proposed by Rohacek), therefore, requires rigorous experiments to prove it, as have been done for *Melipona*.

High humidity may be an important physical factor causing brachyptery in the above case, as suggested by the result of Ohtaki's (1966) experiment. Ohtaki showed that when the larvae of *Sarcophaga peregrina* are kept in a wet condition, they do not pupate, but when they are removed from the wet condition and kept dry, they pupate by releasing ecdysone; this finding is consistent with observations that many kinds of dipterous larvae leave wet environments to pupate. It is possible that in the larvae of some Diptera the hormonal balance is delicately affected by wet or dry condition in which they live, and the development of imaginal structures (wings and so on) could be affected.

In *Leptocera* (*Pteremis*) *fenestralis* (Sphaeroceridae), according to Rohacek (1975), the brachypterous form is very rare in Central Europe, but more frequent in Britain; and in northern Finland the brachyptery is so frequent that the species is recognized as *L.* (*P.*) *fenestralis subaptera* Frey. In this case different photoperiods and temperatures (which affect hormonal balance) available in different geographical areas could be the dominant factors responsible for the wing dimorphism.

At high altitudes in the Eastern African mountains, according to Richards (1954, 1957, 1963), *Leptocera* spp. are brachypterous, and some other genera of Sphaeroceridae are apterous. They live among dead leaves, or in humus, or in burrows. Richards (1957, 1963) thought that these apterous species evolved more or less where they are now found by rather extensive sympatric speciation. It should be pointed out that these species live in essentially the same kind of habitats as the brachypterous individuals of *Limosina* spp. in Europe. It is probable, therefore, that the apterous species have been derived from environmentally induced brachypterous forms of the species which were dimorphic (or polymorphic) for wing development. Evidently, the permanent settlement of these species in wet habitats resulted in the induction of further reduction of wings, leading to the complete loss of them.

4.3.1.19. Lepidoptera

As in other orders, wing reduction is a good indicator of neoteny in Lepidoptera, and as a rule it is confined to the female. The degree of modification (mainly reduction) of other structures associated with wing reduction varies in different groups of Lepidoptera.

Neoteny in the female of Psychidae is, as shown by Saigusa (1961), highly pronounced. Saigusa (1961) classified progressive reduction of structures within the families into four categories. In the most primitive condition (α of Saigusa, which includes *Narychia*, *Melasina*, *Diplodoma*, and so on) pupal and adult legs are nearly as well developed as in the male pupae and adults. In more derived groups (β and γ of Saigusa, which include *Taleporia*, *Solenobia*, *Bacotia*, *Fumea*, *Bruandia*, *Proutia*, and so on) female wings are vestigial or absent; legs are smaller in the female than in the male; tarsal segments are apparently fused (actually undifferentiated), and some of them have lost the ability to walk. In the most derived group (δ of Saigusa, which includes *Clania*, *Nipponopsyche*, *Oiketicoides*, and so on) female wings are completely absent, and legs are represented by small protuberances or absent; the ovipositor is relatively short and the associated internal apophyses are very short. Saigusa (1961) also showed that in the male reduction of the mouthparts, ocelli, epiphysis of the front tibia, the terminal spur of the middle tibia, and wing veins has occurred in parallel with the reduction of wings and legs in the female.

In the above evolutionary sequence of reduction of structures, however, the ovipositor and associated internal apophyses have become long or very long in some female psychid moths (β and γ above), such development representing a case of compensatory growth that accompanies the reduction of other structures (wings and legs). The compensatory growth of legs, which occurs in some free-living female Lepidoptera with reduced wings, does not occur in the female psychids that live inside a case.

Although nothing is known directly about the endocrinology of the psychid moths, the kind of environment in which the female larvae develop (i.e., inside the case) presumably induces abnormally high titers of JH during the larval development, which in turn would result in more limited action of ED; consequently the development of imaginal structures (wings, legs, and so on) would become more or less arrested.[*] At the same time, the abundant supply of JH would expedite the maturation of eggs[**] precociously, before the completion of the normal differentiation of adult structures (wings, legs, and so forth), and hence the neotenous female adults.

Downes (1964) showed that brachyptery combined with physogastry (enlarged abdomen containing excessively developed ovaries) is common in the arctic Lepidoptera, and recorded 11 such species (*Aspilates orciferarius, Psodos coracina*, and so on). He also showed examples of reduction of eyes in several species of the arctic Lepidoptera, and pointed out that the larval life of Lepidoptera in the arctic is relatively long.

Incompletely differentiated imaginal structures in the arctic Lepidoptera probably denote inhibition of development by JH, which is overproduced in the arctic by the effect of cold temperature, the sole major external factor that could cause such abnormal production in that geographical area. The physogastry that accompanies wing reduction can also be attributable to the excessive action of JH, which could cause precocious maturation of ovaries. Further, when JH persists for a long time, one result would be a prolonged larval life; this interpretation is supported by the experimental fact that the administration of JH to the larvae of the silk worm (*Bombyx mori*) prolongs the larval period (Akai, 1971).

Orgyia thyellina is known to be seasonally dimorphic. According to Cretschmar (1928), the first-generation females of this species have normal

---

[*]This interpretation is supported by some experiments with other Lepidoptera. Thus, normally, in Lepidoptera, metamorphosis usually occurs in the absence of JH (Williams, 1959, 1961; Nijhout and Williams, 1974; and others). When the JH titer is increased in old larvae metamorphosis becomes incomplete, as seen in some experiments. For instance, Williams (1961) found that implantation of CA into the pupa of the cecropia moth inhibits imaginal differentiation. Similarly, Sehnal (1968) found that implantation of CA into old larvae of *Galleria* results in the suppression of imaginal structures including wings. Some other studies (Riddiford, 1970, 1971, 1972; Sehnal and Schneiderman, 1973; Truman et al., 1974; and others) show that artificially produced high titers of JH interfere with imaginal differentiation.

[**]In normal development of Lepidoptera the gonadotropic action of JH becomes evident just prior to or after emergence of the adult (Endo, 1970, 1972, 1973; de Wilde and Loof, 1973; Karlinsky and Srihari, 1973; Sroka and Gilbert, 1974; Herman, 1973, 1975).

wings; in the second generation only one-third of them spread the wings, and the rest either retain pupal wings or do not spread the wings fully; in the third generation the majority of them does not spread the wings. Goldschmidt (1940) found that if they are bred at high temperatures (ca. 25°C), the wings in the second and third generations (of the same species) spread out. At any rate, the development of wings in later generations becomes slightly arrested, most probably by the effect of changing environmental factors (temperature and photoperiod). The process of female wing development in *Orgyia antigua* was described in detail by Paul (1937). Seasonal polymorphism is also known to occur in the aquatic moth *Acentropus niveus*; the fully winged female occurs only during late summer (Hackman, 1966).

As Hackman (1966) showed, the Lepidoptera with reduced wings tend to occur at high mountain altitudes, in colder seasons of the temperate zone at lower altitudes, in oceanic islands, and in xerothermic habitats in mountains and deserts. For numerous examples of wing reduction associated with those habitats refer to Hackman (1966). As a general rule, wing reduction is confined to the female sex in Lepidoptera; in fact, in the European Lepidoptera wing reduction occurs only in the female (Eggers, 1939). The reduction of other structures associated with the wing reduction includes the tympanal organ in the female and the mouthparts in both sexes.

Hackman (1966) classified the female wing reduction into two types: a) the heavy egg-filled type, which may be provided with strongly reduced legs (*Orgyia* group of Lymantriidae, Psychina, some Geometridae); and b) the mobile type with a less heavy abdomen and comparatively well-developed legs and sometimes with good running and jumping abilities (a geometrid, *Operophtera brumata*, a tortricid, *Sphaleroptera alpicolana*). Eggers (1939) reviewed the cases of negative correlation between the size of wings and that of ovaries in European Lepidoptera, and discussed its adaptive significance. Rensch (1959) regarded the enlargement of ovaries in the Lepidoptera with reduced wings as a case of material compensation (p. 224); this material compensation results, as repeatedly pointed out throughout this work, from the abnormally heightened titer of JH during development, which has the gonadotropic function. The tendency for the legs to become relatively well developed in some of these Lepidoptera (b above) represents another aspect of material compensation that accompanies the wing reduction.

As is well known, many butterfly species are seasonally dimorphic with respect (mainly) to color pattern. The seasonal dimorphism in *Polygonia c-aureum*, occurring in Japan, has been shown to be environmentally and hormonally controlled. Day length and temperature are the major extrinsic factors determining the seasonal forms of this species. The reproductively active summer form can be produced experimentally by a long day length at a high temperature, and the autumn form (which diapauses) can be produced by a short day at a low temperature (Hidaka and Aida, 1963; Fukuda and Endo,

1966). Fukuda and Endo (1966) found that a long day length at a high temperature activates the brain-corpus cardiacum (CC) system, and that the neuroendocrine material released from the CC determines the wing pattern of the summer form; cardiectomy brought about the change in seasonal forms, but not allatectomy. Further, the result of this study indicated that the brain-CC system is not likely to be activated in producing the autumn form. In another experiment, however, Endo (1972) produced evidence indicating that the stimulation of CA by NSC in the brain is connected with the development of the summer form, and that the NSC do not stimulate the CA in producing the autumn form of this species.

Thus, the factor determining the seasonal morphs has not been definitely identified. What is clear, however, is that the secretion or nonsecretion of the neurosecretory material, which is influenced by external factors (photoperiod and temperature), plays the decisive role in determining the seasonal morphs, which differ in color pattern and reproductive capability. Numerous cases of seasonal dimorphism in butterfly species recently discussed by Shapiro (1976) must be environmentally and hormonally controlled.

4.3.1.20. Strepsiptera

In Strepsiptera, the larvae hatch within the body of the mother and emerge, in large numbers, through the apertures of brood canals. At this stage the young larvae bear resemblance to the triangulin larvae of *Meloe*. The first stage larvae, upon entering the secondary hosts (larvae of some Hymenoptera and some Homoptera), become endoparasites.

The females of the species of Mengenillidae leave the secondary host and live free. They retain three pairs of legs and have the unpaired brood canal between the seventh and eighth abdominal sterna where the primary gonopore opens in other orders of insects (Kinzelbach, 1971). However, the brood canal is not continuous with the oviduct, which is absent in the Strepsiptera. The wingless females of Mengenillidae thus do not differ conspicuously from the larvae in other holometabolous orders, and the larviform female adults can be regarded as neotenous and paedogenetic.

The first instar larvae of the other Strepsiptera, after entering the secondary host, remain there permanently and undergo ecdysis, thereby losing legs and becoming apodous maggot-like larvae. As shown by Lauterbach (1954), the internal organs, including the female gonad in the fertilized female of *Stylops*, undergo nearly total degeneration, and eggs become scattered in the body cavity. Simultaneously, however, several pairs of brood canals are formed on the anterior abdominal segment. Thus, during postembryonic development of these Strepsiptera the process of degeneration is accompanied by the development of the unique (new) structures (brood canals), and the resultant female adults are not larviform. Therefore, the developmental process can be called halmatometamorphosis.

The male larvae undergo pupation, the pupae are enclosed in the exuviae of two preceding instars (in *Stylops*, Richards and Davies, 1964), and eventually winged males emerge. In the adult male of Strepsiptera, the fore wings are greatly reduced, although membranous hind wings are well developed and are functional flight organs. As Kinzelbach (1971) has shown, various degrees of retention of larval features are recognized in the fully winged adults. The juvenile (larval) features are: the trochanterofemur of the first and second pairs of legs, which are comparable with those of the secondary larvae; the apparent reduction of sternal articulation of the mesothoracic legs, which represents the larval condition; the less differentiated antennae that occupy the larval position; the highly reduced mandibles in some groups, highly simplified maxillae, peculiar compound eyes in which ommatidia are widely spaced from one to another, and some internal organs (some nerves, and so on) that can be interpreted as paedo-genetic. The retention of this array of juvenile features in the males of Strepsiptera is an extremely good example of metathetely that occurs in nature.

Although nothing is known about the endocrinology of this order, it is probable that, in both the male and the female, high titers of JH persist during most of their postembryonic development. In the female, the persistence of high titers of JH appears to result in the production of some unique adult structures (brood canals), as if compensating for the loss of certain structures (legs, and so on); this compensatory development of structures is discussed on p. 220. The retention of juvenile features in the male presumably results from a more limited action of ED (than usual) as a consequence of overproduction of JH.

4.3.1.21. Other Holometabolous Orders

Among the Trichoptera, the female with greatly reduced wings occurs in the winter generation of *Dolophilodes* (= *Trentonius*) *distincta* (Ross, 1944); this wing reduction must be environmentally (cold temperature) and physiologically induced. *Agrypnia pagetana* var. *hyperborea* and others are normally macropterous, but are brachypterous in cold areas and at high altitudes (La Greca, 1954). In the female of *Enoicycla pusilla* both fore wings and hind wings are completely lacking (Schmid, 1951). Less marked wing reduction occurs in both sexes of some other Trichoptera, such as *Baicalina reducta* and *Thaumastes dipterus* (Schmid, personal communication).

Among the Mecoptera, permanent aptery occurs in both sexes of *Apterobittacus apterus* (Applegarth, 1939) and *Apteropanorpa tasmanica* (G. Byers, personal communication). *Panorpodes paradoxa* is macropterous at lower altitudes in Japan, but the female is brachypterous at higher altitudes (H. Ando and A. Mutuura, personal communication). In the genus *Boreus* the female is apterous, and wings are modified into an accessory copulatory organ in the male. According to Byers (personal communication), modification of wings occurs also in *Hesperoboreus* spp. from Western North America and *Anomalobittacus* sp. from South Africa.

In Neuroptera, wing reduction is extremely rare. In the female of *Psectra diptera* the hind wings are vestigial and the male is micropterous (Killington, 1946). For wing reduction in the Siphonaptera see p. 225.

4.3.1.22. Collembola

The Collembola are primarily apterous. Therefore, the neoteny in this order can be recognized in the reduction of other structures. The degree of neoteny in Collembola as a whole can be surmised by comparing corresponding structures of this order and the Protura, which are the plesiomorphic sister-group of the Collembola.

The Protura have a nine-segmented abdomen when they hatch from the egg; the number of segments increases by repeated splitting of the penultimate abdominal segment during postembryonic development, until a twelve-segmented abdomen is formed in the adult (anamorphosis). In Collembola, however, the number of abdominal segments remains six throughout embryonic, postembryonic, and adult stages. It is obvious, therefore, that the process of addition of segments or anamorphosis has been completely eliminated in this order, and the six-segmented condition of the abdomen can be regarded as neotenous. Neoteny is also evident in the extremely poorly developed external genitalia, the greatly reduced (undifferentiated) tracheal system, and the lack of tibiotarsal articulation in the prothoracic legs. Certain structures absent in the Protura, however, occur in Collembola. Lateral eyes, each composed of the maximum eight ommatidia, are such characters. The modification of appendages into a jumping organ can also be considered a compensatory development.

Neoteny is evident again in the progressive reduction of certain structures within the order. For instance, the regression of lateral eyes—leading sometimes to their total loss—occurred in many groups (Thibaud, 1976). Parallel reduction of the furcula also occurred in all families of Poduromorpha and Isotomidae Anurophorinae (Cassagnau, 1954). The marked trend of reduction of many structures in this order is probably due to abnormal functioning of the endocrine system. However, nothing is apparently known about the endocrinology of this order, although the anatomy of endocrine organs in some Collembola has been studied.

4.3.2. Myriapoda

4.3.2.1. Endocrine Organs and the Hormonal Control Mechanism of Metamorphosis

In the Chilopoda, the structures called the "cerebral glands" have, on each side, a nervous connection with the NSC in the frontal lobes of the protocerebrum; axons of the NSC form a nervous connection with the cerebral glands. Joly and

Descamps (1968) showed that the major difference in these structures in different groups of the Chilopoda (Geophilomorpha, Scolopendromorpha, and Lithobiomorpha) lies in the location of the glands relative to the protocerebrum. Extirpation of the cerebral glands in *Lithobius* (Joly, 1961, 1966) and *Scolopendra* (Joly, 1962) caused a significant rise in the percentage increase in molting, and implantation of the same glands in *Lithobius* (Joly, 1966) lowered the percentage of molting. Decapitation and implantation of the same glands in *Lithobius* by Scheffel (1969) also yielded essentially the same results. From these findings it can be inferred that an inhibitory hormone is released through the frontal lobes-cerebral glands system.

Further, according to Descamps (1975), the NSC in this system of *Lithobius* have an inhibitory effect on the spermatogenetic cycle. However, Herbaut (1976) found that the destruction of this endocrine complex leads to the degeneration of most of the oocytes at the end of previtellogenesis, and further that the implantation of the cerebral glands immediately after the operation induces the resumption of growth of oocytes in about 50% of the animals operated on. This experiment, therefore, indicates that the hormone has, like the JH in insects, the dual function of promoting oocyte growth and suppressing somatic differentiation; and the hormone may well be the same as a juvenile hormone in insects.

The cerebral glands also occur in the Diplopoda. Histology and anatomy (often ultrastructural) of these organs have been studied by Gabe (1954), Seifert (1971), Seifert and El-Hifnawi (1971, 1972a), Sahli and Petit (1972), Juberthie-Jupeau (1973a), Juberthie and Juberthie-Jupeau (1974), and Sahli (1974, a summary). The ultrastructure of the cerebral glands in *Geophilus* was studied by Ernst (1971), in *Scutigerella* by Juberthie-Jupeau (1973b) and in *Scutigera* by Rosenberg (1976).

Scheffel (1969) found that the *Lymphstränge* (which surround salivary glands) have the function of stimulating the molting process; when that structure was destroyed, molting was delayed. Because of its function, Seifert and Rosenberg (1974) called the structure in *Lithobius* the *glandula ecdysalis*, and pointed out that the *Kragendrüse* in *Polyxenus* described by Seifert and El-Hifnawi (1972b) and an endocrine gland in the head of *Scutigera* described by Rosenberg (1973) correspond, in all structural details, to the *glandula ecdysalis* in *Lithobius*. Further, injection of ecdysone into *Lithobius* larvae caused an increase in the percentage of molting (Joly, 1964), and the application of exogenous ecdysterone induced apolysis (Scheffel, 1969; Scheffel et al., 1974). All these facts indicate that the gland induces molting by secreting an ecdysone.

Pars intercerebralis of the brain in *Lithobius*, unlike many insects, has no neural connection with a neurohemal organ; destruction of this area in the second instar larvae showed no influence on molting (Scheffel, 1969). However, this area stimulates oogenesis (Herbaut, 1975).

Scheffel (1969) pointed out that the inhibitory hormone, like JH in insects, influences the process of release of the molting hormone. Scheffel (1969) further showed that morphological abnormalities comparable with prothetely and meta-thetely in insects can be induced experimentally by causing hormonal disturbance. All these facts strongly suggest that in *Lithobius* metamorphosis (anamorphosis) may well be regulated by ecdysone (ED) and juvenile hormone (JH), as in insects. Apparently, nothing is known directly about the external factors influencing hormonal activity in the Myriapoda.

4.3.2.2. Anamorphosis and Neoteny

In the Lithobiomorpha and Notostigmophora of the Chilopoda, Diplopoda, Pauropoda, and Symphyla, postembryonic development proceeds by anamor-phosis, i.e., the increase in the number of abdominal segments (often bearing appendages) occurs by splitting of the preanal zone of proliferation (teloblastic proliferation); one or more segments are formed at each molt. The number of antennal segments also increases during postembryonic development (Scheffel, 1969). When neoteny occurs, therefore, it should be evident in the decrease of the definitive numbers of abdominal and antennal segments.

Neoteny has apparently occurred within the Diplopoda, and is seen in the reduction in number of segments and loss of ocelli. According to Demange (1974), the number of molts during postembryonic development is rather similar in most groups of diplopods (eight in Polydesmoidea and Penicillata, nine in Craspedosomoidea, eight, ten, or twelve in Spirobolidae and Julidae). The number of segments added at each molt is one, three, four, three, two in Craspedosomidae, two, three, two, one in Polydesmidae, and only one segment at each molt in the Penicillata. On the other hand, in the Juliformia the number of segments added at each molt is high and variable, although in Spirobolidae the large number of segments added at each molt is less variable.

The relatively small definitive number of segments acquired (as compared with those in the Juliformia) in Polydesmidae (about 20), Craspedosomidae (about 30 or more), and Penicillata (about a dozen) is due mainly to the rela-tively smaller number of segments added at each molt, and also to the number of molts they undergo; such development can be regarded as neoteny. Further, Demange (1968, 1974) pointed out that the reduction in number of segments (viz., neoteny) can be induced by environmental effects (seasons, years).

The above findings appear to indicate that the ancestors of the above oligomerous diplopods have been derived from Julida-like ancestors with the potential to produce, depending on environment, variable numbers of segments at each molt. It is imaginable that the individuals with fewer segments thus produced were adaptively advantageous in certain environments, and eventually gave rise to these oligomerous diplopod groups. The alteration of hormonal balance that occurred in the new environments certainly must have played an important role in inducing the neoteny in these groups of diplopods. The fact that in all groups of the Diplopoda sexual excitation (*reveil*) intervenes visibly in the third stage larvae (Demange, 1974) may suggest relatively early maturation of gonads, which could facilitate neoteny.

Neoteny is also evident in the absence of ocelli in some diplopods, such as Polydesmidae, with small numbers of segments. According to Demange (1974), in the diplopods with ocelli (*Pachybolus ligulatus*, Julidae, and so on) the ocelli are absent in the embryo and in very young larvae. Demange (1974) therefore regards the absence of ocelli in adult Polydesmidae as a character of the embryo and very young larvae, and hence neotenous.

The Pauropoda are closely related to the Diplopoda. They live in moist places under logs and stones, in interstices of soil, and so forth, and are therefore minute in body size, not exceeding 2 mm (Snodgrass, 1952). Typically, in the Pauropoda the body consists of 12 segments and 9 pairs of legs. Many other external structures are also reduced. For instance, the collum of diplopods with the well-developed tergal plate is represented, in the Pauropoda, dorsally by the inconspicuous fold behind the head, and ventrally by a pair of small papillae, which represent vestigial legs; and the mouthparts consist of only a pair of mandibles and an underlip structure. The Pauropoda also have no circulatory system and no organs of respiration (Snodgrass, 1952). All these regressed conditions of structures must result from developmental arrest, as is evident in the highly abortive developmental processes of various structures in *Pauropus silvaticus* shown by Tiegs (1947). The first instar larva of *Pauropus silvaticus* has six body segments in front of the pygidium (anal segment), and there are four immature stages during which segments and legs are added (Tiegs, 1947). It should be recalled that in the oligomerous diplopods (Polydesmidae, Penicillata) molting takes place eight times. Clearly, therefore, the reduced conditions of many structures in the Pauropoda represent the neoteny that results from a further contraction of developmental stages. Snodgrass (1952), however, maintained that the simplified structures in the Pauropoda were primitive. Such interpretation presupposes that the structural evolution is one-way toward increased differentiation, which is fallacious.

4.3.3. Arachnida

4.3.3.1. Endocrine Organs and Probable Hormonal Control of Metamorphosis

In several species of Araneida, as shown by Legendre (1958), the endocrine system consists of two pairs of organs (of Schneider). The first pair joins the aorta and is comparable, by its embryological origin, to the CA in insects. The smaller second pair of organs is formed of secretory cells and is comparable to the CC in insects. The two pairs of organs have, as shown by Gabe (1955, 1966), intimate connections with some NSC in the brain. The close resemblance, in their topographic relationships, to the NSC-CC-CA system in insects, however, does not necessarily mean that their functions are exactly comparable to those of the comparable organs in insects. Legendre (1958) also showed putative endocrine "anterior organs" composed of loose aggregations of cells in the anterior prosoma, and interpreted them as homologues of the prothoracic glands in insects.

In the Opiliones, according to Meewis and Naisse (1960), the endocrine system consists of the paraganglionic plates, which are presumably neurohemal organs, and the presumed molt gland situated in the cephalothorax. In *Heterometrus swammerdami* (Scorpiones), what Habibulla (1961) called the "blind anterior organs" may play a role in the molting process. Further, there exists a system (sympathetic nervous system) comparable to the CA-CC-Frontal ganglion in insects (Habibulla, 1970).

In *Ornithodoros* (Acari) two areas of neurosecretory activity are present in the brain (Gabe, 1955, 1966), and a ligation experiment on *Ornithodoros turicata* by Cox (1960) indicated the presence of a molting factor. Further, Hughes (1962) in *Acarus siro* and Dhanda (1967) in *Hyalomma dromedarii* found correlation between the NSC activity and molting.

Injections of ecdysterone into *Araneus*, *Dugesilla*, and *Limulus* by Krishnakumaran and Schneiderman (1968, 1970) and into *Pisaura* by Bonaric (1976), as well as injection of α and β ecdysones into *Limulus* by Jegla et at. (1972), resulted in stimulation of molting, showing that the tissues of these arachnids are sensitive to the ecdysone and ecdysterone. Wright (1969) in *Dermacentor albipictus* and Sannasi and Subramonian (1972) in *Rhipicephalus sanguineus* found that when ecdysone and its analogues are applied externally to larvae, diapause can be terminated, as in some insects. All these findings suggest that ecdysone probably occurs in Arachnida.

Schneiderman and Gilbert (1958) failed to see activity of JH in *Limulus*. The inhibitory hormone in the Arachnida, therefore, still remains to be identified. Yet, it is probable that in the Arachnida, as in insects, termination of each developmental stage by ecdysis takes place by altering the balance between ecdysone and an inhibitory hormone similar to JH in insects.

4.3.3.2. Cases of Neoteny

In the Arachnida, anamorphosis does not occur, and no wings are formed (although winglike structures occur in some families of Oribatei). When neoteny occurs, therefore, the arrest of development of structures is less evident superficially than in the other classes of the Arthropoda. Deficiency in chaetotaxy is, therefore, often an important criterion for the recognition of neoteny.

Thus, in Mesostigmata, according to Lindquist (1965), there are two types of setal deficiency: 1) the chaetotaxy of the larva is deficient and the deficiencies are retained in the adult; and 2) the larval chaetotaxy is complete (holotrichous), but some of the setae usually added to the larval component in the protonymph and deutonymph are suppressed. Lindquist (1965) showed how, in *Hoploseius tenuis*, the deficiencies of the setal pattern occur in conformity with these general tendencies; the deficiencies that resulted represented a localized neoteny. No reduction in number of developmental stages occurs; therefore the case can be compared to metathetely in insects.

The deficiency in setal pattern in some Pseudoscorpiones can also be considered as a localized neoteny. For instance, the chaetae of the posterior row of the cephalothorax in the adult of *Chthonius orthodactylus* are identical to those of the tritonymph of *C. ischnocheleus*, and in some other species of *Chthonius* the protonymphal chaetotaxy of the first and second tergites may be retained in the adult (Gabutt and Vachon, 1963). Similar cases of retention of nymphal trichobothrial pattern in the adult of some other pseudoscorpions were noted by Weygoldt (1969).

4.3.3.3. Acceleration, Halmatometamorphosis, and Neoteny

In parasitic mites, acceleration in development occurs by elimination of a nymphal stage or stages, and this may result in halmatometamorphosis and neoteny.

According to Fain (1965), all free-living species of Ereynetidae are oviparous, and their life cycle consists of a larva, three nymphal stages (proto-, deuto-, and tritonymph) and an adult. However, in some parasitic ereynetids acceleration of development occurs by omission of some developmental stages. Thus, in the species living in the nasal cavities of frogs and toads, eggs contain completely developed "pupae," and the active tritonymphal stage does not occur. Further, in the species living in the nasal cavities of birds and mammals (Speleognathinae) adults emerge from the larvae, and no clearly defined nymphal stages intervene; in these cases nymphal stages are represented by sclerites of the pharynx and membranous sacs (Fain, 1972).

Similarily, in Podapolipidae, which parasitize various insects (Coleoptera, Orthoptera), the female adults emerge from the larval exuviae (Regenfuss, 1968; Baker and Eickwort, 1975; Eickwort, 1976), and the active nymphal stages are eliminated, although the unusual cuticular sac enclosing the pharate adult can be interpreted as a highly degenerate nymphal cuticle (Baker and Eickwort, 1975). The female adult emerging from the larva is provided with different numbers of pairs of legs (zero to four) in different genera of Podapolipidae. Thus, *Chrysomelobia* retains, as in the related family Tarsonemidae, four pairs of legs. In this case development has been accelerated,* without affecting the normal structural organization of the adult.

*Accelerated as compared with the free-living mites with three active nymphal stages, but not accelerated as compared with other groups of the Tarsonemoidea.

In *Dorsipes, Tarsopolipus*, and others, the female adult emerges from the larval exuvium with three pairs of legs. Presumably, in these cases elimination or near total elimination of nymphal stages must occur as in *Chrysomelobia*, and the six-legged female mites thus produced are essentially larval in structural organization, and hence neotenous. Assuming that the number of molts these mites undergo is the same as that which *Chrysomelobia* undergoes, the cases may be regarded as metathetely within the Tarsonemoidea, although the case represents neoteny by acceleration within the Acari.

In some podapolipids the number of legs in the female adults ranges from two pairs to zero. Such reduction occurs in *Tetrapolipus* (to two pairs), in *Podapolipoides, Podapolipus*, and *Locustacrus* (to one pair), and in *Archipolipus*, in which no legs occur (Regenfuss, 1968, 1973). Reduction of other structures, such as tergites, is also pronounced in the podapolipids with one or two pairs of legs. Compensatory constructive development of structures also has occurred in adaptation to the parasitic life of these mites. This includes attachment lobes, the enlarged empodial sucking disc on the ambulacrum in many *Eutarsopolipus* species, and the Y-shaped lobes behind the first pair of legs. Thus, the production of such highly reduced numbers of legs, which does not occur in any free-living mites, and the concomitant production of (compensatory) adaptive structures in these mites illustrate a case of halmatometamorphosis.

Among the male podapolipids, *Chrysomelobia* and a few other genera have four pairs of legs, while the males in the remaining genera have three pairs of legs (Rengenfuss, 1973). In some genera the adult male was found to hatch directly from the egg, thereby skipping the larval stage, although there have been records indicating the emergence of the male from the larva (Regenfuss, 1968). It is probably legitimate to regard the males of the podapolipids with three pairs of legs hatching from the egg as a case of neoteny by an unusual means of acceleration (within the Tarsonemoidea).

As shown by Fain (1962), *Bakterocoptes cynopteri* (Bakterocoptinae, Teinocoptidae) is a parasite of bats, with adult mites six-legged in both sexes. Adult males of this species closely resemble larvae and develop within proto-nymphs (which are also six-legged and very similar to the male adults), and the later nymphal stages have been eliminated from the life cycle. Clearly, therefore, differentiation of structures beyond the larval stage has been largely arrested in this species; such a developmental process can perhaps be called paedogenetic neoteny. In the female of the same species the three pairs of legs are highly regressed and terminate in suckers.

In the Scorpiones, which are acknowledged to be one of the most primitive groups within the Arachnida, the number of nymphal stages is seven to nine. It appears probable, therefore, that such features as the obliteration of tagmosis and abdominal segmentation in many Acari is due, in part, to the abbreviation of developmental stages, and the Acari as a whole can be regarded as more neotenous than the Scorpiones and some other related orders. Within the parasitic mites, as already seen, acceleration of development by elimination of developmental stages has gone further and resulted in neoteny and halmatometamorphosis.

4.3.4. Crustacea

4.3.4.1. Endocrine Organs and Hormonal Regulation of Molting and Metamorphosis

Although the hormonal control mechanism of postmetamorphic moltings* has received considerable attention, mechanisms of young larval development and metamorphosis are very incompletely understood. Furthermore, the material for endocrinological studies has been confined largely to the Malacostraca, especially the Decapoda. An androgenic hormone occurs widely in the Crustacea.

Two major hormones, an inhibitory hormone and a molting hormone (predominantly β-ecdysone) have been shown to be involved in postmetamorphic moltings in the Malacostraca. The NSC situated in the ventral proximal corner of the medula terminalis ganglion of the eyestalk (in Brachyura) form the X-organ, and the axons of the NSC and elsewhere form the sinus gland (Passano, 1960). This X-organ–sinus gland system is where the inhibitory hormone is produced; many experiments involving removal and implantation of this system have often indicated the presence of the inhibitory hormone in this area. Thus, the X-organ–sinus gland system in the Malacostraca is comparable to the NSC–CC–CA system in the Insecta.

The Y-organ in the second antennal or second maxillary segment is apparently the site of production of ecdysone, although no direct proof for the Y-organ as the actual site of production of the ED is as yet available. Extirpation of the Y-glands during the intermolt stage or premolt stage prevents molting, and after reimplantation the animals molt; these facts indicate that ecdysone is a molting hormone. The activity of the Y-organ is controlled by neurohormones from the X-organ–sinus gland complex. Further, elevated titers of ED induce a series of biochemical and morphological changes during proecdysis (initiation of molting including apolysis). Thus, the Y-organ is comparable to the PG in insects (see Willig, 1973, for a summary).

With regard to the endocrine control mechanism of larval development and metamorphosis in the Malacostraca, the eyestalk extirpation experiments with three genera of crabs (*Callinectes, Sesarma,* and *Rhithropanopeus*) by Costlow (1963, 1966a,b) indicate that: 1) the adult X-organ-sinus gland system thought to control molting does not function during zoeal development, and the in-

---

*Many decapods molt after metamorphosis into the adult before the puberty molt, and some of them molt even after the puberty molt. In these respects decapods differ from pterygote insects.

hibitory hormone produced from this system becomes activated only within the megalops stage; 2) the hormone for molting and the hormone for morphogenesis are different; and 3) possibly during the young larval stage only the Y-organ is at work to cause molting.

With regard to the external factors influencing larval development, interaction between salinity and temperature has been shown to affect the time of metamorphosis to the first crab of *Callinectes* (Costlow, 1967), although photoperiods varying from 0 to 24 hr of light but of equal intensity showed (in *Sesarma* and *Rhitropanopeus*) no significant variation in time of metamorphosis, nor was the survival of the larvae affected (Costlow and Bookhout, 1962).

The hormonal control mechanism of molting and metamorphosis in the Entomostraca also remains incompletely known. In the Branchiopoda including Cladocera (Sterba, 1957; Van den Bosch and de Aguilar, 1972) and Anostraca (Menon, 1962; Hentschel, 1964; Kulakovsky, 1976) no well-defined storage-release organ occurs, although groups of NSC occur in the brain, and the sinus gland may be present in *Streptophalus* (Menon, 1962). Presumably, these cells perform an endocrine function by their cyclic activity.

In various cirripeds, Barnes and Gonar (1958a,b) and McGregor (1967) have demonstrated the existence of probable NSC in the central nervous system. Tighe-Ford and Vaile (1972) demonstrated that injection of crustecdysone into *Balanus balanoides* results in increased molting activity, showing that at least the tissue of this species is sensitive to the crustecdysone. Davis and Costlow (1974) showed that eyestalk extracts of *Uca pugilator* significantly increase the time to ecdysis when injected into *Balanus improvisus* at D developmental stage, suggesting that a molt-inhibiting hormone may be produced in this species.*   Although storage and release organs comparable to those in the Malacostraca may be absent, the above experiments suggest the existence of a hormonal regulatory system of molting in these crustaceans.

The third morphogenetic hormone in the Crustacea is the androgenic hormone produced from the androgenic gland, which is formed only in the male; the gland is usually located at the (terminal) ejaculatory region of the spermiduct, although it occurs near the testicules or at the apex of the latter in some isopods (Charniaux-Cotton, 1972). The androgenic gland is known to occur in all superorders, orders, and most suborders of the Malacostraca, the exception being two small orders of the Peracarida (Charniaux-Cotton, 1972). In *Orchestia gammarella* and some others, according to Charniaux-Cotton (1972), spermatogenesis and secondary sexual differentiation are induced only in the presence of

*This interpretation is apparently contradicted by the studies of Gomez et al. (1973) and Cheung and Nigrelli (1973), which showed that cyprid-adult metamorphosis can be induced by a JH mimic (ZR-512) and by farnesol. However, the inhibitory hormone produced in the eyestalks of *Uca* is probably not the same as JH or the JH mimic.

the androgenic hormone; in its absence female (ovary, and others) differentiation occurs. It is, therefore, reasonable to assume that the genome of all individuals of a given species has information both for the production of the male and for the female. The hormone must activate the genes responsible for the development of the male structures; when the hormone is absent, the male genes are not turned on, and the animals develop into females.

Some experiments have shown that the activity of the androgenic gland is moderated by neurosecretion from the eyestalk; in the absence of the neurosecretion, hyperfunction of the gland results. The spermatogenetic activity is also known to vary in different environmental conditions in some species. Charniaux-Cotton (1972) explained such variation in terms of the following chain of reactions: external factor–neurosecretion–androgenic gland–spermatogenesis.

4.3.4.2. Cases of Neoteny in Crustacea (Except Cirripedia)

In a majority of the Crustacea, development is anamorphic, viz., abdominal segments and segmental appendages are added during postembryonic development by teloblastic proliferation as in some Myriapoda. Neoteny is, therefore, most evident in the reduced number of segments, which may or may not bear appendages. The reduced conditions of structures in the adult are attained either by elimination of a developmental stage or stages (acceleration), or without abbreviation of developmental stages.

Mild neoteny affecting only certain structures such as appendages in a population or populations within a species sometimes occurs. For instance, as Margalef (1949) pointed out, in some forms (= subspecies) of *Diacyclops bicuspidatus* the antenna has 14 segments, instead of the typical 17, and the individuals with such reduced antennal segments represent ecological and geographical races; e.g., *D. b.* var. *lubbocki* occurs in salt water in Central Europe. According to Margalef (1949), a similar reduction of structures associated with certain geographical distributions and ecological habitats also occurs in other cyclopids (*Diacyclops bisetosus*, *Megacyclops viridis*) and in the species of *Gammarus* (Amphipoda). In cyclopids, according to Kiefer (1928), reduction in the number of antennal segments is a prevalent evolutionary tendency. However, diversification of furcal setae has accompanied the reduction in number of segments.

In the terrestrial species of the family Talitridae (Amphipoda) the most probable mechanism of speciation is, according to Hurley (1959), neoteny. Several terrestrial species of *Orchestia*, unlike most supralittoral species, show gradation in size and type of the male gnathopod. These intermediate conditions of structures represent premature arrest of development of the ancestral structures. It is imaginable that the change in habitat induced alteration in the hormonal milieu, which could have resulted in the production of the neotenous terrestrial species. A constructive (compensatory) evolution that accompanied the reduction of structures is the modification of the male gnathopod adapted for grasping and carrying the female in coitus that has occurred in most terrestrial species of *Orchestia*.

In some species of *Emerita* (Anomura, Hippidae) the males are very small, and Efford (1967) attributed this size to precocious sexual maturation which results in neoteny; the neotenous male retains certain larval (megalopa) characters. Serban (1960) pointed out that the extremely small body size (often less than 0.3 mm) and the reduced condition of structures in Harpacticoida (Copepoda) are due to abbreviation of developmental stages, associated with their phreatic biotope or the interstitial biotope of the seashore.

A continuous series of reduction of structures in a presumed phylogenetic sequence has been recognized for certain groups of Crustacea. An outstanding case of such neoteny (phylogenetic neoteny) is seen in the evolution of the Conchostraca-Cladocera complex. In the Conchostraca the number of trunk limbs is variable, being 10 to 32 pairs; of these 1 to 16 are postgenital. In the Conchostracha the number of naupliar stages is known to be five (Kästner, 1967). After the nauplius stage, molting occurs many times. In all Cladocera, except for two species, on the other hand, the naupliar stages are passed within the egg; they hatch as juveniles with six (or five) pairs of functional trunk limbs, and no more trunk limbs are added during subsequent growth and development (Anderson, 1973).

In a conchostracan species, *Cyclestheria hislopi*, embryonalization of naupliar stages occurs, and the young conchostracan hatches from the egg with six pairs of functional trunk limbs, as in most cladocerans. Thereafter, however, larval development continues in the usual way, with the addition of further trunk segments and limbs. It is apparent, therefore, that the Cladocera arose from a *Cyclestheria*-like ancestor by suppression of anamorphosis (acceleration), which resulted in neoteny. The neotenous origin of the Cladocera has been discussed by Eriksson (1934), Margalef (1949), de Beer (1958), Anderson (1973), and some earlier workers.

Gurney (1942) hypothesized the evolutionary process through which the grade of organization of the Copepoda came into existence. Throughout the Copepoda, according to Gurney, there is a stage at which three pairs of thoracic appendages occur and the abdomen remains unsegmented; he pointed out that this grade of development of structures precisely corresponds to the protozoea in the Decapoda. According to him, both the protozoea and adult Decapoda the caudal fork bears six pairs of setae as in the Copepoda, and the antenna of the penaid protozoea bears remarkable resemblance to that of the Copepoda. Based on these and other similarities, Gurney suggested that the Copepoda arose by suppression of development from larval forms having the general characters of the decapod protozoea (neoteny by acceleration).

4.3.4.3. Halmatometamorphosis in the Copepoda

There are normally six naupliar instars in the cyclopoid copepods (Elgmork and Langeland, 1970). Thus, in a free-swimming copepod such as *Cyclops*, for instance, the larva hatched from the egg is a typical nauplius, which is followed by a second naupliar instar and four metanaupliar instars. In the metanaupliar instars, the body lengthens, and finally three pairs of appendages are added beyond the mandible (including the maxilliped and two pairs of legs). At the next molt, the larva begins to assume a form with adult structures called the "copepodid." After six copepodid instars it becomes the sexually mature adult; during this period the number of thoracic legs added varies in different copepods, three to five pairs of them being added (Kästner, 1967).

In parasitic copepods, metamorphosis occurs during or after the copepodid stage; the naupliar and metanaupliar stages may be passed within the egg, and the number of these stages may be contracted in these parasitic forms. The degree of metamorphosis associated with parasitism (halmatometamorphosis) varies in different parasitic copepods.

Salmincola californiensis (Lernaeopodidae), described by Kabata (1973), requires a single host, the socky salmon (*Onchorphynchus nerka*). Beyond the naupliar stage the life cycle consists of the free-swimming copepodid stage, four chalimus stages, which are modified copepodid stages (attached to the host), and the adult. Major structural changes during the life cycle (in both sexes) are: the evagination of the filament from the anterior margin of the head upon contact of the copepodid with the host; transfer of the attachment by the frontal filament to the tips of the second maxillae, and degeneration of trunk appendages during the chalimus stages, which eventually become lost; and further development and elaboration of the second maxillae and maxillipeds. The life cycle and structural reorganization in *Achtheres ambloplitis* is similar to that of *Salmincola* (see Snodgrass, 1956). Thus, the magnitude of metamorphic changes (halmatometamorphosis) associated with ectoparasitism in these copepods is relatively small.

The case of profound halmatometamorphosis is seen in the life cycle of the female of *Lernaeocera branchialis* (Lernaeopodidae), which requires two hosts (summarized by Snodgrass, 1956). When the copepodid larva (in both sexes) molts in the gills of the flounder, it becomes the chalimus and is fixed to the host by a filament secreted by a gland in the head. The chalimus goes through four instars; with each molt the chalimus undergoes no conspicuous change in form and structure, and gradually approaches the adult structures, which are attained at the fourth instar after the copepodid stage. The male undergoes no further transformation. The female, still not sexually mature, however, begins her metamorphosis into the final egg-producing stage upon reaching the second host, which should be a cod. The curious transformation of the female in the second host involves the production of large branching hornlike structures of the head as anchoring devices and lengthening of the abdomen, particularly the genital segment, which grows out in a twisted wormlike form; then the abdomen becomes long and straight (penella stage) and eventually swells into a great, elongate twisted bag.

In *Phrixocephalus cincinnatus* (Lernaeoceridae), described by Kabata (1969), metamorphosis of the female in the definitive host results in the production of two pairs of lateral processes, a pair of ventral processes, and the anterior and posterior lobes on the head, which together form antlers anchoring the parasite in the eye of the host *Atherestes stemias.*

In *Cymbasoma rugidum* (Malaquin, 1901; Snodgrass, 1956) the nauplius has the usual three pairs of appendages, and the mandibles are provided with re-curved hooks. The nauplius, once in contact with the host (serpulid worm, *Salmacina dysteri*), casts off its cuticle and its appendages and forces its soft nude body into the host. Within the latter it becomes an oval mass of undiffer-entiated cells. In this form the parasite traverses the coelom of the host and makes its way into the ventral blood vessel. Here it secretes new cuticles and anteriorly two armlike processes, which serve as food (blood)-absorbing organs. As the larva grows, the nutritive arms increase in length, a rostrum is formed in front, and the posterior part of the body becomes encircled with spines. The organs of the future adult now gradually develop within the cuticle of the larva. Eventually, an adult with one pair of antennae and four pairs of swimming legs emerges. Thus, the adaptive changes within the host, as Snodgrass (1956) pointed out, certainly have no counterparts in the presumed free-swimming ancestor of this species. Yet, the adult that emerges after such a profound halmatometamorphosis is a normal free-swimming crustacean.

4.3.4.4. Acceleration in the Origin and Evolution of the Cirripedia

In the generalized cirripeds such as the Ascothoracica the body consists of the head bearing five pairs of appendages, the six-segmented thorax with six pairs of appendages, and the five-segmented abdomen without appendages. This struc-tural organization of the body is essentially the same as that in more generalized copepods. As is generally believed, therefore, the Cirripedia presumably arose from a copepod-like ancestor. In the order Ascothoracica, however, this level of organization is attained at the cyprid stage by acceleration in which the cope-podid stages of the Copepoda have been eliminated from ontogeny. Within the Cirripedia, again, acceleration has often occurred in parasitic forms by contraction of naupliar stages, which are passed partly or wholly within the egg, and the larva that emerges from the egg may be a cypris. Beyond the cypris stage, the mode and degree of metamorphosis vary greatly in different sexes (and hermaphrodites) and groups as discussed below. In the Ascothoracica structures become least modified beyond this stage; therefore they are not included in the discussion.

4.3.4.5. Neoteny and Halmatometamorphosis in the Cirripedia

The mode of abnormal metamorphosis resulting in neoteny and halmatometamorphosis differs in different orders of the Cirripedia, and in different sexes or in the male and the hermaphrodite of the same species. Three different modes of sexual differentiation occurs in the Cirripedia, viz., some cirripeds remain hermaphroditic, some cirripeds differentiate into the male and the hermaphrodite, and some into the male and the female. Judging from the known function of the androgenic hormone in the Malacostraca (p. 204), the action of a similar (or the same) hormone may well be involved in the differentiation of the male in the Cirripedia.

The Thoracica. The adult encolosed in the calcareous shell is considered to be a hermaphrodite. In *Balanus* the metamorphosis of the cypris larvae into barnacles is not excessive, although highly complex changes similar to those occurring in the metamorphosis of holometabolous insects (degeneration of larval structures and construction of imaginal structures) occur before the emergence of young adults (Doochin, 1951; Bernard and Lane, 1962; Walley, 1964; Costlow, 1969).

In some species of the Thoracica the complemental male has been found, in addition to the hermaphrodite. Darwin (1854) thought that the males in *Ibla* and *Scapellum* are produced by developmental arrest (neoteny). Veillet (1961) found that in *Scapellum scapellum* sexes (male and hermaphrodite) are indistinguishable during the stage of cypris larva. In the hermaphroditic larva, however, metamorphosis of the eyes occurs, i.e., the naupliar eyes are replaced by imaginal eyes. However, in the male metamorphosis of eyes does not occur or proceeds very slightly, and the male remains small. Clearly, therefore, the males of these cirripeds remain essentially larval (cyprid) and hence neotenous, as Darwin (1854) maintained.

The Acrothoracica. Sexes are always separate in this order (Newman et al., 1969). In *Berndtia purpurea*, which bores into two kinds of coral, Utinomi (1961) found that the oldest immature male is a legless pupa (and not a typical cypris), as in *Cryptophilanus minutus* studied by Darwin (1854). This pupa develops into the adult male, which is only slightly more than a gametogenic sac (halmatometamorphosis). The adult female of this species differs only slightly from the cypris larva.

The Rhizocephala. Sexes (male and female) are apparently always separate in this order. Reinhard (1942) showed that in *Peltogaster* the male is nothing but a sexually mature cypris larva. He found that the males fix themselves to the opening of the mantle of the female, and expel material which consists of spermatogenic elements and "nurse cells." The material migrates through the cavity of the mantle to enter the so-called testis, which Reinhard found to be nothing but a sperm receptacle. Ichikawa and Yanagimachi (1958, 1960) found, in *Peltogasterella*, *Peltogaster*, and *Sacculina*, that the male is essentially a cypris larva and that the "testis" is the sperm receptacle. Thus, in these rhizocephalans the cypris male apparently does not undergo further meta-morphosis, and therefore they are neotenous. Reinhard and Evans (1951) found the cypris male of *Mycetomorpha*, and further showed evidence for the entry of the male into the mantle; those males entering the mantle metamorphose into mere spermatophores (halmatometamorphosis) within the cavity of the mantle. Veillet (1962) observed spermatogenesis in the colleteric glands (ovi-ducts) in *Sylon*, which had been thought to reproduce parthenogenetically. Yanagimachi and Fujimaki (1967) suggested that the "testis" in *Thompsonia* is a mere mass of cypris cells which have been enjected into the female receptacle.

The female within the mantle consists mainly of the reproductive organs. During the endoparasitic stage the female rhizocephalans undergo even more drastic metamorphosis (halmatometamorphosis). The best-known case of such profound halmatometamorphosis is that of *Sacculina*, described below (based on Snodgrass, 1956).

The free-swimming cypris female larva of *Sacculina*, upon finding a young crab, attaches itself to the latter by its antennules (first antennae), which are provided with a suction cup, detaches the thorax along with legs and antennae, and expels most of the internal tissues, leaving only a mass of cells containing reproductive elements. The body of the larva separates off from the shell and contracts to a sac walled by the ectoderm. From this sac is formed a larva called the "kentrogen" with a dart at the anterior end. The dart pierces the integument of the crab and penetrates into the body of the latter. The soft tissue of the larva enters the crab's body through a minute channel of the dart. Inside the crab the parasite (larval tissues) become a small body consisting of a mass of cells enclosed in the ectodermal epithelium, and becomes attached to the ventral side of the crab's intestine (rhizoids), through which nourishment is absorbed. The parasite at this stage consists of the reproductive organ (ovary) and the food-absorbing fungus-like structures.

At the next molt of the crab the body of *Sacculina* containing the reproduc-tive cells emerges and becomes external. The external parasite consists of the ovary, the sperm receptacle, and the mantle that encloses the peripheral brood chamber.

Finally, as shown by Bocquet-Verdine and Parent (1972), *Boschmaella ballani* is the parasite of a cirriped species, *Balanus improvisus*. The larvae of this parasitic species emerge as blind cyprids, which cannot swim, and they im-mediately reinfest. Bocquet-Verdine and Parent further showed evidence indicating asexual reproduction in this species. Yanagimachi (1961) found that in *Peltogasterella* the adult females on one host usually produce larvae of the same type, either large or small and that the small larvae are prospective females, while the large larvae become males. Yanagimachi further showed that in the metaphase plate of the maturation division of the large male-producing eggs, 15 bivalent chromosomes occur; while in the small female-producing eggs, a univalent chromosome is present besides the 15 bivalent chromosomes.

4.3.4.6. Halmatometamorphosis in the Epicaridea (Isopoda)

All species of the suborder Epicaridea are parasitic on other crustaceans, and their life histories often involve two hosts. Consequently, they often undergo profound metamorphosis and are often sexually dimorphic. An example cited by Snodgrass (1956) is that of *Danalia curvata*, in which the animal is first a functional male and then a functional female. The young larvae of this species, upon leaving the brood pouch of the mother, become typical free-swimming isopod larvae, called the "microniscus." The microniscus larva may adopt a copepod as a temporary host. After several molts, the larva assumes a different form, called the "cryptoniscus," in which the isopod characters become less evident; the body becomes more elongate, eyes have been developed, and the appendages are retained. Within the body is a pair of large hermatophroditic sex organs, each of which contains in its anterior end a small ovary and in its posterior part a large testis. The testes rapidly develop and the larva becomes a functional male. Next, the male cryptoniscus, after finishing insemination of the female already parasitized on the crab, undergoes a radical degenerative metamorphosis, whereby the body becomes a small cylindrical sac. The testis also degenerates, and the ovaries begin to develop, so that the larva changes functionally into a female. For further instances of drastic metamorphosis in the Epicaridea see Snodgrass (1956), Green (1961), and Kästner (1967). The androgenic hormone certainly must play a role in the sex reversal during the metamorphosis of *Danalia*. (See Legrand and Juchault, 1972, for a summary of hormonal control of sexuality in the Isopoda.)

4.4. GENERAL DISCUSSION

4.4.1. Polymorphism Through Abnormal Metamorphosis

The end result of most cases of abnormal metamorphosis is polymorphism at the adult stage.* Now, the question is, how can the polymorphism produced

---

*The term polymorphism also applies to the discontinuous morphs exhibited in ontogenetic sequence (such as the larva, pupa, and adult in holometabolous insects).

through abnormal metamorphosis be accounted for in terms of the existing concepts of polymorphism in evolutionary biology?

The term polymorphism usually refers to Ford's (1961, 1965, and earlier) concept of genetic polymorphism, which is defined as "the occurrence together in the same area of 2 or more discontinuous forms of a species in such a proportion that the rarest of them cannot be maintained by recurrent mutation." This definition connotes the segregation of alleles at a single locus to produce distinct morphs. It excludes geographical variation and continuous variation controlled, for instance, by polygenes. Many attempts have been made to explain polymorphism in terms of the presence or absence of a single allele or alleles, apparently with some success.

The difficulty that is encountered in applying the above method to the analysis of the polymorphism produced through abnormal metamorphosis (in arthropods) is that the genetic mechanism (involving major genes), when present, may manifest itself only under certain environmental conditions. Therefore, the term "environmentally cued genetic polymorphism"* can be applied to such cases of incomplete penetrance of the effect of a gene. Based on these definitions, a review follows of the cases of genetic polymorphism (of the two kinds) revealed in this study.

The wing dimorphism (macroptery and brachyptery) in some species of Carabidae and Curculionidae (Coleoptera) is genetically determined, as shown by some crossing experiments; the inheritance of wing morphs takes place in a simple Mendelian fashion (p. 180). In nature, the wing reduction (which results in the wing dimorphism) in Carabidae apparently occurs regardless of seasons and geographical areas, indicating that the wing morphs are not physiologically induced. Therefore, it is the occurrence of wing morphs in some beetles where Ford's concept of genetic polymorphism may apply.

In all other groups of arthropods undergoing abnormal metamorphosis, however, the kind of good penetrance of the genetic factor seen in some Coleoptera does not occur. The determination of castes (queen and worker) in the stingless bee genus, *Melipona*, has been shown to be genetically determined (double heterozygosity for the production of the queen). However, the activation of the genes for queen development occurs only in a proper hormonal milieu, which is, in turn, induced only when the larvae take optimal amounts of food; in this case the production of the abnormal morph (worker) is usually physiologically induced, and does not require a definite genetic mechanism. Thus, the caste determination in *Melipona* illustrates a well-analyzed (and the only) case of the environmentally cued genetic polymorphism defined earlier. Dis-

*This appears to correspond to what Clark (1976) regarded as intermediate conditions of polymorphism between strictly genetically determined polymorphism (where the environmental effect is of little importance, if any, as in the determination of blood groups, and so on) and polyphenism (environmentally cued polymorphism of Clark).

continuous wing morphs occur also in some groups of parasitic wasps (*Gelis* spp.), and the different wing morphs may be genetically biased as in *Melipona*; the expression of the presumed genetic mechanism is, again, dependent on food intake (and the consequent hormonal balance). A similar case of wing dimorphism is that of Sphaeroceridae (p. 190). In all of these holometabolous insects the genetic mechanism (or a presumed genetic mechanism) for the determination of wing morphs usually does not manifest itself, and one morph (which is usually the physiologically induced one) tends to predominate over the other in a given environment. As discussed shortly, these cases of discontinuous wing morphs may not have a definite Mendelian genetic basis.

Among hemimetabolous insects, the wing dimorphism in water striders, *Gerris* spp., has been analyzed in terms of the segregation of alleles at a single gene locus, with (only) dubious results (p. 176). Here, again, even if a genetic mechanism exists, the effect of environmental factors (photoperiod and temperature) on hormonal activity is (presumably) so strong that only one morph may emerge in a given season and area. Furthermore, the apparent genetic mechanism involved with the production of two distinct morphs in these cases may well be spurious. For instance, in *Gerris odontogaster*, the wing rudiments in the last (fifth) nymphal instar are quite small in all individuals. In prospective macropterous individuals (which emerge mainly in the second generation) the rudiments grow very much during the last molt into adults, while in the micropterous individuals (which predominate in the first generation) the rudiments simply do not grow in size at the corresponding period of development, and ovaries become mature at the same time (in the female). It appears highly probable that at the time micropterous individuals are developing under the influence of lengthening days, overproduction of JH occurs (especially toward the end of nymphal development), and the excessive JH thus produced suppresses further development of wings (in size),while simultaneously the same hormone expedites maturation of ovaries in the female (p. 173).

The above discussion indicates that the discontinuous wing morphs in *Gerris* spp., apparently determined genetically (because of the distinct sizes), can be environmentally and physiologically conditioned morphs. Therefore, similarly discontinuous wing morphs in some hemimetabolous insects (such as some auchenorrhynchous Homoptera) and some holometabolous insects already mentioned may also represent environmentally and physiologically induced polyphenism, discussed below.

The third kind of polymorphism is polyphenism of Mayr (1963), in which two or more distinct morphs occur together in a population or within a species (at different times and/or in different areas), and the differences between morphs are not due to genetic differences. In insects (and presumably also in other arthropods) most cases of polymorphism resulting from abnormal metamorphosis represent environmentally and hormonally induced polyphenism.

Thus, for instance, in *Calotermes flavicollis* and other termites any individual (regardless of sex) can develop into any one of the different castes from any immature stage (Fig. 4.2), and CA activity (JH production), which directly affects the developmental pathways into different castes, is influenced by the pheromonal stimulus and some other external factors. Such highly labile morphogenesis leads us to believe that the production of different castes (morphs) is not due to the genetic difference in different individuals. In caste determination (queen and worker) in the honey bee, the different quality of food taken by young larvae induces different hormonal balances (between JH and ED), which lead to the production of the two different castes; thus all larvae have the potential to be either queen or worker. Solitary and gregarious phases of locusts are determined primarily by different population densities and consequent absence or presence of pheromonal stimulus, which, in turn, induces different balances between JH and ED toward the end of postembryonic development and during the adult stage (Fig. 4.3). In cecidomyid flies, paedogenetic (and parthenogenetic) neoteny occurs in the female, and the production of the paedogenetic generation (in the parthenogenetic female) is dependent on diverse environmental factors, not on a special genetic factor or factors. Under some environmental conditions CA become activated to secrete excess JH; the latter in turn suppresses imaginal differentiation and expedites the maturation of ovaries precociously, and hence the paedogenetic female.

In many of the other cases of polymorphism discussed earlier no direct evidence for hormonal control (involved with morph determination) is as yet available. Yet, the fact that in these cases the production of different morphs is directly dependent on environmental conditions and the fact that the similar environmentally induced polymorphism in related insects is known to be hormonally regulated polyphenism lead us to believe that these cases without direct evidence probably represent polyphenism. The association of a swollen abdomen containing ripe eggs with the short-winged morph in female insects is also a good indication of hormonally induced neoteny without a special genetic basis (p. 142).

It is now clear that the applicability of the concept of genetic polymorphism (in the two forms) to the analysis of polymorphism is quite limited. Polyphenism is obviously a widespread phenomenon in arthropods undergoing abnormal metamorphosis. In polyphenism no particular gene or genes are responsible for the production of a morph; interaction of many genes results in polyphenism. In the discussion on the reaction norm of the genotype that results in various modes of metamorphosis (below), therefore, the concept of genetic polymorphism is disregarded.

4.4.2. Reaction Norm of the Genotype During Various Modes of Metamorphosis

The phenotype at a given time reveals only some of the potentials of a genotype, and phenotypic expression varies in different environments. In the study of structural evolution, therefore, it is of great importance to understand the range of reaction norm of the genotype, and it is in various modes of metamorphosis that a great range in the reaction norm of the genotype manifests itself.

The range of reaction norm of the genotype was a great concern of Goldschmidt (1940) in postulating his theory of macroevolution, since a great range of reaction norm of the genotype constitutes the material basis of macroevolution. Examples exhibiting a wide range of reaction norm of the genotype cited by Goldschmidt were extensive (1940, pp. 252-308), and included such examples as the body size of caterpillars, which can be more variable in experimental conditions than in nature; the modifiability of the asymmetrical abdomen of the hermit crab (*Pagurus*) into straight abdomen when kept outside the shell; seasonal dimorphism of many butterfly species, which can be produced experimentally (see p. 192); phenocopy produced by the effect of heat and cold; and so on. Goldschmidt (1940) thought that these examples indicate that under definite environmental conditions, development may be easily changed, typically in such a way that the order of magnitude of the shift is on the level of higher categorical differences.

More significantly, Goldschmidt was quite well aware of the hormonal action affecting the norm of reactivity of the genotype, and described numerous examples (Goldschmidt, 1940, pp. 271-91). Some of those examples included: neoteny in salamanders (see p. 228), some gobiid fishes in which the fins have been transformed into legs for climbing trees, artificially induced prothetely in the lymantriid moth, hypermetamorphosis in *Sitaris* beetle (see p. 181), different races of dogs and humans, and so on. Goldschmidt attributed these transformations to endocrine malfunction, which may have an immense effect of a macroevolutionary order. Further, according to Goldschmidt, all these macroevolutionary changes arose solely by a small genetic change affecting the hormanal condition. Understandably, however, he was not aware of the effect of environmental factors on the hormonal activity, which is the major concern of this study (p. 137). Goldschmidt discussed further the sexual norm of reactivity, and presented examples in which sexual alternatives in the developmental process may result in differences of a macroevolutionary order of magnitude. He also discussed aspects of regulation in development; he defined regulation as a purposive response of the organism to changed conditions, and regarded regeneration as the most typical form of regulation. Goldschmidt considered regulation as important in the discussion of evolution as it is in experimental embryology. An example of regulation in nature is the fact that in the white races of man the arm of the lever of the heel bone is relatively short and the tendon is long, whereas in the black races the arm of the lever is relatively long and the tendon is short. What Goldschmidt called "regulation" is therefore the same as the "material compensation" of Rensch (1959), which is discussed shortly.

When, as the result of this study shows, the cases of profound metamorphosis (halmatometamorphosis) that occur in parasitic animals are taken into consideration, the range of reaction norm of the genotype is far greater than that recognized by Goldschmidt. Furthermore, there appear to be certain regularities in the reaction norm of the genotype associated with different behaviors and requirements for living, and with the different habitats in which they live, as shown below.

4.4.2.1. Invertebrates Living in Interstitial Environments

When, as shown by Swedmark (1964), invertebrates live in highly restricted spaces such as interstitial spaces on the seashore, they tend to become diminutive in body size, and structures become regressed. An example cited by Swedmark was *Psammodriloides fauveli* (Psammodrilidae, Polychaeta). The body size of this species is $\frac{1}{25}$ of *Psammodrillus balanoglossoides* of the same family. In the former, several important features (pharyngeal apparatus, reversible diaphragmal sac, nephridia) present in the latter are lacking; otherwise the adult organization of the former coincides with a juvenile stage of the latter. Similar examples discussed earlier in this chapter are the pauropods, which measure a maximum of 2 mm in body size, and Harpacticoida (Copepoda), which often measure less than 0.3 mm.

In the above examples a developmental stage or stages are omitted from their ontogeny (acceleration), and they become neotenous adults. Apparently, compensatory development of structures (which are either absent or less-developed in the nonneotenous ancestors) does not occur in these neotenous animals. The reaction norm of the genotype in these animals, therefore, can be described as a substantial suppression of genes, which is presumably effected by an inhibitory hormone. For similar examples see Swedmark (1964).

4.4.2.2. Some Social Insects (Ants and Termites)

Both ants and termites live in colonies. Their worker and soldier castes are produced by abbreviation of the developmental process (prothetely); in these cases that the action of some genes forming adult structures is presumably suppressed. The production of the soldier in termites results from overproduction of JH, and the production of the worker ants apparently results from premature exhaustion of JH. It is apparent, therefore, that a significant alteration in the titer of JH (either increase or decrease) during development results in a switch of regulatory genes, which may, in turn, result in the conspicuous enlargement of the head in these castes, and the development of some adaptive structures (such as the "squirt gun" on the nasute head in the soldier of some termites). For the pheromonal stimuli causing caste differentiation in the termite colony see p. 148.

4.4.2.3. Holometabolous Insects and Frogs

Holometabolous insects undergo three distinct developmental stages (larva, pupa, and adult) in association with changing requirements for food and habitat in their life cycle, and the change in hormonal balance that accompanies such changes has been well known (p. 140). Williams and Kafatos (1971) assumed the presence of three major sets of genes: those for the larva, pupa, and adult; they proposed that changing titers of JH at the three stages (larva, pupa, and adult) of development enable a switch of regulatory genes, which, in turn, enables the expression of the three sets of genes, and hence the complete metamorphosis.* Another well-known case is the metamorphosis of the aquatic tadpole into the terrestrial frog; a sharp increase in the thyroxine titer (for the development of frog structures) and a sharp decline in the prolactin titer (for the development of tadpole structures) occur during this transition in environment.

4.4.2.4. Caenogenesis

Caenogenesis represents the reaction norm of the genotype to environmental requirements during immature stages. The caenogenetic structures are usually not carried over into the adult stage. The development of some conspicuous structures in the juvenile castes in ants and termites can also be regarded as caenogenetic, although they are definitive structures in these insects. The hormonal regulatory mechanism of caenogenesis may, as seen in the production of juvenile castes (above), be highly variable.

Examples of caenogenesis in larval holometabolous insects are: serially developed ambulatory protuberances in *Xenos* (Strepsiptera), dorsal gills on the eighth and tenth abdominal segments in *Panorpa* (Mecoptera), the anal gills in larval Diptera and Trichoptera, paratergal projections in some larval Neuroptera, conspicuous ventral thoracic processes and strongly narrowed posterior portion of the abdomen in *Eucolia* (Hymenoptera), and so on. Cases of caenogenesis in the tadpole include: the gills in *Hyla rosenbergi* that are modified for clinging to the surface film of the water in which it develops (Noble, 1931), the mouth in the tadpole of a semiaquatic species, *Cyclorampus pinderi*, which is especially adapted for clinging to the rock, and the tails of the same species, which are adapted as respiratory organs (Dent, 1968).

---

*According to this model, JH acts as a corepressor, which activates the master regulatory genes of the pupal and adult gene sets. The operator of the master regulatory gene of the pupa is subject to inhibition by a repressor of the pupal regulatory gene, which is active only in the presence of a high titer of JH. So also the operator of the master regulatory gene of the adult is subject to inhibition by another repressor, the important difference being that this repressor remains active in the presence of a lower titer of JH. Both repressors become inactive in the absence of JH. Thus, JH is involved in the negative control of transcription of master regulatory genes.

Caenogenesis also occurs in other immature stages and in animals that do not undergo profound metamorphosis during development. Examples are embryonic pleuropods in insects, egg burster in insects, plastron-bearing spiracular gills in some pupal insects, the modified spiracles in the hemipupa of the paedogenetic cecidomyids (p. 186). The cases of hypermetamorphosis (p. 180) and allometabolism (p. 171) can also be included in the category of caenogenesis. An outstanding example of caenogenesis in the Vertebrata is the amnion and embryonic membrane. For more examples of caenogenesis see Remane (1956) and de Beer (1958).

4.4.2.5. Free-Living Insects and Other Arthropods in Changing External Environments

Compared with animals living in more specialized habitats (in all of their life or part of their life cycle), the requirement for structural reorganization on the part of animals living freely in changing environments is relatively small, and the only mode of abnormal metamorphosis involved here is neoteny.

Among the physical factors in changing environments, temperature and photoperiod are usually dominant ones, influencing the developmental process of many free-living species of insects and other arthropods. The effect of these factors is subject to change in time and space, and such change may induce seasonal dimorphism and geographically related di- or polymorphism. Examples include some cricket species (p. 159), some auchenorrhynchous Homoptera (p. 164), some water strider species (p. 172), some aphid species (p. 169), some cyclopids (p. 205), and so on. In certain cases, population density (whether crowded or isolated) plays a major role in determining morphs, as seen in locusts (p. 154), some auchenorrhynchous Homoptera (p. 164), some aphids (p. 169), and so forth. In wing morph determination in some parasitic Hymenoptera food plays the major role, as it does in caste determination in social Hymenoptera (bees and ants).

The structural reorganization resulting from neoteny in these arthropods is relatively small. For instance, in Heteroptera neoteny may be evident externally only in the short wing, although in some others more structures become modified (e.g., Psocoptera, some Orthoptera, and others). Prevalent constructive developments of structures in these neotenous insects are the enlargement of legs and ovaries.

In well-investigated cases such as locusts, as already noted, overproduction of JH leads to the production of the neotenous morph. It is probable that similar hormonal control mechanisms are often involved in producing the neotenous morphs in the other free-living insects. The constructive development of legs and ovaries in these neotenous insects is also due to an excessive supply of JH, as discussed again on p. 224. The production of the honey bee worker, however, results from lowered titers of JH during development (p. 182).

4.4.2.6. Case-dwelling Female Psychid Moths and Cave-dwelling Beetles

The first larval instar of the female psychid moth builds a case in which she lives for all or most of the rest of her life. The consequence of such living is the reduction of wings and legs, and compensatory growth of the ovipositor occurs in some female psychid moths; presumably, abnormally high titers of JH produced in such a habitat induce such neoteny (p. 191). In some cavernicolous beetles, the hind wings become lost, and reduction of eyes occurs. At the same time, however, appendages become larger (p. 180). For the characteristic structural transformations of other cavernicolous animals refer to Vandel (1965).

4.4.2.7. Parasitic Arthropods

During adaptation to parasitic environments, especially to endoparasitic ones, selection pressure must be very stringent and requires a drastic reorganization of structures on the part of the parasites, by a drastic alteration in the reaction norm of the genotype. However, the parasites have evidently met this requirement successfully. How far this drastic structural reorganization, viz., abnormal metamorphosis, proceeds depends on their life histories.

In some ectoparasites (some copepods, some mites, female scale insects) and the endoparasitic female Strepsiptera, the young larvae are free-living, and comparable to the larvae of nonparasitic relatives. As they enter the parasitic stage, however, degeneration of some structures (e.g., appendages) occurs, and simultaneously some adaptive structures (e.g., attaching organs in copepods and mites, pore structures in female scale insects, brood canals in female Strepsiptera, and so on) develop. At least in the female scale insects overproduction of JH apparently results in such structural reorganization, and the same may well be true with the female Strepsiptera.

Some ectoparasites, such as Phthiraptera and Siphonaptera, spend all or part of their life on their hosts; the development of certain structures (such as wings) is arrested, and some adaptive structures develop at the same time. In these cases overproduction of JH apparently induces such structural reorganization (p. 226).

Hypermetamorphosis occurs in several orders of insects. The insects undergoing hypermetamorphosis change their structures during their larval development, as they alter their hosts. Eventually, however, they emerge as winged, free-living insects. Most probably, as an insect changes hosts, the blaance between JH and ED also changes, and presumably the altered titers of ED on different hosts, as Willis (1974) suggested, regulate the larval polymorphism (hypermetamorphosis); this interpretation agrees essentially with Goldschmidt (1940), who maintained that hypermetamorphosis in *Sitaris humeralis* (Meloidae, Coleoptera) exhibits a wide range of reactivity of the genotype in response to environmental changes, which is mediated by hormonal regulation.*

The halmatometamorphosis that endoparasitic crustaceans undergo shows a limitless modifiability in the reaction norm of the genotype. Examples include the female of *Lernaeocera branchialis* (Copepoda), which undergoes a curious transformation on the second host, and the female of rhizocephalan cirripeds (e.g., *Sacculina*), which becomes a nonarthropodan (in appearance) animal consisting of the reproductive organs and the fungus-like food-absorbing organs (p. 210). An even more impressive case of halmatometamorphosis is that of *Cymbasoma rugidum* (Copepoda), which becomes a free-swimming crustacean after undergoing a highly degenerative transformation of structures during the endoparasitic stage (p. 208). As is well known, parasitic cestods and trematods undergo transformations comparable to those undergone by these endoparasitic crustaceans.

Nothing is known directly about the hormonal control mechanism involved with the halmatometamorphosis in these endoparasitic crustaceans. Theoretically, however, the drastic modification of the reaction norm of the genotype that occurs during the endoparasitic stage is possible only by extensive switches of regulatory genes, which can be triggered by altered titers of an inhibitory hormone, as in holometabolous insects (p. 217); it is conceivable that the hormonal milieu changes drastically, as an animal enters the endoparasitic stage.

4.4.3. Material Compensation (Compensation of Body Parts) and Its Evolutionary Significance

The preceding discussion shows that when reduction or loss of structures occurs during development, often constructive (and adaptive) development or other structures occurs, as if compensating for the reduction and loss of other structures, and the structures formed in them are often highly elaborate new structures. When neoteny occurs in free-living insects, reduction of certain structures (such as wings) tends to be compensated for primarily by excessive growth of some other preexisting structures (such as legs), and development of some new structures may occur. Following Rensch (1959), such compensatory growth and development of body parts (structures) may be called "material compensation."

As Rensch's (1959) review showed, the phenomenon of material compensation was clearly known to Geoffroy St. Hilaire (1822) as the "*loi de balance.*" According to this law, strongly growing parts of an organism consume so much

*Regarding "hormonal regulation," Goldschmidt was not explicit as to how. At that time, the influence of environmental factors on the hormone titer in insects was unknown, except for the blood meal stimulus in *Rhodnius* (Wigglesworth, 1934).

of the body material that the less strongly growing parts remain small or become more or less reduced. Some of the examples of material compensation cited by Rensch include: stronger hind legs and reduced lumbar ribs in vertebrates, increasing differentiation of extremities with shorter vertebral column in vertebrates, well-developed ovaries in short-winged forms in some insects such as *Metrioptera* (see p. 161) and some Lepidoptera (p. 191). Rensch thought that for evaluation of the importance of the compensatory process, modification of structures after experimental alteration must be studied, and he referred to some regeneration experiments. An example he cited was that if, in decapods (*Alpheus, Carcinus*) with asymmetrical chelae, the more solid "male type" chela is amputated, a smaller "female type" chela regenerates, and in the course of following molts the other chela, left intact, becomes increasingly larger until it becomes a male-type chela.

Similarly, Goldschmidt (1940) thought that the phenomenon of regeneration known in experimental embryology suggests the potential for material compensation (regulation of Goldschmidt) in organisms; he paid special attention to the intrinsic ability and tendency of the embryonic cells to move and unite with other cells in the formation of new tissues adapted to the new mechanical conditions (produced experimentally), and referred to some experiments showing convincing evidence for such potency. For instance, cells of an adult hydroid or sponge, if isolated and completely mixed by straining through fine gauze, come together again and build up a perfect new organism. In the propagation of some Myxomycetae, individual, isolated amoeboid cells come together and build up by appropriate morphogenetic movements the complicated toadstool-like structure, and so on.

Goldschmidt (1940) maintained further that the kind of disruption of continuity observed between embryonic parts in the experiments on regulation may also be produced by a genetic change. He believed that many genetic changes result in a relative shift of the rate of interlocked developmental processes. Such a shift, if produced in early developmental stages, at the time of still-labile determination, may act in the same way as an experimental disruption by operation. The consequence of such a mutation would, in many cases, be upsetting to the developmental mechanism, i.e., lethality; but under proper circumstances a certain amount of regulation (= material compensation) takes place, and the result is some kind of monster! In this case the single genetically produced change of an embryonic feature results in a whole series of changed developmental processes—in other words—in a completely new type of development, which may result in profound material compensation.

The contention of Goldschmidt that a single genetic change can result in a completely new type of development can perhaps be reinterpreted in terms of modern regulatory genetics. Indeed, if "a genetic change by a mutation" of Goldschmidt is replaced by "a change in regulatory genes by a mutation," the consequent developmental process could be roughly what Goldschmidt envisaged (above). In fact, some modern workers (Britten and Davidson, 1971; Wilson, 1975; King and Wilson, 1975; Valentine and Campbell, 1975) contended that when mutation of a few regulatory genes occurs, it can result in an extensive repatterning of gene activity, and hence in a drastically different developmental process leading to the production of entirely novel structures; Valentine and Campbell (1975) contended that the repatterning of the regulatory portion of the genome resulted in the origin of invertebrate phyla.

It should be remembered, however, that the change in the chain of gene actions is initiated by the action of an inhibitory hormone, as indicated in the proposed models concerning gene regulation in Eukaryotes (JH in insects by Williams and Kafatos, 1971: steroid hormones, polypeptide hormones, several plant hormones, and several embryonic inductive agents by Britten and Davidson, 1969, 1971, and by Davidson and Britten, 1973). Therefore, when the titer of an inhibitory hormone in a developing animal becomes significantly altered in some environments, there could result a regulatory gene switch and consequent material compensation, as the model by Williams and Kafatos (p. 217, footnote) most clearly indicates,* and the material compensation may lead to the production of an animal with entirely new structural organization (monster of Goldschmidt). Thus, theoretically, the alteration in the titer of an inhibitory hormone during development could have essentially the same morphogenetic effect (viz., material compensation) as that produced by the mutation of regulatory genes. In fact, as already seen, significant alteration in the titer of JH occurs or probably occurs in inducing material compensation (e.g., in the larva-pupa-adult transformation, in the production of juvenile castes in termites and ants, in the compensatory development of legs in the short-winged insects, and so on); and the material compensation probably occurs through the regulatory gene switch in these cases.

Apart from the probable regulatory gene switch involved with material compensation (whose detail still remains to be elucidated), we have abundantly seen throughout this study that the alteration in JH titer (and consequent balance between JH titer and ED titer) is induced directly by alteration in environmental factors. *This fact of direct influence of environments on the hormonal activity, which results in abnormal metamorphosis (with or without material compensation), is of the utmost importance in considering the possibility of macroevolution, viz., a sudden emergence of a new animal (arthropod) species with a conspicuously different structural organization.* It is easy to imagine that as a group of individuals (of the same species) come to live in a new environment, all or most of them probably undergo similar (and substantial) alterations in hormonal activity, which would lead to a significant alteration in the reaction norm of the

*According to this model, the process of the larva-pupa-adult transformation results from a series of regulatory gene switches, and such a sequence of transformation exhibits a pronounced case of material compensation in onotogeny.

genotype and the subsequent production of novel structural organization in them, and hence the emergence of monsters. The perpetuation (inheritance) of the newly acquired structures becomes possible eventually, as long as enough numbers of individuals (monsters) survive the new environment. Thus, it appears probable that when macroevolution occurred, it often did so without mutation, and that an entirely different new arthropod species sometimes arose as a population.

Goldschmidt (1940) was quite right in believing that the change in hormonal activity can result in the production of entirely new structures, and he was also most probably right in maintaining that a small change (in regulatory genes) can result in the production of an entirely different animal. However, he thought that these changes result only from mutation. Following Goldschmidt (and some modern workers), then, an entirely new structure or structures are usually produced in one individual at a time. The resultant "hopeful monster" would, as Mayr (1963) pointed out, have difficulty in finding a mate. At least in the Arthropoda, however, the chances of mutation affecting the hormonal activity and the chances of mutation affecting the genotype drastically to produce new structures (and hence a new animal species) are probably negligible, considering the inevitable alteration in the hormonal activity and in the consequent reaction norm of the genotype (leading to abnormal metamorphosis) that occurs in large numbers of individuals in new (and extreme) environments; and the animals thus produced (without mutation) are not expected to have great difficulty in finding mates.

The persistent tendency of compensatory growth of the hind legs (and other legs*) in insects with reduced wings must now be explained. Acridologists have noticed that in the solitary phase of locust species with shorter wings the hind femur is relatively larger than in gregarious (fully winged) individuals, during the nymphal and adult stages. Since in this case the wing reduction occurs clearly as a result of suppression by a higher titer of JH (see p. 158), the same hormone is expected to suppress also the growth of the hind femur. Therefore, the excessive development of the latter has remained as an enigma. However, this phenomenon, being conditioned by the increased titer of JH, can also result from a repatterning of the regulatory system in gene action, as in soldier production in termites.

The excessive development of the hind legs (and perhaps other legs) in individuals with reduced wings or in wingless species occurs in several orders at least (Orthoptera, auchenorrhynchous Homoptera, Lepidoptera, Coleoptera, Hymenoptera, Siphonaptera). In these insects with more limited ability to fly or without the ability to fly, the well-developed legs must usually be selectively advantageous, since they must often be the sole organs for locomotion and migration.

*Excessive development of other legs has not been investigated, except for the Hymenoptera, in which all legs are known to be enlarged when the wings are reduced.

The tendency of compensatory growth of the hind legs, being fairly common in insects with reduced wings, could be of some predictive value in the practice of systematic entomology. Here the question remains of whether the hind legs are the only appendages that undergo compensatory growth. It is probable that often the other legs and perhaps also antennae could undergo parallel compensatory growth, since it was found by the author and others (Matsuda, 1961a,b, 1962a,b, 1963a,b; Matsuda and Rohlf, 1961; Kumar, 1966; Mukerji, 1972) that the growth of all these appendages is under the control of one physiogenetic factor in various groups of Heteroptera, and this consistent penetrance of the factor into phenotypes (their hypothesis 1), in turn, indicated that selection apparently operates on the growth of all these appendages as a unit. As already seen, the enlargement of all legs occurs in the wingless or wing-reduced Hymenoptera.

Material compensation in insects is by no means confined to the development of wings and legs, and to some other cases shown in this study. Okada (1963) analyzed various morphological characters in Drosophilidae in terms of material compensation. In the taxonomic study on Tortricidae (Lepidoptera) in progress, Dr. A. Mutuura (personal communication) has found the consistent reduction of various genitalic structures and concomitant excessive development of other genitalic parts. Human evolution may perhaps be accounted for in terms of material compensation resulting from neoteny (retardation in development combined with specialization of some facial structures).

The persistent tendency for the ovary to develop excessively (by precocious maturation of eggs) in the insects with reduced wings (in at least several orders) can also be regarded as a case of material compensation, although in this case no genetic change is involved. As has been repeatedly pointed out, this tendency most probably results from excessive titers of JH which suppresses the development of wings and some other structures while simultaneously expediting oogenesis by facilitating yolk deposition. In these insects the number of eggs produced is greater and the period of time necessary for egg maturation is usually shorter (than in fully winged morphs). These attributes are certainly important ecological strategies for adaptation in changing environments (r-selection).

4.4.4. Abnormal Metamorphosis and the Origin of Taxa in the Arthropoda

The preceding discussions show that abnormal metamorphosis often results in conspicuous compensatory development of new structures and compensatory growth of preexisting structures, while some preexisting structures becoming reduced or lost (material compensation). In certain cases, however, the end result of abnormal metamorphosis may simply be highly reduced conditions of structures without compensatory development and growth. A new structure produced through abnormal metamorphosis means initiation of homology of structures when it is inherited in descendants; and when a structure becomes lost by extreme reduction, this means either an interruption in the continuity of a (homologous) structure in phylogeny, or an end of the continuity of the (homologous) structure in phylogeny. The question is therefore whether and to what extent the degree of structural reorganization achieved through abnormal metamorphosis constitutes the basis for the recognition of lower and higher taxa within the Arthropoda.

When abnormal metamorphosis (neoteny, production of juvenile castes, halmatometamorphosis) occurs in only one sex of an animal species, the unique structural organization thus achieved does not constitute the basis for the recognition of a new higher taxon, since in the other sex structures remain normal, and they are comparable with relatives. For instance, no matter how conspicuous female paedogenetic neoteny and halmatometamorphosis may be in some insects (e.g., cecidomyid flies, psychid moths, scale insects, Strepsiptera), they can be placed in taxonomic positions (family and order) based on their males. Similarly, no matter how great the metamorphic changes may be in the females of some parasitic cirripeds, their males are essentially cyprid larvae, a fact that identifies them as crustaceans and cirripeds. No confusion would arise in identifying worker ants as long as they occurr together with their queens, which are normal Hymenoptera. It should be remembered, however, that classification and understanding of phylogenetic relationships (hence origins) of lower taxa may be based on the sex undergoing abnormal metamorphosis (e.g., scale insects).

In the cases of seasonal di- and polymorphism occurring in some insects (e.g., wing dimorphism in *Gerris* spp. and some auchenorrhynchous Homoptera, wing pattern dimorphism in some butterfly species, and so on) and other arthropods, one or more morphs are usually neotenous. The different morphs in these cases, however, should not present a serious taxonomic problem when their natures are clearly understood.

When the individuals produced through abnormal metamorphosis become isolated in certain geographical areas and/or in some extreme environments (such as parasitic environments), such a population may eventually give rise to animals representing new taxa, as discussed below.

The order Siphonaptera (fleas) is often regarded as having been derived from a Diptera-like ancestor, and such a hypothesis is based partly on the fact that transitory wing rudiments* occur, as in Diptera, on the mesothorax during the pupal

*The identity of the structures as wing rudiments has been widely accepted since their discovery by Sharif (1935), although Poenicke (1969), who found the rudiments in eight genera of fleas, denied such homologization. The rudiments clearly arise from the posterolateral ends of the mesonotum, but they later become pleural structures covering spiracles (Poenicke, 1969), and their definitive pleural position is comparable with the definitive pleural position of the subalare bearing wings (in many insects), which are actually the detached posterolateral portions of the tergum (notum) of wing-bearing segments (see Matsuda, 1970); in the fleas, the rudiments do not develop into wings. Therefore, Poenicke's denial of the presence of wing rudiments is unjustified.

stage; in Diptera, the rudiments develop fully into wings. The presence of the wing rudiments indicates, in turn, that the absence of the rudiments in some fleas and their degeneration in others are secondary conditions that occurred as their ancestor acquired the parasitic habit, and further that the genetic factor for wing development has not been lost, at least in some fleas. As discussed below, the absence of adult wings could be hormonally controlled.

The breeding cycle of the rabbit flea (*Spilopsyllus cuniculi*) is known to be linked to the hormonal cycle of the host (rabbit). Rothschild and Ford (1964) and Rothschild (1965) contended further that the hormones under the control of the pituitary gland of the rabbit influence the release of JH from the CA (or probably also from NSC in the brain) of the female flea, and that the JH thus released, in turn, stimulates the development of ovaries. It is probable, therefore, that the activity of the larval CA is also affected by the host's blood, and the kind of hormonal balance (between JH and ED) during larval and pupal stages inhibits further differentiation of the wing rudiments. Enlargement of the hind legs (for jumping) represents a case of material compensation, apparently conditioned by an elevated JH titer.

Judging from close anatomical similarities, there is little doubt that the order Phthiraptera (which includes Mallophaga and Anoplura) arose from a Psocoptera-like ancestor. In the Psocoptera, as already seen (p. 162), wing polymorphism (polyphenism) can readily be induced by altering temperature and population density during rearing; these alterations in turn most probably induce alteration in hormonal balance (between JH and ED) before wing polymorphism results. It is easy to imagine that the psocopterous ancestral individuals settled down on the vertebrate host (or hosts) where the necessary stimulus for the induction of proper hormonal balance (presumably an elevated titer of JH) for aptery was always available. Since aptery was definitely advantageous in the new parasitic environment, they were allowed to stay as they were.* In the meantime (or simultaneously), compensatory (and adaptive) development occurred, involving modification of the generalized mouthparts (present in Psocoptera), consolidation of thoracic segments, and a concomitant shift in position of many structures (see Matsuda, 1970), and so on. All these specializations resulting from neoteny have made the members of this order distinct from the Psocoptera.

In the parasitic mite family Podapolipidae, the female adult that has emerged from the larva is provided with different numbers of pairs of legs (zero to four) in different genera. In *Chrysomelobia* the female retains four pairs of legs as in

---

*This is superficially akin to the "Baldwin effect." In the context of this work, however, alteration in hormonal milieu directly induces alteration in the reaction norm of the genotype, so that aptery can occur; this situation may be comparable to the genetic assimilation of Waddington (1961), in which a phenotype initially produced in response to certain environmental stimuli is taken over by the genotype. In the Baldwin effect (see Mayr, 1963, p. 610), however, mutation occurs later in such a population in producing the favored phenotype.

the related family Tarsonemidae. Some genera (*Dorsipes, Tarsopolipus*, and so on) with three pairs of legs in the female are neotenous, being comparable to the larva with three pairs of legs. Many other genera with two or one pair of legs or without legs in the female are not comparable to any stage of any species; in these podapolipids compensatory development of some adaptive structures occurs (p. 201), and the metamorphosis through which such mites are produced may be called halmatometamorphosis. It is clear that different genera of Podapolipidae arose primarily through different degrees of abnormal metamorphosis, ranging from neoteny to halmatometamorphosis of varying degrees in the female.*

In many instances of origin of taxa through abnormal metamorphosis, however, compensatory development of structures did not occur, or was limited; the neotenous adults thus produced often retain juvenile features of structures, and some imaginal structures may never develop. A good example is the origin of the Collembola from the Protura-like ancestor. In this case the suppression of anamorphosis (during postembryonic development) has obviously resulted in highly reduced or undifferentiated conditions of many structures in the adult Collembola, and compensatory development of structures has apparently been limited (p. 196).

The phylogenetic position of the Grylloblattodea remains unclear; they have been considered to be related to the Orthoptera, or to the Blattaria, or else to the Dermaptera. The members of this small order live under stones and in similar habitats in the alpine zone of the Palaearctic and Nearctic regions. They are wingless, many structures remain juvenile, and little (if any) compensatory development of structures has taken place in these insects (p. 151). Yet, the array of these juvenile and hence neotenous features in the adult makes this order unique enough to be recognized as a distinct order.

Within the Blattaria-Isoptera complex, many structures present in the adult Blattaria remain less developed or undeveloped in the primary reproductives of the Isoptera; apparently the development of conspicuous new structures did not occur, or has been very limited. What makes the two orders truly distinct morphologically is the caenogenetic development in the soldier caste of the latter (enlarged head, defense organ in some groups). The order Phasmida presumably arose from a wing polymorphic ancestor, and within the order evolution has proceeded primarily through prothetely, as discussed on p. 152.

Among the noninsectan arthropods, the Pauropoda (p. 199), some diplopod groups (p. 198), the Copepoda (p. 207), and the Cladocera (p. 206) arose through neoteny without conspicuous compensatory development of structures; a substantial abbreviation of developmental stages (acceleration) has accompanied the neotenic origin of these groups.

Nemobius yezoensis (Gryllidae, Orthoptera) occurs in the northern half of Japan and has short wings. It has been found, however, that the wings in this

*The male in most genera of Podapolipidae is neotenous by acceleration (p. 202).

species can be developed fully by a photoperiodic treatment; the species remains brachypterous as long as certain photoperiods prevail in certain seasons in Northern Japan where this species occurs. The apterous species of Sphaeroceridae (Diptera) living in high mountains in West Africa have presumably been derived from the wing dimorphic ancestor by environmental influence (p. 000). Mani (1962) showed that in the zone above the timber line of Nowthwest Himalayas nearly 50% of the insects are wingless, and the frequency rises to 60% at elevations above 4.000 m. These facts indicate the abundance of wingless (lower) taxa in this area.

The hormonally induced neoteny in salamanders parallels the paedogenetic neoteny in insects. In these animals the immediate agent inducing metamorphosis is thyroxine, released from the thyroid gland, and the activity of the gland is influenced, directly or indirectly, by such exogenous factors as the iodine content in water, food, cold temperature, and so on. The neoteny in salamanders results either from malfunction of the thyroid gland, or from lack of sufficient sensitivity of tissues to the hormone (thyroxine) that leads to the failure in metamorphosis. The degree of failure of metamorphosis (hence neoteny) differs in different species and groups of salamanders, as shown by Dent (1968).

Some salamanders, such as *Ambystoma tigrinum*, are occasionally neotenous (larval), and the metamorphic failure probably results from hypofunction of the thyroid gland; such neoteny is comparable to seasonal dimorphism of insects in which one of the morphs is neotenous (*Euscelis* spp., *Gerris* spp., and so on), and to the locust spp. in which the solitary phase is neotenous.

Some salamanders (Plethodontidae, *Siredon* of Ambystomidae, and so forth) are consistently neotenous in the natural habitat, but metamorphosis can be induced experimentally by thyroxine treatment; these cases are comparable to a brachypterous species, *Nemobius yezoensis*, in which macroptery can be induced by a photoperiodic treatment.

Some salamander species (Sirenidae, Proteidae, Cryptobranchidae, Amphiumidae) are permanently neotenous, and metamorphosis cannot be induced by the thyroxin treatment. Apparently, in these cases tissues have lost their sensitivity to the hormone, and it is at this level that a significant genetic change has occurred. These salamanders are probably comparable to permanently apterous insects.

In many cases of the phenotypes acquired through abnormal metamorphosis, the question remains of how permanently such phenotypes (and hence new taxa) have been fixed genetically. Waddington's (1961) concept of genetic assimilation, which Mayr (1963, 1976, p. 320) called threshold selection, is attractive in this connection (see p. 226, footnote). It is pointless, however, to speculate whether the genotypes of the arthropods living in highly specialized habitats (such as parasitic milieu) have been fixed, since such animals can no longer complete their life cycles in their original habitats.

4.5. SUMMARY

Hormonal Regulations of Normal and Abnormal Metamorphoses in Insects. A general scheme of the balance that steers the postembryonic developmental process in insects is that: ED (ecdysone), in the presence of a high titer of JH (juvenile hormone), apparently induces larval molt; in the presence of less JH, ED induces pupal molt; and in the absence of JH, ED induces imaginal molt. In the Hemimetabola the imaginal molt and differentiation are triggered by ED in the absence of JH. Abnormal metamorphosis results from derangement of the normal hormonal balance, which more or less limits or enhances the action of ED. In this scheme JH is a status quo hormone.

Diverse external factors influence the activity of neurosecretory cells (NSC). Therefore, in certain environments abnormal activity of NSC could occur, and the consequence could be abnormal balance of the two hormones (ED and JH), which might lead to abnormal metamorphosis.

In a majority of insects, JH has the function of stimulating oogenesis, besides its molt-inhibiting role. Therefore, when JH is abnormally high during development, female neoteny often results. In some insects this gonadotropic function has been lost, and this has enabled the development of juvenile castes in termites, and enabled phasmids to evolve by prothetely.

When abnormally high titers of JH occur during late postembryonic development, the result is a more or less limited action of ED (reprogramming of the hormonal action), and hence often neoteny.

Alteration in the titer of JH during postembryonic development may induce a regulatory gene switch and lead to material compensation (p. 221) whereby certain preexisting structures become excessively developed and some new structures may be formed. JH also has some other minor morphogenetic functions.

Tissue sensitivity to JH action is genetically determined and variable in different individuals, species, and groups of insects. Such differences affect the degree of abnormal metamorphosis they undergo.

Blattaria and Mantodea. In Blattaria neoteny is evident in the reduction of wings and arrested development in some other structures, and the neoteny is either metathetely or mild prothetely. Apparently, no conspicuous compensatory development accompanies the reduction of some structures. In Eremiaphilidae (Mantodea) inhabiting the desert in North Africa reduction of wings and tarsal segments occurs.

Isoptera. In a primitive termite, *Calotermes flavicollis*, a colony consists of the primary reproductives, neotenous supplementary reproductives, and essentially nymphal soldiers. Early developmental stages in termites consist of the "larvae," and nymphs with wing rudiments that follow "larvae"; pseudergates are individuals that have regressed from the nymphal stage by molting, or are derived from "larvae."

In termites ED is probably the main oocyte-stimulating hormone, and this probably is the reason why the soldier caste can be formed in the presence of a high titer of JH. Supplementary reproductives are formed from several earlier developmental stages by premature action of ED at the time CA are inactive. Therefore, they represent prothetely. The pheromone of the primary reproductives presumably stimulates the CA to produce more JH, which inhibits the production of reproductives. Soldiers also may produce a pheromone.

Seasons affect the activity of CA in the embryo, and this may determine, in part, the eventual developmental pathways. There is no genetic mechanism determining different castes, since the latter are so freely determined by external factors, which influence CA activity.

Development of some adaptive structures (such as the defense organ in the nasute head) in the soldier caste of some groups of termites is essentially caenogenetic, and represents a case of material compenation apparently triggered by overproduction of JH.

Many adult structures have become reduced in evolution of termites, apparently reflecting the increased effect of JH. However, the trend of compensatory development of structures is apparently absent.

Zoraptera, Ephemeroptera, and Grylloblattodea. In Zoraptera apterous adults without ocelli are neotenous. In some Ephemeroptera the reduction of hind wings is highly pronounced. Grylloblattodea are wingless, and development of many other structures is arrested.

Plecoptera, Phasmida, and Embioptera. Microptery occurs in the male of six species of the Swedish stoneflies, and its occurrence is associated with earlier emergence in the spring (than the female). Wing polymorphism in other stoneflies is environmentally induced.

In Phasmida, the gonadotropic function of JH has been lost, and the latter function is performed by a factor produced from the NSC in the brain. These functional alterations of the hormones have enabled the phasmids to evolve by prothetely.

In Embioptera, all females are wingless. Brachypterous and apterous males occur in some species inhabiting arid regions.

Orthoptera. Solitary locusts are neotenous by metathetely. Lack of pheromonal stimulus induces overproduction of JH, which, in turn, limits the action of ED. The consequences are insufficient differentiation of some imaginal structures (such as wings), retention of some juvenile features, precocious maturation of ovaries, and excessive development of the hind femur. The enlargement of the hind femur in solitary locusts represents a compensatory development. In some wingless species of Orthoptera this phenomenon has apparently been permanently fixed.

In the gregarious phase, a pheromone helps to maintain the normal hormonal balance in producing the fully winged (gregarious) locusts; in the solitary phase overproduction of JH occurs in the absence of the pheromonal stimulus, hence producing neotenous (solitary) locusts. Photoperiod and temperature also influence the phase differentiation.

In Gryllidae also, wing polymorphism is dependent on population density; in crowded conditions the incidence of the macropterous form increases. Photoperiod also has been shown to determine wing morphs. In the northern half of Japan, the brachyptery in *Nemobius vezoensis* is conditioned by certain locally prevailing photoperiods, and represents a case of geographically confined neoteny; macroptery can be induced by photoperiodic treatment in this species. There is evidence indicating that different photoperiods induce different activities of the NSC in the brain, eventually leading to the production of wing morphs in *Gryllodes sigillatus*. In *Metrioptera roeselii*, reversal of brachyptery to macroptery can occur at the cost of loss of fecundity.

Neoteny in brachypterous forms of Gryllidae is evident also in the reduction of some internal structures, such as the male accessory glands. In some Orthoptera (Stenopelmatidae, Prophalangopsidae), arrest of development of some external structures, such as the male and female genitalia, also occurs.

There is a tendency for the wing reduction to occur more frequently in the famale than in the male in Orthoptera.

Dermaptera.   Within a population of *Anisolabis maritime* different individuals tend to undergo different numbers of molts, with different degrees of differentiation of imaginal structures. Such variations are clearly induced by hormonal distrubance, as shown experimentally, and represent highly labile continuous polymorphism.

Various degrees of wing reduction occur in some groups of Dermaptera.

Psocoptera and Thysanoptera.   In many species of Psocoptera, various degrees of neoteny occur, affecting some external structures (including wings). When they are reared in isolation, wing reduction frequently occurs; when reared in crowded conditions, they become macropterous.   However, the effect of crowding disappears at certain higher temperatures.

In Thysanoptera, long days and crowded conditions induce macroptery, and short days inhibit the production of the macropterous form. Kinds of food also influence the development of wings.

Auchenorrhynchous Homoptera.   Development of wings, the shape of the aedeagus, and other structures in *Euscelis plebejus* are affected by photoperiodism, and hence they are seasonally dimorphic; seasonal dimorphism induced primarily by different photoperiods occurs also in other species of *Euscelis* in Europe. Wing dimorphism in *Nilaparvata lugens* is dependent on population density; they tend to be brachypterous in isolation and macropterous in crowded conditions. In the brachypterous form of some Japanese species of Araeopidae the hind legs are, as in solitary locusts, larger than in the macropterous form; other structural modifications also parallel those occurring in the solitary locusts.   In some fulgoroids the quantity of food and the degree of crowding play an important role in determining the wing diamorphism.

Sternorrhynchous Homoptera.   In the wingless morph (female) of aphids the degree of differentiation of various structures is much less than in the winged morph; hence they are neotenous.   JH and JH analogues have a juvenilizing effect when applied to older nymphs; the development of wings is arrested. Judging from the pattern of development of CA in *Brevicoryne brassicae*, the time available for the action of ED is more limited in the wingless than in the winged form; hence neoteny occurs.   In *Megoura viciae*, wing morphs are determined prenatally, and the hormonal balance (between JH and ED) may not be the underlying mechanism of wing morphs; an effect of the mother's JH on the endocrine activity of the developing young within the mother is unlikely.

Diverse external factors are known to affect polymorphism in aphids.

In many species of scale insects the female undergoes fewer molts than the male, and wings are never developed.   Degeneration of legs occurs in varying degrees in different groups of the female scale insects.   Compensatory development of structures that follows the degeneration of the structures comprises various types of pores, ostioles, and so on.   Thus the metamorphosis of the female scale insects can be called halmatometamorphosis; a heightened titer of JH during development is apparently involved with the halmatometamorphosis.

Allometabolism in aleurodids is characterized by the initial simplification of the body and suppression of wings, which are followed by the adaptive modification of the legs first into mobile then to sessile habits.

Heteroptera.   Neoteny in Heteroptera is evident externally only in wing reduction and in the loss of ocelli that may accompany the wing reduction. Usually the neoteny in this suborder is metathetely, and rerely prothetely.

Effects of photoperiod and temperature on wing morph determination are evident in the different wing morphs that occur in association with different seasons and geographical distribution of some species of *Gerris* in Europe. Some circumstantial evidence indicates that overproduction of JH is probably involved with the wing reduction in *Gerris* spp.   At least in most of these species wing polymorphism is not genetically determined.   *Geeris najas* is permanently apterous in highly localized areas in Finland, although the species may be brachypterous in the U.S.S.R.   Wing polymorphism in *Pyrrhocoris apterus* also is induced by environmental factors (photoperiod, temperature, and food). A hormonal theory of wing reduction in Heteroptera and the results of some crossing experiments on the wing morph determination are reviewed.

Coleoptera.   In some Coleoptera neoteny occurs only in the female.   However, wing reduction (neoteny) occurs in both sexes, and the reduction of the hind wings is far more common than the reduction of the elytra.   During post-embryonic development of *Lampyris noctiluca*, the CA and prothoracic glands become nonfunctional at certain critical periods, and the androgenic hormone induces the male differentiation. In the female of the same species, the activity of the endocrine glands is normal, and the winglessness is apparently genetically determined. In *Micromalthus debilis*, five kinds of adults occur, and three kinds of larviform females represent paedogenetic neoteny. Some crossing experiments (in Curculionidae and Carabidae) show clearly that wing dimorphism is genetically determined, the short-winged form being dominant. In Carabidae, wing reduction occurs regardless of geographical area and season. Therefore, wing reduction is probably not induced physiologically, but probably is determined genetically. The mechanism of increased contraction in developmental stages in some cave-dwelling beetles is discussed.

Hymenoptera. For the production of the queen bee (*Apis mellifera*), intake of the royal jelly and a higher titer of JH (than in the worker) are necessary conditions. Lowered titers of JH and the consequent, prolonged action of ED are apparently the necessary conditions for the production of the worker bee. In stingless bees, the queen emerges only after the optimal intake of food. The proper amount of food taken induces the proper hormonal milieu, enabling the expression of the genetic mechanism (double heterozygosity) for the development of the queen. Wing dimorphism in *Gelis* spp. may also be genetically determined. In bumble bees (*Bombus* spp.), the quantity of food is an important factor for differentiation into the queen and the worker. A high titer of JH toward the end of the larval development may be a necessary condition for the production of the queen in *B. hypnorum*, and in *B. terrestris*, caste determination is fixed in the first instar larva.

The worker ants are produced by prothetely, apparently through precocious action of ED. Diverse extrinsic factor are known to be responsible for caste determination in ants.

In several groups of parasitic wasps, wing dimorphism is determined physiologically, by the quantity and quality of food. In Hymenoptera reduction of eyes and ocelli and enlargement of legs accompany the wing reduction (material compensation).

Diptera. Neoteny is far more common in the female than in the male, and the degree of neoteny varies considerably in different Diptera. Paedogenetic female neoteny in Cecidomyidae is determined by diverse environmental factors, which induce the overproduction of JH, the excess JH suppresses imaginal differentiation and normal meiosis of the egg, hence producing diploid parthenogenetic neotenous (paedogenetic) females. Completely apterous female neoteny occurs in the marine chironomid genus *Clunio* and related genera. Metathetelous female neoteny occurs also in some other Nematocera. In all of these cases overproduction of JH is apparently involved, and apparently low temperature is often an important exogenous factor inducing the hormonal disturbance.

Wing reduction of various degrees occurs in more than 20 families of Diptera living in more or less unusual habitats.

Many dipterous larvae are known to leave wet environments to pupate; and there is evidence that ED is released for pupation only when the larvae are kept dry. Wing dimorphism, aptery,and speciation in sphaerocerid flies are discussed.

Lepidoptera. In Lepidoptera, neoteny affecting wings and some other structures usually occurs only in the female. In the female of psychid moths all degrees of arrest of development of structures (affecting mainly wings and legs) occur, presumably because of the overproduction of JH during larval development and more limited action of ED. In some species compensatory development of the ovipositor occurs. Brachyptery, combined with excessively developed ovaries, is common in the arctic moths, and this phenomenon may be attributed to the overproduction of JH (induced by cold temperatures). Brachypterous moths tend to occur at high mountain altitudes, in colder seasons of the temperate zones at lower altitudes, and in some unusual environments (e.g., aquatic environments). A few cases of seasonal dimorphism in wing development are discussed. The seasonal dimorphism in color pattern in *Polygonia c-aureum* has been shown to be environmentally and hormonally induced.

Strepsiptera, other holometabolous orders, and Collembola. The females of most species of Strepsiptera are endoparasitic, and remain larviform as adults; legs and internal organs degenerate during development. A conspicuous compensatory development in these female Strepsiptera is the brood canals through which larvae emerge. The male adult Strepsiptera, though winged and free-living, retain some juvenile features.

Examples of wing reduction in Trichoptera, Mecoptera, and Neuroptera are shown.

Neotenous features in Collembola (as compared with the plesiomorphic sister-group, Protura) include: the six-segmented abdomen in their embryos, nymphs and adults (due to suppression of anamorphosis), the extremely poorly developed male and female genitalia, the greatly reduced tracheal system, and the lack of tibiotarsal articulation in the front legs. Reduction of some other structures (due to neoteny) within the order also occurs.

Myriapoda. The endocrine action regulating molting is apparently similar to that in insects. Paired cerebral glands having a nervous connection with NSC in the frontal lobes of the brain apparently produce an inhibitory hormone (of molting), which also has the function of stimulating oogenesis, as in insects; the glands are therefore comparable to CA in insects. A structure called the *glandula ecdysalis* produces a molting hormone, which is presumably an ecdysone; the glands are therefore comparable to the prothoracic glands in insects. In some myriapods postembryonic development is characterized by anamorphosis, and neoteny in these myriapods occurs by suppression of the anamorphosis (in varying degrees), which results in a decrease in the number of abdominal and antennal segments. Arrest of development of some other external structures and of internal structures also occurs. An increased trend of neoteny is recognized in the phylogenetic sequence within the Diplopoda. The structural organization in pauropods represents an extreme neoteny, in which body size remains diminutive and arrest of development of many structures occurs.

Arachnida.   In araneida, an endocrine system comparable to the NSC–CC–CA system in insects is present, although the function of the system remains obscure. In Acari neurosecretory activity occurs in the brain. Application of ecdysone and ecdysterone to various arachnids has been found to be effective in promoting molt. A mild degree of neoteny in free-living arachnids can be recognized in deficiency in chaetotaxy.

In parasitic mites acceleration occurs by eliminating developmental stages (nymphal stages in the female). The acceleration results in the production of adults with structural organization comparable to that in free-living mites, or neotenous adults comparable to larvae, or adults not comparable to any stage of free-living mites (halmatometamorphosis). Constructive (adaptive) structures such as attaching organs occur in parasitic mites.

Crustacea.   Postmetamorphic moltings in Malacostraca are regulated by an inhibitory hormone produced from the X-organ–sinus gland, and ecdysone presumably produced from the Y-organ. Little is known, however, regarding the endocrine control mechanism of larval development and metamorphosis, in both the Malacostraca and Entomostraca. An androgenic hormone occurs in most groups of the Malacostraca. Experiments have shown that the eyestalk hormone (NSC) moderates the activity of the androgenic gland, and the activity of the NSC is influenced by environmental factors. In a majority of the Crustacea postembryonic development proceeds by anamorphosis. Therefore, neoteny is evident externally in the reduced numbers of abdominal and antennal segments, and in the reduction of some other structures. Cases of geographically and ecologically conditioned neoteny are discussed.

Phylogenetic neoteny is seen in the Conchostraca-Cladocera complex; the Cladocera apparently arose by suppression of anamorphosis (acceleration) in the *Cyclastheria hislopi* (Conchostraca)-like ancestor. The Copepoda might have been derived from the protozoea in the Decapoda by developmental arrest, and the Cirripedia might have been derived from a more generalized copepod by elimination of the copepodid stages (acceleration). Halmatometamorphosis in some parasitic copepods is discussed. The degree of halmatometamorphosis in endoparasitic copepods is far greater than that in the ectoparasitic copepods. The mode of abnormal metamorphosis resulting in neoteny and halmatometamorphosis differs in different orders of the Cirripedia. Usually, the male remains essentially as a cypris larva structurally, and is therefore neotenous. However, in the female and the hermaphrodite profound halmatometamorphosis occurs. During the endoparasitic stage of the female Rhizocephala halmatometamorphosis is quite conspicuous. Halmatometamorphosis also occurs in some parasitic isopods (e.g., Epicaridea). *Cymbasoma rugidum* emerges as a free-swimming adult copepod, after undergoing a profound halmatometamorphosis during the endoparasitic stage.

General discussion. The polymorphism resulting from abnormal metamorphosis can be explained in terms of the concept of genetic polymorphism only to a very limited degree. A great majority of the polymorphism resulting from abnormal metamorphosis represents polyphenism, in which the outstanding effect of an immediately identifiable gene or genes is absent. It is in abnormal metamorphosis that a great range of the reaction norm of the genotype can be seen, which constitutes the material basis of macroevolution. The pattern of the reaction norm of the genotype was shown to vary, depending on the different behaviors and physiological requirements intrinsic to animals, and on the different habitats in which they live (interstitial environment, social behavior, holometaboly, caenogenesis, free living and parasitic living, cave and case dwellings). Various patterns of the reaction norm of the genotype show that when reduction or loss of structures occurs during development, often compensatory (and adaptive) development of other structures occurs; such a compensatory development is called material compensation. Material compensation is the important morphogenetic mechanism through which an animal with an entirely new structural organization (hopeful monster) can emerge during evolution.

The fact that hormonal activity can readily be modified by extrinsic factors is of the utmost importance in considering the possibility of macroevolution in arthropods and perhaps in some other animal groups, since altered hormonal control of morphogenesis could result in the production of new structures in a group of individuals (population) in which they come to live. If enough numbers of these individuals survive the new environment, the acquired structural organization could be perpetuated.

Switch of regulatory genes triggered by alteration in the titer of an inhibitory hormone (e.g., JH) is apparently the underlying mechanism of material compensation, and hence macroevolution. Two persistent tendencies of compensatory development in insects are excessive development of ovaries and of hind legs (perhaps other legs as well) in the individuals and species with reduced wings or without wings. These phenomena can be attributed to the abnormally increased titer of JH during postembryonic development. Important consequences of material compensation are production of a new structure, which means the initiation of a new homology of the structure (when it is inherited in descendants), and interruption or termination of homology of a structure in a given phylogeny.

Many cases of origin of higher and lower taxa within the Arthropoda are discussed, in terms of the concept of material compensation and the reduction of structures alone. The neoteny in salamanders parallels that in insects.

ACKNOWLEDGMENTS

Portions of the manuscript on insects were read by the following taxonomists: G. W. Byers (Lawrence, Kansas), J. M. Campbell (Ottawa, Ontario), D. K. McE.

Kevan (St. Anne de Bellevue, Quebec), S. Masaki (Hirosaki, Japan), C. D. Michener (Lawrence, Kansas), F. Schmid (Ottawa, Ontario), and A. R. Soponis (Ottawa, Ontario). Portions of the manuscript were read by the following geneticists and endocrinologists: G. P. Fontana (Ottawa, Ontario), D. F. Hales (N.S.W. Australia), E. S. Merritt (Ottawa, Ontario), G. F. Steel (Waterloo, Ontario), and J. H. Willis (Urbana, Illinois). Section 4.4 was read by E. Munroe (Ottawa, Ontario). The portion on the noninsectan arthropods was read by E. R. Bousfield (Ottawa, Ontario), E. E. Lindquist (Ottawa, Ontario), and C. T. Shih. Finally, a large portion of the semifinal copy of the manuscript was read by J. Arnold (Ottawa, Ontario). To all these biologists I am sincerely grateful for the information and suggestions they provided.

REFERENCES

Adams, T. S. 1974. The role of juvenile hormone in housefly ovarian morphogenesis. *J. Insect Physiol.* **16**: 349-60.

Adams, T. S., P. I. Grugel, G. Ittycheriah, G. Olstad and J. M. Caldwell. 1975. Interaction of the ring gland, ovaries, and juvenile hormone with brain neurosecretory cells in *Musca domestica*. *J. Insect Physiol.* **21**: 1027-43.

Adiyodi, K. G. and R. G. Adiyodi. 1974. Control mechanism in cockroach reproduction. *J. Sci. Ind. Res.* **33**(7): 343-58.

Akai, H. 1971. Induction of prolonged larval period by the juvenile hormone in *Bombyx mori* L. (Lepidoptera Bombycidae). *Appl. Entomol. Zool.* **6**(3): 138-39.

Alexander, R. D. 1961. Aggressiveness, territoriality, and sexual behaviour in field crickets (Orthoptera: Gryllidae). *Behaviour* **17**: 130-223.

Alexander, R. D. 1968. Life cyle origins, speciation, and related phenomena in crickets. *Q. Rev. Biol.* **43**: 1-41.

Ammar, E. D. 1973. Factors related to the two wing forms in *Javesella pellucida* (Fab.) (Homoptera: Delphacidae). *Z. Ang. Entomol.* **74**: 211-16.

Anderson, N. M. 1973. Seasonal polymorphism and developmental changes in organs of flight and reproduction in bivoltine pondskaters (Hem. Gerridae). *Entomol. Scand.* **4**: 1-20.

Applebaum, S. W., B. Baccah and R. Leiserowitz. 1975. Effect of juvenile hormone and β-ecdysone on wing determination in the aphid, *Myzus persicae. J. Insect Physiol.* **21**: 1279-81.

Applegarth, A. G. 1939. The larva of *Apterobittacus apterus* MacLachlan (Mecoptera: Panorpidae). *Microentomology* **4**: 109-20.

Badonnel, A. 1948. L'effect de groupe chez *Psyllipsocus ranburi* Sél.-Long. *Bull. Soc. Zool. Fr.* **73**: 80-83.

Badonnel, A. 1949. Sur le déterminisme de l'effet de groupe chez *Psyllipsocus ramburi* (Psocoptère). *C. R. Acad. Sci Paris* D **228**: 517-19.

Badonnel, A. 1959. Développement des ailes de *Psyllipsocus ramburi* Sélys-Longchamps: Essai d'interprétation. *Bull. Soc. Zool. Fr.* **84**: 91-98.

Badonnel, A. 1970. Sur les glandes ecdysiales des Psocoptères. *Bull. Soc. Zool. Fr.* **95**: 861-68.

Baehr, J. C. 1973. Contrôle neuroendocrine du fonctionnement du corpus allatum chez *Rhodnius prolixus. J. Insect Physiol.* **19**: 1041-55.

Baker, T. C. and G. C. Eickwort. 1975. Development and bionomies of *Chrysomelobia labidomerae* (Acari Tarsonemina: Podapolipidae), a parasite of the milkweed leaf beetle (Coleoptera: Chrysomelidae). *Can. Entomol.* **106**: 627–38.

Barnes, H. and J. J. Gonar. 1958a. Neurosecretory cells in some cirripeds. *Nature (Lond.)* **181**: 194.

Barnes, H. and J. J. Gonar. 1958b. Neurosecretory cells in the cirripeds *Pollicipes polymerus* J. B. Sowerby. *J. Mar. Res.* **17**: 81–102.

Beier, M. 1968a. Mantodea (Fangheuschrecken), pp. 1–47. *In* J. G. Helmcke et al. (eds.) *Handbuch Zool.* IV Band 2 Hälfte 2 Teil, 12. W. de Gruyter, Berlin.

Beier, M. 1968b. Phasmida, pp. 1–56. *In* J. G. Helmcke et al. (eds.) *Handbuch Zool.* IV Band 2 Hälfte 2 Teil, 10. W. de Gruyter, Berlin.

Bernard, F. J. and C. E. Lane. 1962. Early settlement and metamorphosis of the barnacle *Balanus amphitrite niveus*. *J. Morphol.* **110**: 19–39.

Berthold, G. 1973. Der Einflusz der Corpora allata auf die Pigmentierung von *Carausius morosus* Br. *W. Roux'Arch.* **173**: 249–62.

Bezzi, M. 1916. Riduzione e scomparsa delle ali negli insetti ditteri. *Natura (Pavia)* **7**: 85–182.

Bocquet-Verdine, J. and J. Parent. 1972. Le parasitisme multiple du cerripède operculé *Balanus improvisus* Darwin par le rhizocéphale *Boschmaella balani* (J. Bocquet-Verdine). *Arch. Zool. Exp. Gén.* **112**: 239–44.

Bodenstein, D. 1953. Studies on the humoral mechanisms in growth and metamorphosis of the cockroach, *Periplaneta americana*. II. The function of the prothoracic gland and the corpus cardiacum. *J. Exp. Zool.* **123**: 413–34.

Bonaric, J.-C. 1976. Etude préliminaire de l'action de l'écdystèrone sur le cycle du mue de *Pisaura mirabilis* Cl. (Araneae-Pisauridae). *C. R. Acad. Sci. Paris* D **282**: 477–79.

Bonnemaison, L. E. 1951. Contribution à l'étude des facteurs provoquant l'apparition des formes ailées et sexuées chez les Aphidinae. *Ann. Epiphyt.* **2**: 1–380.

Brauns, A. 1939. Morphologische und physiologische Untersuchungen zum Halterenproblem unter besonderer Berücksichtigung brachypterer Arten. *Zool. Jahrb. Allg. Zool. Physiol.* **59**: 245–390.

Brian, M. V. 1959. The neurosecretory cells of the brain, the corpora cardiaca and the corpora allata during caste determination in ants, pp. 167–71. *In* I. Hrdy (ed.) *The Ontogany of Insects* (Acta Symp. Evol. Insect). Czechoslovak Acad. Sci., Prague.

Brian, M. V. 1965. Caste differentiation in social insects. *Symp. Zool. Soc. Lond.* **14**: 13–38.

Brian, M. V. 1974. Caste differentiation in *Myrmica rubra*. The role of hormones. J. Insect Physiol. **20**: 1351–65.

Brian, M. V. 1976. Endrocrine control over differentiation in a myrmicine ant, pp. 63–70. *In* M. Lüscher (ed.) *Phase and caste differentiation*. Pergamon Press, Oxford.

Brinck, P. 1949. Studies on Swedish stoneflies. *Opusc. Entomol. Suppl.* #11: 1–250.

Brinkhurst, R. O. 1959. Alarmy polymorphism in the Gerroidea (Hemiptera-Heteroptera). *J. Anim. Ecol.* **28**: 211–30.

Brinkhurst, R. O. 1963. Observations on wing-polymorphism in the Heteroptera. *Proc. R. Entomol. Soc. Lond.* **38**: 15–22.

Britten, R. J. and E. H. Davidson. 1969. Gene regulation for higher cells. *Science (Wash., D.C.)* **165**: 349–57.

Britten, R. J. and E. H. Davidson. 1969. Gene regulation for higher cells. *Science (Wash.,* and a speculation on the origins of evolutionary novelty. *Q. Rev. Biol.* **46**: 111–33.

Brossut, R. 1973. Evolution du système glandulaire exocrine céphalique des Blattaria et des Isoptera. *Int. J. Insect Morphol. Embryol.* **2**: 35–54.

Byers, G. W. 1969. Evolution of wing reduction in craneflies (Diptera: Tipulidae). *Evolution* **23**: 346–54.

Campbell, J. M. 1973. New species and records of neotropical Staphylinidae. I. Subtribe Timeparthenina. *Pap. Avul. Zool.* **27 (7):** 83-94.

Cassagnau, P. 1954. Sur un rudiment de furca chez les Neanurinae et sur quelques espèces de ce groupe. *Bull. Soc. Hist. Nat. Toulouse* **89:** 27-34.

Cassier, P. 1966. Variabilité des éffets du groupement (éffets immédiats et transmis) sur *Locusta migratoria migratorioides* (R. et F.). *Bull. Biol. Fr. Belg.* **100:** 135-70.

Cassier, P. and G. Delorme-Joulie. 1976. La différenciation imaginale du tégument chez le criquet pélérin, *Schistocerca gregaria* Forsk: III. Les différences phasaires et leur déterminisme. *Insectes Sociaux* **23:** 179-98.

Cassier, P. and M. A. Fain-Maurel. 1970. Contrôle plurifactoriel de l'évolution des glandes ventrales chez *Locusta migratoria* L. données expérimentales et infrastructurales. *J. Insect Physiol.* **16:** 301-18.

Charniaux-Cotton, H. 1972. Recherches récentes sur la différenciation sexuelle et l'activité génitale chez divers crustaces supérieurs, pp. 127-78. *In* E. Wolff (ed.) *Hormones et différenciation sexuelles chez les invertébres.* Gordon and Breach, Paris.

Cheung, P. J. and R. F. Nigrelli. 1973. The development of barnacles from ciprids in pre-heated seawaters, with or without farnesol. *Amer. Zool.* **13:** 467-68.

Chopard, L. 1949a. Ordres des Dictyoptères, pp. 355-407. *In* P. P. Grassé (ed.) *Traité de Zoologie.* Vol. IX. Masson et Cie, Paris.

Chopard, L. 1949b. Ordre des Chéleutoptères, pp. 594-616. *In* P. P. Grassé (ed.) *Traité de Zoologie.* Vol. IX. Masson et Cie, Paris.

Chopard, L. 1949c. Ordre des Dermaptéroides, pp. 745-70. *In* P. P. Grassé (ed.) *Traité de Zoologie.* Vol. IX: Masson et Cie, Paris.

Church, N. S. 1955. Hormones and the termination and reinduction of diapause in *Cephus cinctus* Nort. (Hymenoptera-Cephidae). *Can. J. Zool.* **33:** 339-69.

Clark, W. C. 1976. The environment and the genotype in polymorphism. *Zool. J. Linn. Soc.* **58:** 255-62.

Clausen, C. P. 1940. *Entomophagous Insects.* McGraw-Hill, New York.

Clouthier, C. and J. M. Perron. 1975. Polymorphism and sensitivity in the potato aphid. *Macrosiphum euphorbiae* (Homoptera: Aphididae). *Entomol. Exp. Appl.* **18:** 457-64.

Cornwell, P. B. 1968. *The Cockroach.* The Pentokil Library, London.

Costlow, J. D., Jr. 1963. The effect of eyestalk extirpation on metamorphosis of megalops of the blue crab, *Callinectes sapidus* Rathbun. *Gen. Comp. Endocrinol.* **3:** 120-23.

Costlow, J. D., Jr. 1966a. The effect of eyestalk extirpation on larval development of the crab, *Sesarma reticulatum* Say, pp. 209-24. *In* H. Barnes (ed.) *Some Contemporary Studies in Marine Science.* Allen and Unwin, London.

Costlow, J. D., Jr. 1966b. The effect of eyestalk extirpation on larval development of the mud crab, *Rhithropanopeus harrisii* (Gould). *Gen. Comp. Endocrinol.* **7:** 255-74.

Costlow, J. D., Jr. 1967. The effect of salinity and temperature on survival and metamorphosis of megalops of the blue crab, *Callinectes sapidus* Rathbun. *Helgolaender Wiss. Meeresuntersuch.* **15:** 84-97.

Costlow, J. D., Jr. 1969. Metamorphosis in crustaceans, pp. 3-41. *In* W. Etkin and L. I. Gilbert (eds.) *Metamorphosis.* Appleton-Century-Crofts, New York.

Costlow, J. D., Jr. and C. G. Bookhout. 1962. The larval development of *Sesarma reticulatum* Say reared in the laboratory. *Crustaceana* **4:** 281-94.

Cousin, G. 1938. La néotenie chez *Gryllus campestris* et ses hybrides. *Bull. Biol. Fr. Belg.* **72:** 79-117.

Cox, B. L. 1960. Hormonal involvement in the molting process in the soft tick, *Ornithodorus turicata* Dugès. *Anat. Rec.* **137:** 347.

Cretschmar, M. 1928. Das Verhalten der Chromosome bei der Spermatogenese von *Orgyia*-Bastarden. *Z. Zellforsch. Mikrosk. Anat.* **7:** 290-399.

Darlington, P. J. 1936. Variation and atrophy of flying wings of some carabid beetles. *Ann. Entomol. Soc. Amer.* **29**: 136–76.

Darlington, P. J. 1943. Carabidae of mountains and islands: Data on the evolution of island faunas, and on atrophy of wings. *Ecol. Monogr.* **13**: 39–61.

Darlington, P. J. 1971. The carabid beetles of New Guinea. Part IV. General considerations; analysis and history of fauna; taxonomic supplement. *Bull. Mus. Comp. Zool. Harv.* **142 (2)**: 129–337.

Darnhofer-Demar, B. 1973. Zur Populationsdynamik einer univoltinen Population von *Gerris lacustris* (Heteroptera, Gerridae). *Zool. Anz.* **190**: 180–204.

Darwin. C. 1854. *A Monograph on the Subclass Cirripedia, with Figures of All Species.* Ray Soc. Publ., London.

Davidson, E. H. and R. J. Britten. 1973. Organization, transcription, and regulation in the animal genome. *Q. Rev. Biol.* **48**: 565–612.

Davis, C. W. and J. D. Costlow. 1974. Evidence for a molt inhibiting hormone in the barnacle *Balanus improvisus* (Crustacea, Cirripedia). *J. Comp. Physiol.* **93**: 85–91.

Davydova, E. D. 1967. Effects of ablation and implantation of corpora allata and cardiaca on wing development and metamorphosis in *Lampyris noctiluca* L. (Coleoptera, Lampyridae) (in Russian). *Dokl. Acad. Nauk S.S.S.R.* **172**: 1218–21.

De Beer, G. 1958. *Embryos and Ancestors.* Oxford University Press, Oxford.

De Kort, C. A. D. 1969. Hormones and the structural and biochemical properties of the flight muscles in the Colorado beetle. *Meded. Landb.-hoogsch.* (Wageningen) **69(2)**: 1–63.

Dehn, M. Von. 1963. Hemmung der Flügelbildung durch Farnesol bei der schwarzen Bohenenlaus, *Dordalis fabae* Scop. *Narturwissenschaften* **50**: 578–79.

Deleurance, S. 1975. Le jeu des corps allates dans la contraction des cycles évolutifs larvaire chez les Coléoptères cavernicoles. *C. R. Acad. Sci. Paris* D **28**: 1389–92.

Deleurance, S. and M. P. Charpin. 1970. Sur la physiologie endocrine des Coléoptères cavernicoles de la famille des Catopidae (sous famille des Catopinae et des Bathyscinae). *C. R. Acad. Sci. Paris* D **270**: 2359–61.

Demange, J.-M. 1968. La réduction métamérique chez les Chilopodes et les Diplopodes Chilognathes (Myriapodes). *Bull. Mus. Nat. Hist. Nat. Paris* **(2)40**: 532–38.

Demange, J.-M. 1974. Reflexions sur le développement de quelques diplopodes. *Symp. Zool. Soc. Lond.* **32**: 273–87.

Denis, R. 1949. Ordre des Zoraptères, pp. 545–55. *In* P. P. Grassé (ed.) *Traité de Zoologie.* Vol. IX. Masson et Cie, Paris.

Denno, R. F. 1976. Ecological significance of wing polymorphism in Fulgoroidea which inhabit tidal salt marshes. *Ecol. Entomol.* **1**: 257–66.

Dent, J. N. 1968. Survey of amphibian metamorphosis, pp. 271–311. *In* W. Etkin and L. I. Gilbert (eds.) *Metamorphosis.* Appleton-Century-Crofts, London.

Descamps, M. 1975. Etude du contrôle endocrinien du cycle spermatogénétique chez *Lithobius forficatus* (Myriapoda Chilopoda). Rôle du complex cellules neurosécrétrices des lobes frontaux du protocérébron-glandes cérébrales. *Gen. Comp. Endocrinol.* **25**: 346–57.

Despax, R. 1949. Ordre des Plécoptères, pp. 557–86. *In* P. P. Grassé (ed.) *Traité de Zoologie.* Vol. IX. Masson et Cie, Paris.

De Wilde, J. 1969. Diapause and seasonal synchronization in the adult Colorado Beetle (*Leptinotarsa decemlineata* Say). *Symp. Exp. Biol.* **23**: 263–83.

De Wilde, J. 1976. Juvenile hormone and caste differentiation in the honey bee (*Apis mellifera* L.), pp. 5–20. *In* M. Lüscher (ed.) *Phase and Caste Differentiation in Insects.* Pergamon Press, Oxford.

De Wilde, J. and A. Loof. 1973. Reproduction-endocrine control, pp. 97–157. *In* M. Rockstein (ed.) *The Physiology of Insecta.* Academic Press, New York, London.

Dhanda, V. 1967. Changes in neurosecretory activity at different stages in adult *Hyalomma dromedarii* Koch, 1844. *Nature (Lond.)* **214**: 508–9.

Doane, W. W. 1973. Role of hormones in insect development, pp. 291–497. *In* S. J. Counce and C. H. Waddington (eds.) *Development Systems.* Vol. II. Academic Press, New York, London.

Doochin, H. D. 1951. The morphology of *Balanus improvisus* Darwin and *Balanus amphitrite niveus* Darwin during initial attachement and metamorphosis. *Bull. Marine Sci. Gulf Caribbean* **1**: 15–39.

Downes, J. A. 1964. Arctic insects and their environment. *Can. Entomol.* **96**: 279–307.

Downes, J. A. 1965. Adaptation of insects in the Arctic. *Annu. Rev. Entomol.* **10**: 257–74.

Efford, I. E. 1967. Neoteny in sand crabs of the genus *Emerita* (Anomura, Hippidae). *Crustaceana* **13**: 6–93.

Eggers, F. 1939. Phyletische Korrelation bei der Flügelreduktion von Lepidopteren. *Verh. VII Int. Kongr. Entomol.* 1934: 694–711.

Eickwort, G. C. 1976. A new species of *Chrysomelobia* (Acari: Trasonemina: Podapolipidae) from North America and the taxonomic position of the genus. *Can. Entomol.* **107**: 613–26.

Ekblom, T. 1927/8. Vererbungsbiologishce Studien über Hemiptera-Heteroptera. I. *Gerris asper* Fieb. *Hereditas* **10**: 333–59.

Ekblom, T. 1949. Neue Untersuchungen über die Flügeldimorphismus bei *Gerris asper* Fieb. *Notulae Entomol.* **29**: 49–64.

Ekblom, T. 1950. Über den Flügelpolymorphismus bei *Gerris odontogaster* Zett. *Notulae Entomol.* **30**: 41–49.

Elgmork, K. and A. L. Langeland. 1970. The naupliar instars in Cyclopoida. *Crustaceana* **18**: 277–82.

Elliott, H. J. 1975. Corpus allatum and ovarian growth in a polymorphic paedogenetic insect. *Nature (Lond.)* **257**: 390–91.

Emerson, A. E. 1961. Vestigial characters of termites and process of regressive evolution. *Evolution* **15**: 115–31.

Emerson, A. E. 1971. Tertiary fossil species of Rhinotermitidae (Isoptera), phylogeny of genera and reciprocal phylogeny of associated Flagellata (Protozoa) and the Staphylinidae (Coleoptera). *Bull. Amer. Mus. Nat. Hist.* **146**(3): 245–303.

Endo, K. 1970. Relation between ovarian development and activity of the corpora allata in seasonal forms of the butterfly, *Polygonia c-aureum* L. *Dev. Growth Differ.* **11**: 297–304.

Endo, K. 1972. Activation of the corpora allata in relation to ovarian development in the seasonal forms of the butterfly, *Polygonia c-aureum* L. *Dev. Growth Differ.* **14**: 263–74.

Endo, K. 1973. Hormonal regulation of mating in the butterfly, *Polygonia c-aureum* L. *Dev. Growth Differ.* **15**: 1–10.

Engelmann, F. 1970. *The Physiology of Insect Reproduction.* Pergamon Press, Oxford.

Eriksson, S. 1934. Studien über die Fangapparate der Branchiopoden, nebst einigen phylogenetischen Bemerkungen. *Zool. Bidr.* **15**: 23–287.

Ernst, A. 1971. Licht- und elektronenmikroskopische Untersuchungen zur Neurosekretion bei *Geophilus longicornis* Leach, unter besonderer Berücksichtigung der Neurohämalorgane. *Z. Wiss. Zool.* **182**: 62–130.

Fain, A. 1962. Les acariens psoriques parasites des chauves-souris. *Bull. Ann. Soc. R. Entomol. Belg.* **98**: 404–12.

Fain, A. 1965. Adaptation to parasitism in mites. *Acarologia* **11**: 429–49.

Fain, A. 1972. Développement postembryonnaire chez les Acariens de la sous-famille Speleognathinae. *Acarologia* **13**: 607-14.

Fain-Maurel, M.-A. and M. P. Cassier. 1969. Etude infrastructurale des corpora allata de *Locusta migratoria migratorioides* (R. et F.), phase solitaire au cours de la maturation sexuelle des cycles ovariens. *C. R. Acad. Sci. Paris* D **268**: 2721-23.

Ford, E. B. 1961. The theory of genetic polymorphism, *In Insect polymorphism. Symp. R. Entomol. Soc. Lond.* **1**: 11-19.

Ford, E. B. 1965. *The Genetic Polymorphism*. Faber and Faber, London.

Fukuda, S. and K. Endo. 1966. Hormonal control of the development of seasonal forms in the butterfly, *Polygonia c-aureum* L. *Proc. Jap. Acad.* **42**: 1082-87.

Fuzeau-Braesch, S. 1960. Etude biologique et biochemique de la pigmentation d'un insecte: *Gryllus bimaculatus* de Geer (Gryllidae-Orthoptère). *Bull. Biol. Fr. Belg.* **96(4)**: 526-627.

Fuzeau-Braesch, S. 1961. Variations dans le longueur des ailes en fonction de l'éffet de groupe chez quelques espèces de gryllides. *Bull. Soc. Zool. Fr.* **86**: 785-88.

Gabe, M. 1954. Emplacement et connexions des cellules neurosécrétrices chez quelques diplopodes. *C. R. Acad. Sci. Paris* D **239**: 828-30.

Gabe, M. 1955. Données histologiques sur la neurosécrétion chez les Arachnides. *Arch. Anat. Microsc. Morphol. Exp.* **44**: 351-83.

Gabe, M. 1966. *Neurosecretion*. Pergamon Press, Oxford.

Gabutt, P. D. and M. Vachon. 1963. The external morphology and life history of the pseudoscorpion *Chthronius ischnocheles* (Hermann). *Proc. Zool. Soc. Lond.* **140**: 75-98.

Geoffroy St. Hilaire, E. 1822. *Philosophie Anatomique*. Paris.

Ghouri, A. S. K. and J. E. McFarlane. 1958. Occurrence of macropterous form of *Gryllodes sigillatus* (Walker) (Orthoptera: Gryllidae) in laboratory culture. *Can. J. Zool.* **30**: 837-38.

Gillott, C. and C.-M. Yin. 1972. Morphology and histology of the endocrine glands of *Zootermopsis angusticollis* Hagen (Isoptera). *Can. J. Zool.* **50**: 1537-45.

Gillott, C. and T. Friedel, 1976. Development of accessory reproductive glands and its control by the corpora allatum in adult male *Melanoplus sanguinipes*. *J. Insect Physiol.* **22**: 365-72.

Girardie, A. and S. Granier. 1973. Système endocrine et physiologie de la diapause imaginale chez le criquet égyptien *Anacridium aegyptium*. *J. Insect Physiol.* **19**: 2341-58.

Goldschmidt, R. 1940. *The Material Basis of Evolution*. Yale University Press, London.

Gomez, E. D., D. J. Faulkner, W. A. Newman and C. Ireland. 1973. Juvenile hormone mimics: Effect on cirriped crustacean metamorphosis. *Science (Wash., D.C.)* **179**: 813-14.

Green, J. 1961. *A Biology of Crustacea*. Aspects of Zoology series. H. F. & G. Witherby Ltd., London.

Gurney, R. 1942. *Larvae of Decapod Crustacea*. Ray Soc., London.

Guthrie, D. M. 1959. Polymorphism in the surface water bugs (Hemiptera-Heteroptera: Gerroidea). *J. Anim. Ecol.* **28**: 141-52.

Guthrie, D. M. and A. R. Tindall. 1968. *Biology of the Cockroach*. G. Witherby Ltd., London.

Habibulla, M. 1961. Secretory structures associated with the neurosecretory system in the immature scorpion, *Hetrometrus swammerdami*. *Q. J. Microsc. Sci.* **102**: 475-79.

Habibulla, M. 1970. Neurosecretion in the scorpion, *Heterometrus swammerdami*. *J. Morphol.* **131**: 1-18.

Hackman, W. 1964. On reduction and loss of wings in Diptera. *Notulae Entomol.* **44**: 73-93.

Hackman, W. 1966. On wing reduction and loss of wings in Lepidoptera. *Notulae Entomol.* **66**: 1-16.

Hagedorn, H. H., J. D. O'Connor, M. S. Fuchs, B. Sage, D. A. Schlaeger and M. K. Bohm. 1975. The ovary as a source of ecdysone in an adult mosquito. *Proc. Nat. Acad. Sci. U.S.A.* **72**(8): 3255-59.

Hales, D. F. 1976. Juvenile hormone and aphid polymorphism, pp. 105-15. *In* M. Lüscher (ed.) *Phase and Caste Determination.* Pergamon Press, Oxford.

Hartmann, R. 1971. Der Einflusz endokriner Faktoren auf die männlichen Akzessorischen Drüsen und die Ovarien bei der Keulenheuschrecke *Gomphocerus rufus* L. (Orthoptera, Acrididae). *Z. Vgl. Physiol.* **74**: 190-216.

Hashimoto, H. 1957. Peculiar mode of emergence in the marine chironomid *Clunio* (Diptera, Chironomidae). *Sci. Rep. Tokyo Kyoiku Daigaku* **B 8**: 217-26.

Hashimoto, H. 1971. Females of *Clunio* (in Japanese). *Makunagi No.* **6**: 1-17. Osaka Pref. Univ., Sakai, Japan.

Henry, L. 1937. Biological notes on *Timema californica* Scudder. *Pan Pac. Entomol.* **13**: 137-41.

Hentschel, E. 1964. Zum neurosekretorischen System der Anostraca, Crustacea (*Artemia salina* Leach und *Chirocephalus grubei* Dybowsei). *Zool. Anz.* **170**: 187-90.

Herbaut, C. 1976. Etude expérimentale de la régulation endocrinienne de l'ovogénèse chez *Lithobius forficatus* L. (Myriapoda Chilopode). Rôle de la pars intercerebralis. *Gen. Comp. Endocrinol.* **27**: 34-42.

Herbaut, C. 1976. Etude expérimentale de la régulation endocrinienne de l'ovagénèse chez *Lithobius forficatus* L. (Myriapoda Chilopode). Rôle du complexe cellules néurosécrétrices protocérébrales-glandes cérébrales. *Gen. Comp. Endocrinol.* **28**: 264-76.

Herman, W. S. 1973. The endocrine basis of reproductive inactivity in monarch butterflies overwintering in Central California. *J. Insect Physiol.* **19**: 1883-87.

Herman, W. S. 1975. Endocrine regulation of posteclosion of the male and female reproductive glands in monarch butterflies. *Gen. Comp. Endocrinol.* **26**: 534-40.

Hidaka, T. and S. Aida. 1963. Day length as the main factor of seasonal form determination in *Polygonia c-aureum* (Lepidoptera, Nymphalidae) (in Japanese). *Zool. Mag.* **72**: 77-83.

Highnam, K. C. and P. T. Haskell. 1964. The endocrine system of isolated *Locusta* and *Schistocerca* in relation to oocyte growth, and the effects of living upon maturation. *J. Insect Physiol.* **10**: 849-64.

Hille Ris Lambers, D. 1966. Polymorphism in Aphididae. *Annu. Rev. Entomol.* **11**: 47-78.

Hinton, H. E. 1946. Concealed phase in the metamorphosis of insects. *Nature (Lond.)* **157**: 552-53.

Honék, A. 1974. Wing polymorphism in *Pyrrhocoris apterus* (L.) (Heteroptera: Pyrrhocoridae): Influence of photoperiod. *Vestn. Česk. Spol. Zool.* **38**: 241-42.

Honék, A. 1976. Factors influencing the wing-polymorphism in *Pyrrhocoris apterus* (Heteroptera, Pyrrhocoridae). *Zool. Jahrb. Syst.* **103**: 1-22.

Hrdy, I. 1976. The influence of juvenile hormone analogues on caste development in termites, p. 71. *In* M. Lüscher (ed.) *Phase and Caste Determination in Insects.* Pergamon Press, Oxford.

Hughes, T. E. 1962. Some aspects of molting in the mite, *Acarus siro. Gen. Comp. Endocrinol.* **2**: 609-10.

Hurley, D. E. 1959. Notes on the ecology and environmental adaptations of the terrestrial Amphipoda. *Pac. Sci.* **13**(2): 107-29.

Hynes, H. B. N. 1941. The taxonomy and ecology of the nymphs of British Plecoptera, with notes on the adults and eggs. *Trans. R. Entomol. Soc. Lond.* **91**: 459-557.

Ichikawa, A. and R. Yanagimachi. 1958. Studies on the sexual organization of the Rhizocephala. I. The nature of the "testis" of *Peltogasterella socialis* Krüger. *Annot. Zool. Jap.* **31**: 82-96.

Ichikawa, A. and R. Yanagimachi. 1960. Studies on the sexual organization of the Rhizocephala. II. The reproductive function of the larval (cypris) males of *Peltogaster* and *Sacculina*. *Annot. Zool. Jap.* **33**: 43–56.

Jackson, D. J. 1928. The inheritance of long and short wings in the weevil, *Sitona hispidula*, with a discussion of wing reduction among beetles. *Trans. R. Soc. Edinburgh* **55**: 665–735.

Järvinen, O. and K. Vepsäläinen, 1976. Wing dimorphism as an adaptive strategy in waterstriders (*Gerris*). *Hereditas* **84**: 61–68.

Jegla, T. C., J. D. Costlow and J. Alspaugh, 1972. Effect of ecdysones and some synthetic analogs on horseshoe crab larvae. *Gen. Comp. Endocrinol.* **19(1)**: 159–68.

Johno, S. 1963. Analysis of the density effect as a determining factor of the wing forms in the brown planthopper, *Nilaparvata lugens*. *Jap. J. Appl. Zool. Entomol.* **7**: 45–48.

Johnson, B. and P. R. Birks. 1960. Studies on wing polymorphism in aphids. I. The developmental process involved in the production of the different forms. *Entomol. Exp. Appl.* **3**: 327–39.

Johnson, R. A. and L. Hill. 1973a. The activity of the corpora allata in the fourth and fifth larval instars of the migratory locust. *J. Insect Physiol.* **14**: 1–20.

Johnson, R. A. and L. Hill. 1973b. Quantitative studies on the activity of the corpora allata in adult male *Locusta* and *Schistocera*. *J. Insect Physiol.* **19**: 2459–69.

Johnson, R. A. and L. Hill. 1975. Activity of the corpora allata in the adult female migratory locust. *J. Insect Physiol.* **21**: 1517–19.

Joly, L. and P. Joly. 1974. Comparaison de la phase grégaire et de la phase solitaire de *Locusta migratoria migratorioides* (Orthoptère) du point de vue de la teneur de leur haemolymphe en hormone juvénile. *C. R. Acad. Sci. Paris* D **279**: 1007–9.

Joly, P. 1972. Environmental regulation of endocrine activity. *Gen. Comp. Endocrinol. Suppl.* **3**: 459–65.

Joly, R. 1961. Déclenchement expérimental de la mue chez *Lithobius forficatus* L. (Myriapode Chilopode). *C. R. Acad. Sci. Paris* D **252**: 1673–75.

Joly, R. 1962. Les glandes cérébrales, organes inhibiteurs de la mue chez les Myriapodes Chilopodes. *C. R. Acad. Sci. Paris* D **254**: 1679–81.

Joly, R. 1964. Action de L'ecdysone sur le cycle due mue de *Lithobius forficatus* L. (Myriapode, Chilopode). *Soc. Biol. Lille, Séance du 14 Février 1964*, pp. 548–50.

Joly, R. 1966. Étude expérimentale du cycle de mue et de la régulation endocrine chez les Myriapodes Chilopodes. *Gen. Comp. Endocrinol.* **6**: 519–33.

Joly, R. and M. Descamps. 1968. Etude comparative du complexe endocrine céphalique chez les Myriapodes Chilopodes. *Gen. Comp. Endocrinol.* **10**: 364–75.

Juberthie-Jupeau, L. 1973a. Étude ultrastructurale des organes paraesophagous chez un diplopode Oniscomorphe *Loboglomeris pyrenaica* Latzel. *C. R. Acad. Sci. Paris* D **276**: 169–72.

Juberthie-Jupeau, L. 1973b. Étude ultrastructurale de l'organe neurohémal céphalique chez un Symphyle *Scutigerella silvatica* (Myriapode). *C. R. Acad. Sci. Paris* D **276**: 1577–80.

Juberthie, C. and L. Juberthie-Jupeau. 1974. Étude ultrastructurale de l'organe neuro-haemal cérébral de *Spelaeoglomeris doderoi* Silvestri, Myriapode diplopode cavernicole. *Symp. Zool. Soc. Lond.* **32**: 190–210.

Jucci, C. and A. Springhetti. 1952. Evolution of seminal vesicles in Isoptera. *Trans. 9th Int. Congr. Entomol.* **1**: 130–32.

Judy, K. J. and L. I. Gilbert. 1970. Effect of juvenile hormone and molting hormone on rectal pad development in *Hyalophora cecropia* (L.). *J. Morphol.* **131**: 301–14.

Kabata, Z. 1969. *Phrixocephalus cincinnatus* Wilson, 1908 (Copepoda: Lernaeoceridae): Morphology, metamorphosis, and host-parasite relationship. *J. Fish. Res. Board Canada* 26: 921-34.

Kabata, Z. 1973. Life cycle of *Salmincola californiensis* (Dana 1852) (Copepoda: Lernaeopodidae). *J. Fish. Res. Board Canada* 30(7): 881-903.

Kästner, A. 1967. Pp. 849-1242. *In: Lehrbuch der Speziellen Zoologie.* Band I, *Wirbellose (Crustacea)*, 2 Teil. G. Fischer Verlag, Stuttgart.

Kaiser, P. 1969. Welche Bedingungen steuern den Generationswechsel der Gallmücke *Heteropeza* (Diptera: Itonididae)? *Zool. Jahrb. Physiol.* 75: 17-40.

Kaiser, P. 1974. Über die "Entwicklungsumkehr" der Imagolarven von *Heteropeza pygmaea* Winnertz und deren hormonale Regelung. *Zool. Jahrb. Physiol.* 78: 199-218.

Kalmus, H. 1945. Correlations between flight and vision, and particularly between wing and ocelli in insects. *Proc. R. Entomol. Soc. Lond.* A 20(7-9): 84-96.

Kambysellis, M. P. and W. B. Heed. 1974. Juvenile hormone induces ovarian development in diapausing cave-dwelling *Drosophila* species. *J. Insect Physiol.* 20: 1779-86.

Kamm, J. A. 1972. Environmental influence on reproduction, diapause, and morph determination of *Anaphothrips obscurus* (Thysanoptera: Thripidae). *Env. Entomol.* 1: 16-19.

Karlinsky, A. and T. Srihari. 1973. La glande prothoracique au cours du développement postembryonnaire chez *Pieris brassicae* L. (Lépidoptère). *Bull. Soc. Zool. Fr.* 98: 243-62.

Kennedy, J. S. 1961. Insect polymorphism. *Symp. R. Entomol. Soc. Lond.* 1: 80-102.

Kerr, W. E. 1950. Evolution of the mechanism of caste determination in the genus *Melipona*. *Evolution* 4: 7-13.

Kerr, W. E. and R. A. Nielsen. 1966. Evidence that genetically determined *Melipona* queens can become workers. *Genetics* 79: 73-84.

Kerr, W. E., Y. Akahira and C. A. De Camargo. 1975. Genetic control of juvenile hormone production in *Melipona quadrifasciata* (Apidae). *Genetics* 81: 749-56.

Kevan, D. K. McE. 1973. The place of classical taxonomy in modern systematic entomology, with particular reference to Orthopteroid insects. *Can. Entomol.* 105: 1211-22.

Kiefer, F. 1928. Über Morphologie und Systematik der Süsswasser-Cyclopiden. *Zool. Jahrb. Syst.* 54: 495-556.

Killington, F. J. 1946. On *Psectra diptera* (Burm.) (Neur., Hemerobiidae), including an account of its life history. *Entomol. Month. Mag.* 62: 161-76.

King, M.-C. and A. C. Wilson. 1975. Evolution at two levels in humans and chimpanzees. *Science (Wash., D.C.)* 188: 107-16.

Kinzelbach, R. 1971. Morphologische Befunde an Fächerflüglern und ihre phylogenetischen Bedeutung. *Zoologica* 119: 1-256.

Kisimoto, R. 1956. Factors determining the wing form of adult, with special reference to the effect of crowding during the larval period of the brown planthopper, *Nilaparvata lugens* Stål. *Oyokontyu* 12: 105-11.

Kisimoto, R. 1957. Studies on the polymorphism in the planthoppers (Homoptera, Araeopidae) III. Difference in several morphological and physiological characters (in Japanese). *Appl. Zool. Entomol.* 1(3): 164-72.

Kistner, D. H. 1958. The evolution of Pygostenini (Coleoptera Staphylinidae). *Ann. Mus. R. Congo Belg. Tervuren ser. 8 Zool.* 68: 1-198.

Kitzmiller, J. B. 1950. The time interval between determination and differentiation of wings, ocelli and wing muscles in the aphid *Macrosiphum sanborni* (Gillette). *Amer. Nat.* 84: 23-50.

Klier, E. 1956. Zur Konstruktionsmorphologie des männlichen Geschlechts-apparates der Psocopteren. *Zool. Jahrb. Anat.* **75**: 207–86.

Klingstedt, H. 1939. Die Uvarovsche Theorie der Wanderheuschreckenphasen und ihre Bedeutung für die Zoologie. *Notulae Entomol.* **19**: 1–16.

Köppä, P. 1970. Studies on the thrips (Thysanoptera) species most commonly occurring on cereals in Finland. *Ann. Agric. Fenn.* **9**: 191–265.

Krishnakumaran, A. and H. A. Schneiderman. 1968. Chemical control of molting in arthropods. *Nature (Lond.)* **220**: 601–3.

Krishnakumaran, A. and H. A. Schneiderman. 1970. Control of molting in mandibulate and chelicerate arthropods by ecdysones. *Biol. Bull. (Woods Hole)* **139**: 520–38.

Kuhr, R. J. and H. S. Cleere. 1973. Toxic effects of synthetic juvenile hormones on several aphid species. *J. Econ. Entomol.* **66**: 1019–22.

Kulakovsky, E. E. 1976. Neurosecretory cells and their cycles in the brain of *Artemia salina* (in Russian). *Zool. Zh.* **55**: 354–61.

Kumar, R. 1966. Contribution to the biology, immature stages, and relative growth of some Australian bugs of superfamily Coreoidea (Hemiptera: Heteroptera). *Austr. J. Zool.* **14**: 895–991.

La Greca, M. 1946. Osservazioni sul brachitterismo degli ortotteri in rapporto al sesso. *Monit. Zool. Ital.* **55**: 138–41.

La Greca, M. 1954. Riduzione e scomparsa delle ali negli insetti pterigoti. *Arch. Zool. Ital.* **39**: 361–440.

Lanzrein, B. and M. Lüscher. 1970. Experimentelle Untersuchungen über die Degeneration der prothorakaldrüsen nach der Adulthäutung bei der Schabe *Nauphoeta cinerea*. *Rev. Suisse Zool.* **77**: 616–20.

Larsen, O. 1931. Beitrag zur Kenntnis der Pterygopolymorphismus bei den Wasserhemipteren. *Lunds. Univ. Årsk. N.F. Avd. 2 Bd.* **27**(8): 1–30.

Laufer, H. and H. Greenwood. 1969. The effects of juvenile hormone on larvae of the dipteran, *Chironomus thummi*. *Amer. Zool.* **9**: 603.

Lauterbach, G. 1954. Begattung und Larvengeburt bei der Strepsipteren. Zugleich ein Beitrag zur Anatomie der Stylops-Weibchen. *Z. Parasitenk.* **16**: 255–97.

Lebrun, D. 1967a. La détermination des castes du termite à cou jaune (*Calotermes flavicollis* Fabr.). *Bull. Biol. Fr. Belg.* **101**(3): 139–217.

Lebrun, D. 1967b. Nouvelles recherches sur le déterminisme de *Calotermes flavicollis*. *Ann. Soc. Entomol. Fr. (N.S.)* **3**: 667–71.

Lebrun, D. 1970. Intercastes expérimentaux de *Calotermes flavicollis* Fabr. *Insectes Sociaux* **17**: 159–76.

Lees, A. D. 1961. Clonal polymorphism in aphids. *Symp. R. Entomol. Soc. Lond.* **1**: 68–79.

Lees, A. D. 1967. The production of the apterous and alate forms in the aphid *Megoura viciae* Buckton, with special reference to the role of crowding. *J. Insect Physiol.* **13**: 289–318.

Lefeuvre, J. C. 1971. Hormone juvénile et polymorphisme alaire chez les Blattaria (Insecte, Dictyoptère). *Arch. Zool. Exp. Gén.* **112**: 653–66.

Legendre, R. 1958. Contribution à l'étude du système nerveux des Aranéides. *Ann. Biol.* **34**: 194–223.

Legrand, J.-J. and P. Juchault. 1972. Le contrôle humoral de la sexualité chez les crustaces isopodes gonochoriques, pp. 179–218. *In* E. Wolff (ed.) *Hormones et Différenciation Sexuelle Chez les Invertébrés*. Gordon and Breach, Paris.

Lengendorf, D. 1937. Beitrag zur Kenntnis und Systematik der bisher unbekannten palaearktishcen Lycoriidae (Sciaridae) bei den Flügellosigkeit oder Flügelrückbildung und Reduktion der Palpen gleichzeitig auftritt. *Drechenianan* **95B**: 30–36.

Lengendorf, E. 1949. Rückbildungen bei Trauermycken-Imagines (Dipt., Sciaridae). *Entomon* **4**: 115–17.

Levita, B. 1962. Contribution à l'étude du mécanisme d'un effet de groupe chez un insecte orthoptère: *Gryllus bimaculatus* de Geer. *Bull. Soc. Zool. Fr.* **87**: 197–221.

Light, S. F. 1942. The determination of castes of social insects. *Q. Rev. Biol.* **17**: 312–26.

Light, S. F. 1943. The determination of castes of social insects (continued). *Q. Rev. Biol.* **18**: 46–63.

Light, S. F. 1944. Experimental studies on ectohormonal control of the development of supplementary reproductives in the termite genus *Zootermopsis* (formerly *Termopsis*). *Univ. Calif. Publ. Zool.* **43**: 413–54.

Lindquist, E. D. 1965. An unusual new species of *Hoploseius* Berlese (Acarina-Blattisociidae) from Mexico. *Can. Entomol.* **97**: 1121–31.

Lindroth, C. 1946. Inheritance of wing dimorphism in *Pterostichus anthracinus* L. *Hereditas* **32**: 37–40.

Lindroth, C. H. 1949. Die Fennoskandischen Carabidae. III. *Göteborgs Vetenskaps-och Vitterhets-samhälles Handlingar 6. följd Ser. B.* **4**: 1–911.

Lindroth, C. 1957. *The Faunal Connection Between Europe and North America.* Almqvist and Wiksell/Gebers, Stockholm, and Wiley, New York.

Loher, W. 1960. The chemical acceleration of the maturation process and its hormonal control in the male desert locust. *Proc. R. Entomol. Soc. Lond. B.* **153**: 380–97.

Lüscher, M. 1958. Experimentelle Erzeugung von Soldaten bei der Termite *Kalotermes flavicollis* (Fabr.) *Naturwissenschaften* **45**: 69–70.

Lüscher, M. 1961. Social control of polymorphism in termites. *Symp. R. Entomol. Soc. Lond.* **1**: 57–67.

Lüscher, M. 1963. Functions of the corpora allata in the development of termites. *Proc. 16th Int. Congr. Zool.* **4**: 244–50.

Lüscher, M. 1969. Die Bedeutung des Juvenilhormons für die Differenzierung des Soldaten bei der Termite *Kalotermes flavicollis. Proc. 6th Congr. IUSSI, Bern 1969:* 165–70.

Lüscher, M. 1974. Kasten und Kastendifferenzierung bei niederen Termiten, pp. 694–739. *In* G. H. Schmidt (ed.) *Sozialpolymorphismus bei Insekten.* Wissenschaftliche Verlagsges, Stuttgart.

Lüscher, M. 1975. Pheromones and polymorphism in bees and termites. *Symp. Intern. Union Stud. Soc. Ins.* **1975**: 123–41.

Lüscher, M. 1976a. Introduction, pp. 1–4. *In* M. Lüscher (ed.) *Phase and Caste Determination in Insects.* Pergamon Press, Oxford.

Lüscher, M. 1976b. Evidence for an endocrine control of caste determination in higher termites, pp. 91–103. *In* M. Lüscher (ed.) *Phase and Caste Determination.* Pergamon Press, Oxford.

Lüscher, M. and A. Springhetti. 1960. Untersuchungen über die Bedeutung der Corpora allata für die Differenzierung der Kasten bei der Termite *Kalotermes flavicollis* (Fabr.). *J. Insect Physiol.* **5**: 190–212.

Malaquin, A. 1901. Le parasitisme évolutif des Monstrillides (Crustacés Copépodes). *Arch. Zool. Exp. Gén. ser. 3*, **9**: 81–232.

Mani, M. S. 1962. *High Altitude Entomology.* Methuen, London.

Margalef, R. 1949. Importancia de la neotenia en la evolucion de los crustaceos de agua dulce. *P. Inst. Biol. Appl.* **6**: 41–51.

Masaki, S. 1966. Photoperiodism and geographic variation in the nymphal growth of *Teleogryllus yezoensis* (Ohmachi and Matsuura) (Orthoptera: Gryllidae). *Kontyû* **34**: 277–88.

Masaki, S. and N. Oyama. 1963. Photoperiodic control of growth and wing development in *Nemobius yezoensis* Shiraki. *Kontyû* 31: 16–26.

Mathad, S. B. and J. E. McFarlane. 1968. Two effects of photoperiod on wing development in *Gryllodes sigillatus* (Walk.). *Can. J. Zool.* 46: 57–60.

Mathad, S. B. and J. E. McFarlane. 1970. Histological studies of the neuroendocrine system of *Gryllodes sigillatus* (Walk.) in relation to wing development. *Indian J. Exp. Biol.* 8: 179–81.

Matsuda, R. 1960. Morphology, evolution and a classification of the Gerridae (Hemiptera-Heteroptera). *Univ. Kansas Sci. Bull.* 41: 25–632.

Matsuda, R. 1961a. Studies of relative growth in Gerridae (1–3). *Ann. Entomol. Soc. Amer.* 54: 578–98.

Matsuda, R. 1961b. Studies of relative growth of Gerridae (IV). *J. Kansas Entomol. Soc.* 34: 5–17.

Matsuda, R. 1962a. Studies of relative growth of Gerridae (VI). *Univ. Kansas Sci. Bull.* 43(4): 113–39.

Matsuda, R. 1962b. Relative growth of appendages of some species of Heteroptera. *Kontyû* 30: 152–59.

Matsuda, R. 1963a. Evolution of relative growth in Arthropoda. *Z. Wiss. Zool.* 169(1–2): 64–81.

Matsuda, R. 1963b. A study of relative growth of leg and antennal segments in two species of *Orthotylus* (Miridae: Heteroptera). *Proc. R. Entomol. Soc. Lond.* A 38: 86–89.

Matsuda, R. 1970. Morphology and evolution of the insect thorax. *Mem. Entomol. Soc. Can. No. 76.*

Matsuda, R. 1976. *Morphology and Evolution of the Insect Abdomen.* Pergamon Press, Oxford.

Matsuda, R. and J. Rohlf. 1961. Studies of relative growth in Gerridae (5). Comparison of two populations. *Growth* 25: 211–17.

Mayr, E. 1963. *Animal Species and Evolution.* Harvard University Press, Cambridge, Mass.

Mayr, E. 1976. *Evolution and Diversity of Life.* Belknap Press, Cambridge, Mass., London.

McFarlane, J. E. 1966. Studies on group effect in crickets. III. Wing development of *Gryllodes sigillatus* (Walk). *Can. J. Zool.* 44: 1017–21.

McGregor, D. B. 1967. The neurosecretory cells of barnacles. *J. Exp. Mar. Biol. Ecol.* 1: 154–67.

McKenzie, H. L. 1967. *Mealybugs of California.* University of California Press, Berkeley.

Meewis, H. H. and J. Naisse. 1960. Phénomènes neurosécrétrices et glandes endocrines chez les Opilions. *C. R. Acad. Sci. Paris* D 245: 858–60.

Menon, M. 1962. Neurosecretory system of *Streptocephalus* sp. (Anostraca: Branchiopoda). III. *Intern. Symp. Neurosecretion, Bristol* 1962: 412–15.

Metwally, M. M. and F. Sehnal. 1973. Effects of juvenile hormone analogues on the metamorphosis of beetles *Trogoderma granarium* (Dermestidae), and *Caryedon gonagra* (Bruchidae). *Biol. Bull. (Woods Hole)* 144: 368–82.

Michener, C. D. 1974. *The Social Behaviour of the Bees.* Belknap Press, Cambridge, Mass.

Miller, M. 1969. Caste determination in the lower termites, pp. 283–310. *In* K. Krishna and F. M. Weesner (eds.) *Biology of Termites I.* Academic Press, New York, London.

Mitis, H. Von. 1937. Ökologie und Larvenentwicklung der mitteleuropäischen *Gerris* Arten (Heteroptera). *Zool. Jahrb. Syst.* 69: 337–72.

Mittler, T. E. 1973. Aphid polymorphism as affected by diet. Perspectives in aphid biology. *Bull. Entomol. Soc. New Zealand* 2: 65–75.

Mittler, T. E. and O. R. W. Sutherland. 1969. Dietary influences on aphid polymorphism. *Entomol. Exp. Appl.* 12: 705–13.

Mjeni, A. M. and P. E. Morrison. 1976. Juvenile hormone analogue and egg development in the blowfly, *Phormia regina* (Meig.). *Gen. Comp. Endocrinol.* **28:** 17-23.

Mochida, O. 1973. The characters of the two wing-forms of *Javesella pellucida* (F.) (Homoptera: Delphacidae), with special reference to reproduction. *Trans. R. Entomol. Soc. Lond.* **125(2):** 177-225.

Mockford, E. L. 1965. Polymorphism in the Psocoptera. A review. *Proc. North Central Branch, Entomol. Soc. Amer.* **20:** 82-86.

Mordue, W., K. C. Highnam, L. Hill and A. J. Luntz. 1969. Environmental effects upon endocrine-mediated processes in locusts. *Mem. Soc. Endocrinol. No.* **18:** 111-36.

Mouton, J. 1969. Données histochimiques sur la neurosécrétion de *Carausius morosus* (Phasmides-Orthoptères). *Ann. Endocrinol.* **30:** 839-51.

Mouton, J. 1970. Étude histophysiologique et expérimentale de la fonction neuroendocrine au niveau de la chaine nerveuse de *Carausius morosus* (Phasmides). *Ann. Biol.* **9(7-8):** 9-10.

Mouton, J. 1971. Influence de la neurosécretion sur la reproduction du phasme *Carausius morosus* (Cheleuthoptère). *Ann. Endocrinol.* **32:** 709-10.

Mukerji, M. K. 1972. A study of allometric growth in five species of Miridae (Miridae: Hemiptera). *Can. Entomol.* **104:** 1223-28.

Müller, H. J. 1954. Der Saisondimorphismus bei Zikaden der Gattung *Euscelis* Brüllé. *Beitr. Entomol.* **4:** 1-56.

Müller, H. J. 1957. Die Wirkung exogener Faktoren auf die zyklische Formbildung der Insekten, insbesondere der Gattung *Euscelis* (Hom. Auchenorrhyncha). *Zool. Jahrb. Syst. Ökol.* **85:** 317-430.

Müller, H. J. 1964. Über die Wirkung verschiedener Spektralbereiche bei der photoperiodischen Induktion der Saisonformen von *Euscelis plebejus* Fall. (Homoptera: Jassidae). *Zool. Jahrb. Physiol.* **70:** 411-26.

Nagel, R. H. 1934. Metathetely in larvae of the confused flour beetle (*Tribolium confusum* Duval). *Ann. Entomol. Soc. Amer.* **27:** 425-28.

Naisse, J. 1966a. Contrôle endocrinien de la différenciation sexuelle chez *Lampyris noctiluca* (Coléoptère Lampyridae) II. *Gen. Comp. Endocrinol.* **7:** 85-104.

Naisse, J. 1966b. Contrôle endocrinien de la différenciation sexuelle chez *Lampyris noctiluca* III. *Gen. Comp. Endocrinol.* **7:** 105-10.

Naisse, J. 1966c. Contrôle endocrinien de la différenciation sexuelle chez *Lampryis noctiluca* I. *Arch. Biol. Liège* **77:** 139-201.

Naisse, J. 1969. Rôle des neurohormones dans la différenciation sexuelle de *Lampyris noctiluca*. *J. Insect Physiol.* **15:** 877-92.

Neumann, D. and H. J. Dordel. 1972. *Clunio marinus* (Chironomidae). Schlüpfen und Kopulation, p. 11. *Encycl. Cinematographica E 1798/1971.* Göttingen.

Newman, W. A., V. A. Zullo and T. H. Withers. 1969. Cirripedia, pp. R206-R295. *In* R. C. Moore (ed.) *Treatise on Invertebrate Paleontology*, Part R, Arthropoda 4. University of Kansas Press, Lawrence, Kansas.

Nijhout, F. and C. M. Williams. 1974. Control of moulting and metamorphosis in the tobacco hornworm, *Manduca sexta* (L.): Cessation of juvenile hormone secretion as a trigger for pupation. *J. Exp. Biol.* **61:** 493-501.

Nikolei, E. 1961. Vergleichende Untersuchungen zur Fortpflanzung heterogener Gallmücken unter experimentellen Bedingungen. *Z. Morphol. Ökol. Tiere* **50:** 281-329.

Noble, G. K. 1931. *Biology of the Amphibia.* McGraw-Hill, New York.

Noirot, C. 1957. Neurosécretion et sexualité chez le termite à cou jaune *Calotermes flavicollis.* *C.R. Acad. Sci. Paris* D. **245:** 743-45.

Noirot, C. 1969. Formation of castes in the higher termites, pp. 311-50. *In* K. Krishna and F. M. Weesner (eds.) *Biology of Termites.* Vol. I. Academic Press, New York, London.

Novák, V. J. A.   1965.   The question of evolution of the juvenile hormone. *Acta Entomol. Bohemoslov.* **62(3):** 165–70.

Novák, V. J. A.   1975.   *Insect Hormones.* Chapman and Hill Press, London.

Oertel, R.   1924.   Studien über Rudimentation, ausgeführt an den Flügelrudimenten der Gattung *Carabus. Z. Morphol. Ökol. Tiere* **1:** 38–120.

Ohtaki, T.   1966.   On the delayed pupation of the fleshfly, *Sarcophaga peregrina* Robineau-Desvoidy. *Jap. J. Med. Sci. Biol.* **19(2):** 97–104.

Okada, T.   1963.   Cladogenetic differentiation of Drosophilidae in relation to material compensation. *Mushi* **37:** 70–100.

Ozeki, K.   1958.   Effect of corpus allatum hormone on development of male genital organs of the earwig, *Anisolabis maritima. Sci. Papr. Coll. Gen. Ed. Univ. Tokyo.* **8(1):** 69–75.

Ozeki, K.   1959.   Further studies of the effects of the corpus allatum hormone on the development of the genital organs in males of the earwig, *Anisolabis maritima. Sci. Papr. Coll. Gen. Ed. Univ. Tokyo* **9(1):** 127–34.

Palmen, E.   1944.   Die anemohydrochore Ausbreitung der Insekten als zoogeographische Faktor. *Ann. Zool. Soc. Zool.-Bot. Fenn. Vanamo* **10(1):** 1–262.

Pantel, J.   1917.   A proposito de un *Anisolabis* alado, contribucion al estudio de los organo valadores y de los esclerites toracicos en los Dermapteros; datos para la interpretacion del macropterismo excepcional. *Mem. Acad. Ci. Barcelona* **14(1):** 1–160.

Passano, L. M.   1960.   Molting and its control, pp. 473–536. *In* T. H. Waterman (ed.) *The Physiology of the Crustacea.* Vol. I. Academic Press, New York, London.

Paul, H.   1937.   Transplantation und Regeneration der Flügel zur Untersuchung ihrer Formbildung bei einem Schmetterling mit Geschlecht-dimorphismus. *Orgyia antiqua* L. *W. Roux' Arch. Entwicklmech.* **136:** 64–111.

Pener, M. P., A. Girardie and P. Joly.   1972.   Neurosecretory and corpus allatum controlled effects on mating behaviour and color change in adult *Locusta migratoria migratorioides* males. *Gen. Comp. Endocrinol.* **19:** 494–508.

Perez, Y., M. Verdier and M. P. Pener.   1971.   The effect of photoperiod on male sexual behaviour in a North Adriatic strain of the migratory locust. *Entomol. Exp. Appl.* **14:** 243–50.

Pesson, P.   1951a.   Ordre des Thysanoptera, pp. 1805–69. *In* P. P. Grassé (ed.) *Traité de Zoologie.* Vol. X. Masson et Cie, Paris.

Pesson, P.   1951b.   Ordre des Homoptères, pp. 1390–1656. *In* P. P. Grassé (ed.), *Traité de Zoologie.* Vol. X. Masson et Cie, Paris.

Pflugfelder, O.   1936.   Vergleichend-anatomische, experimetelle und embryologische Untersuchungen über das Nervensystem und die Sinnesorgane der Insekten. *Zoologica (Stuttg.)* **34(5,6):** 1–56, Taf. 1–8; 57–102, Taf. 14–25.

Pflugfelder, O.   1938.   Weitere experimentelle Untersuchungen über die Funktion der Corpora allata von *Dixippus morasus* Br. *Z. Wiss. Zool.* **151:** 149–91.

Plateaux, L.   1971.   Sur le polymorphisme social de la fourmi *Leptothorax nylanderi* (Förster). II. *Ann. Sci. Nat. Zool. Biol. Anim. 12 ser.* **13:** 1–90.

Poenicke, W.   1969.   Über die postlarvale Entwicklung von Flöhen (Insecta, Siphonaptera), unter besonderer Berücksichtigung der sogenannten Flügelanlagen. *Z. Morphol. Tiere* **65:** 143–86.

Pohlhammer, K.   1969a.   Die Reaktion der larvalen Ovarien der heterogenen Gallmücke *Heteropeza pygmaea* Winnertz 1846 (Diptera Cecidomyiidae) auf Farnesyl-methyläther. *Zool. Anz.* **182:** 272–75.

Pohlhammer, K.   1969b.   Hormonale Steuerung der Ovarentwicklung bei einer heterogenen Gallmücke. *Naturwissenschaften* **56:** 39.

Pohlhammer, K. and H. Pohlhammer.   1970.   Beeinflussung des Geschlechtes der durch

Pädogenese entstehenden Junglarven von heterogenen Gallmücken. *Naturwissenschaften* 57: 677.

Pohlhammer, K. and K. Treiblmayr. 1973. Juvenilhormon als gonadotroper Faktor bei den Larven der heterogenen Gallmücke. *Heteropeza pygmaea* Winnertz 1846. *Zool. Jahrb. Physiol.* 77: 145-52.

Poisson, R. 1924. Contribution à l'étude des Hémiptères aquatiques. *Bull. Biol.* 58: 49-305.

Ramme, W. 1931. Verlust und Herabsetzung der Fruchtbarkeit bei macropteren Individuen sonst brachypterer Orthopterenarten. *Biol. Zentralbl.* 51: 533-40.

Regenfuss, H. 1968. Untersuchungen zur Morphologie, Systematik und Ökologie der Podapolipidae (Acarina, Tarsonemini). *Z. Wiss. Zool.* 177: 183-282.

Regenfuss, H. 1973. Beinreduktion und Verlagerung des Kopulationsapparates in der Milbenfamilie Podapolipidae, ein Beispiel für Verhaltengesteuerte Evolution morphologischer Strukturen. *Z. Zool. Syst. Evolutionsforsch.* 11: 173-95.

Reid, J. A. 1941. The thorax of wingless and short winged Hymenoptera. *Trans. R. Entomol. Soc. Lond.* 91: 367-446.

Reinhard, E. G. 1942. The reproductive role of the complemental males of *Peltogaster. J. Morphol.* 70: 389-402.

Reinhard, E. G. and J. T. Evans. 1951. The spermiogenic nature of the "mantle bodies" in the aberrant rhizocephalid, *Mycetomorpha. J. Morphol.* 89: 59-69.

Remane, A. 1956. *Die Grundlagen des natürlichen Systems der vergleichenden Anatomie und der Phylogenetik.* Teest and Portig K-G., Leipzig.

Rembold, H. 1973. Biochemie der Kastenbildung bei der Honigbiene. *Naturw. Rundschau* 26: 95-120.

Rembold, H. 1976. The role of determination in caste formation in the honey bee, pp. 21-34. *In* M. Lüscher (ed.) *Phase and Caste Determination in Insects.* Pergamon Press, Oxford.

Rensch, B. 1959. *Evolution above the Species Level.* Methuen Press, London.

Richards, O. W. 1954. New wingless species of Diptera, Sphaeroceridae (Borboridae) from Ethiopia. *J. Linn. Soc. Lond. (Zool.)* 42: 387-91.

Richards, O. W. 1957. On apterous and brachypterous Sphaeroceridae from Mt. Elgon in the collection of the Musée royal du Congo belge, Tervuren. *Rev. Zool. Bot. Afr.* 55(3-4): 374-88.

Richards, O. W. 1963. The genus *Mesaptilotus* Richards (Diptera: Sphaeroceridae) with description of new species. *Trans. R. Entomol. Soc. Lond.* 115: 165-79.

Richards, O. W. and R. G. Davies. 1964. A. D. Imms' *A General Textbook of Entomology.* Methuen, London.

Riddiford, L. M. 1970. Effects of juvenile hormone on the programming of postembryonic development in eggs of the silkworm, *Hyalophora cecropia. Dev. Biol.* 22: 249-63.

Riddiford, L. M. 1971. Juvenile hormone and insect embryogenesis. *Mitt. Schweiz. Entomol. Ges.* 44: 177-86.

Riddiford, L. M. 1972. Juvenile hormone in relation to the larval-pupal transformation of the cecropia silkworm. *Biol. Bull. (Woods Hole)* 142: 310-25.

Riddiford, L. M. and J. W. Truman. 1972. Delayed effects of juvenile hormone on insect metamorphosis are caused by the corpus allatum. *Nature (Lond.)* 237: 458.

Robeau, R. M. and S. B. Vinson. 1976. Effects of juvenile hormone analogues on caste differentiation in the imported fire ant, *Solenopsis invicta. J. Georgia Entomol. Soc.* 11: 198-203.

Rohacek, J. 1975. Die Flügelpolymorphie bei der europäischen Sphaeroceridenarten und Taxonomie der *Limosina heteroneura*-Gruppe (Diptera). *Acta Entomol. Bohemoslov.* 72: 196-207.

Rohdendorf, E. B. and F. Sehnal. 1972. The induction of ovarian dysfunctions in *Thermobia domestica* by Cecropia juvenile hormones. *Experientia* 28: 1099-1101.

Röseler, P.-F. 1970. Unterschiede in der Kastendetermination zwischen den Hummelarten *Bombus hypnorum* and *Bombus terrestris*. *Z. Naturf.* 25b: 543-48.

Röseler, P.-F. and I. Röseler. 1974. Morphologische und physiologische Differenzierung der Kasten bei den Hummelarten *Bombus hypnorum* (L.) und *Bombus terrestris* (L.). *Zool. Jahrb. Physiol.* 78: 175-98.

Rosenberg, J. 1973. Topographie und Ultrastruktur der endokrinen Kopfdrüsen (Glandulae capitis) von (*Scutigera: Notostigmophora*). *Z. Morphol. Tiere* 79: 311-21.

Rosenberg, J. 1976. Die Ultrastruktur der Gabeschen Organe ("Cerebraldrüsen") von *Scutigera coleoptrata* (Chilopoda: Notostigmophora). *Zool. Beitr.* 22: 281-306.

Ross, E. S. 1970. Biosystematics of the Embioptera. *Annu. Rev. Entomol.* 15: 157-71.

Ross, H. H. 1944. The caddis flies, or Trichoptera of Illinois. *Illinois Nat. Hist. Surv.* 23(1): 1-326.

Rothschild, M. 1965. The rabbit flea and hormones. *Endeavour* 24(9): 162-66.

Rothschild, M. and B. Ford. 1964. Breeding of the rabbit flea (*Spilopsyllus cuniculi*) Dale. *Nature (Lond.)* 201: 103-4.

Rüschkamp, F. 1927. Der Flugapparat der Käfer. *Zoologica (Stuttg.)* 75: 1-88.

Saeki, H. 1966. The effect of the population density on the occurrence of the macropterous form in a cricket *Scapsipedus aspersus* Walker (Orthoptera: Gryllidae) (in Japanese). *Jap. J. Ecol.* 16: 104.

Sahli, F. 1974. Sur les organes neurohémaux et endocrines des Myriapodes Diplopodes. *Symp. Zool. Soc. Lond.* 32: 217-30.

Sahli, M. F. and J. Petit. 1972. Observations sur l'ultrastructure des organes de Gabe des Polydesmidae et Iulidae (Diplopoda). *C. R. Acad. Sci. Paris* D 275: 2017-20.

Saigusa, T. 1961. On some basic concepts of the evolution of psychid moths from the points of view of the comparative ethology and morphology (in Japanese). *Tyo to Ga* 12: 120-43.

Salt, G. 1941. The effects of hosts upon their insect parasites. *Biol. Rev.* 16: 239-64.

Salt, G. 1952. Trimorphism in the ichneumonid parasite *Gelis corruptor*. *Q. J. Microsc. Sci.* 93: 453-74.

Sannasi, A. and S. T. Subramonian. 1972. Hormonal rupture of larval diapause in the tick *Rhipicephalus sanguineus* (Lat.). *Experientia* 28: 666-67.

Scheffel, H. 1969. Untersuchungen über die hormonale Regulation von Häutung und Anamorphose von *Lithobius forficatus* (L.) (Myriapoda, Chilopoda). *Zool. Jahrb. Physiol.* 74: 436-505.

Scheffel, H., C. Wilke and W. Pollak, 1974. Die Wirkung von exogenen Ecdysteron auf Larven des Chilopoden *Lithobius forficatus*. *Acta Entomol. Bohemoslov.* 71: 233-38.

Schmid, F. 1951. Le groupe de *Enoicyla* (Trichopt. Limnoph.). *Tijdschr. Entomol.* 94: 207-26.

Schmieder, R. G. 1936. The polymorphic forms of *Melittobia chalybii* Ashmead and the determining factors involved in their production (Hymenoptera: Chalcidoidea, Eulophidae). *Biol. Bull. (Woods Hole)* 65: 338-54.

Schneiderman, H. A. and L. I. Gilbert. 1958. Substances with juvenile hormone activity in Crustacea and other invertebrates. *Biol. Bull. (Woods Hole)* 115: 530-35.

Scott, A. C. 1936. Haploidy and aberrant spermatogenesis in a coleopteran, *Micromalthus debilis* Le Conte. *J. Morphol.* 59: 485-509.

Scott, A. C. 1938. Paedogenesis in the Coleoptera. *Z. Morphol. Ecol. Tiere* 33: 633-53.

Scott, A. 1941. Reversal of sex production in *Micromalthus*. *Biol. Bull. (Woods Hole)* 81: 420-31.

Seevers, C. H.   1957.   A monograph on the termitophilous Staphylinidae (Coleoptera). *Fieldiana* **40**: 1–334.

Seifert, G.   1971.   Ein bisher unbekannte Neurohämalorgan von *Craspedosoma rawlinsii* Leach (Diplopoda, Nematophora). *Z. Morphol. Tiere* **70**: 128–40.

Seifert, G. and E. El-Hifnawi.   1971.   Histologische und elktronenmikroskopische Untersuchungen über die Cerebraldrüse von *Polyxenus lagurus* (L.) (Diplopoda, Penicillata). *Z. Zellforsch.* **118**: 410–27.

Seifert, G. and E. El-Hifnawi.   1972a.   Die Ultrastruktur des Neurohämalorgans am Nervus protocerebralis von *Polyxenus lagurus* (L.) (Diplopoda, Penicillata). *Z. Morphol. Tiere* **71**: 116–27.

Seifert, G. and E. El-Hifnawi.   1972b.   Eine bisher unbekannte endokrine Drüse von *Polyxenus lagurus* (L.) (Diplopoda, Penicillata). *Experientia* **28**: 74–76.

Seifert, G. and J. Rosenberg.   1974.   Elektronmikroskopische Untersuchungen der Häutungsdrüsen Lymphstränge von *Lithobius forficatus* L. (Chilopoda). *Z. Morphol. Tiere* **78**: 263–79.

Sehnal, F.   1968.   Influence of the corpora allatum on the development of internal organs in *Galleria mellonella* L. *J. Insect Physiol.* **14**: 73–85.

Sehnal, F. and H. A. Schneiderman.   1973.   Action of the corpora allata and of juvenilizing substances on the larval-pupal transformation of *Galleria mellonella* (Lepidoptera). *Acta Entomol. Bohemoslov.* **70**: 289–302.

Sellier, M. R.   1949.   Diapause larvaire et macroptèrisme chez *Gryllus campestris* (Ins. Orth.). *C. R. Acad. Sci. Paris* D **228**: 2055–56.

Sellier, M. R.   1954.   Recherches sur la morphogénèse et le polymorphisme alaire chez les Orthoptères Gryllides. *Ann. Sci. Nat. (Zool.) 11 ser.* **16**: 595–735.

Serban, M.   1960.   La néotenie et le probème de la taille chez les copépodes. *Crustaceana* **1**: 77–83.

Shapiro, A. M.   1976.   Seasonal polymorphism, pp. 259–332. *In* T. Dobzhansky, M. R. Hecht and W. C. Steere (eds.) *Evolutionary Biology*, Vol. IX. Appleton-Century-Crofts, New York.

Sharif, M.   1935.   On the presence of wing buds in the pupa of Aphaniptera. *Parasitology* **27**: 461–64.

Shull, A. F.   1937.   The production of intermediate winged aphids with special reference to the problem of embryonic determination. *Biol. Bull. (Woods Hole)* **72**: 259–87.

Silvestri, F.   1905.   Descrizione di un nuovo genere di Rhipiphoridae. *Redia* **3**: 315–24.

Slama, K.   1964a.   Die Einwirkung des Juvenilhormons auf die Epidermiszellen der Flügelanlagen bei Künstlich beschleunigter und verzögerter Metamophose von *Pyrrhocoris apterus* L. *Zool. Jahrb. Physiol.* **70**: 427–54.

Slama, K.   1964b.   Physiology of sawfly metamorphosis. 2. Hormonal activity during diapause and development. *Acta Soc. Entomol. Cech.* **61**: 210–19.

Slama, K.   1975.   Some old concepts and new findings on hormonal control of insect morphogensis. *J. Insect Physiol.* **21**: 921–55.

Slama, K. and C. M. Williams.   1965.   Juvenile hormone activity for the bug *Pyrrhocoris apterus*. *Proc. Nat. Acad. Sci.* **54**: 411–14.

Slama, K., M. Romanuk and F. Sorm.   1974.   *Insect Hormones and Bioanalogues*. Springer-Verlag, New York.

Smithers, C. N.   1972.   The classification and phylogeny of the Psocoptera. *Mem. Aust. Mus.* **14**: 1–349.

Snodgrass, R. E.   1952.   *A Textbook of Arthropod Anatomy*. Comstock Publ. Assoc., Ithaca, New York.

Snodgrass, R. E.   1956.   Crustacean metamorphosis. *Smithson. Misc. Collect.* **131**: 1–78.

Socha, R. and F. Sehnal. 1973. Inhibition of insect development by simultaneous action of prothoracic gland hormone and juvenile hormone. *J. Insect Physiol.* 19: 1449–53.

Southwood, T. R. E. 1961. A hormonal theory of the mechanism of wing polymorphism in Heteroptera. *Proc. R. Entomol. Soc. Lond.* (A) 36(4–6): 63–66.

Springhetti, A. 1964. Sulla strutura delle vesicole seminali delle Termiti. *Atti. Accad. Naz. Ital. Entomol. Rendiconti* 11: 212–19.

Springhetti, A. 1971. II controllo dei reali sulla differenziazione degli alati in *Kalotermes flavicollis* Fabr. (Isoptera). *Bull. Zool.* 38: 101–10.

Springhetti, A. 1974. The influence of farnesenic acid ethyl ester on the differentiation of *Kalotermes flavicollis* Fabr. (Isoptera) soldiers. *Experientia* 30: 1197–98.

Sroka, P. and L. I. Gilbert. 1974. The timing of juvenile hormone release for ovarian maturation in *Manduca sexta. J. Insect Physiol.* 20: 1173–80.

Staal, G. B. 1961. *In* Insect polymorphism. *Symp. R. Entomol. Soc. Lond.* 1: 88.

Staal, G. B. and J. de Wilde. 1962. Endocrine influences on the development of phase characters in *Locusta. Coll. Int. CNRS No.* 114: 89–102.

Steel, C. G. H. 1976a. Photoperiodic regulation of neurosecretory cells controlling polymorphism in the aphid *Megoura viciae. Gen. Comp. Endocrinol.* 29: 265–66.

Steel, C. G. H. 1976b. Neurosecretory control of polymorphism in aphids, pp. 117–30. *In* M. Lüscher (ed.) *Phase and Caste Determination in Insects.* Pergamon Press, Oxford.

Stein, W. 1973. Zur Vererbung des Flügeldimorphismus bei *Apion virens* Herbst (Col. Curculionidae). *Z. Ang. Entomol.* 74: 62–63.

Sterba, G. 1957. Die neurosekretorischen Zellgruppen einiger Cladoceren (*Daphnia pulex* und *magna, Streptocephalus vetulus*). *Zool. Jahrb. Anat.* 76: 303–10.

Swedmark, P. 1964. The interstitial fauna of marine sand. *Biol. Rev.* 39: 1–42.

Szöllosi, A. 1975. Imaginal differentiation of the spermiduct in acridids: Effects of juvenile hormone. *Acrida* 4: 205–16.

Tanaka, S. 1976. Wing polymorphism, egg production and adult longevity in *Pteronemobius taprobanensis* Walker (Orthoptera, Gryllidae). *Kontyû* 44: 327–33.

Thibaud, J.-M. 1976. Structure et régession de l'appareil oculaire chez les insectes Collemboles. *Rev. Ecol. Biol. Sol.* 13: 173–90.

Thomas, A. 1964. Etude expérimentale relative du contrôle endocrine de l'ovogenèse chez *Gryllus domesticus* L. *Bull. Soc. Zool. Fr.* 89: 835–54.

Thomas, E. S. and R. D. Alexander. 1962. Systematic and behavioral studies on the meadow grasshopper of the *Orchelimum concinnum* group (Orthoptera Tettigoniidae). *Occas. Papr. No.* 626. Univ. Mich. Mus. Zool., Ann Arbor, Michigan.

Tiegs, O. W. 1947. The development and affinities of the Pauropoda, based on a study of *Pauropus silvaticus. Q. J. Microsc. Sci.* 88: 165–336.

Tighe-Ford, D. J. and D. C. Vaile, 1972. The action of crustecdyson on the cirriped *Balanus balanoides* (L.). *J. Exp. Mar. Biol. Ecol.* 9: 19–28.

Troisi, S. J. and L. M. Riddiford. 1974. Juvenile hormone effects on metamorphosis and reproduction of fire ant *Solenopsis invicta. Env. Entomol.* 3(1): 112–16.

Truman, J. W. 1970. The eclosion hormone: Its release by the brain, and its action on the central nervous system of silkworm. *Amer. Zool.* 10: 511–12.

Truman, J. W., L. M. Riddiford, and L. Safranek. 1974. Temporal patterns of response to ecdysone and juvenile hormone in the epidermis of the tobacco hornworn. *Manduca sexta. Dev. Biol.* 39: 247–62.

Turner, B. D. 1974. Altitudial variations in the size of females of two species of Psocoptera on mango trees in Jamaica. *Entomol. Month. Mag.* 109: 247–49.

Ulrich, H. 1936. Experimentelle Untersuchungen über den Generationswechsel der heterogenen Cecidomyide *Oligarces* paradoxus. *Z. Induk. Abstamm. Vererbl.* 71: 1–60.

Ulrich, H. 1938. Untersuchungen über Morphologie und Physiologie des Generationswechsels von *Oligarces paradoxus. Verh. 7 Int. Kongr. Entomol. Berlin* 1938: 955–74.

Ulrich, H. 1940. Über den Generationswechsel und seine Bedingungen. *Naturwissenschaften* 78: 569–76.

Usinger, R. and R. Matsuda. 1959. Classification of the Aradidae (Hemiptera-Heteroptera). *Brit. Mus. Monogr.* Brit. Mus., London.

Utinomi, H. 1961. Studies on the Cirripedia Acrothoracica III. Development of the female and male of *Berndtia purpurea Utinomi. Publ. Seto. Mar. Biol. Lab.* 9(2): 167–98.

Uvarov, B. 1966. *Grasshoppers and Locusts.* Cambridge University Press, Cambridge.

Valentine, J. W. and C. A. Campbell. 1975. Genetic regulation and the fossil record. *Amer. Sci.* 673: 674–80.

Vandel, A. 1965. *Biospeleology.* Pergamon Press, Oxford.

Van den Bosch, A. and P. de Aguilar. 1972. Les caractéristiques tinctoriales des cellules neurosécrétrices chez *Daphnia pulex* (Crustacea: Cladocera). *Gen. Comp. Endocrinol.* 18: 140–45.

Veillet, A. 1961. Sur la métamorphose et le déterminisme du sexe du cirripède *Scapellum scapellum* Leach. *C. R. Acad. Sci. Paris* D 253: 3087–88.

Veillet, A. 1962. Sur la sexualité de *Sylon hipplites* M. Sars. Cirripède parasite de crévettes. *C. R. Acad. Sci. Paris* D. 254: 176–77.

Velthuis, H. H. W. and R. M. Velthuis-Kluppel. 1967. Caste differentiation in a stingless bee, *Melipona quadrifasciata* Lep., influenced by juvenile hormone application. *Koninkl. Neder. Acad. Wetnesch. Pro. ser. C.* 78(1): 81–94.

Vepsäläinen, K. 1971a. The role of gradually changing day length in determination of wing length, alary dimorphism and diapause in a *Gerris odontogaster* (Zett.) population (Gerridae, Heteroptera) in South Finland. *Ann. Acad. Sci. Fenn. Ser. A, IV. Biologica* 183: 1–25.

Vepsäläinen, K. 1971b. The role of photoperiodism and genetic switch in alary polymorphism in *Gerris* (Het. Gerridae) (a preliminary report). *Acta Entomol. Fenn.* 28: 101–2.

Vepsäläinen, K. 1974a. The life cycles and wing lengths of Finnish *Gerris* Fabr. species (Heteroptera, Gerridae). *Acta Zool. Fenn.* 141: 1–73.

Vepsäläinen, K. 1974b. The wing lengths, reproductive stages and habitats of Hungarian *Gerris* Fabr. species (Heteroptera, Gerridae). *Ann. Acad. Sci. Fenn. Ser. A, IV.* 212: 1–18.

Vepsäläinen, K. 1974c. Determination of wing length and diapause in water-striders (*Gerris.* Fabr., Heteroptera). *Hereditas* 77: 163–76.

Vinson, S. B. and R. Robeau. 1974. Insect growth regulator effect, on colonies of the imported fire ant. *J. Econ. Entomol.* 67: 584–87.

Vinson, S. B., R. Robeau, and L. Dzuik. 1974. Bioassay and activity of several insect growth regulators on the imported fire ant. *J. Econ. Entomol.* 67: 325–28.

Waddington, C. H. 1961. Genetic assimilation, pp. 257–94. *In* E. W. Caspari and J. M. Thoday (eds.) *Advances in Genetics* Vol X. Academic Press, New York, London.

Walley, L. J. 1964. Histolysis and phagocytosis in the metamorphosis of *Balanus balanoides. Nature (Lond.)* 201: 314–15.

Wanyonyi, K. 1974. The influence of the juvenile hormone analogue ZR 512 (Zoecon) on caste development in *Zootermopsis nevadensis* (Hagen) (Isoptera). *Insectes Sociaux* 1: 35–44.

Weber, H. 1931. Lebensweise und Umweltbeziehungen von *Trialeurodes vaporariorum* (Westwood) (Homoptera-Aleurodiina). *Z. Morphol. Ökol. Tiere* 23: 575–753.

Weber, H. 1934. Die postembryonale Entwicklung der Aleurodiden (Hemiptera: Homoptera). *Z. Morphol. Ökol. Tiere* 29: 268–305.

Wesenberg-Lund, C. 1943. *Biologie der Süsswasserinsekten.* Verlag-J. Springer, Berlin, Vienna.

Weygoldt, P. 1969. *The Biology of Pseudoscorpions.* Harvard University Press, Cambridge, Mass.

White, D. 1965. Changes in size of the corpus allatum in a polymorphic insects. *Nature (Lond.)* 208(5012): 807.

White, D. F. 1968. Postnatal treatment of the cabbage aphid with a synthetic juvenile hormone. *J. Insect Physiol.* 14: 901-12.

White, D. F. 1971. Corpus allatum activity associated with development of wingbuds in cabbage aphid embryos and larvae. *J. Insect Physiol.* 17: 761-73.

White, D. F. and J. B. Gregory. 1972. Juvenile hormone and wing development during the last larval stage in aphids. *J. Insect Physiol.* 18: 1599-1619.

Wigglesworth, V. B. 1936. The function of the corpus allatum in the growth and reproduction of *Rhodnius prolixus. Q. J. Microsc. Sci.* 79: 91-119.

Wigglesworth, V. B. 1952. Hormone balance and the control of metamorphosis in *Rhodnius. J. Exp. Biol.* 29: 620-31.

Wigglesworth, V. B. 1955. The breakdown of the thoracic gland in the adult insect, *Rhodnius prolixus. J. Exp. Biol.* 32: 485-91.

Wigglesworth, V. B. 1961. Insect polymorphism—A tentative synthesis. *Symp. R. Soc. Lond.* 1: 103-13.

Williams, C. 1959. The juvenile hormone I. Endocrine activity of the corpora allata of the adult *Cecropia* silkworm. *Biol. Bull. (Woods Hole)* 116: 323-38.

Williams, C. M. 1961. The juvenile hormone II. Its role in the endocrine control of molting, pupation, and adult development in the *Cecropia* silkworn. *Biol. Bull. (Woods Hole)* 121: 572-85.

Williams, C. M. 1969. Photoperiodism and the endocrine aspects of insect diapause. *Symp. Exp. Biol.* 23: 285-300.

Williams, C. M. and F. C. Kafatos. 1971. Theoretical aspects of the action of juvenile hormone. *Mitt. Schweiz. Entomol. Ges.* 44(1-2): 151-62.

Willig, A. A. 1973. Die Rolle der Ecdysone in Häutungszyklus der Crustaceen. *Fortsch. Zool.* 11: 55-74.

Willis, J. H. 1974. Morphogenetic action of insect hormones. *Annu. Rev. Entomol.* 19: 97-115.

Wilson, A. C. 1975. Evolutionary importance of gene regulation. *Stadler Symp.* 7: 117-33.

Wilson, E. O. 1971. *The Insect Societies.* Harvard University Press, Cambridge, Mass.

Wirtz, P. 1973. Differentiation in the honeybee larva. *Meded. Landbouwhoogsch. Wageningen* 73(5): 1-155.

Wood, E. A. and K. J. Starks. 1975. Incidence of paedogenesis in the greenbug. *Env. Entomol.* 4: 1001-2.

Wright, J. E. 1969. Hormonal termination of larval diapause in *Dermacentor albipictus. Science (Wash., D.C.)* 163: 390-91.

Wyatt, I. J. 1961. Pupal paedogenesis in the Cecidomyidae (Diptera). I. *Proc. R. Entomol. Soc. Lond.* (A) 36(10-12): 133-43.

Wyatt, I. J. 1963. Pupal paedogenesis in the Cecidomyidae (Diptera). II. *Proc. R. Entomol. Soc. Lond.* (A) 38(7-9): 136-44.

Wyatt, I. J. 1964. Immature stages of Lestremiinae (Diptera: Cecidomyidae) infesting cultivated mushrooms. *Trans. R. Entomol. Soc. Lond.* 116: 15-37.

Wyatt, I. J. 1967. Pupal paedogenesis in the Cecidomyidae (Diptera). 3. A reclassification of the Heteropezini. *Trans. R. Entomol. Soc. Lond.* 119(3): 71-98.

Wygodzinsky, P. W. 1966. A monograph of the Emesinae (Reduviidae, Hemiptera). *Bull. Amer. Mus. Nat. Hist.* 133: 1-614.

Yanagimachi, R. 1961. Studies on the sexual organization of the Rhizocephala III. The mode of sex-determination in *Peltogasterella. Biol. Bull. (Woods Hole)* 120: 272-83.

Yanagimachi, R. and N. Fujimaki. 1967. Studies on the sexual organization of the Rhizocephala IV. *Annot. Zool. Jap.* 4(2): 98-104.

SECTION III
Sense Organs

5

Evolution of Antennae, Their Sensilla and the Mechanism of Scent Detection in Arthropoda[1]

P. S. CALLAHAN

5.1. INTRODUCTION

In his philosophical and evolutionary treatise *The Phenomenon of Man* (1959), Teilhard has summarized the contents of that evolutionary Pandora's box that scholars must develop these insights if they are to make headway in the study of evolution and mankind. Teilhard maintains that gradual acquisition of these insights "covers and punctuates the whole history of the struggle of the mind."

This essay is a short and tremulous attempt to elaborate on Teilhard's seventh "sense." I shall attempt to relate that sense to what I have discovered concerning the insect communication system. The last sense outlined in his treatise is, in my opinion, the most important of his insights. Since Teilhard placed it in an emphatic position at the end of his list, he may well have shared my opinion. His seventh sense, as defined in *The Phenomenon of Man*, is "a sense of the organic, discovering physical links and structural unity under the superficial juxtaposition of successions and collectivities." In this single elegant sentence Teilhard has summarized the contents of that evolutionary Pandora's box labeled "heredity vs. environment." He tells us in his "Teilhardian" style (he

[1]DEDICATION: This chapter is dedicated to Dr. Ayodhya P. Gupta, our editor, for encouraging the publication of my ideas, and to Dr. Eleanor H. Slifer, whose work pointed the way for all morphologists interested in the insect antenna.

also liked to coin words) that mankind must develop this seventh sense in order to evolve toward perfection. He suggests that we may gain this important sense by discovering the physical links, presumably from the environment, that relate to, or control, structural unity (biological form). He visualizes the physical phenomenon as being hidden under a superficial mixture of successions (generations) and collectivities (populations). He labels these integrated phenomena of physics and morphology the "organic," and implies that we cannot understand evolution without developing our seventh "organic sense." No truer statement of the problems of modern evolutionary thought could be made.

The age-old and sometimes vicious conflict (Koestler, 1973) between neo-darwinism and neo-lamarckism has its roots in educational overspecialization, which has reinforced our ignorance of Teilhard's organic sense. My reader may protest, however, that with the discovery of Muller's high energy mutations we already understand the physical link between the environment and structure. It is only necessary that we continue and concentrate our research on these high energy mutagenic radiations.

I believe that our single-minded infatuation with the effects of high energy radiations on organic life has led us down a walled arroyo to the dried up hard pan of a dead sea, while the real sea of life overflows on the plateau above. That the sea of life is infrared (IR) should be obvious, as demonstrated by our concern with the ozone layer, which keeps away the sea of high energy radiation. Studying the relationship between environment and evolution and ignoring the IR sea within which we dwell is comparable to an ichthyologist's ignoring the water wherein the fish swim.

One should not, of course, ignore the effects of high energy radiation on biological systems, any more than the low energy effects. It is merely my contention that a thorough review of the literature will demonstrate that low energy phenomena are ignored.

One can plead, of course, that the low energy IR portion of the spectrum is extremely difficult to handle experimentally. This contention only emphasizes the mathematical and theoretical impoverishment of modern biological research. A little theory is good for the soul and should not be left to the physicists alone. Any biologist who postulated such an elegant contradiction as black holes, as have the physicists, would probably be intellectually lynched by his biological peers. It is the intrepidness of the physicists that gives me the courage to state that perhaps, as Albert Szent-Gyorgyi (1960) suggests, organic life does exist in a metastable state.

It is my contention, I dare not say theory, that this organic metastable state is controlled and programmed by energy in the IR portion of the spectrum. I shall even venture further into the mysterious realm of organic, solid state physics and proclaim my belief that naturally occurring low energy IR masers (micrometer amplification of stimulated emission radiation) are the controlling environmental messengers of an elegant molecule-to-cell organic wavelength control system. If this thesis is true, then the implications to evolutionary theory are worthy of thought. With this in mind we may begin with a common definition of the organic metastable state.

5.2. METASTABLE STATE

The biologist is not, in most cases, familiar with the physical subtleties of metastable systems. The following paragraphs are ordered to give the reader a basic insight into the metastable state. This is a necessary prerequisite to understanding the relationship between the nonlinear maserlike wavelengths, and insect behavior and evolution.

Webster's Third Unabridged Dictionary defines metastable state as "a state of precarious stability; specifically, such a state of an atom, which, though excited, cannot emit radiation without a further supply of energy." As everyone knows, an atom is thought to consist of a central nucleus surrounded by electrons in different levels of orbit—almost like our sun and its orbiting planets. Each different layer of electrons is located at a distinct energy level. In the normal state of equilibrium, all of the electrons stay put in their "natural" levels of orbit, as do our planets. Figure 5.1 (top) shows a hypothetical three-level energy state for a large group (called a population) of atoms. They are all in their normal equilibrium state. When energies from another source, such as an electrical discharge or light, impinge upon this population of atoms, the electrons orbiting each nucleus are raised from one energy level to a higher energy level. This phenomenon is called *absorption* by the atom or molecule. We might think of absorption as being the effect of an outside energy source that drives a significant population of little people from the first floor, called ground state, up to the third-floor level. As more and more of the little electron people reach the third floor, it finally gives way, and they drop through the third floor to the second floor. In doing so, they give off heat and electrical energy, which is lost to the system. In some populations of atoms or molecules (combined atoms), the little electron people fall through the second floor, so that they reach the ground state without any further detectable energies being given off. However, in other atomic or molecular combinations, particularly in the gaseous state [remember, a scent (pheromone) is a complex gas], the second floor (energy level) is of such a character—like a strong oak floor—that many, many electron people pile up on the floor, and are held there for a much longer period of time before they break through the strong oak flooring and fall back to the ground level state. If the second-floor energy level holds enough of the electron people for a long time, they pile up until the huge population that finally falls to the ground floor level gives off a tremendous burst of electromagnetic energy. While falling back to ground level, this huge temporarily

Hypothetical 3-Level State

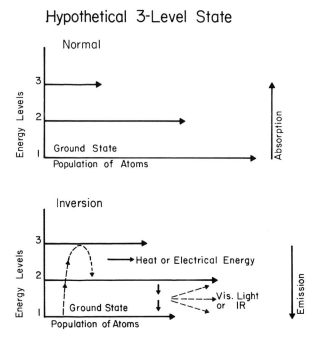

Fig. 5.1. Energy levels of a population of atoms in a three-level state. At equilibrium (top drawing) the largest population of atoms rests at ground state with all electrons at their own energy levels. A stimulated population of atoms may "store" electrons at one energy level, after rising to a higher level (absorption), and then drop to ground state, emitting light in various regions of the spectrum (fluorescence, bottom drawing).

"stored" population emits either visible, ultraviolet (UV), or IR light frequencies. These frequencies are in different portions of the electromagnetic spectrum and depend on the makeup of the gas atoms or molecules involved.

Physicists call the second-floor storage level the *metastable state.* They have shown that to get lasing action, you must first have *fluorescent* frequencies given off as the electrons fall from the second floor metastable state to the ground state. If the emitted fluorescent light is red, green, or other visible frequencies, we can detect these visible light energies with our eyes. There must be, however, as Townes (the inventor of the laser) points out, many fluorescent IR energies that we cannot detect with our eyes; we need complex IR sensors to detect them. This is especially true of the far IR, where radiations are very weak in terms of electron-volt energy. The mirrors in a laser are spaced so that they put all of these fluorescing electrons in phase; they come out marching in synchronization, and thus are amplified (Fig. 5.2).

It becomes obvious to me from reading physics that it takes a lot of energy going into a system to get visible fluorescence, which is fairly high-energy

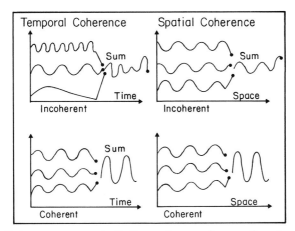

Fig. 5.2. Radiation may possess both time coherence (temporal) and space coherence (spatial). Radiation that is incoherent is out of phase, so waves do not sum in time and space (top). Radiation that is in phase does sum, so there is an amplification of the wave (bottom).

radiation, out. However, it does not take nearly as much energy going into a system to get low electron-volt IR fluorescent energy out. Of course, we could not see such nebulous, glowing IR fluorescing radiations because our eye is not tuned to them.

Physicists call the energy going into a system, e.g., the electrical discharge of a helium-neon laser, the "pumping" energy. In the case of the ruby solid state laser, which puts out a red coherent beam, the energy going in comes from a high-intensity visible flash lamp, which causes the chrome impurities in the ruby to fluoresce.

My studies of the mechanisms of visible and microwave lasers convinced me that very low quanta of energy from the sun, or even from the stars, moon, or blackbody radiation of night, "pump" scent molecules so that they flouresce in the far-IR region of the spectrum. In other words, I believed that the gas (scent) does not need to be contained in a tube and pumped by a high electrical discharge, but that scent is, in essence, a free-floating fluorescence pumped by the *natural* day, night, sun, and star light, or the blackbody energies, of our spaceship Earth. Scent, in my mind, is a fleeting-floating world of vapors that luminesce in many, many different IR colors, which can be collected and amplified by a scent organ such as the insect antenna. The natural sensilla are tuned as a resonating system to these IR frequencies. Accordingly, I coined the term "maserlike frequencies" (micrometer amplification of stimulated emission radiation) for the scent IR colors that we could not detect until recently.

Until four years ago, the presumed existence of my maserlike wavelengths was based entirely on the mathematics of antenna design, as applied to the shape of the sensilla sensors. Experimental verification of my predictions is contained in several publications (Callahan, 1975a,b, 1977a,b,c).   A characteristic of nonlinear (laser or maser) radiation from organic dye systems is that a change in concentration of the lased dye produces a change in wavelength (Webb et al., 1974). The higher the concentration of dye is in the solvent

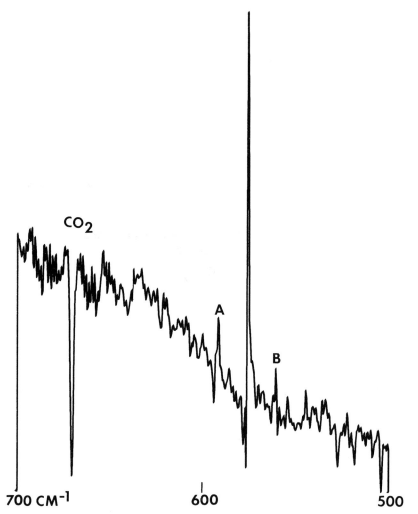

Fig. 5.3.   Solid state organic (*T. ni* pheromone-beeswax) far IR maser line at 17.39 μm (26°C) after one-half hour in source beam and modulated at 50 to 60 Hz with MSC 300 dark blue filter over strobe light.   Both main line and side bands A and B have a narrow band width (less than 1.5 cm^{-1}), a characteristic of all phase-modulated laser systems. From Callahan (1977a).

carrier, the longer the wavelength. This is called "concentration tuning" by laser physicists. There is also a temperature wavelength shift. Blackbody physics predicts that the higher the temperature the shorter the wavelength. Such is not the case for nonlinear laser optics. The thermal shift of a solid state GaAs diode laser is from 9100 Å at room temperature to 8400 Å at the cold temperature of liquid nitrogen, the reverse of blackbody radiation. As Stehling (1966) points out, this is an interesting feature of such nonlinear semiconductor devices.

I "doped" beeswax with 100 μg of cabbage looper pheromone [(Z)-7-dodecen-1-ol acetate], and by modulating it at the same vibration frequency (50 Hz) as the cabbage looper moth antenna produced nonlinear maser radiation at 17.39 μm from the pheromone-doped wax (Fig. 5.3, Callahan 1977a). Decreasing the temperature shifted the frequency to shorter wavelengths (16.81 μm). This phenomenon of wavelength shift, plus the side bands evident in the emission (Fig. 5.3), are irrefutable proof that low energy IR organic maser wavelengths are real and no longer a theoretical probability. If the reader keeps in mind that those wavelengths were predicted and produced from a known pheromone based on an analysis of the form of the antenna sensilla sensor, then it becomes understandable that wavelength and form are inexorably related. As is the case in efficient man-made communication systems, a wavelength-antenna-waveguide system is evident, the only difference being, of course, that man-made systems operate in the long-wave inorganic radio region of the spectrum, whereas natural systems work in the organic IR region. If one cannot accept the existence of such an IR organic metastable system, then it is best that he omit this chapter, as it will have no meaning to him. For those willing to believe that my discovery of the first solid state organic metastable maser may have meaning, the application of that meaning to evolutionary theory should begin with the phylogenetic tree of life.

5.3. PHYLUM ARTICULATA

Evolutionists agree that the group Arthropoda is a valid category. It is usual in the literature to classify Arthropoda as the highest category (phylum). However, I prefer, as do Sharov (1966), Tiegs (1947), and Snodgrass (1952), to consider Arthropoda as a subphylum of a higher category called Articulata. Accordingly, in this chapter I shall follow the phylogenetic lineage suggested by Sharov (1966) (Fig. 5.4).

Most zoologists are unanimous that the annelids are the original ancestors of the arthropods (see also Weygoldt, Chapter 3). Articulata includes all creatures that are segmented and wormlike, and that have secondary cavities called coelomic sacs, either as an embryo or throughout life. The series of paired internal pouches develop along the entire body length of the embryo and larvae of

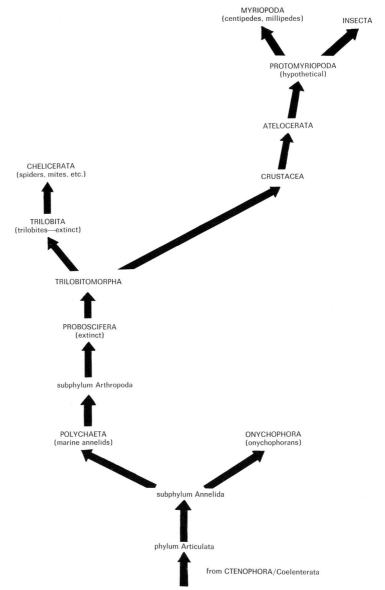

Fig. 5.4.   Phylogenetic tree showing descent of insects from presumable coelenterate ancestors. From Sharov (1966).

annelids. Embryologists point out that the coelomic sacs are not as distinctly formed in the arthropod embryos and larvae, and may even be completely absent.

The phylum Articulata, as given by Sharov (1966), consists of four subphyla, of which the Annelida are the first and Arthropoda the fourth (see Table 5.1). Included under the arthropods as superclasses are the Chelicerata—spiders and related animals—and Crustacea—crabs, crayfish, and so on. Because the annelids are considered the most primitive of the four subphyla, they are looked upon by almost all insect evolutionists as the original ancestors of organisms in the other subphyla.

There is a great difference of opinion as to the ancestors of the segmented Articulata. Many modern insect evolutionists consider some unknown ctenophore-like animal to be their direct ancestor. Sharov thinks that the Ctenophora have evolved toward a more mobile form of life. There are some present-day ctenophores that creep along the ocean floor. Most embryologists see ctenophores as a transitional form between organisms that have radial symmetry and those with bilateral symmetry.

If annelids evolved from some ctenophore-like ancestor, what then connects ctenophores to the more advanced annelids, and the annelids to the more mobile arthropods? The most primitive of the annelids belong to the class Polychaeta ("many-spined"). The polychaetes are marine worms with creeping locomotion and a well-defined head bearing eyes and *antennae.* Fossil poly-chaetes, from the family Spintheridae, have been taken from the late pre-

Table 5.1. The Phylum Articulata (From Sharov, 1966).

Phylum: Articulata
 Subphylum 1. Annelida
    (polychaetes, earthworms, leeches)
 Subphylum 2. Stelechopoda
    (tardigrades, or "water bears")
 Subphylum 3. Malacopoda
    (Onychophora)
 Subphylum 4. Arthropoda
  Superclass a. Proboscifera
    (prosciferans)
  Superclass b. Trilobitomorpha
    (trilobites)
  Superclass c. Chelicerata
    (scorpions, spiders, mites)
  Superclass d. Crustacea
    (crayfish, crabs, shrimp, etc.)
  Superclass e. Atelocerata
    (myriopods—centipedes and millipedes—and insects)

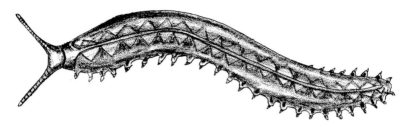

Fig. 5.5.   A present-day onychophoran, genus *Macroperipatus.* This species is found in leaf mold in the tropics.   Originally, they were thought to be a missing link: between annelids and arthropods; but they are now considered a dead-end divergent branch of Annelida.

Cambrian period.   Present-day species are segmented and are covered with numerous stiff setae, or spines.   They swim or crawl around on the ocean floor; some are predatory.   Among certain groups of these bristly worms certain structures resemble those found on arthropods.   This is particularly true of what is called the "oral lip," which resembles the labrum of insects.   Some polychaetes also have faceted compound eyes like those in insects.   Such similarities indicate that the polychaetes may be in the line between annelids and arthropods.

Which is the nearest living organism to the arthropods that resembles annelids in its structure?   Most evolutionists are agreed that it is a strange wormlike group called the Onychophora; the *Peripatus,* or "velvet worm," is probably the best-known example.   The Onychophora were originally considered by some evolutionists to be a sort of "missing link" between the annelids and the arthropods.   Sharov includes them as a class under his third subphylum, Malacopoda (Table 5.1).   There are about 65 existing species of Onychophora, which live in humid tropical areas of the world.   They resemble slugs with legs, but have much in common with arthropods.   Unlike the case of insects, the external segmentation was lost and the proboscis was reduced, and finally disappeared, in these animals.   Like insects, however, they have segmented antennae, a mouth on the lower surface of the body, and an exoskeleton made of chitin.   The jaws and antennae, as in arthropods, have evolved from leg appendages.   The circulatory system is much like that of arthropods.   Despite all these similarities and the earlier belief that they are the "missing link," most evolutionists now regard Onychophora as an independent branch that diverged along the line between annelids and arthropods and shares with them a common ancestor (Fig. 5.5) (see also Manton, Chapter 7).

5.4. PRIMITIVE ARTHROPODS

The most primitive, extant arthropods are peculiar-looking organisms placed in a superclass called Proboscifera, so named for their long snouts.   Compound eyes on long stalks protruded from the heads of these strange creatures. Many evolutionists think they were in the line of descent of the ancestors of the Crustacea. The Proboscifera are believed to have also given rise to the now extinct fossil group called trilobites, which led to spiders. Insect evolutionists disagree as to whether insects are descended directly from crustacean ancestors or from the trilobites. Sharov (1966) supports the idea that insects come from the primitive myriopods (see also Bergström, Chapter 1).

The crustaceans (Fig. 5.6) are considered the forerunners of the first group of primitive myriopods in the superclass Atelocerata (Fig. 5.4). The crustacean ancestors of the Atelocerata probably lived in shallow water and in humid regions where they gradually crossed over to living on dry land, after evolving a tracheal system for breathing, malpighian tubules for excretion, and fat body for food storage.

The primitive myriopods did not have a fixed number of body segments. Like crustaceans and insects they had mandibles and *segmented antennae*. Along the evolutionary line from crustaceans, which have *two pairs of antennae*, to the atelocerates there was a reduction and loss of the *second pair of antennae*. Modern myriopods, like insects, have only one pair of antennae, but—distinctly unlike insects—have legs on every segment of the body (some millipedes have over 400 legs). They are land animals, which survive in deserts and jungles, and in hot and cold climates.

According to Sharov (1966) the first atelocerates probably appeared during the Devonian period. They evolved along one branch into our modern millipedes and centipedes; a second branch presumably evolved into the insects. These ancient hypothetical myriopods have been named *Protomyriopoda* (Fig. 5.4).

Most evolutionists believe that the primitive wingless insects (Apterygota) came from this ancient ancestor of the myriopods. These insects include springtails, silverfish, and bristletails. Another branch developed into winged insects, the Pterygota. This group includes all other insect orders.

In my opinion the phylogenetic tree, as presented here (Fig. 5.4), from the Ctenophora through the annelids to the insects, is based on very convincing evidence. Some future evolutionists, however, might uncover new fossil evidence

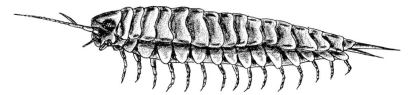

Fig. 5.6.   Reconstruction of the supposed ancestor of Atelocerata. This crustacean probably inhabited shallow water and was the forerunner of the first myriopods. Similarity of microstructure of compound eye, musculature, and joints of mandibles, and likeness of coxal appendages relate crustaceans to Atelocerata. From Sharov (1966).

and give a clearer picture of what is now accepted as a problematic but logical phylogenetic tree of the arthropods.

5.5. ORIGIN OF THE ANTENNA AND THEIR SENSILLA IN ARTHROPODA

There is no evidence that the ancient ctenophore ancestors of the Polychaeta developed antenna-like structures. We must hypothesize that the segmented antenna evolved somewhere along the line between primitive annelids and the more modern marine annelids (Polychaeta). Since modern Polychaeta swim or crawl along the ocean floor, we may postulate that the segmented antenna evolved from the locomotor organs. As Sharov (1966) points out, certain morphological structures of the polychaetes are also characteristic of arthropods. The prostomium of *Nereis vereus*, for instance, bears a pair of tentacles, a pair of palps, and two pairs of simple eyes. Some polychaetes have more complex compound eyes. According to present theory, as the polychaetes passed from creeping organisms to the more mobile swimming or burrowing predacious forms, the aboral processes moved forward to fuse with the developing prostomium (head). The fossil polychaetes of the family Spintheridae had no prostomium. According to Sharov, the unpaired apical tentacle of the aboral complex evolved into the medial tentacle of the prostomium, but retained traces of its paired origin in two separate roots of nerve fibers running from the tentacle to the brain. The apical tentacle, despite its paired nerve roots, did not evolve into the antennae. In the species *Chrysopetaltum debile* (Sharov, 1966), the antennae are quite similar to the notopodia of the body segments. Since the notopodium is the dorsal branch of the locomotory parapodium of the third segment, we must presume that the setae (chaetae)-covered dorsal notopodium moved forward to fuse with the aboral portion of the head and evolved into the forward-directed antenna. The notopodial chaetae eventually evolved into sensilla.

The evolution of the lobopod arthropods was toward increased segmentation, whereas in the Onychophora (Fig. 5.5) external segmentation was lost, so they cannot be considered as part of the evolutionary lineage of arthropods. In Onychophora the proboscis also disappeared, whereas in the primitive arthropods it was retained for grasping prey.

In Onychophora the aboral complex was also reduced, although the segmented antennae were retained. The aboral complex of primitive arthropods remained much as in its primitive annelid ancestors.

The most primitive arthropod may have belonged to the superclass Proboscifera. An example is *Opabinia regalis* Walcott (Fig. 5.7), in which the eversible probosis is similar to an analogous organ in polychaetes. With Proboscifera however, we still have not reached a change in the evolution of the locomotory process (third segment) into a segmented aboral antenna. Sharov

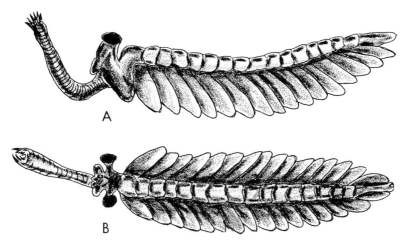

A

B

Fig. 5.7. *Opabinia regalis*, a fossil probosciferan from the Middle-Cambrian period, over 600 million years ago (A, side view; B, top view). Sharov considers strange Proboscifera to be ancestors of other arthropods. This group is characterized by a long, protrudable proboscis, with grasping spines at the tip, and a two-segmented head with compound, stalked eyes. Side lobes, called paratergal lobes, were apparently used as paddles to propel the animal. Under each lobe was a leg with gill filaments. From Walcott (1912) and a drawing by Sharov (1966).

(1966) states that "on the dorsal surface of the head there was probably an aboral complex, as shown in the reconstruction of the animal." In other words, the aboral complex cannot be observed, for a certainty, in the fossil remains. However, in the evolution from Proboscifera, by way of Dicephalosomita, to the more advanced Arthropoda (Trilobitomorpha), the segmented antenna suddenly appears. Sharov (1966) states: "Evolution from such Dicephalosomita to Trilobitomorpha consisted in reduction of the proboscis, transfer of the labrum to the ventral side of the head, addition of 2 trunk segments to the head tagma, *transformation of the limbs of the third segment into tactile organs-antennae*—and appearance of outgrowths on the inner side of the prebasal joints of the legs—endites, gill outgrowths being maintained in nearly all joints of the legs."

The flagellum-like antenna of the Trilobita also developed from locomotor-type appendages. Presumably in the Trilobita (Fig. 5.8), and also the crustaceans and atelocerates, the antennae lost all locomotor function and became *tactile* organs. To quote Sharov (1966) directly: "As a result they were transformed into multi-articulate flagellum-like appendages or appendages of a different type in which as shown by Imms (1939)—only some basal joints are true joints and the greater part of the flagellum-like antenna developed by secondary subdivision into rings of the distal joint of the antenna (Imms, 1940)."

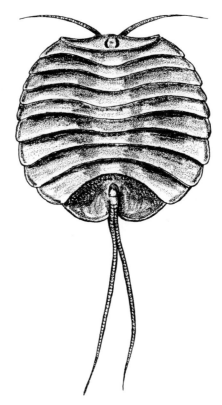

Fig. 5.8. *Cheloniellon calmani*, an Upper Devonian trilobite, seen from above. Extinct trilobites, which were marine arthropods, are known through almost 4,000 species. From Sharov, after Broili (1932).

Størmer (1944, 1951) has demonstrated that the bases of the antennae of Trilobitomorpha were covered by the labrum. The margins of the labrum had lateral recesses through which the antennal bases protruded. The fourth segment (behind the third antennal segment of Trilobitomorpha) bears the first set of walking appendages. According to this phylogeny the walking appendages of the fourth segment of the Trilobitomorpha were transformed into a second pair of antennae in the Crustacea and into chelicerae in the Chelicerata. In the Atelocerata they were reduced, and they disappeared altogether in the Protomyriopoda. Table 5.2 is a summary (after Sharov, 1966) of the head and first body segment homologies of the Atelocerata.

Finally, there is valid phylogenetic evidence that insect antennae evolved from the locomotory appendages of the third segment of the most primitive annelid form.

Table 5.2. Head Segment Homologies in Various Representatives of
Articulata (From Sharov, 1966).

| No. of segment | Annelida | Onychophora | Proboscifera | | Trilobitomorpha | Chelicerata | Crustacea | Atelocerata |
			Dicephalosomita	Pycnogonida				
1	Prost	Palp	Lbr	(Lbr)	Lbr	Lbr	Lbr	Lbr
2	1st Perist	(Praejaw)	Oc	Chf	Oc	Oc	Oc	Oc
3	2nd Perist	Jaw	L_1	Palp	Ant	(Pchl)	1Ant	Ant
4	Ppd_1	Slp	L_2	Ovg	L_1	Chl	2Ant	(Prmnd)

Ant = antennal; chf = cheliphoral; chl = cheliceral; jaw = of jaws; L_1 and L_2 = 1st and 2nd
pair of legs; lbr = labral; oc = ocular; ovg = ovigenous leg; palp = palpar; *(pch1) = precheliceral; 1st and 2nd perist = 1st and 2nd peristomal; ppd = 1st parapodial segment; (prae-jaw) =
pre-jaw; *(prmnd) = premandibular; prost = prostomial; slp = oral papillae (segment of slime).
*Parentheses indicate segments that have undergone reduction.

5.6. MONOPHYLY VS. POLYPHYLY IN ARTHROPODA

According to the Sharov hypothesis, the first annelids were lobopodial and
passed from a creeping to a swimming or burrowing way of life. The lobopodia
became the ventral segment of the parapodia (neuropodia), and the dorsal
part, the notopodia, formed from ridges or plates bearing rows of dorsal chaetae.
The parapodia function as both locomotor and respiratory organs. We must
assume that the setae (chaetae) served not only for traction and propulsion
through mud or water but also as a filter for debris in the water currents directed
into the gill clefts. Since they are located directly behind the gill plate, the
physical placement indicates a respiratory as well as a locomotory function in
the Polychaeta.

The sequence of the displacement of these gill-like locomotor appendages
forward into the preoral region of the head seems to have varied among the
various classes (Trilobita, Chelicerata, and Crustacea). This raises some interesting questions as to whether or not polyphyly occurred in the group characterized as arthropods. Cisne (1974) believes that polyphyly is demonstratable in
the arthropods, and that evolution of the group occurred along two parallel
lines. As he points out: "The relationship of trilobites, perhaps the most familiar
of extinct arthropods, is a critical question because trilobites are among the most
primitive arthropods known. Their relationship to other groups may help
delineate the major branches into which arthropods first diverged."

Cisne (1974) regards Trilobitmorpha, where branching is believed to have occurred (Fig. 5.4), as a sort of "taxonomic wastebasket." He considers that arthropodization took place at least twice and perhaps three times. Groups of metamerically segmented ancestral animals may have independently solved the same adaptive problems by separately developing a jointed, segmented exoskeleton (polyphyly). The works of Tiegs and Manton (1958) and Manton (1964) have challenged the prevailing belief in monophyly in the group Arthropoda (see also Anderson, Chapter 2, and Manton, Chapter 7). Cisne (1974) proposes that the phylum Arthropoda (the subphylum of Sharov) actually consists of two parallel phyla. One group includes the Trilobita, Chelicerata, and Crustacea as three parallel lines with common arthropod ancestors (the TCC group); the other, a separate line, is the Uniramia (which includes the Atelocerata of Sharov). According to this phylogeny, the trilobites show the characteristics of both Crustacea and Chelicerata in their primitive organization, but none of the characteristics of the Uniramia, from which the Myriopoda and Insecta evolved. Cisne states: "Crustacea can be derived from the condition in trilobites by fusion of an additional segment to the head posteriorly (Trilobita T, crustacean 2nd maxillary), movement of an additional segment to a preoral position in the adult (trilobite C, crustacean 2nd antennae) and further differentiation of head segments with reduction and specialization of what would be the 1st postoral limb (trilobite C, crustacean mandible) to a gnathobasic mandible." Cisne (1974) points out that nauplius larvae are similar to the fossil trilobite larvae in the genus *Triarthrus* in that the second antenna (the homolog of the first postoral limb E, in the trilobites) is also postoral and bears a finger-like enditic process that extends to the side of the mouth cavity. If, as indicated by the evidence presented by Tiegs and Manton (1958) and Manton (1964) and summarized by Cisne (1974), this phylogeny is correct, then the antennae and their sensilla might have evolved more than once along parallel lines. In view of what I have discovered concerning the mechanism of insect olfaction, I consider this highly unlikely—but not impossible! My reasoning, of course, is not paleontological in nature, but rather based on the mechanism of scent detection in arthropods and upon certain environmental parameters. If the generalized phylogeny of insects from water-dwelling annelid ancestors is correct, then the development of the antennal sensilla must be considered in reference to a mechanism common to a liquid as well as a gaseous environment. *In other words, we must examine antennal evolution in terms of the gradual transformance of a common olfactory mechanism workable in both a liquid and a gaseous medium and in both cases utilizing the H_2O molecules in the system.*

Kristensen (1975) disagrees with Cisne (1974) that polyphyly occurred in Arthropoda. Nor does he believe in the five-fold independent evolution of the five classes of Hexapoda (Insecta) as suggested by Manton (1972) (polyphyly in the lower categories of hexapods). She (Manton, 1972, 1973) proposed that the Diplura, Collembola, Protura, Thysanura, and Pterygota all evolved independently from an ancestor that was soft-bodied, multilegged and lobopod. Sharov (1966) also separated the Apterygota from the Pterygota. Manton (1972) has stated that "since functional continuity must have been maintained through all evolutionary stages, and since the leg mechanisms of the hexapod classes are mutally exclusive, these classes cannot have given rise to one another and must have evolved in parallel from ancestors with lobopodial limbs and little trunk sclerotization." As Kristensen (1975) points out, however, the only conclusion that can be legitimately drawn from the mutual exclusiveness of leg mechanisms is that the five taxa are monophyletic. Polyphyly is based on convergence, not divergence. I do not understand how "mutual exclusiveness" can lead to convergence as implied in a five-fold independent evolutionary process for hexapods. Mutually exclusive characteristics are divergent and thus monophyletic.

Since the antennal sensilla and the "system" wherein they operate apparently evolved from the setae (chaetae) on the dorsal portion of the notopodia of ancestral parapodial forms, we must assume that these ancestral chaetae were in contact with a flowing liquid medium, and that their locomotor beginnings assured a considerable amount of movement. We may speculate that currents of water, carrying various complex molecules, were drawn across chaetae surfaces into the gills, and further that oftentimes, during locomotory movements, the notopodial chaetae tapped or brushed against the substrate that supported the organism.

We may further speculate that throughout the evolutionary eons the processes of movement (tapping) and current flow selected for, or conversely worked against selecting, certain configurations of sensilla. In order to understand how such a mechanism could develop (since the antenna is a "molecule-seeking" organ), we may postulate that the constant tapping, vibrations, and rubbing behavior of arthropods has meaning within a molecular context. All of my experimental evidence leads to the conclusion that the tapping engaged in by insects, e.g., ants, etc., the antennal vibrations, and the rubbing of antenna and feet with legs or wings is a part of a system that was selected for, along with the sensilla forms, in the evolutionary development of the insect antennae.

It is not possible at this time to make a really detailed comparative study of the morphology of the antennal sensilla. There are huge gaps in our knowledge of the morphology of the antennae of various orders in the class Insecta and arthropod groups. Based on my own observations, however, and also on the meticulous work of Eleanor Slifer and others, it is possible to make a cursory examination of antennal divergence from the lowest to the highest orders of insects (Table 5.3).

5.6.1. Evidence of Monophyly from Insecta Sensilla

The antennal parameters compared (Table 5.3) are sensilla type, mean length of sensilla trichodea, array characteristic (arrangement), grooming, and antennal

Table 5.3. Type, Size, Array, Vibration Frequency and Grooming Parameters in Five Representative Insect Orders.

Order	Collembola	Orthoptera	Dermaptera	Anoplura	Lepidoptera	Hymenoptera
Type	Thick wall T. Thin wall T.	Thick wall T. Thin wall T. Basiconica Coeloconica	Thick wall T. Thin wall T. Basiconica Coeloconica	Thick wall T. Thin wall T. Basiconica Tuft organ Pits	Thick wall T. Thin wall T. Basiconica Coeloconica Styloconica Chaetica Auricillicum	Thick wall T. Thin wall T. Basiconica Coeloconica Styloconica Placodea Pyramid Chaetica Globular Slots and pits Sicula
Mean length of trichodea (μm)	12 to 130	19 to 150	20 to 40	20 to 50	20 to 150	20 to 120
Array	None	None	Even spacing	Even spacing	Even spacing Log-periodic	Even spacing Zig-zag spacing Log-periodic spacing
Grooming	Mouth alone	Mouth and legs	Mouth and claws	Claw alone	Legs, special organs	Legs, special organs, wings
Frequency	Waves slowly	Waves 2 to 8 Hz	Waves 2 to 12 Hz	Revolves	Waves and vibrates 12 to 100 Hz	Waves and vibrates 40 to 500 Hz

These parameters were taken as averages from the literature and from personal observations. It is obviously an extremely simplified and generalized attempt. There is tremendous overlapping, and huge gaps exist in the literature of antenna sensilla morphology.

vibration frequency. Orders were chosen at approximately even intervals from the lower (Collembola) to the higher (Hymenoptera) phylogenetic categories.

In general, we may state with some assurance that among the lower orders, the sensilla of the antenna are not of varied form. The Collembola (personal observations, Figs. 5.9, 5.10) exhibit only thick- and thin-walled trichodea types. The species *Ctenolepisma lineata pilifera* (Thysanura) exhibits thick- and thin-walled trichodea, short, thin-walled basiconica, and coeloconica, as do the grasshoppers (Slifer, 1955; Slifer and Sekhon, 1970). In general, the thick- and thin-walled trichodea, basiconica, and coeloconica are the main forms demonstrable up through the order Dermaptera.

Modifications of these four basic types begin to occur at the center of the order groupings. The Anoplura, for instance, show a modification of the basiconica called the tuft organ (Miller, 1970). At the lepidopteran level the number of types has increased to seven distinctive forms. Also in these higher orders the surface texturing of the sensilla becomes much more complex. Among the Coleoptera and up to and including the Hymenoptera one is apt to find corrugated, helical, terraced, equiangular, and smooth type sensilla surfacing. At the highest category among the social Hymenoptera, the sensory sensilla types have increased to at least 12 (and perhaps more) basic forms (Callahan, 1975b).

The behavioral parameters also become more complex from the lowest to highest orders. At the Collembola level, antennal movement is very restricted. Most of the Collembola species I have observed do not vibrate their antenna. Waving is a better terminology for the low frequent motion of the antenna of all the orders below the Dermaptera. Higher orders all superimpose high frequency antennal vibrations on the slower antenna waving. Furthermore, among the lower orders most grooming is done with the mouth, or mouth and legs alone. Among the higher orders we begin to observe the wings and claws being utilized in grooming behavior, but more importantly many species have devel-

Fig. 5.9. *Lepidocyreus cyaneus* (Entomobryidae, Collembola) inhabits moist turf around greenhouses in Florida. Note primitive condition where scales extend out from body and cover entire antenna.

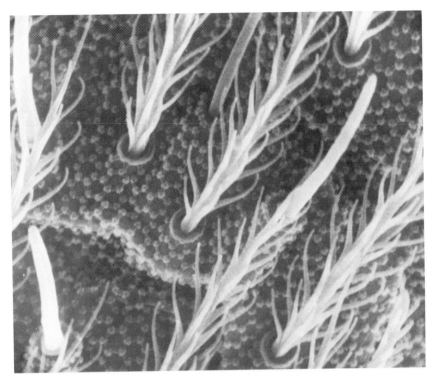

Fig. 5.10.   Tip of antenna of *Lepidocyreus cyaneus* (Entomobryidae, Collembola). Note surface texture and branched scales.   At lower left is a thin-walled short trichodea and at right a longer thick-walled trichodea.   They are almost lost among branched scales (8 X 10 print—24,600X).

oped special organs and behavior patterns of grooming (Callahan, 1969; Callahan and Carlysle, 1971; Goldman et al., 1972; Jander, 1966).

I have never observed any species at any other level that at some time or another does not tap the substrate. All insects apparently exhibit tapping behavior. It is this well-know behavioral characteristic that first led early entomologists to call the antennae "feelers."

Very little correlation can be made between lower and higher orders and the length of sensilla. Trichodea are apt to range from 12 to 150 μm in length. One is as apt to find long trichodea on the lower orders, such as Collembola (Table 5.3), as on the higher and more complex orders, such as the social Hymenoptera.

The considerable divergence of antennal sensilla form, and the increasing complexity of behavioral parameters associated with the sensory mechanism, leads me to postulate a gradual divergence in the class Insecta from the lowest Apterygota to highest Pterygota. The lineage (Sharov, 1966) that shows the Trilobita as having evolved with one pair of antenna and the Crustacea with two pairs, from a common ancestor the Trilobitomorpha (Fig. 5.4), leads me to speculate that divergence, and thus monophyly, is an evolutionary characteristic of the arthropoda in general. Sharov (1966) also points out that to accept the view that the atelocerates originated from Onychophora means accepting the polyphyletic origin of arthropods and the explanation of many fundamental features common to all arthropods as a manifestation of convergence. I personally favor monophyly based on such "antennal system" divergence. I must caution, however, that polyphyly may have occurred between the Apterygota and the Pterygota. I certainly have not looked at enough of the lower orders to make a valid judgment based on sensilla form and sensory mechanisms.

In the last part of this chapter I shall attempt to explain how the IR environment may have operated as a selective mechanism in the evolution of the insect sensilla and the insect communication system. *My ideas are based on well-documented paleontological evidence that demonstrates that the antenna of insects evolved from the locomotory parapodia of the third segment; as such these evolutionary precursor organs were subject to the current flows of the liquid substrate. They also maintained mechanical contact, from which tapping might evolve, with the ocean floor substrate.*

5.7. ORIGIN OF GROOMING IN ARTHROPODS AND INSECTS

We assume that the arthropods evolved from ancestral Coelenterata that exhibited a benthic mode of life (Fig. 5.11). The ancient Ctenophora descended from the upper water and developed bilateral symmetry, and the ability to creep along the muddy ocean floor. The earliest annelids utilized the developing parapodia for crawling, and some of their descendants later evolved burrowing and swimming forms.

The transition forms between the Annelida and Crustacea oscillated back and forth between crawling and swimming. The Proboscifera and Trilobitomorpha were mainly crawling organisms. Most evolutionists believe that the tran-

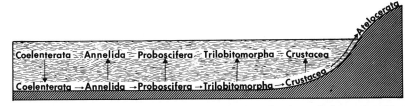

Fig. 5.11. Diagram of presumed direction of evolution of arthropods from a benthic to a pelagic way of life. Transition to pelagic life of Atelocerata can be regarded as a split from the main direction of evolution from Coelenterata to Crustacea. Transitions to swimming forms occurred in early higher systematic groups. From Sharov (1966, p. 236).

sition from crawling to swimming occurred in each of the superclasses, but did not in turn lead to any new higher groups. Eventually, the Crustacea, which exhibit modern swimming forms, evolved from groups that crawled or ran across the ocean bottom. The later Atelocerate emerged from the bottom on to the land. I favor this hypothesis for the simple reason that a crawling or running bottom-inhabiting organism would be far more likely to pick up debris on its notopodial chaetae than would a free-swimming form. It would therefore likely develop the habit of rubbing the chaetae-covered notopodial appendages, either against some suitable substrate or against another body organ in order to dislodge the debris. A free-swimming organism would be much less likely to develop the rubbing or tapping behavior, since presumably the free flow of water across the notopodia and gill cleft would keep the spines farily clean.

Grooming is a behavioral activity found throughout the orders of insects. It occurs in the lower Apterygota, such as the Thysanura (and also the Myriapoda) as well as in the higher orders of Diptera and Hymenoptera. The reader is referred to the work of Jander (1966) for an excellent study of significant grooming behavior throughout Insecta.

Jander (1966) points out that there is an unmistakable relationship between the general level of grooming activity and general habits. She found that highly active forms, such as the Hymenoptera, generally exhibit a markedly high frequency of grooming.

The utility of grooming behavior as a mechansim for keeping the antenna and proboscis sensilla clear of debris has been experimentally demonstrated by Callahan and Carlysle (1971) and Goldman et al. (1972). Despite the necessity of maintaining clear sensory surfaces, grooming behavior is equally important to the communication system of the insect for a totally different reason.

Jander's work has demonstrated that, with few exceptions, the Myriapoda and Thysanura groom the antenna and all of the legs with the mouthparts. She regards this type of grooming as the primordial mode among the arthropod. In the Hemimetabola, the Odonata and Hemiptera exhibit a derived form (apomorphic) of grooming that utilizes exclusively the rubbing actions of the legs. Among Coleoptera, grooming of the antenna and first two pairs of legs is usually by the primordial method of using the mouthparts. Interestingly, this method is often replaced by other unique and specialized methods in certain species of beetles. Neuroptera typically groom all of the legs with the mouthparts, but groom the antenna by rubbing one antenna at a time between the raised forelegs.

Higher orders of insects such as the Lepidoptera, Hymenoptera, and Diptera have special grooming organs on the forelegs. Callahan and Carlysle (1971) described the morphology of the epiphysis on the forelegs of *Heliothis zea*, and the complex grooming activity of that moth. Goldman et al. (1972) described the morphology of the tibial combs of *Aedes aegypti* (L.) and related it to the proboscis cleaning of the mosquito. It is obvious that grooming, as is the case of antenna vibrations, is an integral part of the insect communication system. The organs with the sensory sensilla are the ones that are groomed in each and every case where the phenomenon has been studied in detail.

I (Callahan 1967, 1975a, 1977a) have demonstrated that a thin molecular layer of molecules (the cabbage looper pheromone) coated on a wax substrate and rubbed will emit a low energy maser wavelength in the 17-μm IR region (Fig. 5.12). Rubbing the waxy substrate of the thin molecular layer increases the amplitude (output energy) of the IR emission; the electret charge on the waxy substrate is further supplemented by the rubbing charge. The insect antenna is an electret (Callahan 1967, 1977a). (An electret is analogous to a magnet and for that reason the terms ends in "et.") Just as a magnet will line up *inorganic* metal particles in a magnetic field, an electret will line up charged or polarizable *organic* particles in its electret field (Callahan, 1967; Perlman, 1973). As in the case of a magnetic field, the electret field is permanent. Insect waxes make excellent electrets (Fridkin and Zheludev, 1960).

It is not necessary in this thesis to go into the origin of the waxy insect exoskeleton. It obviously evolved as a necessary prerequisite in an organism

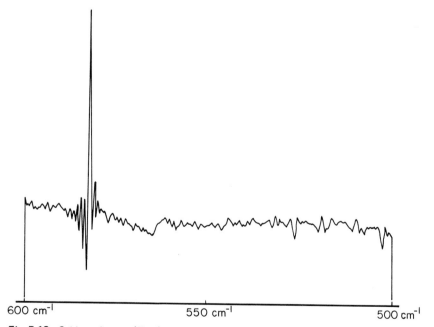

600 cm^{-1} 550 cm^{-1} 500 cm^{-1}

Fig. 5.12. Cabbage looper (*T. ni*) pheromone maser emission line from a single monomolecular layer coated on a wax-coated aluminum mirror. This line is emitted at 17.27 μm after 7 min of blackbody radiation and is between water rotation lines 5C and 6A. This monolayer has been rubbed on an aluminum mirror and modulated at 55 Hz to give a high intensity, narrow band maserlike infrared line.

moving from a benthic to a pelagic mode of life. The water balance mechanism of the waxy epicuticle is discussed in detail by Neville (1975).

In summary, my thesis is that the notopodial chaetae gradually evolved and formed over the eons into the preoral antennal sensilla, and during the process were constantly contacting a solid substrate which coated the evolving organ with debris. This stimulated the rubbing and tapping behavior in the various groups. It was inevitable that the developing chaetae (sensilla) should at times become coated with thin layers of molecules from the environment, whether from a solid (the ocean floor) or a watery substrate. In the process the constant rubbing, cleaning, and tapping was continued in the pelagic forms. A considerable charge was constantly being superimposed on the permanent electret charge of the organ. This charge attracted organic molecules to the surface of the spines in much the same manner that a magnet attracts metal filings. The narrow band, weak, maser emissions from these environmental molecules would thus be brought into close proximity to the evolving sensilla as the organic molecules were attracted to the antenna. Sensilla "tuning" would occur during the selection process, dependent upon the various lengths of sensilla that resonated to the molecular emissions. In this latter respect the phenomenon of organic "concentration tuning" as discovered by laser physicists is of great significance to the entire molecular-waveguide-vibration communication system.

5.8. EVOLUTION OF INSECT COMMUNICATION SYSTEM

5.8.1 Nonlinear Radiation and Concentration Tuning

There are three types of laser—solid, liquid, and gaseous. In general they all operate on the same principles; that is, *they take advantage of the very small but always present nonlinear response of materials to light.* Every molecular substance known to man has some energy level region where it will exhibit a minute nonlinear response to some portion of the electromagnetic spectrum. In most molecular substances, whether solid, liquid, or gaseous, it was, until lately, impossible to measure such small nonlinear responses. Laser design is a quest for maximizing and amplifying the nonlinear response of solids, liquids, and gases to light. A linear response is, of course, a response where the amplitude (signal strength) of the radiation coming from a system is proportional to the energy entering into the system, e.g., increasing the voltage to the filament of a light bulb and increasing the output of light. This is not true of nonlinear emissions, where both the frequency and the amplitude may be quite disportionate to the stimulating frequency. A nonlinear laser emission is narrow band (highly monochromatic) and of amplitude (signal strength) above the stimulating emission, or the molecular noise (static) from the molecules of other vibrating substances in the same frequency portion of the electromagnetic spectrum.

My thesis that the insect sensilla are electromagnetic resonators (waveguide-antenna) to narrow-band IR radiation is based on Maxwell's equations as applied to dielectric antennae-waveguide design. A dielectric is an insulative substance, e.g., beeswax, not a metallic substance. At high frequencies dielectric materials may be utilized for antenna-waveguide systems. Just as glass dielectrics (lenses) can collect and guide visible waves, "elongated lenses" (spines) can collect and guide the longer IR waves. For a dielectric waveguide-antenna to work, however, the IR electromagnetic energy must be narrow band (highly monochromatic) and of enough energy output (amplitude) to be above all of the random noise (static) from the vibrating molecules that fill the atmosphere or watery environment.

My thesis, that trace amounts of molecules in the gaseous atmosphere, e.g., pheromones, plant scents, CO_2, and so on, or in liquid media (water) absorb radiation and are stimulated to emit nonlinear (maser)* radiation, is based on antennal design as applied to the sensilla and not on the chemist's selection rules for absorption and emission as applied to the physical chemistry of molecular reactions. At the present state of the art the chemist's selection rules are not sufficient to predict such nonlinear radiation from all of the complex organic molecules utilized by insects in their communication system.

I am certain that the insect sensillum has evolved over the eons as an organ that can "lock in" on molecular nonlinear emissions and handle them in such a manner as to amplify and focus the radiation to the pore kettles and dendrite endings where detection takes place.

Concentration tuning is a term applied by physicists to the phenomenon of wavelength shift in organic dye lasers. A laser may be thought of in exactly the same way as any other electronic oscillator; the only difference is that the wavelengths are generated in the UV, visible, or IR portion of the spectrum instead of the radio region.

The trace organic molecules in a dye laser are mixed in a solvent, usually water, ethyl alcohol, glycerol, or acetone, and stimulated to fluoresce by "pumping" them with an electric discharge or flash of visible light. As described at the beginning of this chapter the electrons are "pumped" to a higher metastable state, where they are "stored" until they fall to ground level, giving off electromagnetic energy. The organic molecules must have enough degrees of freedom in the solvent to oscillate. The higher the concentration and the warmer the temperature, the longer the wavelength is; the lower the concentration and the cooler the temperature, the shorter the wavelength. Of course, there are limits of concentration above and below the optimal where the system will not operate. The rest of the system consists of a resonant cavity (spaced mirrors) in

---

*I use *m*aser instead of *l*aser for micrometer amplification by stimulated emission of radiation.

which the emitting wavelengths obtain enough feedback from the system to sustain and amplify the molecular oscillations—in much the same way as the sound box under vibrating violin strings does.

None of the gases or liquids used in these laser systems works as a "pure" substance. They are always complex mixtures. In almost all cases either nitrogen, ammonia, water, or water vapor is a part of the mixture.

Organic molecules attracted to a sensillum by its electret charge form a thin layer or "monolayer" between the base and tip of the sensillum. The length and diameter of the sensillum, plus the tapering, curvature, and surface texture, are analogous to the resonant box of the string instrument. It forms a resonant open waveguide that "collects" the electromagnetic waves riding on its surface. Every single one of the design parameters of a man-made, dielectric, open resonator, rod waveguide are applicable to the tapered, curved, thin-layer configuration of the insect antenna sensillum.

In summary, the insect has evolved an open resonator, waxy rod that takes advantage of the very small but always present nonlinear response of gaseous molecules to light.

5.8.2 Antenna Vibrations and Phase-Amplitude Modulation

This is not the place for a long discourse on the physics of phase and amplitude modulation. Essentially, modulation, as applied to electronic systems, is variation of the amplitude (strength), frequency, or phase (waves in or out of unison) of an electromagnetic signal in a periodic or intermittent manner. The so-called carrier frequency of radiation has superimposed upon it the vibrations of the modulator.

Insects wave or vibrate their antennae (Table 5.3) when searching the environment for molecular signals. As in the case of rubbing (grooming) this is of necessity a part of the communication system. It is an important part and should not be ignored. All insects, except Collembola, have a Johnston's organ at the base of the antenna. In other words, they are equipped with an organ for recording the exact vibrations of their own antennae. The fact that the Collembola do not have a Johnston's organ favors the ideas about polyphyly in lower orders as presented by Manton (1972) and Sharov (1966). Apparently Collembola are so restricted in their environment that distant searching is not required, and they depend on low amplitude waving and rubbing alone for signal amplification.

Phase modulation is a technique utilized lately by astronomers to amplify weak IR and microwave signals from outer space. Essentially, it consists of vibrating a collecting mirror at audio frequencies. The phase-amplitude content of the IR portion of the spectrum lies in the region from low audio (1 or 2 Hz) up to 600 Hz.

Put more simply, the collected molecules on the waveguide sensilla are vibrated in an audio range (depending on the species) and the molecules "shaken" in such a manner that the weak nonlinear radiation is amplified. Conversely, molecules coating a substrate are modulated by shaking and tapping the substrate and causing them to emit their nonlinear energy at a much higher amplitude (amplification).

I (Callahan 1975a, 1977a,b,c) have demonstrated that the far-IR emission from the female cabbage looper sex scent (pheromone) can be amplified by phase-amplitude modulating the gas or a thin layer of the molecules at the same frequency as that of the vibrating male antenna (50 to 55 Hz) (Figs. 5.13, 5.14).

The phenomenon of insects tapping with their antenna is recorded throughout the entomological literature. Flying insects vibrate their antenna in order to modulate the airborne stream of gas molecules, but upon landing the insects often switch to a form of "tapping" modulation. This tapping behavior is most often referred to as antennal palpation by insect behaviorists. The word palpation implies actually touching the surface while the antenna vibrations (Hz) are in motion.

An excellent description of palpation behavior is given by Hendry et al. (1973). It has been demonstrated that certain parasitic wasps locate their host larvae while utilizing palpation behavior. Dr. P. D. Greany has been good enough to furnish me with a photograph (Fig. 5.15) of the braconid wasp parasite, *Orgilus lepidus* (Muesebeck), palpating a piece of filter paper impregnated with heptanoic acid. Hendry et al. (1973) demonstrated that heptanoic acid, extracted from the frass on the surface of a potato containing the larvae of the potato tuberworm, *Phthorimaea operculella* (Zell.), elicited palpation of the impregnated spot by the wasp's antenna. The chemical heptonoic acid serves as an identification signal indicating the presence of the host larvae in the potato.

Wheeler (1960) gives examples of tapping behavior in numerous species of ants. Forel (1904) describes behavior of ants in the genus *Camponotus* as including, besides "rapid antennary vibrations," "the butting of the head" and "striking the nest with the gaster." Dante, in the *Divine Comedy* (Purgatoria Canto XXVI), wrote of ants, "even so within their dark batallions one ant rubs muzzle with another, per chance to spy out their way and their fortune." We may presume that the ants, mentioned by Wheeler, Forel, and Dante, are conveying some sort of message to one another. They are, we suspect, using a make-and-break vibration system to communicate.

In one of his pioneering papers on pheromone communciation of the cabbage looper moth, Shorey (1964) describes in detail the approach to the female by the male and the copulatory behavior of the species. He demonstrated that after approaching the female through the air the male landed behind and palpated the female pheromone gland with his vibrating antenna.

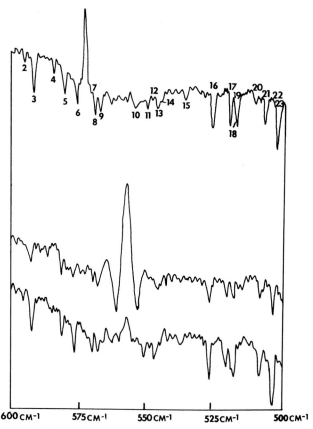

Figs. 5.13–5.14. (Top) Cabbage looper (*T. ni*) pheromone *maserlike emission* line from a single monomolecular layer coated on a polished aluminum mirror. This line is emitted after 10 min of blackbody radiation and is at 17.45 μm between water rotation lines 6A and 7A (Bureau of Standards). Aluminum mirror and pheromone rubbed but not modulated. (Bottom) A thick "monolayer" of pheromone on a gold-coated front-surface mirror. Low fluorescence on a steady mirror at 18.05 μm is shown. (Middle) Same thick monolayer fluorescing at 18.05 μm but audio frequency modulated by vibrating mirror at same frequency as cabbage looper moth antenna (55 Hz). A ca. 10-fold increase in energy is obtained by modulating the pheromone.

Pheromone molecules emitted from the gland would be in much higher concentration that those evaporated into the air. The molecules would be expected to coat over the external membrane of the gland while evaporation from the surface.

If modulation of the IR emissions ("carrier waves") from the pheromone molecule is a true factor of the insect communication system, then the modulation frequency should ride the "carrier" IR wavelengths as they shift frequency with concentration and temperature of the pheromone. The highest output and

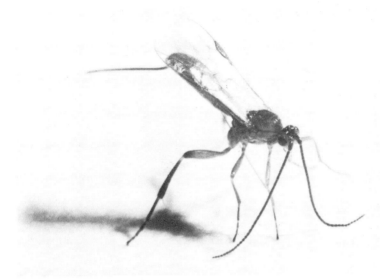

Fig. 5.15. Parasitic wasp, *Orgilus lepidus* (Muesebeck), in midst of tapping (palpating) a piece of filter paper impregnated with heptanoic acid extracted from frass at entrance hold of larvae of potato tuberworm. Acid in frass tells wasp that there is a tuberworm feeding inside potato. From Hendry et al. (1973).

longest wavelength would be expected from the highest concentration right at the gland, provided that the concentration is not so great as to instigate cutoff due to "swamping" of the thin layer of molecules coating the abdominal gland.

Figure 5.16 shows the 50 Hz cabbage looper modulation frequency riding the 17-μm region IR carrier wavelength from the cabbage looper pheromone (Z)-7-dodecen-1-ol acetate (Callahan, 1977c).

As the spectra presented in this paper and published in detail elsewhere demonstrate (Callahan, 1975a,b, 1977a,b,c), it is possible to obtain low energy nonlinear wavelengths (coherent) from insect sex scents. The coherent non-linear emissions match the wavelength criteria of the sensilla and are far above the background noise (static) of any other molecular emitters in the atmosphere.

The fact that I could predict and locate the nonlinear radiation, utilizing the same system as the insect, is irrefutable proof that insect sensilla have evolved into a nonlinear resonant system for "looking" at coded narrow-band IR frequencies from insect and host plant molecular attractants.

5.8.3. Nonlinear Radiation and the Atmosphere

The earth is essentially a water planet. Without the magic substance water there is no organic life. Water is the universal solvent, the precursor of all living sys-

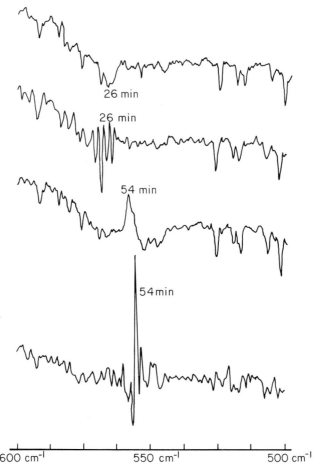

Fig. 5.16.   Top two spectra show a "thick" layer of cabbage looper pheromone absorbing energy after 26 min of radiation (17.61 μm) wavelength). Modulation was accomplished by gently tapping pheromone wax-coated mirror. Bottom two spectra show pheromone after 54 min (18.03 μm) when it has moved to a longer wavelength. A live cabbage looper moth vibrating at 55 Hz was held against the mirror. At the proper modulation frequency IR line emits like a narrow band laser (maserlike). Lengths of cabbage looper sensilla fit emitted lines.

tems.   Keeping this fact in mind, it is indeed strange that present day-research treats insect semiochemicals as if they existed "unto themselves."

Among the hundreds of papers that appear yearly on pheromones or host plant scents there is little concern for the simple and obvious fact that once the scent leaves the gland or plant it becomes a part of the complex system surrounding the earth called the atmosphere. As I have demonstrated, it is impossible to cause the emission of nonlinear coherent IR radiation from the cabbage looper pheromone if I purge the Fourier transform spectrophotometer system, by which I measure the emission, of water vapor (Callahan, 1975a).

The earth's atmosphere is composed of 78% nitrogen, 21% oxygen, and about 1% argon. The rest of the air is a complex mixture of trace molecules, the most important of which are CO_2 methane, and ammonia (Table 5.4).

Is it a coincidence that in almost all laser systems, these very same gases are used as coadditives to the gas that is lased? A CO_2 gas-dynamic laser, for instance, takes advantage of the nonlinear interaction between CO_2 and the vibrational energy of nitrogen. In order to get maximum output from a gas-dynamic laser, water vapor is added as a catalyst and the mixture heated and blown from a nozzle so that it quickly expands and cools (Anderson, 1976). It is essentially the same thermodynamic process, only at a far higher temperature, as that which takes place on the release of a pheromone or host scent from a surface or gland.

If we had eyes that could see on a submicroscopic level, we would observe that the atmosphere is composed of approximately 2.7×10^{19} molecules per cubic centimeter, and moving at an average velocity slightly higher than the speed of sound. All of the molecules in this complex mixture are colliding with neighboring molecules at a rate of 10^{10} collisions per second. It is these complex vibrational and rotational collisions that are important to the laser physicist. Nonlinear laser radiations are produced by transitions from one energy level to another, which in turn are brought about by collisons of the molecules and absorption of incoming "pumping" electromagnetic radiation from an outside source. It is for this reason that merely blowing the pheromone increases the nonlinear output (as in electrophysiological studies).

The complexity of laser design is not in principle, but rather in locating, for

Table 5.4. Trace Gases in the Atmosphere.

Formula	Gas	Source	Abundance (4.2 Km)	Lifetime
N_2O	Nitrous oxide	Anaerobic decay	0.24 ppmv	130 years
CH_4	Methane	Anaerobic decay	1.50 ppmv	3 to 7 years
NH_3	Ammonia	Anaerobic decay	6.00 ppmv	10 days
C_2H_4	Ethylene	Anaerobic decay	0.20 ppbv	1 day
O_3	Ozone	Sunlight + O_2	3.43 mm STP	–
CH_3Cl	Methyl chloride	Marine biological decay	0.50 ppbv	1 year
CO_2	Carbon dioxide	Plant respiration	330.00 ppmv	–
CO	Carbon monoxide	Cumbustion	0.07 ppmv	–
SO_2	Sulfur dioxide	Combustion	2.00 ppbv	10 days

From Wang et al. (1976).

various mixtures, these complex transitions from lower to higher energy levels, and then figuring out ways to store them in a metastable state so that they can emit into a resonant cavity system. In the case of the insect the open resonant cavity is the surface of the proper-length sensilla, or pit.

The cabbage looper pheromone, tomato plant scent, or human breath become low energy, gas-dynamic systems the moment the organic gases are released into the atmosphere. Any change in concentration or in temperature is going to shift the nonlinear emissions, as I have already demonstrated (concentration tuning, Callahan, 1975a,b, 1977a,b,c).

Furthermore, since the semiochemicals are released into the atmosphere, any one, or more than one, of the atmospheric constituents (Table 5.4) may be a part of the system.

Another difficulty in the case of pheromones is the fact that these specialized sex attractants are not usually "pure" substances but complex mixtures (Klun et al., 1973; Butler et al., 1977).

Ammonia and nitrogen are coadditives of numerous laser systems. It is probably not unreasonable to consider that once a pheromone leaves the scent gland, certain atmospheric gases are "catalysts" or "collision components" of the free-floating scent. If so, then we can well understand the widly varying results among pheromone field studies by entomologists. A field, for instance, newly fertilized might create an atmospheric environment either "over or under-charged" with the necessary ammonia or nitrogen concentration in the atmosphere. In summary, the atmosphere over a field of corn or cabbage is as much a part of the nonlinear scent system as is the attractant scent itself.

5.8.4. Nonlinear Radiation and the Atmospheric Windows

Although nonlinear emissions are monochromatic and of an amplitude above any background emission (static), nevertheless any such molecular nonlinear-waveguide-antenna system would be very likely to evolve in a region of maximum efficiency. This being the case, an empirical search for such emissions should concentrate in the IR water vapor windows or partial windows (Fig. 5.17). The main window of course, is the 7 to 14 μm window.

For the nonphysically oriented, an atmospheric window is precisely what the name implies—it is a region of the electromagnetic spectrum where the water vapor or other major gases, e.g., CO_2, do not absorb electromagnetic radiation. Other than the visible region, where the window is obvious to our eyes, the major IR regions are described in detail at four different altitudes (called path lengths) by Traub and Stier (1976). The transmission spectrum for terrestrial atmosphere plotted in their publication shows numerous microwindows in all regions of the IR spectrum. These are small regions between the numerous water absorption lines where no absorption occurs. The cabbage looper phero-

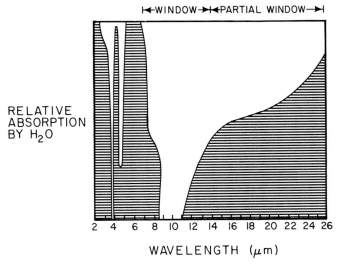

Fig. 5.17. Infrared water vapor windows in the 3, 5, and 7 to 14 µm regions and partial window from 14 to 26 µm.

mone, stimulated by blue and black light, emits a log-periodic series of nonlinear coherent IR wavelengths in the 17 µm microwindow between water absorption bands 3B (16.90 µm) and 10A (18.03 µm). *It is no coincidence that the male cabbage looper moth flies to a candle flame and dies in the flame.* The exact same lines are emitted by the stimulated hydrocarbons around and above the flickering (modulated) candle flame (Callahan, 1977b). The wax or petroleum candle is a man-made mimic emitter of hundreds of such lines in the 15 to 30 µm partial IR water vapor window.

As Wang *et al.* (1976) have so aptly warned, anthropogenic gas may well alter our climate by plugging the atmospheric windows. Excesses of gases such as N_2O, CH_4, NH_3, HNO_3, C_2H_4, SO_2, CCl_2F_2, CCl_3F, CH_3Cl, and CCl_4 may at present be increasing at an accelerated rate. The problem of excess CO_2, which has strong absorption in the middle of the 3 to 5 and 5 to 30 µm regions, and the so-called greenhouse effect are so well publicized as to need no explanation in this work. The same may be said for the present anthropogenic problem of deodorants and the ozone layer. One may be pessimistic and predict that we will all die of cancer smelling nice—such are the problems of olfaction!

Although pollution and anthropogenic gases are important for the short run, my thesis involves the evolution of the atmosphere over the eons, and what I believe to be the direct effect of the always changing atmosphere on the evolution of living plants and creatures.

There can be little argument with the fact that the atmospheric windows, and also the trace molecules in the atmosphere, have varied considerably over the eons. Since the nonlinearity of my IR signals depends on mixtures of gases, or in the case of sea creatures on the concentration of trace molecules in the water, it is obvious that "closing" and "opening" atmospheric windows puts tremendous selective pressure on the evolving system. Just as atmospheric CO_2 from high volcanic activity can partially close the 7 to 14 μm window, volcanic dust and SO_2 can plug the blue visible window in the sky and water.

What I have been saying is that the speciation of insects has progressed in direct proportion to the changes in the atmosphere. Those changes are dependent on complex molecular mixing, and so also are the nonlinear molecular emissions that control the communication system of the insect. Change one and you change the other, and that is the meaning of the word evolution!

5.8.5 Nonlinear Radiation, the Arthropod Antenna and Monophyly

The best description of a fossil crustacean is of the 300-million-year-old species *Lepidocaris rhyniensis* from the Rhynie Chert (Green, 1961). The species has two pairs of antennae (see Fig. 1.20 in Bergström, Chapter 1). The posterior pair is long, branched, and curved with a fringe of long hairs. The front pair is unfringed and much shorter. Since *Lepidocaris rhyniensis* was not a crawling species, Green (1961) believes the fringed posterior antennae were very likely utilized as swimming organs. The front shorter pair, called antennules in Crustacea, is the sensory antennae. Both were movable. Snow (1972) has demonstrated that in the hermit crab, *Porgurus alaskensis*, the anterior antennules are chemoreceptive organs. A large amount of electrophysiological and ultrastructural evidence has accumulated that the antennules are the primary chemoreceptive organs of crustaceans (Snow, 1972). Crustaceans exhibit four varieties of antennal movement, flicking, pointing, wiping, and withdrawing. Maynard and Dingle (1963) believe that all these types occur in a wide variety of crustaceans. If we consider flicking as a modified form of tapping, then we see that these behavior antennal parameters are shared by insects and crustaceans. It is interesting to pick up a sand crab from a beach and blow one's breath on its antennules. They are stimulated to vibrate at a high frequency (personal observation). A vibrating antennule submerged in water would probably be slowed to a flicking pattern of movement.

It is a simple fact that in order to take advantage of the small, but always present, nonlinear response of molecules to light, exactly the same physical design principles apply whether or not it is a gas or liquid laser system. It becomes obvious then that in the transition of an organism from a benthic to a pelagic form of life, exactly the same overall form (generalized morphology) and behavioral parameters, rubbing, vibrating, and so on would serve the same system. The changes over the evolutionary eons are not in physical-chemical principles but in the wavelengths of emissions, which are in turn dependent upon the ever changing atmospheric mixtures and watery constituents and the resulting change in form of the sensilla programmed through the tremendous selective pressure of the liquid and gaseous environmental windows.

The important environmental parameters that react in the evolution of the communication system are: 1) the incoming radiation from the sun, sky, and night-light which "pumps" the organic molecules and lifts them to a nonlinear metastable state; 2) the changing atmospheric and water windows at any one period in the formation of our planet; 3) the many organic molecular oscillators arising from evolving plants and animals; 4) the mixing of such organic molecules in the air; and, finally, 5) the amount of the trace molecules in the air or rivers, lakes, and oceans, CO_2, NH_3, and so on, at any one time in the atmospheric and watery history of the planet.

It is axiomatic of my thesis that taste (the feet or probosis of an insect on molecular surfaces or submerged in liquids), olfaction (the antenna submerged in a gaseous atmosphere), or chemoreception of a water-dwelling organism (Crustacea, and so forth) are all variations, dependent upon the media, of exactly the same nonlinear physical-chemical molecular system.

My thesis is in no way that of Wright (1966), or any other researcher, for it states that the dielectric waveguide-antenna sensilla of arthropods, regardless of the environmental media, are resonant "collectors and amplifiers" of molecular nonlinear wavelengths from scent semiochemicals. I further believe the mammals (human) and other living creatures have an olfaction system that operates with exactly the same physical-chemical principles! Scents drawn into the blackbody nose cavity are mixed as trace elements into the mucous "carrier" media and nonlinear coded IR wavelengths modulated and amplified by the vibrations of the olfactory cilia (Callahan, 1975a). Thus could a salmon or sea turtle follow a nonlinear molecular trail through the water, or a polar bear sniff the scent of a sea lion's nonlinear windborne trail from miles across the ice cap. The arctic scent would work with extremely high efficiency because of cooling as the scent moved through the arctic air. Gas-dynamic lasers depend on the thermodynamic cooling of warmed gases!

For an excellent review of the importance of solar activity on climate the reader is referred to Huntington (1945). A review of the effects of volcanic activity on the insects of the Florissant lake basin is given in my book (Callahan, 1972). The tremendous amount of debris and atmospheric gases generated by volcanic activity has been documented numerous times within living memory. It is recorded that the destruction of St. Pierre on the Island of Martinique, Lesser Antilles (May 8, 1902), killed over 40,000 persons. On that day a deadly flood of fire and ashes descended from the sky (Morris, 1902). Morris wrote: "It is believed that Mount Pelee threw off a great gasp of some exceedingly heavy and noxious gas, something akin to firelamp, which settled upon the city and rendered the inhabitants insensible."

It is interesting that animals can sense such impending natural disasters. He further wrote: "Wild animals disappeared from the vicinity of Mount Pelee. Even snakes, which at ordinary times are found in great numbers near the volcano crawled away. Birds ceased singing and left the trees that shaded the sides of Pelee."

One can well imagine what effect constant volcanic activity, such as that which has occurred during numerous geological periods, has on the water vapor atmospheric windows, and even upon the clear visible window of sea water. Since the pumping radiations for the nonlinear IR wavelengths are mainly in the visible, blacklight, and near IR, there would likely be great shifts in the nonlinear IR scent emissions. The cabbage looper pheromone, for example, would be completely quenched by such volcanic hot gases. Also, the resultant closing of the black light and blue windows (which pass the pumping radiation) would cut off the emissions.

More subtle selections would occur during less traumatic atmospheric shifts. A slowly evolving ice age would shift the average night or day temperature, and thus shift the wavelength of a semiochemical a fraction of a micrometer. I have already shown experimentally (Callahan, 1977a) that a decrease of a few degrees centigrade will shift the nonlinear emission of the cabbage looper pheromone practically a whole micrometer from 17.39 μm at 26°C to 16.81 μm at 20°C. A gradual cooling of the atmosphere would change the thermodynamic properties of the semiochemical mixtures and thus its wavelength of nonlinear IR radiation. There is no doubt at all in my mind that this is exactly what has happened over the eons. Populations of insects or crustaceans with sensilla of one length are, over the eons, selected out for populations with shorter sensilla by a cold climate. Is it by chance alone that one is more apt to find small species with average short sensilla in the arctic and large, or giant, species with longer average sensilla in the tropics? Why do the giant moths and beetles occur in the tropical latitudes? Of course the same thermodynamic nonlinear shifts could occur among body hormone or cellular molecules and be equally responsible for selection of small insect species in cold climates. This would not hold for mammals that have a controlled body temperature.

Two very important predictions of my thesis are: 1) small insects having short sensilla will resonate to short nonlinear IR wavelengths (below 16 to 20 μm) and the modulation frequencies will be high, 200 to 500 Hz (gnats, blackflies, and so on), and any coded harmonics far apart in the IR spectrum; and 2) large insects (saturnids, large beetles, and so on) having long sensilla will resonate to longer nonlinear IR wavelengths (above 10 μm) and the modulation frequencies will be low, 2 to 200 Hz, and the coded harmonics close together (Callahan, 1977c).

We may come to understand from my theories and experimental data (presented over the last 20 years) that olfaction and taste are nonlinear, low energy, IR systems. These nonlinear wavelengths have shifted, appeared, and disappeared with the ever changing windows in the water and the atmosphere. Thus has evolution progressed. The vibrating spines of arthropods are selected for or against in a constant changing pattern. Such a system must by its very nature lead to a tremendous diversity of species—a monophyly, of divergence! The system ensures tremendous speciation, for who among my readers would deny that arthropods are a diverse and probably increasing group of organisms?

I leave my reader with the thought that just as physicists have utilized high energy, inorganic, nonlinear systems in the design of inorganic radio and light lasers, so also has nature evolved a system for taking advantage of the nonlinear, IR emissions of organic molecules. Nature did it long before man! Like the great philosopher, Teilhard de Chardin, I refuse to believe that man, with all his technology, is smarter than the Creator!

5.9 SUMMARY

Teilhard de Chardin (*The Phenomenon of Man*, 1959) has outlined seven "senses" that scholars must develop if they are to make headway in the study of evolution. His seventh sense is a sense of the "organic," which is based on physical links with the environment. In order to understand these physical links, a definition of the metastable state is given. Life exists in a metastable state.

The relationship of the metastable state to evolutionary theory is developed by tracing the evolution of Arthropoda from the primitive annelids to the modern insects (Sharov, 1966). The phylogenetic tree as presented (Fig. 5.4), from the Ctenophora to the insects, is based on very convincing evidence.

The origin of the antenna and its sensilla in Arthropoda is traced, and phylogenetic evidence presented that the insect antenna evolved from the locomotory appendages of the third segment of the most primitive annelid form (Sharov, 1966).

I favor monophyly in the group Arthropoda, based on the mechanism of olfaction and my belief in a gradual transformance of a common olfactory mechanism workable in both a liquid and a gaseous medium, and in both media utilizing the H_2O molecule in the system. Evidence for monophyly in insect sensilla is given by comparison of the sensilla sizes, types, and arrangements and also the vibration frequency and grooming parameters.

My ideas about monophyly and evolution are based on well-documented paleontological evidence which demonstrates that the antennae of insects evolved from the locomotory parapodia of the third segment. As such, these evolutionary precursor organs were subject to the current flows of the liquid substrate. They also maintained mechanical contact, from which tapping might evolve, with the ocean floor substrate.

The origin of grooming in arthropods is traced. The higher orders have more specialized grooming organs and behavior.

My theory on the evolution of the insect communication system is presented.

It is demonstrated, by presenting the maserlike emissions predicted from pheromones (Callahan, 1975a,b, 1977a,b,c), that insects (as does man with lasers) take advantage of the very small but always present nonlinear response of materials to light. Insects phase-amplitude modulate molecular emissions by vibrating their antennae at appropriate modulation frequencies.

Atmospheric trace gases affect the emission and absorption properties of the IR radiation from insect semiochemicals. The primary transmissions of IR wavelengths are through the water vapor atmospheric windows. The changing atmosphere over the eons, opening and closing of atmospheric windows, changing concentrations of trace molecules in the atmosphere, and temperature shifts at the earth's surface have created an environmental selective pressure that probably contributed to a gradual divergence of insect forms and thus monophyly in Arthropoda. Concentration tuning and temperature-dependent shifts of wavelengths in the changing atmosphere were among the most important selection mechanisms in the continual speciation of the group Arthropoda.

REFERENCES

Anderson, J. D. 1976. *Gasdynamic Lasers–An Introduction*. Academic Press, New York, London.

Broili, F. 1932. Ein neuer Crustacee aus den rheinischen Unterdevon. *S. Ber. Bayer. Akad. Wiss.* 1: 27–38.

Butler, L. I., J. E. Halfhill, L. M. McDonough and B. A. Butt. 1977. Sex attractant of the alfalfa looper, *Autographa californica* and the celery looper, *Anagrapha falcifera* (Lepidoptera: Noctuidae). *J. Chem. Ecol.* 3(1): 65–70.

Callahan, P. S. 1967. Insect molecular bioelectronics: A theoretical and experimental study of insect sensilla as tubular waveguides, with particular emphasis on their dielectric and thermoelectret properties. *Misc. Publ. Entomol. Soc. Amer.* 5(7): 315–47.

Callahan, P. S. 1969. The exoskeleton of the corn earworm moth,*Heliothis zea*, Lepidoptera: Noctuidae, with special reference to the sensilla as polytubular dielectric arrays. *Exp. Stn. Res. Bull. Univ. Georgia*, 54: 5–105.

Callahan, P. S. 1972. *The Evolution of Insects*. Holiday House, New York.

Callahan, P. S. 1973. Studies on the shootborer, *Hypsipyla grandella* (Zeller) (Lep., Pyralidae). XIX. The antenna of insects as an electromagnetic sensory organ. *Turrialba* 23(3): 263–74.

Callahan, P. S. 1975a. The insect antenna as a dielectric array for the detection of infrared radiation from molecules. *Proc. 1st Int. Conf. Biomed. Transducers (Paris)* Part 1: 133–38.

Callahan, P. S. 1975b. Insect antennae with special reference to the mechanism of scent detection and the evolution of the sensilla. *Int. J. Insect Morphol. Embryol.* 4(5): 381–430.

Callahan, P. S. 1977a. Solid state organic (pheromone-beeswax) far infrared maser. *Appl. Opt.* 16(6): 1557–62.

Callahan, P. S. 1977b. The moth and the candle–The candle flame as a sexual mimic of the coded infrared wavelengths from a moth sex scent (pheromone). *Appl. Opt.* 16(12): 3089–97.

Callahan, P. S. 1977c. Tapping modulation of the far infrared (17 μm region) emission from the cabbage looper pheromone (sex scent). *Appl. Opt.* 16(12): 3098–3102.

Callahan, P. S. and T. C. Carlysle. 1971. A function of the epiphysis on the foreleg of the corn earworm moth, *Heliothis zea*. *Ann. Entomol. Soc. Amer.* **64(1)**: 309-11.

Cisne, J. L. 1974. Trilobites and the origin of Arthropods. *Science (Wash., D.C.)* **186(4185)**: 13-18.

Forel, A. 1904. *Ants and Some Other Insects*. Open Court Publ., Chicago.

Fridkin, V. M. and I. S. Zheludev. 1960. *Photoelectrets and the Electrophotographic Process*. Consultant Bureau, New York.

Goldman, L. J., P. S. Callahan and T. C. Carlysle. 1972. Tibial combs and proboscis cleaning in mosquitos. *Ann. Entomol. Soc. Amer.* **65(6)**: 1299-1302.

Green, J. 1961. *A. Biology of Crustacea*. Quadrangle Books, Chicago.

Hendry, L. B., P. D. Greany and R. J. Gill. 1973. Kairomone mediated host-finding behavior in the parasitic wasp *Orgilus lepidus*. *Entomol. Exp. Appl.* **16**: 471-77.

Huntington, E. 1945. *Mainsprings of Civilization*. John Wiley and Sons, Inc., New York.

Imms, A. D. 1939. On the antennal musculature in insects and other arthropods. *Q. J. Microsc. Sci.* **81**: 273-320.

Imms, A. D. 1940. On growth processes in the antenna of insects. *Q. J. Microsc. Sci.* **81**: 285-593.

Jander, U. von. 1966. Untersuchungen zur Stammesgeschichte von Putzbewegungen von Tracheaten. *Z. Tierphychol.* **23**: 799-844.

Klun, J. A., O. L. Chapman, K. C. Mattes, P. W. Wojtkowski, M. Beroza and E. P. Sonnet. 1973. Insect sex pheromones: Minor amounts of opposite geometrical isomer critical to attraction. *Science (Wash., D.C.)* **181**: 661-31.

Koestler, A. 1973. *The Case of the Midwife Toad*. Vintage Books-Random House, New York.

Kristensen, N. P. 1975. The phylogeny of hexapod "orders." A critical review of recent accounts. *Z. Zool. Syst. Evolutionsforsch.* **13(1)**: 1-44.

Manton, S. M. 1964. Mandibular mechanisms and evolution in Arthropods. *Phil. Trans.* **247(B)**: 1-183.

Manton, S. M. 1972. The evolution of arthropodan locomotory mechanisms. Part 10. Locomotory habits, morphology and evolution of the hexapod classes. *Zool. J. Linn. Soc.* **51**: 203-400.

Manton, S. M. 1973. Arthropod phylogeny—A modern synthesis. *J. Zool. Proc. Zool. Soc. Lond.* **171**: 111-30.

Maynard, D. M. and H. Dingle. 1963. An effect of eyestalk abolation on antennule function in the spiny lobster *Panuleris argus*. *Z. Vgl. Physiol.* **46**: 515-40.

Miller, F. H. 1970. Scanning electron microscopy of *Solenopotes capillatus* Enderlein (Anoplura: Linognathidae). *J. N.Y. Entomol. Soc.* **78(3)**: 139-45.

Morris, C. 1902. *The Volcano's Deadly Work*. W. E. Scull, Washington.

Neville, A. C. 1975. *Biology of the Arthropod Cuticle*. Springer-Verlag, New York.

Perlman, M. M. (ed.). 1973. *Electrets-Charge Storage and Transport in Dielectrics*. The Electrochemical Soc. Inc., Princeton, New Jersey.

Sharov, A. G. 1966. *Basic Arthropodan Stock with Special Reference to Insects*. Pergamon Press, Oxford.

Shorey, H. H. 1964. Sex pheromone of noctuid moths. II. Mating behavior of *Trichoplusia ni* (Lepidoptera: Noctuidae) with special reference to the role of the sex pheromone. *Ann. Entomol. Soc. Amer.* **57(3)**: 371-77.

Slifer, E. H. 1955. The distribution of permeable sensory pegs on the body of the grasshopper (Orthoptera: Acrididae). *Entomol. News* **66**: 1-5.

Slifer, E. H. and S. S. Sekhon. 1970. Sense organs of a thysanuran, *Ctenolepisma lineata pilifera*, with special reference to those on the antennal flagellum (Thysanura, Lepismatidae). *J. Morphol.* **132(1)**: 1-25.

Snodgrass, R. E. 1952. *A Textbook of Arthropod Anatomy*. Comstock Publ. Co., Ithaca, New York.

Snow, P. J. 1972. The antennular activities of the hermit crab, *Pogurus alaskensis* (Benedict). *J. Exp. Biol.* **58**: 745-50.

Stehling, K. R. 1966. *Lasers and Their Applications*. World Publications, New York.

Størmer, L. 1944. On the relationships and phylogeny of fossil and recent Arachnomorpha. *Skrift. Vid.-Akad. Oslo. I. Math.-Nat. Kl.* **(5)**: 1-158.

Størmer, L. 1951. Studies on trilobite morphology. Part III. The ventral cephalic structures with remarks on the zoological position of the trilobites. *Norsk. Geol. Tidsskr.* **29**: 108-58.

Szent-Gyorgyi, A. 1960. *Introduction to a Submolecular Biology*. Academic Press, New York, London.

Teilhard de Chardin, P. 1959. *The Phenomenon of Man*. Harper and Row Publishers, New York.

Tiegs, W. O. 1947. The development and affinities of Pauropoda, based on a study of *Pauropus silvaticus*. *Q. J. Microsc. Sci.* **88**: 165-336.

Tiegs, W. O. and S. M. Manton. 1958. The evolution of the Arthropoda. *Biol. Rev. (Cambridge)* **33**: 255-337.

Traub, W. A. and M. T. Stier. 1976. Theoretical atmospheric transmission in the mid- and far-infrared at four altitudes. *Appl. Opt.* **15(2)**: 364-78.

Wang, W. C., Y. L. Yung, A. A. Locis, T. Mo and J. E. Hansen. 1976. Greenhouse effects due to man-made perturbations of trace gases. *Science (Wash., D.C.)* **194(4266)**: 685-90.

Webb, J. P., F. G. Webster and B. E. Plourde. 1974. Sixteen new infrared laser dyes excited by a simple, linear flashlamp. *Eastman Organic Chem. Bull.* **46(3)**: 1-8.

Wheeler, W. M. 1960. *Ants—Their Structure, Development and Behavior*. Columbia University Press, New York.

Walcott, C. D. 1912. Cambrian geology and paleontology. No. 6. Middle Cambrian Branchiopoda, Malacostraca, Trilobita, Merostomata. *Smithson. Misc. Collect.* **57**: 145-228.

Wright, R. H. 1966. Odour and molecular vibration. *Nature (Lond.)* **209(5023)**: 571-73.

6

Eye Structure and the Monophyly of the Arthropoda

H. F. PAULUS

6.1 INTRODUCTION

Eyes are widely distributed in Arthropoda. In the Trilobita, there are lateral composite eyes and four median eyes (Sharov, 1966). Among Chelicerata, Xiphosura (*Limulus*) have a pair of facetted eyes and one pair of median eyes. Simple-lens eyes are found in other Arachnida, whereas compound and median eyes are the rule in the Mandibulata. These photoreceptors have played an important role in the discussion of the mono- or polyphyletic origin of the Arthropoda, but this discussion has normally been very superficial. Mostly it has been claimed that because composite eyes are distributed in all the main groups of Arthropoda, they must be the basic type; thus the Arthropoda are monophyletic. These eyes are also taken to be a consequence of convergent evolution (Tiegs and Manton, 1958). However, we will see in the following account that facetted or compound eyes evidently are the basic type of eyes in all groups of Arthropoda, indicating that the latter are monophyletic. But for refuting the idea of polyphyletic origin it is necessary to have a look at this question in a more detailed and critical way.

In order to prove monophyly of Arthropoda, one has to show that all members of this group can be traced back to a common ancestor. This can be accomplished by the synapomorphic scheme of Hennig (1950, 1966). It is possible to reconstruct a dendrogram with the aid of common derived characters (synapomorphies). The main difficulty in this approach is the correct recognition and evaluation of the synapomorphic characters in the members of the various groups of arthropods. These characters must be separated from primitive plesiomorphic characters, and above all from convergences and parallelisms. There-fore, one has to look for homologies in analyzing the forms and functions of the characters. In the following text, I will first compare the different eye structures in the various groups of Arthropoda to show the similarities and dissimilarities. Dissimilarities may be due to the fact that a particular function was resolved during evolution in different ways. Such structures are called the product of analogy. Dissimilarities may also result from secondary functional changes. Similarities likewise may be the result of nearly identical solutions of a particular function by different groups (convergences).

In discussing the evolution of eyes in different arthropod groups, it must be said that proof of convergence in the development of composite eyes or the ommatidia does not automatically imply polyphyly among those groups. On the other hand, homology of eye characters does indicate monophyly. The reliability of homologies is based on the complexity of characters compared. Facetted eyes are unquestionably highly complicated structures, so that it is possible to find homologies to determine mono- or polyphyly in the Arthro-poda. There are three possible interpretations of the evolution of eyes in various groups of Arthropoda:

1. The first arthropod had a relatively unspecialized, small lateral eye from which developed independently the facetted eyes in the Chelicerata, Crustacea, Myriapoda, and Insecta. These facetted eyes were the product of parallelism, but the basis is homologous. In this case we presume the monophyly of the Arthropoda, but with such simple characters it is not possible to be certain.

2. The facetted eyes evolved completely independently in various arthropod groups, indicating polyphyly (in the sense of Tiegs and Manton, 1958; Manton, 1973). Similarities or identical construction of the ommatidia are the consequence of functional forces.

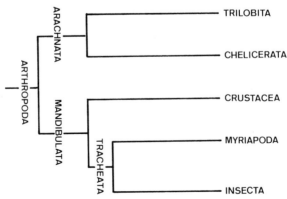

Fig. 6.1.   Dendrogram showing phylogenetic relationships of various arthropod groups.

3. The first arthropod already had a facetted eye, which was modified or specialized in the evolution of the different arthropod groups, indicating monophyly.

I favor the third interpretation. Within the different arthropod groups one can find many modifications, some of them being remarkable cases of parallelisms; nearly all of them can be traced back to a typical facetted eye. A monophyletic relationship between various groups is shown in Fig. 6.1, which also supports the opinions of other monophyletists (Hanström, 1926a,b; Snodgrass, 1935; Remane, 1959; Siewing, 1960; Sharov, 1966). This proposed dendrogram is supported by a comparative study of the different eye structures in the Arthropoda.

6.2. EYES (LATERAL) OF ONYCHOPHORA

The Onychophora show a mixture of annelidan and arthropodan characters. Most authors agree that Onychophora are not the precursors of the Arthropoda, but an early branch of the prearthropodan groups. The eyes of the Onychophora also are a mixture of a polychaete and a possibly primitive arthropodan eye (Eakin and Westfall, 1965). For this reason, I do not think that their eye structure can contribute anything to our discussion of the mono- or polyphyly of the Arthropoda. The retinula cells belong to the rhabdomeric type described by Eakin (1963, 1968, 1972), but one can also find rudiments of cilia as in most photoreceptor cells in Polychaeta. One further difference is found in the construction of the lens system. It is formed from a homogenous secretion from a subcornegenous cell layer. In most respects however, the onychophoran eye is very similar to the polychaete eye. For further details one is referred to Eakin and Westfall (1965).

6.3. EYE TYPES IN THE ARTHROPODA

Lateral eyes occur in two completely different structural and functional types: 1) the simple-lens eye with a cup-shaped retina (all Arachnida, some aberrant eyes in fleas, lice, and insect larvae); 2) the typical facetted eye, composed of many ommatidia. Very similar to the simple-lens eyes are the dorsal or median eyes. The difference between these two eyes is not in structure but in innervation from the brain. The important functional difference between simple-lens eyes and facetted eyes is in the different optical systems. In the former, light falls from a single, mostly strongly biconvex lens on different parts of the cup-shaped retina, depending on the angle of incidence (Fig. 6.3). In the facetted eye, small arrays of light rays come through separate systems of lenses (cornea and crystalline cone) to a separated retinula, the rhabdom. Each ommatidium forms an image of a point of the surrounding area. The sum of all ommatidia forms the whole image. The effect of both systems is surely identical, depending of course on the optical system and the neuronal connections to the brain. Both eye types are different solutions to the problem of forming sharp images. Innervation of the lateral eyes comes from lateral parts of the protocerebrum. The optic lobes in most Chelicerata, primitive Crustacea, all Myriapoda, and some insects with reduced eyes consist of two neuropiles, the outer lamina ganglionaris and the medulla externa. In higher Crustacea and all insects except those mentioned is added a third neuropile, the medulla interna or lobula plate. In some higher Crustacea one can find a fourth neuropile, the medulla terminalis. The neuronal structures of these neuropiles are highly complicated, but in certain respects are very similar in all groups (Hanström, 1924; Elofsson and Dahl, 1970; Hamori and Horridge, 1966; Ribi, 1974, 1975; Strausfeld, 1976; Campos-Ortega and Strausfeld, 1972, 1973; Nässel, 1975).

6.3.1. Median Eyes (Frontal Ocelli) and Frontal Organs in Arthropoda

All Arthropoda except the Myriapoda have the so-called median eyes or frontal ocelli. In most cases, they vary from simple-lens eyes with a few rhabdomeres to many rhabdoms arranged in many small retinulae of two or three, occasionally more, rhabdomeres. A vitreous body is present only in Arachnida. A similar body was described for *Cloeon* (Ephemeroptera) by Hesse (1901). The median eyes in Crustacea are called nauplius eyes. These eyes are the only photoreceptors in the larvae. But in many species they persist even in the adult stage. In insects, they are called frontal eyes or ocelli. At least in most Crustacea and in Collembola (Entognatha), there are other simple eyes, known as frontal organs; these are innervated by a special center in the protocerebrum, the nauplius eye neuropilem or the ocelli center in insects (Hanström, 1940). But there are numerous unidentified organs that are often called frontal organs, and are related to these photoreceptors. As shown by Dahl (1958, 1965) and in particular by Eloffson (1963–1970) in different crustacean groups these are not frontal organs at all. The confusion about the nature of these frontal organs is due to the fact that people call all unidentified small organs in front of the brain frontal organs, such as X- and Y-organs in Crustacea, and cerebral glands in Myriapoda and primitive insects, which have no role in photoreception. Elofsson (1965, 1966a) established, following Hanström (1926b), the following criteria for recognition of typical frontal organs: they are photoreceptors which are innervated by the same center as the nauplius eye. The often-mentioned glandular function is due to the confusion with real glands. According to Elofsson (1965, 1966a), the Crustacea have the following types of photoreceptors besides the facetted eyes: 1) median frontal organs (= ventral frontal organs), 2) dorsal frontal organs (= lateral frontal organs), and 3) nauplius eyes. Originally, all the organs were in pairs. And finally there are four nauplius eyes (Paulus, 1972b, 1973). The original basic number must have been eight ocelli-like structures (see below, Fig. 6.31).

6.4. EYES (LATERAL FACETTED) OF THE TRILOBITA

Many fossil trilobites had large composed eyes, although nothing is known about the fine structure. However, during the last few years some interesting details have become available as a result of special techniques. These details are those of the dioptric system (Clarkson and Levi-Setti, 1975). There are two different types of composite eyes within the Trilobita. On the basis of the arrangement of the single facets, they are called either holochroal or schizochroal. The primitive and most common type is the holochroal. In this eye the single facets are very narrow, polygonal lenses, forming an image, as in most of the Crustacea and Insecta. I prefer to call this type the facetted eye. In the schizochroal type are bound isolated lenses, forming round semicircular facets. I prefer to call such eyes composite eyes. Such eyes are present only in the family Phacopidae, and must have been derived from the normal holochroal type (Müller, 1960). Clarkson and Levi-Setti (1975) think that this eye type is modified as a consequence of paedomorphosis. Towe (1973) considers this eye type as an aggregation of isolated lens eyes, comparable to the lateral eyes of the Myriapoda.

Unfortunately, only the schizochroal type is well known. These eyes have large convex lenses, isolated by dark pigmented rings. These rings surround the lens cylindrically, forming a so-called sublensar alveolus (Clarkson, 1967). The distal surface of each lens is protected by a membrane, that continues into the cuticula and a so-called sclera as an "intrascleral membrane." This membrane also forms a cylinder, which surrounds the lens and the sublensar alveolus. Sometimes there are found some rudiments of sublensar structures, possibly as crystalline cones. A further remarkable detail is a calcium crystal, lying just on the lens surface. The functional significance of these structures was discussed by Clarkson and Levi-Setti (1975). According to these authors, the trilobitan eye was a strong photoreceptor and able to see very sharp images, although with only a very short rhabdom. Nothing is known about the fine structure of the retina.

6.5. EYES OF CHELICERATA

Chelicerata have either lateral facetted eyes (Merostomata, Eurypterida) or lateral ocelli (Arachnida) and four (Pantopoda) or two (all remaining Chelicerata) externally visible median eyes. For a phylogenetic discussion it is necessary to separate these two eye types because they are of completely different origin (as in Mandibulata). In the following sections these different eyes are compared in the various chelicerate orders to show their phylogenetic significance.

6.5.1. Lateral Facetted Eye of *Limulus* (Merostomata: Xiphosura)

Limulus, one of the few recent representatives of the fossil Merostomata, has a pair of well-developed lateral facetted eyes. Many works are available on the morphology of these eyes (Demoll, 1914; Miller, 1957; Lasansky, 1967; Fahrenbach, 1968, 1970, 1971). The relatively large lateral eyes consist of numerous units, which do not form a polygonal facetted eye, but show a schizochroal arrangement as in the phacopid eye, an aggregate of many isolated but very narrow, circular lenses. The corneae are not dome-shaped but completely flat. The lens is formed by a large cuticular cone (Fig. 6.7). In this respect, the lens is comparable to the so-called exocone type of dioptric system in insects. Of course the two types are not homologous. The epidermis surrounds both the regions between the corneae and the cuticular cones. Just beneath the cone, there are about 100 high prismatic hyaline cells. Some of these cells send roots to the basal part of the ommatidium, forming supporting structures comparable to the cone cell roots of the Mandibulata. These roots pass between the retinula cells. The epidermal cells surrounding the cone are filled with pigment granules and are called distal pigment cells (Fahrenbach, 1968). The rhabdom is normally composed of 10–13 retinula cells, but the number varies between 4 and 20 (Fahrenbach, 1968). The rhabdom is a starlike monolayered aggregate of these retinula cells (Fig. 6.7B). A specific character of the *Limulus* ommatidium is the so-called excentric cell, which has a large dendritic process in the center of the rhabdom. This dendritic process has no microvillous border. Similar central retinula cells, but with small microvillous borders, can be found in the ommatidia in the thysanurans, *Lepisma*, *Thermobia* (Zygentoma) (Paulus, 1974, 1975), and in *Entomobrya* (Collembola) (Paulus, 1977). Besides these distal pigment layers, there are proximal pigment cells and interretinular pigment cells similar to the secondary pigment cells in Mandibulata. On the caudal margin of each facetted eye is situated a supplementary rudimentary lateral eye, which is invisible externally (Fig. 6.31B). The ultrastructural analyses of Fahrenbach (1970) show that this eye is in fact a degenerate part of the lateral eye composed of the same type of ommatidia and numerous active neurosecretory cells. A small nerve bundle from this degenerate eye connects it, or it along with the lobus opticus of the normal eye, to the protocerebrum. I am certain that this rudimentary eye is a part of the facetted eye. Nothing or nearly nothing is known about the function, and ontogenetic or postembryonic development. Possibly, it is an early larval eye, because during embryonic development of the lateral eyes a strong shifting of the anlage occurs (Johannson, 1933), which continues during the first and second larval stages. During this, the facetted eye is enlarged. It is possible that this rudimentary eye is the first part of the eye that later degenerates, as happens in the development of the holometabolic insects, where before pupation the larval eyes degenerate. A second possibility is that in earlier evolutionary time the Merostomata had larger eyes. Thus, the *Limulus* eye is a secondary one modified from the normal holochroal type.

6.5.1.1. Median Eyes of Limulus

There is one pair of median eyes (Demoll, 1914; Jones et al., 1971; Fahrenbach, 1971). Each eye contains about 130–150 sensory cells with an irregular arrangement of their rhabdomeres. Between them are some arhabdomeric cells (Jones et al., 1971) with processes in the distal region of the retina cells. There is no other lens than the corneal one.

Just beneath the median eye inside the prosoma is situated an organ sending axons in two bundles along the nerves of the median eyes. From its structure and position, Demoll (1914, 1917) thought it must be a photoreceptor called the "endoparietal eye." Millecchia et al. (1966) also have shown in electrophysiological experiments its photosensitive nature. From these data, I believe that these paired organs are rudimentary median eyes. The nerves from the organs run to the same center in the middle part of the protocerebrum as those from the functional median eyes. Thus it is reasonable to speculate that *Limulus* (and possibly the Merostomata) had ancestors with four external median eyes. This theory is very important, since the prearthropodan stock can be derived from animals with four median eyes.

6.5.1.2. Ventral Eyes of Limulus

A strange pair of light-sensitive organs is located on the ventral part of the prosoma just above the mouth (Fig. 6.31B). For a long time nothing was known about the real nature of these sense organs. Patten (1888) and Johannson (1933) thought that they were olfactory organs. But from the ultrastructural details (Clark et al., 1969) and the type of innervation, there is no doubt that they are photoreceptors (Demoll, 1914; Hanström, 1926a,b). Some of the differences of opinion about these organs are explainable because they are active only during the earlier instars. In older instars, they are suppressed, and replaced by olfactory sense organs (Hanström, 1926a,b, 1928). The sensory cells distally contain typical rhabdomeres, forming irregular rhabdoms. The rudimentary character of these organs is seen in the adults, where many sensory cells are present along the nerve bundle, and partly immediately on the protocerebral surface. During the early postembryonic stages, these organs develop as derivatives of the two optic ganglia on each side of the basal optic lobes. Later, these organs migrate to the ventral part just above the mouth. The phylogenetic significance of these photoreceptors is not known. However, there are two possibilities: 1) since these organs are innervated from the two optical ganglia, they could be early precursors of the facetted eyes; 2) they develop in the place where, in the embryonic stages, develop the median eyes; and furthermore, the innervation and the ultrastructure in adult stages are similar, suggesting that they are probably median eyes; if this interpretation is correct, then I believe the ventral eye to be homologous with one of the frontal organs of the Mandibulata (Elofsson, 1963, 1966a,b; Paulus 1972b) (Fig. 6.31B).

6.5.2. Eyes (Lateral Facetted and Median) of the Eurypterida

This only known fossil group of Chelicerata had a pair of facetted eyes, possibly similar to those of the Xiphosura (*Limulus*) and one pair of externally visible

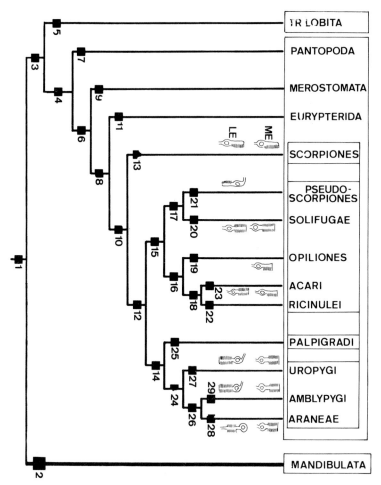

Fig. 6.2.   Phylogenetic dendrogram of Arachnata (= Tribobita + Chelicerata).   Black squares referred to by numbers are only some of the characters considered as synapomorphies. Some of the proposed characters are not quite proved and might be convergences (e.g., no. 10: book-lungs, trichobothria, or no. 14).   Characters: 1. Characters of Arthropoda; facet eyes, at least four median eyes. 2. Mandible, mandibulate ommatidium. 3. See contribution of Weygoldt (this book). 4. Lacking antennae, chelicerae. 5. Glabella. 6. Extremities on prosoma (3 + 4); eight ostia (?), tendency for fusion or reduction to two median eyes. 7. Lateral eyes lacking; inverse retina in four median eyes. 8. Two median eyes (instead of four). 9. Excentric cell in ommatidium (?). 10. Lateral eyes dispersed into five lenses; ontogenetic development of median eyes to so-called converted eyes; trichobothria (reduced in Opiliones and Solifugae (?); slit organs; book-lungs; malpighian tubules; genital opening unpaired; suction pump for fluid food. 11. One opisthosomal segment lacking. 12. Retina of lateral eyes with rhabdomere network; lyriform slit organs. 13. Pairs of ostia; combs (third segment). 14. Metasoma with a contact-sensitive flagellum (reduced in Amblypygi and Araneae).   From Weygoldt (1972). 15. Lateral eyes reduced to two, one, or zero

median eyes. Some authors (Snodgrass, 1952) have suggested a small pitlike organ just above the mouth as the ventral eye like that of *Limulus*. However, this is unconfirmed.

6.5.3. Eyes (Median) of Pantopoda

The recent Pantopoda have four large eyes on an eye hill, which are true median eyes (Sokolow, 1911; Wirén, 1918). Lateral eyes are absent even in all known fossil Pantopoda (Sharov, 1966). If we consider the possession of four median eyes as primitive, then Pantopoda would be more primitive than all terrestrial Chelicerata, and possibly also the Xiphosura (Merostomata) and Eurypterida. This consideration is important for the systematic position within the Chelicerata. As the Pantopoda definitely belong to Chelicerata (Helfer and Schlottke, 1935; Firstman, 1973) the possession of two median eyes seems to be a synapomorphic condition in the Merostomata, Eurypterida, and Arachnida (presuming that the reduction of four median eyes to two happened only once); this further implies that the Pantopoda must have separated from the prechelicerate group before branching into Merostomata, Eurypterida, and Arachnida (Fig. 6.2). The retina of these median eyes is inverse, meaning the rhabdom is located proximally far from the vitreous body.

6.5.4. Lateral Eyes of Arachnida

All groups of the terrestrial Arachnida have only simple-lens eyes, often called ocelli. Lateral facetted eyes are always absent, even in known fossil Arachnida. There are up to five relatively large, circular lateral eyes. Five eyes are present only in Scorpiones and Uropygi—in some respects the two most primitive groups of the Arachnida. The number of lateral eyes in the other groups varies from none to three. In the works of Widmann (1908) and Scheuring (1913, 1914) we can see the great diversity of the arachnid eyes. The distribution of everted or

---

lenses; tubelike tracheae. 16. Two leg-pairs used and modified as touch organs; two pairs of ostia; progoneate. 17. Two lateral eyes on each side of head; two-segmented scissors-like chelicerae with ventrolaterally inserted terminal segment; tracheae with stigma in third and fourth opisthosomal segments. 18. Larvae with three pairs of legs; preoral space enclosed by labrum and fused coxae of pedipalps. 19. Lateral eyes lacking. 20. Lateral eyes minute. 21. Median eyes lacking. 22. Totally blind; third leg modified as male gonopodium. 23. Gnathosoma. 24. Jack-knife–like chelicerae; chelicerae two-segmented; two pairs of book-lungs in second and third opisthosomal segments. 25. Totally blind. 26. Three lateral eyes on each side of head; postcerebral sucking pump; labellum; petiolus; flagellum of metasoma reduced. 27. Typical mating behavior, from Weygoldt (1972); preoral space modified as "camarostome." 28. Chelicerae with poison glands; opisthosoma with spinning orifices; male copulatory organs in pedipalpi. 29. Flagellum-like legs. *ME* = median eye; *LE* = lateral eye.

inverted retinae is very complicated, and Demoll's (1917) evolutionary ideas are in many respects erroneous because of misinterpretation and unwarranted ideas on the evolution of Xiphosura. *Limulus* was considered by Versluys and Demoll (1921) a secondary marine chelicerate. Demoll (1917) considered the position of the rhabdom (inverse or everse, distally or proximally) and the type of innervation. Unfortunately, he did not separate the median and lateral eyes. Some of the errors of the earlier workers have now been corrected by electron microscopic work (EM). However, in most of this EM work the type of axons is not considered, and much further work in Arachnida has to be done.

Median and lateral eyes differ in embryological development, their position and type of innervation. Median eyes develop from an inverted invagination of the epidermis, producing a three-layered anlage. From the distal layer develops a vitreous body consisting of hyaline cells; the median layer forms the retina; the basal one forms the so-called postretinal membrane, becoming later mostly pigment cells. Since during later embryonic development, the retina cells turn over 180°, the initial external surface becomes the internal surface. For this reason the median ocellus is called an inverted eye. However, the later retina is in all cases everse, meaning that the rhabdom is located distally just beneath the vitreous body. The lateral eyes develop as simple invaginations of the epidermis. These eyes very seldom have a vitreous body (except in some mites). In some cases, the retina is inverse although there is no inverted epidermis layer. In such cases either a secondary transformation of the retinula cells must occur, or a complete 180° turnover. Correlated with the inverse position of the rhabdomeres is always the development of a tapetum. The inverse retina seems to be different in the various groups of Arachnida, a point which is not elaborated sufficiently. Clear inverse rhabdomeres seem to be present in Pseudoscorpiones. The sensory tips of the cells are located proximally to the lens, the axons running just beneath the lens, laterally out of the eye cup (Scheuring, 1913). In Uropygi and Amblypygi, these retinula cells look very similar, except that the axons run from the middle of the cells. But this must be reinvestigated. In Araneae especially, this point has been corrected. The retinula cells have their nuclei distally, the rhabdomeres proximally; but the axons come out of the cell proximally, penetrating the tapetum layer (Land, 1969, 1972). Everse retinae are to be found in Scorpiones, Solifugae (Scheuring, 1913), and Acari (Wachmann, 1975; Mills, 1974; McEnroe, 1969).

6.5.4.1. Lateral Eyes of Scorpiones

At least since the time of Scheuring's (1913) work, the ocelli or the lateral eyes of the Scorpiones have been interpreted as a dispersed *Limulus* facetted eye. The fine structure of one ocellus seems to confirm this. The retina is composed of several isolated rhabdoms, forming different retinulae. One retinula is composed of 5 to 10 retinula cells, forming a starlike rhabdom as in the ommatidium of *Limulus.* The excentric cell dendrite is absent (Bedini, 1967; Fleissner and Schliwa, 1977). There are numerous pigment cells between these retinulae. A tapetum is absent; therefore, the retina is everse (Fig. 6.3C, D).

6.5.4.2. Lateral Eyes of Other Arachnida

Palpigradi and Ricinulei are completely blind, and the lateral eyes are absent in Opiliones and most Acari. In all other groups, the retinula cells form a network of connected rhabdomeres. In higher Araneae, we find many cells having two rhabdomeres, arranged parallel to one another (Baccetti and Bedini, 1964). In contrast to the Scorpiones, typical closed rhabdoms are missing. I described above some of the differences between the orders. The distribution of everse or inverse retinae is shown in Fig. 6.2. Homann (1950, 1952, 1971, 1975) has worked extensively on the different eye structures in Araneae. From his work one could show the significance of many different eye types in systematic considerations. The secondary eyes in spiders are homologous with the lateral eyes. Some EM work has been done on Lycosidae (Baccetti and Bedini, 1964; Melamed and Trujillo-Cenoz, 1966) and Salticidae (Eakin and Brandenburger, 1971). But all this work cannot be used for phylogenetic conclusions until more details of the other groups are known. All we can say now is that the various groups, during their occupation of land, have modified and adapted their eyes in response to the requirements of their respective ecological zones.

6.5.5. Median Eyes of Arachnida

In addition to the lateral simple eyes, all groups have a pair of median eyes, except the Pseudoscorpiones, and the totally blind Ricinulei and Palpigradi. In some Acari, this pair is fused into one. Rudimentary median eyes ("endoparietal eyes") or ventral eyes have never been found. In contrast to those of the Mandibulata, these median eyes are often the most important ones for orientation ("primary eyes"). In some higher Araneae (Salticidae), they are one of the most efficient eye types of the Arthropoda. In Opiliones and Solifugae (the lateral eyes of the latter group are very minute) they are the only photoreceptors. Retinulae with closed isolated rhabdoms are distributed in Scorpiones, Uropygi, Amblypygi, and Opiliones, whereas the retina in the Solifugae and Araneae is composed of a network of rhabdomeres as in the lateral eyes. The closed rhabdoms of the retinulae are often starlike (Scheuring, 1913; Bedini, 1967; Curtis, 1969, 1970). But in Opiliones the rhabdom structure is much different from that in the *Limulus* ommatidium (Curtis, 1969, 1970). There are five cells per rhabdom in *Euscorpius* and four in *Mitopus* (Opiliones).

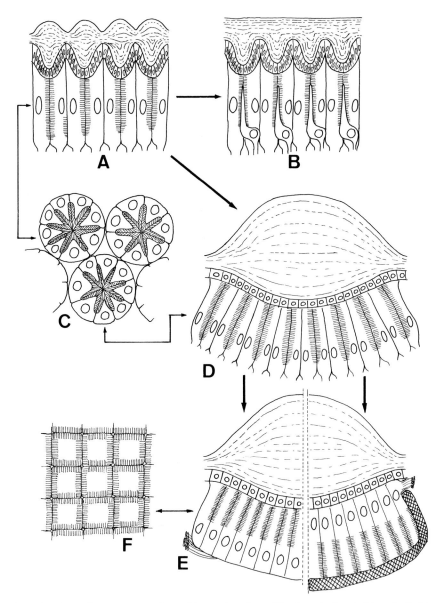

Fig. 6.3. Lateral eye types and possible modes of modifications in Arachnida. A, Facet eye type in possible ancestor of Arthropoda; B, ommatidia in *Limulus*; C, transverse section through retina of eye in (A) and (D) to show similarity; D, lens eye type of Scorpiones—this eye is supposed to have developed from the facet eye by fusion of corneae; E, lens eye type in Arachnida: right—inverse retina, left—everse retina; F, transverse section through rhabdomere region of (E) to show netlike connections of rhabdomeres.

6.5.6. Evolution of Chelicerate Lateral Eyes

Like the Trilobita, the fossil and extant Merostomata (Xiphosura) and the Eurypterida have lateral facetted eyes; it is quite certain that the early chelicerate groups had such eyes. However, it is difficult to say whether they were constructed as in the recent Limulida. The dioptric system of the Trilobita, especially of the schizochroal type of eye, is very different from that of the Chelicerata; but that of the holochroal (facetted) one seems to be similar. In *Limulus*, the arrangement of the ommatidia is different from that in the typical holochroal type because the ommatidia are circular and somewhat isolated. It is not known whether that is the case in the fossil Merostomata too. The fact that *Limulus* has some rudimentary eyes (lateral and median ones) seems to show that its progenitors probably had better-organized eyes in earlier times. The so-called excentric cell could be a special character of the *Limulus* ommatidium, since in Arachnida there is no such comparable structure. The relative lack of special characters in the *Limulus* ommatidium makes it difficult to homologize it with the facetted eyes in other Arthropoda. For a discussion of the evolution or modification of the lateral eyes within the Chelicerata, only the form of the rhabdom and arrangement of the microvilli offer some possibilities. To derive the arachnid lateral eyes from the facetted eye of the Merostomata (*Limulus*) or the Eurypterida one could postulate as follows: 1) Since in Arachnida there are up to five lenses on each prosomal side, it is reasonable to assume that the first arachnid ancestor probably had its eyes reduced to five lenses. Thus the condition in the present Arachnida is a synapomorphic situation (Fig. 6.2, character 10). Each lens corresponds to a facet in the *Limulus* eye. In this case the facetted eye is dispersed and reduced to five isolated lenses. 2) The five lenses correspond to the whole facetted eye, which disintegrated into five parts, each part with many ommatidia that persisted; the single cornea was modified into a fused lens.

Although both situations are found in different insect facetted eyes (see later), the first possibility does not seem to be true in Arachnida, since there is no evidence that arachnid eyes are comparable to one single *Limulus* ommatidium. The second possibility, however, fits exactly the eyes of Scorpiones. The retina of these eyes consists of numerous, isolated, closed rhabdoms, with starlike arrangement of the rhabdomeres (Bedini, 1967; Fleissner and Schliwa, 1977). Only the excentric cell is lacking, which I believe to be a special character of the *Limulus* eye. The retina of the remaining Arachnida is strongly modified. There are never any isolated closed rhabdoms with more than two cells. Although the evidence for the second possibility is lacking, it is more likely to have been the case. I think the eyes in the remaining Arachnida are a further modification of the situation in Scorpiones, and are comparable to those in the males of Coccidae (Insecta) (Paulus, unpublished). This type of eye has developed from a normal facetted eye (see below, Figs. 6.25, 6.26, 6.27). As in the male coccids, the arachnid retina consists of rhabdomeres forming a uniformly distributed network of microvilli. The rhabdomeres were originally present on each side of the retinula cell. In Acari and some Araneae, we find only two rhabdomeres on opposite sides of each cell, forming a small rhabdom with the neighboring rhabdomere (Baccetti and Bedini, 1964; Schroer, 1974, 1976) (Fig. 6.4). This arrangement of the retina with four-sided rhabdomeres in each cell seems to me a good synapomorphic situation in the Arachnida, excluding the Scorpiones (character 12 in Fig. 6.2). This means that the monophyly of the Arachnida as a whole is not provable on the basis of the lateral eyes. The same is true with the other characters (no. 10 in Fig. 6.2), which are very evident but not provable because of the strong possibility of convergences. All given characters in Fig. 6.2 are directly correlated with the terrestrial mode of life. The synapomorphic characters seem to be the unpaired genital openings and the sucking pump. Some authors treat Scorpiones separately from the remaining Arachnida. Since some fossil Eurypterida are very similar in general shape to Scorpiones, some authors consider them as a sistergroup of the Scorpiones (Sharov, 1966; Kraus, 1976). This suggests that the recent scorpions and the remaining Arachnida colonized land independently. If so, reduction and modification of the lateral eyes too must have happened independently. This view is supported by the

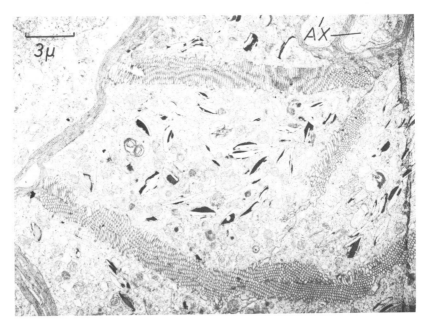

Fig. 6.4. Transverse section through proximal inverse retina of lateral eye of *Tarantula marginemaculata* (Amblypygi). *AX* = axon.

fact that there are fossil marine scorpions, some of them with numerous lateral lenses (*Mesophonus*) (Størmer, 1963). If the synapomorphies of the Arachnida are real homologies and not convergence, then these fossil scorpions must be precursors of the Arachnida, and not of the Scorpiones alone, if they are not scorpion-like Eurypterida. Then the present Scorpiones as a taxon are a para-phyletic group according to Hennig (1950, 1966). In other words, from the marine scorpion-like precursors have developed the recent Scorpiones and the remaining Arachnida. I believe one should also consider the combs in the fossil marine scorpions as characters from which the synapomorphic combs of recent Scorpiones possibly have been derived. The polyphyletic origin of the Arachnida also should be examined. But the characters given in Fig. 6.2 seem to support monophyly. As mentioned above, the lateral eyes do not provide good characters to clear up the evolutionary pathways. But the network-like distribu-tion of the rhabdomeres is a good synapomorphic feature of the Arachnida, excluding the Scorpiones. It is possible that an inverse retina also is a synapo-morphic feature of this group. The number of lateral lenses seems to support the idea of sistergroups. The Pseudoscorpiones, Solifugae, and primitive Acari have two lateral eyes on each side of the prosoma. This could be a synapomor-phic condition in these groups, or at least in the Pseudoscorpiones and Solifugae, which are considered as sistergroups because of certain characters (no. 17, Fig. 6.2). It is possible that the precursors of the Pseudoscorpiones + Solifugae, and Opiliones + Acari + Ricinulei had only two pairs of lateral eyes, which are reduced in Opiliones, Ricinulei, and most Acari. It is difficult to tell about the mono- or polyphyletic origin of Acari (van der Hammen, 1970, 1973).

According to mating behavior (Weygoldt, 1972), the Schizopeltida + Uropygi are a monophyletic group. They also have the primitive number of lateral eyes. The possession of three pairs of lateral eyes in Amblypygi and Araneae is, to my mind, a further proof of their sistergroup relationship, besides other characters (no. 26 in Fig. 6.2).

6.5.7. Homology of Chelicerate Median Eyes

Median eyes can be found in all arthropodan groups. In Merostomata, Eury-pterida, and terrestrial Arachnida we always find one pair of externally visible lenses. It is reasonable to assume that the primitive number of median eyes was four. Only the Pantopoda have four well-developed median eyes. Since this group belongs to the Chelicerata (Størmer, 1944; Raw, 1957) they should be considered as the most primitive representatives in this respect. The same might be true of the Merostomata ancestors. In Merostomata, as in *Limulus*, we find the so-called endoparietal eye, which is a pair of rudimentary median eyes. It is quite probable that these ancestors too had four median eyes. However, there are, as far as I know, only two visible median eyes in all known Merostomata

Eurypterida and Arachnida (fossil and recent), so that reduction to two must have happened before the separation into Merostomata and other groups. For this reason, the Pantopoda could be a sistergroup of the remaining Chelicerata. In addition to character no. 7 given in Fig. 6.2, the inverse retina can be a further synapomorphic feature. In the other Chelicerata, it is always everse. Hanström (1926a,b) thought the median eyes of Arachnida homologous with the ventral eye of *Limulus*. He based this idea on the type of innervation of the median eyes in scorpions, which is very similar to that of the *Limulus* ventral eye. But Johannson (1933) refuted this idea.

As shown above, the monophyly of the entire Arachnida is very probable, and can also be supported by the special mode of embryonic development of the median eyes. The development results in the so-called inverted eye, with its everse retina despite the reversal of the epidermal layer. Another character is the vitreous body, known only in terrestrial Arachnida. According to Johannson (1933) and Weygoldt (1975), the canal-like invagination of the developing median eyes in *Limulus* is homologous with the eye-hole of the Arachnida. This peculiar mode of development of the *Limulus* median eyes from an anlage near the mouth supports Lauterbach's (1973) ideas for the lateral head duplication of the early arthropod ancestors. The margins of the scutum or carapace in *Limulus* are rudiments of this duplication. The embryonic mode of development recapitulates this evolution of the lateral duplication (Weygoldt, 1975, p. 193). The embryonic mode of median-eye development is secondary (or better tertiary), simplified by removing the anlage dorsally without changing the mode of reversal of the epidermis. The phylogenetic significance of this will be considered with the discussion of the frontal organs of the Mandibulata.

6.6. EYES (LATERAL FACETTED) OF CRUSTACEA

There are many histological and EM works dealing with the eye structures of Crustacea. Some of the review articles are by Debaisieux (1944), Bullock and Horridge (1966), Waterman (1966), and Wolken (1971). Unfortunately, only the Decapoda have been well investigated. For phylogenetic discussions the eyes of the primitive groups should be better studied. The question of mono- or polyphyly of the Mandibulata as a whole can be perhaps better understood by a detailed comparison of their eyes. The astonishing similarity in the ommatidium structure in Crustacea and Insecta is well known (Grenacher, 1879; Hesse, 1901). This similarity has been interpreted both as homology (Siewing, 1960; Paulus, 1972a,b,c, 1974) and convergence (Korschelt and Heider, 1890; Tiegs and Manton, 1958). For this reason, it is necessary first to show that this typical ommatidium has developed only once within Crustacea and Insecta. Only in these two cases is the homology probable and the ommatidium shows the ground-plan characters of the Mandibulata. The ancestors of the Crustacea had facetted eyes even when these eyes were lacking in the most primitive representative, the Cephalocarida (Sanders, 1957, 1963; Lauterbach, 1972). But well-developed facetted eyes occur in Anostraca, Phyllopoda, and most other Crustacea.

6.6.1. Eyes of Anostraca

Recent works on anostracan eyes are those of Debaisieux (1944), Elofsson and Odselius (1975), and Paulus (in prep.) on *Artemia* and *Tanymastix*. The omma-tidium consists of an eucone, tetrapartite crystalline cone. The cornea is weakly curved and not thickened. Although neither Debaisieux nor Elofsson and Odselius mention the two corneagen cells, there is no doubt that these cells are present (Paulus, in prep.). They are present between the distal semper cells of the crys-talline cone and the cornea. Their proximal parts envelop the cone in the distal part (Fig. 6.5). Debaisieux called these cells "cellules épidermiques juxta-cristal-lines" because of their position beside the cone. As the two distal pigment cells of the other Crustacea are lacking, Elofsson and Odselius seem to be uncertain whether these are pigment cells or corneageous cells. However, if one considers the criterion of the absence of pigment granules and their special position be-tween the cornea and the crystalline cone, and therefore and function of build-ing the cornea, it is very clear that they are corneagenous cells.

The rhabdom is composed of six retinular cells, five of them forming a closed rhabdom. A sixth cell has a short microvillous border either distally in *Artemia* and *Tanymastix* (Debaisieux, 1944; Elofsson and Odselius, 1975; Paulus, un-published) or proximally in *Artemia* (Paulus, unpublished). The discrepancy in *Artemia* is possibly due to different species studied. Elofsson and Odselius found in *Artemia* an unlayered simple rhabdom. The *Artemia* species I studied had a small but clearly visible layering in the rhabdom (Fig. 6.5). Well-layered rhab-dom is also found in *Tanymastix* (Fig. 6.5). From the known data, the Ano-straca have an ommatidium, consisting of an unthickened cornea with two cor-neagenous cells, four Semper cells forming an eucone crystalline cone, and a clearly layered rhabdom of six retinula cells and some basal pigment cells.

6.6.2. Eyes of Phyllopoda

In this group *Triops* and *Lepidurus* (Notostraca) (Wenke, 1908; Debaisieux, 1944), *Daphnia* (Röhlich and Törö, 1965; Waterman, 1966; Güldner and Wolf 1970), *Leptodora* (Wolken and Gallik 1965) (Cladocera), and *Limnadia* (Concho-straca) (Nowikoff, 1905) have been studied by light and electron microscopy. In *Triops* and *Lepidurus*, we find ommatidia that are very similar to those of the Anostraca. The "cellules épidermiques juxta-cristallines" of Debaisieux (1944) are surely the two corneagenous cells, as in *Artemia*, *Tanymastix*, and other Crustacea. But a further very important difference is found in the number of

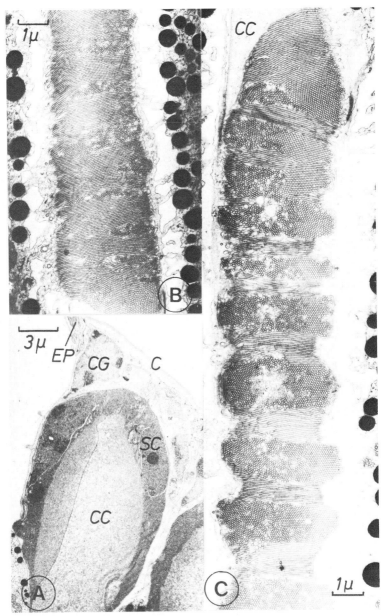

Fig. 6.5.   EM sections through ommatidium of Anostraca. A, Longitudinal section through distal part of crystalline cone of *Artemia salina*, with two corneagenous cells clearly displayed (*CG*); B, longitudinal section through rhabdom of *Artemia* to show slightly layered microvilli; C, longitudinal section through rhabdom of *Tanymastix lacunae* to show strongly tiered microvilli.   *C* = cornea; *CC* = crystalline cone; *CG* = corneagenous cell; *SC* = Semper cell.

retinula cells. The rhabdom consists of eight cells, five of which are the so-called primary, and the remaining three secondary retinula cells, all forming a closed rhabdom. It is not clear though whether the rhabdom is layered or not. Debaisieux (1944) considered it to be layered in the distal region.

From the data in older works, Dahl (1963) suggested a crystalline cone of five cells and five retinula cells in Conchostraca and Cladocera. However, the new EM works on *Daphnia* show that even in this genus there are four Semper cells and eight retinula cells. According to Nowikoff (1905), *Limnadia* has a pentapartite crystalline cone; the rhabdom is composed of five retinula cells and two so-called supporting cells. But as in *Daphnia*, I believe it is made up of at least seven or eight retinula cells. The ommatidium of *Leptodora* (Cladocera) consists of a pentapartite crystalline cone and an unlayered rhabdom of five retinula cells (Wolken and Gallik, 1965).

6.6.3. Eyes of Ostracoda

The systematic position of Ostracoda is quite uncertain. According to Dahl (1963), they have an intermediate position between his Maxillopoda and Malacostraca (see below, Fig. 6.32). Most of the species have only nauplius eyes. In Cypridinidae we find a small rudimentary facetted eye. Some species have a bipartite crystalline cone and a rhabdom of seven retinula cells (Waterman, 1961).

6.6.4. Eyes of Malacostraca

The crustacean ommatidium normally given in textbooks is that of the Decapoda, which have, in most cases, well-developed, stalked facetted eyes. The decapod ommatidium conforms, in most cases, with that of the typical crustacean eye. Even that of the Leptostraca (*Nebalia*) has a tetrapartite eucone crystalline cone, two typical corneagenous cells, and a retina of seven cells, forming a layered closed rhabdom.

The facetted eye of the Stomatopoda (*Squilla*) is similar to the type in higher Crustacea (Schiff and Gervasio, 1969; Perrelet et al., 1971; Schönenberger, 1977). The only difference is that there are eight typical retinula cells. The eyes of many Decapoda (Eguchi and Waterman, 1966; Rutherford and Horridge, 1965; Krebs, 1972) are well known. We find a special eye type in *Panulirus* (Eguchi and Waterman, 1966; Meyer-Rochow, 1975b). In this species, the normal fused layered rhabdom in the distal region becomes strongly armed in the proximal region. The type of neuronal arrangement of the retinula axons is in certain respects similar to the neuronal superposition eye of Diptera (Meyer-Rochow, 1975b). Although the ommatidia of most of the Malacostraca are identical, we find modifications in the Syncarida. According to Hanström (1934, 1935), the crystalline cone consists of two parts and a layered rhabdom of seven cells. Greater modifications are found in the Amphipoda. Their ommatidia consist

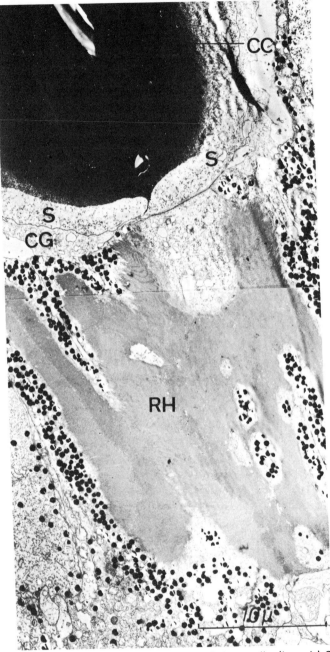

Fig. 6.6.   Longitudinal section through ommatidium of *Oniscus asellus* (Isopoda).Crystalline cone is composed of only two parts; retinula consists of 16 cells. *C* = cornea; *CC* = crystalline cone; *CG* = corneagenous cell; *RH* = rhabdom; *S* = Semper cell.

of bipartite crystalline cones and only five retinula cells, which form an unlayered rhabdom (Debaisieux, 1944; Donner, 1971). But in one species Schatz (1929) found some ommatidia with a tripartite or even a tetrapartite cone. Like the facetted eyes of Amphipoda, those of Isopoda are unstalked. The Isopoda are a heterogeneous group, and in many respects very interesting. For example, they contain many terrestrial species. Nonetheless, we have a nearly continuous series of purely aquatic to semiaquatic, to terrestrial forms; there are no principal dif ferences in eye structure, except in the strongly biconvex cornea in terrestrial forms (Thiele, 1971). There are EM studies of *Ligea* (Edwards, 1969), *Porcellio* (Tuurala and Lehtinen, 1964, 1971; Nemanic, 1975; Paulus, unpublished), and *Oniscus* (Paulus, unpublished), and light microscopic ones on different species (de Lattin, 1939; Plabody, 1939; Tuurala and Lehtinen, 1964; Debaisieux, 1944). De Lattin investigated the different reductional stages of cave isopodan eyes. In most cases, the isopod ommatidium consists of two corneagenous cells, a bipartite crystalline cone, and seven or eight retinula cells, forming a closed unlayered, short rhabdom (Fig. 6.6). In some cases, there is a single ommatidium with a tetrapartite (de Lattin, 1939) cone. *Oniscus asellus* has a giant rhabdom, consisting of 16 retinula cells. Perhaps this is a double retinula.

6.6.5. Number of Retinula and Semper Cells in Crustacea

The crustacean ommatidium shows as much variability, and modifications in its composition, as found in insects. The number of the crystalline cone cells varies from two to four. But in all cases, this modification seems to be a secondary one because in most representatives of the different higher taxa four Semper cells are typical. The same is true of the varying number of retinula cells. We find five or six cells in Anostraca, Maxillopoda, and Amphipoda. And as seems to be the case with Semper cells, this is a secondary modification or reduction from seven or eight. In most cases, these groups with modified ommatidia are living in strongly modified biotopes, and have corresponding modified habitats. From the distribution of the typical character combinations of the ommatidium within the different crustacea (see below, Fig. 6.32), I am convinced that the basic crustacean eye consisted of two corneagenous cells, a tetrapartite eucone crystalline cone, and a retinula of seven or eight cells forming a layered closed rhabdom.

6.6.6. Median (Nauplius Eyes) and Frontal Eyes in Crustacea

The median eyes of Crustacea are called nauplius eyes. The frontal organs are simple eyes innervated by the same center in the protocerebrum. The original number of these photoreceptors is present in primitive Crustacea. The genus *Triops* (Notostraca) has four nauplius eyes and the two pairs of frontal eyes. Four nauplius eyes are found in most Phyllopoda (Elofsson, 1966b) (Fig. 6.31C, below). Three median or nauplius eyes is the most common situation in Crustacea, except the Phyllopoda (Fig. 6.31D). Even the Anostraca have three median eyes. The number of the present frontal organs varies widely, frequently as a consequence of parallel reduction (Elofsson, 1963, 1965, 1966a,b). According to Elofsson, there is a great difference between the nauplius eyes of the Malacostraca and those of non-Malacostraca. In the non-Malacostraca, the sensory cells are inverse; those in Malacostraca everse. This difference led Elofsson to believe that eyes in the two groups developed independently. In Copepoda, the structural modifications are very complicated. On the basis of embryological investigation it seems that the nauplius eyes in this group are homologous with the lateral eye cup, suggesting an eventual lack of facetted eyes and the dorsal frontal organs. Elofsson ignores the fact that it is very easy to displace anlagen in the embryo. The spectacular example from Diptera makes this clear. The frontal eyes of insects normally develop from a special area of the epidermis by simple invagination. But in higher Diptera (e.g., in *Drosophila*), they develop from a common eye-antenna imaginal disc. In this case, no one would consider that because of different embryological origin, these frontal ocelli of *Drosophila* are not homologous with those of other insects. The difference between inverse or everse retinae is not always found. Rasmussen (1971) showed in *Artemia* that the retina of the nauplius eyes is everse. There are many accounts of the fine structure of nauplius eyes (Dudley, 1968; Elofsson, 1966b, 1969; Fahrenbach, 1964; Krebs and Schaten, 1976; Martin, 1971, 1976; Ong, 1970; Rasmussen, 1971; Umminger, 1968; Wolken, 1967). All these investigations establish the identity of these eyes. Differences are found especially in the number of the retinula cells. In many Isopoda where nauplius eyes previously were unknown, Martin (1971, 1976) reported three nauplius eyes and a pair of the dorsal frontal organs in many species. The sensory cells of these receptors are incorporated in all cases in the protocerebrum. They resemble the ocelli in the Collembola (Paulus, 1972b). Very peculiar nauplius eyes are found in female copepods. The genus *Copilia* has two lateral eyes, which are half the size of the body. According to Vaissière (1961), these two eyes are strongly modified lateral parts of the tripartite nauplius eye. The median part is strongly reduced. Functionally, each of these eyes is like a single ommatidium. One eye has a biconvex lens in the frontal region of the body and a small lens far from it. This latter lens is analogous to a crystalline cone but formed by one single cell. Beneath it is a L-shaped retinula, consisting of five isolated rhabdomeres (Wolken and Florida, 1969).

6.7. EYES (LATERAL) OF MYRIAPODA

Myriapoda, the sistergroup of the Insecta, usually have two groups of up to 40 relatively large circular lenses of different sizes. They are frequently reduced or even absent, as in Pauropoda, Symphyla, some Chilopoda (Geophilidae), and some Diplopoda (Polydesmidae). The arrangement of these lenses is never as narrow as in a facetted eye. Only in the Scutigeromorpha (Notostigmophora) do we find a typical facetted eye with about 250 facets. The genus *Scutigera* often plays an important role in phylogenetic discussions of the Mandibulata; because this genus is the only representative of the Myriapoda with a facetted eye, it is often considered primitive (Siewing, 1960), or in some cases as a secondary development (Adensamer, 1894: "Pseudofazettenauge"; Paulus, 1974). Indeed, the myriapodan eyes do not fit in the ommatidial scheme of the Mandibulata because of their completely different structures. This difference is one of the main arguments to support the fact that the ommatidia of Crustacea and those of the Insecta must be analogues and consequently the Mandibulata are at least diphyletic. Unfortunately, there are only a few detailed EM works on myriapodan species (*Lithobius:* Joly and Herband, 1968; Joly, 1969; Bähr, 1971, 1972, 1974; *Polybothrus:* Bedini, 1968); both genera are representatives of the Chilopoda (Lithobiidae). Older accounts are by Graber (1880), Grenacher (1880), Willem (1892), Heymons (1901), Hesse (1901), and Hanström (1934) on Chilopoda and by Grenacher (1880), Hesse (1901), and Bedini (1970) on Diplopoda (*Glomeris, Julus*).

Grenacher (1880) divided the myriapodan eyes into four types, two of which are shown in Fig. 6.8: 1) simple biconvex lens, without vitreous body (*Lithobius, Glomeris*); 2) lens with an internal chitinous corneal cone ("exocone"), without vitreous body (*Julus*); 3) simple biconvex lens, with a vitreous body (*Scolopendra*); and 4) simple biconvex lens, with a large crystalline conelike body, consisting of many parts (*Scutigera*, Fig. 6.9).

Obviously, these differences are not correlated with systematics. They are different adaptations to perceive light in the dioptric lens system. In all species, the retina is always many-layered, and contains up to 100 retina cells (Bedini, 1968). The ommatidium of *Scutigera* will be described in the next section. The difference from the single eye in the Arachnida is in the monoaxonal arrangement of all the rhabdomeres in the myriapodan eye; in Arachnida there is polyaxonal arrangement of the rhabdomeres or rhabdoms. In this respect, the eye of *Lithobius* is very similar to the stemma of some insect larvae, especially those of Coleopteroidea. Because the stemma is clearly derived from ommatidia, as will be shown later, the same I believe is possible with the myriapodan eye. There is no vitreous body in *Scolopendra* (Paulus, unpublished), although Heymons (1901) suggested one. I found instead about 1,000 retinula cells. A detailed comparison of different myriapodan eye types is not possible unless new investigations are made, especially on Diplopoda.

6.7.1. Lateral Facetted Eye of *Scutigera*

The Scutigeromorpha or Notostigmophora are interesting because of some remarkable characters. They have a row of dorsomedian stigma, somewhat resembling similar organs in Arachnida. The palpus of the second maxilla is leglike; they seem to be more primitive in this respect than all other Chilopoda. They also have a facetted eye with about 200-250 units, arranged as narrow, hexagonal facets. *Scutigera* is the single representative in the Myriapoda with such eyes. Snodgrass (1952) and Hennig (1958) for this reason consider this group to be the most primitive in Chilopoda. Grenacher (1880), Adensamer (1894), Hemenway (1900), Hesse (1901), Hanström (1934), and Miller (1957) have studied the eye in *Scutigera*. But there has never been any real effort to reconstruct the ommatidium; the nature of the crystalline conelike vitreous body is not known. I have studied (Paulus, unpublished) the rhabdom and the structure of the dioptric system in this species. The ommatidium is shown in Figs. 6.7 and 6.9. The cornea is biconvex, and the surface nearly smooth. Just beneath the lens is the very large crystalline conelike vitreous body, consisting of numerous confluent parts without any visible nuclei. The number of these parts is not fixed, and is variable even within the same eye. One segment of the vitreous body consists of a hyaline substance, and along the cell membranes are dense cytoplasmatic zones, and sometimes small spheres. The latter are called rudimentary nuclei. In electron micrographs cone-segments appear to be extracellular secretions of

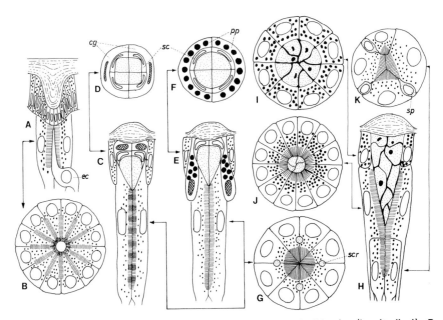

Fig. 6.7. Different types of ommatidia in Arthropoda. A, *Limulus* (longitudinal); B *Limulus* (transverse); C, D, Crustacea; E. F, Insecta; G, transverse section through rhabdom of (C) and (D); H–K, *Scutigera coleoptrata* (Myriapoda, Chilopoda); transverse sections in (J–K) are referred to levels in (H). *CG* = corneagenous cell; *EC* = excentric cell; *PP* = primary pigment cell; *SC* = Semper cell; *SCR* = cone cell root; *SP* = supporting cell.

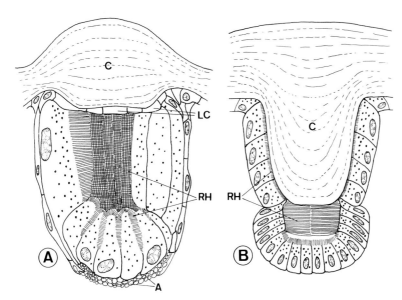

Fig. 6.8. Two types of lateral lens eyes in Myriapoda. A, *Lithobius forficatus*, simplified from Bähr (1974); B, *Julus* (Diplopoda), from Hesse (1901). *A* = axons; *C* = cornea; *LC* = lentigenous cell; *RH* = rhabdom.

the distal pigment cells surrounding the distal zone of the vitreous body (Figs. 6.7, 6.9). I believe that these cells secrete both the crystalline cone-segments and the cornea above them during the last ecdysis. The number of these pigmented cone cells varies from 8 to 12 or 13. Each cell produces more than one segment during its lifetime. The cone is functionally comparable with the pseudocone of some insects, but it is not homologous structurally. The rhabdom is two-layered. The distal retinula cells extend far around the cone laterally. Hemenway (1900) correctly observed the number to be 10-12, but it varies according to the region of the eye. For example, in the center of the eye there are about 7-10 retinula cells, and in the lateral parts (dorsally) 9-23, forming a circular ring of rhabdomeres around the cone tip (Figs. 6.7, 6.9). The proximal layer consists, in all cases, of four retinula cells, forming a triangular rhabdom (Figs. 6.7 and 6.9). A fourth cell is present somewhat excentrically, forming a small microvillous border. Between these four proximal retinula cells are to be seen three or four smaller cells, possibly functioning as supporting cells. In addition to the distal pigment cells, there are other distal and basal pigment cells. As seen from these data, the ommatidium of *Scutigera* is much different from those of Crustacea and Insecta. There are neither typical corneagenous cells, nor a tetrapartite crystalline cone. The retinula consists of 11-16 cells, forming a closed rhabdom in two layers. On comparison with the eye of *Lith-*

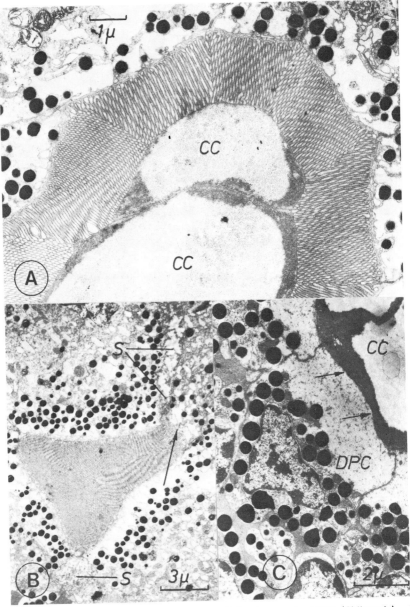

Fig. 6.9. EM sections through ommatidium of *Scutigera coleoptrata* (Chilopoda). A, Transverse section through proximal part of vitreous body with surrounding distal rhabdom, consisting of about 10-12 retinula cells; B, transverse section through proximal rhabdom. There are four rhabdomeres, one of them very small (arrow). Note small cells possibly supporting cells (*S*). C, Transverse section through distal part of vitreous body to show distal pigments cells, which produce segments of vitreous body.

obius, the *Scutigera* ommatidium seems to be derived from the former. Thus, the facetted eye of *Scutigera* is not homologous with those of insects or cray-fishes. The evolution of this eye type will be discussed in Section 6.10.3.

6.8. EYES (FACETTED) OF INSECTA

As in Crustacea, the ommatidium of insects consists of exactly the same number of cells. There are a biconvex cornea, two primary pigment cells, four Semper cells forming a crystalline cone, and normally eight retinula cells. Only the number of the secondary pigment cells varies. There are many works on the structure and function of the insect ommatidium, some of them included in Bullock and Horridge (1965), Wehner (1972), Horridge (1975a), or Snyder and Menzel (1975). In spite of the great diversity in composition of the dioptric system and the rhabdom, the number of participating cells is constant in an astonishing manner. Only in Hymenoptera (as so far investigated) do we find nine retinula cells; and in some butterflies there are up to twelve. But in nearly all other insects, there are eight, so that the basic number for insects seems to be eight. Modifications, therefore, are secondary, and in most cases easy to demonstrate. The only principal difference from the crustacean ommatidium is the presence of two primary pigment cells, instead of two corneagenous cells (Fig. 6.7C,D and E, F). But their homology was proved and demonstrated by Hesse (1901) and Paulus (1972a, 1974). As in Crustacea, it is important to investigate the eye structure in the primitive insect groups (apterygote insects) for the phylogenetic discussion. Hesse (1901) thought that their eyes represent transitional stages between the myriapodan eyes and those of other insects. If this were true, it would mean an independent evolution of the typical mandibulate ommatidium. This in turn would imply that the ommatidium described above in Crustacea and Insecta is a case of convergence. Since I am convinced that the typical ommatidia s. str. are homologous (Paulus, 1974), they must have evolved in their typical form in the ancestors of the Mandibulata. The eyes of Myriapoda are secondarily modified as discussed later. One way to show the wrong interpretation of the evolution of the ommatidium by Hesse (1901) and Tiegs and Manton (1958) is to explain the real construction of the apterygote insect eye. In the following account I will demonstrate that even these primitive insects have the typical insect ommatidium.

6.8.1. Degenerate Facetted Eyes of Collembola

Collembola are the only representatives of the Entognatha with eyes. They usually have two groups of up to eight relatively large circular lenses, characteristic of the species in size and arrangement. From the investigations of Hesse (1901), Paulus (1970a,b, 1972a,b, 1974, 1975, 1977), and Barra (1971), we know that these eyes are typical ommatidia, at least originally completely identical with those of higher insects. The composition of the rhabdom is quite diverse in the different colembolan groups, even within one species. But the number of retinula cells is remarkabley constant, and as in Crustacea or other Insecta, nearly always eight (Fig. 6.10). Hesse (1901) considered the two primary pigment cells as corneagenous cells. But in the suborders Symphypleona and Arthropoleona (*Entomobrya*) there are two typical primary pigment cells. These two cells in the other species do not have pigment granules or any other cytoplasmatic structures. For this reason I consider them as degenerate, which indicates that in earlier times they were primary pigment cells. This modification is shown in Fig. 6.11. The crystalline cone is always formed by four Semper cells. In Arthropleona the cone is a large, extracellular, homogenous ball (ecteucone). The cone in Symphypleona is intracellular, and therefore tetrapartite as in the other insects (enteucone). After a greater reduction of the dioptric system, as seen in some cave species of the Hypogastruridae or Poduridae (*Anurida maritima*), the crystalline cone is lost (Thibaud, 1967; Thibaud and Massoud, 1973; Paulus, 1974). But even in such cases, the four Semper cells and the two degenerate primary pigment cells are present. This acone condition surely is secondary (Fig. 6.12). As mentioned before, the rhabdom always consists of eight cells, which is arranged in Poduromorpha (Cassagnau, 1970) as two-layered. The other Collembola have a single-layered rhabdom as in most other insects. From these data, it

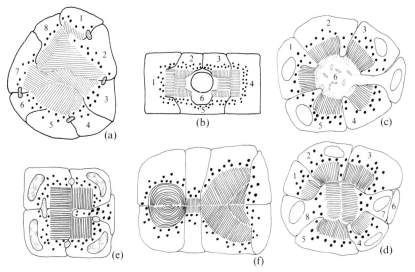

Fig. 6.10. Transverse sections through different rhabdom types in Collembola: a, Irregular type in *Orchesella*; b, extreme bilateral type in secondary eyes in *Entomobrya muscorum*; c, open-rhabdom type in distal region of primary eye in *Entomobrya muscorum*; d, as in (c) but proximal region; e, *Podura aquatica*; f, *Allacma fusca*. Modified from Paulus (1972 a, 1974, 1977).

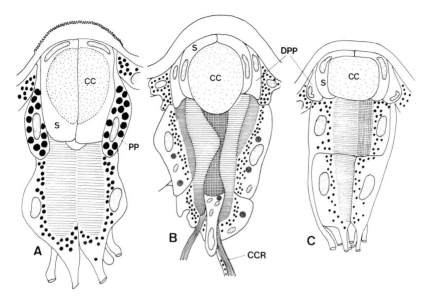

Fig. 6.11. Different types of ommatidia in Collembola: A, Normal insectan ommatidial type in Symphypleona (*Allacma, Dicyrtomina*); B, monolayered type with dispersed primary pigment cells and an enteucone crystalline cone in Arthropleona-Entomobryomorpha (*Orchesella*); C, two-layered type similar to (B) in *Podura* or *Neanura* (Poduromorpha). Modified from Paulus (1972a, 1975). *CC* = crystalline cone; *CCR* = cone cell root; *DPP* = degenerated primary pigment cell; *PP* = primary pigment cell; *S* = Semper cell.

appears that the ancestors of the Collembola had typical ommatidia as the higher insects, and that the collembolan eye is a dispersed facetted eye. Therefore, this eye has nothing in common with the recent myriapodan eye.

6.8.2. Lateral Facetted Eyes of Zygentoma

As in Collembola, some thysanurans (silverfish and firebrats) possess a reduced compound eye, which consists of 12 ommatidia in *Lepisma, Thermobia*, and *Ctenolepisma. Tricholepidion gertschii*, the only living representative of the family Lepidotrichidae, has an eye with 40–50 ommatidia. Their structures were examined by Hesse (1901), Hanström (1940), and Elofsson (1970). EM studies are available for *Lepisma, Thermobia*, and *Ctenolepisma* (Brandenburg, 1960; Paulus, 1972a, 1974, 1975; Meyer-Rochow, unpublished). The structure of the ommatidium is shown in Fig. 6.13. The Zygentoma have, like most insects, a biconvex cornea, four Semper cells (which form in *Lepisma* an acone, and in other genera an enteucone crystalline cone), and two primary pigment cells. In some cases, the pigment granules of the primary pigment cells are seen to have migrated from these cells to an extracellular space. This pigment displacement

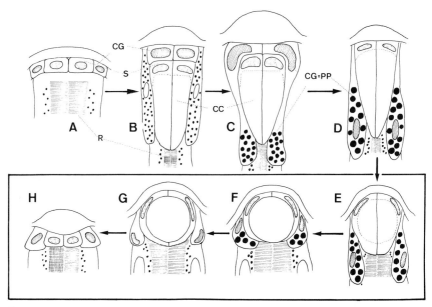

Fig. 6.12. Homologies of corneagenous and primary pigment cells from a premandibulate ancestor (A) to Crustacea (B), to Insecta (C) and (D). A, Suggested protomandibulate ommatidium with two corneagenous and four cone cells, indicated differently by drawing of nucleus; B, typical crustacean; C, Archaeognatha (*Machilis, Dilta*) with two corneagenous cells, developed also as pigment cells; D, pterygote insect of eucone type—primary pigment cells are connected tenously with cornea in imago; E–H, progressive reduction seen in Collembola: E, symphypleonid collembolan (Sminthuridae) similar to D; F, *Entomobrya*, retaining primary pigment cells; G, arthropleonid collembolan, with reduced unpigmented primary pigment cells; H, *Anurida maritima* with reduced crystalline cone but retaining all cells. From Paulus (1974). *CC* = crystalline cone; *CG* = corneagenous cell; *PP* = primary pigment cell; *S* = Semper cell.

simulates the existence of more primary pigment cells as described by Elofsson (1973) in *Lepisma.* But in *Thermobia* and *Tricholepidion* there are only two pigment cells. The retinula consists of eight cells, forming a two-layered rhabdom. Distally, there are four cells, one of them with a very small microvillous border in the center, and an excentric nucleus as in *Entomobrya* (Collembola) (Paulus, 1977). There are three cells in the proximal region, forming a triangular rhabdom, as seen in cross sections.

6.8.3. Lateral Facetted Eyes of Archaeognatha (Thysanura: Machilidae)

The Machilidae (jumping bristletails) are the only apterygote insects with fully developed facetted eyes, which are large and meet on the dorsal midline of the head, a good synapomorphic feature in the Archaeognatha. It is reasonable to

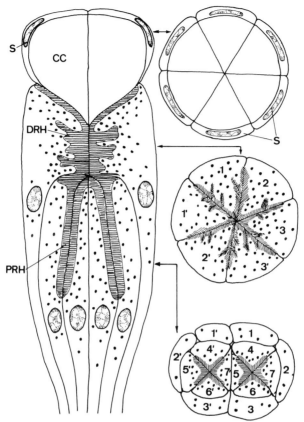

Fig. 6.13. *Thermobia domestica* (Zygentoma) ommatidium in longitudinal section with transverse sections at levels indicated. Note secondary extruded pigment grains from primary pigment cells. From Paulus (1974). *CC* = crystalline cone; *EP* = extruded pigment grains; *PP* = primary pigment cell; *S* = Semper cell.

assume that the ommatidium in this group is preserved unchanged from the ancestors of all insects. Therefore, the knowledge of their eyes is of particular interest. As we will see, the ommatidial structures are in certain respects a transition from the Crustacean ommatidium. General accounts of the eyes in Archaeognatha are available by Hesse (1901) and Hanström (1940). EM studies are by Paulus (1972a, 1975) and Meyer-Rochow (1972a). A more detailed analysis has been conducted by Paulus (1977, in prep.) on the ommatidia of *Machilis* and *Dilta*. I also have obtained some unpublished data on *Allomachilis* (from Horridge in Canberra).

For our discussion the construction of the dioptric apparatus of some interest. The cornea is sligh1y thickened, and covered, as in most insects, with nipples

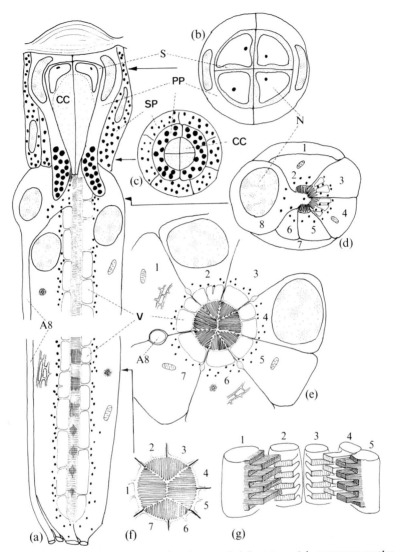

Fig. 6.14. Ommatidium of *Machilis* (Archaeognatha) (not to scale), transverse sections at levels indicated. Accessory or secondary pigment cells are omitted in (b). Note rhabdomere 8 in (d), bilateral symmetry in (f) and crustacean pattern of rhabdom detailed in (g). From Paulus (1975). *A8* = axon of the eighth sensory cell; *CC* = crystalline cone; *N* = nucleus; *pp* = primary pigment cell; *SP* = secondary pigment cell; *S* = Semper cell; *V* = vacuoles.

(Bernhard et al., 1970). Directly beneath the lens, two large cells enclosing the cone have nuclei situated (at least in part) between the cornea and the crystalline cone, and the cytoplasm extending for some distance proximally (Fig. 6.14). According to Hesse (1901), these are typical corneagenous cells. But from the EM studies, it is evident that these proximal extensions around the tip of the cone contain pigment grains, which are slightly larger than those in retinula cells. These two cells are, therefore, primary pigment cells. But from their position and function, these two cells represent an intermediate stage between Crustacea and Insecta. In one respect, they are corneagenous cells, producing the new cornea during the many permanent ecdyses, and in another, they are iris pigment cells like the primary pigment cells of other insects. The crystalline cone, which tapers sharply, consists of four parts formed by four cone cells with nuclei arranged in four quadrants between the cone and lens. The elongated rhabdom consists of seven retinula cells, arranged as in a typical apposition fused-rhabdom eye. But within one complex eye we find in some ommatidia an eighth cell, which forms a small rhabdomere either distally just beneath the tip of the cone (in *Machilis, Dilta:* Paulus, 1975, and unpublished) or basally (in *Allomachilis:* Meyer-Rochow, 1972a). The rhabdom is layered or tiered at least in the basal region, which means that, as in Crustacea, there is a periodic change in the direction of the rhabdomeres. The extension of this layering within one rhabdom is not constant (Fig. 6.15).

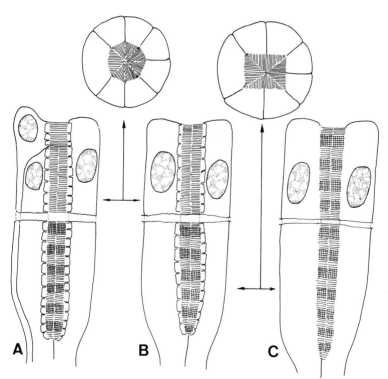

Fig. 6.15. Rhabdom types in *Machilis* in longitudinal sections and transverse sections indicated at corresponding levels. A, Type with eight retinula cells and only proximal part layered; B, same as in (A) but seven cells; C, same as in (B) but entire rhabdom layered.

6.8.4. Lateral Facetted Eyes of Pterygota

Since the adaptive radiation in pterygotes has led to great diversity of groups and species with many new adaptations and changes in modes of life, the lateral facetted eyes have undergone modifications to adapt to the different demands on light resolution and form and color perception. Consequently, numerous types of apposition, superposition, or clear-zone eyes have evolved. Rhabdoms and the dioptric apparatus changed very often. The different types of crystalline cones are shown in Fig. 6.18, and rhabdom cross sections of different insect groups in Fig. 6.17. A review article on the function of the superposition or clear-zone eyes is given by Horridge (1975b,c). Snyder and McIntyre (1975) have worked on the light-guide theories of the rhabdom, and Waterman (1975), Menzel (1975), and Wehner et al. (1975), and Wehner (1976) on the perception of polarized light. In spite of the great diversity in ommatidial structure one unit of the compound eye is remarkably consistent in its number of cells. The

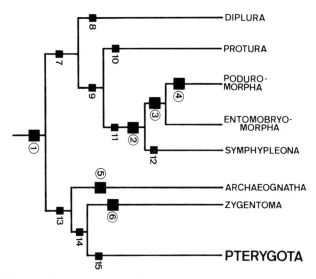

Fig. 6.16. Phylogenetic dendrogram of apterygote insects. Synapomorphies given by Hennig (1969) are indicated as small squares with uncircled numbers; synapomorphies of eyes are indicated with large squares and circled numbers. 1, Typical insect ommatidium with primary pigment cells; 2, dispersed facet eye with up to eight ommatidia; 3, ecteucone crystalline cone type; 4, retinula two-layered; 5, facet eyes meeting dorsally; 6, retinula two-layered, distal central cell with small rhabdomeres; 7, entognathous; 8, lacking eyes; 9, antennae reduced to four segments, unpaired dactylus, lacking cerci; 10, lacking antennae, blind (independently from Diplura); 11, antennae four-segmented, six abdominal segments, tibiotarsus, abdominal legs specialized; 12, fusion of abdominal segments; 13, antennae with scape and pedicel and a flagellum; 14, dicondylous mandible, sperm: 9 + 9 + 2; 15., wings. Adapted from Hennig (1969) and original data.

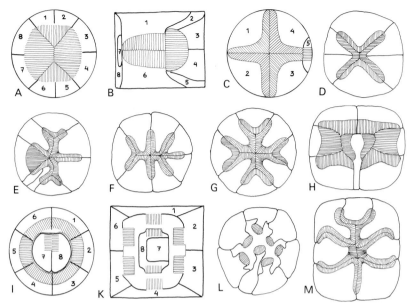

Fig. 6.17. Different transverse sections through rhabdoms of pterygote insects. A, *Apis*, *Gryllus*; B, *Ishunura* (Odonata)—from Ninomiya et al. (1969); C, *Blaberus* (Dictyoptera)—from Wolken and Gupta (1961); D, *Dytiscus* (Coleoptera)—from Horridge (1969b); E, *Archichauliodes* (Megaloptera)—from Walcott and Horridge (1971); F, *Ripsemus* (Col., Scarabaeidae)—from Horridge and Giddings (1971b); G, *Ephestia* (Lepidoptera)—from Fischer and Horstmann (1971); H, *Sartallus* (Col., Staphylinidae)—from Meyer-Rochow (1972b); I, *Aedes* (Diptera)—from Brammer (1970); K, *Gerris* (Heteroptera)—from Schneider and Langer (1969); L, *Drosophila* (Diptera); M, *Atelophlebia* (Ephemeroptera)—from Horridge (1975b).

crystalline cone, whether acone, eucone, or pseudocone, is always composed of four segments, produced by four Semper cells. Even if the cone is replaced by the extension of the cornea, as in some acone or especially exocone types, the four Semper cells are present. The same is true of the two primary pigment cells. Basicly, the rhabdom is composed of eight retinula cells, forming the rhabdom normally in one layer. Sometimes there are nine cells (Hymenoptera: *Formica* or *Apis:* Menzel, 1972; Skrzipek and Skrzipek, 1971, 1973; Grundler, 1974) or as in some Lepidoptera from eight to twelve (*Ephestia:* Horridge and Giddings, 1971a; Fischer and Horstmann, 1971; *Galleria:* Stone and Koopowitz, 1976). The number of secondary pigment cells is variable.

6.8.4.1. Types of Crystalline Cones in Pterygota

Figure 6.18 shows the types of cones known since the time of Grenacher (1879). Eltringham (1933) added the exocone type for cases where the cone is replaced

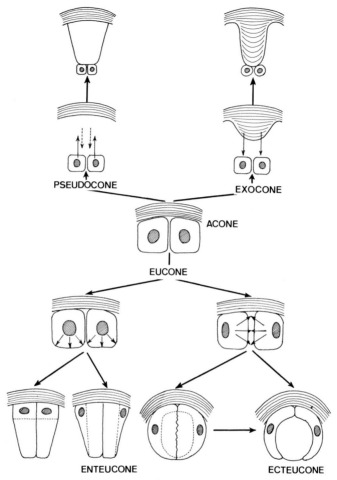

PSEUDOCONE

EXOCONE

ACONE

EUCONE

ENTEUCONE

ECTEUCONE

Fig. 6.18. Cone types in insects. Crystalline cone is very variable in different insect forms. But in all cases four cone cells are present. From Paulus (1972a).

by a corneal cone. This type is distributed in some Coleoptera (*Dascillus* [Paulus, unpublished] , Elateroidea, and Cantharoidea), and was misinterpreted by Meixner (1935) as a pseudocone type. New EM accounts have now confirmed (Horridge, 1969a, on *Photuris*) its exocone nature. The different types, however, are not separated distinctly in various insects. Within the Insecta the basic type of cone is without any doubt the eucone type. As I have shown (Paulus, 1972a), this type is most widely distributed, and especially so in most apterygote insects with eyes, and also in the primitive pterygote groups (Blattodea, Ephemeroptera, Odonata, and Plecoptera). Derived from this original type is the acone type found in Dermaptera, Heteroptera, some Coleoptera (Staphylinidae: Grenacher, 1879; Meyer-Rochow, 1972b; Tenebrionidae: Eckert, 1968; Cerambycidae: Grenacher, 1879), and primitive Diptera (Tipulidae: Dietrich, 1909; *Aedes:* Brammer, 1970; *Chironomus:* Paulus, unpublished.) From this acone type is derived in one direction the pseudocone one (in higher Diptera) and in another direction the exocone one, replacing the lost cone by an extracellular fluid or by a chitinous corneal cone. As mentioned before, the four Semper cells remain in all these cases. Modifications in the number of cone segments as described in *Apis* (Skrzipek and Skrzipek, 1971) or in *Dilta* (Paulus, unpublished) are aberrations in embryonic development. Only in the larvae of Trichoptera and Lepidoptera do we find a fixed modified number of cone cells. They always have three parts (Paulus and Schmidt, 1978).

6.8.4.2. Rhabdoms of Pterygota

The different rhabdom forms are mostly not interpreted at this time either in a functional or in a phylogenetic way. I believe the radially symmetrical and simple, bilaterally symmetrical rhabdoms are the most primitive in insects; similarly the one-layered arrangement of the eight retinula cells also is the basic type in insects, and even in Mandibulata. Sometimes a cell (many times called the eighth) is present on the proximal or distal part of the rhabdom and is called the rudimentary or dgenerate cell. But it seems to me that this cell always has a distinct function, and the smallness is an adaptation. In *Apis*, this cell perceives the polarized light, along with UV (Menzel and Snyder, 1974; Gribakin, 1972, 1975; Helversen and Edrich, 1974; Wehner et al., 1975). Only in a few cases can the arrangement of the rhabdomeres be correlated with function. The open rhabdom of the higher Diptera is correlated with the special mode of direction-vision and resolution. The separated rhabdomeres 1–6 are connected separately in the lamina, in a manner described as neuronal superposition (Kirschfeld, 1973). But the rhabdom of the primitive Diptera (*Aedes:* Brammer, 1970) consists of the rhabdomeres 1–6, forming a closed ring; even the rhabdomeres 7 and 8 are connected with this ring (Fig. 6.17). It is hard to explain the evolution of the neuronal superposition in Diptera, since it is different from all other insects. The primitive Diptera (Nematocera) might reveal some transitional forms.

Snyder's (1975) suggestion of rhabdom function applies only to rhabdoms with a uniform rhabdom volume. Cross sections of many holometabolan rhabdoms (Fig. 6.17) show a cross-shaped or cruciform figure, which is possibly the basic one for the Holometabola. Horridge and Giddings (1971b) called it the neuropteran type. We find this type in Coleoptera (*Dytiscus:* Horridge, 1969b; *Gyrinus:* Wachmann and Schroer, 1975; Burghause, 1976; *Anoplognathus:* Meyer-Rochow and Horridge, 1975), Lepidoptera (*Ephestia:* Horridge and Giddings, 1971a; Fischer and Horstmann, 1971), Neuroptera (*Chrysopa:* Horridge,

1976; *Ascalaphus:* Schneider and Langer, 1975), Megaloptera (*Archichauliodes:* Walcott and Horridge, 1971), and primitive Hymenoptera (*Perga* larva: Meyer-Rochow, 1973). But within these groups, this basic rhabdom was modified and simplified secondarily. Some very peculiar rhabdoms are found in some Ephemeroptera (Meyer-Rochow, 1971; Horridge, 1976; Wolberg-Buchholz, 1976).

6.8.5. Larval Eyes in Insects

In holometabolic insects, the larvae have special eyes, called stemmata. They are not uniformly constructed, according to Hesse (1901). Some authors see in them a recapitulation of the lateral eyes of the Myriapoda (Demoll, 1917; Snodgrass, 1935; Weber, 1933; Kaestner, 1972). Since some stemmata are similar to the myriapodan eye, Kaestner called the latter also stemmata. Consideration of the structure of the stemmata is significant in the discussion of the evolution of the facetted insect eye from the multilayered chilopodan-like eyes (Hesse, 1901; Tiegs and Manton, 1958). This mode of evolution is presented in nearly all textbooks of entomology (Weber, 1933, 1954; Eidmann-Kühlhorn 1970; Chapman, 1969). The stemmata are always considered as precursors of the ommatidia, and are believed to have had this simple structure in the insect precursor since the myriapodan time. That means the evolution of the insect eye started with the myriapodan eye (Tiegs and Manton, 1958). This mode of evolution suggests convergent evolution of the crustacean and insect ommatidia (Fig. 6.1). However, it could be demonstrated that the insect larval stemma is the result of ommatidial modification; this would serve as a good model for the evolution to the myriapodan eye, and would thus suggest the monophyletic origin of the mandibulate ommatidium. The eyes of the Myriapoda are secondarily modified from this type of ommatidium in the same manner as the secondary modification of the holometabolan larval eyes, as a consequence of nearly identical selection pressure.

6.8.5.1. Larval Facetted Eyes of Panorpa (Mecoptera)

The larvae of most Mecoptera have a small facetted eye on each side of the head, consisting of 30–35 ommatidia. In this respect, they are the only Holometabola with an original plesiomorphic facetted eye. Similar large facetted eyes in the larvae of some Culicidae (*Chaoborus* = *Corethra*) (Diptera) are actually prematurely developed adult eyes from the imaginal disc (Constantineanu, 1930). These larvae have larval eyes, in addition to the imaginal facetted eyes. As shown by Bierbrodt (1942) and Paulus (unpublished) these eyes are slightly modified, typical insect ommatidia. Knowledge of these ommatidia, therefore, is very important for a discussion of the evolution of the insect larval and myriapodan eyes. The cornea is slightly biconvex, and the surface smooth. The eucone crystalline cone is nearly spherical, consisting of four Semper cells. Around this cone and just beneath the cornea, are two primary pigment cells. The form the corneal cuticule in the distal region during ecdysis. In the proximal part of these two cells are pigment granules, slightly more than in the other pigmented cells. The rhabdom is composed of eight reginula cells, arranged in two layers. In the distal layer are four cells, which form a cruciform rhabdom (Fig. 6.19). The proximal region has four cells, forming a rectangular rhabdom. Further details are to be seen in Fig. 6.19. The form and extension of the cone cell roots is unusual. As in most insects, they extend between the retinula cells in the distal region. In the proximal rhabdom region, one of the four disappears. The other three become quite voluminous, containing many pigment granules, in addition to the normal cell organelles. These *Panorpa* ommatidia are comparable to typical insect ommatidia. The only difference is the two-layered retina, which occurs in a similar form in some Collembola and Zygentoma (Thysanura: Lepismatidae). As is the case in the latter groups, in *Panorpa* also this is a consequence of re-

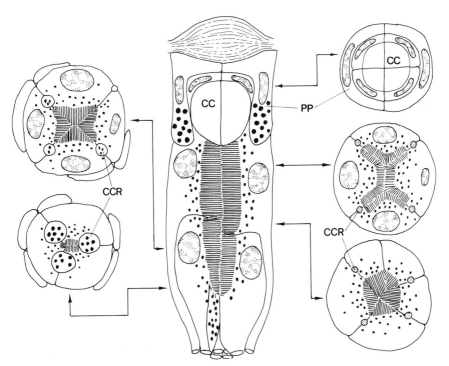

Fig. 6.19. Schematic reconstruction of larval ommatidium of *Panorpa communis* (Mecoptera). Transverse sections at levels indicated. Retinula is two-layered, dioptric apparatus normally, as in typical adult insects. *CC* = crystalline cone; *CCR* = cone cell root; *PP* = primary pigment cell.

duction of the eyes. Because facetted eyes are lacking in all other holometabolous larvae, *Panorpa*, and possibly the Mecoptera, are in this respect the most primitive forms in Holometabola. Before pupation, the larva loses this small facetted eye, which does not increase during the larval period. The larval ommatidium is possibly much different from that of the imago. Therefore, the adult makes a new complex from a special part of the epidermis near the larval eye. In this respect, the Mecoptera are typical Holometabola, in which the larval eye is always reduced in the pupa, and a new one formed from the imaginal disc of the eye.

Presumably, the described ommatidial structure in *Panorpa* is identical with that of most other Mecoptera larvae. However, modifications can be found in species with a special mode of life in which their eyes are reduced. One such larva is that of *Boreus hiemalis*. This larva, found in deep moss pillows, has only three small lenses on each side of the head. In these eyes, the crystalline cones are completely reduced, but the corresponding cells seem to be present. One retinula consists of eight cells, forming a well-organized, closed rhabdom, just beneath the cornea (Fig. 6.20). Externally, there often are two additional rudimentary ommatidial rhabdoms; one of them is near the small lens group, but the other is completely integrated in the basal optical nerve near the brain, or sometimes partly integrated in it. The rhabdoms of both rudimentary eyes also are well organized, consisting of eight retinula cells (Paulus, in prep.).

6.8.5.2. Larval Ommatidia of Trichoptera and Lepidoptera

As Hesse (1901) has stated, the lepidopteran larval eye must be very similar to that of *Lepisma* and the Collembola. In all textbooks these two eye types are always compared to prove Hesse's theory that the insect facetted eye evolved from the myriapodan ancestors. But as shown here there are great differences between these eyes, and the similarities come from parallel modes of modification (Paulus and Schmidt, 1978). The stemmata of Trichoptera and Lepidoptera are completely identical. In Trichoptera larvae we always find a large pigmented spot on each side of the head. This spot contains in *Rhyacophila* and *Hydropsyche* seven small ommatidia, and normally six in the other groups. But in most cases, there are no corneal lenses, the exceptions being the terrestrial *Enoycila* larva, in which one of the ommatidia has a small biconvex lens. The caterpillars of the Lepidoptera normally have six isolated circular biconvex lenses, widely distributed on each side of the head. In Micropterygidae, Tillyard (1923) described five small lenses arranged as in Trichoptera. This eye type is sometimes called a facetted eye (Hinton, 1958), although it is questionable. Trichopteran (Phryganaeidae) larval eyes have been studied by Pankrath (1890) and Hesse (1901). Our recent work (Paulus and Schmidt, 1978) deals with representatives of Hydropsychidae, Rhyacophilidae, Philopotamidae, Limnophilidae,

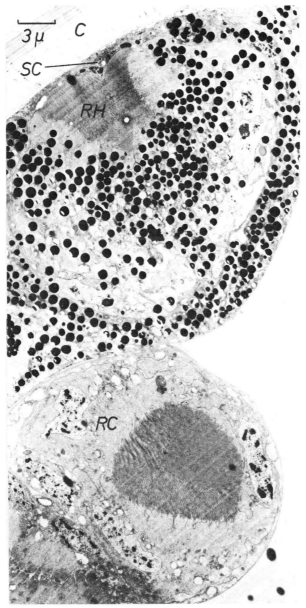

Fig. 6.20. Longitudinal section through rudimentary eyes of *Boreus hiemalis* (Mecoptera). Crystalline cones are reduced. Retinula is relatively small. Note completely extruded second eye from cuticle. *C* = cornea; *RC* = retinula cell; *RH* = rhabdom; *SC* = Semper cell.

and Sericostomatidae. Works on caterpillars are those of Pankrath (1890) (*Gastropacha*), Hesse (1901) (*Smerinthus, Arctia*), Busselmann (1935) (*Ephestia*), Dethier (1942, 1943) (*Isia, Calpodes, Pieris*), and Paulus and Schmidt (1978) (*Gastropacha, Operophthera*). Philogène (1975) gave some SEM pictures of the corneae of some species.

In all species, the eucone crystalline cone in both stemma types consists of three segments of the three Semper cells. In most cases, this cone is partly enveloped by three very large pigmented cells, called by Pankrath and Hesse "Mantelzellen" (mantel or enveloping cells). From their structure and position, they are the primary pigment cells, whose size varies from species to species. In the caterpillars, they envelop practically the whole stemma. In some Trichoptera, they are either normal in size or very small. But in all cases, there are three such cells. The retinula always consists of seven cells, which arrange the rhabdom in two layers (Fig. 6.21). The distal zone is composed ot three, the proximal one of four rhabdomeres, which form the rhabdom differently. We find a remarkable

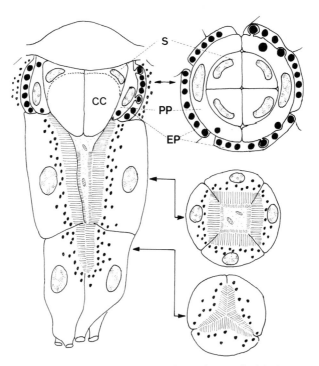

Fig. 6.21.   Longitudinal section of larval ommatidium (stemma) of *Sericostoma*, transverse sections at levels indicated. This ommatidium is a double stemma. Cone consists of six parts; two distal rhabdoms (*DRH*) are fused, but proximal (*PRH*) ones are separated. Numbers indicate single retinula cells. *CC* = crystalline cone; *DRH* = distal rhabdom; *PRH* = proximal rhabdom; *S* = Semper cell.

situation within the Limnephiloidea; representatives of this group have only six stemmata, one of which (the most anterior one) is apparently a double-stemma. It is made up of six primary pigment cells, and a sexpartite crystalline cone. The large retinula contains 14 cells, with six cells in the distal and eight cells in the proximal region. The fusion of the rhabdom is not complete in the proximal region, there being two separate rhabdoms, each with four rhabdomeres (Fig. 6.21). This fusion of stemmata is advanced in the genus *Philopotamus* (Hydropsychoidea). Externally, there appear to be only two stemmata. However, in sections we observe that each crystalline cone is fused with those of several stemmata. The number of the segments is not easy to see. The retinula consists of a strongly armed rhabdom with at least 20 or 21 cells in one observed case, and about 14 cells in the other eye. That means that there are at least five fused stemmata in each of the two isolated large new stemmata. The two-layered rhabdom is not clearly seen. But there seems to be a multilayered arrangement of the rhabdomeres.

These data are interesting in many respects. The dioptric system is composed of six cells: three primary pigment cells and three semper cells. Compared with the normal ommatidium, then, it is clear that only one Semper cell changed into a primary pigment cell. This is supported by the abberant cone number in *Hydropsyche:* there is one crystalline cone with four parts, and there are only two primary pigment cells. This is an atavism, since it occurs at the evolutionary level of the Amphiesmenoptera (Trichoptera + Lepidoptera). This change in cell arrangement of the dioptric apparatus, and the special arrangement of the seven retinula cells are good synapomorphies in both these groups. They are sister-groups (Amphiesmenoptera of Hennig, 1969) and the special, modified ommatidium must have been derived from a common ancestor.

6.8.5.3. Larval Eyes of Tenthredinidae (Hymenoptera)

The larvae of the Symphyta have on each side of the head one very large lens within a large pigmented spot. Hesse (1901), Corneli (1924), and Meyer-Rochow (1973) have studied the eyes in this family. The structure of this lens eye is similar to the single-lens eye of the scorpions or the frontal eyes of insects. Under a single strongly biconvex lens and a lentigenous cell layer are found many isolated retinulae-rhabdoms. One of these rhabdoms is composed of eight retinula cells, which form a cruciform figure in cross section. These eight cells are arranged in one single layer. Crystalline cones are absent.

As shown in the dipteran larval eye (see next section), this single-lens eye of hymenopteran larvae is the result of the fusion of the entire facetted eye, possibly similar to the situation in the scorpions. The evolutionary developmental mode must have been the fusion of all corneae into one single lens, and complete reduction of the single crystalline cones. I would like to call this eye type the "unicorneal composite eye." In Holometabola, this eye type is present in all larvae of Hymenoptera with eyes. Similar eyes are found in some larvae of Diptera, although the latter evolved independently.

6.8.5.4. Larval Eyes of Diptera

As shown in the comprehensive work of Constantineanu (1930) on different eye types in culicomorph dipteran larvae, these eyes are very different from one another. Each eye type described in Trichoptera, Lepidoptera, or even Hymenoptera is found also in different dipteran larvae. There are no new, especially EM, works except White's (1967) on *Aedes*. Within the different primitive Diptera (Nematocera) one finds the various modifications and changes in the larval ommatidia. Four such modifications are shown in Fig. 6.22. A typical ommatidium is not present. In all cases we find fused ommatidia, either with their fused crystalline cones or with reduced cones (Fig. 6.22, no. 1, *Chironomus*). In

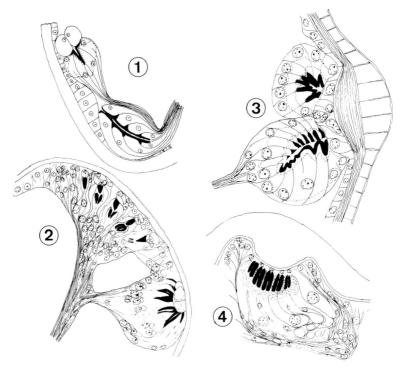

Fig. 6.22. Some larval stemmata of nematoceran Diptera to show different ways of reduction and modification of ommatidia. 1, *Chironomus:* one stemma normal as ommatidium, second multilayered, lacking a cone; 2, *Culex:* fusion of many cones to one (right part)— left upper part is developing imaginal disc; 3, *Eulalia* (Stratioymidae): two multilayered fused stemmata under one lens; 4, *Tipula:* a possibly unicorneal composite eye. From Constantineanu (1930).

Culex, it is easy to see that the real larval eye is a precociously differentiated part of the eye anlage. Several retinulae are grouped together under a single multipartite crystalline cone, each consisting of about eight retinula cells, forming a rhabdom like that of the adult (Fig. 6.22, no. 2). In *Tipula*, the situation looks very similar, but the cone is reduced (Fig. 6.22, no. 4). This eye type shows some similarity with the unicorneal composite eye of the sawfly larva. Figure 6.22, no. 3 shows the *Eulalia* (Stratiomyidae) type of eye, which is very similar to that of the Coleoptera or Neuroptera. As seen in *Culex*, the larval facetted eye is the differentiated imaginal disc of the eye during the last larval instar. The onset of development of the imaginal eyes is very different in various species. In *Chaoborus* it starts during the early larval instar, whereas in Tipulidae it starts during the pupal stage.

6.8.5.5. Larval Eyes of Neuropteroidea and Coleopteroidea

The stemmata of these two groups are poorly known. From the few histological accounts, it seems that in both groups the eyes are very similar. Coleoptera and Megaloptera have up to six stemmata on each side of the head, (Neuroptera) and Raphidioptera up to seven, and the Strepsiptera five. Stemmata in Coleoptera are described for *Dytiscus* (Grenacher, 1879; Hesse, 1901; Günther, 1912), *Acilius* (Patten, 1888), *Gyrinus* (Bott, 1928), *Cicindela* (Friedrichs, 1931), and *Nebria* (Paulus, unpublished), in Megaloptera for *Sialis* (Grenacher, 1879; Hesse, 1901), and in Neuroptera for *Myrmeleon* and *Euryleon* (Hesse, 1901; Doflein, 1916; Jockusch, 1967) and *Protohermes* (Yamamota et al., 1975). As far as I know, the larval eyes of Raphidioptera are not investigated. In adult Strepsiptera, we find a composite eye of typical stemma units, as in *Myrmeleon*. I refer to this eye type as the "stemmataran facetted eye." The structure of this eye has been investigated by Strohm (1910), Rösch (1913), Kinzelbach (1967), and Wachmann (1972a,b). Kinzelbach (1971) calls this eye the "paedomorphic complex eye." The stemmata of the two superorders have a strongly biconvex lens, formed by a number of corneagenous, or better lentigenous cells. Just beneath this layer is a multilayered, monaxonically arranged rhabdom, consisting of numerous retinula cells. This rhabdom is strongly branched, like that of the trichopteran, *Philopotamus*. (Paulus and Schmidt, 1978) or the dipteran, *Eulalia* (Constantineanu, 1930). For this reason, it is likely that the coleopteran stemma evolved by the fusion of several ommatidia with reduced crystalline cones. But in the eyes of *Myrmeleon*, *Euryleon*, and *Sialis* these cones are possibly preserved from the ancestors of these two superorders. As in *Sialis*, the cone consists of eight segments; the stemma could be the fusion of two ommatidia. Unfortunately, the number of retinula cells is not known. In *Myrmeleon*, the cone consists of seven parts, but the retina contains about 30–40 sensory cells. Yamamoto and Toh (1975) described in *Protohermes* a cone consisting of

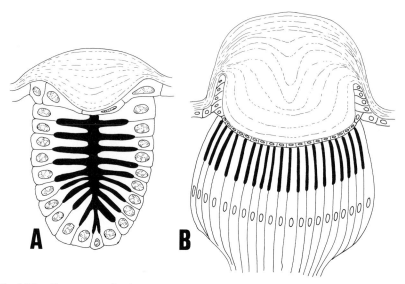

Fig. 6.23. Two types of coleopteran stemmata. A, Multilayered monorhabdomere type of most Adephaga (here *Gyrinus* or *Nebria*); B, secondary enlarged multirhabdomeral type of *Cicindela* larva with a net-retina as in higher Arachnida. A, original; B, from Friedrichs (1931).

10-20 cells, and a retina of 200-300 cells. In these cases, I suppose, the number of sensory cells multiplied because of the diurnal predatory mode of life. This has happened in *Cicindela* larvae, with as many as 1,000 sensory cells (Friedrichs, 1931) (Fig. 6.23).

The eyes of different groups of Coleoptera or Neuropteroidea are not known. As in the hymenopteran larvae, there is only one large single lens in the larvae of Cantharoidea, Elateroidea and many Dryopoidea; this lens is, possibly, a unicorneal complex eye.

6.8.5.6. Nymphal Eyes of Hemimetabola

Normally, we find the adult facetted eye even in the nymphal stages of the different groups. During ecdyses, this eye grows by adding rows of new, but identical, ommatidia (Bodenstein, 1953; Edwards, 1969; Meinertzhagen, 1973). In Odonata, the lamina is reorganized before adult ecdysis (Lew, 1933; Lerum, 1968; Mouze, 1972). But in some higher hemimetabolan forms the young instars have eyes different from those of the adults. Such eyes are found in the nymphs of Aphidina, Coccina, and Aleurodina. Some accounts of the structure of these larval eyes are given by Pflugfelder (1936). In Aphidina and Aleurodina, these larval eyes are prematurely developed ommatidia, with some differences

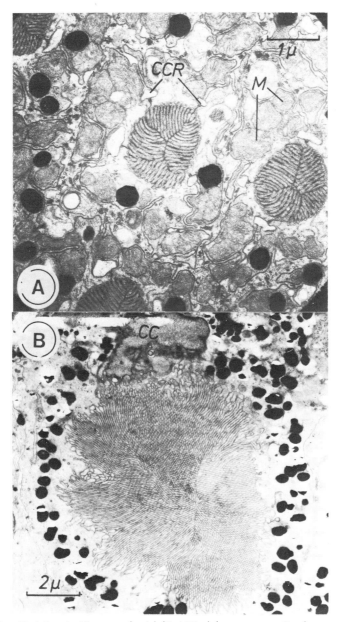

Fig. 6.24.   Rhabdom in *Myzus* sp. (male) (Aphidina) (transverse sections) to compare persisting larval eye rhabdom in (B) with that of imago in (A).

from the later adult ones. The difference consists in the extension of the dioptric system and shortening of the rhabdom to make the eye more light-sensitive. In cross section, the rhabdom of the larval ommatidium is much thicker (Fig. 6.24). These larval eyes are not reduced at adult emergence as in Holometabola, and persist beside the adult facetted eye. They are not integrated in the adult eye, owing to the difference in structure and function. A special type of larval eye is found in the genus *Pseudococcus* (Coccina). This eye is completely modified and has very strange photoreceptor cells—the only such example in the Arthropoda. As this eye has some similarity to the adult eye of that species, it is described in the next section.

6.8.6. Atypical Lateral Eyes in Adult Insects

Adult Siphonaptera, Mallophaga, Anoplura, some Psocoptera, and Coccina have large unicorneal eyes that in some species are similar to the median eyes. But from the type of lateral innervation, they undoubtedly are lateral eyes derived from facetted ones. The evolutionary development is not known, but possibly comparable with the evolution of some larval eye types (e.g., Tenthredinidae). In Mallophaga, there are two rudimentary ommatidia under one single lens (Wundrig, 1936). In Anoplura, the unicorneal eye is composed like the unicorneal complex eye of hymenopteran larvae or scorpions (Stöwe, 1943). In Siphonaptera, the eye is similar to the monaxonical multilayered larvel eye of some Coleoptera. Possibly, this eye is transferred from the larval stage to the imago. The present-day larvae are blind. The rhabdom was investigated by Wachmann (1972a).

Remarkable eye structures are found in some Coccina males. In addition to the persisting larval eyes, they have either typical facetted eyes (in Monophlebinae and Ortheziinae) or a pair of dorsal and a pair of ventral giant unicorneal eyes. The genus *Steingelia* has more such lenses, with a total of 14. None of them is homologous with the frontal ocelli as shown by Pflugfelder (1937). From histological accounts, these eyes seem to be constructed like the ocelli of insects (Krecker, 1909; Marshall, 1935; Srivastava and Sinha, 1966; Ramachandran, 1963). The eye of the *Pseudococcus* male is reconstructed schematically in Fig. 6.25. There is nothing peculiar in the retinulae arrangement of this eye. However, this type of retina is very similar to that of Arachnida (except Scorpiones). The rhabdomeres show a network of sensory borders, uniformly distributed in the eye cup. The sensory part is around the cell border. But this border is remarkable in that it does not consist of microvilli as in Arthropoda, but flat discs (Fig. 6.26). These discs are piled up like the discs in the vertebrate rod. One sensory cell of the *Pseudococcus* eye is reconstructed in Fig. 6.27. Very similar photoreceptor cells are found in frogs (e.g., *Rana pipiens*: Wolken, 1971). But in arthropods this type is completely new. The

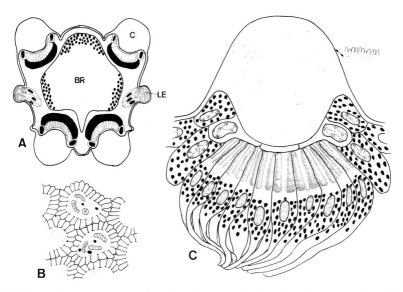

Fig. 6.25. Schematic reconstruction of big lens eye of male of *Pseudococcus* sp. (Coccina). A, Frontal section to show two lateral larval eyes and four large imaginal lens eyes; C, Longitudinal section through one eye; B, transverse section through some of retinula cells. Note Completely different type of membrane enlargement instead of microvilli. *BR* = *C* = cornea; *LE* = larval eye.

cilium rudiments, such as are found in the vertebrate rod cell, are absent. I believe that the *Pseudococcus* disc can be derived from one microvillus by flattening. The same type of photoreceptor cell can be found in the larval eye of this species. Here, the discs are much more extended than in the adult eye. Nothing is known about the causes and functional significance of such strong modifications and changes of the eye structure. Since primitive Coccina have facetted eyes, these unusual lens eyes must have been derived from them. The possibility of such modifications demonstrates how easily great changes in organ structure can occur in the evolution of groups. However, one must not conclude from this that comparable diversity of structural modifications of other organs also has occurred as reported by Tiegs and Manton (1958) and Manton (1973).

6.8.7. Evolution of Larval Eyes in Insects

As stated earlier, the insect larval eyes are greatly different organs, and cannot be derived from those of the Myriapoda, although Hesse (1901), Demoll (1917), Snodgrass (1935), Weber (1933), and Tiegs and Manton (1958) have suggested this. Another argument against such a theory is the fact that all primitive insects have facetted eyes even in larval stages (Fig. 6.28). The main argument against

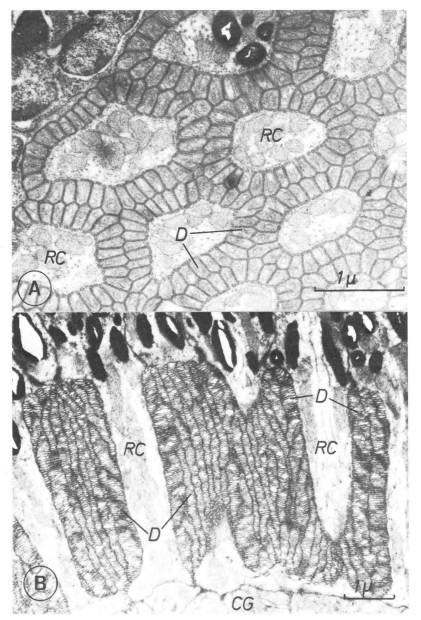

Fig. 6.26. Retina of imaginal lens eye in male *Pseudococcus*. A, Retina is composed of numerous cells with a surrounding disc border instead of microvilli; B. longitudinal section through retina. Discs are piled up as in rods of some Amphibia.

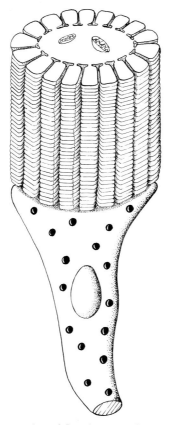

Fig. 6.27. Schematic reconstruction of *Pseudococcus* photosensory cell with discs instead of microvilli (original).

this opinion is that all larval eyes can be shown to be derived from typical insect ommatidia.   This derivation has occurred in Holometabola at least five times independently, and mostly in different ways.

The Holometabola are certainly a monophyletic group, implying a common ancestor for all representatives, and possession of at least one synapomorphic character (Hennig, 1950, 1969; Kristensen, 1975).   Such a synapomorphic character is the degeneration of the larval eyes before adult ecdysis.   Larval eyes in Hemimetabola always persist until the adult stage.   Highly modified larval eyes appear only in the sistergroup of Holometabola, the Paraneoptera.   But these eyes developed independently or parallel to larval eyes in Holometabola.   From the fact that in one group of the Holometabola (Mecoptera) we find a typical small larval facetted eye with only slightly modified ommatidia, it is reasonable

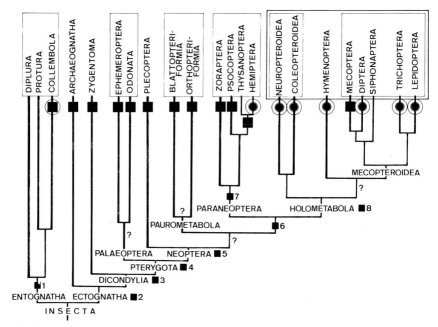

Fig. 6.28. Dendogram of Insecta, composed from data in Hennig (1969), to show distribution of larval eyes (circles) in different orders. Typical larval eyes are present only in some higher Hemiptera besides Holometabola. It is impossible that these eyes are related to the Myriapod type of eye. Characters: 1 = entognathous; 2 = flagellate antennae; 3 = dicondylous mandible; 4 = wings; 5 = wings able to fold back; 6 = larvae lacking frontal ocelli; 7 = rhynchote mouthparts; 8 = holometaboly.

to assume that the holometabolan group probably has had such eyes in the larva. These eyes must have been modified in such a way that they could not be integrated with the adult eye; they degenerate in the prepupal stage, and a new one is formed from the imaginal eye anlage or disc, as in Mecoptera. Possibly, the *Panorpa* facetted larval eye represents exactly this situation. The ommatidium had a normal dioptric apparatus modified for a greater light reception at the cost of resolution. It had two primary pigment cells and four Semper cells, which form a eucone crystalline cone. The retinula originally was one-layered, for it is single-tiered in the unicorneal complex of Hymenoptera, and even in the rudimentary eye of *Boreus* or in some different eye types of Diptera. The two-layerd retinula of *Panorpa* and of the Amphiesmenoptera (Trichoptera + Lepidoptera) must be convergent as a consequence of special modifications. From this stage of evolution, the larval eyes evolved in the three supposedly monophyletic groups of Holometabola: Coleopteroidea + Neuropteroidea, Hymenoptera, and Mecopteroidea (Antliophora + Amphiesmenoptera) (Hennig's, 1969, nomenclature).

Let us consider first the development in the Hymenoptera. As described earlier, the unicorneal complex eye develop by the fusion of all corneae and cone reduction. The retinulae of the original facetted eye remained unchanged. In a truly analogous way the eyes of scorpions, other Arachnida, and the lens eyes of the Anoplura developed. Unfortunately, this mode of eye evolution in Hymenoptera does not reflect the phylogeny of this group and their sistergroups because this type of modification is an autapomorphic character (Hennig, 1969; Kristensen, 1975; Königsmann, 1976).

Within the Mecopteroidea, Mecoptera are the most primitive. Possibly, the eruciform larval type and its mode of life is also primitive in Mecoptera, and possibly in Mecopteroidea as a whole. Chen (1946) thought that the most primitive larval type, even in Holometabola, is aquatic. The sistergroup of the Mecoptera (Diptera, according to Hennig, 1969) started with the small larval facetted eye—a modification by degeneration as seen in *Boreus*. Culicidae and Chironomidae demonstrate how this degeneration occurred; the most commonly found degeneration in these groups is the retardation of the eye development in the early embryonic stage. The different species of these groups show different stages of this development. In most cases, the ommatidia fuse or are grouped together with one single cornea. Some of these modifications are present also in Diptera (e.g., *Tipula, Culex,* or *Chironomus*).

In Amphiesmenoptera (Trichoptera + Lepidoptera), we find the third principal mode of evolution of larval eyes. The first step was a decrease in the number of ommatidia. The remaining seven ommatidia first stayed together as in Trichoptera or Zeugloptera (*Micropteryx*). In Leipdoptera, they moved to special positions on each side of the head. Before this event, however, a transformation of the cells of the dioptric system must have occurred. One of the cone cells became a primary pigment cell. This transformation occurred in the ancestor of the Amphiesmenoptera because this special kind of structure was found in all the species investigated.

The larvae of Coleoptera have up to six and those of Neuropteroidea up to seven isolated lenses. The structure of the stemma is strongly modified from the typical ommatidium. This type of multilayered monaxonal stemma possibly had already been modified in a common ancestor of these two superorders. This would mean that this special eye type, which closely resembles that in the Myriapoda, is a synapomorphic character of Coleopteroidea and Neuropteroidea. In some cases, there is a crystalline cone, but always modified (as in *Myrmeleon*) with seven parts, and in *Sialis* (Megaloptera) with eight. The vitreous body, described in *Dytiscus* by Günther (1912), was not observed by Hesse (1901). As far as is known, no crystalline cone is found in any Coleoptera. The evolutionary development of this larval eye type is possibly very similar to those in some Diptera and Trichoptera (*Philopotamus*), although somewhat modified. The first step was a decrease in the facet number, then a fusion of some ommatidia, and

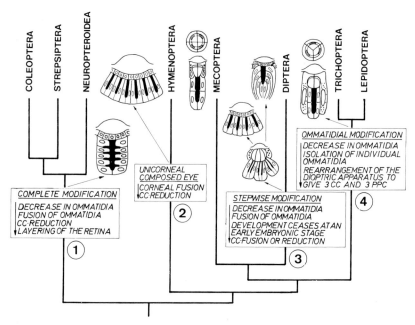

Fig. 6.29. Distribution and evolutionary parthway of holometabolan larval eyes. There are at least 4 evolutionary radiations in different orders.

finally a new arrangement of the retinula cells by layering. The crystalline cones too either fused or were reduced. The larval eyes of the Strepsiptera fit into this scheme. For this reason their close relationship with the Coleoptera is supported. It is remarkable that in this group the adults have facetted eyes consisting of these larval eyes; I have called this eye the "stemmataran facetted eye."

In summary, three important points are to be emphasized: 1) the different larval eyes of insects are derived in all cases from typical insect ommatidia, and are most decidedly not derived from the myriapodan eye by recapitulation or atavisms; 2) the similarity with the myriapodan eye is the consequence of parallel evolution by modification of the typical original ommatidia; 3) as will be discussed later, the myriapodan eye has evolved in exactly the same manner from ommatidia as the insect larval eye. These conclusions are summarized in Fig. 6.29.

6.8.8. Frontal Ocelli and the Dorsal Frontal Organs of Insects

Normally, insects have up to three frontal ocelli: a median and two lateral ones. In higher insects, each eye has a strongly biconvex lens, and a one-layered retina, consisting of many isolated closed rhabdoms, or sometimes, as in Arachnida, a network of connected rhabdomeres. The small rhabdoms are often separated by pigment cells. The structural and functional details are summarized by Goodman (1970). For our discussion some structural details of the primitive insects are of interest. First, it must be said that in Myriapoda median eyes and the frontal organ are always lacking. The latter organ in the frontal part of the head of *Scutigera*, described as a median eye by Knoll(1974), is surely no photoreceptor. The so-called frontal organs in Myriapoda (cerebral glands, organ of Gabe a.o.: Hörberg, 1931, Fahlander, 1938; Palm, 1955; Gabe, 1967; Prabhu, 1961) are neurohaemal organs (Joly, 1969; Seifert and El-Hifnawi, 1971).

Although Archaeognatha and Zygentoma (*Tricholepidion*) have three frontal ocelli and no frontal organs, the following situation is generally found in Collembola, according to Marlier (1941) and Paulus (1972b, 1973) (Fig. 6.30): 1) four frontal ocelli: the median ocellus found in other insects is paired in these primitive insects; 2) one pair of ocelli besides the four frontal ocelli—these two ocelli, located on the dorso-median part of the head, are innervated from the same center as the frontal ocelli, and these organs are homologous to the dorsal frontal organs of the Crustacea (Paulus, 1972b).

The Collembola have, besides the facetted eyes, six ocelli, (Fig. 6.31E). These ocelli are structurally very primitive. They have no lens; the retina is composed of two to 12 retinula cells; in many cases they have migrated under the epidermis. In the genus *Tomocerus*, the dorsal frontal organ has migrated deep into the head, lying on the oesophagus. The sensory rhabdoms containing cells are partly incorporated in the brain as in isopods. This organ was known for a long time as "Nabert's organ" (Barra, 1969; Paulus, 1972b) (Fig. 6.30). In the section Poduromorpha of the Collembola these dorsal ocelli always seem to be absent. The ventral frontal organs are absent in Collembola as in all other insects. The *Lepisma* frontal organ described by Holmgren (1916) contains rudimentary rhabdoms (Pipa et al., 1964). A similar organ is found in *Thermobia* (Watson, 1963). Because the ocelli are absent in these genera, I believe the so-called frontal organs of *Lepisma* or *Thermobia* are the migrated ocelli under the epidermis; in *Tricholepidion*, there are three ocelli, but the frontal organ is lacking (Elofsson, 1970). Such internal ocelli could be found in some moths (Dickens and Eaton, 1973). The others, called frontal organs in Zygentoma (Thysanura: Lepismatidae) and Archaeognatha (Hanström, 1940; Watson, 1963; Chaudonneret, 1950; de Lerma, 1951; Bart, 1963; Bitsch, 1963a,b), are possibly neurohaemal organs. For further discussion of the homology of these organs see Paulus (1972b). The insects have basically four ocelli, and at least the dorsal pair of frontal organs as photoreceptor structures. As in most Crustacea, the four ocelli have been reduced to three and fused with the median ocellus. Its double nature can be seen in many insects by its duplicated nerve roots.

As described in section 6.5.7, the basic number of median eyes seems to be four in Chelicerata also. This basic number is present in Pantopoda. *Limulus*

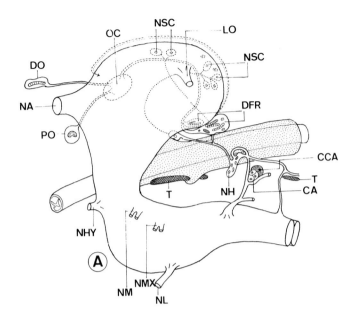

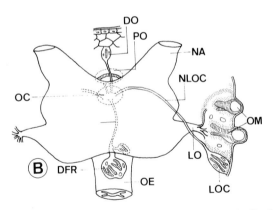

Fig. 6.30. Brain of *Tomocerus longicornis* (Collembola) to show distribution of different ocelli and the dorsal frontal organ. A, Lateral view: dorsal frontal organ has penetrated deep into the head, now lying just on the oesophagus (known as "Nabert's organ" in this genus)—note two median ocelli; B, dorsal view: lateral ocelli are sitting next to composite eyes. From Paulus (1972b). *CA* = corpora allata; *CCA* = corpora cardiaca; *DFR* = dorsal frontal organ (Nabert's organ); *DO* = distal median ocellus; *LO* = lobus opticus; *LOC* = lateral ocellus; *NA* = n. antennalis; *NH* = neurohaematic organ; *NHY* = n. hypopharyngealis; *NLOC* = nerve of lateral ocellus; *NL* = n. labialis; *NM* = n. mandibularis; *NMX* = n. maxillaris; *NSC* = neurosecretory cells; *OC* = ocellar center; *OE* = oesophagus, *OM* = ommatidium; *PO* = proximal median ocellus; *T* = tentorium.

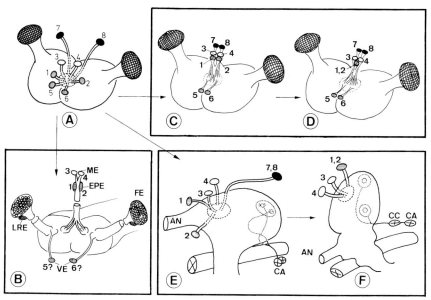

Fig. 6.31. Distribution and possible evolutionary relationships of median eyes (nauplius eyes, ocelli) and frontal organs in Arthropoda. A, Protarthropodan ancestor with eight medial photoreceptors besides facet eyes; B, different eye types in *Limulus*. Ocelli 1-4 are homologous to same numbered in C-F, and ventral eyes (5, 6) are possibly homologous to ventral frontal organs. C, Distribution of all eight eyes—four nauplius eyes (1-4) and dorsal (7, 8) and ventral (5, 6) frontal organs in Phyllopoda; D, distribution of these eyes in Malacostraca. Nauplius eyes are reduced to three (possible fusion of 1 and 2). E, In Collembola all eyes are present, except ventral frontal organs; F, in higher insects only three ocelli are present. *AN* = n. antennalis; *CA* = corpora allata; *CC* = corpora cardiaca; *EPE* = endoparietal eye; *FE* = facet eye; *LRE* = lateral rudimentary eye; *ME* = median eyes; *VE* = ventral eye.

too has four median eyes, two of them known as endoparietal rudimentary eyes. In other Chelicerata, we always find one pair of external median eyes, although occasionally they are reduced (e.g., in Pseudoscorpiones). The Eurypterida have median eyes as in *Limulus*. These median eyes are, without any doubt, homologous within the Chelicerata. At least in Arachnida, the peculiar embryological development (converted eyes) must be considered as a good synapomorphy for this terrestrial group. The strange development of the median eye in *Limulus* is possibly a recapitulation of the head duplicature, as suggested by Lauterbach (1972) and a key event in arthropodan evolution (Johannson, 1933; Weygoldt, 1975). This interpretation includes the theoretical fact that the Chelicerata as well as the Trilobita had these median eyes before developing this lateral head duplicature. Indeed, it would seem that the fossil Dicephalosomita (Sharov, 1966) with the genus *Opabinia* from the middle Cambrian had four median eyes, if we accept the reconstruction of Walcott (1912) or Sharov (1966). It is not inconceivable that these four eyes could lead back to four similar prostomial eyes of a polychaete-like arthropodan ancestor.

6.9. HOMOLOGY OF MEDIAN EYES IN ARTHROPODA

As stated above, all groups of Arthropoda originally had, besides the lateral facetted eyes, two pairs of median eyes. The homology within Chelicerata as well as in Mandibulata seems certain. Any lingering doubt about it comes from Myriapoda, which lack these eyes. But this absence may be explained by the strong reduction and modification of the lateral eyes. Possibly, Myriapoda had no median eyes for the same reason as holometabolic insect larvae. Unfortunately, the specific structures of these organs are insignificant for them to be regarded as clear synapomorphy. However, the basic number four and the innervation from the protocerebrum in all Arthropoda seem to establish that all these organs are homologous. The presence of these organs in Collembola is a very strong argument in favor of homology of the mandibulate frontal organs (Fig. 6.31).

6.10. EYE STRUCTURES AND ARTHROPOD PHYLOGENY

We will now consider the significance of certain eye structures in arthropod phylogeny.

6.10.1. Evolution of the Rhabdom

Photoreceptor cells are characterized by typical membrane specializations, which are of two different types. According to Eakin (1963, 1968) these two types are distinguishable by different origins: 1) the ciliary type in which the photoreceptor organelle is derived from the plasma membrane of a modified cilium: 2) the rhabdomeric type, in which this organelle is derived directly from the cell membrane, independently of cilia. Eakin sees in this difference two main evolutionary lines, principally corresponding to the Protostomia and Deuterostomia. The rhabdomeric type is found especially in Protostomia, the ciliary type in Deuterostomia. Vanfleteren and Coomans (1976) provide a reviewed interpretation of the Eakin hypothesis. According to them, there are many photoreceptor cells with cilia or rudiments of cilia even in Protostomia. They are widely distributed in Annelida (Bocquet, 1972; Dorsett and Hyde, 1968; Fischer and Brökelmann, 1966; Hermans, 1969; Hermans and Eakin, 1974; Kerneis, 1971; Whittle and Golding, 1974). Even in the eyes of Onychophora we always find rudiments of cilia (Eakin and Westfall, 1965). Therefore, the ciliary type seems to be the original one, from which the rhabdomeric type is derived. This has happened several times in Protostomia, but in Deuterostomia nearly all photo-receptor organelles are the strongly modified ciliary membrane. But the development of a typical arthropodan rhabdomere, characterized by a complete, parallel row of microvilli, possibly happened only once. In recent years, rudimentary cilia have been reported in arthropodan photoreceptor cells. For example, Home (1972) in *Adalia* (Coleoptera), Wachmann and Hennig (1974) in *Megachile* (Hymenoptera), and Juberthie and Muñoz-Cuevas (1973) in *Ischyropsalis* (Opiliones) found centrioles, which they interpreted as rudiments of cilia. But this sporadic occurrence in such highly evolved groups leads me to believe that these centrioles have nothing to do with Eakin's ciliary type. One can find such centrioles in nearly all cell types, especially in the epidermis. According to Fahrenbach (1968), such centrioles are occasionally observed in the distal pigment cells in the ommatidium of *Limulus*. Muñoz-Cuevas (1975) considers these centrioles in *Ischyropsalis* as a ciliary induction of the photoreceptor organelle.

The typical arthropodan photoreceptor cell has a very regular microvillous border, called a rhabdomere. Several such rhabdomeres fuse together to form a closed light-guide system, called the rhabdom. Many such rhabdoms form the retina, which is of three different compositions: 1) Many retinula cells are distributed equally under one common dioptric system; they form a network of microvillous borders (Fig. 6.4); the single retinula cell has four or only two microvillous borders; such retinas are distributed in most Arachnida, in the nauplius eyes of Crustacea, and the ocelli of insects. 2) A series of retinula cells forms around a central axis, composing a closed rhabdom; such rhabdoms are found under a lens, and many such rhabdoms form a facetted eye; derived from this by secondary fusion of the separated lenses into one single lens is the eye with a retina composed of many rhabdoms that are not connected to one another (Fig. 6.3); such eyes are found in Scorpiones, in the unicorneal composed eyes of fleas, Anoplura, and Hymenopteran larvae. 3) Similar to (2), but with rhabdomeres that do not form a closed rhabdom, but stay isolated, this type is called the open rhabdom; such rhabdoms are found in higher Diptera (Wada, 1974), Heteroptera (Schneider and Langer, 1969; Walcott, 1971; Shelton and Lawrence 1974), Coleoptera (possibly in most Cucujiformia: Wachmann, 1977), and even Collembola (*Entomobrya*: Paulus, 1977); it consists of eight rhabdomere, two of them centrally situated (Fig. 6.11).

The form of the rhabdom is certainly correlated with its function, but only a few facts about this are really known. The correlation of the distribution angles of the different rhabdomeres within one rhabdom with the perception of polarized light (Shaw, 1969; Snyder, 1973) is generally mentioned. According to Laughlin et al. (1975), each rhabdomere is sensitive to polarized light because of the dichroitic absorption possibility of the microvillous membranes, provided all microvilli are parallel to one another. But the consequence of this dichroitic absorption of the photosensitive molecules is not only the ability of polarized light perception but also the ability to reach the highest light sensitivity (Menzel, 1975). This last ability is possibly the final stage in the evolution of such rhabdomeres, and the ability of polarized light perception is only a secondary effect. On this basis, I suggest three possibilities of rhabdom formation: 1) All microvilli of one cell are completely parallel to each other along the entire extension of the border to attain high light sensitivity; the polarized light sensitivity is not important for the species; there is no selection against it. 2) The polarized light sensitivity is important for the species for orientation (e.g., for the bee); in this case, at least one cell, maybe all cells, are exactly oriented at specific angles to one another (frequently at 90°) with highly parallel-oriented microvilli. 3) polarized light sensitivity is not only not important but is disruptive of orientation.

There is a selection pressure to eliminate the polarization figures. One way of accomplishing this is to produce discs instead of microvilli as in vertebrates, or in the males of *Pseudococcus* (Fig. 6.27), but without losing a high light sensitivity. Another possibility for eliminating the polarization sensitivity is to change the direction of the microvillous border within one single cell. Twisting the rhabdomere along the longitudinal axis of the retinula cell is possibly one way to accomplish this (Snyder and McIntyre, 1975). Such twisted rhabdomeres are indeed found in *Apis* (Grundler, 1974; Wehner et al., 1975) or *Myrmecia* (Menzel, 1975). To retain the polarization sensitivity a short ninth cell is present on the base of the rhabdom, which is not or is nearly not twisted. Probably only this single cell is the polarization receptor, as shown in the ethological experiments by Helversen and Edrich (1974) for the bee. Another way to eliminate or decrease this sensitivity is to have different microvilli directions in one cell. The microvillous border may be circular as in *Photuris* (Horridge, 1969a), *Dicyrtomina* (Paulus, 1975), *Gyrinus* (Wachmann and Schroer, 1975), or Aphidina larval eyes (Fig. 6.27). It may be angular in different ways as in most Arachnida (Schroer, 1974) (Fig. 6.4), or the central retinula cells in many Coleoptera (Wachmann, 1977). The very peculiar rhabdoms in some Ephemeroptera may be explained in this way. There are, as seen from these few examples, many ways to increase or decrease the polarization sensitivity, which has been realized independently and in parallel in Arthropoda. The fused rhabdom is known in two different types: one, most common in insects, myriapods, and a few crustaceans, is the contiguous one. All microvilli borders extend from the tip of the cell to the end without any interruption; the second type is the so-called layered or tiered rhabdom (Fig. 6.5). The alternate layers are visible in longitudinal sections. The microvilli borders of one retinula cell are periodically interrupted during the latter's longitudinal extension. These periodic spaces are filled by microvilli of the neighboring cells. This rhabdom type is typical of Crustacea, but is also found in some insects. It has been reported in Archaeognatha (Meyer-Rochow, 1972a; Paulus, 1972a, 1975), some Coleoptera, and Lepidoptera

(Meyer-Rochow, 1972a; Butler et al., 1970; Kolb, 1977). According to Shaw (1969), and Waterman (1975), the polarization perception of this type is much better than of the contiguous rhabdom. But Horridge (1975b) stated that these many layers only increase the number of reflecting boundaries to increase the light efficiency.

Possibly the tiered rhabdom is the basic one for Mandibulata, even though the proof is weak. Contrary to the belief of Elofsson and Odselius (1975) and Elofsson (1976) that the tiered rhabdom is typical only for Malacostraca, I found it in *Artemia* and *Tanymastix* (Anostraca) (Fig. 6.5). Lack of a tiered rhabdom was enough for Elofsson to suggest that the two rhabdom types developed independently. To ensure that the tiered rhabdom is the basic one in Crustacea the rhabdoms of more phyllopods, especially of *Triops* should be investigated. From Debaisieux's (1944) work, *Triops* has such a rhabdom, at least in the proximal region. But this should be confirmed by EM study. If the tiered rhabdom is the basic one in Crustacea, then the tiered rhabdom of the Archaeognatha in insects is possibly preserved from the last common ancestor of insects and crustaceans. But the possibility of convergence cannot be excluded, since such rhabdoms are distributed in some higher insects.

6.10.2. Homology of Ommatidia in Arthropoda

An ommatidium is the facet unit in a facetted eye. It consists of a lens system and of a monaxonal light-sensitive cell arrangement. But this definition is not useful for recognizing homologies or analogies. According to this functional definition we find ommatidia in all groups with facetted eyes. But the real question that interests us is whether all these facetted eyes or better the ommatidia, are homologous. Before answering this, it is necessary to define homology. Homologous structures are derived genetically from the last common ancestor that had exactly the same type of structure, or a modification of it. The stages of modification must be shown in series of changes (criterion of continuity of Remane, 1956). To recognize homologies, the structures considered must have a level of complexity of enough characters to ensure that the agreement is not trivial (criterion of specific quality). This criterion is important for separating special characters considered as homologous from those that developed in parallel from simple common characters. Parallelisms are the product of convergent developments from a homologous basis. For reconstruction of evolution or phylogeny, it is very important to separate parallelisms from homologies, and of course analogies. Monophyly (in the sense of Hennig, 1966) can be established only with special homologies or synapomorphies. A taxon is monophyletic if all its members are derived from one single, common ancestor. That means parallelisms cannot prove the monophyly of a taxon because the common structures of both are in most cases simpler, primitive characters, the so-called plesiomorphies (for detailed discussions, see Hennig, 1965, 1966; Schlee, 1971; Brundin, 1972, 1976). To separate synapomorphies of a taxon from parallelisms or homoiologies it must be shown for a special character complex that it has developed even in this synapomorphic feature in its ancestor group. Of course, such synapomorphic characters may be modified during the evolution of a taxon, or in some cases completely reduced.

To prove the monophyly of the group under consideration, one has to show the changes or at least the possibility of modification. Returning to the question of homology of the ommatidia, we have to look at the complexity of characters of one ommatidium. Are ommatidial characters complex enough to prove their mono- or polyphyly? Theoretically, one proven synapomorphic feature would prove the monophyly of a group. Unfortunately, there are only few proven synapomorphic characters. The possibility of convergence in many characters is serious. If we can prove the ommatidium as a synapomorphic character complex at least for the Mandibulata, then this group would surely be monophyletic. If we compare the ommatidia of the recent representatives with facetted eyes, we find both similarities and differences (Fig. 6.7). The most characteristic difference between Chelicerata (*Limulus*) and Mandibulata is the complete absence of a crystalline cone in the Chelicerata. Mandibulata have a crystalline cone with the same number of cells. If one were to build a functional photoreceptor in arthropods with a minimal number of cells, one would need a chitinous lens and a retina with sensory cells, whose density and arrangement would depend on what that receptor would have to perform. The *Limulus* eye fits this minimal requirement, since it has these elements and almost nothing more. Supposedly, a photoreceptor that evolved in any arthropod independently was constructed in this way. The *Limulus* eye is constructed so simply that it cannot provide a good basis for homologization in looking for synapomorphies. The common characters, the lens, a lentigenous layer, pigment cells, even the rhabdom formation, are too simple. The ommatidia of Crustacea and Insecta are very similar. As a matter of fact, the similarity is so great that the cell in the crustacean ommatidium can be matched with a corresponding cell in the insect ommatidium. In both groups we find a cornea lens made originally of the same two cells, the corneageous cells, a eucone crystalline cone made up of the four Semper cells, and the retinula forming a rhabdom of eight cells. In insects, the two corneagenous cells are transformed in the two primary pigment cells. Besides these two cells, there are some other pigmented cells, called the secondary pigment cells, distal or proximal pigment cells. Since the number and arrangement, and, possibly, the types of pigments (Struwe et al., 1975) vary a great deal, it is not possible to consider them for phylogenetic discussion. In some cases, the number of other elements varies too, often correlated with a special mode of life. We will come back to this variability later.

The ommatidium of *Scutigera* differs considerably from that of Crustacea and Insecta. The dioptric apparatus consists of a crystalline conelike structure made up of many segments; there are also corneagenous and pigment cells. The retinula is composed of more than eight cells arranged in two layers (Figs. 6.7, 6.9). The single lateral eyes of the remaining Myriapoda do not fit in any way the mandibulate ommatidium. Three possible conclusions may be drawn: 1) The ommatidia of the Crustacea and Insecta are convergent; the similarity is the consequence of a parallel evolution from a simple precursor structure, which we find unchanged in Myriapoda even today; the ommatidia are not a synapomorphic feature in the Mandibulata. 2) The ommatidia of the Crustacea and Insecta are analogous; the similarity is the consequence of a completely independent evolutionary development; the nearly identical structures are a consequence of a functional necessity to make an ommatidium in this way and not in any other. 3) The ommatidia are homologous, which would mean that the last common ancestor of the Mandibulata had exactly this type of mandibulate ommatidium; this eye therefore is a synapomorphy and with this a good proof of the monophyly of the Mandibulata.

Point 1 does not prove a polyphyly of the Mandibulata, but only the polyphyletic origin of the eyes. But the monophyly cannot be proved with the eye structure because the eye structure of the precursor was too simple. Point 2 means the polyphyly of the Mandibulata as supported by Tiegs and Manton (1958) or Manton (1973). Point 3 proves the monophyly of the Mandibulata. Because the Myriapoda are the sistergroup of the Insecta (Hennig, 1969) their eyes must have been modified secondarily. The simplicity of the myriapodan eye has nothing to do with primitiveness.

In order to support the monophyletic view, the homology of the ommatidia and their elements must be proved, and the secondary modification of the myriapodan eye be made plausible. It must be shown that within each group in Crustacea and Insecta the ommatidium in its special form is monophyletic. Only then can we say that this type of ommatidia must have been present even in the last common ancestor of the Mandibulata. This kind of proof is necessary to exclude the theory in point 1 (parallelism).

6.10.2.1. Monophyly of Crustacean Ommatidium

Figure 6.32 shows the distribution of the typical crustacean ommatidium within the different orders. Presuming that this dendrogram is correct in its main features—at least the expressed monophyly of the Malacostraca is in agreement with Dahl (1963) and Siewing (1963a,b)—it is seen that the typical ommatidium is represented in nearly all groups. The most important groups in this respect are the primitive taxa. Unfortunately, the most primitive Crustacea, the Cephalocarida (Hessler, 1964; Sanders, 1963; Lauterbach, 1974), are totally blind. This blind state does not support Lauterbach's opinion that the Cephalocarida have retained their present mode of life unchanged from the ancestors of the Crustacea. If we agree that the ancestors of the Crustacea had ommatidia or a facetted eye,

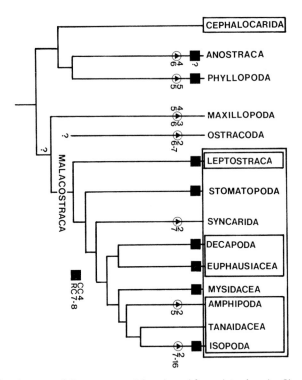

Fig. 6.32. Dendrogram of Crustacea, mainly adapted from data given by Siewing (1963b). Black squares show distribution of typical crustacean ommatidium with four cone cells and seven to eight retinula cells. Circled triangles represent modifications of ommatidia: upper number means number of cone cells; lower, number of retinula cells. It is clear that in all main groups the typical ommatidium is present. It is therefore justifiable to take this type of ommatidium as the original one in Crustacea.

then the recent mode of life in Cephalocarida must be a secondary development. Typical ommatidia are present in *Triops* and in *Daphnia* (Phyllopoda). The modification in the number of Semper cells and the retinula cells also is secondary. The Anostraca have typical ommatidia too, at least in the dioptric system. The apparent lack of the two corneagenous cells is clearly disproved by Paulus, unpublished) (Fig. 6.5). Even within the highly modified Maxillopoda of Dahl (1956) (I do not think that this is a monophyletic group), we find some representatives with nearly typical ommatidia. The number of retinula cells is reduced to five or six. But as in *Daphnia*, where the earlier authors reported five or six cells, a new EM study has to confirm this. Possibly, there are at least in some species eight cells. But these modified features are easily explained by the highly changed morphology and mode of life within these groups.

Within Malacostraca, the most primitive groups, the Leptostraca (*Nebalia*) and Stomatopoda (*Squilla*) have typical crustacean ommatidia. Modifications concerning the crystalline cone can be found in Syncarida, Amphipoda, and Isopoda. The cone in these groups is composed of only two segments. But at least in some Isopoda de Lattin (1939) found some species with a tetrapartite cone. For this reason I believe that this modification also is secondary. The typical tiered or layered rhabdom was possibly present in the ancestor of the Crustacea as discussed before. I also believe that the last ancestors of the Crustacea had a facetted eye, consisting of ommatidia with possibly an unthickened cornea made by two corneagenous cells, a tetrapartite eucone crystalline cone, and a retinula of six to eight (possibly eight) retinula cells which formed a closed, layered rhabdom. It is important to consider three questions at this point, the first two being: 1) Whether the ancestors of Crustacea had stalked eyes, and 2) What the functional basis is of the evolution of such eyes. Looking at most Malacostraca, it seems that stalked eyes are correlated with the possession of a carapace. But stalked eyes are present also in Anostraca, which do not have a carapace. In Phyllopoda we find a carapace but no stalked eyes. However, in Phyllopoda the eyes are located in a chamber that is considered to have invaginated secondarily. Even the embryonic development shows that these sessile eyes are probably derived from stalked eyes. How easily such stalked eyes can become sessile is shown in a mutant of *Artemia*, which has sessile eyes instead of stalked ones. According to Dahl (1963) and Lauterbach (1974), the crustacean ancestors did not have a carapace. Lauterbach (1974) also reported that the carapace is surely lacking in Cephalocarida and in Copepoda, but possibly also in Anostraca, Mystacocarida, and Branchiura. But if stalked eyes have evolved in correlation with a carapace, then the carapace must be reduced secondarily in Anostraca. This, however, we do not know. The Copepoda, possibly neotenic forms, do not have facetted eyes. Dahl (1976), on the other hand, suggests the possibility of parallel development of the carapace. The third question is that of the structure of the optic lobe. In the entomostracan orders there are present only two optic neuropiles without a chiasma, whereas the Malacostraca have three neuropiles, each with a chiasma (Hanström, 1928, Dahl, 1963; Elofsson and Dahl, 1970). There are only two neuropiles in Xiphosura (*Limulus*) and in Myriapoda, and possibly also in Collembola. From this it seems that the three neuropiles in Crustacea and Insecta developed independently. But at least in insects it can be shown that a reduction of eyes is always correlated with a reduction of the medulla interna (e.g., in nearly all insect larvae). This feature in Collembola and Zygentoma seems also to simplify the neuronal connections. In Archaeognatha we find the three normal insectan ganglia with chiasmata. The two optic ganglia of the Myriapoda are possibly a consequence of the reduction of their eyes. Even in *Scutigera* there are two ganglia. According to Hanström (1928), there are no chiasmata in Myriapoda. This situation in the Entomostraca may be explained in two ways: 1) that this situation, which we find also in Myriapoda, is the original situation; 2) that a secondary simplification of the optic ganglion occurred. The latter must be correlated with a modification of the ontogenetic mode of optic ganglia formation, since the difference in the neuronal connections between Eutomostraca and Malacostraca is based on a different mode of development (Elofsson and Dahl, 1970; Meinertzhagen, 1973).

6.10.2.2. Monophyly of Insect Ommatidium

Hesse (1901) and Tiegs and Manton (1958) believed that the ommatidium is the product of the Myriapoda-Insecta evolution. Hesse (1901) proposed a developmental series from the multilayered coneless eye of the Diplopoda or Chilopoda, to the two-layered rhabdom with a crystalline cone in *Scutigera*, to finally the two-layered rhabdoms of Collembola and Zygentoma, and holometabolic insect larvae (especially those of caterpillars). The eyes of Collembola, of Zygentoma, and of Archaeognatha have two corneagenous cells, which developed into primary pigment cells in Pterygota. Hesse (1901) did not separate the crustacean and myriapodan eyes as two corneagenous cells are not present in Myriapoda. Paulus (1970a, 1972a, 1974, 1975) has demonstrated that the evolutionary pathways proposed by Hesse, and accepted by Tiegs and Manton (1958), could not be right. Collembola have eyes with two primary cells, four Semper cells, forming originally a tetrapartite eucone crystalline cone, and originally a one-layered retinula (in Entomobryomorpha and Symphypleona), consisting of eight cells. In Archaeognatha, too, we find the typical ommatidium of the Pterygota. Only the Zygentoma have a two-layered retina, but the dioptric system is very similar to that of Pterygota. This contradicts Hesse (1901) and Tiegs and Manton (1958). There can be no doubt that the last insect ancestor had the typical insect ommatidium, consisting of two primary pigment cells, four Semper cells, and eight retinula cells. The possession of two primary pigment cells is a good synapomorphic feature in Insecta. This proves the monophyly of the insects. Contrary to Manton's (1973) view, the Collembola (and with them the Entognatha) clearly belong to the Insects (Figs. 6.16, 6.28).

6.10.3. Problem of Myriapodan Eye

As stated earlier, the typical ommatidium must have been present not only in the ancestor of Crustacea but also that of Insecta. There would be no doubt about the homology of eyes in those two groups if the eyes of the Myriapoda also were constructed in the same way. But the myriapodan eyes are much different from a typical ommatidium. Even the so-called ommatidium of the *Scutigera* facetted eye does not fit the definition of the mandibulate ommatidium. Since the detailed similarity of the insect and crustacean ommatidia is so great, it cannot be due to convergence; and the eyes of Myriapoda must have been modified secondarily. This is supported by many other homologies. For ex-

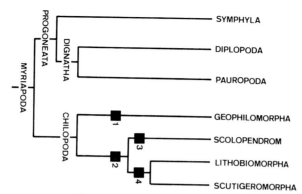

Fig. 6.33. Phylogenetic dendrogram of Myriapoda to show derived position of Scutigeromorpha. Names of groups characterize possible synapomorphies. Characters in Chilopoda: 1 = blind, reduction of Tömösvary organ; 2 = heterosegments; 3 = ? (I do not know a good synapomorphy); 4 = anamorphic, 15 pairs of legs.

ample, the mandible is a good synapomorphic feature of the Mandibulata, in spite of Manton's (1973) contrary opinion. The theory of the "whole-limb mandible" has been refuted convincingly by Lauterbach (1972). The same is true of the contention that the Diplopoda were originally dignathous and had only five head segments. Diplopoda originally had six head segments, those of the two maxilla being reduced during embryogeny as shown by Dohle (1964, 1974). But how can one explain the secondary modification of the myriapodan eye? How can one explain the facetted eye of *Scutigera*? Unfortunately, there is no recent representative with typical ommatidia or even transitional forms. In order to explain the evolutionary mode of modification, we have to look for models that show us the possibility of such a modification. As mentioned before, such a model is presented by the evolution of the holometabolic larval eye. In both groups, a change in mode of life was the starting point for the modification of their facetted eyes.

The ancestors of Myriapoda had facetted eyes, as known from fossil material (e.g., *Pleurojulus*), which modified because these ancestors had just come out of the water and had no good protection against evaporation from their cuticle. For this reason, they could be active only by night. In insect larvae, this eye reduction possibly started with a wood-boring mode of life. The result of these two different modes of life was a remarkably identical eye, at least in some insect larvae (e.g., Coleoptera). This parallel evolution can be proved very easily. As shown before, the insect larval eye is definitely derived from the normal insect ommatidium. All reductions and modifications start with a diminution of the facetted eye (as in *Panorpa*) and in decreasing the number of ommatidia. In insect larvae, as well as in some adult insects we can reconstruct at least three possible methods of eye modification (Fig. 6.34): 1) Reduction of the remaining

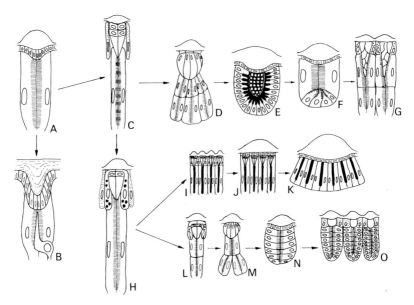

Fig. 6.34. Evolutionary pathways to arthropod ommatidia and facet eyes. A, Simple protarthropodan ommatidium; B, ommatidium of *Limulus*; C, ommatidium of Protmandibulata as seen today unchanged in Crustacea. From this type there developed in Myriapoda by reduction different myriapodan eyes. D, Intermediate stage in a myriapodan ancestor; E, *Scolopendra*; F, *Lithobius*; G, "Pseudofacet" eye of *Scutigera*. From (C) type insect type of ommatidium in (H) developed. Different ways of modification of this type of ommatidia are shown in (I–K): I, normal facet eye; J = intermediate stage to K; unicorneal composite eye as in hymenopteran larvae; L, larval ommatidium of Mecoptera; M, double ommatidium of Trichoptera; N, multilayered stemma of Coleoptera; O, "Pseudofacet" stemmatic type of facet eye in adult Strepsiptera.

ommatidia to isolated lenses: these remaining ommatidia are either unchanged in structure as in Collembola, or slightly modified as in the larvae of Trichoptera and Lepidoptera; the first step of modification in most cases is the formation of a two-layered rhabdom, with distribution of the eight cells in two layers. 2) Fusion of ommatidia: such cases are found frequently in the larvae of Trichoptera and Diptera, which have double stemma showing complete fusion at least of the crystalline cone; in cases where the two ommatidia or stemmata fuse, the retinula cells very often distribute in many layers, as seen in *Philopotamus* with its giant stemma a fusion of three or four ommatidia; often the crystalline cone no longer develops; such eyes are present in the larvae of Coleoptera and some Neuropteroidea. 3) Fusion of all corneae into one single giant lens without modification of the retina, but with reduction of the now superfluous crystalline cones: in this way had developed the so-called unicorneal composite eye; such eyes are present in all larvae of Hymenoptera and some Diptera; Werringloer (1932) has demonstrated the development of such an eye in the females of different species in the Dorylinae (Formicidae); in one known case, the crystalline cones are not reduced (in *Ampelisca*: Amphipoda) (Demoll, 1917).

The lateral eyes are structurally nearly identical with those of the Coleoptera larvae. They have a multilayered retinula, no crystalline cone, and a multicellular lentigenous epidermis layer. It seems very likely that the myriapodan eye developed by the second method just described. Since all recent myriapodan eyes were formed in this way, this modification must have occurred in the ancestor of the Myriapoda, whereas the ancestor of the Tracheata had an ommatidium. But we cannot exclude the possibility that this mode of ommatidial modification happened independently in the two groups of Myriapoda. In insect larvae we find very similar stemma types, not only in Coleopteroidea + Neuropteroidea branch but also in some Diptera. For this reason, we cannot be sure that the myriapodan type of eyes is a synapomorphy of this group. Possibly, the known fossil Diplopoda with facetted eyes or at least composed eyes with over 1,000 units had normal mandibulate ommatidia. But there is nothing known about fossil Chilopoda with facetted eyes (Kraus, 1974).

The facetted eye of *Scutigera* is nothing but a further-developed chilopodan eye, that is, a secondarily made facetted eye from the modified ommatidia formed as a result of the reduction of the primary facetted eye. Because of the irreversibility of evolution (Dollo's law), this highly evolved Scutigeromorpha (Fig. 6.33) could not go back to reactivate the lost ommatidia, and developed the myriapodan ocelli. They developed from these the "pseudofacetted eye," as stated correctly by Adensamer (1894). To correct the optical conditions, they made a crystalline conelike vitreous body, which is formed, in contrast to those of the Crustacea or Insecta, of numerous distal pigmental cells (Figs. 6.7, 6.9). There are parallel cases of such secondary facetted eyes in insects. The Strepsiptera have a facetted eye and not typical ommatidia. I have referred to this type of eye as the "stemmataran facetted eye".

6.10.4. Evolution of Arthropodan Facetted Eye

The question of the evolution of the ommatidium can throw light on the question of the evolution of the Arthropoda or at least the Mandibulate. If the ommatidium can be shown as a monophyletic structure, this could suggest the monophyly of the Arthropoda or Mandibulata. If the ommatidium can be shown to have evolved independently, it does not automatically mean the polyphyly of the Arthropoda. Even monophyletic groups can develop eyes independently or convergently in parallel.

In early evolution, the ancestors of the Arthropoda possibly had eyes similar to these of Polychaeta. They had some isolated lens eyes without fixed numbers of elements. During the evolution of the different arthropodan characters, especially after building a chitinous cuticle, these eyes had a chitinous lens, an epidermal lentigenous layer, and beneath it a cuplike retina. By composing and increasing many of these eyes, they developed a composite eye to perfect movement vision and form resolution. By decreasing the opening angles they soon developed a facetted eye. As mentioned before, this type of facetted eye or its ommatidium has only those cells which it necessarily uses and no more. There are no special characters that could form the basis for homologization. For this reason, I think the xiphosuran facetted eye and that of the Mandibulata cannot be homologized with certainty, even if the homology seems probable. I cannot exclude the possibility that both facetted eye types evolved independently from simple precursor structures. In this case, the facetted eye in its special form would be the result of a parallel evolution. But from the fossil distribution of a facetted eye in the different groups (Petrunkewitch, 1955; Størmer, 1944), I think it is more probable that the last ancestor of the Arthropoda had the facetted eye. Therefore, this eye type is homologous within the Arthropoda (Siewing, 1960). Within the Chelicerata this facetted eye has been reduced at least twice: in the Pantopoda, and in the Arachnida during occupation of the land. The Arachnida modified the facetted eye by forming a unicorneal composed eye as the first step (seen in Scorpiones). From eye structure alone, it cannot be stated whether this modification happened only once. Possibly, the scorpions colonized land independently from other Arachnida. The eyes of scorpions and of the remaining Arachnida differ in their retina arrangement (Fig. 6.3), but this difference can be easily explained by further evolution of the retina from this scorpion-type of retina. The possible synapomorphies can be seen in Fig. 6.2.

The facetted eyes of the Crustacea and Insecta are undoubtedly homologous. The typical ommatidial structure, with its special characters which can be developed only within a facetted eye and not as a single structure, as mentioned before, is surely monophyletic. The argument that this ommatidium is the result of functional forces can be refuted by the numerous modifications which prove that the ommatidium can be structured completely differently and still be functional. The fixation of the number of cells has possibly been accidental, like the fixation of seven cervical vertebrae in mammals. There is possibly no selection pressure today to make just seven (Steiner, 1955; Osche, 1966, 1973). Tiegs and Manton (1958) see in these detailed similarities the result of fantastic convergences. They cite Grenacher (1879) and especially Hesse (1901) in support of their views. Because Tiegs and Manton have attempted to prove the polyphyly in Arthopoda on the basis of Hesse's (1901) work on the insect ommatidium, I think it is necessary to provide a different interpretation: 1) According to Grenacher (1879), the primitive crystalline cone in insects is the acone one. But primitive and simple are not necessarily the same. If the acone type really is the original one in insects, then Crustacea have developed the eucone type independently from the eucone one in insects. But as shown before, the eucone crystalline cone is distributed in all primitive insects. For this reason, the eucone type must have been present at the beginning of insect as well as at the beginning of crustacean evolution. 2) The second important argument is the evolutionary development of the insect ommatidium from the isolated two-layered myriapodan eye as still seen in Collembola and Zygentoma; but these groups also have the typical insect ommatidia, and therefore they have secondarily reduced facetted eyes (Paulus, 1972a, 1975). The same is true of the eye of *Lepisma*, which can be proved by the *Tricholepidion* facetted eye (Elofsson, 1970). 3) The holometabolic insect larvae recapitulate the myriapodan eye. The holometabolic insect larvae recapitulate the myriapodan eye. This is the third main argument of Tiegs and Manton. As shown earlier, the insect larval eyes had developed many times from typical ommatidia. The similarity of such eyes with those of Myriapoda is the result of convergence.

With the refutation of all arguments that the mandibulate ommatidium is diphyletic, the arguments in favor of monophyly of at least Mandibulata seem more plausible. This is strongly supported by the homology of the ommatidium, at least those of Crustacea and Insecta. The ommatidium is for this reason a good synapomorphic character which proves the monophyly. The eyes of the Myriapoda are secondarily simplified photoreceptors, which in one case changed secondarily again to a new facetted eye (Fig. 6.34).

6.11. SUMMARY

The phylogeny of the Arthropoda is discussed, especially in light of the important works of Tiegs and Manton (1958) and Manton (1973). Whereas these authors think that arthropods are polyphyletic, most other contributors, especially the European ones, think they are monophyletic. A polyphyletic origin implies that all important arthropodan characters evolved many times independently: the head with its particular number of segments, the special sense organs, legs, mouthparts, and others. To prove monophyly one must show that all recent arthropods are descended from one common ancestor who had the basic arthropodan characters. One of these basic characters is the photoreceptive organs: lateral facetted eyes and median eyes. Such sense organs can be found in all fossil and recent major arthropod groups. But to prove their monophyletic origin one must compare them in detail to show the differences, the similarities, and ways of modification. Important are such similarities which are not convergences, or parallelisms. The only recent chelicerates with lateral facetted eyes are the Xiphosura (*Limulus*). All terrestrial Arachnida have lateral ocelli (up to five on each side of the prosoma). The eyes of Scorpiones are derived from the *Limulus* type of eye by diminution and separation of the complex into five parts, and fusion of the remaining ommatidial facets into five big lenses. Under each lens, we find numerous isolated ommatidial rhabdoms. The other Arachnida have a further modification of the situation in scorpions.

The isolated rhabdoms are fused to a uniformly distributed network of micro-villi. This kind of retina seems to be a good synapomorphy for the Arachnida (besides Scorpiones). The phylogeny of the Chelicerata as a whole is discussed (see dendrogram, Fig. 6.2). It is shown that the basic number of median eyes in Chelicerata is four. The kind of embryonic development of these eyes (converted eyes) is a good synapomorphy of the Arachnida (including Scorpiones).

It is shown that the Mandibulata (Crustacea, Myriapoda, Insecta) are monophyletic. The ommatidia within Crustacea and Insecta have evolved once. One can show a basic mandibulate ommatidium consisting of a cornea made by two cells (corneagenous cells), a tetrapartite eucone crystalline cone, and a retinula of eight cells. The basic type of rhabdom in Crustacea is a tiered rhabdom, possibly that of insects as well (*Machilis*). Within these two groups these eyes were modified during evolution. To prove this, it was shown that the basic subgroups in Crustacea (Phyllopoda, Anostraca) and in Insecta (Collembola, Thysanura) have typical ommatidia.

It is shown that the myriapodan eyes are modified mandibulate ommatidia. The insect larval eyes are a good model for this modification. The eyes of the holometabolic insects are in many respects very similar to the myriapodan ones. The insect larval eyes have modified to stemmata from normal adult ommatidia at least in four or five different ways (Fig. 6.29). From new ultrastructural data it is shown that the *Scutigera* facetted eye is a secondary one, therefore called a "pseudofacetted eye."

The basic number of median eyes in Mandibulata is four as well. Besides them, we find two pairs of frontal organs, which are median eyes. A total number of eight is found in Phyllopoda. In Insecta, the Collembola have four median eyes (frontal ocelli) and the dorsal pair of frontal organs.

From the given data it can be shown that the mandibulate ommatidium has evolved only once. The similarity is shown to correspond cell by cell. The same is true of the median eyes and the frontal organs, which are reduced in Myriapoda. All these corresponding organs prove the monophyly at least of the Mandibulata. The monophyly of Chelicerata + Mandibulata is very likely because of the basic number of four median eyes, but from the lateral ommatidia it cannot be proved with certainty. The eyes of *Limulus* are too simply constructed to have enough characters for a good synapomorphy.

ACKNOWLEDGMENTS

This research was supported by "Deutsche Forschungsgemeinschaft." For technical help I thank Annette Gudat. For many discussions and critical remarks, I thank Prof. Dr. G. Osche, Prof. Dr. P. Weygoldt, and Dr. Otto von Helversen. I also thank Prof. Dr. A. P. Gupta and Dr. C. M. Bate for reading and correcting the English manuscript.

REFERENCES

Adensamer, T. 1894. Zur Kenntnis der Anatomie und Histologie von *Scutigera coleoptrata*. *Verh. Zool. Bot. Ges. Wien* **43**: 573-78.

Baccetti, B. and C. Bedini. 1964. Research on the structure and physiology of the eyes of a *Lycosid* spider. I. Microscopic and ultramicroscopic structure. *Arch. Ital. Biol.* **102**: 97-122.

Bähr, R. R. 1971. Die Ultrastruktur der Photorezeptoren von *Lithobius forficatus* (Chilopoda). *Z. Zellforsch.* **116**: 70-93.

Bähr, R. R. 1972. Licht- und dunkeladaptive Änderungen der Sehzellen von *Lithobius forficatus* (Chilopoda). *Cytobiologie* **6**: 214-33.

Bähr, R. R. 1974. Contribution to the morphology of chilopod eyes. *Symp. Zool. Soc. Lond. No.* **32**: 383-404.

Barra, J. A. 1969. Les photorécépteurs des Collemboles. Nouvelles formations à structure rhabdomérique propre au genre *Tomocerus* (Insectes, Collemboles). *C.R. Acad. Sci. Paris* **268**: 2088-90.

Barra, J. A. 1971. Les photorécépteurs des Collemboles, étude ultrastructurale. I. L'appareil dioptrique. *Z. Zellforsch.* **117**: 322-53.

Bart, A. 1963. Données histologiques et expérimentales sur le système neurosecreteur de l'insecte Apterygote *Petrobius maritimus*. *Gen. Comp. Endocrinol.* **3**: 298-411.

Bedini, C. 1967. The fine structure of the eye of *Euscorpius carpathicus*. *Arch. Ital. Biol.* **105**: 361-78.

Bedini, C. 1968. The ultrastructure of the eye of a centipede *Polybothrus fasciatus*. *Monit. Zool. Ital.* **2**: 31-47.

Bedini, C. 1970. The fine structure of the eye in *Glomeris* (Diplopoda). *Monit. Zool. Ital.* **4**: 201-19.

Bernhard, G. C., G. Gemne and J. Sailström. 1970. Comparative ultrastructure of corneal surface topography in insects with aspects on phylogenesis and function. *Z. Vgl. Physiol.* **67**: 1-25.

Bierbrodt, E. 1942. Der Larvenkopf von *Panorpa communis* und seine Verwandlung, mit besonderer berücksichtigung des Gehirns und der Augen. *Zool. Jahrb. Anat.* **68 (1)**: 49-136.

Bitsch, J. 1963a. Le complex nerveux hypocérébral et les corpora allata des Machilides. *C.R. Acad. Sci. Paris* **(D) 254**: 1501-3.

Bitsch, J. 1963b. Morphologie céphalique des Machilides (Ins., Thysanura). *Ann. Sci. Nat. Zool. Paris (12 sér.)* **5**: 501-706.

Bocquet, M. 1972. L'infrastructure de l'organe photorécepteur des Syllidae (Annélides). *C.R. Acad. Sci. Paris* **(D) 274**: 1689-92.

Bodenstein, D. 1953. Postembryonic development, pp. 822-65. *In* K. D. Roeder (ed.) *Insect Physiology*. John Wiley, New York.

Bott, H. R. 1928. Beiträge zur Kenntnis von *Gyrinus natator*. I. Lebensweise und Entwicklung. I. Der Sehapparat. *Z. Morphol. Ökol Tiere* **10**: 207-306.

Brammer, P. R. 1970. The ultrastructure of the compound eye of a mosquito *Aedes aegypti* (Diptera). *J. Exp. Zool.* **175**: 181-96.

Brandenburg, J. 1960. Die Feinstruktur des Seitenauges von *Lepisma saccharina* (Thysanura). *Zool. Beitr.* (N.F.) **5**: 291-300.

Brundin, L. 1972. Evolution, causal biology, and classification. *Zool. Scr.* **1**: 107-20.

Brundin, L. 1976. A neocomian chironomid and Podonominae-Aphroteniinae (Diptera) in the light of its phylogenetics and biogeography. *Zool. Scr.* **5**: 139-60.

Bullock, T. H. and G. A. Horridge. 1965. *Structure and Function in the Nervous System of Invertebrates, I, II.* Freeman and Co., San Francisco.

Burghause, F. 1976. Adaptationserscheinungen in den Komplexaugen von *Gyrinus natator* (Coleoptera). *Int. J. Insect Morphol. Embryol.* **5**: 335-48.

Busselmann, A. 1935. Bau und Entwicklung der Raupenocellen der Mehlmotte *Ephestia*. *Z. Morphol. Ökol. Tiere* **29**: 218-28.

Butler, L., R. Roppel and J. Zeigler. 1970. Post-emergence maturation of the eye of the adult Black Carpet Beetle *Attagenus megatoma*. *J. Morphol.* **130**: 103-27.

Campos-Ortega, J. A. and N. J. Strausfeld. 1972. The columnar organization of the second synaptic region of the visual system of *Musca domestica*. *Z. Zellforsch.* **124**: 561-85.

Campos-Ortega, J. A. and N. J. Strausfeld. 1973. Synaptic connections of intrinsic cells and bascet arborizations in the external plexiform layer of the fly's eye. *Brain Res.* **59**: 119-36.

Cassagnau, P. 1970. La phylogenie des Collemboles à la lumière des structures endocrines rétrocérébrales. I. *Simp. Int. Zoofilog. Univ. Salamanca* **1970**: 333-49.

Chapman, R. F. 1969. *The Insects: Structure and Function.* American Elsevier, Publishing Company, New York.

Chen, S. H. 1946. Evolution of the insect larva. *Trans. R. Entomol. Soc. Lond.* **97**: 381-404.

Chaudonneret, J. 1950. La morphologie céphalique de *Thermobia domestica* (Thysanoure). *Ann. Sci. Nat. Zool.* **11**: 145-278.

Clark, A. W., R. Millecchia and A. Mauro. 1969. The ventral photoreceptor cells of *Limulus*. 1. The microanatomy. *J. Gen. Physiol.* **54**: 289-309.

Clarkson, E. N. K. 1967. Fine structure of the eye in two species of *Phacops* (Trilobita). *Palaeontology* **10 (4)**: 603-16.

Clarkson, E. N. K. and R. Levi-Setti. 1975. Trilobite eyes and the optics of des Cartes and Huygens. *Nature (Lond.)* **254**: 663-67.

Constantineanu, M. J. 1930. Der Aufbau der Sehorgane bei den im Süßwasser lebenden Dipterenlarven und bei Puppen und Imagines von *Culex*. *Zool. Jahrb. Anat.* **52**: 253-346.

Corneli, W. 1924. Von dem Aufbau des Sehorgans der Blattwespenlarven und der Entwicklung des Netzauges. *Zool. Jahrb. Anat.* **46**: 573-605.

Curtis, D. J. 1969. The fine structure of photoreceptors in *Mitopus morio* (Phalangida). *J. Cell Sci.* **4**: 327-51.

Curtis, D. J. 1970. Comparative aspects of the fine structure of the eyes of Phalangida (Arachnida) and certain correlations with habitat. *J. Zool. (Lond.)* **160**: 231-65.

Dahl, E. 1956. Some crustacean relationships, pp. 138-47. *In* K. G. Wingstrand (ed.) *B. Hanström: Zool. Papers in Honour of his Sixty-fifth Birthday, Nov. 20th, 1956.* Lund.

Dahl, E. 1958. The ontogeny and comparative anatomy of some protocerebral sense organs in notostracan phyllopods. *Q. J. Microsc. Sci.* **100**: 445-62.

Dahl, E. 1963. Main evolutionary lines among recent Crustacea, pp. 1-15. *In* H. B. Whittington and W. D. J. Rolfe (eds.) *Phylogeny and evolution of Crustacea.* Mus. Comp. Zool. Spec. Publ. Cambridge, Massachusetts.

Dahl, E. 1965. Frontal organs and protocerebral neurosecretory systems in Crustacea and Insecta. *Gen. Comp. Endocrinol.* **5**: 614-17.

Dahl, E. 1976. Structural plans as functional models exemplified by the Crustacea Malacostraca. *Zool. Scr.* **5**: 163-66.

Debaisieux, P. 1944. Les yeux des Crustacées: Structures, dévelopment, réactions à l'énclairement. *Cellule* **50**: 5-122.

Demoll, R. 1914. Die Augen von *Limulus*. *Zool. Jahrb. Anat.* **38**: 443-64.

Demoll, R. 1917. *Die Sinnesorgane der Arthropoden, ihr Bau und ihre Funktion.* View, Braunschweig.

Dethier, V. G. 1942. The dioptric apparatus of lateral ocelli. I. *J. Cell. Comp. Physiol.* **19**: 301-313.

Dethier, V. G. 1943. The dioptric apparatus of lateral ocelli. II. *J. Cell. Comp. Physiol.* 22: 115-26.

Dickens, J. C. and J. L. Eaton. 1973. External ocelli in Lepidoptera previously considered to be anocellate. *Nature (Lond.)* 242: 205-6.

Dietrich, W. 1909. Die Fazettenaugen der Dipteren. *Z. Wiss. Zool.* 92: 465-539.

Doflein, F. 1916. *Der Ameisenlöwe.* G. Fischer Verl., Jena.

Dohle, W. 1964. Die Embryonalentwicklung von *Glomeris marginata* im Vergleich zur Entwicklung anderer Diplopoden. *Zool. Jahrb. Anat.* 81: 241-310.

Dohle, W. 1974. The segmentation of the germ band of Diplopoda compared with other classes of Arthropods. *Symp. Zool. Soc. Lond.* 31: 143-61.

Donner, K. O. 1971. On vision in *Pontoporeia affinis* and *P. femorator* (Crustacea, Amphipoda). *Soc. Sci. Famica Comm. Biol.* 41: 1-17.

Dorsett, D. A. and R. Hyde. 1968. The fine structure of the lens and photoreceptors of *Nereis virens. Z. Zellforsch.* 85: 243-55.

Dudley, P. L. 1968. The fine structure and development of the nauplius eye of the copepod *Doropygus. Cellule* 68: 7-42.

Eakin, R. M. 1963. Lines of evolution of photoreceptors, pp. 393-425. *In* D. Mazia and A. Tyler (eds.) *General Physiology of Cell Specialization.* McGraw-Hill, New York.

Eakin, R. M. 1968. Evolution of photoreceptors, pp. 194-242. *In* T. Dobzhanski, M. K. Hecht and W. Steere (eds.) *Evolutionary Biology II.* Plenum Press, New York, London.

Eakin, R. M. 1972. Structure of invertebrate photoreceptors, pp. 626-684. *In* H. Autrum, R. Jung, W. R. Lowenstein, D. M. McKay and L. H. Teuber (eds.) *Handbook of Sensory Physiology*, Vol. VII. Springer, Berlin, New York.

Eakin, R. M. and J. L. Brandenburger. 1971. Fine structure of the eyes of jumping spiders. *J. Ultrastruct. Res.* 37: 618-63.

Eakin, R. M. and J. A. Westfall. 1965. Fine structure of the eyes of *Peripatus* (Onychophora). *Z. Zellforsch.* 68: 278-300.

Eckert M. 1968. Hell-Dunkel-Adaptation in aconen Appositionsaugen der Insekten. *Zool. Jahrb. Physiol.* 74: 102-20.

Edwards, A. S. 1969. The structure of the eye of *Ligea oceanica. Tissue Cell* 1: 217-28.

Eguchi, E. and T. H. Waterman. 1966. Fine structure patterns in crustacean rhabdoms, pp. 105-124. In C. G. Bernhard (ed.), *The Functional Organization of the Compound Eye.* Wenner-Gren Center Int. Symp. Ser. 7. Pergamon Press, Oxford.

Eidmann, H. and F. Kühlhorn. 1970. *Lehrbuch der Entomologie.* Parey, Hamburg.

Elofsson, R. 1963. The nauplius eye and frontal organs in Decapoda. *Sarsia* 12: 1-68.

Elofsson, R. 1965. The nauplius eye and frontal organs in Malcaostraca (Crustacea). *Sarsia* 19: 1-54.

Elofsson, R. 1966a. The nauplius eye and frontal organs in non-Malacostraca. *Sarsia* 25: 1-128.

Elofsson, R. 1966b. Some aspects of the fine structure of the nauplius eye of *Pandalus borealis* (Decapoda). *Acta Univ. Lund, N. S.* 28: 1-16.

Elofsson, R. 1969. The ultrastructure of the nauplius eye of *Sapphirina* (Copepoda). *Z. Zellforsch.* 100: 376-401.

Elofsson, R. 1970. Brain and eyes of Zygentoma. *Entomol. Scan.* 1: 1-20.

Elofsson, R. 1973. A peculiar kind of pigment cell in the compound eye of *Lepisma saccharina. Entomol. Scand.* 4: 87-90.

Elofsson, R. 1976. Rhabdom adaptation and its phylogenetic significance. *Zool. Scr.* 5: 97-101.

Elofsson, R. and E. Dahl. 1970. The optic neuropiles and chiasmata of Crustacea. *Z. Zellforsch.* 107: 343-60.

Elofsson, R. and R. Odselius. 1975. The anostracan rhabdom and the basement membrane. An ultrastructural study of the *Artemia* compound eye. *Acta Zool. (Stockh.)* **56**: 141-53.

Eltringham, H. 1933. *The Senses of Insects.* Methuen's Biol. Monogr., London.

Fahlander, K. 1938. Beiträge zur Anatomie und systematischen Einteilung der Chilopoda. *Zool. Bidr. Upps.* **17**: 1-148.

Fahrenbach, W. H. 1964. The fine structure of a nauplius eye. *Z. Zellforsch.* **62**: 182-97.

Fahrenbach, W. H. 1968. The morphology of the eyes of *Limulus*. I. Cornea and epidermis of the compound eye. *Z. Zellforsch.* **87**: 278-90.

Fahrenbach, W. H. 1970. The morphology of the eyes of *Limulus*. III. The lateral rudimentary eye. *Z. Zellforsch.* **105**: 303-16.

Fahrenbach, W. H. 1971. The morphology of the *Limulus* visual system. IV. The lateral optic nerve. *Z. Zellforsch.* **114**: 532-45.

Firstman, B. 1973. The relationship of the chelicerate arterial system to the evolution of the endosternite. *Z. Arachnol.* **1**: 1-54.

Fischer, A. and J. Brökelmann. 1966. Das Auge von *Platynereis dumerilii* (Polychaeta). *Z. Zellforsch.* **71**: 217-44.

Fischer, A. and G. Horstmann. 1971. Der Feinbau des Auges der Mehlmotte *Ephestia kuehniella* (Lepidoptera). *Z. Zellforsch.* **116**: 275-304.

Fleissner, G. and M. Schliwa. 1977. Neurosecretory fibres in the median eyes of the scorpion *Androctonus australis*. *Cell Tissue Res.* **178**: 189-98.

Friedrichs, H. F. 1931. Beiträge zur Morphologie und Physiologie der Sehorgane der Cicindeliden (Col.). *Z. Morphol. Ökol. Tiere* **21**: 1-172.

Gabe, M. 1967. *Neurosecretion.* Gauthier-Villars, Paris.

Graber, V. 1880. Über das unicorneale Tracheatenauge. *Arch. Mikrosk. Anat.* **17**: 58-93.

Grenacher, H. 1879. *Untersuchungen über das Sehorgan der Arthropoden.* Göttingen.

Grenacher, H. 1880. Über die Augen einiger Myriapoden. *Arch. Mikrosk. Anat.* **18**: 415-67.

Gribakin, F. G. 1972. The distribution of the long wave photoreceptors in the compound eye of the honey bee as revealed by selective osmic staining. *Vision Res.* **12**: 1125-34.

Gribakin, F. G. 1975. Functional morphology of the compound eye of the bee, pp. 154-178. *In* G. A. Horridge (ed.) *The Compound Eye and Vision of Insects.* Clarendon Press, Oxford.

Grundler, O. J. 1974. EM-Untersuchungen am Auge der Honigbiene (*Apis mellifica*). I. Untersuchungen zur Morphologie and Anordnung der neun Retinulazellen in Ommatidien verschiedener Augenbereiche. *Cytobiologie* **9**: 203-20.

Güldner, F. H. and J. R. Wolf. 1970. Über die Ultrastructur des Komplexauges von *Daphnia pulex*. *Z. Zellforsch.* **104**: 259-74.

Goodman, L. J. 1970. The structure and function of the insect dorsal ocellus. *Adv. Insect Physiol.* **7**: 97-195.

Günther, K. 1912. Die Sehorgane der Larve und Imago von *Dytiscus marginalis*. *Z. Wiss. Zool.* **100**: 60-115.

Hammen, L. van der. 1970. La phylogenèse des Opilioacarides et leurs affinités avec les autres Acariens. *Acarologia* **12**: 465-73.

Hammen, L. van der. 1973. Classification and phylogeny of mites. *Proc. 3rd Int. Congr. Acarol.* **1971**: 275-81.

Hamori, J. and G. A. Horridge. 1966. The lobster optic lamina. I-III. *J. Cell Sci.* **1**: 249-74.

Hanström, B. 1924. Untersuchungen über das Gehirn, insbesondere die Sehganglien der Crustaceen. *Ark. Zool.* **16(10)**: 1-119.

Hanström, B. 1926a. Das Nervensystem und die Sinnesorgane von *Limulus*. *Acta Univ. Lund (Adv. 2)* **22**: 1-79.

Hanström, B. 1926b. Eine genetische Studie über die Augen und Sehzentren von Turbellarien, Anneliden und Arthropoden. *K. Svenska Vetensky Akad. Handl.* (3) 4 (1): 1-176.

Hanström, B. 1928. *Vergleichende Anatomie des Nervensystems der wirbellosen Tiere.* Springer, Berlin.

Hanström, B. 1934. Bemerkungen über das Komplexauge der Scutigeriden. *Acta Univ. Lund (Adv. 2, N. F.)* 30 (6): 1-14.

Hanström, B. 1935. Fortgesetzte Untersuchungen über das Araneengehirn. *Zool. Jahrb. Anat.* 59: 455-78.

Hanström, B. 1940. Inkretorische Organe, Sinnesorgane und Nervensystem des Kopfes einiger niederer insektenordnungen. *K. Svenska Vetonsky Akad. Handl.* (3) 18 (8): 1-266.

Helfer, H. and E. Schlottke. 1935. Pantopoda. *Bronn's Klassen und Ordnungen d. Tierreichs* 5/4/2: 1-312.

Helversen, O. v. and W. Edrich. 1974. Der Polarisationsempfänger im Bienenauge: ein Ultraviolettempfänger. *J. Comp. Physiol.* 94: 33-47.

Hemenway, J. 1900. The structure of the eye of *Scutigera forceps. Biol. Bull.* (*Woods Hole*) 1: 205-213.

Hennig, W. 1950. *Grundzüge einer Theorie der phylogenetischen Systematik.* Deutscher Zentralverlag, Berlin.

Hennig, W. 1958. *Taschenbuch der Zoologie*, Bd. 3: *Wirbellose II (Gliedertiere).* Leipzig.

Hennig, W. 1965. Phylogenetic systematics. *Annu. Rev. Entomol.* 10: 97-116.

Hennig, W. 1966. *Phylogenetic Systematics.* Illinois University Press, Urbana, Chicago, London.

Hennig, W. 1969. *Die Stammesgeschichte der Insekten.* Verl. von W. Kramer, Frankfurt.

Hermans, C. D. 1969. Fine structure of the segmental ocelli of *Armandia brevis* (Polychaeta). *Z. Zellforsch.* 96: 361-71.

Hermans, C. D. and R. M. Eakin. 1974. Fine structure of the eyes of an alciopid *Vanadis tagensis* (Annelida). *Z. Morphol. Tiere* 79: 245-67.

Hesse, R. 1901. Untersuchungen über Organe der Lichtempfindung bei niederen Tieren. VII. Von den Arthropodenaugen. *Z. Wiss. Zool.* 70: 347-473.

Hessler, R. A. 1964. The Cephalocarida. *Mem. Conn. Acad. Arts Sci.* 16: 1-97.

Heymons, R. 1901. Die Entwicklungsgeschichte der Scolopender. *Zoologica (Stuttg.)* 33: 1-244.

Hinton, H. E. 1958. The phylogeny of the Panorpoid orders. *Annu. Rev. Entomol.* 3: 181-206.

Holmgren, N. 1916. Zur vergleichenden Anatomie des Gehirns von Polychaeten, Onychophoren, Xiphosuren, Arachniden, Crustaceen, Myriapoden und Insekten. *K. Svenska Vetensky Akad. Handl.* 56: 1-303.

Homann, H. 1950. Die Nebenaugen der Spinnen. I. *Zool. Jahrb. Anat.* 71: 56-144.

Homann, H. 1952. Die Nebenaugen der Spinnen. II. *Zool. Jahrb. Anat.* 72: 345-65.

Homann, H. 1971. Die Augen der Araneen: Anatomie, Ontogenie und ihre Bedeutung für die Systematik. *Z. Morphol. Tiere* 69: 201-72.

Homann, H. 1975. Die Stellung der Thomisidae und der Philodromidae im System der Araneae. *Z. Morphol. Tiere* 80: 181-202.

Home, E. M. 1972. Centrioles and associated structures in the retinula cells of insect eyes. *Tissue Cell* 4: 227-34.

Hörberg, T. 1931. Studien über den komparativen Bau des Gehirns von *Scutigera. Acta Univ. Lund. 2 (N. F.)* 27: 1-24.

Horridge, G. A. 1969a. The eye of the firefly *Photuris. Proc. R. Soc.* B 171: 445-63.

Horridge, G. A. 1969b. The eye of *Dytiscus* (Col.) *Tissue Cell* 1: 425-42.

Horridge, G. A. (ed.). 1975a. *The Compound Eye and Vision of Insects.* Clarendon Press, Oxford.

Horridge, G. A. 1975b. Optical mechanisms of clear-zone eyes, pp. 225-98. *In* G. A. Horridge (ed.) *The Compound Eye and Vision of Insects.* Clarendon Press, Oxford.

Horridge, G. A. 1975c. Arthropod receptor optics, pp. 459-78. *In* A. W. Snyder and R. Menzel (eds.) *Photoreceptor Optics.* Springer, Berlin, New York.

Horridge, G. A. 1976. The ommatidium of the dorsal eye of *Cloeon* as a specialization for photoreisomerization. *Proc. R. Soc. Lond.* B **193**: 17-29.

Horridge, G. A. and C. Giddings. 1971a. The retina of *Ephestia* (Lep.). *Proc. R. Soc. Lond.* B **179**: 87.

Horridge, G. A. and C. Giddings. 1971b. Movement on dark and light adaptation in beetle eyes of the neuropteran type. *Proc. R. Soc. Lond.* B **179**: 73-86.

Jockusch, B. 1967. Bau und Function eines larvalen Insektenauges. Untersuchungen am Ameisenlöwe. *Z. Vgl. Physiol.* **56**: 171-98.

Johannson, G. 1933. Beiträge zur Kenntnis der Morphologie und Entwicklung des Gehirns von *Limulus polyphemus. Acta Zool.* (*Stockh.*) **14**: 1-100.

Joly, R. 1969. Sur l'ultrastructure de l'oeil de *Lithobius forficatus. C.R. Acad. Sci. Paris* D **268**: 3180-82.

Joly R. and C. Herband. 1968. Sur la régénération oculaire chez *Lithobius forficatus* (Chilopode). *Arch. Zool. Exp. Gén.* **109**: 591-612.

Jones, C., J. Nolte, and J. E. Brown. 1971. The anatomy of the median ocellus of *Limulus. Z. Zellforsch.* **118**: 297-309.

Juberthie, C. and A. Muñoz-Cuevas. 1973. Présence de centriole dans la cellule visuelle de l'embryon d'*Ischyropsalis luteipes* (Opiliones). *C.R. Acad. Sci. Paris* **276**: 2537-39.

Kaestner, A. 1972. *Lehrbuch der Spexiellen Zoologie* I/3/A: Insecta. G. Fischer, Stuttgart.

Kerneis, A. 1971. Étude histologique et ultrastructurales des organes photorécepteurs du panache de *Potamilla* (Annelide). *C.R. Acad. Sci. Paris* **273**: 372-75.

Kinzelbach, R. 1967. Zur Kopfmorphologie der Fächerflügler (Strepsiptera). *Zool. Jahrb. Anat.* **84**: 559-684.

Kinzelbach, R. 1971. Morphologische Befunde an Fächerflüglern und ihre phylogenetische Bedeutung. *Zoologica* (*Stuttg.*) **41**: 1-256.

Kirschfeld, K. 1973. Das neurale Superpositionsauge. *Fortschr. Zool.* **21**: 229-57.

Knoll, H. J. 1974. Untersuchungen zur Entwicklungsgeschichte von *Scutigera coleoptrata* (Chilopoda). *Zool. Jahrb. Anat.* **92**: 47-132.

Kolb, G. 1977. The structure of the eye of *Pieris brassicae* (Lep). *Zoomorphologie* **87**: 123-46.

Königsmann, E. 1976. Das phylogenetische System der Hymenoptera. I. *Dtsch. Entomol. Z.* (*N. F.*) **23**: 253-79.

Korschelt, E. and K. Heider. 1890. *Lehrbuch der vergleichenden Entwicklungsgeschichte der wirbellosen Thiere.* Spez. Teil 1., Jena.

Kraus, O. 1974. On the morphology of palaeozoic diplopods. *Symp. Zool. Soc. Lond.* **32**: 13-22.

Kraus, O. 1976. Zur phylogenetischen Stellung und Evolution der Chelicerata. *Entomol. German.* **3**: 1-12.

Krebs, W. 1972. The fine structure of the retinula of the compound eye of *Astacus fluviatilis. Z. Zellforsch.* **133**: 399-414.

Krebs, W. and B. Schaten. 1976. The lateral photoreceptor of the barnacle, *Balanus eburnus. Cell Tissue Res.* **168**: 193-207.

Krecker, J. 1909. The eyes of *Dactylopius* (Coccina). *Z. Wiss. Zool.* **93**: 73-89.

Kristensen, N. P. 1975. The phylogeny of hexapod "orders". A critical review of recent accounts. *Z. Zool. Syst. Evolutionsforsch.* **13**: 1-44.

Land, M. F. 1969. Structure of the principal eyes of jumping spiders in relation to visual optics. *J. Exp. Biol.* **51**: 443-70.

Land, M. F. 1972. Mechanisms of orientation and pattern recognition by jumping spiders (Salticidae), pp. 231-247. *In* A. Wehner (ed.) *Information Processing in the Visual Systems of Arthropods.* Springer, Berlin, New York.

Lasansky, A. 1967. Cell junctions in ommatidia of *Limulus. J. Cell Biol.* **33**: 365-84.

Lattin, G. de. 1939. Untersuchungen an Isopodenaugen. *Zool. Jahrb. Anat.* **65**: 417-68.

Laughlin, S. B, R. Menzel and A. W. Snyder. 1975. Membranes, dichroism and receptor sensitivity, pp. 237-259. *In* A. W. Snyder and R. Menzel (eds.) *Photoreceptor Optics.* Springer, Berlin, New York.

Lauterbach, K. E. 1972. Über die Herkunft des Carapax der Crustaceen. *Zool. Beitr. N. F.* **20**: 273-327.

Lauterbach, K. E. 1973. Schlusselereignisse in der Evolution der Stammgruppe der Euarthropoda. *Zool. Beitr. N. F.* **19**: 251-99.

Lauterbach, K. E. 1974. Über die Herkunft des Carapax der Crustaceen. *Zool. Beitr. N. F.* **20**: 273-327.

Lerma, B. de. 1951. Note originali e critiche sulla morfologia comparata degli organi frontali degli Arthropodi. *Ann. 1st. Mus. Zool. Univ. Napoli* **3**: 1-25.

Lerum, J. E. 1968. The postembryonic development of the compound eyes and optic ganglia in dragon flies. *Proc. Iowa Acad. Sci.* **75**: 416-32.

Lew, G. T. W. 1933. Head characters of the Odonata. *Entomol. Amer.* **14**: 41-97.

Manton, S. M. 1973. Arthropod phylogeny–A modern synthesis. *J. Zool. Proc. Zool. Soc. (Lond.)* **171**: 111-30.

Marlier, G. 1941. Recherches sur les organes photorécepteurs des insectes Aptilotes. *Ann. Soc. R. Zool. Belg.* **72**: 204-36.

Marshall, W. S. 1935. The development and structure of the eyes, ocelli of the female Black Scale *Saisetia oleae. J. Morphol.* **57**: 12-36.

Martin, G. 1971. Étude préliminaire d'une structure photosensible dans la région centrale du protocérébron de *Porcellio dilatatus* (Crustacé). *C.R. Acad. Sci. Paris* **272**: 269-71.

Martin, G. 1976. Mise en évidence et étude ultrastructurelle des valles médians chez les Crustacés Isopodes. *Ann. Sci. Nat. (12 sér.)* **18**: 405-36.

McEnroe, W. 1969. Eyes of the female two-spotted Spider Mite *Tetranychus urticae.* I. Morphology. *Ann. Entomol. Soc. Amer.* **62**: 461-66.

Meinertzhagen, J. A. 1973. Development of the compound eye and optic lobe in insects, pp. 51-104. *In* D. Young (ed.) *Developmental Neurobiology of Arthropods.* Cambridge University Press, Cambridge.

Meixner, J. 1935. Coleoptera, pp. 1035-1348. *In* W. Kükenthal and T. Krumbach (eds.) *Handbuch Zoologie*, Vol. 4/2. W. de Gruyter, Berlin.

Melamed, J. and O. Trujillo-Cenoz. 1966. The fine structure of the visual system of Lycosidae. I. Retina and optic nerve. *Z. Zellforsch.* **74**: 12-31.

Menzel, R. 1972. Feinstrucktur des Komplexauges der roten Waldameise *Formica polyctena.* (Hymenoptera). *Z. Zellforsch.* **127**: 356-73.

Menzel, R. 1975. Polarization sensitivity in insect eyes with fused rhabdoms, pp. 372-387. *In.* A. W. Snyder and R. Menzel (eds.) *Photoreceptor Optics.* Springer, Berlin, New York.

Menzel, R. and A. W. Snyder. 1974. Polarised light detection in the bee *Apis mellifera. J. Comp. Physiol.* **88**: 247-70.

Meyer-Rochow, V. B. 1971. Fixerung von Insecktenorganen mit Hilfe eines Netzmittels: Das Dorsalauge der Eintagsfliege *Atalophlebia costalis. Mikrokosmos* **60**: 348-52.

Meyer-Rochow, V. B. 1972a. A crustacean-like organization of insects-rhabdoms. *Cytobiologie* **4**: 241-58.

Meyer-Rochow, V. B. 1972b. The eyes of *Creophilus erythrocephalus* and *Sartallus signatus*. *Z. Zellforsch.* **133**: 59–86.

Meyer-Rochow, V. B. 1973. Structure and function of the larval eye of sawfly *Perga* (Hymenoptera). *J. Insect Physiol.* **20**: 1565–91.

Meyer-Rochow, V. B. 1975a. The dioptric system in beetle compound eyes, pp. 299–313. *In* G. A. Horridge (ed.) *The Compound Eye and Vision in Instects.* Clarendon Press, Oxford.

Meyer-Rochow, V. B. 1975b. Axonal wiring and polarisation sensitivity in eye of the rock lobster. *Nature (Lond.)* **254**: 522–23.

Meyer-Rochow, V. B. and G. A. Horridge. 1975. The eye of *Anoplognathus* (Col., Scarabaeidae). *Proc. R. Soc. Lond.* **B 188**: 1–30.

Mickoleit, G. 1973. Über den Ovipositor der Neuropteroidea und Coleoptera und seine phylogenetische Bedeutung. *Z. Morphol. Tiere* **74**: 37–64.

Millecchia, R., J. Bradbury and A. Mauro. 1966. Simple photoreceptors of *Limulus polyphemus*. *Science (Wash., D.C.)* **154**: 1199–1201.

Miller, W. H. 1957. Morphology of the ommatidia of the compound eye of *Limulus*. *J. Biophys. Biochem. Cytol.* **3**: 421–28.

Mills, L. R. 1974. Structure of the visual system of the two-spotted spider-mite, *Tetranychus urticae*. *J. Insect Physiol.* **20**: 795–808.

Mouze, M. 1972. Croissance et metamorphose de l'appareil visuel des Aeschnidae. *Int. J. Insect Morphol. Embryol.* **1**: 181–200.

Müller, A. H. 1960. *Lehrbuch der Paläontologie*, Bd. 2/2. G. Fischer, Jena.

Muñoz-Cuevas, M. A. 1975. Modèle ciliaire de dévelopement du photorécepteur chez l'opilion *Ischyropsalis luteipes*. *C.R. Acad. Sci. Paris* **280**: 725–27.

Nässel, D. R. 1975. The organization of the lamina ganglionaris of the prawn, *Pandalus borealis* (Crustacea). *Cell Tissue Res.* **163**: 445–64.

Nemanic, P. 1975. Fine structure of the compound eye of *Porcellio scaber*. *Tissue Cell* **7**: 453–68.

Ninomiya, N., Y. Tominaga and M. Kuwabara. 1969. The fine structure of the compound eye of a damsel fly. *Z. Zellforsch.* **98**: 17–32.

Nowikoff, M. 1905. Über die Augen und Frontalorgane der Branchiopoda *Z. Wiss. Zool.* **79**: 432–64.

Ong, J. E. 1970. The micromorpholoy of the nauplius eye of the estmarine calanoid copepod *Sulcanus conflictus* (Crustacea). *Tissue Cell* **2**: 589–610.

Osche, G. 1966. Grundzüge der allgemeinen Phylogenetik, pp. 817–906. *In* L. v. Bertalanffy and F. Gessner (eds.) *Handbuch Biologie* 3 (2). *Akad. Verlagsgesell.*, Athenaion, Frankfurt.

Osche, G. 1973. Das Homologisieren als eine grundlegende Methode der Phylogenetik. *Aufsätze u. Reden Senckenberg. Naturf. Ges. Frankfurt* **24**: 155–65.

Palm, N. B. 1955. Neurosecretory cells and associated structures in *Lithobius*. *Ark. Zool.* **(2) 9**: 115–20.

Pankrath, O. 1890. Das Auge der Raupen und Phryganidenlarven. *Z. Wiss. Zool.* **49**: 690–708.

Patten, W. 1888. Studies on the eyes of Arthropodes. II. Eyes of *Acilius*. *J. Morphol.* **2**: 97–190.

Paulus, H. F. 1970a. Zur Feinstruktur des zusammengestetzten Auges von *Orchesella*. *Z. Naturforsch.* **25b**: 380–81.

Paulus, H. F. 1970b. Das Komplexauge von *Podura aquatica*, ein primitives Doppelauge. *Naturwissenschaften* **57**: 502.

Paulus, H. F. 1972a. Zum Feinbau der Komplexaugen einiger Collembolen. Eine vergleichend-anatomische Untersuchung (Insecta). *Zool. Jahrb. Anat.* **89**: 1–116.

Paulus, H. F. 1972b. Die Feinstruktur der Stirnaugen einiger Collembolen (Insecta, Entognatha) und ihre Bedeutung für die Stammesgeschichte der Insekten. *Z. Zool. Syst. Evolutionsforsch.* **10**: 81-122.

Paulus, H. F. 1972c. The ultrastructure of the photosensible elements in the eyes of Collembola and their orientations, pp. 55-59. *In* R. Wehner (ed). *Information Processing in the Visual Systems of Arthropods.* Springer, Berlin, New York.

Paulus, H. F. 1973. Die Feinstruktur der Stirnaugen einiger Collembolen und ihre Bedeutung für die Stammesgeschichte der Mandibulata. *Verh. Dtsch. Zool. Ges.* **66**: 56-60.

Paulus, H. F. 1974. Die phylogenetische Bedeutung der Ommatidien der apterygoten Insekten (Collembola, Archaeognatha, Zygentoma). *Pedobiologia* **14**: 123-33.

Paulus, H. F. 1975. The compound eye of apterygote insects, pp. 1-20. *In* G. A. Horridge (ed.) *The Compound Eye and Vision of Insects.* Clarendon Press, Oxford.

Paulus, H. F. 1977. Das Doppelauge von *Entomobrya muscorum* (Collembola). *Zoomorphologie* **87**: 277-93.

Paulus, H. F. and M. Schmidt. 1978. Evolutionswege zum Larvalauge der Insekten. I. Die Augen der Trichoptera und Lepidoptera. *Z. Zool. Syst. Evolutionsforsch.* **16**: (in press).

Perrelet, A., L. Orci and F. Baumann. 1971. Evidence for granulolysis in the retina cells of a stomatopod crustacean. *Squilla mantis. J. Cell Biol.* **48**: 684-88.

Petrunkewitch, A. 1955. Arachnida, pp. 42-162. *In* R. C. Moore (ed.) *Treatise on Invertebrate Palaeontology* (Arthropod. 2). The University Press of Kansas, Lawrence, Kansas.

Pflugfelder, O. 1936. Bau und morphologische Bedeutung der sogenannten Ozellen der Schildlausmännchen. *Zool. Anz.* **114**: 49-55.

Pflugfelder, O. 1937. Vergleichend-anatomische, experimentelle und embryologische Untersuchungen über das Nervensystem und die Sinnesorgane der Rhynchoten. *Zoologica (Stuttg.)* **34**: 1-102.

Philogène, B. J. R. 1975. Observations sur la structure des ocelles larvaires (Stemmata) de certaines Lepidoptères. *Can. Entomol.* **107**: 1073-80.

Pipa, R. L., R. S. Nishioka and H. A. Bern. 1964. Thysanuran median frontal organ: Its structural resemblance to photoreceptors. *Science (Wash., D.C.)* **145**: 829-31.

Plabody, E. B. 1939. Pigmentary responses in the isopod, *Idothea. J. Exp. Zool.* **82**: 47-83.

Prabhu, V. K. K. 1961. The structure of cerebral glands and connective bodies of *Jonospeltis splendidus. Z. Zellforsch.* **54**: 717-33.

Ramachandran, S. 1963. Structure and development of the compound eye in the male *Drosicha* (Homoptera). *Proc. R. Entomol. Soc. Lond.* **(A) 38**: 23-31.

Rasmussen, S. 1971. Die Feinstruktur des Mittelauges und des ventralen Frontalorganes von *Artemia salina.* (Crustacea, Anostraca). *Z. Zellforsch.* **117**: 576-96.

Raw, F. 1957. Origin of chelicerates. *J. Palaeontol.* **31**: 139-92.

Remane, A. 1956. *Die Grundlagen des natürlichen Systems der vergleichenden Anatomie und Phylogenetik.* 2. Aufl. Akad. Verlagsgesellsch. Geest u. Portig, Leipzig.

Remane, A. 1959. Die Geschichte der Tiere, pp. 340-422. *In* G. Heberer (ed.) *Die Evolution der Organismen.* G. Fischer, Stuttgart.

Ribi, W. A. 1974. Neurons in the first synaptic region of the bee, *Apis mellifica. Cell Tissue Res.* **148**: 277-86.

Ribi, W. A. 1975. The neurons of the first optic ganglion of the bee (*Apis mellifera*). *Adv. Anat. Embryol. Cell Biol.* **50 (4)**: 5-43.

Röhlich, P. and J. Törö. 1965. Fine structure of the compound eye of *Daphnia* in normal dark and strongly light-adapted state, pp. 175-86. *In* O. W. Rohen (ed.), *The Structure of the Eye, II. Symp.* Schattauer, Stuttgart.

Rösch, P. 1913. Beiträge zur Kenntnis der Entwicklungsgeschichte der Strepsiptera. *Jenaer Z. Naturwiss.* **50:** 97-146.

Rutherford, D. J. and G. A. Horridge. 1965. The rhabdom of the lobster eye (Crustacea). *Q. J. Microsc. Sci.* **106:** 119-30.

Sanders, H. L. 1957. The Cephalocarida and crustacean phylogeny. *Syst. Zool.* **6:** 112-29.

Sanders, H. L. 1963. The Cephalocarida. Functional morphology, larval development, comparative external morphology. *Mem. Conn. Acad. Arts Sci.* **15:** 1-80.

Schatz, E. 1929. Bau und Entwicklung der Augen von *Gammarus. Z. Wiss. Zool.* **135:** 539-73.

Scheuring, L. 1913. Die Augen der Arachnoideen. 1. *Zool. Jahrb. Anat.* **33:** 335-636.

Scheuring, L. 1914. Die Augen der Arachnoideen. 2. *Zool. Jahrb. Anat.* **37:** 369-464.

Schiff, H. and A. Gervasio. 1969. Functional morphology of the *Squilla* retina. *Publ. Staz. Zool. Napoli* **37:** 610-29.

Schlee, D. 1971. Die Rekonstruktion der Phylogenese mit Hennig's Prinzip. *Aufsätze u. Reden Senckenberg. Naturf. Ges. Frankfurt* **20:** 1-62.

Schneider, L. and H. Langer. 1969. Die Struktur des Rhabdoms im Doppelauge des Wasserläufers *Gerris lacustris. Z. Zellforsch.* **99:** 538-59.

Schneider, L. and H. Langer. 1975. EM investigations on the structure of the photoreceptor cells in the compound eye of *Ascalaphus macaronius* (Ins., Neuroptera), pp. 410-12. *In* A. W. Snyder and R. Menzel (eds.), *Photoreceptor Optics.* Springer, Berlin, New York.

Schönenberger, N. 1977. The fine structure of the compound eye of *Squilla mantis* (Stomatopoda). *Cell Tissue Res.* **176:** 205-33.

Schroer, W. D. 1974. Zum Mechanismus der Analyse polarisierten Lichtes bei *Agelena gracilens* (Araneae). *Z. Morphol. Tiere* **79:** 215-31.

Schroer, W. D. 1976. Polarisationsempfindlichkeit rhabdomerialer Systeme in den Hauptaugen der Trichterspinne *Agelena gracilens* (Araneae). *Entomol. German.* **3:** 88-92.

Seifert, G. and E. S. El-Hifnawi. 1971. Histologische und elektronenmikroskopische Untersuchungen über die Cerebraldrüse von *Polyxenus largurus* (Diplopoda). *Z. Zellforsch.* **118:** 410-27.

Sharov, A. G. 1966. *Basic Arthropodan Stock.* Pergamon Press, Oxford.

Shaw, S. R. 1969. Sense-cell structure and interspecies comparisons of polarized light absorption in arthropod compound eyes. *Vision Res.* **9:** 1031-40.

Shelton, P. M. J. and P. A. Lawrence. 1974. Structure and development of ommatidia in *Oncopeltus fasciatus* (Heteroptera). *J. Embryol. Exp. Morphol.* **32:** 337-53.

Siewing, R. 1960. Zum Problem der Polyphylie der Arthropoda. *Z. Wiss. Zool.* **164:** 238-70.

Siewing, R. 1963a. Zum Problem der Arthropodenkopfsegmentierung. *Zool. Anz.* **170:** 429-68.

Siewing, R. 1963b. Studies in malacostracan morphology: Results and problems, pp. 85-103. *In* W. D. J. Rolfe (ed.) *Phylogeny and Evolution of Crustacea.* Mus. Comp. Zool. Spec. Publ., Cambridge, Mass.

Skrzipek, K. H. and H. Skrzipek. 1971. Die Morphologie der Bienenretina (*Apis mellifica*) in elektronemikroskopischer und lichtmikroskopischer Sicht. *Z. Zellforsch.* **119:** 552-76.

Skrzipek, K. H. and H. Skrzipek. 1973. Die Anordnung der Ommatidien in der Retina der Biene (*Apis mellifica*). *Z. Zellforsch.* **139:** 567-82.

Snodgrass, R. E. 1935. *Principles of Insect Morphology.* McGraw-Hill, New York.

Snodgrass, R. E. 1952. *A Textbook of Arthropod Anatomy.* Comstock, Ithaca, New York.

Snyder, A. W. 1973. Polarization sensitivity of individual retinula cells. *J. Comp. Physiol.* **83:** 331-60.

Snyder, A. W. 1975. Photoreceptor optics–Theoretical principles, pp. 38-55. *In* A. W. Snyder and R. Menzel (eds.) *Photoreceptor Optics.* Springer, Berlin, New York.

Snyder, A. W. and P. McIntyre. 1975. Polarisation sensitivity of twisted fused rhabdoms, pp. 388-91. *In* A. W. Snyder and R. Menzel (eds.) *Photoreceptor Optics.* Springer, Berlin, New York.

Snyder, A. W. and R. Menzel (eds.). 1975. *Photoreceptor Optics.* Springer, Berlin, New York.

Sokolow, I. 1911. Über den Bau der Pantopodenaugen. *Z. Wiss. Zool.* **98**: 339-80.

Srivastava, W. S. and P. K. Sinha. 1966. The eyes of *Centrococcus insolitus* (Coccoidea). *Entomologist* **99**: 247-53, 261-68.

Steiner, H. 1955. Die Bedeutung der Zufallszahlen in der Stammesgeschichte der Tiere. *Nat. Volk (Frankfurt)* **85**: 133-43.

Stone, G. C. and H. Koopowitz. 1976. Ultrastructure of the visual system of the wax-moth *Galleria.* I. The retina. *Cell Tissue Res.* **174**: 519-31.

Størmer, L. 1944. On the relationships and phylogeny of fossil and recent Arachnomorpha. *Skrift. Vid.-Akad. Oslo, I. Math.-Nat. Kl.* **1944 (5)**: 1-158.

Størmer, L. 1963. *Gigantoscorpio willsi*, a new scorpion from the lower Carboniferous of Scotland and its associated preying microorganisms. *Skrift. Vid.-Akad. Oslo, I. Math.-Nat. Kl. (N.S.)* **8**: 1-171.

Stöwe, E. 1943. Der Kopf von *Trimenopon jenningsi* (Mallophaga). *Zool. Jahrb. Anat.* **68**: 137-226.

Strausfeld, N. J. 1976. *Atlas of an Insect Brain.* Springer, Berlin, New York.

Strohm, L. 1910. Die zusammengesetzen Augen der Männchen von *Xenos rossii* (Strepsiptera). *Zool. Anz.* **36**: 156-59.

Struwe, G., E. Hallberg and R. Elofsson. 1975. The physical and morphological properties of the pigment screen in the compound eye of a shrimp (Crustacea). *J. Comp. Physiol.* **97**: 257-70.

Thibaud, J. M. 1967. Structure et regression de l'appareil visuel chez les Hypogastruridae (Collemboles). *Ann. Spéléol.* **22 (2)**: 407-16.

Thibaud, J. M. and Z. Massoud. 1973. Étude de la regression des cornéules chez les insectes Collemboles. *Ann. Spéléol.* **28**: 159-66.

Thiele, H. 1971. Über die Facettenaugen von land- und wasserbewohnenden Crustaceen. *Z. Morphol. Tiere* **69**: 9-22.

Tiegs, O. W. and S. M. Manton. 1958. The evolution of the Arthropoda. *Biol. Rev. (Cambridge)* **33**: 255-337.

Towe, K. M. 1973. Trilobite eyes: Calcified lenses *in vivo. Science (Wash., D.C.)* **179**: 1007-09.

Tuurala, O. and A. Lehtinen. 1964. Über die photomechanischen Erscheinungen in den Augen zweier Asselarten, *Oniscus asellus* und *Porcellio scaber. Ann. Acad. Sci. Fenn.* **A 4: 77**: 1-9.

Tuurala, O. and A. Lehtinen. 1971. Über die Einwirkung von Licht und Dunkel auf die Feinstruktur der Lichtsinneszellen der Assel *Oniscus asellus. Suomal. Tiedeakad. Toim.* **(A) 117**: 1-8.

Umminger, B. L. 1968. Polarotaxis in copepods. II. The ultrastructural basis. *Biol. Bull. (Woods Hole)* **135**: 252-61.

Vaissière, R. 1961. Morphologie et histologie comparées des yeux des Crustacées Copepodes. *Arch. Zool. Exp. Gén.* **100**: 1-126.

Vanfleteren, J. R. and A. Coomans. 1976. Photoreceptor evolution and phylogeny. *Z. Zool. Syst. Evolutionsforsch.* **14**: 157-69.

Versluys, J. and R. Demoll. 1921. Die Verwandtschaft der Merostomata mit den Arachnida und den anderen Abteilungen der Arthropoda. *Proc. K. Acad. Wetens. Amsterdam* **23**: 739-65.

Wachmann, E.   1972a.   Das Auge des Hühnerflohs *Ceratophyllus gallinae* (Siphonaptera). *Z. Morphol. Tiere* **73**: 315–24.

Wachmann, E.   1972b.   Zum Feinbau des Komplexauges von *Stylops* (Strepsiptera). *Z. Zellforsch.* **123**: 411–24.

Wachmann, E.   1975.   Feinstruktur der Lateralaugen einer räuberischen Milbe (*Microcaeculus*) (Acari). *Entomol. German.* **1**: 300–7.

Wachmann, E.   1977.   Vergleichende Analyse der feinstrukturellen Organisation offener Rhabdome in den Augen der Cucujiformia (Insecta, Coleoptera). *Zoomorphologie* **88**: 95–131.

Wachmann, E. and A. Hennig.   1974.   Centriolen in der Entwicklung der Retinulazellen von *Megachile rotundata* (Hymenoptera). *Z. Morphol.* **77**: 337–44.

Wachmann, E. and W. D. Schroer.   1975.   Zur Morphologie des Dorsal- und Ventralauges des Taumelkäfers *Gyrinus substriatus* (Gyrinidae). *Zoomorphologie* **82**: 43–61.

Wachmann, E., J. Haupt, S. Richter and Y. Coineau.   1974.   Die Medianaugen von *Microcaeculus* (Acari). *Z. Morphol. Tiere* **79**: 199–213.

Wada, S.   1974.   Spezielle randzonale Ommatidien der Fliegen (Diptera: Brachycera): Architektur und Verteilung in den Komplexaugen. *Z. Morphol. Tiere* **77**: 87–125.

Walcott, B.   1971.   Cell movement on light adaptation in the retina of *Lethocerus* (Hemiptera). *Z. Vgl. Physiol.* **74**: 1–16.

Walcott, C. D.   1912.   Cambrian geology and palaeontology. No. 6 – Middle Cambrian Branchipoda, Malacostraca, Trilobita and Merostomata. *Smithson. Misc. Collect.* **57**: 145–228.

Walcott, R. and G. A. Horridge.   1971.   The compound eye of *Archichauliodes* (Megaloptera). *Proc. R. Soc. Lond.* B **179**: 65–72.

Waterman, T. H.   1961.   Light sensitivity and vision, pp. 1–64. *In* T. H. Waterman (ed.) *The Physiology of Crustacea*, Vol. II. Academic Press, New York, London.

Waterman, T. H.   1966.   Polarotaxis and primary photoreceptor events in Crustacea, pp. 493–511. *In* C. G. Bernhard (ed.) *The Functional Organization of the Compound Eye.* Pergamon Press, Oxford.

Waterman, T. H.   1975.   The optics of polarization sensitivity, pp. 339–71. *In* A. W. Snyder and R. Menzel (eds.) *Photoreceptor Optics.* Springer, Berlin, New York.

Waterman, T. H., H. R. Fernandez and T. H. Goldsmith.   1969.   Dichroism of photoreceptive pigment in rhabdoms of the crayfish *Oronectes*. *Gen. Physiol.* **51**: 415–32.

Watson, J. A. L.   1963.   The cephalic endocrine system in *Thermobia domestica* (Thysanura). *J. Morphol.* **113**: 359–69.

Weber, H.   1933.   *Lehrbuch der Entomologie.* G. Fischer, Jena.

Weber, H.   1954.   *Grundriss der Insektenkunde.* G. Fischer, Stuttgart.

Wehner, R. (ed.).   1972.   *Information Processing in the Visual Systems of Arthropods.* Springer, Berlin, New York.

Wehner, R.   1976.   Polarized-light navigation by insects. *Sci. Amer.* **235** (1): 106–15.

Wehner, R., G. D. Bernard and E. Geiger.   1975.   Twisted and non-twisted rhabdoms and their significance for polarization detection in the bee. *J. Comp. Physiol.* A **104**: 225–46.

Wenke, W.   1908.   Die Augen von *Apus productus* (Phyllopoda). *Z. Wiss. Zool.* **91**: 236–65.

Werringloer, A.   1932.   Die Sehorgane und Schzentren der Dorylinae nebst Untersuchungen über die Fazettenaugen der Formicidae. *Z. Wiss. Zool.* **14**: 432–524.

Weygoldt, P.   1972.   Geißelskorpione and Geißelspinnen. *Z. Freunde Kölner Zoo (Köln)* **15**: 95–107.

Weygoldt, P.   1975.   Untersuchungen zur Embryologie und Morphologie der Geißelspinne *Tarantula marginemaculata* (Amblypygi). *Zoomorphologie* **82**: 137–99.

White, R. H.   1967.   The effect of light and light deprivation upon the ultrastructure of the larval mosquito eye. II. The rhabdom. *J. Exp. Zool.* **116**: 405–26.

Whittle, A. C. and D. W. Golding. 1974. The fine structure of prostomial photoreceptors in *Eulalia viridis* (Annelida). *Cell Tissue Res.* **154:** 379-98.

Widmann, E. 1908. Über den feineren Bau der Augen einiger Spinnen. *Z. Wiss. Zool.* **90:** 258-312.

Willem, V. 1892. Les ocelles de *Lithobius* et de *Polyxenus* (Myriapoda). *Bull. Seances Soc. R. Malacolog. Belg.* **27:** 1-12.

Wirén, E. 1918. Zur Morphologie und Phylogenie der Pantopoden. *Zool. Bidr. Upps.* **6:** 41-181.

Wolberg-Buchholz, K. 1976. The dorsal eye of *Cloeon dipterum. Z. Naturforsch.* **31c:** 335-36.

Wolken, J. J. 1967. The eye of the crustacean *Copilia. J. Gen. Physiol.* **50:** 2481.

Wolken, J. J. 1971. *Invertebrate Photoreceptors.* Academic Press, New York, London.

Wolken, J. J. and R. G. Florida. 1969. The eye structure and optical system of the crustacean copepod *Copilia. J. Cell Biol.* **40:** 279-85.

Wolken, J. J. and G. J. Gallik. 1965. The compound eye of a crustacean, *Leptodora kindii. J. Cell Biol.* **26:** 968-73.

Wolken, J. J. and D. D. Gupta. 1961. Photoreceptor structures. The retinal cells of the cockroach eye. IV. *J. Biophys. Biochem. Cytol.* **9:** 720-44.

Wundrig, A. 1936. Die Sehorgane der Mallophagen, nebst vergleichenden Untersuchungen an Lioceliden und Anopluren. *Zool. Jahrb. Anat.* **62:** 45-110.

Yamamoto, K. and Y. Toh. 1975. The fine structure of the lateral ocellus of the Dobsonfly larva. *J. Morphol.* **146:** 415-30.

SECTION IV

Anatomy, Morphology, and Physiology

7

Functional Morphology and the Evolution of the Hexapod Classes

S. M. MANTON

7.1. INTRODUCTION

In 1949 a great palaeontologist, D. M. S. Watson, reviewed the various mechanisms of evolution that clearly have been operative in the animal kingdom. The idea that adaptation of species to specific environmental conditions must lead to slow evolutionary change is a view which has been expressed over the centuries. The close fit of animals and plants to their special circumstances of life is an important matter. But there are instances in which it is very difficult to believe that adaptation to habitat conditions has in fact played an important part in evolution. The vertebrate fossil record shows examples where, over some 20 million generations, steady changes have taken place that have increased the efficiency of the animals, without restricting them to one particular habitat, but instead allowing them to live well under a variety of conditions. These changes are called advances; they suit life in general, and are not adaptations to some particular set of environmental conditions.

It is this concept of persistent habits over vast periods of evolutionary time, habits that have been as constant as any morphological feature, which has largely been overlooked by invertebrate zoologists, although the concept and its evolutionary importance are equally applicable to invertebrates.

A vertebrate illustration from Watson's work concerns the plesiosaurs, a group extending in time from the Lower Lias to Upper Cretaceous. At the beginning of their reign, there were longer-necked forms with no more than 35 neck vertebrae and shorter-necked forms with no fewer than 27. Steadily, the longer-necked forms increased their number of neck vertebrae to 76, and their heads became smaller, while the shorter-necked forms reduced their neck vertebrae down to 11, and increased the size of their heads. Both forms lived in the same environment with the same food potential. The long-necked plesiosaurs, (e.g., *Elasmosaurus*) could rapidly move the neck and small head sideways and snatch fish by guile, while the short-necked, large-headed forms (e.g., *Trinacromerium*) caught their prey by speed of swimming. Many anatomical features of the body and head facilitated these differences of habit.

Habit divergencies in land arthropods are also intimately associated with body shape and structure, the latter fitting the animals to no one set of environmental conditions, but instead to better living in a variety of circumstances by various means. A second illustration of persistent habits can be seen by splitting open a decaying log in South Africa, where, under and within it, in the daytime, may be found: onychophorans, millipedes, centipedes, pauropods, symphylans, collembolans, diplurans, pterygotes, and perhaps thysanurans and proturans. The animals are not directly adapted to living in a log; instead, the structure of each is associated with divergent habits which fit them to live better, either in the log or outside it in a variety of other habitats (see below).

Watson further pointed out that many of the great changes which are seen to occur in the evolution of large groups of animals are of a kind unlike those which bring about the differentiation of species and of the subspecies contained within them. Again, this great generalization, based upon vertebrate palaeontology, is vindicated by its applicability to the arthropods. As will be shown below, the uniramian classes are based upon structural differences associated with habit advances, while adaptive radiations (e.g., in malacostracan crustaceans or pterygote insects) are associated with species differentiation, and related species do not usually share the same locations. Adaptive radiations appear to be associated with detailed adaptation to particular circumstances, in contrast to the two forms of plesiosaurs referred to above living together in the same places.

The potential habits that may become divergent one from another are many. Not all may be implemented today, but a number are shown, even by different arthropodan groups in the Cambrian, and today by Onychophora, four classes of myriapods and five classes of hexapods. The two contrasting types of evolution, by adaptive radiation and by advances leading to better living anywhere, cannot be expected always to have been quite distinct. There must also have been intermediates, but this does not mask the reality of the advances that do not lead to adaptation to particular circumstances. We have no direct knowledge of the initial origin of the large arthropodan groups. If their supposed phylogenies are to be put on paper in a diagrammatic manner, it is best that it be done by diverging lines, as in Fig. 7.1. The lines may be likened to the main stems of bushes, and below the "soil" level we have no information. It is not permissible to join up such main stems to form evolutionary trees on inadequate evidence. For many years our understanding of the evolution of arthropods has had shady

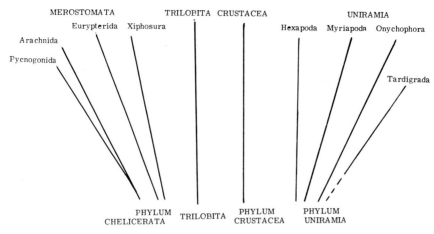

Fig. 7.1. Grouping of arthropodan taxa on the basis of comparative jaws and trunk limbs. On the right, phylum Uniramia bite with tip of a whole limb and trunk lacks biramous limbs. Three subphyla, Onychophora, Myriapoda and Hexapoda, have probably evolved independently from a primitive stock of early terrestrial Uniramia. In the middle, phylum Crustacea uses mandibular gnathobases, primitively by a rolling movement giving grinding of food. Biramous trunk limbs are present, and exopod may carry swimming setae but never a close series of flat lamellae. On the left, phylum Chelicerata also uses gnathobases, but bites by primitive adduction in the transverse plane, not by rolling. Opisthosomal limbs of *Limulus* appear to be basically biramous with many flattened gill lamellae carried by the outer branch. Phylum Chelicerata and phylum Crustacea appear to be entirely distinct. Trilobita are also entirely distinct from Crustacea, outer rami of trilobite limbs carrying flattened, probably respiratory lamellae. Trilobita show many primitive features and perhaps stand nearest to Chelicerata and early merostomes but far removed from phylum Crustacea. Pycnogonida and Tardigrada are also tentatively placed where their relationships appear to lie.

foundations, based on partially understood palaeontology and on speculation, the latter unbridled because of a lack of factual information of the right kind. The situation today is very different, owing to three lines of advancement in knowledge. New and trustworthy descriptions of extinct animals are available, and of particular interest are the very diversified Middle Cambrian arthropods and the Ordovician ostracods. Secondly, the field of comparative and functional embryology indicates with certainty many arthropodan relationships and lack of relationships. Thirdly, the comparative studies on functional morphology of arthropods, now covering a very large field, show with certainty the functional significance of many features, and structural assemblies that hitherto have been credited either with purely speculative significance or with no significance at all. Yet their understanding shows the evolutionary paths which must have been followed in reaching the present form of many taxa, and also demonstrates their relationships with one another.

Many arthropodan features, such as metameric segmentation and all that appertains thereto, do not indicate monophyly of the group. Metameric segmentation has been evolved more than once in animals. There is a similar basic plan to trunk musculature in all arthropods, showing similar functions within each category of muscles (Manton, 1973), although there are plenty of examples of muscles which change their functions and their insertions. The basic plan is dependent upon a trunk with some mobility of one segment on another, each possessing some degree of surface sclerotization or calcification, not on monophyletic evolution.

The subectodermal layer of connective tissue fibres, which may be very thin or represented by a basement membrane in some arthropods while thick in others, is associated with forming direct sites of muscle insertions or with connections to the tonofibrils passing from the muscle to the cuticle. All along the body, at least in ontogenetic stages, there are transverse intersegmental tendons formed by connective tissue. Their development from first initiation in the ectoderm was described by Manton (1928). As ontogeny proceeds, muscle rudiments become associated with the tendons, as do the myotomes and connective tissue in vertebrates. Segmental tendons provide a scaffolding for creeping muscle rudiments; parts of the tendon systems may remain unclothed as compact skeletal structures. Transverse tendons may sink forward into the preceding segment or backward into the following segment, or may grow extensions in both directions. Elaborations may be great; some tendons separate in part or entirely from the body wall in the adult and grow internally by their own formative cells retained from the ectoderm. The elaborations of tendons are in accordance with functional needs in both head, prosoma, cephalon, and trunk. Where segments are immobile on one another, fusion of segmental tendons may occur, giving the endosternite of *Limulus* or the corresponding structures in the copepod cephalothorax, or the tendons may disappear. Again, basic similarity in formation of tendon systems is due to metameric segmentation, not to monophyly of arthropods. Apodemes are of frequent occurrence and appear just where there is functional need for them. They are often, but not necessarily, segmentally homologous or serial in arrangement.

Transverse tendons have been described for a trilobite by Cisne (1975), and Whittington (1960) has figured the trilobite apodemes on which the enrollment muscles presumably inserted on each tergal arch. It is surprising that Cisne did not see these muscles, which should have been much more conspicuous than the slender muscles of the basic plan that he reconstructed. The presence of intersegmental or segmental tendons is dependent upon metameric segmentation, not upon a supposed monophyly in arthropods, but the tendon can be simpler in the less-advanced taxa.

How many of the protagonists of monophyly in arthropods have considered the real effects of metameric segmentation? That myotomes and developing muscles in vertebrates utilize connective tissue for their insertions does not mean monophyly of vertebrates and arthropods. The resemblances between skeleto-muscular metamerism in the arms of an ophiuroid, a vertebrate, and an arthropod does not add up to monophyly of all three.

There are other arthropodan features that are not explicable in a concept of monophyletic derivation of arthropodan groups. On the contrary, the evidence for the polyphyletic derivation of major arthropodan taxa seems inescapable at the present time. This subject has been reviewed in recent years, and Fig. 7.1 is repeated here to indicate the conclusions reached (Manton, 1964, 1974, 1977; Anderson, 1973; Manton and Anderson, in press).

The recognition of the phylum Uniramia (Manton, 1972) as a homogeneous group, encompassing present-day Onychophora, Myriapoda, and Hexapoda, rests on sure foundations. The relationship was first recognized, in all its strength, by Tiegs (1944, 1947), as a result of his embryological studies of myriapods and insects. His argument would have been more complete had he had access at the time to accounts of onychophoran embryology other than those of the last century, a shortcoming remedied by Manton (1949) and Anderson (1973).

The phylum Uniramia possesses: 1) uniramous limbs, 2) whole-limb jaws biting with the tip, and 3) a unified type of embryonic development, which differs from those of Crustacea and Chelicerata in a decisive manner (Anderson, 1973, Manton and Anderson, in press).

A second phylum, the Crustacea, contrasts in: 1) the biramous limbs, modi-fied in numerous ways and sometimes secondarily uniramous; 2) jaws or man-dibles used for biting, grinding, or scraping, formed by a gnathobase (or proximal endite), the more distal part of the limb in the adult providing a biramous or uniramous palp, or disappearing entirely; 3) a contrasting type of ontogenetic development; with 4) a nauplius larva often present, as in no other class or phylum, and a corresponding stage present when embryonic development is long.

The Chelicerata form a third phylum, in which: 1) there are probably basically biramous limbs, which are usually secondarily reduced to the protopod and endopod alone; 2) the jaws are formed by gnathobases, but they are moved and used in a manner contrasting with crustacean gnathobases (Manton, 1964, 1977); 3) the embryonic development is unlike that of Crustacea (Anderson, 1973, Manton and Anderson, in press), and there is no nauplius larva.

The extinct Trilobita showed no major change in their basic organization throughout their entire reign of some 250 million years. The cephalon con-trasted with the crustacean head and with the chelicerate prosoma, and the fossil record shows no intermediate types of animals, even in the Cambrian. The record shows instead many diverse types of arthropods with no clear connection or relationship with one another. As early as the Cambrian, the trilobite leg was biramous, similar on cephalon, thorax, and pygidium and of a type quite distinct from that of the Crustacea, none of which possesses the lamellae born by the outer ramus as in trilobites, which are probably respiratory in function but also may have served for swimming (see also Bergström, Chapter 1)* The trilobite gnathobases on all postoral limbs are unarticulated with the trunk, quite unlike the wide articulated coxae of the merostome prosoma and some of the limbs on the trunk of *Sydneyia*. Trilobite limbs are structurally and functionally different from those of both Crustacea and Chelicerata. Trilobite ontogenetic development, through a long series of larval stages, is well known and quite distinct from the ontogenetic development of the other principal taxa. Nothing in the least like the nauplius larva of crustaceans or the larva of *Limulus*, lacking a terminal spine, or the protonymphon larva of pycnogonids is present in any trilobites, and again these differences are decisive.

If only the three phyla, the Crustacea, the Chelicerata, and the Uniramia, were known, the trilobites might claim the status of a fourth phylum, but we are also faced with the growing number of Cambrian arthropods, well-preserved and well-restored, that show no intermediates between one another or types of animals leading on towards the trilobites, or *Aglaspis*, recently shown to be no chelicerate. Some Middle Cambrian arthropods, such as *Branchiocaris* (Briggs, 1976), possessed a simpler cephalon than any crustacean; a long series of flattened limbs extended posteriorly to the caudal furca, a bivalved carapace was present. Others possessed a large caudal furca, a large bivalved carapace, and many limbless segments posteriorly, such as *Protocaris*, where no limbs can be seen. Only *Canadaspis* possessed crustacean-like head limbs, but the antennule was minute and the maxilla 2 resembled the eight thoracic legs. But even at this early date there are remarkable structural advances. The limbs of *Canadaspis* end in strong claws, and the cusps on the mandibular incisor process are enormous, suggesting carnivorous or scavenging feeding on large prey, living or dead. There is no caudal furca, and presumably little swimming; but the stout limbless abdomen with large, strong unarticulated spines on the seventh abdominal segment suggests a habit of shallow ploughing, the body being driven into the surface of the substratum by abdominal sculling movements. The large unarticulated spines appear to have had long muscles serving them. Thus, in spite of some crustacean resemblances there were striking differences from modern Leptostracaca, presumably

---

*In reply to the hypothesis that trilobites used the outer ramus lamellae as digging organs, much could be said. Briefly, mechanical strength, joints and suitably placed musculature are essential for mechanical digging. These properties are possessed by the digging endopods of *Limulus*; trilobite endopods could have been similarly used. The blunter limb-tips in trilobites and *Limulus* than in myriapods is compatible with digging in the two former.

Secondly, trilobite outer rami and their lamellae are shown to lie dorsal to the succeeding endopods in the fine photographs of *Olenoides* by Whittington (1975). In such a position they could not dig, but they might have caused water currents which could have wafted away particles scraped up by endopods.

Thirdly, the exact dimensions of the outer ramus lamellae and of the respiratory exites of various crustaceans show that the lamellae could have housed a respiratory blood circulation (Tiegs and Manton, 1958).

the nearest relatives of *Canadaspis*. And the differences do not lead toward the rest of the Crustacea.

Thus, at present the three phyla, Uniramia, Crustacea, and Chelicerata, are each constant in their contrast with the other two, and the great group Trilobita must at present remain of uncertain status, along with a large number of smaller extinct groups. The classifying on paper of these early arthropods into higher categories, such as Trilobitoidea, Trilobitomorpha, Merostomoidea, and so on merely confuses the picture and does not clarify arthropodan evolution.*

The embryological evidence undoubtedly indicates a quite different origin of the Uniramia from the Crustacea. Uniramian ontogeny could have been descended from that of some yolky-egged annelid, but not from that of any extant polychaete. Crustacean ontogeny, on the other hand, could not have come from that of any known annelid; but what these extinct ancestors may have been, we do not know (Anderson, 1973; Manton and Anderson, in press). The lobopodium of the Onychophora shows a limb dependent upon haemocoelic fluid and muscles for its functioning (Manton, 1967, 1977). If an annelidan ancestor of the Uniramia possessed a lobopodium, in contrast to the functionally quite different, often biramous, parapodium of living annelids, this ancestor might be expected to possess a haemocoel to work the lobopodium. We are thus led in a speculative manner to the possibility of a haemocoelic, metamerically segmented worm, call it what we will, which contrasted with the modern annelids, and which disappeared as its immense evolutionary potential was absorbed into uniramian evolution.

7.2. HABITS AND THE EVOLUTION OF ONYCHOPHORA

The importance of habits that have persisted over long periods of time and which are associated with evolutionary progress has been noted above. Such habits lead to better living anywhere in a wide variety of circumstances. These comments are indisputable, but it has taken a long time for the realization that they are equally applicable in principle to many major lines of arthropodan evolution. The concept that divergence of habits is correlated with structural evolution which fits animals to no particular niche, will be considered in detail below concerning the hexapod classes. First a short summary of habits and structural differentiation of the myriapods and onychophorans is appropriate.

It appears reasonable to suggest (see below) that present-day uniramians have evolved from soft-bodied ancestors, possibly some with and some without incipient head capsules when they emerged from the water (Fig. 7.33), and whose success on land depended on various contrasting habits. It is possible that already armored uniramians also left the water, and gave rise to the myriapod-like Arthropleurida. *Arthropleura* was gigantic in the Carboniferous, and

*See also Whittington, H. B. 1978. Early Arthropods. Syst. Assn. Reports (in press).

Eoarthropleura was already in existence on land in the Lower Devonian (Rolfe, 1969; Størmer, 1976). The Arthropleurida are further considered in Section 7.7. p. 440 and Manton (1977). How many landings of uniramian arthropods there may have been we do not know, and the evidence concerning unarmored Uniramia in the sea is almost a closed book. The Middle Cambrian *Aysheaia* was probably on this line.

A first requirement on reaching land would have been the gaining of shelter, protecting the animals from osmotic uptake of dew and rain, which could blow them up to bursting. Land planarians with damp skins and no protection against excessive osmotic uptake are not found in wet woodland but under stones and other cover in drier open country (Pantin, 1950).

The next requirement would have been the ability to withstand drying up, thus permitting a wider range of habitats. A dry cuticle, rendered hydrofuge in a variety of ways, is a frequent feature of terrestrial uniramians, but even this may not be enough to conserve water.

The cuticle of onychophorans is remarkably hydrofuge, yet these animals dry out in the room atmosphere about twice as fast as a wet-skinned earthworm (Manton, 1937). The lack of control of water loss by onychophorans shows that the need to conserve water is not their highest priority. The thin, dry cuticle is very hard to wet, and is perforated by innumerable minute pits from whose bottom some 60 or more narrow, mostly unbranching tracheae arise and pass to all parts of the body, ending inside tissue cells. It seems to be more advantageous to Onychophora to possess tracheae about 2 μm wide, which will not be deformed by haemocoelic pressure changes associated with the advantageous ability to deform the body shape momentarily and extremely without harm (Fig. 7.2a). Onychophorans achieve shelter by deforming the body so that they can penetrate narrow crevices and reach more commodious spaces into which predators cannot follow. The possession of fewer, larger, branching tracheae in hexapods is not obligatory. Restriction in the number of spiracles promotes conservation of water, but the centipede, *Craterostigmus*, using extensive protractor movements of the head, caused by hydrostatic pressure of haemolymph, possesses the onychophoran-like type of minute tracheae, which cannot be damaged by changes in internal hydrostatic pressure. The ventral ganglionic chain of this centipede is invested by an unusually thick sheath of connective tissue, a probable protection against hydrostatic pressure changes (Manton, 1965). The pressure protracting the head of *Craterostigmus* is generated by muscles at a distance, there being no head protractors.

It may be concluded that the evolution of the Onychophora on land is associated firstly with the dry, hydrofuge cuticle preventing osmotic uptake of dew or rain; secondly with the large number of structural features associated with the ability to distort the body extremely. Among these features are: the many spiracular pits, giving origin to narrow and numerous unbranched tracheae, and

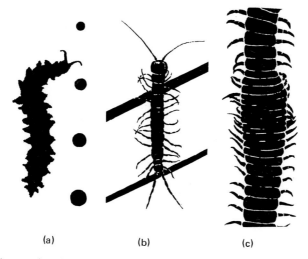

(a) (b) (c)

Fig. 7.2. Diagrams based on photographs of live animals to show their abilities to alter body shape momentarily. a, An onychophoran in a chamber containing dry air and below a damp dark chamber with access via several-sized round holes. All these holes were passable to the animal shown, legs in squeezing through smallest hole being withdrawn, so offering no projection. b, The centipede, *Lithobius*, running across slits in a card. Animal can go through these slits; head in particular is seen to be much wider than slits. Much complex morphology permits this centipede to maintain head flattening and leave space for manipulation of food in such shallow crevices. c, A burrowing centipede, *Orya*, pushing against a glass plate, with a group of segments becoming short, wide, and thick, as occurs during their normal earthworm-like burrowing. A continuity of surface armor is maintained at all shapes of a segment, as a result of facilitating morphology.

the thick connective tissue skeleton onto which the muscles are inserted, many muscles and connective tissue fibres are arranged in a criss-cross lattice, set at an angle to the long axis of the body, as in coelenterate mesogloea, an essential arrangement where connective tissue fibres are incapable of stretching. If those fibres stretched, there would be no skeletal rigidity for the muscles. The trunk musculature is unstriated, if this were not so, the wide range in shape changes would not be possible. The animal can traverse a space of but one-ninth of the transverse sectional area of the body when resting (Fig. 7.2a). There is no stiff tranverse mandibular tendon which would impede such shape changes; instead, each jaw is supported by a strong longitudinal apodeme, which carries massive jaw muscles and does not hinder body distortions (see Fig. 7.7c). There is an isolating hydrostatic-muscular mechanism for the leg, so that the legs can move independently and be flattened, leaving no projections, as does not occur in other uniramians, which thus facilitates crevice penetration (Manton, 1967, 1977).

The habit of greatest evolutionary significance to ancestral Onychophora, and adopted early in their terrestrial existence, must have been associated with the facilitating features just noted, and with others, such as a three-segmented head with jaws on the second segment in the form of short, wide little-modified legs, the terminal claws being enlarged, forming cutting paired blades on each jaw. A simple series of gaits is used by Onychophora, as is characteristic of the Uniramia, in contrast to the other arthropodan phyla (see Fig. 7.4). But the onychophoran gaits are unexploited. Neither speedy nor strong leg movements are compatible with the fundamental habit of deforming the body extremely and slowly and achieving shelter without pushing employing unstriated muscles. Further development of the onychophoran type of gaits is found in the myriapods and hexapods and always associated with special needs and proficiencies.

7.3. HABITS, FACILITATING MORPHOLOGY, LIMB MOVEMENTS, AND EVOLUTION OF THE MYRIAPOD CLASSES

The Myriapoda and Hexapoda, in contrast to the Onychophora, do not need the slow movements characteristic of the latter because the presence of sclerites on the body surface and apodemes, or differentiated tendons within, means that usually muscle contractions at variable speeds will not cause changes in internal hydrostatic pressure. The ability to perform strong movements by the body or legs in some taxa and alternatively, but of necessity, that of performing weaker movements in other taxa are both dependent upon: striated muscles; jointing of body and legs; good articulations at the joints; and either rigidity- or flexibility-promoting trunk morphology, depending upon habits and needs of particular taxa. Advanced types of gaits are used, which could have originated from the simple onychophoran gait series.

Support of muscles, in Onychophora dependent upon haemocoelic pressure and the connective tissue sheath, changes with the advent of sclerites, apodemes, segmental and intersegmental tendons, and other elaborations of the thin subectodermal connective tissue layer. Tonofibrils passing through the ectodermal cells link the muscles with their sites of tension on the cuticle. The skeletal function of the haemocoel of a soft-bodied animal gives way in myriapods and hexapods to that of cuticular and tendon systems whose action is not controlled by fluid pressure. Some muscles pull on the flexible and not on the stiff parts of the cuticle, while the larger segmental tendons become separated entirely or in part from the subectodermal layer. The physiological functions of the haemocoel persist in the myriapods and hexapods, but the mechanical ones wane, except in special cases, such as the earthworm-like burrowing by geophilomorph centipedes, where muscles, working on the incompressible haemocoel, cause local thickening by a group of segments, which thereby exert the burrowing heave against the soil (Fig. 7.2c). Other examples are the hydrostatic extension of the collembolan jumping organ (see below) and the protrusion of the head of the centipede, *Craterostigmus*, during feeding, as noted above, movements caused indirectly by muscles acting on the hydrostatic skeleton.

The joints, articulations, and type of differentiated exoskeleton permitting the slow, strong movements used in burrowing by Diplopoda differ in a mutually exclusive manner from the morphology facilitating fast running in Chilopoda. Muscles providing strength have large transverse sectional areas and are short, wide, and few in number (see extrinsic muscles of a diplopod leg, Fig. 7.3a). The unavoidable weakness of fast-moving muscles is compensated for by their greater numbers in speedy runners. A strongly moving millipede leg has but two extrinsic leg muscles, while *Scutigera*, the fleetest of all centipedes, has 34 such muscles, and many are long, so that a maximal effect is caused by a minimal degree of muscle shortening, all approaches to isometric muscular contraction being physiologically advantageous (Fig. 7.3b).

The myriapods and hexapods have exploited simple, presumably primitive, gaits such as seen in Onychophora (Fig. 7.4a) to form extensive series of gaits (Fig. 7.4c), not found in Crustacea or Arachnida. From the left end of the onychophoran series advancement could have produced the slow, strong movements (Fig. 7.4b) with long relative duration of backstroke (heavy lines on the figures); and from the right-hand end the fastest gait patterns may have originated, with very rapid stepping (short-pace duration) and small relative duration of backstroke, resulting in as few as three legs out of 40 being in the propulsive phase at one moment (Fig. 7.4c). By contrast, the strong, slow diplopod movements (left side of Fig. 7.4b) have many simultaneously propulsive legs. The upper zig-zag lines (gait "diagrams") show the movement of the legs relative to the head, and the lower gait "stills" show views from above at one moment in time during the execution of the gait. The phase difference between successive legs changes in an obligatory manner with change in the pattern of the gait (the relative durations of forward and backward strokes).

Many morphological features facilitate the execution of the various types of gait. No one myriapod or hexapod implements the full range of observable gaits. The stability requirements of the hexapods restrict their performances, usually to gaits providing at least three legs supporting the body at all times and arranged about the center of gravity (see Fig. 7.26).

The diplopods exercise various modes of burrowing (Manton, 1954, 1961), and the morphology permits the achievement of strong movements by slow gaits with many or all of the legs propulsive at one moment. Facilitating features include: the rigid cuticle investing the whole body; the diplo-segments, usually with incompressible joints between them—a maximum number of legs are thereby accommodated on the body, and their total thrust applied as a burrowing force; the firm coxa-body articulation; the musculature of body and legs; the morphology of the head end or the dorsal surface widened by keels,

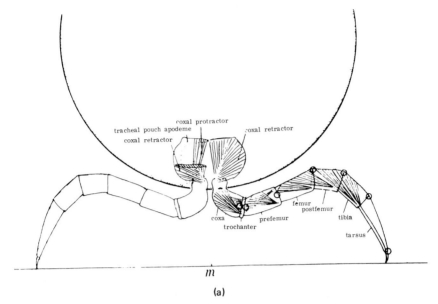

(a)

(b)

used in "flat-back" pushing between layered decaying leaves; and some 40 other features (Manton, 1954, 1958, 1961, 1977). The diplopods could have evolved from early terrestrial uniramians that sought shelter by pushing into the substratum, and adopted a diet of decaying vegetation eaten in quantity. The evolution of the various types of diplopodan morphology would follow the adoption and development of this habit.

By contrast, the Symphyla and Chilopoda usually limit their sclerites to the sternal and tergal series, leaving a flexible pleuron bearing few pleurites. Both seek shelter in existing crevices; body flattening (see Figs. 7.21c, 7.2b) and lateral flexibility are useful to both, and they have achieved fleetness but by different means. Symphyla feed on decaying leaves and wood; small size and flexibility-promoting morphology enable twisting and turning into small existing crevices, but unlike the Onychophora, without body deformation other than flexures, and unlike the Diplopoda, without pushing. Symphyla are fleet for their size and achieve this fleetness by very rapid stepping and slowish pattern gaits. This surprising combination is necessitated by trunk flexibility being a higher priority than an ability to use fast-pattern gaits, which would need potential trunk rigidity. There are structural features facilitating a rapid forward swing of the leg (Manton, 1966), and an inability to reduce further the duration of this stroke appears to be the limiting factor controlling their speed, again as in no other arthropod. The extreme trunk flexibility is a major achievement, involving separation of intercalary from principal tergites, in the presence of extra tergites on some segments and a freely folding ventral surface (Manton, 1966).

The Chilopoda show the opposite in gait patterns, and in the fastest Scolopendromorpha there are very few legs in contact with the ground at one moment (Fig. 7.4c). This arrangement can be achieved only by the presence of elaborate skeleto-musculature controlling yawing. Trunk undulations reduce speed and efficiency, and fastest stepping and fastest gait patterns are possible only where antiundulation mechanisms are well developed. These measures include tergites of alternate length (Fig. 7.5); long sectors of muscles crossing more than one joint; tergites being each linked indirectly by muscles with up to five sternites (Fig. 7.5); well-formed leg-rocking mechanisms (see below); and remarkable coxa-body and coxa-trochanter joints. Such running abilities serve the pursuit of live prey, even the catching of spiders and flies, and escape from predators.

Fig. 7.3. Extrinsic leg muscles (a) of a iuliform diplopod with short, wide muscles suited for strong, slow movements, contrasted with (b) the ventral extrinsic muscles of a fleet scolopendromorph centipede inserting on sternite. Muscles are long and numerous and cross to other side of the body. In addition, there are more leg extrinsic muscles inserting on tergite and elsewhere. Longer, more numerous, muscles provide speed of movement, each muscle contributing only a small force. From Manton (1958, 1965).

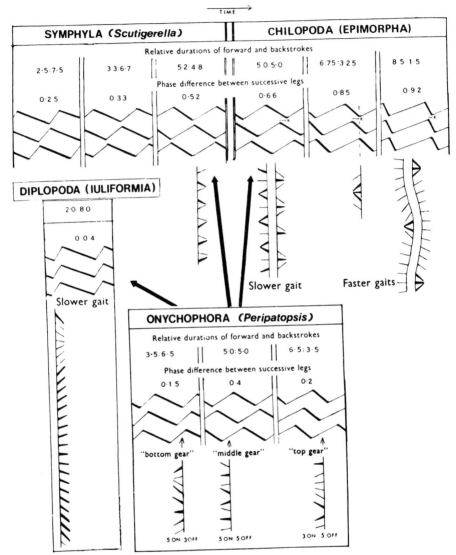

Fig. 7.4.   Diagrams showing typical uniramian gaits. Each zig-zag line from left to right represents movements of one leg; thick parts of lines indicate propulsive back stroke with limb tip on ground, and thin lines show recovery forward swing. Only a few successive legs are drawn, sufficient to indicate phase differences between them. Heavy arrowed lines indicate manner in which simple onychophoran-type gaits might have been advanced to give various types of gait suiting speed or strength. Only a few examples are shown; a fuller summary of uniramian gaits is given in Manton (1973), together with contrasting arachnid gaits. From Manton (1973).

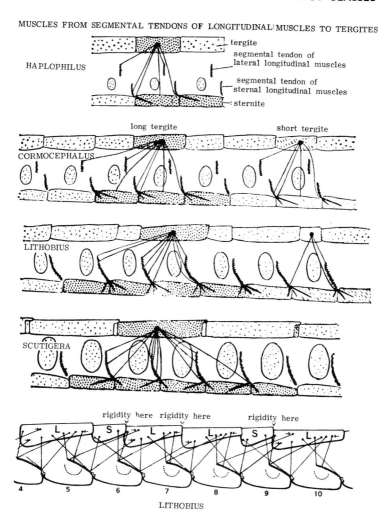

Fig. 7.5. Diagrams showing how muscle-tendon arrangements in chilopods, together with alternating tergite lengths, provide increasing facultative trunk rigidity from (a) *Haplophilus* with tergites of equal length, to (b) a scolopendromorph centipede with alternating long and short tergites, one of each being drawn on left and right; (c) *Lithobius*, with greater degree of tergite heteronomy; and (d) *Scutigera*, where short tergites are so short as to be overlapped by long tergites. Diagrams show how one long tergite becomes linked by tendons and muscles (simple lines) to progressively more sternites. (e) shows a lateral view of *Lithobius*; leg-bearing segments are marked below; lines represent muscles that have shifted their insertions off short tergites and onto long tergites. Two successive long tergites over legs 7 and 8 have acquired muscles from in front and behind and thus between them form most stable part of the body, as demonstrated by cinematography. For full details, see Manton (1965). From Manton (1965).

The morphology and performance of fast-running centipedes would have arisen by an initial adoption of a habit of speedy running and carnivorous feeding, contrasting with those of both Diplopoda and Symphyla, leading to entirely different head and trunk morphology.

The minute Pauropoda have adopted fast-patterned gaits, accompanied by trunk morphology in some ways more extremely modified toward the execution of these gaits than that of the Chilopoda. The tergites and musculature of the pauropod trunk provide rigidity and the flexibility-promoting muscles, so well-formed in the Symphyla (see Fig. 7.22a), are absent (Fig. 7.22b). The latter category of muscles is needed for certain chilopod activities, and is present along with the rigidity-promoting musculature (*dvma, dvmp, dvtr*, Fig. 7.22b).

In all, we now have a functional understanding of many onychophoran and myriapodan features, only some of which can be touched upon here. This knowledge indicates the paths of evolutionary advancement that must have led to the existence of these features today. They include: the nature of the cuticle, whether conspicuously rigid, flexible, or elastic; the types of tracheal systems; the presence of single segments; presence of diplo-segments; presence of keels extending the dorsal surface, giving cover to legs exerting a burrowing thrust; presence of intersegmental joints that are flexible, but lack articulations; presence of others that are flexible and constructed with engineering precision in the hard parts; presence of intersegmental joints that are immobile—there are both intersegmental and intrasegmental zones of particular flexibility or rigidity; presence of sclerites that are segmentally united into a cylinder; sclerites separated by arthrodial membrane and movable on one another; the tergite and sternite series of sclerites separated, or not, by an expanse of flexible pleuron; intercalary tergites and sternites separated from the principal tergites and sternites, completely or in part; extra tergites present on some segments that enhance the flexibility of the trunk—subdivision of the principal tergites in an already heteronomous series secondarily confers the same advantage in the centipede, *Craterostigmus*; tergites alternating in length; long tergites over legs 7 and 8; simple sternites; extra sternites; sliding sternites; tilted sternites; sternal carpophagous structures (certain Geophilomorpha); pleurites varying in position, in numbers, and in connections; pleurites with fixed margins or with flexible margins whose furling and unfurling movements, dependent upon particular cuticular structure, change their shape momentarily, as required in maintaining an intact surface armor during extensive shape changes of a segment in earthworm-like burrowing Geophilomorpha (Fig. 7.2c), and similar, but lesser, changes in position of tergite margins; movable and fixed pleurites, the pleurites moving in different ways in the several taxa where they serve different purposes; spiraling mechanisms of various kinds; the form of the internal trunk apodemes and tendon systems; limb morphology suiting strength or speed or exceptional flexibility in movements; the leg-rocking mechanisms, differing from those of the hexapod classes; segment numbers, which are controlled by known factors, resulting in long- or short-bodied arthropods—even loading on the legs during running is a mechanical and physiological advantage and acquired by suitable combinations of segment numbers with practicable gaits (e.g., the first instars of *Lithobius*, cannot use the fast-patterned gaits of the adult, but only the slower patterns that give even loading on the smaller number of legs; when subsequent moults have added the optimal number of legs, then even loading can be combined with fast-pattern gaits; but further increase in leg number in adult *Lithobius* and *Scutigera* would be no advantage since it would bring uneven loading on the legs; Manton, 1953). The functional components of the muscular system of the Uniramia are comprehended for the first time; the presence or absence of certain muscle categories in certain taxa is meaningful, as are the shifts in muscle insertions from one location to another (Fig. 7.5), so essential for the faculative rigidity of anamorphic centipedes; but shifts of muscle insertions, correlated with other functions, are clear in various myriapodan taxa.

A logical case can be made for the supposition that present-day Oncyhophora and the four myriapodan classes have evolved from early uniramian terrestrial ancestors, each pursuing the habit divergencies mentioned above. With these habits are associated the many structural features whose functions we now appreciate for the first time, the full list of features being much longer than the sample mentioned above. A functional interpretation of the trunk of Onychophora and the myriapod classes indicates a past divergence from ancestors with little sclerotization of the trunk region, legs, and leg-bases (see Section 7.5). The difference between the morphology and associated capabilities of these taxa do not suggest that any one could have descended from any other or from the Arthropleurida, if functional continuity has existed, as indeed it must. The hexapod classes possess a different set of trunk characteristics from those of myriapods, which will be considered in outline below. But it is convenient first to summarize the fundamental differences between the heads of myriapods and hexapods as demonstrated by recent functional studies of head structure.

7.4. HEADS, MANDIBLES, TENTORIAL APODEMES, AND FEEDING IN MYRIAPODS AND HEXAPODS

Head morphology and the mode of action of the mandibles and associated structures of the four myriapod and five hexapod classes have been as misunderstood as those of the trunk and legs. Mandibular morphologies and mechanisms are not easy to comprehend. A selection from the original drawings, as needed for this purpose, is included here, but lacks being printed in three colors (Manton, 1964, 1977).

The hexapod head is trignathan (Fig. 7.6) with mandibles, maxillae 1, and labium (maxillae 2) situated behind the mouth. The myriapods are basically

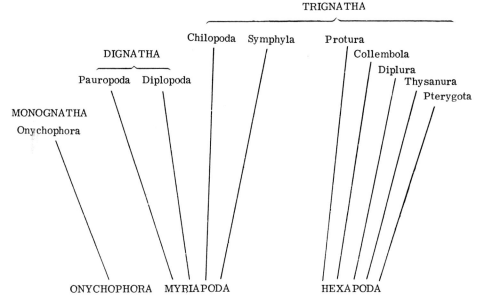

Fig. 7.6. Diagram showing three subphyla of Uniramia and their superimposed grades of organization, Mono-, Di- and Trignatha, which have no taxonomic significance.

dignathan, but with posterior additions to the head, to various extents, in three of the four classes. The Diplopoda possess a gnathochilarium behind the mandibles formed by maxillae 1 and part of the collum segment added during ontogeny. Symphyla possess maxillae 1 and no true labium; the fused maxillae 2 form a united structure with different connections and movements from those of the hexapod labium. The Chilopoda possess maxillae 1, leglike maxillae 2 united at their bases and the poison claw (first trunk) segment closely associated with the posterior part of the head. The pauropods are simply dignathan. In spite of these postmandibular differences in essentially dignathan heads, the mandibles and the skeleto-musculature that works them are basically similar in all myriapods and contrast with the hexapod classes. The monognathan Onychophora possess entognathous mandibles and a long, sclerotized jaw apodeme (Fig. 7.7c), but these mandibles are on the second head segment and not on the fourth, as in myriapods and hexapods.

The mandibles, tentorial apodemes, and their modes of action are uniform throughout the myriapod classes; they may be either ectognathous or entognathous, but in essential they differ decisively from those of hexapods. Myriapodan mandibles are segmented, while those of hexapods are not. The distal segment, or gnathal lobe of the myriapod mandible, bites against its fellow, forming a pair of crushing or cutting organs (Figs. 7.7a, 7.8). Adduction takes

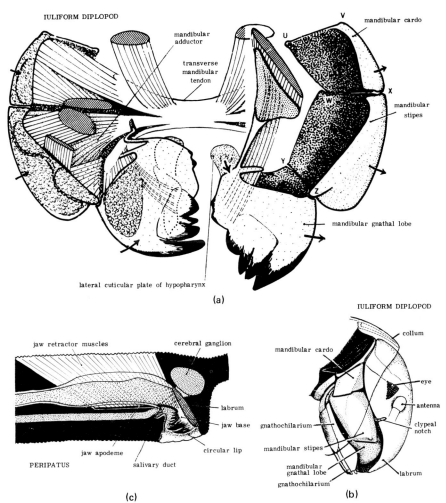

IULIFORM DIPLOPOD

mandibular cardo

mandibular
adductor

transverse
mandibular
tendon

V

U

X

mandibular
stipes

W

Y

Z

mandibular gnathal lobe

lateral cuticular plate of hypopharynx

(a)

IULIFORM DIPLOPOD

jaw retractor muscles

cerebral ganglion

mandibular cardo

collum

eye

antenna

labrum

jaw base

gnathochilarium

clypeal
notch

jaw apodeme

circular lip

mandibular stipes

PERIPATUS

salivary duct

mandibular
gnathal lobe

labrum

gnathochilarium

(c)

(b)

Fig. 7.7. a and b, An iuliform diplopod. a, Head in side view to show three-segmented mandible forming much of side wall of head, labrum in front and gnathochilarium behind mandible. Clypeal notch gives origin to anterior tentorial apodeme. b, Preparation showing mandibles, transverse mandibular tendon, and some muscles viewed from in front. Mandibular cardo is hinged to cranium at *U–V*. Mandibular adductor muscles from cavity of mandible are removed on apparent right where gnathal lobe is fully abducted as a result of a thrust by anterior tentorial apodeme at point marked by arrow. Tendon from mandibular gnathal lobe carries a large sclerite, which gives origin to main muscle, here cut. On apparent left, mandibular adductor muscles from cavities of mandibular segments converge onto transverse mandibular tendon shown in white. c, Diagrammatic longitudinal half of head of an onychophoran to show jaw bearing a pair of cutting blades surrounded by circular lip. Longitudinal jaw apodeme leaves base of jaw and carries massive muscles indicated diagrammatically. a and b, from Manton (1964).

place in the transverse plane of the head and is caused by variously arranged strong adductor muscles, the largest and strongest pulling from a tendon or apodeme leaving the gnathal lobe and inserting widely on the cranium. Other adductor muscles leave the cavities of the proximal segments in diplopods and symphylans and insert on transverse mandibular tendons. These muscles are directly adductor, not promotor-remotor as in branchiopod crustaceans and machilid thysanurans, where mandibular tendons are also present. When a jointed mandible, as in diplopods, forms much of the side wall of the head (Fig. 7.7a), there is no possible arrangement of muscles that could directly abduct the mandibles. Instead, such mandibles are pushed by a thrust exerted on the side of each gnathal lobe by the anterior tentorial apodeme (arrow in Figs. 7.7b, 7.8). This apodeme is a hollow cuticular bar of complex shape and jointed in part, which swings from the clypeal notch, its point of intucking from the surface cuticle (Fig. 7.8a). Here a short link of flexible, uncalcified cuticle unites the rigid apodeme with the head capsule. Large muscles from the apodeme pass in various directions to insert on the cranium and thus swing the apodeme within the head (Fig. 7.8b).

In Symphyla, the mandible is two-segmented and articulates with the head at one point instead of by a long hingelike articulation as in diplopods (Fig. 7.9a). The tentorial apodeme presses on the stiff gnathal lobe apodeme bearing the prinicpal gnathal lobe adductor muscles (Fig. 7.9b). There are differences in detail, but the principles of mandibular movement in Symphyla and Diplopoda are the same.

The Chilopoda (Fig. 7.10) and Pauropoda possess entognathous mandibles. The strong articulation between head and mandible seen in diplopods and symphylans is represented by a long flexible suspensory ligament from the base of the mandible to the floor of a sunken gnathal pouch. Each pouch is formed by the downgrowth during ontogeny of a pleural fold situated at the side of the head. A short pleural fold exists in *Petrobius* (see Fig. 7.12a), where it covers the base of the mandible and forms part of the elaborate boxing in of this organ. In entognathous myriapods the fold grows so as to enclose a gnathal pouch on either side of the smaller preoral cavity. In Onychophora the fold grows round the labrum from an initial lateral position and encloses the whole preoral cavity as a round mobile lip. Each gnathal pouch in Chilopoda covers most of the mandible, the mandibular biting tips projecting into the preoral space. The mandible is now a piercing, protrusible organ, the base executing wide excursions within the gnathal pouch (double-headed line in Fig. 7.10). The distal part of the mandible is supported as it moves, so that neat cutting by the distal tips can be effected when the mandible is protruded.

The segmentation of the chilopod mandible (Fig. 7.10b, c) is more complex than in diplopods, but as in the latter, permits flexure of the mandible on its long axis. Intrinsic mandibular muscles are also concerned in flexures of the mandible (see *Scutigera*, Fig. 7.11c). Mandibular adduction is still caused by

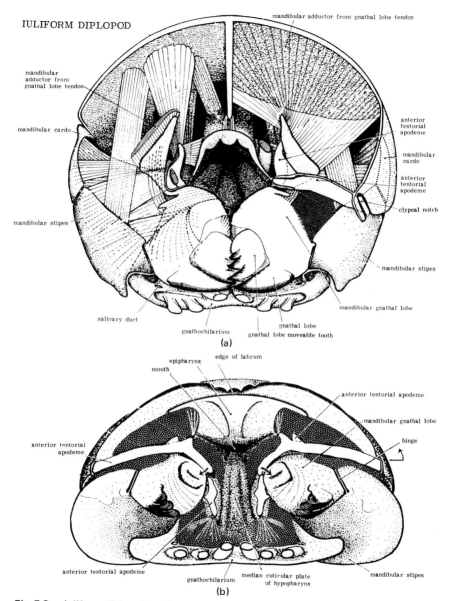

IULIFORM DIPLOPOD

mandibular adductor from gnathal lobe tendon

mandibular adductor from gnathal lobe tendon

mandibular cardo

mandibular stipes

salivary duct

gnathochilarium

gnathal lobe

gnathal lobe moveable tooth

anterior tentorial apodeme

mandibular cardo

anterior tentorial apodeme

clypeal notch

mandibular stipes

mandibular gnathal lobe

(a)

epipharynx

mouth

edge of labrum

anterior tentorial apodeme

anterior tentorial apodeme

mandibular gnathal lobe

hinge

anterior tentorial apodeme

gnathochilarium

median cuticular plate of hypopharynx

mandibular stipes

(b)

Fig. 7.8.   Iuliform diplopod head structure continued. a, Frontal view with labrum and anterior cuticular head wall cut back to show anterior tentorial apodeme, white, passing in from clypeal notch. Gnathal lobes are fully adducted and precral cavity passes into pharynx above. b, View looking into preoral cavity, floored by ganthochilarium, with mandibles abducted and labrum raised extremely to show anterior tentorial apodemes, white. Arrows show direction of thrust from anterior tentorial apodemes on either side abducting mandibles. From Manton (1964).

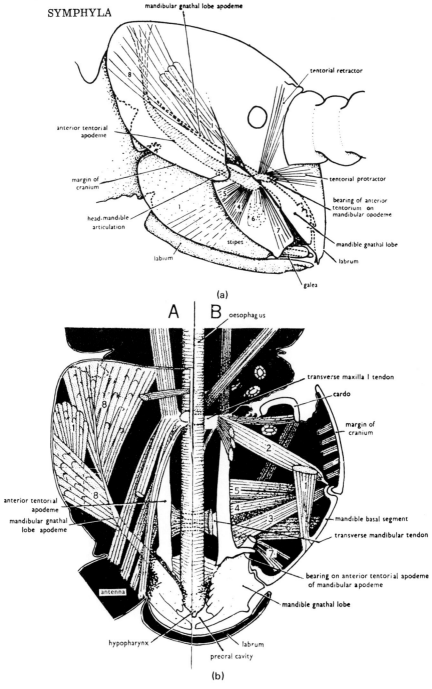

SYMPHYLA

(a)

(b)

direct muscles from the mandible to the transverse mandibular tendon. But muscles from a tendon corresponding with that from the gnathal lobe in diplopods and symphylans, inserting on the cranium, are the largest and most important adductors (shown for *Scutigera*, muscle 20, Fig. 7.11b, d). The tentorial apodeme on either side forms a fulcrum about which a mandibular process impinges, thus controlling the effects of tension by the mandibular muscles. Abduction is still caused by the swing of the anterior tentorial tentorial apodemes, but the ensuing movement in chilopods is a complex see-saw about the transverse mandibular tendon. Protrusion and depression of the distal part of the mandibles are followed by retraction and levation, as the mandibles bite together. The basal, proximal ends of the mandibles deep in the gnathal pouches execute wide excursions in the opposite directions from those at the tips. The food is not only cut up but pulled deeply into the preoral cavity, where the mandibles release the food near the mouth.

The belief that arthropods "stuff food into their mouths" (Cisne, 1974) is not true. All they do is prepare food and bring it near enough to the mouth opening for esophageal or pharyngeal muscular suction to cause swallowing. Neither can one agree that the Phyllocarida "gnaw on relatively large food items seized directly with the mandibles." Which phyllocarid? one may ask. The group is a heterogeneous collection of animals.

When a centipede's prey is large and immobilized by the bite of the poison claw, a lesion may be made by the cusped base of these claws that permits the whole head of the centipede to be inserted into the prey, whose soft parts are manipulated by the mandibles. Smaller food can be swallowed whole. The complex movements of the huge mandibles of *Scutigera* (Fig. 7.11) within a gnathal pouch sunk to the posterior limit of the cranium, are a peak of evolutionary perfection, producing a centipede capable of cutting a hard, large spider, into pieces that are swallowed. In the opposite direction, minute mandibles and small heads in geophilomorph centipedes are suited to easy penetration of prey; several shore centipedes may feed on the same barnacle at the same time at night, after an initial lesion has been achieved.

Fig. 7.9. Symphylan, *Scutigerella immaculata*, jaw mechanism. a, Two-segmented mandible articulates at one point only with cranium; basal segment bears marked muscles 4-7. Gnathal lobe, or distal segment, lies in preoral cavity; from gnathal lobe an apodeme passes deeply into head as shown in (a) and (b) and from it massive adductor muscles insert on cranium as shown on left side of (b). Anterior tentorial apodeme curls round mandibular apodeme and causes mandibular abduction. Posterior to mandible, maxillary stipes and so-called labium (maxilla 2) complete contours of head. b, Frontal views of partial reconstructions of head, level *A* being more superficial than level *B*. Short transverse mandibular and maxillary tendons are present, and in level *B* manner of overlap of mandibular apodeme by anterior tentorial apodeme is shown. Principles of mandibular adduction and abduction are similar to those in diplopods; for details see Manton (1964, 1977). From Manton (1964).

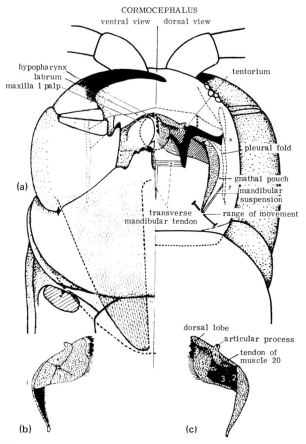

Fig. 7.10. (a) Head of a scolopendromorph centipedea, to show entognathy, mandibular mechanism, and anterior tentorial apodeme (black), ventral view of left and dorsal view on right, drawn as if head capsule were transparent and musculature omitted. External face of mandible is mottled, hatched area indicates internal cavity of mandible; (b) and (c) show opposite faces of an isolated mandible, five mandibular segments being marked. For further description see Manton (1965, 1977). Anterior tentorial apodeme swings from point *X*, corresponding with diplopod clypeal notch, and abducts mandible. From Manton (1965).

The pauropod mandible is so small that cuticular flexibility is sufficient to eliminate the need for mandibular jointing. The tentorial apodemes are well formed and presumably function much as in other myriapods, although their action has not been studied in the living animals.

An understanding of the essentials concerning the skeleto-musculature of myriapod mandibles and their modes of action in ectognathous and entognathous classes is necessary for comprehending the basic differences between myriapod and hexapod heads.

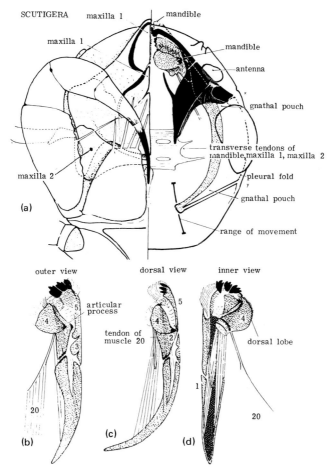

SCUTIGERA   maxilla 1   mandible

maxilla 1

mandible

antenna

gnathal pouch

transverse tendons of
mandible, maxilla 1, maxilla 2

maxilla 2

pleural fold

gnathal pouch

(a)

range of movement

outer view   dorsal view   inner view

articular
process

5

tendon of
muscle 20

dorsal lobe

4

5

4

4

3

2

1

20

20

(b)

(c)

(d)

Fig. 7.11.   Centipede, *Scutigera coleoptrata*, drawn as in Fig. 7.10 to show very large size of mandible and very stout anterior tentorial apodeme that abducts it. For further description, see Manton (1965, 1977). From Manton (1965).

Hexapod mandibles bite with the tip, as in all uniramians, in contrast to the crustaceans, chelicerates, and trilobites, which bite with a gnathobase; but hexapod mandibles are unjointed, and do not bite together, primitively, by transverse adductor-abductor movements as in myriapods. The basic hexapod mandibular movement, with some embellishment, is seen in the Machilidae. Here the axis of movement of the mandible on the head is vertical (see heavy line in Fig. 7.12a), and the movement is equivalent to the promotor-remotor swing of a walking leg. The remotor swing rolls the molar areas across one another and scrapes the distal cusps on the substratum, dislodging unicellular algae. The further collection of these algal cells can be regarded as a specializa-

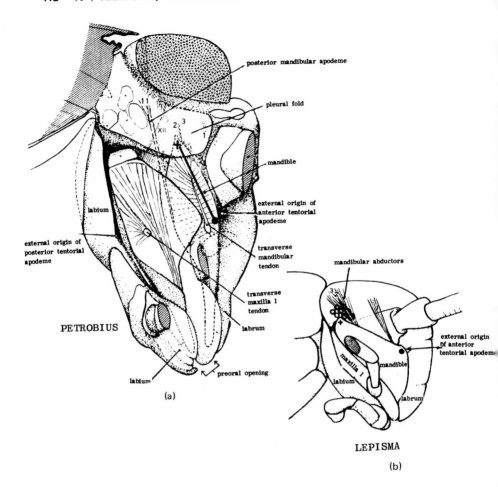

PETROBIUS

posterior mandibular apodeme

pleural fold

mandible

external origin of
anterior tentorial
apodeme

transverse
mandibular
tendon

transverse
maxilla 1
tendon

labrum

labium

external origin of
posterior tentorial
apodeme

labium

preoral opening

(a)

mandibular abductors

external origin
of anterior
tentorial apodeme

maxilla 1

mandible

labium

labrum

LEPISMA

(b)

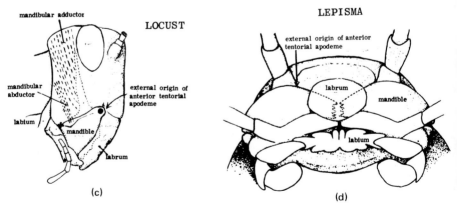

LOCUST

mandibular adductor

mandibular
abductor

labium

mandible

labrum

external origin of
anterior tentorial
apodeme

(c)

LEPISMA

external origin of anterior
tentorial apodeme

labrum

mandible

labium

(d)

tion, not appertaining to the basic mandibular mechanism. The distal parts of the elongated mandible of *Petrobius* are boxed in by other head limbs. Maxillary gland secretions pass into the enclosed space, and algal cells are sucked up to the grinding molar areas with this fluid and thence to the mouth, the long and narrow oesophagus exerting the suction. The promotor movement parts the molar areas.

The promotor-remotor mandibular muscles of *Petrobius*, arising from the anterior and posterior mandibular margins, pass to the cranium and anterior tentorial apodemes (Fig. 7.13a), while muscles from the concavities of the mandibles converge onto the transverse mandibular tendon, as in anostracan crustaceans, and contribute force to the rolling movement, as in Fig. 7.18b. These muscles are not adductor muscles, and neither is the transverse tendon, an adductor muscle tendon. A strong remotor pull on the mandible is exerted by an apodeme from its posterior margin whose muscles insert on the cranium above (Fig. 7.13a, right side). Strong promotor muscles leave the anterior mandibular margin and insert on the almost rigid anterior tentorial apodeme (Fig. 7.13a, left side).

The hexapods contrast with the myriapods in possessing two pairs of tentorial apodemes. Both are approximately rigid and do not cause mandibular movements. The anterior tentorial apodemes provide sites of insertion of mandibular muscles, and the posterior tentorial apodemes carry maxilla and labial muscles. The anterior tentorial apodeme (Fig. 7.12a, 7.13a) arises from a point corresponding with the clypeal notch of a myriapod, but there is no flexible link with the surface cuticle allowing any swinging within the head.

Thus, the mandible of *Petrobius* is fundamentally different from those of the Myriapoda in structure, movements, and musculature. *Petrobius* shows a partial and superficial similarity to the mandibles of *Chirocephalus* and *Anaspides* among the Crustacea, but only because all use the promotor-remotor swing of a walking leg, which is the simplest way in which a mandible can be used, apart

Fig. 7.12. To show positions of axes of swing of mandible on the head in various hexapods. a, Side view of head of thysanuran, *Petrobius*; the axis of rolling mandibular movement lies between cross, marking dorsal ball-and-socket articulation, and black spot. These two points are similar in *Lepisma* in (b) and locust in (c). In (a) mandible of *Petrobius* is considerably boxed in by maxilla 1; and long intucking giving origin to posterior tentorial apodeme is shown by dark line anterior to labium. b, Similar view of *Lepisma* to show change in positions of cross and black spot, resulting in mandibular biting almost in transverse plane of head, postaxial part of mandible being enlarged while preaxial part is reduced (see Fig. 7.14a); c, similar view of locust where change in position of mandibular axis of swing is even more extreme (see also Fig. 7.15); d, ventral view of head of *Lepisma* to show narrow labrum, no wider mandibular gape being possible because of position of mandibular adductor apodeme and muscles from mandibular cavity to anterior tentorial apodeme (see Fig. 7.14a). For further description, see Manton (1964, 1977). From Manton (1964).

PETROBIUS

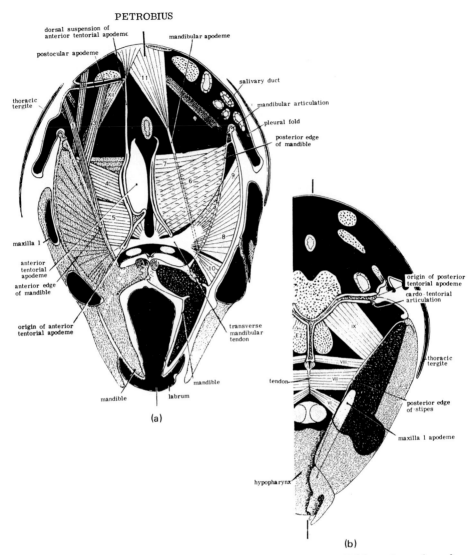

(a)

(b)

Fig. 7.13. To show mandibular mechanism of *Petrobius*. a, Superficial anterior surface of mandible, mottled, is drawn on left, and a section through mandible, a little more posterior in level, on right. Transverse mandibular tendon is shown on both sides and bears muscles that give rolling movement of mandibles; anterior tentorial apodeme is shaped differently at two levels; mandibular adductor apodeme arises from posterior margin of mandible. b, Section through a more posterior level of head behind that of (a) to show dorsal part of in-tucking, forming posterior tentorial apodeme (see Fig. 7.12a), transverse connection between the two apodemes, and median downgrowth united with transverse maxilla 1 tendon. For further description, see Manton (1964, 1977). From Manton (1964).

from the backward and forward slicing action of the onychophran jaw (Fig. 7.7c). The myriapods contrast in using the adductor-abductor movement of a walking leg, but this movement in the walking leg usually takes place distal to the coxa and not from the very base of the limb, although a few small coxal adductor and abductor muscles can be found in some hexapod classes, but not in myriapods. The jointed myriapodan mandible does not employ a primitive locomotory limb movement but one suiting the unique jointing of the mandibles. A swinging anterior tentorial apodeme is also a unique feature. Myriapodan mandibular movements, jointing and muscles, and the swinging anterior tentorial apodemes form a tightly integrated complex that has evolved in this group of arthropods alone, in contrast to all others. The mandible of the hexapod, *Petrobius*, is fundamentally different from myriapod mandibles and from those of all the primarily aquatic arthropods.

The posterior tentorial apodemes of *Petrobius* arise from paired, long vertical intuckings situated anterior to the labium (Figs. 7.12a, 7.13b). These apodemes are elaborate in shape, and the pair is united by a hollow transverse connective above the maxilla 1 cardo; a median downgrowth from this connective united with the transverse maxilla 1 tendon. The anterior tentorial apodemes are linked by stabilizing muscles to the dorsal cranial surface and by a downgrowth on either side to the transverse mandibular tendon. The anterior and posterior tentorial apodemes are linked on either side by anteroposterior muscles. The muscular connections of the two pairs of tentorial apodemes in *Petrobius* possibly indicate support for rapid vibratory movements of the mandibles (for details see Manton, 1964).

There has been considerable confusion concerning the existence and nature of tentorial apodemes in myriapods and hexapods. The morphology is easily observable from well-fixed esterwax sections stained by Mallory's triple stain. The apodemes appear red, with no confusion between them and purple muscles and other head structures. Segmental tendons stain bright blue. This clear distinction between tendons and apodemes is not apparent with many other staining techniques (see figures in Manton, 1964, 1977). Reconstruction from sections is an essential procedure for investigation of tentorial apodemes, besides the making of other types of preparation. Neither a true labium nor posterior tentorial apodemes are present in any myriapod.

In *Lepisma* (Fig. 7.14) the anterior pair of tentorial apodemes is fused across the middle line and separated by the shortest of muscles from the posterior pair. In the Pterygota all four apodemes, the anterior and posterior pairs, are fused to form the tentorium, an endoskeletal plate traversing the cranium and fixed to it at four zones, the sites of the apodemal intuckings, and bracing the cranium against the tension exerted on it by the massive mandibular adductor and abductor muscles (Fig. 7.15). The tentorium also supports labial, maxillary, and other muscles.

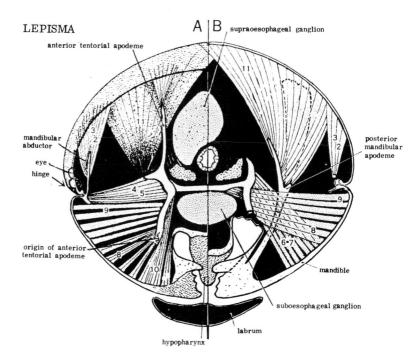

(a)

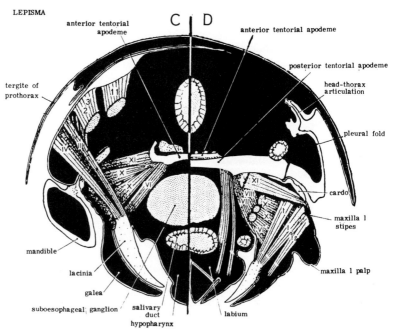

(b)

The formation of a united tentorium is associated with pterygote mandibles which bite together, as in a locust, in the transverse plane of the head. *Lepisma* shows how this condition probably has arisen. Here a shift in the position of the axis of swing of the mandible on the head, compared with *Petrobius*, brings the dorsal end of this axis to a more posterior position, a change carried further in the locust (Figs. 7.12, 7.15). The now roughly horizontal axis of mandibular movement and a reduction of the preaxial part of the mandible result in mandibles biting together strongly in the transverse plane of the head. This is a secondary condition, derivable from the rolling unjointed *Petrobius*-type of mandible and quite unlike the myriapod mandibles, which primarily bite in the transverse plane. In *Lepisma*, a strong adductor apodeme leaves the mandible near its distal end, and thus, although the leverage is great, the gape is small and the labrum narrow (Fig. 7.12b). In the locust, the adductor and abductor apodemes arise from the proximal mandibular rim, and the gape is wide with the labrum spreading almost across the full head width. *Lepisma* can feed on the hard grain of cereals, among other things, and the locust can cut up leaves of all kinds with rapidity. It is probable that the transverse biting of the mandibles of *Lepisma* and of the locust represent alternative modes of solving the same problems.

The hexapod classes comprise the entognathous Collembola, Diplura, and Protura, and the ectognathous Thysanura and Pterygota. A posterior pair of tentorial apodemes is present in all hexapod classes. The anterior pair is lacking only in the Diplura where the minute preoral cavity hardly leaves room for them.

In Collembola (*Tomocerus*, Figs. 7.16, 7.17a) the posterior tentorial apodeme passes across the base of the gnathal pouch and unites with the cuticle of the head by fine fibrils at a point just proximal to the origin of the plural fold and anterior to the labium, a position corresponding with the dorsal or proximal end of the intucking of the origin of the posterior tentorial apodeme in *Petrobius* (Figs. 7.12a, 7.13b). Distally, the pair of apodemes in *Tomocerus* unite and support the hypopharynx. The anterior tentorial apodemes lie more antero-dorsally than the posterior pair, and their posterior tips join with the transverse mandibular tendon. From the complex anterior branches of the anterior tentorial apodeme is formed a cuticular boss which (Fig. 7.16, left side), with the frontal sclerite, restricts the movement of the mandible on each side, controlling its

---

Fig. 7.14.   Mandibular mechanisms of thysanuran, *Lepisma*. a, Transverse section of head viewed from in front, level *A* being close to anterior face of mandible and level *B* being through middle of mandible. Axis of swing of mandible is shown in Fig. 7.12b; preaxial part of mandible is much reduced and carries an abductor tendon and abductor muscles, drawn on left of Fig. 7.14a; adductor apodeme and its muscles are seen on right. Anterior tentorial apodemes (white) are fused. b, Sections at more posterior levels *C* and *D* to show posterior tentorial apodemes fused with one another. Roman numerals mark maxilla 1 muscles. For further description, see Manton (1964, 1977). From Manton (1964).

LOCUST

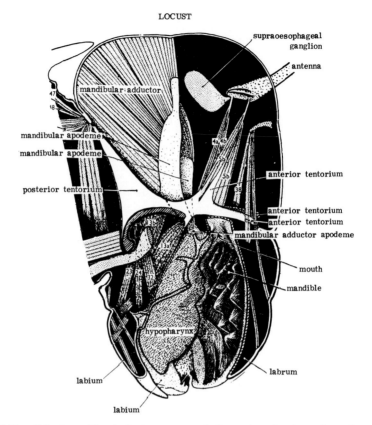

Fig. 7.15. Side view of head of a locust cut sagitally to show fused anterior and posterior tentorial apodemes and mandibular adductor apodemes with their muscles. Axis of swing of mandible on head is shown in Fig. 7.12c. For further description, see Manton (1964, 1977). From Manton (1964).

rolling as fingers control a pencil. The mandibular movements are effected by complex musculature giving rotator and counter-rotator (as in Fig. 7.18b), adductor-abductor, and movements of protrusion and the reverse.

The details of entognathy in the Diplura are quite different from those of Collembola. The large posterior tentorial apodemes are organized much as in Collembola, and possess an arm extending across the base of the gnathal pouch, anchored laterally; anterodistally the pair unites in support of the hypopharynx. The mandibles work in a contrasting manner to those of Collembola. *Campodea* possesses a wide transverse mandibular tendon (Fig. 7.18), and long and short muscles, converging on the tendon, protrude the mandibles and pull them back, besides effecting the rolling movement, as in *Petrobius*. The cutting cusps at the tips of the mandibles are brought together and apart. The control of the move-

TOMOCERUS

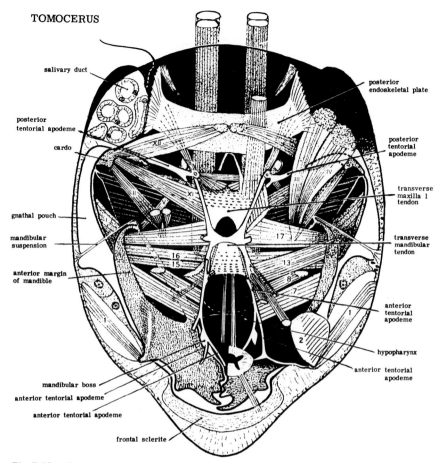

salivary duct

posterior
endoskeletal plate

posterior
tentorial apodeme

cardo

XII

posterior
tentorial
apodeme

transverse
maxilla 1
tendon

gnathal pouch

mandibular
suspension

XI

17

transverse
mandibular
tendon

anterior margin
of mandible

16
15

13

8

anterior
tentorial
apodeme

1

2

hypopharynx

anterior tentorial
apodeme

mandibular boss

anterior tentorial apodeme

anterior tentorial apodeme

frontal sclerite

Fig. 7.16. Reconstruction of head of *Tomocerus* in anterodorsal view. Mandibles are withdrawn into gnathal pouch, upper surface of mandible being mottled on left, while section lies deeper in on right. Behind mandible lies maxilla 1 in each gnathal pouch. Anterior and posterior tentorial apodemes are seen in part, and transverse tendons of mandible and maxilla 1 are shown. Further understanding of head structures can be gained from many sections of head given in Manton (1964, 1977), skeletal features being shown in color, more details than can be shown by one uncolored figure here.

ment is not by "fingers" as in Collembola, but is probably effected by the long tendon and muscle from the prostheca, which inserts on the posterior wall of the cranium. The gape is small. The mandibular musculature in Collembola and Diplura is conspicuously different.

Maxilla 1 in entognathous hexapods is sunk into the gnathal pouch, and the maxillary muscles insert on the posterior tentorial apodeme. No myriapod maxilla 1 is entognathous (compare Figs. 7.10, 7.11 with 7.16, 7.18). Entog-

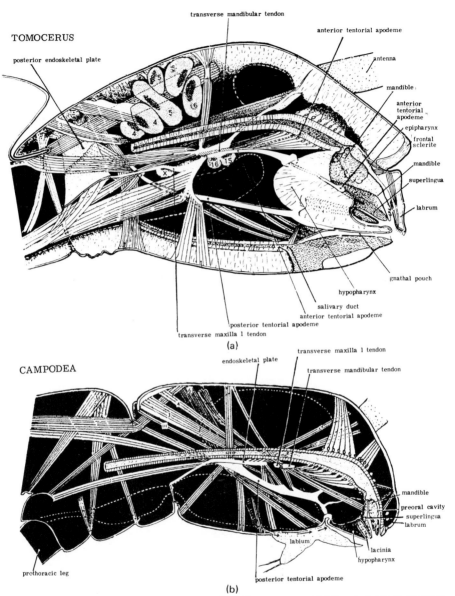

Fig. 7.17. Longitudinal halves of heads of (a) *Tomocerus* and (b) *Campodea* to show great differences between details of entognathy in Collembola and Diplura. Anterior and posterior tentorial apodemes are white; only the posterior pair is present in *Campodea*. For further details and explanation see Fig. 7.18 and Manton (1964, 1977) where color is used to show endoskeletal structures. From Manton (1964).

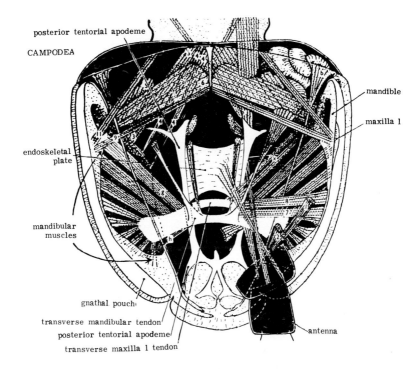

Fig. 7.18.   Reconstruction of head of *Campodea* to show entognathy and mandibular mechanism. a, Anterodorsal view, levels shown being more superficial on right. Mandible fills deep gnathal pouch, mandibular and maxilla 1 tendons are wide, and only posterior tentorial apodemes are present. b, Transverse section (thick) through mandibles within gnathal pouches on either side and transverse mandibular tendon. Muscles between each mandible and tendon cause rolling mandibular movement as in *Petrobius*. Edge of maxilla 1 is cut beside each mandible. For further figures and description see Manton (1964, 1977). From Manton (1964).

nathy in the Protura differs in details from entognathy of Collembola and Diplura.

The ectognathous hexapod classes (Thysanura and Pterygota) should not be regarded as more advanced in the essentials of their head structure than the entognathous classes. On the contrary, entognathy must have been independently evolved in each class of hexapod showing this feature. Yet all could have descended from a condition much as in the Machildae, with simple rolling mandibles and separate anterior and posterior tentorial apodemes, but without the hydraulic mechanisms for sucking up minute particles.

The advantages to entognathous hexapods of protrusible, piercing, or neatly cutting minute mandibular blades are similar to those of entognathous Chilopoda, but the mechanisms of mandibular movement in the two are quite different. Those present in the ectognathous Chilopoda and in the Machilidae are recognizable in modified form in the respective entognathous classes of the two subphyla.

Entognathy is a widespread, convergently acquired phenomenon (Fig. 7.19), being shown by some ectoparasitic and free-living copepods with stylet-like

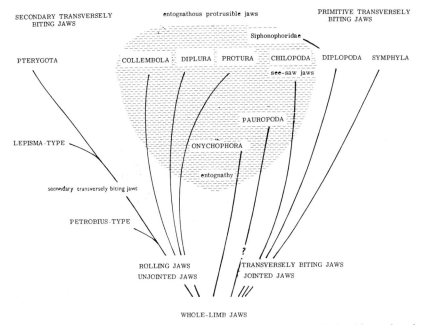

Fig. 7.19.   Diagram showing conclusions reached concerning interrelationships and evolution of jaw mechanisms of Onychophora, Myriapoda, apterygote classes, and Pterygota. Shaded area indicates convergent evolution of entognathy and protrusible mandibles. Pauropoda possess an unsegmented mandible, but there is no evidence as to whether it has been derived from a primitively segmented or an unsegmented mandible. From Manton (1964).

mandibles, maxillae 1 remaining outside the trunk, formed by fusion of the labrum with the paragnaths; by onychophorans; by siphonophoran diplopods; as well as by chilopods and pauropods. Entognathy is no indication of affinity, and the classification of hexapod classes into Entognatha and Ectognatha consequently is of no taxonomic value.

If the mandibular muscles of the three entognathous classes are compared in detail with those of *Petrobius*, a considerable measure of homology is evident in spite of superficial differences. The origin of the different entognathous hexapods presumably lies among early arthropods with the essentials of the machilid mandibular and tentorial apodeme organization. The ectognathous hexapods show no evidence of having been descended from any entognathous group. The entognathous hexapods cannot be regarded as simple primitive "insects." They are just as advanced along their own lines as are the Pterygota.

The entognathous hexapods probably evolved their mandibular arrangements as suggested above, but it is unlikely that their ancestors were hexapodous at that time. It is shown below that the leg-base mechanisms of all five hexapod classes are so different, one from another, that it is inconceivable that any one type of leg-base gave rise to any other. It is probable that all are descended from arthropods with trunk sclerotization scanty enough to have permitted independent evolution of each type in a multilegged state. Each leg-base mechanism is associated with particular habits, which must have been as important in the evolution of the hexapod classes as they have been shown to have been for the myriapod classes (see below).

The lack of validity of the Ectognatha as a taxonomic unit is apparent on the evidence of both heads and limb-bases. The biting by the mandibles of *Lepisma* and of the locust have probably been independently acquired from mandibles with the basic assets of that of *Petrobius* without the hydraulic embellishments for collecting fine particles. Both *Lepisma* and the locust bite secondarily in the transverse plane, but have done so on different lines and probably independently and in parallel. The Thysanura appear to form a valid class contrasting absolutely with the Pterygota in leg-base mechanisms (see below), as in the utilization of jumping gaits and associated morphology, which contrasts with all other Uniramia (Manton, 1972). These latter features cannot be regarded as primitive.

Thus, we see profound differences between myriapod heads, with their: jointed mandibles, primitively and not secondarily biting in the transverse plane; the swinging anterior tentorial apodemes; no posterior tentorial apodemes; and no true labium; and the hexapod heads with: basically rolling, unjointed mandibles; two pairs of nonswinging tentorial apodemes; and a labium. A claim for a symphylan origin of the hexapods is quite untenable on head, trunk, and limb structure and on movements. (More figures than shown above are required for better comprehension of the mandibular mechanisms; see Manton, 1964, 1965, summary in 1977.)

7.5. LEGS, LEG-BASE MECHANISMS, PLEURITES, AND THEIR FUNCTIONS AND HOMOLOGIES IN MYRIAPODS AND HEXAPODS

It is necessary to consider the skeloto-musculature and movements of legs and leg-bases in a little detail because they are relevant to phylogenetic considerations of the hexapods. Only the Scutigeromorpha among myriapods and adult Pterygota are plantigrade (Fig. 7.20). The other myriapods and the wingless hexapods are unguligrade, standing on the tarsal claws. The distal leg joints of unguligrade limbs usually lack extensor muscles, since their dorsally placed hinge joints allow only flexors to operate. Extension of distal leg-joints, lacking extensors, is accomplished by proximal depressor muscles, intrinsic and extrinsic, when the limb tip is on the ground. The dorsal hinge articulations between podomeres and the dorsal leg cuticle provide the only incompressible face of the leg. When extension of a leg takes place during the latter part of a propulsive backstroke, a slight but beautifully designed basal rock brings the entire dorsal face of the telopod to a more anterior position, and this facilitates leg extension

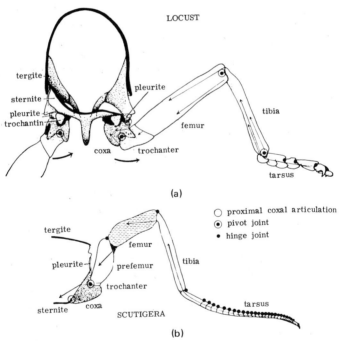

Fig. 7.20.   Diagrammatic comparisons of prothorax of the locust and segment of the only plantigrade centipede, *Scutigera*.   Hinge and pivot joints along legs are marked.   Locust pleurite is fixed and there is no leg rocking, while pleurite of *Scutigera* is mobile and leg rocking takes place.   From Manton (1966).

by the proximal depressor muscles. Working models illustrate these features clearly.

Leg-rocking mechanisms differ in the myriapod and hexapod classes. The types of rocking mechanism appear to be mutually exclusive. The functions of pleurites is often intimately bound up with the rocking mechanisms.

Myriapod leg-rocking is similar in chilopods, pauropods, and symphylans. In all, the proximal ventral rim of the coxa articulates closely at one point with the sternite; the dorsal rim of the coxa extending up the flanks is pulled alternately anteriorly and posteriorly, deforming the flexible pleuron (Fig. 7.21c). In Symphyla a single pleurite dorsal to the coxa articulates with it and bears the rocking muscles (Fig. 7.22a, *tcx*, *tep*). Two pleurites in Chilopoda are separated from the coxa (Fig. 7.21c), the katopleure giving origin to rocking muscles *pct.1* and *tcx*, as shown. There are others, such as *rot.tr.p.* from the trochanter and coxa, which assist the forward rock of the dorsal face during the propulsive backstroke. In the minute Pauropoda (Fig. 7.22b), muscles *ret.rot.co.b.* achieve the double purpose of leg-rocking and effecting the promotor-remotor swing of the leg. The wide coxal base in the Chilopoda, expanded up the flanks, provides leverage for the rocking movements in these larger animals. As is to be expected, the total musculature causing the forward rock of the dorsal face of the leg during the propulsive backstroke is stronger than the muscles causing the opposite recovery rock.

The Diplopoda are a little different owing to the frequent ventral insertion of the coxa on the sternite, as in Iuliformia. Here the horizontal axis of swing is fixed by tight articulations. Figure 7.21a shows a promotor-remotor coxal swing in ventral view, and in Fig. 7.21b a diagrammatic lateral view is given of three such coxae seen end-on during the promotor-remotor swing. Here the rock and the promotor-remotor swing are one and the same mechanism. As the coxa swings forward, the ventral surface becomes more anterior than the mid-dorsal surface, and conversely on the remotor swing. A posterior displacement of the ventral leg-face means the equivalent of a forward displacement of the dorsal surface, as is normal for the remotor swing, the dorsal face being in effect forwardly displaced during the propulsive stroke.

In the Pterygota where the tarsus is plantigrade there is no rocking mechanism, which might be mechanically difficult and is rendered needless by the presence of an abundance of antagonistic muscles at the leg-joints. Most joints between the podomeres of other classes are represented by pivot joints, which allow both flexor and extensor muscles. The presence of distal hinge joints in myriapod and apterygote hexapod classes cannot reflect any inabilities, but instead must mean that their leg extension, effected by their leg-rocking and other mechanisms, is sufficient or suitable for their needs.

In the unguligrade apterygote hexapods, leg-rocking is engineered differently from class to class. In Diplura and Protura the coxal margin is articulated with

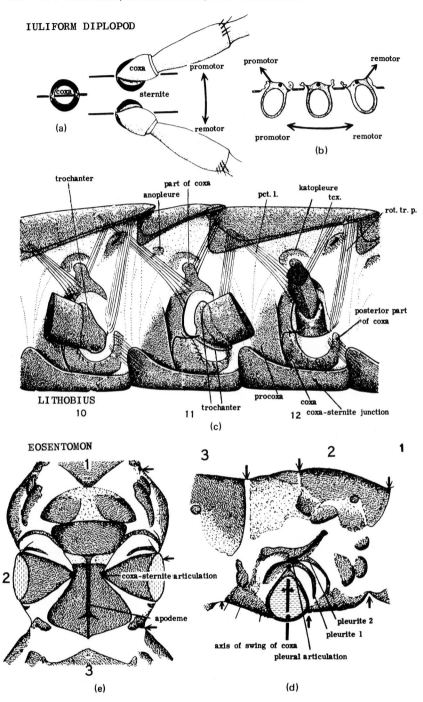

IULIFORM DIPLOPOD

(a)

(b)

LITHOBIUS

(c)

EOSENTOMON

(e)

(d)

the sternite, a feature facilitating trunk flexibility. There is, however, no coxal rocking in the Diplura (Fig. 7.24a). Instead, an intrinsic rocking of the leg takes place at the trochanter-femur joint in a unique manner, with muscle and joint modifications that rock the leg and transmit the movement strongly to the more distal podomeres. The Protura also have a unique and different method of achieving leg-rocking. The dorsal rim of the coxa is linked to a pleurite, which in turn slides against another pleurite, thus implementing a rock (Fig. 7.21d).

The Collembola lack all coxa-body articulations. Each pendant coxa is slung from the tendinous endoskeleton and tergite above in an elaborate manner (Fig. 7.30b). In the absence of any articulations, the coxa can move in any direction, stability being provided within.

The variety of leg-rocking mechanisms may now be left and consideration given to the homologies between pleurites in general and to the functional morphology of the leg-base in Thysanura and Pterygota in particular.

In the Pterygota the dorsal rim of the coxa articulates with the body, while the ventral rim usually is freely surrounded by flexible membrane, in contrast to the myriapods and apterygote classes. The dorsal coxal rim articulates with a fixed pleurite strongly anchored to both tergite and sternite and incapable of movement (Fig. 7.20a). The promotor-remotor swing takes place between the coxa and the fixed pleurite, with some support from a pleurite, the trochantin, situated in the arthrodial membrane anterior to the coxal base. The procoxal pleurite in Chilopoda similarly supports the coxa anteriorly, and in all arthropods it is the anterior side of the leg that is in greatest need of support. The strides taken by plantigrade pterygote legs are shorter than those taken by most unguligrade hexapods.

The thysanuran coxa also is articulated dorsally with a pleurite, but the pleurites are mobile, not fixed. They are horizontal in position in Machilidae and dorsoventral in Lepismatidae (Fig. 7.23b). In the Machilidae, the promotor-remotor swing of the leg, remarkably and uniquely takes place between pleurite and body, the pleurite on the remotor swing deforming the pleuron, the leg going with it. Leg joints are well stabilized and strong and, unlike other hexapod classes, jumping gaits are employed. The unique movements of lepismatid pleurites and legs are considered in Section 7.6.

Fig. 7.21. Three different methods of achieving leg rocking and extension of distal leg joints lacking extensor muscles. a, Diplopod "built in" leg rocking mechanism showing swing of coxa on body in ventral view; b, longitudinal section through three successive diagrammatic coxae, promotor-remotor swing moving dorsal and ventral faces of coxae as shown; c, pleuron of *Lithobius* with legs cut short; leg 10 in promotor position, leg 11 in remotor position, and levated position is shown by leg 12. Muscles *pct. 1* and *tcx* from dorsal part of coxa rock it alternately about coxa-sternite junction. d and e, Proturan, *Eosentomon*, lateral view in (d) and ventral in (e). Leg is cut short. Sliding of pleurite 1 against pleurite 2 swings dorsal edge of coxa alternately forward and backward about its sternal articulation. For further description, see Manton (1972, 1977). From Manton (1972).

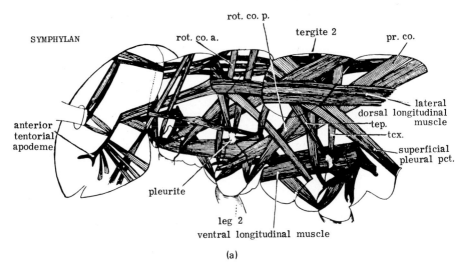

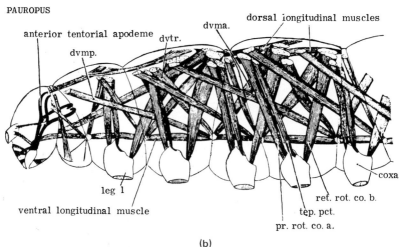

Fig. 7.22. Trunk musculature of (a) a symphylan, *Hanseniella agilis*, and (b) a pauropod, *Pauropus sylvaticus*, to show flexibility-promoting superficial pleural muscles *pct* in (a) and deep rigidity–promoting muscles *dvmp*, *dvma*, *dvtr* in (b). Symphylan trunk is conspicuously flexible, while that of *Pauropus* is not. For further description, see Manton (1966, 1977); muscle drawings from Tiegs (1940, 1947). Interpretation after Manton (1966).

The variety in the morphologies of the coxa-body junction in the hexapod classes is so great that no one type could be conceived of having had the potentiality of giving rise to any other, and all contrast with the myriapod solutions to the problem. A coxa, articulating ventrally with a sternite and movable at its dorsal rim, and a coxa articulated dorsally and free ventrally, are mutually

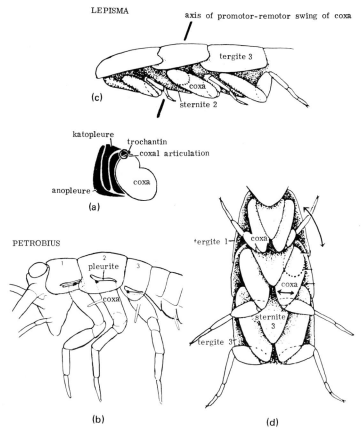

Fig. 7.23. Thysanura, Lepismatidae. a, Pleurites situated anterior to coxa, coxal articulation being marked by a ring. Three pleurites overlap one another slightly. b, Lateral view of thorax of *Petrobius* to show positions of pleurites, here situated dorsal to coxa; coxa-pleurite articulation is marked by a black spot. Paratergal lobes cover bases of coxae. c, Lateral, and, d, ventral views of *Lepisma*. For description, see text and further description in Manton (1972, 1977). From Manton (1972).

exclusive. The one could not be transformed into the other without passing through a nonfunctional stage or an intermediate stage lacking the mechanical efficiency of that which went before, an unlikely transition. The type of coxal articulation in Thysanura and Pterygota could have arisen as alternative modes of sclerotization in early uniramian ancestors with incipient stiffening of cuticle on body and legs. But it is recognized that the specialized uses of the legs of pterygotes introduce particular features in some groups, such as the extra coxal articulation with the sternite, additional to the principal dorsal pleural articulation, in beetles capable of wedge-burrowing as well as running (Evans, 1977).

The coxa-body junctions and the leg-rocking mechanisms in myriapods and hexapods probably have evolved in parallel from little-sclerotized ancestors. This means that there never could have been a hexapodous "ancestral insect" with the potentiality of giving rise to all five classes. Their class characteristics must have been established separately and in divergent manners from ancestral arthropods lacking much trunk and limb sclerotization and probably lacking differentiation of podomeres (see also below concerning the hexapodous state).

The only two hexapod classes with a leg-base feature in common are the Thysanura and Pterygota, but pleurites of the former are mobile, while those of the latter are not. These surely are mutually exclusive features. Fixed pleurites are an essential prerequisite for the evolution of flight and flight muscles. These at their simplest, as Tiegs (1955) has shown, are modified muscles used for walking by nonflying nymphs. The mobile pleurites of the Thysanura are associated with their unique jumping gaits (see Section 7.6), and their thoracic musculature is quite different.

The legs and pleurites of *Lepisma* have been considered to be primitive among hexapods. This viewpoint, made in ignorance of the mode of action of both machilid and lepismatid legs and leg-bases, cannot be accepted. In *Lepisma*, the coxa is long, backwardly directed, and housed in an almost horizontal groove between paratergal lobe and the expanded sternite (Fig. 7.23c,d). In most arthropods promotor-remotor and levator-depressor movements take place at right angles to one another, at the coxa-body and coxa-trochanter joints (Figs. 7.21c, 7.24a); the latter movement occurs also at more distal joints along the leg. In *Lepisma*, the two movements are combined and sited at the coxa-body joint. The amplitude is small and the rapidity great. The coxa moves upward and outward in its grove and then downward and toward the middle line. On the depressor-adductor movement the posterior rim of the coxa deforms the flexible pleuron, as occurs during the backstroke of many myriapods and apterygote hexapods. Three vertical, imbricating, and well-musculated pleurites strongly support this leg movement, shuffling over one another on the recovery leg movement and opening out strongly, supporting the very rapid adductor-remotor coxal movement. The coxa articulates anterodorsally with the first adjacent pleurite the trochantin, but there is no second coxal articulation. With such a leg-base, *Lepisma* can execute the almost horizontal and wide-angled movements of the telopod (Fig. 7.32b).

The almost vibratory action of the *Lepisma* coxa is not shown by the Machilidae. Here, the jump by the legs takes place almost in the transverse plane of the body. The coxa is articulated with the anterior end of the horizontally elongated pleurite over legs 2 and 3. A pleural apodeme arises from the middle of the three vertical pleurites in *Lipisma*, the so-called katopleure, and from the single horizontal pleurite over legs 2 and 3 of *Machilis*, but from the coxa of leg 1 and not from either of the two horizontal pleurites here. The apodeme carries

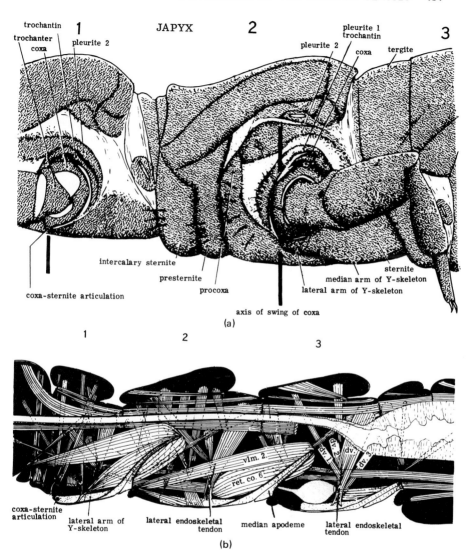

Fig. 7.24. *Japyx.* a, Lateral view of pro- and mesothorax, more sclerotized parts of cuticle mottled; b, longitudinal half of body to show Y-skeleton with spine apodeme, its appended muscles, coxa-apodeme articulation, and salient features of thoracic musculature; see text and full description in Manton (1972, 1977). From Manton (1972).

the long muscles from the trochanter, which cross two joints. They are mainly depressors in *Petrobius* but levators in *Lepisma*. The rest of the thysanuran leg morphology and movements are considered in Section 7.7.3.

The jumping gaits of the Machilidae and Lepismatidae differ from those of all other uniramians. *Lepisma* depends on the small rapid coxal movements, which are strongly supported by the unique array of anterior pleurites and by the deformability of the pleuron at the posterior coxal base, as is conspicuous in Chilopoda, Diplura, and the Machilidae (Figs. 7.21c, 7.24a). The great running speed of *Lepisma* is dependent upon the very lightly built telopod, which swings through an enormous angle (Fig. 7.32b). Leg morphology is even more advanced than that of the machilid thoracic legs, but there is nothing in other hexapods that in the least resembles the abdominal high jumping morphology and musculature of the Machilidae (Sections 7.6., 7.7.3.).

How can the pleurites or leg structure of either *Machilis* or *Lepisma* be considered primitive? The idea stems from a complete lack of understanding of the functions of the parts or of the supreme achievements of the entire mechanisms.

The problem of pleurite homologies and the evolution of pleurites have hardly been approached on a functional basis in entomological literature. Consequently, existing assumptions concerning pleurite homologies between different classes inspire little confidence. The three pleurites of *Lepisma* are termed anopleure, katopleure, and trochantin. The so-called trochantin bears the principal and only articulation of the coxa on the body (there is no second articulation, Manton, 1972), yet in pterygotes the trochantin never carries the principal coxal articulation and usually bears none at all. A much larger fixed pleurite bears the principal coxal articulation, so what justification in there for naming the pleurite bearing the coxal articulation in *Lepisma* a trochantin? One suspects an entomological "escape reaction", having named the other two pleurites katopleure and anopleure. But since no case can be made for common ancestry of myriapods and hexapods, the so-called anopleure and katopleure of *Lepisma* can have no homology with the one pleurite in Symphyla, two in Chilopoda, two quite different pleurites in Protura, and be expected to be homologous with any particular pleurite in Machilidae? The Machilidae possess one pleurite over meso- and metathoracic legs but two over the prothoracic legs, so what reasonable homologies do they have? The myriapodan procoxa is not considered, although it has obvious resemblance to the lepismatid pleurites, much more so than the chilopodan katopleure and anopleure.

Furthermore, the three lepismatid pleurites cannot represent the remains of complete rings that may be supposed to have encircled the leg-base. Such structures, if present, would be a hindrance to lepismatid leg movements and would prevent the posterior coxal margin from being pulled into the pleuron during the remotor-adductor movement. Similarly, the single metathoracic pleurite of *Tomocerus*, forming an anterior crescent at the leg-base, is in a functionally appropriate position. A narrow complete ring would not provide suitable support for the metathoracic leg.

The positions of pleurites, horizontal in Machilidae and vertical in Lepismatidae, are associated with differences in the jumping techniques of the two groups. Pleurite numbers and positions are of functional significance. The pleural apodemes of the two Thysanura are also of functional significance, although their origins are not identical, arising from the so-called katopleure of *Lepisma*, and in *Petrobius* from the only pleurite over legs 2 and 3, but from the coxa of the prothoracic legs. These apodemes, as already noted, carry the long extrinsic muscles from the trochanter.

It is difficult to see any real resemblances, either structural or functional, between the leg-bases and pleurites of Thysanura and Pterygota, nearer than some little-sclerotized uniramian with incipient trunk stiffening of cuticle, tending toward dorsal rather than ventral coxa-body attachments. Limb podomeres may not have been differentiated at such an early stage of body sclerotization. Thereafter the Thysanura, dependent upon mobile pleurites and jumping gaits, must have followed an opposite evolutionary path to the ancestral pterygotes with fixed pleurites, running or walking gaits, plantigrade tarsus, and absence of leg rocking, all measures conferring considerable stability, as would be necessary for the subsequent evolution of flight. Without such features there would have been no effective evolution of flight. (See further in Manton, 1972, 1977.)

In pterygotes, the thorax is fully invested with sclerites, fused to one another and supporting the main pleurite bearing the coxal articulation: the apodemal phragma at either end of each thoracic segment; the tergites; sternites; and small sclerites at the wing bases. This investment differs from that of the Geophilomorpha, where the armor is also complete, but the pleurite edges here are not fixed, or fused, but supported by discontinuous cones of heavy sclerotization set in flexible endocuticle (Blower, 1951). Thus the marginal zones can curl inward, so maintaining a continuous external armor at all moments during extensive shape changes of a segment (Fig. 7.2c). Diplopods possess an entire segmental or displo-segmental armor, fused into a rigid cylinder in the Iuliformia. Some mobility of sclerites is present in many diplopodan orders, but their pleurites are entirely different in size, number, and positions from those of other myriapods and from the hexapods. The position, size, and number of pleurites accords with functional needs in each class, and there is no real correspondence between pleurites of different classes.

A few other pleurites deserve mention. The procoxa supports the coxal base anteriorly in chilopods and is stouter and more extensive than the metacoxa behind the leg-base (Fig. 7.21c). The metacoxa may be absent. This is understandable, since there is greater stress on the anterior side of the leg, as shown by the anterior pivot-joint articulations being stronger than the posterior ones, which are often absent in joints of fast-running legs (Fig. 7.21c).

The dipluran thorax is very flexible, with free lateral bending at the intersegments and within the segments. Pleurites here in positions corresponding with pro- and metacoxa are present, but supported by fusion with the presternite and sternite respectively (Fig. 7.24a). The body can flex laterally so that the intercalary sternite and its lateral extension entirely cover the procoxal sclerite, achieving greater flexibility than dorsally by pleurite 1 and 2 (Fig. 7.24a). There is no leg-rocking at the coxa-body junction.

In *Craterostigmus*, a scolopendromorph centipede, the many measures that have secondarily been evolved providing extra trunk flexibility have given the need for fusion of coxa and procoxa and the formation of a vertical apodeme at the junction of the two. The metacoxa is absent. The whole provides different stability measures from those of the Diplura.

The Protura show marked intrathoracic flexibility; extra transverse lines of flexibility are provided by lesser sclerotization, achieving the same end as in the Diplura, but by different means. The two pleurites above the coxa serve leg-rocking, while those of the Diplure do not. The number of pleurites above the coxa in the two classes is only roughly similar or dissimilar, according to views about pleurite limits.

It may be concluded that theories concerning homologies between pleurites in classes distantly or unrelated to each other are usually invalid, as is the existence of subcoxal segments (see above and below).

7.6. GAITS AND EVOLUTION OF THE HEXAPOD CLASSES

The advantages of hexapodous running are such that it has been adopted by arthropods as diverse as some prawns and crabs, certain arachnids in different orders, and four hexapod classes (Fig. 7.25). The utilization of many legs, when they are short, gives no mechanical interference of one leg by the next (see *Peripatus*, Fig. 7.25a); but when the legs are many and long, the fields of movement overlap considerably, gaits have to be executed very exactly, and only a limited range may be practicable at all, if mechanical interference is to be avoided (Fig. 7.25b). When locomotory legs are few and well fanned out anteroposteriorly, the fields of movements can be nearly or entirely separated even if legs are long, and this is the principal advantage of using only three pairs of legs (Fig. 7.25d–h). There are also limitations in that hexapodous gaits are usually serviceable only when at least three legs are propulsive at any one moment, and arranged about the center of gravity, thus maintaining stability. Only certain gaits are useful, but the range is greater than in the arthropods with many and long legs.

The similarities between the gaits of the five hexapod classes are due to the above limitations imposed by the use of only three pairs of legs, but these similarities do not mask the differences between the gaits of the five classes. These class differences in performance are correlated with different morphologies and are readily seen in the tracks (Fig. 7.27). But first the basic similarities must be considered shortly.

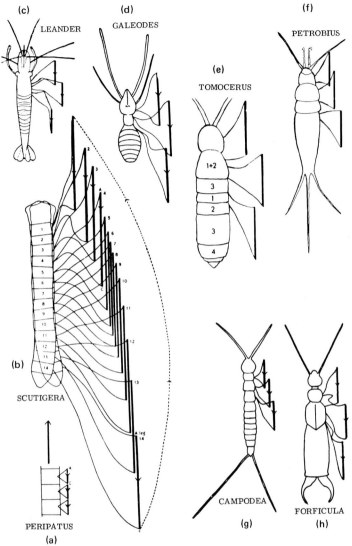

Fig. 7.25. Showing range of movement during walking of legs of various arthropods; thin lines show position of each leg at beginning and end of backstroke relative to leg base, and thick vertical lines show limb tips on ground during propulsive backstroke relative to leg base. a, *Peripatus*; fields of movement of successive legs do not overlap. b, In long-legged centipede, *Scutigera*, there is extreme overlap of fields of movement. Reduction of ambulatory legs to three pairs in prawn *Leander* (c), arachnid *Galeodes* (d), collembolan *Tomocerus* (e), thysanuran *Petrobius* (f) (walking on a ceiling), dipluran *Campodea* (g) and pterygote *Forficula* (h) results in little overlap of fields of movement combined with long legs, but the placing of legs on the ground varies in the several taxa. From Manton (1953, 1972).

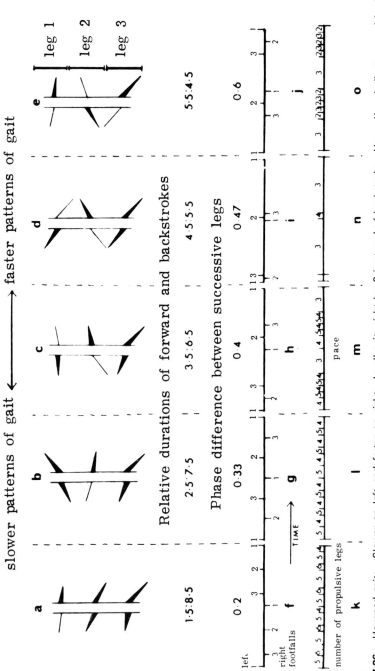

slower patterns of gait ←———→ faster patterns of gait

Relative durations of forward and backstrokes

Phase difference between successive legs

FIG. 7.26. Hexapod gaits. Slower on left and faster on right. In all gaits right leg 3 is at end of backstroke. Heavy lines indicate propulsive legs; thin lines indicate legs off ground, performing recovery forward stroke. Below "gait stills" *a* to *e* are entered order of footfalls in *f* to *j*, and below again in *k* to *o* are given number of propulsive legs at successive moments throughout one pace. For further description, see Manton (1972, 1977). From Manton (1972).

All hexapods use a range of gaits, such as illustrated in Fig. 7.26, providing different speeds of locomotion. Pace durations also vary, and when shortest, so is the duration of the backstroke relative to the forward stroke of the leg. The phase difference between successive legs is adjusted with every change in the "pattern" of the gait, as shown. The faster gaits are seen on the right and the slower on the left. Only gait e, with a backstroke less than 0.5 of the pace duration, possesses short moments with only two of the six legs supporting the body, moments of instability that are carried over by the velocity of movement of some small hexapods. There is no limit to the progressive slowing of gait patterns beyond that shown on the left side of Fig. 7.26. Only gait b possesses footfalls at even time intervals, and therefore even loading on the legs at all times, a physiological and mechanical advantage, which cannot be maintained when speed of running is a higher priority (cf. even loading on legs of anamorphic Chilopoda, Manton, 1953, 1977, and above). The time interval between the footfall of leg $n + 1$ and raising of n, marked k, promotes stability and is almost absent in the fastest gaits. This is not obligatory, since k is large in the fast running of arachnids. It is seen in Fig. 7.26 how the order of footfalls changes from one of alternation between the two sides of the body, on the right, to three successive footfalls on one side of the body followed by three on the other, as on the left. This very striking difference is not a fundamental one, but just the expression of the utilization of progressively different gait patterns.

The class differences in the use of hexapodous gaits, superimposed on the basic similarities inherent in their use, are important evidences of parallel evolution of hexapody in the five classes.

Comparisons between the tracks made by four classes of hexapods are shown in Fig. 7.27. The track of the collembolan *Tomocerus* (Fig. 7.27a), gait (4.4:5.6), is made up of groups of footprints, staggered on the two sides because legs are moved in opposite-phase relationship. Each footprint is indicated by the number representing the mark left by each tarsus. Each group consists of a transverse row of close marks, legs 2 being placed on the ground level with and lateral to leg 1 and simultaneously with the raising of leg 1, time interval k being almost or quite absent (Fig. 7.26). Leg 3 is placed on the ground between the footprints of legs 1 and 2, at same transverse level and simultaneously with the raising of leg 2. The effective lengths of the leg is shown by their fields of movement in Fig. 7.25e; leg 3 does not work at a distance from the body markedly different from those of legs 1 and 2 (cf. Fig. 7.25f–h).

The tracks of the dipluran, *Campodea* (Fig. 7.27b), gait (6.5:4.5) (Fig. 7.26e), contrast in that the footprints of each group are forwardly staggered and arranged in a different order. Leg 2 oversteps leg 1 so that its footfall occurs laterally and in front of that of leg 1, and the footfall of leg 3 similarly is placed lateral to and in front of leg 2. The change in leg support is approximately similtaneous, as in Collembola, time interval k being nonexistent. These leg

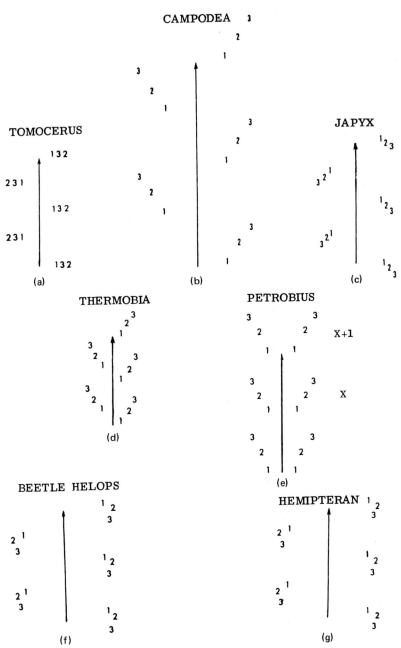

Fig. 7.27. Footfalls, recorded on smoked paper, of various hexapods and tracks here scaled to common segment volume so that widths of track and stride lengths are comparable from one taxon to another. Median lines show two stride lengths, and numbers show dispositions of footfalls of legs 1, 2 and 3. Differences in tracks are constant for each taxon. See Manton (1972, 1977). From Manton (1972).

movements are possible because of the great difference in leg lengths, as shown in Fig. 7.25g. There are short unstable moments. The tracks of *Japyx* shows backwardly staggered groups of footprints arranged in the same order as in *Campodea*. This is due to the shortness of japygid legs, associated with their strong burrowing activities.

The Thysanura employ jumping gaits as in no other hexapods, but this is not evident in the tracks. The order and placing of footfalls in each group shows some resemblance to those of *Campodea*. Legs of *Petrobius* move in similar phase, and so the groups of footprints in the two sides do not alternate as in *Lepisma* or *Thermobia*, which jump with the legs in opposite phase. The stride lengths taken by *Petrobius* are much shorter than those of *Campodea*, and those of *Lepisma* are shorter still, the legs being shorter. But the pace durations of the Lepismatidae are extremely short, resulting in very swift locomotion.

Jumping or hopping by any animal depends on a strong thrust being exerted against the ground at the beginning of the backstroke of the leg, not by a steady thrust throughout the backstroke, as in walking. Unstable moments of thysanuran jumping gaits, when only two legs support the body, are carried over by the impetus of the jump. In Thysanura, a series of jumps replaces the walking-running of other hexapods. In *Petrobius* legs 1 jump together on footfalls 1 in groups of footprints $x + 1$ (Fig. 7.27e); then legs 3 jump together on footfalls 3 in group x, followed by legs 2 jumping on footfalls 2 in groups of footprints $x + 1$. Neither the thysanuran walking tracks nor the fields of movement of their legs are in the least like those of Pterygota; there is some resemblance to the tracks of Campodeiidae but with less overstepping (Figs. 7.25f–h, 7.27b), but note the different ways in which apparently similar footprints are made.

The Pterygota are typically plantigrade, unlike the other hexapods. The fields of movement of the legs of an earwig are shown in Fig. 7.25h for comparison with other hexapod classes. The tracks of a beetle, gait (5.0:5.0), and a hemipteran, gait (4.0:6.0), Fig. 7.27f, g, differ from the tracks of all other hexapod classes in the arrangement of footprints with each group. Leg 2 is placed on the ground lateral to and a little behind the footprint of leg 1, while leg 3 is placed just behind the footprints of legs 1 and 2 but between them in distance from the middle line of the track. These triangular groups alternate on the two sides of the track, the paired legs being in opposite phase. The gaits are not fast in pattern, and neither pterygote runs very rapidly; in contrast to *Campodea*, and the Thysanura, the pterygote strides are short. It is probable that the plantigrade stance of pterygotes hinders the taking of long strides, which are easier when the tarsus is unguligrade.

The Protura are quadrupedal in their slow walking, the first pair of legs not being used. The gait pattern is slow, (2.0:8.0), as is its execution, and there is a short time interval k during which legs 2 and 3 are simultaneously on the ground, the limb tips being close to one another. Such gaits could easily have been derived from hexapods with a gait pattern (2.0:8.0) and phase difference between successive legs of 0.25, giving footfalls at even intervals of time, a short time interval k, and thus an easy start for stable quadrupedal gaits to carry on without legs 1. No tracks have been directly recorded on smoked paper from these minute animals, but tracks would consist of closely placed double marks in each group, alternating on the two sides.

An appreciation of the differences between the hexapodous features common to all hexapods, accruing from the use of only three pairs of legs, and the superimposed features in which each of the five hexapod classes uses its legs differently, supports the view that the hexapodous state has been evolved independently five times. The evidence from locomotory mechanisms accords with the conclusions reached concerning the five different types of leg-bases in the hexapod classes and with that concerning mandibles and tentorial apodemes already considered above. (Further details concerning locomotory mechanisms of the hexapods are given in Manton, 1972, summarized in 1977, and the contrasts with hexapodous arachnids in 1973).

Each class of hexapods uses its three pairs of legs in a distinctive manner with different leg actions, gaits, and fields of movement, much of which is readily appreciated from the tracks; also the leg-base mechanisms of each class are decisively different. It may therefore be suggested that the condition in each class has arisen independently, from multilegged ancestors. Indeed no other postulate carries with it any credibility as to ancestral stages. Multilegged ancestors in which trunk sclerotization had only just begun would have had the plasticity for evolution of the various types of leg-base. Podomere differentiation may have been only just starting, because it could not have proceeded far in advance of sclerotization at the leg-base, which is so different in each class. But a head capsule may have been established earlier on a plan contrasting with that of the myriapods and comprising the beginnings of a rolling ectognathous mandible, maxillae and labium, and two pairs of tentorial apodemes.

7.7. HABITS, STRUCTURE, AND THE EVOLUTION OF THE HEXAPOD CLASSES

Some of the correlations between the habits of the Uniramia and the evolution of their distinctive structures and accomplishments have been outlined. The fundamental differences between the jointed, transversely biting mandibles of the myriapods and the rolling, unjointed mandibles of the hexapods have been noted. Hexapod modifications of their basic type of mandible are many, and suit different ways of life, particularly in the Pterygota. But there appears to be no transition from the myriapod to the hexapod types of jaw, and both must have been established early in terrestrial life.

The Carboniferous period was one of arthropodan abundance, and possibly a less competitive era than the present. But long before then habit diversification must have set in among early terrestrial Uniramia. The seeking of cover giving protection from osmotic uptake of water, desiccation, and predators must have been the same for ancestral myriapods and hexapods. Ancestral hexapods must have possessed an incipient head of the composition already noted, with a labium, two pairs of tentorial apodemes, and a rolling mandible, contrasting with the myriapod head capsule, with one pair of tentorial apodemes, supporting a different type of mandible. The head differences between the two subphyla must have contrasted with one another long before full sclerotization was achieved, and probably before hexapody had been evolved in parallel from ancestors with little trunk cuticular stiffening and many legs. Whether the similarity in position, but not function, of the anterior tentorial apodeme in myriapods and hexapods represents some community in stock dating from a time when incipient mandibles with no particular sclerotization were present we do not know, but the Onychophora show how an apodeme can be very serviceable to a soft mandible provided with terminal cusps, here situated on a different head segment from that of myriapod and hexapod mandibles. The completely different paths of further advancement of Onychophora, Myriapoda, and Hexapoda are summarized below in Fig. 7.33.

The outstanding hexapod proficiencies are: speedy locomotion by jumping in Collembola, achieved in a hydrostatic manner by a terminal springing organ; and that achieved by rapid jumping gaits in Thysanura (both being unique achievements in Uniramia); and locomotion by flight in the pterygote adult state (the Protura and Diplura are crevice specialists in soil and litter). The hexapod achievements are quite different from those of myriapods, and include in pterygotes all the functional advantages of exploiting a more elaborate life cycle than exists in the myriapods, with consequent diversification of feeding and general habits of life.

The differentiation of a thorax would depend on the acquisition of the hexapodous state because paired longish legs moving in opposite phase, unless closely placed, would tend to throw the body into yawing flexures, as in Chilopoda. Such undulations are undesirable because they reduce the potential speed of locomotion and waste energy in unwanted transverse movements. Differentiated limb-bearing segments, if few in number, need to be placed close together and just behind the head where they can support the feeding movements.

7.7.1. Collembola

The Collembola are in many ways the most interesting of hexapod classes because they show unequivocally how they have been derived from soft-bodied ancestors independently of other hexapods.

All Collembola are small, but the larger surface- and litter-living species, about 5 mm in length, are those which walk by normal hexapod gaits, and execute high jumping escape reactions, in any direction and repeated at random time intervals; which, with the body coloration, foil predators. The animals tire easily. Other Collembola have become smaller, and have penetrated deeper into soil and decaying wood. They show progressive stages of degeneration of the jumping organ; alteration in its muscular connections, which are clearly secondary; shortening of the legs; and a secondary increase in trunk flexibility. The small species with relatively shorter legs than the surface and litter livers could not have initiated the evolution of a thorax because leg length would have been too small. The jumping habit is the one with which the evolution of distinctive collembolan structure has been associated.

The jumping organ is shown folded at rest in the abdominal groove, where it is held by the hamula, and extended after the jump (Fig. 7.28a,b). Both the jumping organ and hamula appear to be derivatives of paired legs, a terminal pair having fused to form the manubrium and leaving free the distal paired dentes which strike the ground. Onychophoran paired legs move in opposite or in similar phase at different moments. Some species have little more than 10 trunk segments, while others have over 40. A short, soft-bodied arthropod could have pushed with a terminal leg-pair moving in unison, much as the median process at the posterior end of a staphylinid beetle larva assists forward locomotion. Pushing or stepping may have become jumping, thus originating the jumping organ, but only from a short-bodied arthropod. The collembolan abdomen is six-segmented; an effective jumping organ could not work on the end of a long abdomen. It may be concluded that the Collembola are primitively shorter than other hexapods.

The pressure that causes the jumping organ to strike the ground is hydrostatic and generated by special trunk musculature. The thorax and anterior abdominal segments can be held in a rigid manner; the tergo-pleural arches envelop the body, leaving little flexible pleuron (Fig. 7.28d); intersegmental arthrodial membrane is scanty; pro- and mesothoracic tergites are fused; the flexible parts of the trunk surface are firmly held by connective tissue ties and by muscles which prevent any outbillowing of flexible zones of cuticle under momentary increases in hydrostatic pressure that would reduce the effective hydrostatic blast to the jumping organ. As many as 10 ties similar to q and e

---

Fig. 7.28. Collembolan jumping organ (a) to (c) and thoracic legs (d) to (f). a, Lateral view of abdomen with jumping organ extended; b, same with jumping organ at rest in abdominal groove; c, ventral view of abdomen, jumping organ extended, main flexure taking place about horizontal dotted line; d, lateral view of thorax, stippled areas showing extent of unstiffened arthrodial membrane, above left and at coxa–trochanter joint; zones *a* and *b* on coxae are not separated on leg 1, they are completely separated on leg 2 and united on mesial face on leg 3. *pl* denotes pleurite of metathorax. e, Transverse section through metathorax near anterior edge of limb where zones *a* and *b* are separated from one another by arthrodial membrane, and showing origin of some of some extrinsic leg muscles, segmental endoskeleton in black; f, more posterior level through legs 3, zones *a* and *b* united on mesial side of coxa; pleurite of metathorax is cut, and some substantial endoskeletal ties *p* and *r* are shown. For further description, see Manton (1972, 1977). From Manton (1972).

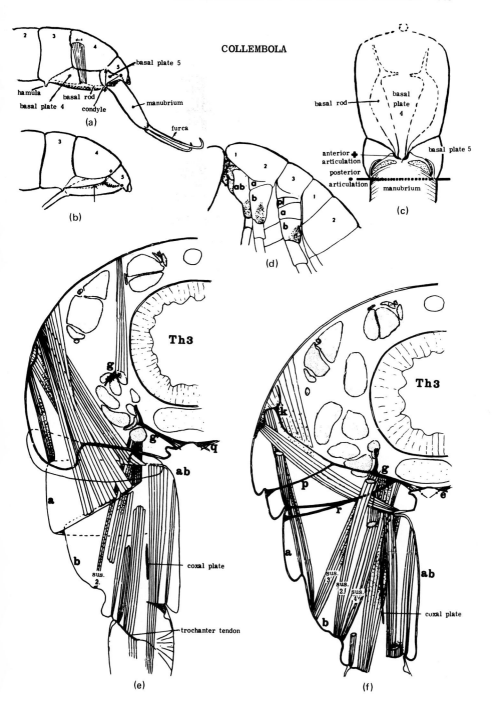

COLLEMBOLA

in Figs. 7.28e,f hold each paired segmental endoskeletal unit to the cuticle on either side; there are stouter, more solid, ties such as p and r in Fig. 7.28f; and there are large tendons inserting at either end on the cuticle, such as m and n on Fig. 7.29a, which maintain the shape of the abdominal furrow housing the resting jumping organ; the series of superficial pleural muscles shown on the same figure hold the flexible pleuron all along the body; and muscles *sus.2-sus.4* hold the arthrodial membrane at the distal end of the coxa (Fig. 7.28e,f). A small longitudinal shortening of the body caused by the complete cylinder of longitudinal muscles, the dorsal, lateral (see below), and sternal series, and very large deep obliques (Fig. 7.29b-d) drives the haemolymph into the jumping organ of much smaller diameter than the trunk, and effects both the release of the organ from the hamula and the strike on the ground. In addition, there are facilitating elaborations within the jumping organ (Manton, 1972) and stabilizing measures at the leg joints (see below).

The manubrium is elaborately hinged to the sixth abdominal segment (Fig. 7.28c). The restoration of the jumping organ to its resting position is done by muscles. The anterolateral point marked by a cross (Fig. 7.28a) is folded deeply into the median groove (Fig. 7.28b), so that it lies as shown by the arrow. The movement is made possible by basal plates 4 and 5 and lines of flexible cuticle, which allow complex folding (Fig. 7.28a-c). The long muscle 1 from the lateral longitudinal system originates at the proximal anterolateral rim of the manubrium and arises not from rigid manubrium cuticle, but from an intucking of swinging flexible cuticle, which enables the tonofibrils of this muscle to maintain constant alignment with the muscle fibres during extensive manubrial movements, a provision as effective as that of the arcuate sclerite in the spider's femur-patella joint and many other comparable devices (Manton, 1972, Fig. 12c; 1977, Fig. 10.5).

The musculature causing the slight longitudinal shortening of the body is seen in Fig. 7.28a-d, except for the dorsal series shown elsewhere (Manton, 1977, Fig. 9.2c). Alone among hexapods, a lateral longitudinal system is present, which, with the dorsals and sternals, forms a muscular cylinder around the body several layers thick. Lateral longitudinals, as separate entities, are present in Onychophora and in Geophilomorpha where shape changes of the body are conspicuous. In diplopods, the muscular cylinder is broken up into separate paratergal (lateral) and dorsal longitudinals and many oblique muscles of various homologies; the sternal longitudinal series is usually incompatible with spiraling and is often almost or entirely absent. Sternal and lateral longitudinals are close together in Scolopendromorpha and fused in Lithobiomorpha, Scutigeromorpha, Pauropoda, and Symphyla (Fig. 7.22).

The so-called subcoxal segments and the segmentation, complete or partial, of collembolan coxae have not been understood prior to the work presently being reviewed. Just above the collembolan prothoracic leg is a zone of flex-

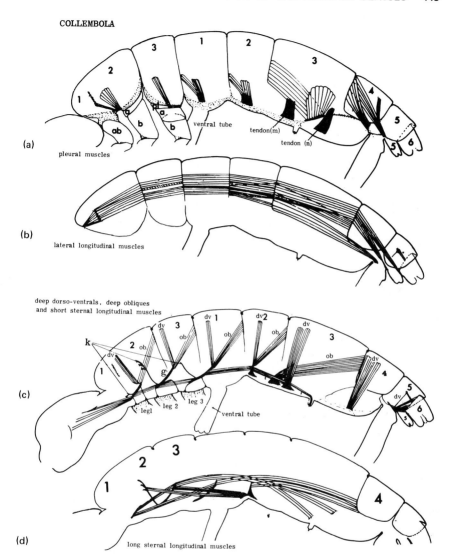

Fig. 7.29. Collembolan trunk musculature in lateral view, tendons simplified and shown in black. a, Superficial pleural muscles; b, lateral longitudinal muscles—sector 1 serves to fold up jumping organ after use; c, deep dorsoventral and deep oblique muscles with short sectors of sternal longitudinal muscles; d, long sectors of sternal longitudinal muscles with segmental thoracic endoskeletal connective tissue skeleton shown diagrammatically in black. From Manton (1972).

ible pleuron (Fig. 7.28d), used in the movements of leg 1, which lacks any coxal subdivisions. The prothoracic extrinsic muscles from leg 1 fan out, as is usual in arthropods, being inserted both in the head and on the mesothoracic tergite. The flexible zone, shown by stippling on Fig. 7.28d, is stabilized by two ties of connective tissue, one above the other, which unite with the pro-thoracic tendinous endoskeleton. The tension on these ties leaves slight billow-ing between them, and this cuticular bulge has been called a "subcoxal segment." The cuticle here is entirely unlike that of a trunk sclerite or of a leg podomere, and appertains to no true segment.

The coxa of leg 1 is entire; that of leg 2 is completely divided into the sections marked *a* and *b* on Fig. 7.28d, and these regions are united on the mesial face only on leg 3, which is larger and more posteriorly directed than the other legs. The base of all coxae is unarticulated to the body; articulations, if present, would slide apart under momentary increases in hydrostatic pressure, and the leg needs to support the body firmly prior to and at hydrostatic jumping, and to land securely.

The coxa is slung from the body in an elaborate manner differing from any-thing found in other Uniramia and summarized in the diagram in Fig. 7.30b. Muscles link the coxa to the tendinous endoskeleton above, which is complex in shape and represented by *G*, and this is strongly linked by tendon *k* to the tergite above. Distally, across the lumen of the coxa, a thin but substantial connective tissue sheet, the coxal tendon, is strongly linked with the anterior and posterior coxal cuticle, and this plate is linked to the complex *G* above and gives origin to tendinous "ropes" anchoring the trochanter and femur and their joints. Muscles originating close to the posterior articulations between trochanter and femur and to the dorsal articulation between femur and tarsus can cause no flexure at these joints, but can hinder longitudinal dislocation of joints by hydrostatic blast.

I return now to coxal subdivision into the regions *a* and *b* (Fig. 7.28d–f). The pendant coxae do not swing widely on the body, as coxae do in many other arthropods. The extrinsic coxal muscles on the mesothorax arise both from the proximal coxal rim and from the proximal edge of coxal "segment" *b*. These more distal muscles gain in length by the coxal subdivision, and the latter provides extra sites of origin. The promotor and remotor muscles do not diverge but converge above the leg to insert on the tergite; cf. the extrinsics from leg 1 (Manton, 1972, Fig. 18). The space needed by the usual fanning forward and backward of these extrinsic muscles is thereby reduced, and the dorsal convergence is a help in accommodation of muscles in the thorax where there is an acute space problem; sufficient length of muscle is provided by the distal origin on coxal "segment" *b*. The same features obtain on the meta-thoracic coxa, but the strength required of the coxa is met by the mesial lack of separation of coxal "segments" *a* and *b*. There is no justification for naming

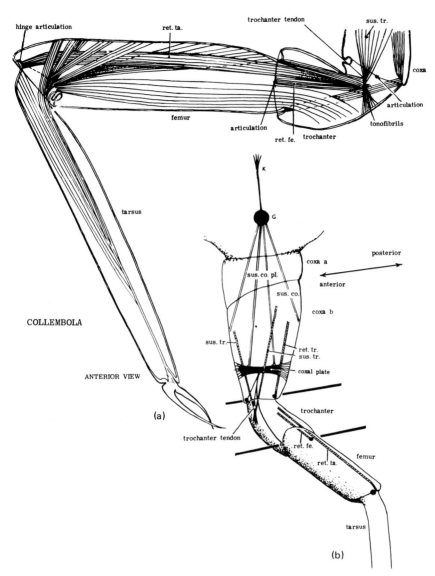

Fig. 7.30. Leg of collembolan, *Tomocerus*. a, Anterior view of leg distal to coxa-trochanter joint; b, diagram of mesothoracic leg to show leg mechanism associated with hydrostatic jumping and need to prevent dislocation at joints. Connective tissue coxal plate gives origin to suspensory muscles inserting on segmental endoskeletal complex above, indicated by *G*, and this is slung from tergite by tendon *k*. Axes of movement at coxa-trochanter and trochanter-femur joints are marked. Muscles *ret. fe.* and *ret. ta.* prevent dislocation at joints; they cannot cause flexures when situated so close to articulation. For further description, see Manton (1972, 1977). From Manton (1972).

these regions as true podomeres. They are just part of the extremely efficient structural organization of the Collembola associated with hydrostatic jumping.

The single pleurite on the metathorax has also been misinterpreted. It supports the large and backwardly sloping metathoracic coxa just where support is required. It encircles the anterior side of the leg-base, fading into the general flexible cuticle at its two ends. It cannot be the remains of a subcoxal complete ring because such a structure would not support the coxal base any more adequately than would complete sclerotized rings at the base of the coxa of *Lepisma* (see above). There are no pleurites supporting the pro- and mesothoracic coxae because they are functionally needless, and if present would not in fact be an asset.

Thus it can be concluded that the Collembola have no subcoxal segments of any kind. The many advances shown by this class, in association with hydrostatic jumping, include one podomere fewer in the legs than in most hexapods, and this results in one fewer joint to be stabilized by methods as utilized at the other leg joints.

Collembolan trunk organization appears to be a direct derivative of a soft-bodied ancestor with musculature arranged much as in Onychophora today and with lobopodia in which jointing may have been little advanced. Such an ancestor must have possessed an incipient head capsule typical of hexapods with the jaws on the fourth and not on the second head segment as in Onychophora. The earliest known collembolan is *Rhyniella praecursor*, from the Devonian, and even at this early date the pro- and mesothoracic tergites are fused, a feature associated with trunk rigidity, useful for hydrostatic jumping.

All the features correlated with collembolan jumping could have evolved along with the persistent use of the mechanical properties of the haemocoel. As already noted, the mechanical properties tend to wane in most hexapods and myriapods, but are fully utilized in geophilomorph centipedes for burrowing where muscles on the site of a burrowing heave and at a distance from it step up the momentary internal pressure.

That Collembola have originated from many-legged, soft-bodied ancestors, independently from other hexapods, and that they have maintained functional continuity throughout their evolution, seems to be inescapable. The view that Collembola are "primitive insects" stems from an entire lack of understanding of the functions carried out by their morphology. We may conclude that the collembolan jumping habit is as long-standing and as fundamental in its effects as are the body distortability of Onychophora, the bulldozing of Diplopoda, the running of Chilopoda, and the twisting and turning of the Symphyla; but the incipient head capsule must have differed from that of Myriapoda, although stiffening of the cuticle may have been scanty. As soon as sclerotization and/or calcification became more advanced, so the contrasting use of hexapod and myriapod mandibles must have become reflected in the jointing and otherwise in these mandibles and in the contrasting leg-base morphologies.

7.7.2. Diplura and Protura

The Diplura, which include the Japygidae and the Campodeiidae, have also been regarded as "primitive insects," but a study of their structure on a functional basis indicates otherwise. The Protura are a very small class.

The habitats of both Campodeiidae and Japygidae typically are soil crevices. *Campodea* can be found in a wide variety of superficial places under light cover, but the greatest abundance in temperate regions are found deep in porous soils and in particular locations, such as the hard soil surfaces under large piles of rotting leaves. Both groups are carnivorous.

Campodea is fleet for its size (see long stride lengths on the tracks in Fig. 7.27b). No pressure is exerted on the soil by the delicate body. The ubiquitous distribution probably depends on fleetness at night in finding new and suitable micro-habitats when the soil becomes too dry or food too scarce. *Japyx* on the other hand, alone among hexapods, walks on a flat, damp surface slowly and with reluctance. It is often content to shelter by stones, particularly the larger species, but the smaller are most agile in penetrating porous soil. A small japyid species from the Solomon Islands would not maintain steady progression sufficient for an adequate speed record to be made. But given some porous soil, at once these animals showed their accomplishments: those of very free thoracic flexibility; feeling by the antennae; and "scrabbling" by the strong legs, shifting soil particles. No pushing by the body surface was observed.

In contrast to the rigidity-promoting morphology of the jumping Collembola, the Diplura and Protura possess a thorax that is very flexible. Tergites and sternites leave the pleuron fully exposed, not boxed in, as in Collembola. Intercalary tergites and sternites (often named differently), good intersegmental flexibility, and organized levels of potential intrasegmental flexures promote the ability to bend in all directions. *Japyx* can stand on the metathoracic legs and turn the head laterally, then posteriorly and across the dorsal surface to face the other side, without movement of the metathoracic limbs. The neck and thoracic flexibility of the Protura, although conspicuous, is not so great. The Campodeiidae with longer legs than Japygidae and the Protura do not show such extreme trunk flexibility; presumably greater rigidity is needed by the Campodeiidae for supporting the longer legs.

The zones of the thoracic intrasegmental flexibility of the pleuron have already been noted and the morphology shown in Figs. 7.21d,e, 7.24a. These zones are differently contrived in Diplura and Protura, as are the leg-rocking mechanisms (Section 7.5 above). The thoracic intersegmental regions of japygids are constricted, and the short legs can be pulled well into the flanks so that they do not project beyond the width of the abdomen, the widest part of the body. Japygids are not the parallel-sided arthropods depicted in some textbooks. The nonmobile dorsolateral pleurite of Diplura, curving round the coxa, stabilizes the dorsal end of the axis of promotor-remotor swing of the leg (Fig. 7.24a).

This arrangement contrasts with that of the Protura and of the myriapods with conspicuously flexible bodies, the Symphyla and Geophilomorpha, where flexibility must have been independently acquired.

The second particular advance of the Japygidae is the telescopic antenna, which can shorten to two-thirds of its resting length, and is used in feeling crevices before a route is selected. Each antennal segment can be pulled into the next by muscles, and blood pressure appears to push it out again, a large sinus being present in each antennal segment, whose degree of extension is presumably controlled by muscles in the wall. Flexibility exists between antennal segments in Campodeiidae, Collembola, and Myriapoda but without alterations in length, and the pterygote antenna can be moved only from the base. The evolution of a telescopic antenna suits the penetration of soil crevices by japygids and is a unique arthropodan achievement.

The third particular advance by japygids concerns leg strength used in "scrabbling" and shifting soil particles. The vertical axis of coxal swing on the body passes through the strong sternal articulation, but only a sclerotized pleurite fold holds in place the dorsal end of this axis (Fig. 7.24a). The coxa-sternite articulation is very strongly supported by the ends of the arms of the "Y-skeleton," an apodeme situated on each thoracic sternite and formed by a surface fold of cuticle (Fig. 7.24a,b). Where the arms unite at the posterior margin of the sternite, the stem of the Y passes posteriorly into the body as a hollow apodeme. The length of this spine apodeme, or "spina," varies in different species; it carries the largest of the coxal remotor muscles on one side and a long sector of the sternal longitudinal muscles on the other. A longer "spina" means extra length to these muscles and is doubtless associated with the exact needs of other particular habitats, habits, or body size. The other sectors of the sternal longitudinal muscles are also shown in Fig. 7.24b. These muscles are often called ventral longitudinals, $vlm.$, when no lateral series is present as a separate entity. The spine apodeme, its muscles, and the anterior ends of the Y-skeleton, supporting the coxal articulation more strongly than in any other uniramian, account for the japygids being able to exert so much force with their legs while all the longitudinal muscles, both dorsal and ventral, hold the body firmly in any useful curvature.

The spine apodeme has been hailed as a primitive hexapodan feature. Much ingenuity has been expended in trying to justify this and to identify the "spina" in other classes. But in fact a similar "spina" has not been found outside the Japygidae. In *Campodea* a cuticular endoskeleton is present (Manton, 1972, 1977), which is united with the median part of the sternite and supports the ventral longitudinal muscles, which are here very simple; but there is no "spina," or indeed a need for one. The apodemes arising from the sternite in Campodeiidae are doubtless comparable structures, and no apodemes from sternites are present in Collembola and Thysanura. The Protura have a median cuticular ridge (Fig. 7.21c) on the sternite but no internally projecting apodeme. In this class and in the Pterygota the so-called spina is not an apodeme, but merely the connective tissue linkage of muscles or tendinous endoskeleton with the posterior margin of the sternite. Connective tissue endoskeleton arises intersegmentally (Manton, 1928) and shifts forward, entirely or in part, to the middle of the thoracic segment in Collembola (shown diagrammatically in black, Fig. 7.29b, and also posteriorly in other arthropods). The myth of the "spina" is deeply seated in the literature. The distinction between the apodeme and tendinous endoskeleton is easily seen in esterwax sections stained with Mallory's tripple stain, showing the apodemes red and connective tissue blue. Thus, the "spina" cannot be regarded as a primitive feature but instead is a highly advanced one associated with particular habits of japygids.

The musculature of the japygid thorax shown in Fig. 7.24b illustrates many points of interest, which contrast with collembolan musculature considered above. The elaborate ventral longitudinal muscles and the spine apodeme give strength to the downward bending of the body, useful for the leg movements. The antagonists are the dorsal longitudinals, also in many sectors, attached to the anterior margin of principal and intercalary tergites and to their faces. The deep dorsoventral muscles in several sectors, $dv.$, $dv1.$, $dv2.$, $dv3.$, arise from their own tendon, but not a ribbon tendon as in Collembola, shown in black in Fig. 7.24b; they pull out the relaxed, contracted dorsal and ventral longitudinal muscles. The complexity of the dorsoventrals is greater than in myriapods and collembolans because of the thoracic flexibility and variety of movement. The deep oblique muscles, $dvmd.$, are slender but as long as can be contrived. They cannot promote as much rigidity of trunk as in Collembola and epimorphic Chilopoda where these muscles are either larger, shorter, or more elaborate; but some holding together of the mobile segments is desirable, and muscles arising nearest to the middle line are the most appropriate. There are further functions associated with the small additional Y-skeleton situated anteriorly on the japygid prothorax (Manton, 1972). Thus the muscular and skeletal systems of the japygid thorax are very far removed from any concept of primitiveness in Diplura which could lead on to the morphology of other classes. Instead, the details show a high level of structural advance suiting particular habits.

The legs of japygids are as remarkable in structure as is the thorax. The very robust coxal base is supported by $dvc.$ muscles, as in the flexible-bodied Geophilomorpha (Manton, 1965). These muscles provide stabilizing strength, but cannot cause coxal rocking. A unique leg-rocking mechanism is present within the leg itself in all Diplura (see Manton, 1972, Figs. 12, 14).

The legs of campodeiids contrast with those of all other hexapods not only in being long, but in markedly increasing in length along the thorax. It is possible for leg 2 to be placed on the ground lateral to and considerably in front of leg 1 before leg 1 is raised. Similarly, leg 3 oversteps leg 2. This degree of overstepping is rare in arthropods. The long legs, relative to segment size, result in long strides and the forwardly staggered groups of footprints shown in Fig. 7.27b. No other hexapods have been observed to do this. The forward staggering of the footprints of the Thysanura is dependent upon jumping gaits in which, at speed, legs 1 leave the ground before the footfalls of legs 2.

Fairly small size in the Diplura seems to be desirable, if their proficiencies are to be used profitably. The great leg length of campodeiids is surprising for a crevice-inhabitant in soil, but advantages provided by speedy running on the surface must outweigh the disadvantages within the soil, provided size is small. It is the larger japygids in India and Australia that shelter by and under stones. There appears to be no information as to whether these species are too large for easy penetration of porous soils.

Thus the Diplura cannot rank as simple "primitive insects" and the "spina" of japygids cannot be considered to be a primitive hexapodan feature. Dipluran habit proficiencies are reflected in their morphology, and this no more leads toward or away from Collembola than any other hexapod class. Dipluran morphology is distinctively their own, facilitating their habits; their mandibular and leg-base mechanisms are unlike those of other classes of hexapods and of myriapods.   Diplura could have originated by independent evolution from multilegged, soft-bodied ancestors.

Various categories of trunk muscles have been mentioned above in giving an outline of collembolan and dipluran trunk and thoracic musculature. It should be appreciated that these categories are constant throughout uniramian taxa and indeed in all arthropodan groups where fusion of segments does not mask the basic plan (except perhaps for the lateral longitudinal muscles of the Uniramia). The similarity in plan of muscles and of segmental tendons is dependent upon metameric segmentation, not on monophyletic evolution of arthropods. The divergent elaborations of both these systems are very great. The function of the muscles in the several categories, together with the muscle identifications, in the uniramian taxa has been given elsewhere (Manton, 1973, Appendix).

The small class Protura differs from other hexapod classes in the head, pleural, and limb-base morphology and functioning. The absence of antennae and the use of only the meso- and metathoracic legs for walking puts them undeniably aside from other hexapods. The functions of the large prothoracic legs are not fully known. They are held forward, with no "elbows" projecting, when closely surrounded by the substratum or decaying matter. Legs 1 move in similar or opposite phase and are uncoordinated with the following walking limbs. But it is possible that these anterior legs are mainly sensory and operate beside the pointed head when moving under cover or deep in soil or litter.

The Protura, seldom over 1 mm in length, also fail to qualify as primitive hexapods in any way. Their type of flexibility, entognathy, leg-rocking, and absence of antennae are entirely individual. If a proturan touches a drop of water in a dish, the cuticle sticks, and the animal, unable to free itself, usually dies, even if the legs stand on damp filter paper. Moisture in the environment of Protura is not usually in the form of droplets, but an inability to weather the results of rain or heavy dew may be responsible in part for this class (by 1932) numbering only 43 described species.

7.7.3. Thysanura

Much of the mandibular, pleural, and leg-base morphology and functioning has already been described in Sections 7.4 and 7.5, and the jumping gaits have been considered briefly in Section 7.6. The replacement of normal hexapod running gaits by a series of jumps is not found in other hexapods. There is no essential difference between running and walking, the terms referring largely to speed and pattern of gait. The Thysanura comprises two families, the Machilidae, or springtails, and the Lepismatidae, including silverfish.

The Machilidae live near the shore, hiding among rocks and under fallen wood. They are larger than most of other apterygotes. High jumping escape reactions, giving high jumps in any direction, can be repeated a few times before the animals tire and run off by the normal jumping gaits. The agility is striking. Hexapod gaits and no jumping are practiced on the underside of rocks, etc. The high escape jumps are executed by entirely different means from those of Collembola, although the abdomen is used for the purpose in both.

Machilid abdominal limbs are short, flat, and fused with their sternites forming coxal plates; their depression against the ground effects the takeoff for high jumps. The abdominal tergo-pleural arches envelop the body, extending laterally as far as the level of the sternite. This tergo-pleural exoskeleton is strongly constructed in such a manner that lateral abdominal flexure is impossible, but the tergites can shuffle over one another dorsally with ease, the abdomen then assuming the prejump position with a dorsal concavity and the terminal filaments elevated. This shuffling together is effected by dorsal longitudinal muscles, which are mostly twisted round one another like ropes (Fig. 7.32). Possibly extra length, facilitating longitudinal shortening, is thereby obtained. A similar feature is shown by the shrimps and prawns that jump through the water by forward flapping of the abdomen and the tail fan.

The details of the high jump have been elucidated by Evans (1975), using high speed cinematography. Energy is stored during the prejump position with cocked abdomen. Two movements follow. Strong depression of the abdomen against the ground, the dorsal concavity being straightened out, is accompanied by thoracic flexure in the opposite direction, causing ventral concavity and a raising of the center of gravity. The two movements spread toward each other. Whether the animal jumps forward, upward, or backward, or turns a circle in the air to land facing in the opposite direction, depends on these flexures.

The strength of the tail beat is caused by complex musculature. Since the exoskeleton prevents lateral flexure of the abdomen, the absence of the superficial oblique muscles, which cause such movements in other classes, is understandable. The deep obliques, the rigidity-promoting muscles in many classes, which strongly shorten the body of Collembola (*ob.*, Fig. 7.29c), are enormously bulky and complex in machilids (Fig. 7.31), as in no other hexapods. The normal single sector, double in chilopods, *dvma.* and *dvmp.* (see Manton, 1965, 1972, 1977), is represented by many large sectors, which loop around one another and insert in various places on four successive segments; muscle 23 is much like muscle *dvmp.* in chilopods, and muscles 24, 25A, 25B, and 26 roughly correspond with muscle *dvma.* and the even more elaborate, looped and twisted muscles of the malacostracan abdomen in the jumpers. The tendinous endoskeleton, shown diagrammatically with the muscles in Fig. 7.31, is forwardly displaced ventrally from the intersegmental level, so that the segmental tendon in ventral longitudinal muscle 15 lies in the middle of the preceding sternite

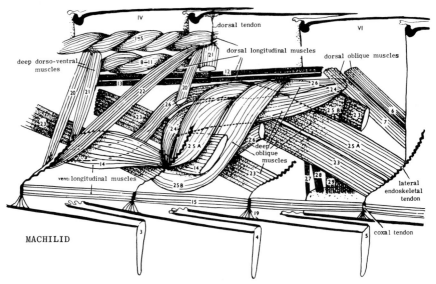

Fig. 7.31. View from sagittal plane of abdominal musculature of machilid, *Petrobius*, which causes high jumping by a strong tail beat against ground. Intersegmental tendons shown diagrammatically in black; dorsal longitudinal muscles 1–5 and 8–11 are twisted like ropes; ventral longitudinal system is duplicated as muscles 14, 15, as are deep dorsoventral muscles 20, 21 in association with great complexity and looping of deep oblique muscles 23–26. For further description, see text and Manton (1977). From Manton (1977).

and attached to it by the short muscle 19. The tendon system is very elaborate. There are two ventral longitudinal muscles, sector 15 the normal one, between the ventral ends of the tendon systems and a bulky more dorsal sector 14 linking the middle of the segmental tendons and doubtless supporting them against the tension exerted by the many deep obliques. These elaborate deep oblique muscles must be responsible for most of the tail beat, but exactly how looped muscles work here or in malacostracan crustaceans is not known. The antagonists to the deep obliques, doing no outside work, are as usual the deep dorso-ventrals, generally a simple muscle dv. (Fig. 7.29c) which is duplicated here as sectors 20 and 21, a feature correlated with the bulk of relaxed muscles to be stretched out. There is possibly an additional intermittent hydrostatic force operating against the coxal plates from the haemocoel, but the deep obliques appear mainly to be responsible for the abdominal tail beat.

Thus, in respects other than the basic structure and movements of the mandible, the Machilidae show features that cannot be regarded as primitive. Even the mandibular mechanism is assisted by an advanced hydraulic system for the collection of scraped-up algal cells. The legs of *Petrobius*, in performing the jumping gaits or in contributing to the takeoff for the high escape jumping, as in all jumping, need to flex a little, and then extend rapidly and with force, the initial thrust at the jump driving the animal forward, as opposed to the steady thrust exerted throughout the backstroke during a normal running or walking gait.

The leg swings rapidly forward after the jump, pausing a little before the next takeoff. It is not surprising to find good provisions for flexure and extension in the leg skeleto-musculature.

The extrinsic muscles are bulky and strong. The smallest are the coxal adductors, which are substantial and probably flex the distal ends of the long coxae together before the jump. The abductors inserting on the tergite are long and bulky. The coxa-trochanter joint (Fig. 7.32a) is very strong, its musculature providing for wide and strong angular movements and the same holds for the femur-tibia joint. There is little movement at the trochanter-femur joint, and the tibia-tarsus joint is strongly held.

At the coxa-trochanter joint, there is abundant arthrodial membrane on either side of the pivot-joint articulations; the morphology giving strong leg extension consists here of a very large apodeme from the trochanter from which long and bulky depressor muscles pass not only into the coxa, but through it to insert on the pleural apodeme (Fig. 7.32a). There is also an elaborate tendon arising from the ventral face of the trochanter supporting levator muscles; the tendon swings on the cuticle, thus enabling a pull to be exerted on the trochanter at an unchanging angle to the muscle fibres.

The Lepismatidae are smaller than most Machilidae and are speedier runners for their size than found in any other apterygote class. Their maximum speeds

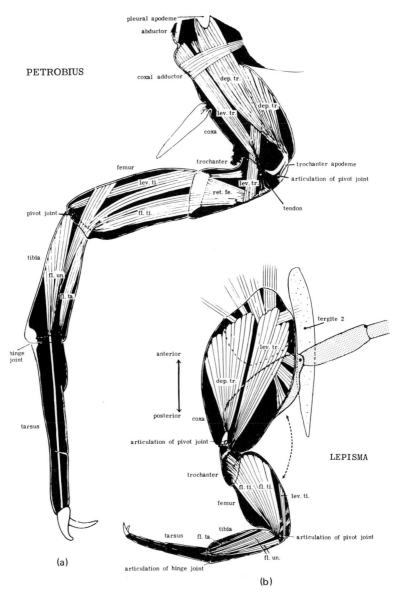

Fig. 7.32.   Legs of Thysanura. a, *Petrobius*, and b, *Lepisma*, to show remarkable jointing and apodemes connected with jumping gaits, which are used differently in the two taxa. Note trochanter apodeme carrying intrinsic and extrinsic depressor muscles and coxa-trochanter joint, as shown by two outlines, femur flexing forward dorsal to coxa as shown, a most unusual accomplishment and associated with jumping gaits. For further description, see Manton (1972, 1977). From Manton (1972).

during fastest sprints have not been recorded. A repeat of their maximum effort is not easily elicited. The limb-base and pleurites, as noted above, are unique and are correlated with the most remarkable jumping gaits and leg action existing within myriapods or hexapods (see above). The Pterygota show examples of jumping as isolated events, as in grasshoppers, fleas, and click beetles. The mechanisms differ and have recently been investigated; often catapult mechanics are employed, but the actions are intermittent and do not give the sustained locomotion of the jumping lepismatid gaits.

The jumping by the legs of *Lepisma* is almost horizontal and effected largely at the coxa-trochanter joint, with strong but important supporting movements by the coxa. The leg is shown in Fig. 7.32b. The trochanter is very small and almost immovable on the femur. The coxa-trochanter joint is narrow and remarkably mobile. An even larger trochanter apodeme than in *Petrobius* penetrates the coxa, and the pivot articulations of the joint lie directly on the wide base of the apodeme, which thus actually turns about this strong fulcrum (as in *Petrobius*, Fig. 7.32a). The extraordinary wide swing of the trochanter-femur on the coxa is made possible by the actual shape of the coxa and trochanter at their junction, whereby the trochanter-femur at the end of the forward stroke comes to lie dorsal to the coxa and not just up to it (Figs. 7.23c,d, 7.32b), a most unusual position. The coxa houses the large depressor muscles, and extrinsic insertions of long trochanter muscles are present. The femur-tibia joint is a strong pivot, the articulations displaced toward the dorsal face so permitting the large ventral emargination at the distal end of the femur. Bulky and wide femur-tibia depressor muscles can cause very strong flexure against the ground, used in jumping, the antagonistic levator muscle being very small and near to the articulation, doing no outside work.

These striking locomotory features of the Thysanura are unique, and the skeleto-musculature associated with them cannot be regarded as primitive, in that their high level of peculiarities, all now understood functionally for the first time, do not lead on to a basic condition of leg morphology in the Pterygota or in any other class.

7.7.4. Pterygota or True "Insects"

Comments have already been made on the pterygote head, mandibular structure, and its mechanism, together with those of the pleuron and leg-base. The first steps toward pterygote evolution, besides hexapody (see above), must have been a general tendency towards greater stability of the body by the formation of pleural rigidity and the plantigrade tarsus. A clawed, plantigrade tarsus or foot is present in the Onychophora, but, contrary to the figures presented in at least one textbook, this foot is directed laterally and not anteriorly, and on easy ground is held up and not used at all, the animal habitually walking on the pad of short, stiff subterminal setae (see the photographs and tracks shown by Manton, 1950, and the many detailed works concerning Onychophora published during the last 90 years). The foot is depressed and used when the going is slippery or rough. If the Onychophora evolved such a foot, presumably it could also have been evolved by the ancestral Pterygota. The immediate result of such an advance would have been easy walking on the surface at many inclinations, but strides would have been short and speeds not spectacular in contrast to unguligrade uniramians.

At the same time the Pterygota would appear to have avoided all the advances found in apterygote classes concerning: speed of running by ordinary and by jumping gaits; protection by high escape jumping, except as a later specialization; the measures providing a great flexibility of thorax and neck; and seeking refuge at an early stage into soil, and other deep shelter, accompanied by small size and short legs. The initial pterygote evolution may be likened to that of primates which remained generalized in limbs and teeth, while other mammals became much more highly modified as carnivores, fast-running herbivores, swimmers, and so on. As competition for survival became more acute, the apterygote classes must have advanced their different habits and facilitating morphology outlined above. Meanwhile, the Pterygota presumably did not compete in: hydrostatic rigidity for mechanical purposes; or in trunk (thoracic) flexibility; or in speed of running; or in using any of the particular apterygote modifications of the basic jaw mechanism. The retention of generalized structure and habits for a long time appears to have been as rewarding to ancestral pterygotes as to the primates.

The potentiality of simple ancestral stability of body and legs must have been great, leading to flight and the evolution of some 750,000 species, in contrast to the much more limited success of the apterygote classes—Collembola with some 1,500 species, Diplura and Thysanura with under 500 each, and Protura with under 100 species so far identified. Each apterygote class has reached a high level of performance and advance in structure along its own individual lines. None of them appears to have been capable of close association with pterygote evolution.

The great success of the Pterygota stems from the evolution of flight in some strong and stable-bodied stock with a dorsal coxal articulation on a fixed pleurite and the conversion of normal locomotory musculature, such as seen in exopterygote nymphs, into flight muscles (Tiegs, 1955). Neither the skeleton nor the thoracic musculature of the Thysanura is in the least suitable for such modifications. There are some ontogenetic resemblances between Pterygota and Thysanura, but they cannot be indicative of affinity closer than uniramians with incipient trunk sclerotization and pleurite formation starting dorsal to the coxa. Thereafter the leg-base and pleurite mechanism of the two classes must have proceeded differently. The mere possession of a pleurite articulated with the dorsal rim of the coxa is no more indicative of affinity than the converse in Diplura, Protura, and Myriapoda when the ventral rim of the coxa articulates with the sternite; they are each fundamentally different from one another in their several ways. If there are few mechanical possibilities, either ontogenetic or in adult structure, such features can have been adopted independently more than once. Pivot-joints and hinge-joints, for example, situated between leg podomeres are the only ways in which simple jointing is mechanically possible, and they have undoubtedly been evolved independently in Uniramia, Crustacea, and Merostomata, resulting in similarities between simple joints existing in arthropods of many kinds. Such joints are no more indicative of monophyly than is metameric segmentation of the trunk. The avoidance by Pterygota of the structural advances shown by the apterygote classes has, in the end, led to the amazingly successful radiation of the Pterygota, exceeding in numbers the rest of the animal kingdom put together.

7.8. CONCLUSIONS

The analysis of habits and facilitating morphology outlined above shows, as already noted, that the most plausible hypothesis which fits the facts appears to be that the five hexapod classes have originated by independent divergence from soft-bodied uniramian ancestors (Fig. 7.33). Each class has progressed by a series of functionally advantageous stages. The Myriapoda represent four parallel evolutions from uniramians with a different type of head capsule from that of ancestral hexapods. There can have been no one ancestral myriapod any more than one ancestral "insect." Probably there were early habit divergencies that were not successful enough to persist to the present time.

The fossil record tells us far less about the evolution of Uniramia than do the studies of functional morphology and embryology. There are pterygotes in plenty in amber and in recent deposits representing modern orders. The Permian has provided the beautifully preserved Monura, a group with conspicuous resemblances to the Machilidae but with less anchylosed head sclerites and longer abdominal limbs. There is no justification for picturing a single evolutionary line with the Monura as ancestral Thysanura, the latter in turn posing as ancestral pterygotes, as depicted in a modern textbook. So much license in the search for simplicity makes havoc of firm evidences in many subjects.

The Carboniferous shows us abundant wings; large gliding flyers; some hexapods with small prothoracic lateral flaps anterior to the large meso- and metathoracic wings. Some could fold the wings, an obvious advantage, but there is no clear evidence of relationship with modern pterygotes. The myriapods are represented among the fossils by well-armored Diplopoda. The softer-bodied chilopods so far have not been found fossilized until a much later date, but this cannot be taken as an indication of later differentiation of the class.

The Devonian *Rhyniella praecursor* is the earliest known hexapod, but this

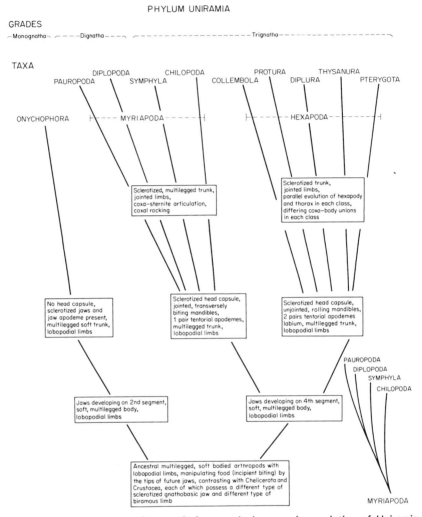

Fig. 7.33. Diagram representing conclusions reached concerning evolution of Uniramia, based upon facts established by studies of comparative functional morphology and embryology; see Manton (1972, 1977) and Anderson (1973). From Manton (1972).

collembolan is already an advanced member of its class with fused pro- and mesothoracic tergites that facilitate the facultative rigidity and jumping habits of Collembola.

The Carboniferous was an era of large size for both vertebrate and invertebrate animals. There was *Arthropleura*, nearly 2 m long, which has been known for many years and was recently redescribed by Rolfe (1969). This myriapod-like animal has now been joined by *Eoarthropleura* (Størmer, 1976), a few cm long, from the Lower Devonian. Both were very similar in form, the tergal surfaces of *Eoarthropleura* being the smoother. The roughness of *Arthropleura* was probably of mechanical significance, increasing cuticular rigidity, as in the larger extant Polydesmoidea. In both arthropleurids well-sclerotized tergites were expanded laterally into keels covering the legs, as in diplopodan Polydesmoidea and some Colobognatha (Manton, 1954, 1961). The longitudinal line of apparent sclerotization between keels and the main tergite was probably a line of strengthening, not a hinge. No land arthropod is known to have hinged keels; such flexibility would not have been a functional asset. Keels are useful for their rigidity, covering and protecting the legs during shallow burrowing.

The sternites and eight-segmented legs of Arthropleurida were well sclerotized. No spiracles have been found. Two pairs of plates, the K and B plates, were ventrally situated and overlapped the edges of the sternites, each other, and the base of the coxa. It has been suggested that these plates may have covered some respiratory device, as lay below the coxal plates of eurypterids. An animal as large as *Arthropleura* presumably must have possessed some respiratory system. The leg-base was associated with or articulated anteriorly to very large and elaborate "rosette plates" in roughly the same position as the chilopodan procoxa, but entirely different in detail from any myriapodan or hexapodan procoxal pleurite, although possibly functionally similar. The head is inadequately known. So large an animal as *Arthropleura* with very many trunk segments and living in Carboniferous forest litter would have used its paired limbs in similar phase, as in Diplopoda. One would not expect speedy running by fast patterned gaits from such morphology, or the body undulations consequent upon the use of such gaits, as drawn for *Eoarthropleura* in its original description.

There is no adequate reason for including the Arthropleurida within the Myriapoda. Presumably the Arthropleurida are uniramian. One can see no way in which their trunk structure could have led to those of the myriapod or hexapod classes. No functionally possible intermediates can be suggested between Arthropleurida and modern uniramians. The Arthropleurida persisted for some 100 million years without significant change in body structure. During that time they appear to represent a line of their own. They may have emerged from the sea with their known degree of sclerotization, and independently from the little-sclerotized, presumed uniramian ancestors of the extant taxa.

It is possible that early terrestrial uniramians adopted every conceivable divergence in habits, but that only some of these habits became associated with facilitating morphology capable of successful competition and persistence to the present day, evolving distinctive structure with the passage of time, as did Watson's plesiosaurs.

The concept of persistent habits that lead to better living anywhere, rather than adaptation to particular circumstances, is seen to be as valid for Uniramia as for vertebrates. Once the several proficiencies of the hexapod classes are appreciated, never again can any apterygote class be considered to be near the ancestry of any other hexapod class. The Pterygota in pursuing early trends toward simple stability and doubtless slow walking, avoided the several lines of advance of the apterygote hexapods and ultimately evolved flight and an adaptive radiation and further advance, both structural and physiological, which has affected all terrestrial animals and plants in a manner that needs no elaboration here.

The view that insects originated in the soil and then invaded the more exposed habitats cannot be endorsed, no matter by what route the animals may be supposed to have become subterranean. It has been mentioned above that the soil and the decaying log-living collembolans and japygids, although very successful below ground, all have legs too short to account for the evolution of a thorax. These hexapods and many others must be secondarily subterranean, as proved by the stages in reduction of collembolan jumping organs and associated muscles among extant species.

The above account is of necessity brief, and the figures shown here are insufficient to maintain the arguments. Further information and a fuller series of figures are to be found in Manton (1950, 1953, 1958, 1961, 1964, 1965, 1966, 1972, 1973, summary in 1977).

The functional account of uniramian structural systems presented here depends in its conclusions on much wider arthropodan studies, which are connected one with another, the first two having been published 50 years ago. Meanwhile, no hasty decisions have been adopted. It has become progressively clear that segmental and intersegmental tendon systems, muscle systems, and nervous systems, which are dependent upon the muscles or mesodermal somites and not vice versa, all follow the metameric segmentation of the animals. The components of all these systems, often complex, are visible most clearly when metameric segmentation is not complicated by fusion of segments or masked by the ventral elaborations taking place during head formation. Many other systems such as arteries, segmental organs and limbs are due to the existence of metameric segmentation. Such segmentation is no indication of monophyly among arthropods, and no systems, such as basic muscular components, or tendons, or limbs, provide any acceptable evidence of monophyly in general, or any reason for supposing that there once was a "primitive insect" that was descended from Symphyla, and many other inadequately considered hypotheses.

7.9. SUMMARY

Vertebrate palaeontology draws a distinction between some evolutionary changes that are associated with persistent habits fitting animals to live better under many circumstances, and adaptations to particular sets of environmental conditions. This concept is shown to be applicable equally forcefully to arthropods.

Habits that have persisted unchanged over vast periods of time have resulted in the basic morphological structure of the Onychophora, the four myriapod and the five hexapod classes. Large numbers of structural features are now understandable on a functional basis, giving better living anywhere. Basic distinctions between myriapods and hexapods concerning head structure, mandibular mechanisms, and different leg-base mechanisms are mutually exclusive in the myriapods and the five hexapod classes.

Entognathy is a convergently acquired phenomenon, widespread among arthropods, and its manifestation in three classes of hexapods appears to be convergent, providing particular assets by different means. Each could have arisen independently from the basic rolling type of hexapod mandible. The apterygote hexapods cannot be regarded as a related taxon, or as primitive hexapods. On the contrary, they are more advanced in their head structure than the basic pterygote type.

The leg morphology, gaits, and leg-base mechanisms of the five hexapod classes are mutually exclusive and superimposed upon the basic assets of hexapody common to all hexapods, be they uniramian, crustacean, or arachnid. But the five uniramian hexapod classes each use their three leg pairs in a distinctive manner.

Collembola appear to have evolved in association with persistent use of the mechanical properties of the haemocoel, since the whole body structure is associated with hydrostatic jumping. The Diplura and Protura contrast in lacking the collembolan rigidity-promoting mechanisms and instead have developed flexibility of trunk in their basic accomplishment.

Only the Thysanura and Pterygota have the coxa articulated dorsally with a pleurite, but this pleurite is mobile in Thysanura and fixed in the Pterygota. The Thysanura display unique jumping gaits and limb morphology that cannot be regarded as primitive.

Pleurite homologies are considered on a functional basis. Subcoxal segments are not found to be valid structures.

The Pterygota avoided the advances in structure and habits displayed by the apterygote and myriapod classes, and instead went in for trunk stability by the plantigrade tarsus and a fixed pleurite. Only such morphology and initial lack of habit differentiation could have led to the evolution of flight and the enormous success and adaptive radiation of the group.

Consideration of the hexapod classes in particular illustrates many things that are true for arthropods as a whole; but it must be emphasized that similarities due to metameric segmentation, serial repetition of limbs, muscles, tendons, nerves, and so on, are not evidences upholding a monophyletic interpretation of arthropodan evolution.

ACKNOWLEDGMENTS

My thanks are due to the many colleagues who have helped me to obtain the living arthropods on which this work has been based. I am grateful for the hospitality which I have enjoyed in the Zoology Department of the British Museum (Natural History), London, and in particular to the librarians and the Keeper of Zoology. I have profitted by the kindness and wisdom of Mr. J. Gordon Blower for reading the manuscript and for his helpful suggestions. To Dr. J. P. Harding I owe thanks for much photography of living animals and for abundant help in preparing duplications of the illustrations. I thank the Royal Society of London for permission to use certain of my drawings from the *Phil. Trans. R. Soc.* B 1964. published here, without color, as Figs. 7.6-7.8, 7.11-7.18. I also thank the Linnean Society of London for permission to use some of my drawings from their *Journal of Zoology* 1952-73 and published here as Figs. 7.9, 7.10, 7.19-7.30.

REFERENCES

Anderson, D. T. 1973. *Embryology and Phylogeny of Annelids and Arthropods.* Pergamon Press, Oxford.

Blower, J. G. 1951. A comparative study of chilopod and diplopod cuticles. *Q. J. Microsc. Sci.* **92:** 141-61.

Briggs, D. E. G. 1976. The arthropod *Branchiocaris* n. gen. Middle Cambrian Burgess Shale, British Columbia. *Geol. Surv. Can.* **264:** 1-28.

Cisne, J. L. 1974. Evolution of the world fauna of aquatic freeliving arthropods. *Evolution* **28:** 337-66.

Cisne, J. L. 1975. Anatomy of *Triarthrus* and the relationships of the Trilobita. *Fossils Strata Oslo* **4:** 45-63.

Evans, M. E. G. 1975. The jump of *Petrobius* (Thysanura, Machilidae). *J. Zool. Lond.* **176:** 49-65.

Evans, M. E. G. 1977. Locomotion in Coleoptera Adephaga, especially Carabidae. *J. Zool. Lond.* **181:** 189-226.

Manton, S. M. 1928. On the embryology of the mysid crustacean, *Hemimysis lamornae. Phil. Trans. R. Soc.* **B. 216:** 363-463.

Manton, S. M. 1937. Studies on the Onychophora II. The feeding, digestion, excretion and food storage of *Peripatopsis. Phil. Trans. R. Soc.* **B. 227:** 411-64.

Manton, S. M. 1949. Studies on the Onychophora VII. The early embryonic stages of *Peripatopsis*, and some general considerations concerning the morphology and phylogeny of Arthropoda. *Phil. Trans. R. Soc.* **B. 223:** 483-580.

Manton, S. M. 1950. The evolution of arthropodan locomotory mechanisms. Part 1. The locomotion of *Peripatus. J. Linn. Soc. (Zool.)* **41:** 529-70.

Manton, S. M. 1953. The evolution of arthropodan locomotory mechanisms. Part 3. The locomotion of Chilopoda and Pauropoda. *J. Linn. Soc. (Zool.)* **42:** 118-66.

Manton, S. M. 1954. The evolution of arthropoda locomotory mechanisms. Part 4. The structure, habits and evolution of the Diplopoda. *J. Linn. Soc. (Zool.)* **42:** 299-368.

Manton, S. M. 1958. The evolution of arthropodan locomotory mechanisms. Part 6. Habits and evolution of the Lysiopetaloidea (Diplopoda), some principles of the leg design in Diplopoda and Chilopoda, and limb structure in Diplopoda. *J. Linn. Soc. (Zool.)* **43:** 487-556.

Manton, S. M. 1961. The evolution of arthropodan locomotory mechanisms. Part 7. Functional requirements and body design in Colobognatha (Diplopoda), together with a comparative account of diplopod burrowing techniques, trunk musculature and segmentation. *J. Linn. Soc. (Zool.)* **44:** 383-461.

Manton, S. M. 1964. Mandibular mechanisms and the evolution of arthropods. *Phil. Trans. R. Soc.* B. **247**: 1-183.

Manton, S. M. 1965. The evolution of arthropodan locomotory mechanisms. Part 8. Functional requirements and body design in Chilopoda, together with a comparative account of their skeleto-muscular systems and an Appendix on a comparison between burrowing forces of the arthropodan haemocoel. *J. Linn. Soc. (Zool.)* **45**: 251-484.

Manton, S. M. 1966. The evolution of arthropodan locomotory mechanisms. Part 9. Functional requirements and body design in Symphyla and Pauropoda and the relationships between Myriapoda and Pterygota. *J. Linn. Soc. (Zool.)* **46**: 103-41.

Manton, S. M. 1967. The polychaete *Spinther* and the origin of the Arthropoda. *J. Nat. Hist.* **1**: 1-22.

Manton, S. M. 1972. The evolution of arthropodan locomotory mechanisms. Part 10. Locomotory habits, morphology and evolution of the hexapod classes. *Zool. J. Linn. Soc.* **51**: 203-400.

Manton, S. M. 1973. The evolution of arthropodan locomotory mechanisms. Part 11. Habits, morphology and evolution of the Uniramia (Onychophora, Myriapoda, Hexapoda) and comparisons with Arachnida, together with a functional review of uniramian musculature. *Zool. J. Linn. Soc.* **53**: 257-375.

Manton, S. M. 1974. Arthropod phylogeny–A modern synthesis. *J. Zool. Lond.* **171**: 111-30.

Manton, S. M. 1977. *The Arthropoda: Habits, Functional Morphology and Evolution.* Oxford University Press, Oxford.

Manton, S. M. and D. T. Anderson. 1978. Polyphyly and the evolution of arthropods. *Systematics Association Symposium* (in press).

Pantin, C. F. A. 1950. Locomotion in British terrestrial nemertines and planarians. *Proc. Linn. Soc. Lond.* **162**: 23-37.

Rolfe, W. D. I. 1969. Phyllocarida, pp. R296-R331. *In* R. C. Moore (ed.) *Treatise on Invertebrate Paleontology. Part R. Arthropoda 4.* Geol. Soc. Amer. and University Press of Kansas, Lawrence, Kansas.

Størmer, L. 1976. Arthropods from the Lower Devonian (Lower Emsian) of Alken an der Mosel, Germany. Part 5: Myriapoda and additional forms, with general remarks on fauna and problems regarding invasion of land by arthropods. *Senckenb. Lethaea* **57**: 87-183.

Tiegs, O. W. 1940. The embryology and affinities of the Symphyla, based on a study of *Hanseniella agilis. Q. J. Microsc. Sci.* **82**: 1-225.

Tiegs, O. W. 1944. The post-embryonic development of *Hanseniella agilis* (Symphyla). *Q. J. Microsc. Sci.* **85**: 191-328.

Tiegs, O. W. 1947. The development and affinities of the Pauropoda, based on a study of *Pauropus sylvaticus. Q. J. Microsc. Sci.* **88**: 165-336.

Tiegs, O. W. 1955. The flight muscles of insects–Their anatomy and histology; with some observations on the structure of striated muscle in general. *Phil. Trans. R. Soc.* B. **238**: 221-348.

Tiegs, O. W. and S. M. Manton. 1958. The evolution of the Arthropoda. *Biol. Rev.* (Cambridge) **33**: 255-337.

Watson, D. M. S. 1949. The mechanism of evolution. *Proc. Linn. Soc. Lond.* **160**: 75-84.

Whittington, H. B. 1960. Trilobita. *Encyclopedia of Science and Technology.* McGraw-Hill, New York.

8

Visceral Anatomy and Arthropod Phylogeny*

K. U. CLARKE

8.1. INTRODUCTION

The use of internal anatomy in studies of arthropod phylogeny has fallen far behind that of external anatomy and of embryology: only in the fields of skeletal-muscular systems (Manton, 1973) and of nervous systems (Bullock and Horridge, 1965) have really significant contributions appeared in recent years. One of the major advances has crystallized out in the proposals of Manton that "Arthropoda" is best considered a grade of organization, and phylogenetically consists of three independent lines of evolution, the Uniramia, the Crustacea, and the Chelicerata. The purpose of this paper is to examine some aspects of internal anatomy and determine if they support this proposition. The internal anatomy to be dealt with here is that group of systems primarily concerned with the passage of materials into, about, and out of the animal—namely, the input systems: the alimentary systems, the respiratory system; the "about" system: the circulatory system; and the output system: the excretory and reproductive systems. This is considered a "natural" grouping of systems, since the properties of any one have an effect on those of any other.

Work on the anatomy of these systems really falls into two categories, the first being early work that established the general internal anatomical structures of the different groups of arthropods and paid attention to the phylogenetic interpretation of their results. Much of this work was done before the functions of the structure were adequately known, and when the techniques, especially of

*The author's decision not to include illustrations in this chapter was based on his belief that descriptions alone are adequate for general understanding of the topics covered. *Editor.*

histology, were substantially less good than those of the present day. Much of this work was brilliant and still forms the basis of present-day thoughts. More recently, and in the second category, much anatomical work has been done in connection with physiological studies and is interpreted in terms of function rather than phylogeny.

In both categories work in one group, such as the Arachnida, was done without reference to studies in the Crustacea or the Uniramia, which results in many difficulties at the present time, since in the terms used different structures often are referred to by the same name. The use of the name nephridia for the coelomic organs in some groups but not in others is an example of a name persisting unjustifiably in arthropods as well as creating confusion within the grade. Throughout this work it has seemed advisable to use a common name for structures whose anatomy clearly shows identity, though in doing so familiar terms in some groups will have to give way to familiar terms in others. There is a great need for students of arthropods to agree on a common nomenclature to be used in comparative discussions ranging across the grade.

8.2. LEVELS OF ANALYSIS OF INTERNAL STRUCTURE

Traditionally, internal anatomical analysis of whole animal structure is by breakdown of the whole animal into systems, each system into organs, each organ into tissues, and each tissue into cells. Recognition of the different cell types demands attention be paid to the constituent organelles, but this level is rarely added to the analytical sequence quoted above. Further analysis to the chemical level has shown the importance of proteins and enzymes in establishing relationships; but again traditional analysis does not usually go down to this level, and its proper evaluation is outside the scope of this paper.

It is important to note that each level of analysis has a set of properties peculiar to that level and either unaffected by or only slightly influenced by changes in properties at other levels (Clarke, 1973). This relative independence of levels means that the morphology of a system or organ can remain unchanged while the physiological functions of its cells can alter. Thus organs can change function without notable change of shape and design; for example, the production of silk by the malpighian tubules in some insects and the acceptance of nitrogen excretion by some coelomic organs. Function therefore is not a reliable criterion of phylogenetic relationship, and judgment must be made on the anatomical patterns of organs and systems rather than similarities and differences in cellular physiology.

At each level of analysis it is possible to distinguish properties that can be assigned to one of the following three grades: 1) those that are inherent in the basic design of all arthropods, e.g., segmental arrangement of coelomic organs; 2) those characteristic of one or more major divisions within the arthropods, e.g., tracheoles in insects, book-lungs in Chelicerata; 3) features that are adaptations to different ways of life and are found widely distributed throughout arthropods wherever the animal has adopted a specialized mode of life requiring this structural adaptation, e.g., silk production, luminous organs, loss of eyes.

The properties peculiar to the different levels of analysis are not of equal importance as indicators of phylogenetic relationships.

At the cellular level, it is to be noted that cell differentiation is an adaptation to do a particular job, and the constraints of their work produce a high degree of resemblance between independently adapted cells; for example, the presence of trachea and tracheoles in Arachnida and Insecta. Fine analyses often show differences, but comparisons at this level are rather rare in the systems to be considered here. Occasionally, the presence or absence of a cell type coincides with phylogenetic boundaries that have been established on other grounds, which then too can be used as an additional characteristic, but can rarely or never stand on their own.

At the tissue level, tissue design is fairly characteristic in arthropods. The absence of a closed circulatory system does not favor the development of massive thicknesses of tissue, and except for the neural tissues, the dominant design is of a single epithelial layer supported by a basement membrane, given mobility by muscle fibres which are either widely spaced or form thin layers, and served by fine nerves. Increase in size is most frequently due to folding, and has little effect on these simple relationships. Tissue design differs little in the major groups of arthropods, but within any one animal it is adapted to many requirements, and in this respect much convergence occurs between them.

At the organ level there is much more help in determining phylogenetic relationships. Diversity in design is considerable, and they are sufficiently complex to have an evolutionary history that aids considerably where problems of convergence may arise.

System design is also very helpful in establishing evolutionary relationships and presents much the same scope and limitations as does organ design. It is sometimes rather more flexible, as for example, the development of the heart in the abdomen of Isopoda where it is correlated with respiratory development of the abdominal limbs; a similar arrangement occurs in *Squilla*. This kind of flexibility at system level emphasizes the need for the subdivision of each level into the three categories of major (1), minor (2), and adaptive (3), mentioned above as a guide to judgment of the importance of the differences and similarities that occur.

8.3. GENERAL PROBLEMS OF STRUCTURAL ANALYSIS

This article will place much emphasis on shape and form of internal structures in assessing their similarities and differences for phylogenetic purposes. Some attention must be paid to the forces that affect these properties. Structures at all levels of analysis are usually multifunctional. One of these functions is often clearly the major one, and it is from this function that the structure takes its name. For example, the tracheal system in insects has the conduction of air from the exterior to the cells as its major function. It also acts as connective tissue binding organs together and plays a part in suspending them in the haemocoele. It has a role at ecdysis associated with the changes in body size that occur at this time. All these functions have left their imprint on the design of the system, whose final pattern results from the interaction of all these requirements. A similar tracheal system developed in some Arachnida does not always have the same subsidiary function, i.e., has little or no importance as a connective tissue and plays no part in size changes at ecdysis. The overall design of the system is quite different in the two groups. In some cases the main function of the structure is not the dominant feature in its design.

Structural features in a system or organ must be designed to do the maximum work load that is to be inflicted upon the animal, and the need for this stress factor in design may not be evident all the time. Some stress requirements, such as egg maturation, may occur only for a limited time in the animal's life cycle, and structures to meet their demands may differentiate and persist only when required. Others, such as respiratory stress, may occur spasmodically throughout life; then the maximum load facility must always be present. Accounts of internal structure should take note of changes throughout the life span of the animal. All too frequently they are limited to adult stages.

Tolerance in structure design is usually documented as variation between normal individuals of the same species. Rarely is there any information on the performance of these different patterns of structural design. It seems likely that, in internal anatomy, many functions can be adequately performed without too rigorous adherence to a preset plan. Deficiencies in the work load and performance capabilities of one organ or structure can often be made good by an increase of those of another. Where there is more than one organ of a particular type present, i.e., coelomic organs, malpighian tubules, it is desirable that all be examined and the system reported on as a whole. Paired organs in bilaterally segmental animals are rarely identical. It is unwise to use too rigorous criteria in deciding questions of identity in questions of internal anatomy until much more is known of the forces that govern structural design and the tolerances that can be permitted for normal functioning.

8.4. SPECIAL STRUCTURAL PROBLEMS IN INTERNAL ARTHROPOD ANATOMY

A number of features in arthropod organization set special problems in using internal anatomy in phylogenetic studies. These are: the early breakdown in embryonic development of the coelomic sacs and tissues; the development of a large, spacious, body cavity, the haemocoele; the presence of an exoskeleton; and growth by moulting and ecdysis.

The early breakdown of the coelomic cavities and tissues in an animal whose developmental mechanics appear to have evolved or were in the process of evolving in animals in which they played or were to play a more prominent part, has resulted in an environment in which the organization of cells into tissues, organs, and systems is not so rigidly dictated. The patterns of origin of organs are more readily diversified as they are now free of the tyranny of the coelomic programme. An example is the formation of gonads that capture the potential reproductive cells together with their "bits" of coelomic epithelium in diverse ways in the Insecta, Crustacea, and Arachnida. It is possible that induction phenomena in embryological development have passed from the now nonexistent coelom-complex to other tissues that have not lost their organization. In this situation induction may affect any cell entering its field regardless of its origin.

If this turns out to be so, then the origin of cells, whether ectodermal, mesodermal, or endodermal, may be of little importance in deciding homologies among organs. The apparent ectodermal origin of the antennal glands in Ostracoda, or the ectodermal origin of the malpighian tubules in insects as compared with their endodermal origin in arachnids, may be of no phylogenetic consequence. This phenomenon could be the basis of Shimkevich's (1908) metorisis.

The large haemocoele and exoskeleton provide, as it were, a box in which there is plenty of room for the organs and systems to develop their own patterns without undue interference with each other, and for their shape and form to be relatively uninfluenced by the external shape of the animal. Problems of organ suspension within the "box" are solved in a different manner from those of organ suspension in animals with a more well-developed coelom. This subject will be touched on later.

The relative independence of the shape and form of internal organs from the animal's external form allows evolution of one to proceed almost independently of that of the other. This allows wide variation of external body form to be coupled with conservative internal design, and similarly considerable variation in internal structure with invariant external body form. Although this uncoupling of external and internal body design occurs in many animal groups, the possession in arthropods of exoskeleton and haemocoele allows it to be taken to the extreme. Great care must be taken in interpreting internal body designs on the basis of external structural relationships. The approach in this paper is to try to interpret internal structure with as little reference to external structure as possible, since it is thought that in this manner the maximum contribution of internal anatomy to arthropod phylogeny can be obtained.

The effect of moulting and ecdysis on internal anatomy has been little studied.

The noncuticular parts of the systems studied are generally supposed to be uncoupled from the cycles of mitosis and growth that affect the cuticular structures. Much work needs to be done on the manner of growth of these noncuticular structures. For example: how does the heart grow in size? The internal anatomy has to cope with the steplike nature of the growth that occurs, and almost all internal organs are affected by the actual mechanics of ecdysis and must have in their design features that enable them to act appropriately to this event.

8.5. GUT

8.5.1. Introduction

A mouth opens into the gut at the anterior end of the body on the ventral surface. A buccal or preoral cavity, bounded anteriorly by the labrum, posteriorly by a labium, and laterally by the base of one or more pairs of appendages, is usually present. Some manipulative preparation of the food occurs here before it is passed through the mouth. The mouth opens to an intucked tube of ectoderm lined with cuticle, the foregut. Posteriorly, the foregut is invaginated into the lumen of the midgut. This invagination often forms a valve, the proventricular or oesophageal valve. The midgut is of endodermal origin; it lacks a cuticular lining, and is the region of the gut where most of the chemical processing of the food takes place. The midgut opens posteriorly into the hindgut, which, like the foregut, is an intucked tube of ectodermal origin lined by cuticle. A valve between the midgut and hindgut occurs sporadically in all groups, but does not seem to be a regular feature in any major arthropod division. The hindgut opens posteriorly on the telson through an anus.

Except in the Onychophora, the muscle associated with the gut in arthropods is striated. In insects these muscles perform relatively slow contractions resembling those of smooth muscle, and whilst their ability to stretch is less than that of smooth muscle, it is greater than that of striated skeletal muscle. An unusual feature of all arthropod guts at the tissue level is the arrangement of the muscle layers in the fore- and hindguts. In both cases the longitudinal muscles lie beneath the circular ones next to the basement membrane of the gut epithelia, while the circular muscles adjoin the haemocoele. In the midgut the circular muscles are next to the basement membrane of the midgut epithelium, while the scanty longitudinal muscles are outside facing the haemocoele.

Extrinsic gut muscles originating from the body wall are inserted into the gut. These muscles are concentrated especially about the anterior region of the foregut and the posterior region of the hindgut. They act to dilate these regions of the gut, and to suspend them from the body wall. The midgut is usually devoid of extrinsic muscles and in many instances is tensioned between the fore- and hindguts. Dorsal and ventral mesenteries suspending the gut of the type found in the Annelida and Chordata are absent.

Movement of materials through the gut is by muscular action and by peristalsis. Ciliary cells are absent from the guts of all arthropods and thus cannot make any contribution to fluid movement within the system. Mucus cells are also frequently said to be absent; however, a mucin-like substance is found in both the Crustacea and the Chelicerata, but is often replaced by other devices for protecting the cells of the midgut epithelium. The absence of these cell types leaves the epithelium free to concentrate on other requirements, so that area for area the arthropod epithelium is more efficient in enzyme production and food absorption than that of many other animal groups.

Apart from its primary function of food processing, the gut plays an important role in many other functions in arthropods: it plays an active role in the distension of the body at ecdysis; it functions in the storage of food material in the cells of the midgut; much of the intermediate metabolism of amino acids, fats, and carbohydrates may be performed in its cells: primitively, nitrogen excretion is associated with the gut; in aquatic arthropods it is important in osmoregulation and in terrestrial animals in water absorption. Although the term digestive system is often used to designate the gut, digestion is only one of its functions. The gut is always designed to accomplish more than this single function; hence, the term gut is used in preference to digestive system.

8.5.2. Types of Gut

The guts of arthropods can be grouped into two major structural types, depending upon the anatomy of the midgut. In one type it is a simple tube, which is characteristic of the Uniramia; in the other it is a tube from which arise numerous caeca, and this is characteristic of the Chelicerata and Crustacea. Caeca do occur in the Uniramia, but are secondary adaptations limited to a few groups of insects. It is unfortunate that caeca occur in the midgut of cockroaches and locusts are so widely used in teaching, as their condition is by no means representative of the Uniramia. In both the Chelicerata and the Crustacea caeca are present in nearly all members of these groups: in the few cases where they are absent, this absence is associated with small size of the animal, whose other features make it reasonable to suppose that they have become secondarily lost, perhaps as an adaptation to the small size of the animal. In large members of the Crustacea, especially the Malacostraca, and in the larger Chelicerata, the Xiphosura and the Scorpiones, some or all of the caeca become much branched to form a large racemose organ referred to by different names, digestive gland, liver, or hepatopancreas, in the different groups. In spite of the similarity in function, these organs are of different origin in the Chelicerata and the Crustacea, and it is thought that clearer thinking on phylogeny will result if they are given different names. In this article they will be called digestive glands in Chelicerata and hepatopancreas in Crustacea.

Other criteria may be used to group the guts into major types. In the Uniramia and the Crustacea a peritrophic membrane is formed in the midgut. This membrane is secreted by the midgut cells and forms a sheath around the food mass, protecting the cells of the midgut epithelium from abrasion by hard particles in the food and often aiding in food movement through the midgut. In Uniramia it is lost in some forms that have adopted a fluid diet. In Crustacea it is reported in a few widely scattered species, and its firm status as a characteristic of the group needs further study. In the Chelicerata a peritrophic membrane has not been described. Its absence from the Arachnida, which are all fluid feeders, will perhaps occasion little surprise, but it is also absent from *Limulus*, where its protective function would appear to be useful. The mucus-like material produced in the *Limulus* gut may be adequate for this function, and in the Chelicerata the membrane was never developed. It was therefore primarily, not secondarily, absent from the fluid-feeding arachnids, and this was perhaps a factor in forcing them to adopt this mode of feeding.

8.5.3. Chelicerata

As a basis for considering the structure and function of the chelicerate gut, a brief description will be given of the gut of *Limulus* as representative of the Merostomata, and of the spider, *Argiope*, as a representative of the Arachnida.

In *Limulus* the mouth, which is situated some way back from the anterior end, opens into a foregut, which runs dorsally for a short distance; the foregut then curves anteriorly and dorsally and finally turns posteriorly, where it opens to the midgut. This junction is marked externally by a circular groove in the gut wall. The midgut is a wide tube with two pairs of ducts, which are asymmetrically placed close to its anterior end which ramify to form a digestive gland. The midgut opens to a short hindgut, which leads to the anus.

The foregut is further differentiated into: a) a short oesophagus which passes dorsally through the vascular ring and gradually expands to form b) a large crop with many longitudinal folds in its walls, leading to c) the gizzard, a short muscular tube lined with ridges bearing flattened teeth, opening to d) a conical proventricular valve that projects into the lumen of the midgut and is lined by a ridged cuticle. The crop and gizzard are surrounded by a well developed circular muscle layer. Extrinsic muscles arising from the endocranium and ventral body wall are inserted into the oesophagus and crop.

The midgut is lined with columnar cells, some of which secrete a mucilaginous substance that is used to coat the fecal pellets. Each midgut caecum branches to form smaller and finer tubes, which eventually end in alveoli. The cells lining the ducts and alveoli are of two types, glandular cells, which secrete the digestive enzymes, and cells that absorb the products of digestion. These absorptive cells are thought also to secrete calcium phosphate, which is mainly derived from the shells of lammellibranch molluscs on which the animal feeds. Regenerative cells occur among the base of the enzyme-secreting and absorptive cells. The products of digestion taken up by the absorptive cells may be synthesized into fats, carbohydrates, and proteins in these cells, or in connective tissue cells scattered among the alveoli. In any case these connective tissue cells store the food reserves of the animal. The ramifications of caeca penetrate among other organs and tissues of the body, forming the massive digestive gland of the animal, and helping to pack the organs firmly in the haemocoele.

In *Argiope* the mouth is anterior and opens into a foregut, which runs dorsally and then turns sharply posteriorly leading back to the midgut. The midgut bears numerous caeca and opens posteriorly to a short hindgut leading to the anus.

The foregut is differentiated into: a) a pharynx directed dorsally and situated immediately behind the mouth—it is characterized by sclerotic plates in its anterior and posterior walls; it opens to b) a slender oesophagus, which arises at right angles from the pharynx and runs posteriorly, and has thickened dorsal and lateral walls; this leads to c) a chamber lined with sclerotic plates and with extrinsic muscles that arise from the dorsal wall of the cephalothorax inserted on its dorsal wall, and muscles from the endosternite inserted on its ventral wall. It forms a powerful sucking organ, usually referred to as a "sucking stomach," and opens posteriorly to a short cuticle-lined tube, which leads to the midgut.

The midgut forms a broad tube in the prosoma, which narrows where it passes through the pedicel, then greatly enlarges to form a tube that turns sharply dorsally and then curves posteriorly toward the end of the prosoma. In the prosoma a large caecum arises from the midgut, which passes anteriorly giving off a number of branches and ending just behind the poison gland. Branches from this caecum pass to the base of each leg. In the opisthosoma a number of pairs of ducts arise from the dilated region of the midgut; each caecum ramifies to end in fine tubules and alveoli. These together form a digestive gland, which is less extensive then in *Limulus* and is derived from posterior caeca that do not occur in that animal.

Posteriorly the midgut narrows to open to a short hindgut. Anterior to this junction a single branched malpighian tubule arises from the dorsal surface of the midgut. The tubule is a specialized organ of nitrogenous excretion producing guanine as the end product of nitrogen metabolism.

The hindgut is short and has a large dorsal sac called the stercoreal pocket, in which feces may be stored.

8.5.3.1. General Survey of the Chelicerate Gut

The merostomate gut as represented by *Limulus* is unique in being the only chelicerate gut known that handles particulate food. Initially the food is very efficiently shredded by the chelate limbs and the gnathobases of the limbs surrounding the mouth.   The absence of a muscular pharynx indicates that there is little difficulty in passing the shredded food to the crop.   Further tituration of the food occurs in the gizzard.   Only small, finely ground particles can enter the fine ducts of the digestive gland, or there may be a danger of clogging the tubes and alveoli.   In the absence of a peritrophic membrane, mucus produced by the midgut cells protects them from damage by the food.

All the Arachnida are fluid feeders.   The mouth is usually guarded by setae, which strain off any solid particles.   Regurgitation of enzymes produced in the midgut, to bring about external digestion of the prey, commonly occurs.   A pumping pharynx is developed immediately following the mouth in all Arachnida, and is clearly essential for fluid movement in or out of the gut.   Only in the Araneae and the Amblypygi is it reinforced by the development of a sucking stomach.   A gizzard is absent, a fact that speaks for the efficiency of the mouth filters, which allow only fluid to enter the gut.   A proventricular valve is not found; there are no special valves guarding the entrance of the midgut.

In the Acari, Pseudoscorpiones, and Opiliones the pharynx has an X-shaped lumen, an indication that when the extrinsic muscles relax, the lumen can be very tightly closed, and if the cuticle lining it has the appropriate elastic properties, very rapid pumping movements can be made.   It is in fact an indication of a powerful pump, and may explain the absence of a sucking stomach in these groups and provide a clue to its absence elsewhere.

A number of caeca arise from the midgut.   In general they are more numerous in the primitive orders of arachnids, a maximum of seven pairs being found in the scorpions.   Apart from this it has not been possible to establish firm correlations between taxonomic divisions and the number of caeca present. This is partly due to the caeca sometimes appearing fused at their bases and the fact that they are often branched, both features making it difficult to establish how many caeca were primitively present in these cases.   The caeca that open to the midgut lying in the prosoma are often of different anatomical and histological structure from those that lie in the opisthosoma.

In the Scorpiones and the Uropygi all the caeca ramify and end in fine tubules opening to alveoli and thus forming a large digestive gland.   In the Amblypygi, Araneae, and Solifugae both simple and ramifying caeca occur.   In the Amblypygi there are four pairs of unbranched caeca in the prosoma and four pairs ramifying to form a digestive gland in the opisthosoma.   In the Araneae the caeca are similar to those described for *Argiope*, but the number of caeca in the prosoma varies and can be correlated with the different families in the spiders.   In the Solifugae four pairs of simple caeca are present in the prosoma, two pairs of which reach the bases of the third and fourth pairs of limbs.   A single pair of ramifying caeca forms a large digestive gland in the opisthosoma.

In the Palpigradia there are a pair of simple caeca in the prosoma and five pairs of caeca in the opisthosoma, the first three pairs being simple and the last two much branched.

In the Acari, Ricinulei, and Opiliones only simple caeca occur. In the Opiliones the midgut forms two chambers, one above the other. Three pairs of caeca arise from the dorsal side of the dorsal chamber, one pair lying in the prosoma, the other two in the opisthosoma.

The unbranched caeca that occur in the Acari and the Pseudoscorpiones may be correlated with small size, but branched caeca occur in the Palpigradi, whose adult size is about 2.8 mm. In these orders in some species the tips of the caeca may fuse to form loops that open to the midgut at each end. In Pseudoscorpiones the posterior caeca open to the midgut by a common duct, and a single ventral unpaired caecum occurs.

The gut diverticula, like the midgut, produce enzymes and absorb the products of digestion. A certain amount of intracellular digestion has been observed in many groups, and in the Araneae waste material may accumulate in the tips of the midgut cells, which break off and pass out of the animal. In a number of arachnids fats, carbohydrates, and proteins have been shown to be stored in the connective tissue cells associated with the gut caeca. These cells may be regarded as forming a loosely organized fat body.

Where malpighian tubules occur in the Arachnida they open into the posterior end of the midgut. They are often branched, the point of branching being distal from their point of entry to the midgut. In this respect they differ from the malpighian tubules of the Uniramia. Where their function has been investigated, they have been shown to excrete the end products of nitrogen metabolism in the form of guanine. There seems to be no evidence of their origin, but they could easily have arisen from midgut caeca that have become specialized for nitrogen excretion. It would be interesting to search for caeca specialization in arachnids that lack morphologically recognizable malpighian tubules.

The hindgut is usually short and does not appear to be specialized for water absorption, although it is possible that the stercoreal pocket, which is found only in the Araneae, has this function. In the Solifugae the hindgut is compressed laterally; there is no indication as to the function of this unusual modification. In the family Hoplopeltidae anal gands are present, which are used to spray an offensive liquid in defense of the animal.

8.5.4. Crustacea

Notions about the gut in Crustacea are very much dominated by investigations into the structure and function of the guts of the Malacostraca, such as *Nephrops*, *Astacus*, *Homarus*, and *Cancer*. A very elaborate filter mechanism occurs in these animals, which is not characteristic of the lower members of the Malacostraca or of other groups of Crustacea. On balance it seems better to give a general survey of the Crustacea gut than to cite particular examples as types.

The mouth opens to a short vertical foregut lined with cuticle, which opens to the midgut, whose long axis is usually set at right angles to that of the foregut.

In the majority of Crustacea the foregut is not differentiated into a pharynx, oesophagus, or crop but is usually referred to as the oesophagus. Oesophageal glands, which consist of small groups of cells in the general epithelium and open into its lumen, are frequently present. These produce a mucus-like substance. As mentioned above, elaborate filter and valve mechanisms may occur. These develop in the posterior foregut and at the foregut-midgut junction. The latter may represent the proventricular valve. Such mechanisms are particularly well developed in the decapod Crustacea.

In the majority of Crustacea the midgut is a long tube of nearly uniform diameter. A peritrophic membrane has been described in *Daphnia*, *Lepas*, and various Malacostraca. It was probably primitively present throughout the Crustacea. Both digestion and absorption have been demonstrated to occur in the midgut of *Artemia*. Excretory granules have been observed to be pinched off from the midgut cells in *Cyclops*. Some differentiation of function in different parts of the midgut has been described. Digestion of the food mass appears to take place in the posterior region and absorption in the anterior region of the midgut.

Caeca are always present on the midgut, but there is much variation in their number and position. A pair of caeca is generally present, opening into the midgut just behind its junction with the foregut. In *Artemia* they are simply a pair of pouches. In the Malacostraca, particularly the Decapoda, they ramify to form fine tubules ending in alveoli forming the hepatopancreas of these animals (Balss, 1926). In the majority of Crustacea they form a pair of simple, unbranched caeca.

A medium dorsal caecum is frequently present at the anterior end and at the posterior end of the midgut. They appear to be simple extensions of the midgut epithelium serving only to increase its surface area.

The midgut caeca are more numerous in the Malacostraca than they are in the lower Crustacea. In the alveoli of the hepatopancreas of definite cycle of cell activity occurs. The regenerative cells are at the tip of alveolus: their daughters function first as absorptive cells, then, as they move along the walls of the alveolus, as enzyme-secreting cells, and finally become fibrillar cells before they are cast out. Their production rate is greater than their death rate; hence the tubules and alveoli increase in size throughout the life of the animal. Phagocytosis has not been observed in the cells of the hepatopancreas. Excretion of dyes from the haemolymph by the hepatopancreas have been demonstrated, and it is perhaps an indication of a more extensive excretory function in this organ. This general description contrasts with that of the Chelicerate digestive gland, emphasizing the need to distinguish between the two which differ in both their origin and their mode of function.

The hindgut is usually short. It may or may not have a caecum and, like the oesophagus, has glands that produce a mucoid-like substance.

Food reserves are said to be stored in cells in the lateral parts of the trunk and in the appendages, but apparently not in connective tissue cells associated with the alveoli of the hepatopancreas.

Little will be said here about the specializations in the foregut of many of the Malacostraca. The short oesophagus opens into an elaborate grinding chamber, the gastric mill. Posteriorly there is an elaborate system of filters, which effectively prevents large, hard particles from entering the caeca of the hepatopancreas. In some cases back projections of the foregut convey the rejected particles right through the lumen of the very short midgut and deposit them directly into the hindgut.

The detailed anatomy of the gastric mill and proventriculus differs in the different species and higher groups of the Malacostraca, but has clearly evolved within the group.

8.5.5. Uniramia

The gut in the Uniramia is best surveyed by examining descriptions of representatives of the three main groups that make up this unit.

8.5.5.1. Onychophora

The description of the gut in this animal is founded almost entirely on the work published by Manton (1937). The mouth lies anteriorly and ventrally and opens to a cuticular lined foregut; this leads posteriorly to a long, tubular midgut, which opens to a short hindgut leading to a posterior anus.

The foregut is differentiated into a) a pharynx lined with thick cuticle and having a well-developed circular muscle coat. Radial muscles lie between these circular fibres; they originate from the cuticular lining and are inserted into the outer connective tissue sheath of the gut. Extrinsic dilator muscles suspend the pharynx from the body wall. The pharynx is followed by b) the oesophagus. Here the cubical epithelium secreting the cuticular lining is surrounded on its haemocoelic side by a thin, dense layer of connective tissue. A space separates this coat from another less dense sheath containing circular and longitudinal muscle fibres. Some strands of connective tissue connect these two coats across the intervening space. The oesophagus opens to c) an oesophageal valve, which is invaginated into the lumen of the midgut when the longitudinal muscles contract; when they are relaxed the foregut is not invaginated into the midgut, both fore- and midgut forming a single straight tube.

The midgut is a soft-walled tube of larger diameter than the oesophagus and extending from about the second to third pair of legs to near the posterior end of the body. The epithelium of the midgut is largely composed of tall columnar cells, some of which are gland cells loaded with secretory granules, which tend to be scarce or absent from the anterior and posterior ends of the midgut; others are storage cells, which contain fats, carbohydrate, and proteins synthesized from the absorbed products of digestion. These storage cells also secrete the peritrophic membrane, and excretory granules which are not uric acid. Apparently uric acid crystals are formed in the lumen of the gut between the cell borders and the peritrophic membrane, perhaps from these excreted granules. A peritrophic membrane is secreted from the entire midgut surface, and is produced about once every 24 hr. It serves to protect midgut epithelium from injury by abrasive particles in the food, contributes greatly to keeping the gut sterile, since it is discharged daily, and acts as a vehicle by which the uric acid crystals, which adhere to its surface, are conveyed to the outside. Regenerative cells occur between the bases of the columnar cells.

There are no caeca, and no mucoid-like substances are produced by the gut. Storage of materials is within the midgut epithelium and not in connective tissue cells associated with it. There are no malpighian tubules, the midgut epithelium functioning as the site of nitrogenous excretion.

The hindgut is a narrow tube opening to the anus.

A pair of salivary glands open into the buccal cavity.

8.5.5.2. Myriapoda

The Myriapoda are a diverse group of animals, and here again it seems best to deal with the gut in a general survey rather than by reference to specific types. In general plan the gut is very similar to that of the Onychophora, consisting of a foregut, a long simple tubular midgut, and a short hindgut.

The length of the foregut is relatively short in the Symphyla, the Pauropoda, and the Diplopoda, but varies in the different families of the Chilopoda. Differentiation of the foregut is not very marked. Immediately behind the mouth extrinsic muscles arising from the head capsule are inserted into the foregut and can bring about active dilation of this region. Contraction of the circular muscles brings about constriction of the gut lumen. This region then is differentiated into a pharynx, but the muscle layers may be poorly developed, and in the Diplopoda a pharynx is said to be nonexistent. Extrinsic muscles may also be inserted into the foregut behind the brain and bring about dilation of this region, which is normally narrow and referred to as the oesophagus. A crop may be present or absent; where present it is a simple widening of the posterior region of the oesophagus. In the chilopod, *Cryptops*, forward-directed bristles in the crop prevent coarse food particles from entering the midgut. Posteriorly the foregut is invaginated into the midgut to form a proventricular valve. In the diplopod, *Spirobolus*, a muscular sphincter occurs at the foregut-midgut junction.

The midgut is a tube lined with columnar epithelium. There is little differentiation in its cells; a few gland cells scattered among the others have been described in *Lithobius*, and regenerative cells are generally present among the bases of the columnar ones. A peritrophic membrane is present and is apparently secreted by the entire midgut epithelium. The midgut is often ensheathed in fatty tissue, which probably also contains the carbohydrate and protein stores of the animal.

The hindgut is rather variable in length and in degree of differentiation. It is quite long in the Symphyla, where it is differentiated into four regions, and in the Diplopoda, where it is differentiated into three. In the chilopod, *Lithobius*, it is quite short and apparently undifferentiated. In the diplopods, *Spirobolus* and *Strongylosoma*, its differentiation into three regions is marked by muscular sphincters. Fecal pellets collect in the middle chamber. The posterior chamber forms a wide eversible sac, which ejects the fecal pellets out through the anus.

A pair of malpighian tubules opens into the hindgut at its junction with the midgut in the Symphyla, the Pauropoda, the Diplopoda, and the Chilopoda. They lack muscles and are incapable of independent movement in the chilopod, *Lithobius*.

In spite of the general similarity of the gut in the Onychophora and the Myriapoda, considerable differences exist between them. The pharynx of Onycophora is clearly quite different from that of the myriapods; the proventricular valve is a permanent feature in myriapod guts; the cells of the midgut do not store the food reserves and have lost at least part of their nitrogenous excretory function to the malpighian tubules developed from the hindgut. There is greater differentiation of the hindgut in Myriapoda than in Onychophora.

8.5.5.3. Insecta

A survey of the gut in such an enormous and diverse group as the insects is outside the scope of this article. Comments will be restricted to the Apterygota and the more primitive orders of the Pterygota.

8.5.5.3.1. Apterygota.   In the Protura, the foregut is a simple tube apparently lacking muscles and only slightly invaginated into the lumen of the midgut at its posterior end. The simple tubular midgut lies well forward in the anterior region of the thorax and abdomen. The epithelium of the midgut is a sheet of single cells; lateral junctions between the cells appear to be lacking, the cells resting on a basement membrane. The inner surfaces of the cells are produced into a number of filaments directed toward the lumen. These filaments are described as cilia, but proof of their possessing the ultrastructural characteristic of cilia is lacking. Certainly they have a very different appearance from the tight epithelium sheet and brush border of other arthropod midguts. A peritrophic membrane is absent, and intracellular digestion is said to occur (Berlese, 1909). Granules of guanine have been reported in the gut, indicating that it is the site of nitrogen excretion. Guanine has been reported elsewhere in arthropods only in the Arachnida. The hindgut forms a narrow tube; immediately following the midgut it broadens posteriorly to form a large extensible colon, and a short rectum leads to the anus. In Berlese's figure the basement membrane of the midgut appears to be continuous with the cuticular lining of the hindgut, the basement membrane of the hindgut forming a kind of cone that comes forward and joins the basement membrane of the midgut some little way anterior to the junction. Within this cone, and just behind the junction, the cells of the hindgut form a ring of uni- or bicellular papillae, a duct cell being apparently associated with each papilla. These are developed at the same site as the malpighian tubules develop in other insects, and are thought by some workers to be precursors of them.

In the Collembola, the buccal cavity is well developed and has extrinsic dilator muscles inserted on its anterior wall. It acts as a pump replacing the pharynx and is often referred to, incorrectly, as the pharynx. The mouth opens to a narrow foregut which first runs vertically and then turns to pass posteriorly through the nerve ring and open to the midgut just behind the head. In its posterior region its cuticle may bear a few denticles, and it is only slightly invaginated into the midgut. In the majority of Collembola the midgut forms a simple broad tube beginning just behind the head and ending in the posterior region of the abdomen. In *Achorutes* and *Protanura*, Mukerji (1932) states that a pair of long caeca arises from the anterior midgut. Granules of unknown nature, but perhaps associated with nitrogen excretion, accumulate in the midgut cell tips, which break off and are shed at ecdysis when regeneration of the epithelium occurs. A peritrophic membrane appears to be lacking. The hindgut may form a simple tube, the posterior end of which widens to form a rectal sac having extrinsic dilator muscles arising from the body wall inserted on its surface. A pyloric valve may be present in some species. There are no malpighian tubules in the majority of Collembola, but Heymons (1897) has described small pockets at the anterior end of the hindgut in *Isotoma*, which may represent them.

In the Diplura, the mouth opens to a narrow foregut, which, in some forms (*Anajapyx*), is very long, extending back to the fourth abdominal segment. The midgut is relatively short with a well-developed ring of markedly columnar cells at its posterior end. A peritrophic membrane does not seem to have been described and is probably lacking. The hindgut in *Japyx* is a simple tube enlarged posteriorly and with extrinsic dilator muscles. Six small caeca arise at the anterior end of the hindgut in *Anajapyx*, and 16 such caeca occur in *Campodea*; they are completely absent in *Japyx*. They resemble short malpighian tubules.

In the Thysanura, the foregut in *Machilis* consists of a well-developed pharynx immediately behind the mouth, opening to a narrow oesophagus, which passes posteriorly to open into the midgut through a slightly developed invagination and simple proventricular valve. In *Lepisma* the foregut is much longer and expands posteriorly to form a well-developed crop leading to a muscular gizzard bearing teeth and bristles on its cuticular lining, and then opening by a simple proventricular valve invaginated into the lumen of the midgut. The midgut forms a long, broad tube with a number of caeca at its anterior end. These are better developed in *Lepisma* than they are in *Machilis*. A peritrophic membrane is present and is formed by secretion from the surface of the entire midgut epithelium. The hindgut in *Machilis* forms a straight tube differentiated into four regions: an ileum immediately following the midgut-hindgut junction; a colon; a more posterior rectal sac characterized by the presence of extrinsic dilator muscle, and, internally, by six longitudinal bands of columnar epithelial cells similar to the rectal pads of Pterygote insects. Well-developed malpighian tubules are present, 12-20 tubules in the various species of Machilidae, 4-8 in the Lepismatidae.

8.5.5.3.2. Pterygota.   In the Odonata the mouth opens to a well-developed pharynx, which leads to a long, narrow oesophagus, passing back to the base of the abdomen, where it expands to form a crop. Posteriorly, the crop opens to a well-developed gizzard, much better developed in the larvae than in the adult, and finally the foregut opens to the midgut through a simple proventricular valve invaginated to the midgut lumen. The midgut is a simple tube, lacking caeca at its anterior end. It is lined by a columnar epithelium; the anterior region appears to be specialized for the formation of the peritrophic membrane. Posteriorly the midgut narrows sharply to open into the hindgut: a strong sphincter muscle is present at this junction. The hindgut is differentiated to form a short, narrow tube, the ileum, which opens to a broader tube characterized by a thick pad of epithelium on its ventral surface. Tillyard (1917) named this region the pre-rectal ampulla. It leads posteriorly to a rectal sac which contains well-developed gills, and from this a short rectum leads to the anus. In the adult, the rectal sac is very short and lacks gills; there is no separation of the hindgut to form ileum and pre-rectal ampulla, but a ventral epithelial pad extends the whole length of the tube. Numerous malpighian tubules are present. They unite near their bases in groups of five or six, which open to the lumen of the hindgut through a short, very narrow tube. In the early instars only three malpighian tubules are present.

In the Plecoptera the mouth leads to a well-developed pharynx. Behind this is a short, narrow oesophagus, which swells to form a relatively huge crop that extends well back into the abdomen. The gizzard is absent or in a very rudimentary form. The midgut is short and has 10 well-developed caeca at its anterior end. The hindgut is a simple tube swollen posteriorly to form a rectal sac. A large number of malpighian tubules are present.

In the Orthoptera the insect gut assumes the form commonly described in textbooks as characteristic of the whole class. The foregut is differentiated into: a) a well-developed muscular pharynx immediately behind the mouth, on which are inserted extrinsic dilator muscles arising from the head capsule and tentorium. It opens to b) a narrow oesophagus, which passes between the nerve ring and widens to form a capacious crop. This leads to c) a powerful gizzard lined with six well-developed teeth and pads of bristles that can completely close the gut lumen and act as a valve; behind this the foregut is invaginated into the lumen of the midgut to form a proventricular valve. The midgut is a simple tube bearing a number of caeca at its anterior end. These caeca are simple extensions of the midgut epithelium and like it are lined by a peritrophic membrane, which is secreted from the surface of the midgut epithelium. The columnar cells of the midgut have regenerative cells between their bases, and may be differentiated to form enzyme-secreting and absorptive cells.

Posteriorly, the midgut opens into the hindgut. The anterior region of the hindgut forms the ileum, which leads to a wider tube, the colon, which opens to a rectal sac. The rectal sac contains six rectal pads, which consist of columnar binucleate cells and are organs of water absorption. A short tube, the rectum, opens from the rectal sac and leads to the anus. At the midgut-hindgut junction are numerous malpighian tubules that excrete nitrogenous waste, which forms uric acid in the lumen of the tubules.

8.5.6. Trilobita

In several species of trilobites the fossil imprints have yielded information about the gut structure in this group of animals. In *Triarthrus* (Cisne, 1973) the mouth, which is directed posteriorly, opens to an oesophagus, which passes forward and then curves dorsally and posteriorly where it expands to form a "stomach." This appears comparable with the gizzard of other arthropod guts, since it possesses a complex musculature suitable for food trituration. No teeth have been found associated with it. Dilator muscles are associated with the oesophagus. Posteriorly the gizzard opens to a wide tube. Lateral to the tube are imprints of a large ramified gland, though the large ducts have not in this species been shown to connect with the gut. There is no indication that the caeca have the form of segmental masses. The hindgut appears to be a simple tube, about the same diameter as the oesophagus, leading back to the anus.

In other trilobites, e.g., *Cryptolithus*, caeca are known to connect with the gut. In a re-examination of *Burgessia bella*, Hughes (1973) reports that the gut shows paired lateral caeca which branch to give bi- or trilobed caeca.

Evidence then clearly supports the relationship of the trilobite gut to the caecate type associated with Chelicerata and Crustacea and not with Uniramia.

8.5.7. Tardigradia

The gut in *Macrobiotus hufelandi* (Cuénot, 1949) consists of a simple foregut passing directly posteriorly from an anterior-placed mouth and leading to a broad tubular midgut. The foregut about halfway along its length is modified to produce a sucking bulb, the pharyngeal bulb. The midgut is a simple tube lacking caeca; it opens to a short hindgut leading to the anus. At the junction of the midgut and hindgut are three large glands thought to be excretory in function and often referred to as malpighian tubules.

8.5.8. Pycnogonida

The mouth, which is situated at the tip of the proboscis, leads to a muscular pharynx, which opens to a thinner oesophagus whose epithelium becomes multi-layered as it approaches the midgut. In this region the cuticular lining of the foregut thins and disappears. Posteriorly the oesophagus invaginates into the lumen of the midgut. While this region appears to be equivalent to the proventricular valve of other arthropods, the absence of a cuticular lining raises the question of whether this really is part of the foregut or not. On strict criteria this multilayered noncuticulate region ought to be considered midgut. The midgut gives rise to a number of caeca, which are long and pass to each appendage. They may extend the entire length of each limb. The lumen of each caecum narrows toward its tip leading to a solid cord of cells. Their structure is basically similar to that of the midgut. Digestion is intracellular: there is no trace of a peritrophic membrane. The midgut opens to the hindgut through a valve similar to that guarding the foregut and the midgut. The hindgut is short. It is possible that cells at its anterior end have an excretory function: this type of cell is absent from the hind end. Muscles arising from the hindgut and attached to the anal appendage function in opening the anus.

8.5.9. Pentastomida

In the Pentastomida the mouth opens to a pharynx having extrinsic dilator muscles on its dorsal side; from this a narrow oesophagus passes back to open into a long tubular midgut, lacking a peritrophic membrane. A narrow hindgut passes back to the anus. There are no malpighian tubules.

8.5.10. Discussion on the Gut

A simple tubular noncaecate midgut is to be found in the Tardigradia, Pentastoma, Onychophora, Myriapoda, and Insecta. The presence of caeca at the anterior end of the midgut in some insects would appear to diminish the importance of the presence or absence of caeca as a phylogenetic character. However, like many features in arthropod comparative anatomy, their presence in insects is a low-grade adaptive feature not comparable with the phylogenetic differences at the grade 1 level. In the apterygote orders Protura and Collembola the foregut is short and the midgut long. In the order Thysanura, in *Machilis* and particularly in *Lepisma*, the foregut has increased greatly in length and the midgut proportionally shortened. This shortening, if uncompensated, would greatly reduce the surface area of the midgut epithelium relative to other organs and tissues. Compensation for this comes from the development of caeca, which form a ring around the anterior end of the midgut. This ring of caeca is confined to the Thysanura and the primitive pterygote orders: in higher orders the gut loses its caeca and returns to the simpler tubular type. In certain special instances caeca may again appear but often not in the same region, and they are clearly specializations developed within the order for particular methods of feeding or for the accommodation of microorganisms.

The caecate midgut is found in the Tribolita, Chelicerata, Pycnogonida, and Crustacea. Its condition in the Crustacea, however, is rather different from that in the other classes.

In the Chelicerata caeca arise from along the length of the midgut, while in Crustacea they are confined to the anterior end; only in the Malacostraca and particularly the Decapoda do caeca arise from the posterior end and never from along the length of the midgut.

In the Crustacea the caeca are, in primitive forms, simple extensions of the midgut epithelium: only in the more advanced Malacostraca do the anterior caeca ramify to form the fine tubules and alveoli of the hepatopancreas. This complex organ is clearly originated within the Crustacea and is not characteristic of the group. Caeca have been described in the Trilobita; in *Burgessia*, Hughes (1973) stated that they appear to be bi- or trilobed; in *Triarthrus*, Cisne (1973) stated that imprints of a large ramifying gland have been found in the fossils. No ducts connected with the midgut have been found, and the gland may not be an early form of hepatopancreas.

In the Chelicerata, *Limulus* and scorpions both show caeca that form tubules and alveoli of a large digestive gland. In the scorpions six pairs of ducts, each with its own ramification of ducts and alveoli, form the digestive gland, while in *Limulus* two pairs of caeca give rise to the digestive gland. Elsewhere in the arachnids both simple and complex caeca may exist or only simple caeca are present. In all cases caeca arise from the whole length of the midgut. It seems likely that the caeca would have had a simple structure in the earliest representative of the group and formed the complex digestive gland in response to the large size of *Limulus*, scorpions, and other arachnids. The simple caeca persist perhaps, in their primitive form, in such animals as *Koenenia* (Palpigradia) where

the midgut is a simple sac with lateral pouches bulging out between the segmental muscles in the opisthosoma. A pair of simple caeca is present in the
prosoma. Elsewhere the simple caeca may be a response to small size, and the
curious arrangement in some Opiliones, where the caeca fuse at their tips to
form loops of midgut, could well repay study in the mechanics of caecate guts.
In the Pycnogonida the midgut caeca are simple and extend into the appendages
almost to their tip. In some of the arachnids the midgut caeca also reach or
enter slightly the bases of the appendages.

The evidence of comparitive anatomy supports the presence of three rather
than two, types of gut in the arthropods: a simple tubular midgut characteristic
of the Uniramia; a midgut with anterior caeca characteristic of the Crustacea;
and a midgut with caeca originating from the whole of its length correlated with
segmentation in the Chelicerata. Initially caeca were of simple design, the digestive gland of the Chelicerata and the hepatopancreas of the Crustacea developing
independently in response to size increase by animals in these groups. Secondarily, simple caeca may occur, and there is need to reexamine caeca of all types
to try to establish satisfactory criteria for distinguishing between primary simple
and secondary simple caeca.

The presence or absence of, its function, and the origin of the peritrophic
membrane is another feature of the arthropod gut that requires reexamination.
In the Uniramia it appears to be absent in the Protura and the Collembola, and is
present in the Onychophora, Myriapoda, and the other insect orders. It is absent
in Pentastoma and Tardigradia. The above-mentioned classes in which it is
absent are all small animals, and Protura are fluid feeders, traditionally reasons
for supposing it is absent, since its protective function seems unnecessary. The
Collembola, however, are not fluid feeders, and the midgut cells in Protura as
figured by Berlese (1909) have inner borders that do not appear compatible with
membrane production and differ widely from the typical brush border of the
Collembola and other groups. It seems possible that the peritrophic membrane
may have been absent from the ancestors of these animals and has developed
within the group of noncaecate guts.

The peritrophic membrane appears to be completely absent from the
Chelicerata. In the arachnids, all of which are fluid feeders, this is again to be
expected, based on the protective functions being unnecessary in fluid-feeding
animals. However, its absence from the Xiphosura is surprising, since here its
protective functions would appear to be required. The protective functions in
the Xiphosura gut are performed by mucoid-like material secreted from the
midgut epithelium, a circumstance surely more primitive than the production of
a peritrophic membrane. It should be noted that intracellular digestion has been
described as taking place in the digestive gland. The uptake of particulate matter
by the midgut cells would be incompatible with the presence of a peritrophic
membrane. It seems not improbable that a peritrophic membrane is absent from

the Chelicerata, that it was never developed in them, and its primary absence is another distinctive feature of the guts of these animals.

A peritrophic membrane is present in the guts of some species of Crustacea. On the present information it seems impossible to judge whether it is widespread and characteristic of the group, or if it is a special adaptation developed in a few isolated cases. It must be remembered that mucus-like substances are to be found in the guts of Crustacea and like similar substances in the Chelicerata must be judged to be a more primitive form of protection than the development of a peritrophic membrane.

The functions of a peritrophic membrane are more than just protection of the gut epithelium from abrasion by particles within the food, and may have developed for different reasons in the different groups. In Onychophora the protective function would appear to be of relatively small importance, since the feeding habits of the animal may well exclude much abrasive material from entering the midgut. Lawrence (1953) has called attention to the possibility of fungal infection of the animal via the midgut as a result of the large number of spores in the forest floor habitats in which the animals live. Their method of feeding would perhaps help to exclude these, but they are likely to occur on the body surfaces of the animal to which they apply their funnel-like lips. The frequent expulsion of the membrane and its contents and the resulting sterility of the gut may be the major force in its development in these animals.

A number of scattered observations in insects and Crustacea indicate that there is considerable movement of fluid along the length of the gut between the peritrophic membrane and the gut epithelium. In a number of cases too the enzyme-producing cells appear to be concentrated in the posterior region of the midgut and the absorptive cells in the anterior region, the opposite way round to that which might be expected. Remember that in many insects the midgut is relatively short. These observations put together suggest that something akin to a counter-current situation can occur in the gut which has developed a peritrophic membrane. The backflow of food in the gut can be against a forward flow of enzymes and products of digestion produced by secretion posteriorly and absorption anteriorly. By analogy with other counter-current systems this would enable processes to be carried out efficiently in a shorter length of gut than is otherwise possible. The direction of flow is set by the direction of food passage in the gut. Such a system would not be very useful in a fluid-feeding animal, though in some circumstances its retention in fluid feeders could be explained on this hypothesis.

Some such circumstances as these may be the reason for the development of the peritrophic membrane in some Crustacea, where intake of water through the rectum and reverse peristalsis have been observed in forms like copepods that lack a peritrophic membrane. A counter-current situation could also be a factor favoring its development in the shorter midgut of insects. It hardly seems likely that this could be an important factor in such long gut forms as the myriapoda, Symphyla, Chilopoda and Diplopoda, all of which possess a peritrophic membrane; perhaps the sterility and protective functions are dominant here.

Clearly much more work needs to be done on the peritrophic membrane. It should be remembered that in many instances where the peritrophic membrane is considered to be absent, this is because it has not been stated to be present rather than that it has been looked for an found not to be present. Tentatively it may be considered that the peritrophic membrane developed within the arthropods. It need not have been a feature of their common ancestor, and it may have evolved in response to different needs in the Uniramia and the Crustacea.

Malpighian tubules are present in the Uniramia and in the Arachnida. They are absent in the Merostomata, Crustacea, Tardigradia, Pentastoma, and Pycnogonida. Their independent origin in the Uniramia and Arachnida is certain.

Malpighian tubules are present in many of the larger arachnids, where they are present as long tubes, often much branched, arising from the posterior end of the midgut. They do not appear to be modified midgut caeca of the lateral series, but new structures. Evidence suggests that primitively the waste end products of nitrogenous metabolism were excreted from the midgut, and, as the malpighian tubules have been shown to be important in the excretion of nitrogenous waste in the form of guanine in arachnids, it does not seem difficult to visualize the malpighian tubules as specialized organs of nitrogenous excretion, concentrating into a suitable region a primitively more generalized property.

In the Uniramia the matter is not quite so simple. Malpighian tubules are absent from the Onychophora and in typical form from the Protura and Collembola. They are also formed from the ectoderm of the hindgut, though this is questioned by Hensen (1932). In Protura there are six simple bicellular papillae, both cells apparently opening to a duct. These papillae are formed from a collar of cells surrounding the hindgut just behind the midgut-hindgut junction. In the Diplura short tubes are present in this position, and in the Thysanura and Pterygota well-developed malpighian tubules occur. These cannot be seen therefore as a simple concentration of a generalized function of an epithelial sheet into a specialized organ, but a new structure to which a function of another organ is to become transferred. The development of the malpighian tubules can be seen to be consistent with the development of the counter-current situation mentioned above. If this developed, the forward flow of material from the hind to the fore region of the midgut would be incompatible with the backward flow of fluid from the mid- to the hindgut. The malpighian tubules may have developed primarily to irrigate the hindgut, whose increasing differentiation indicates its importance. Also correlated with this is the development of water absorption (rectal pads) at the hind end of the hindgut. The small size of the papillae in Protura and the short tubes in Diplura would be adequate for a small amount of irrigation, but could scarcely contribute in any significant way to nitrogen excretion at this time. Once a water circulation system had been established of the type found in present-day insects, transfer of nitrogen excretion from the midgut to the malpighian tubules could readily follow. Malpighian tubules occur in Symphyla, Pauropoda, Diplopoda, and Chilopoda. A single pair is present; they lack muscles in the Chilopoda, and, like the insectan malpighian tubules, produce uric acid. If the malpighian tubules originated in the insects as shown by the series in the Apterygota, then the myriapodan malpighian tubules must represent an independent evolutionary line. The six papillae present in Protura contrast with the two malpighian tubules in Myriapoda and serve to emphasize this possibility. Much more needs to be known about the physiology of the myriapods before useful speculation can be made about the forces involved in the origin of the malpighian tubules.

The transfer of the excretion of nitrogenous waste away from the gut presumably has the advantage of limiting the functions of the gut epithellium with consequent increase in efficiency of the remaining functions. In arthropods that lack malpighian tubules excretion may still largely be through the gut, or partially transferred to other organs such as the coelomic organs or the gills or, indeed, in aquatic animals, almost any surface through which it can be lost. Malpighian tubules are active excretory organs: both water and wastes are passed through by active transfer; they are then superior to organs such as the coelomic organs where the prime process is filtration followed by active resorption of wanted materials from the lumen of the organ. In this system all the haemocoelic fluid has to pass through the organ in the course of time with consequent risk of loss. In the malpighian tubules wastes and minimum water are secreted through the tubules; most of the haemocoelic fluid does not need to leave the haemocoele to be "cleaned." No wonder malpighian tubules are selectively favored in terrestrial environments.

8.6. RESPIRATORY SYSTEM

8.6.1. Introduction

Studies on the respiratory system will be confined to those structures adapted to facilitate the passage of oxygen or air from the environment across the body surface. In many animals these organs are protected in pouches, and in aquatic animals, irrigated by water currents produced by associated appendages. These accessory respiratory structures, while essential for the proper functioning of the system, require for their full evaluation a much broader consideration than can be given in an article dealing with internal anatomy.

Many arthropods of moderate or small size lack specialized respiratory structures, diffusion of gases across their body surfaces being adequate for their requirements. Even when specialized organs are present, a proportion of the gaseous exchange still takes place through the general body surface.

As a preliminary for further discussion a brief account of the respiratory organs found in the different groups will be given.

8.6.2. Chelicerata

8.6.2.1. Merostomata

In the xiphosuran, *Limulus*, the respiratory organs occur on the posterior wall of the platelike appendages of the five posterior segments of the mesosoma. On each appendage the posterior integument is folded into a large number, some 1,500 thin-walled lamellae projecting from the surface. These lamellae lie parallel to each other resembling the pages of a book, and they are given the characteristic name of book-gills. The appendages are fused in the midline and directed posteriorly so that their posterior surface is directed dorsally towards the ventral wall of the animal. The beating of these appendages causes a current of water to flow over the book-gills. The velocity and volume of this current are controlled by the shape of the body-wall of the ventral cavity in which these appendages lie. The coxa of the last pair of legs bears a short spatulate flabellum that is said to be used in cleaning the gills and in sensing the oxygen content of the water current.

In the Eurypterida, Moore (1941) has reexamined the evidence for the presence of respiratory organs in the genus *Slimonia* and concluded that the gills are not borne on the genital operculum or the mesosomal plates, but are represented by a highly vascularized area on the ventral body wall. Waterston (1973), in his recent studies on the lower Devonian *Stylonurus scoticus*, has described a number of gill tracts, highly vascularized areas on the ventral body wall posterior to the base of the gill pouches formed by the appendages of the first, second, third, and fourth mesosomal segments. It is thought that each gill tract was irrigated independently of the others by movement of its associated appendages—a less well-organized system than the integrated movements of the mesosomal appendages of *Limulus*.

8.6.2.2. Arachnida

The respiratory organs of the arachnids may be grouped under two headings: the book-lungs and the tracheae clearly derived from them, and tracheae whose origin does not appear to be from a book-lung.

The Book-lungs. The respiratory organs in the scorpions are contained in paired sclerotized pockets on the ventral wall of the second, third, fourth, and fifth segments of the mesosoma. The anterior dorsal wall of the pocket is folded to form a large number of thin-walled lamellae, lying parallel with one another like the pages of a book. Thin cuticular processes project from the surfaces of the lamellae facing each other. These keep the lamellae apart so that air can circulate freely between them. These lamellae are very similar to those of the book-gills, and these organs are often referred to as book-lungs. The pocket opens through a slitlike spiracle to the exterior. Muscles attached to the dorsal side of the pocket can dilate the atrial chamber, drawing air into it; when the muscles relax, air is expelled as the atrium returns to its normal size.

Lankester (1881) demonstrated the homology between the book-gills of *Limulus* and the book-lungs of scorpions. Embryological studies have shown that the book-lungs of scorpions originate from the posterior basal surface of the developing limb. The book-lung has moved away from the limb base during later development to take up its present position in the adult animal. Størmer (1944, 1963) concluded that during the evolution of the Chelicerata there was a gradual transition of these respiratory structures from the limb base to the ventral surface of the animal. The Eurypterida represent an intermediate stage in this process, where the respiratory organs have become simplified and moved on to the ventral surface to form gill tracts. In this respect the scorpions are closer to the Eurypterida than they are to the Xiphosura, and indeed some of the Devonian and Carboniferous scorpions are thought to have led an aquatic life.

On functional grounds this simplification of book-lungs to form ventral respiratory tracts in animals as large as the Eurypterida seems inefficient and unlikely. It would be more in keeping if the vascularized ventral tracts were the primitive respiratory organs and the book-gills and book-lungs developed from them.

Book-lungs are also present in the Amblypygida and the Uropygida on the third and fourth segments of the opisthosoma (mesosoma) and on similar segments in the Araneae in the listomorph and mygalomorph spiders. In other spiders they occur only on the third segment, being replaced on the fourth segment by tracheae. In two small groups of spiders both book-lungs are replaced by tracheae.

In the Araneae the spiracles lead to a small atrial cavity from which four primary tracheal tubes extend anteriorly: in some cases the tubes do not extend beyond the opisthosoma; in others they penetrate as far as the anterior region of the prosoma. The spiracles may migrate posteriorly and fuse across the midventral line to form a simple opening lying just in front of the spinnerets. In the family Caponiidae the anterior pair of book-lungs is also converted to tracheae and respiration is entirely by this means.

It should be noted that the largest spiders have retained their book-lungs, and tracheae have only completely replaced them in two families of very small spiders.

In a number of arachnid orders book-lungs are entirely absent, and tracheae are the only respiratory organs present. In pseudoscorpions two pairs of spiracles are present leading to tracheae. One pair opens onto the ventral side of the third segment, and the other between the third and fourth segments. Small protrusions from the body wall that occur during development in association with these spiracles are considered to be rudimentary limb buds. In Opiliones there is a single pair of spiracles on the ventral side of the second segment, which leads to a well-developed system of tracheae. In the Solifugae a very well-developed tracheal system is present, which opens to the exterior by three pairs of spiracles. Two pairs of these, on the third and fourth segments, might be equated with those in pseudoscorpions, but the third pair opens on the prosoma just behind the second pair of legs and does not correspond with any similarly placed pair in the arachnids. In the Ricinulei there is a single pair of spiracles situated on the posterior lateral angles of the prosoma. It leads to an atrial cavity from which a very large number of minute sclerotized tubes conduct air to the organs of the prosoma. In the Acari, a well-developed tracheal system is present in many forms. The tracheae open to the exterior through one to four pairs of spiracles situated on different parts of the anterior region of the body. According to Ripper (1931) there is a reasonable similarity between the tracheae of the Ricinulei and the Acari, and both differ from those of other arachnids.

Compared with our knowledge of the tracheae and tracheoles of insects very little is known about the tracheal system in the Arachnida. There is much need for work on this system in noninsectan groups, and one is tempted to quote from Richards (1951): "Accordingly, in determining any tube a trachea it is desirable to have some evidence that the tube does function in respiration."

8.6.3. Crustacea

The available evidence strongly indicates that respiration in primitive Crustacea was accomplished by gaseous exchange across the surface of the body and its appendages. In the primitive Cephalocarida no special respiratory structures have been identified, although the large, flattened pseudo-epipodite at the base of each limb could function as a gill. True epipodites, which are also flattened plates, occur on the limbs of many Crustacea and are often referred to as gills. Where they have a large surface area, thin cuticle, and well-developed blood supply, gaseous exchange may well be their most important function. Evidence that they are specialized regions in this respect is usually lacking. In any case, even where specialized respiratory structures do occur, as in the isopods *Ligia* and *Oniscus*, 50% of the normal respiration still takes place through the general body surface.

In the different groups of Crustacea different regions of the body have become specialized for respiratory exchange.

The carapace is present in many Crustacea and is generally considered as characteristic of the group; its appearance is seen as one of the initial steps in their evolution (Siewing, 1960). However, it is often absent or poorly developed in primitive groups such as the Cephalocarida, the Mystacocarida, and the Notostraca, which, on other grounds, are considered primitive. It may have developed independently in the various Crustacean lines that possess it. The outer wall of the carapace is usually covered with a relatively thick integument, but its inner wall is thin and acts as a gill in the Malacostraca, Myscidacea, Tanaidacea, and Decapoda. The carapace is well vascularized, and well-directed respiratory currents of water are directed over its inner surface.

In the Branchiura it is possible that a restricted dorsal area of the shell-fields is a specialized respiratory region. In the Cirripedia the mantle is thought to be the main site of gaseous exchange. Most other groups lack identifiable specialized respiratory organs except for the Malacostraca, where a well-defined set of gills is present.

In the primitive Malacostraca such as *Nebalia* and *Anaspides* the gills are epipodites of simple platelike form occurring on the thoracic limbs as they do elsewhere in the Crustacea. In the stomatopod, *Squilla*, epipodites are present on the thoracic limbs, but here large branching gills develop on the exopodites of the pleopods. Small hooks on the endopodites of these pleopods join each pair together in the midline, a mechanism reminiscent of that found in *Limulus* but clearly of independent phyletic origin. In the Peracarida the lophograstrid myscids have gills developed as branched and foliaceous epipodites on the thoracic limbs. In the Cumacea they are filamentous, and they are lacking altogether in the Tanaidacea. In the Isopoda the epipodites are lacking, and the abdominal pleopods are modified for respiration. Here the exopodite and endopodite of each pleopod is flattened to form a large lamella, which comes to lie against the ventral surface of the abdomen. In the Oniscoidea the exopodite forms an operculum, covering the respiratory endopodite. In the Porcellionidae and the Armadillididae the surface of the exopodite may invaginate to form an atrial cavity from which finger-like processes branch throughout the exopodite base. These are the pseudotracheae, which are specialized areas of respiratory exchange. They are quite different from the tracheae of the Chelicerata and the Uniramia.

In the Amphipoda simple platelike epipodites on the thoracic limbs act as gills.

In the Eucarida, the Euphausiacea have lost the platelike form of their thoracic epipodites, which have become larger and filamentous and are irrigated by a respiratory current generated by the thoracic limbs.

It is in the Decapoda that gills have reached their maximum development. Primitively four gills were thought to be present on each thoracic limb: a) the pleurobranch, which arises from the thoracic wall just above the limb base;

b) two arthrobranchs, which arise from the membrane between the body wall and the limb base; c) the podobranch, which arises from the dorsal wall of the coxa and is held to be part of the epipodite of other Crustacea; the other part, called the mastigobranch, does not function as a gill but serves to clean them.

Not all these gills are present in any one decapod. Of the 32 possible gills this series proposes, the maximum number found is 24 in the peneid shrimp, *Benthesicymus*. Elsewhere in the Decapoda smaller numbers are present. In addition to their numbers, gills themselves show adaptations to respiratory exchange. In the peneid shrimps the gills have a central axis that bears biserially arranged main branches which are themselves subbranched. The phyllo-branchiate gills, consisting of a central axis bearing flattened platelike branches, are found in many Caridea, Anomura, and Brachyura. A third type, the trichobranchiate gill, consisting of many unbranched filamentous projections, is found in many Macura, some Anomurans, and primitive Brachyura. Gills may be reduced or altogether absent, and respiration may be entirely through the inner surface of the carapace in some terrestrial brachyurans.

8.6.4. Uniramia

The primary respiratory organs of the Uniramia are the tracheoles and the tracheae. They are almost universally present throughout the group. Our knowledge of the tracheolar and tracheal function within the group is completely dominated by our studies on insects. Hardly any studies on function have been published in other groups, and very often anatomical distinctions between tracheoles and tracheae, so clearly drawn in insects, have not been recorded. Hence the precise nature of the structure being considered is often uncertain. As in the arachnids there is need for a great deal of work to be done on noninsectan Uniramia.

8.6.5.1. Onychophora

In the onychophorans the body surface is covered with a large number, up to 75 per segment, of spiracles which lead to a shallow atrial cavity that penetrates the body wall. Closing mechanisms are absent. From the bottom of the atrium a number of fine tubules arise. Each tubule opens into the atrium independently of the others. These tubules have a fine cuticular lining containing chitin that shows fine transverse striations. Some of these tubes can be very long and conduct air to the cells and tissues of the body. Often, where they leave the atrium, they remain together as a thick bundle for a little way before separating to go to their final destinations. Very little seems to be known about their physiology. Nothing seems to be known about the precursors of these tubes. In development they have not appeared by the time the animal is born. Kästner (1968) stated that they first appear in the dorsal region of the animal and shift ventrally during postembryonic growth.

8.6.5.2. Myriapoda

The majority of the Pauropoda lack special respiratory organs. Tracheae have so far only been described in the order Hexamerocerata. Where present the spiracles are situated on the coxae of the walking legs and open to a pair of very short tracheae. The tracheae from the spiracle of the first pair of walking legs supply the tissues and organs of the head. This spiracle is considerably larger than the others.

In the Symphyla there is a single pair of spiracles situated on the head just above the base of the mandibles. The tracheae arising from it supply the head musculature and the nervous system, and a few fine branches pass posteriorly to the most anterior trunk segments.

In the Diplopoda there is a pair of spiracles per trunk segment, two pairs per diplo-segment. They are situated on the sternites anterior and lateral to each pair of walking legs. The spiracles can be opened or closed by a valvular mechanism. The spiracles open to an atrial cavity, which is thin-walled in *Polyzonium* and in some Colobgnatha. In all other Diplopoda the atrial walls are heavily sclerotized and function both as an atrium and as a hollow apophysis for muscle attachment. Bundles of slender tracheae arise from the atrium; usually they are unbranched, do not anastomose, and do not taper, but anastomosis has been recorded in the vicinity of the spinning glands in the Nematomorpha, and branching occurs in the Pselaphognatha. The tracheae are very long in *Glomeris*. In many diplopods they are absent from the sclerotized atria of the first three trunk segments.

In the Chilopoda, except in the Scutigeromorpha, spiracles are situated in the pleural region of the body segments, usually on separate sclerotized plates called stigmatopleurites. They may be present on all the trunk segments, but in *Lithobius* occur only on segments 3, 5, 8, 10, 12, and 14; other patterns may occur in other centipedes. Spiracles at the anterior end of the body are usually larger than those at the posterior end; this is correlated with the more bulky musculature in the anterior region (Manton, 1965). Each spiracle leads to an atrium, which is shaped like a flattened funnel tapering inward. Its wall bears a number of fine cuticular processes, the trichomes, which may function to keep the atrial lumen open during the movements of the animal, or in certain instances enable it to act as a physical gill (Manton, 1965). Muscles associated with the stigmatopleurites do not seem to be part of a closing mechanism but act as part of the body musculature. They may facilitate the ventilation of the atrial cavity.

The tracheae arising from the atrium go to the tissues and organs in their im-

mediate vicinity. No connections between tracheae arising from separate spiracles occur in the Lithobomorpha. If then a spiracular opening were blocked, the
area it supplies would suffer oxygen deprivation, making the role of the trichomes in keeping the atrial cavity open especially important in these forms
(Currey, 1974; Manton, 1965). In *Cryptops* vesicular swellings occur at the
branching points of the tracheae, and transverse connections have become established between tracheae in the same segment. In *Scolopendra* longitudinal
tracheae have developed; in Geophilomorpha longitudinal tracheae extend the
whole length of the body.

Among the Chilopoda, the Scutigeromorpha differ greatly from all other
Myriapoda in their respiratory arrangements. In *Scutigera* the spiracles are slits
opening in the middorsal line on the posterior tergites on all the trunk segments
that have long tergal plates. Each spiracle leads to an atrium, from each side of
which a large number (600) of fine tubules arise. These tubules may branch
once or twice near their bases and end blindly to form a compact kidney-shaped
organ. Air is drawn into the organ, though whether through pressure created by
the heartbeat or by functioning in a manner similar to the tracheoles of insects is
unknown. This type of tracheal organ occurs sporadically in other arthropods
and may be referred to as a tracheal lung.

8.6.5.3. Insecta

8.6.5.3.1. Apterygota.   Among the Apterygota the tracheal system exhibits
various degrees of development.

In the Protura it is absent in the Acerentomidae but present in the Eosentomidae. In *Eosentomon* two pairs of spiracles are present, an anterior pair on
the mesothorax and a posterior pair on the metathorax. Tracheae arising from
the anterior pair supply the mesothorax, prothorax, and head, and those from
the posterior pair, the metathorax and abdominal segments, the latter by a pair
of long wavy tracheae that lie laterally. There is no anastomosis between the
anterior and posterior, or between the left- and right-hand trachea. There is no
indication of any abdominal spiracles.

In the Collembola the majority of species lack tracheae, and respiration is
through the cuticle. Tracheae are present in the species of the family Sminthuridae and those of the rather distantly related family Entomobryidae. In both
cases there is a single pair of spiracles situated between the head and the prothorax. Tracheae arising from the spiracles supply cells and tissues in the head,
thorax, and abdomen. No anastomosis takes place between tracheae from opposite sides of the body, but a major trachea arising from each spiracle supplies
the contralateral prothoracic leg, which does not receive ipsilateral tracheae.

In the Diplura there is great variation in the tracheal system within the order.
In *Campodea* there are three pairs of spiracles: an anterior pair on the anterior

region of the mesothorax, from which tracheae go to the prothorax and the head; a posterior pair on the mesothorax, which supply that segment; and a posterior pair on the metathorax, which supply that segment and send long branches back to the posterior of the abdomen. Although there is some overlap of the tracheal fields, no anastomosis occurs between tracheae from the different spiracles, either longitudinally or transversely. In Japygidae, in *Anajapyx* there are nine pairs of spiracles, two pairs on the thorax and seven pairs on the abdomen; in *Projapyx* there are ten pairs, three on the thorax and seven on the abdomen; in *Japyx solifugus* there are eleven pairs, four on the thorax and seven on the abdomen. The spiracles open to tracheae that are connected longitudinally, but transverse connections are few, and often only one occurs at the posterior end of the abdomen. In *J. solifugus* the four pairs of spiracles on the thorax consist of two pairs on the mesothorax and two pairs on the metathorax. The third pair lying anteriorly on the metathorax is in addition to the three pairs found in *Campodea*. In *Japyx isabella*, *Projapyx*, and *Anajapyx*, there are only two pairs on the thorax, which correspond with the anterior pair on the mesothorax in *Campodea* and *Japyx solifugus*, and a pair on the mesothorax, which corresponds with the third pair found in *J. solifugus* and thus represents a different pair from that on the mesothorax of *Campodea*.

In the Thysanura there are nine pairs of spiracles in the Machilidae, one pair on the mesothorax, one pair on the metathorax, and seven pairs on the second to eighth abdominal segments. There are no spiracles on the first abdominal segment. The first pair is located either on the anterior border or in the membrane between the pro- and mesothorax, the second pair on the anterior border of the metathorax. In position, therefore, they correspond to the positions of the first and third spiracles of *Japyx solifugus* and with other Japygidae.

In the Lepismidae there are ten pairs of spiracles, one pair on the mesothorax, one pair on the metathorax, and eight pairs on the first eight abdominal segment.

In *Machilis* the trachea associated with each spiracle forms an independent system with no anastomosis with its neighbors. The trachea arising from the mesothorax spiracle supplies the head, prothorax, mesothorax, and the metathoracic leg and the first abdominal segment (Oudemans, 1888). The metathoracic spiracle supplies principally the metathoracic leg. Elsewhere where anastomosis is absent, the tracheae arising from particular spiracles are rarely limited in their internal distribution by segmental boundaries. In the Lepismidae anastomosis between tracheae occurs. Longitudinal trunks are well developed, and transverse links are to be found welding the otherwise isolated tracheae into a single system.

8.6.5.3.2. Pterygota.   In the Pterygota there is much variation in the tracheal system and considerable development and adaptation in the different groups for various purposes. In the more primitive orders there are ten pairs of spiracles, two on the thorax and eight on the abdomen. The first pair is usually in the membrane between the pro- and mesothorax, the second pair on the mesothorax, usually in an anterior position, and the remainder are in the first eight abdominal segments. They thus correspond with the arrangement found in the Lepismidae. The spiracles open to a tracheal system in which there are numerous longitudinal and transverse anastomoses welding the whole into a complex efficient respiratory system.

In development the tracheae arising from each spiracle are at first independent from those arising from the other tracheae. Anastomoses develop as the animal grows.

The tracheal system in pterygote insects consists of two quite distinct organs, the tracheoles and the tracheae. The tracheoles are intercellular tubules of small, usually 2 μm, diameter. When at rest they are filled with fluid. When the cells they supply are active, the increased osmotic pressure produced by the breakdown of metabolites causes fluid to be withdrawn and air sucked up to the region of the active cell. When metabolic activity is less, and osmotic forces are reduced, fluid reenters the tracheole under the forces of capillarity, and air is excluded from the "resting" cell. Thus a very efficient system of demand and supply of air to the appropriate cells exists. There has to be a balance between the forces due to changes in osmotic pressure produced by metabolic change and the forces of capillarity, which depend upon a number of features, the most important of which is the diameter of the tracheole tube.

Tracheae are intercellular tubes usually of relatively large diameter compared with the tracheoles. Tracheae are lined with cuticle; a wax layer is present, at least in the larger branches, and, except during embryonic development and in aquatic insects sometimes at moulting, they do not contain fluid. They are tubes held open by a spiral lining, the taenidea, and they conduct air to the tissues. A circulation of air through the major trunks is usually developed in the larger pterygote insects.

In most accounts of the tracheal system in apterygote insects and in the Myriapoda the presence of tracheoles is not remarked upon, and their presence or absence in these cases is uncertain.

8.6.5. Other Groups

Special respiratory organs are absent in the Pycnogonida, Tardigradia, and Pentastoma.

8.6.6. General Discussion of the Respiratory System

It is convenient to open the discussion on the respiratory system by considering the arthropods group by group.

8.6.6.1. Chelicerata

The book-gills of the Xiphosura such as *Limulus* are usually cited as the arche-type of the primitive respiratory system in the Chelicerata, largely on the basis that it is an aquatic respiratory system, that its evolution to the book-lungs of the scorpions has been demonstrated, and it originates on the limbs and migrates to the ventral body wall in scorpions. Recent work in Eurypterida on *Slimonia* and *Stylonurus* shows the presence of a much more primitive type of respiratory organ, but on the ventral body wall, not on the base of any of the limbs. This organ appears to have been an area of integument permeated by a plexus of canals and channels, lying in a position beneath the ventrolateral blood sinuses of the animal. The acceptance of the structures as respiratory organs depends upon their position in the same region and segments as the book-lungs of scor-pions. They would appear strangely inadequate for animals of this size, even if they were extremely sluggish. By the same token it would be strange if, after developing book-gills, these large and apparently fairly active predators were to return to a system of relatively small surfaces rather than develop the book-lung type of respiratory organs evolved elsewhere. It is possible that other integu-mental areas were vascularized, and systems such as the circulation of blood through the marginal artery lying close to the dorsal prosomal carapace, such as is found in *Limulus*, may also have been present and supplemented the ventral respiratory areas. The respiratory organs in Eurypterida indicate a type from which book-lungs could evolve, and it may well be that their position on the ventral surface is also primitive and that their migration to the limb base in *Limulus* and subsequent return in arachnids are secondary adaptive maneuvers.

In any case the respiratory organs develop on the ventral surface of the animal or the ventroposterior surface of the limbs, which contrasts with the Crustacea where they develop on the dorsal side of the limb.

Among the Arachnida the book-lungs of the scorpions correspond in position to the respiratory areas of the Eurypterida and, although invaginated into an atrial cavity, are in structure remarkably like the book-gills of the Xiphosura. In embryo-scorpions the first indications of the book-lungs appear on the base of the limbs and then migrate to and become incorporated into the sternal plates. The limbs are rudimentary and disappear in later embryonic development. The book-lungs where they occur in arachnids are homologous with the book-gills, and the transformation of the latter to the former has been clearly demonstrated.

The transformation of book-lungs into tracheae, which here will be called book-tracheae to distinguish them from the tracheae arising from other struc-tures, has occurred in a number of arachnid orders. The process can be followed in the Araneae. The large and active spiders such as the Mygalidae possess only book-lungs. The majority of spiders have a pair of book-lungs and one pair of book-tracheae. Only in two or three families of very small spiders have the book-lungs been entirely replaced by book-tracheae. It seems unlikely that size and activity are dominant forces in bringing about this change, nor is adaptation to terrestrial life a convincing reason. Spiders that have only book-lungs, the scorpions and the Amblypygi, are unquestionably thoroughgoing terrestrial animals, but do not possess book-tracheae. There is no evidence to suggest that water loss from book-lungs is greater than from book-tracheae. Indeed, in view of the mechanisms of tracheolar function the reverse might be expected to be true.

Tracheae occur in a number of other arachnids in which book-lungs are absent, but in positions in which book-lungs normally occur. These too may be regarded as book-tracheae. The Solifugae have a very well-developed tracheal system, which opens to the surface by two pairs of spiracles on the opisthosoma and in the position of the book-tracheae, but also by a third pair of spiracles on the posterior borders of the prosoma. These must be a new structure.

The Palpigradia have no special respiratory organs and pose another example of the problem of whether this is because they are primitively absent or because they have been lost, the small size of the animal making special respiratory structures unnecessary. In the Ricinulei the existing tracheal system cannot be derived from book-tracheae. It must represent an independent development. Ripper (1931) has associated the tracheal system of ticks with that of the Ricinulei, and the opening of the spiracles situated on the anterior body-region of these animals cannot be derived from book-tracheae.

8.6.6.2. Uniramia

The respiratory organs in the Uniramia are tracheoles and tracheae. They are absent from most of the Pauropoda, Protura, and Collembola where, like the Palpigradia, there is confusion between the possibility that the system is primitive and that which claims secondary loss due to small size. They are, however, present in some members of each group, and a close comparative study may resolve the problem in this case.

A comparison of the tracheal system in the Onychophora, Myriapoda, and Insecta may help to clarify some of the stages in the evolution of the tracheal system within the Uniramia.

In the Onychophora the system consists of numerous isolated tracheae scattered over the body surface, some 75 per segment. Each trachea consists of an external aperture, the spiracle, opening to a short multicellular tube, the atrium, that just penetrates the integument and basement membrane. From this a number of tracheoles convey air to the nearby tissues and organs. Tracheolar function comparable to that found in insects does not seem to have been demonstrated in *Peripatus*, but the dimensions of these tubules indicate that they may well function in a similar manner. There are no anastomoses between tracheoles arising from different tracheae. These tracheae could develop into the type found in other Uniramia by the further growth and development of the atrium. This already has the form of an intucked ectodermal tube, and its further growth into the body, carrying with it the tracheoles, is all that is needed to give an organ very similar to that found in other primitive Uniramia.

In all other Uniramia there is only one pair of spiracles per segment or, in *Scutigera*, one spiracle per segment, on those body segments that bear them. There is no indication whether this condition is due to a reduction in number from one similar to that found in Onychophora, or if only a single trachea or a pair of tracheae per segment was the primitive state for these animals.

If numerous spiracles were primitively present, reduction to a single pair would be favored by the decrease in water loss that would accompany it. Although the tracheal system is clearly an adaptation for air breathing, in the first instance its development would have been favored in a very moist environment, since its design encourages water loss, and mechanisms to reduce this appear later in its evolution. Again, if reduction occurred, why was it that these spiracles rather than any others are the ones selected? In *Scutigera* the tracheae form tracheal lungs that oxygenate the haemolymph flowing past them. Situated in the middorsal line close to or in the main blood flow of the heart, they are in a position to ensure maximum oxygenation of the haemolymph immediately prior to its being distributed directly to the organs and tissues through the complex circulatory system that exists in this animal. *Scutigera* apparently lacks a respiratory pigment, and maximum solution of oxygen in the haemolymph needs the type of adaptation indicated here. Where the tracheae develop to conduct oxygen directly to the tissues, the pleural region is the optimal site for the spiracles. The main organs, heart, ventral nerve cord, and gut are about equidistant from the lateral line, and thus a spiracle in this position gives optimal diffusion distance for the pattern of organs its tracheae will supply. This position of the spiracles is found in most Chilopoda and the vast majority of the Insecta. In Symphyla the spiracles are on the coxae and in the Diplopoda on the sternum in a position anterior and dorsal to the limb articulations. These differences probably reflect adaptive requirements in the groups named.

These conditions would apply in selecting the position for spiracles whether they had to operate on a segment with numerous small tracheal trees or bring them forth on an otherwise tracheless animal.

Kästner (1968) remarked that tracheae develop in the middorsal line in *Peripatus* and migrate from there over the rest of the segment. I have not been able to find any confirmation of this, but I regard its possibility as very important and consider it worthwhile to write a few lines on its implications.

In the primitive Uniramia which lack tracheae gaseous exchange is across the body surface, and the oxygen reaches the cells and tissues through physical solution in the haemolymph. The most important site of exchange is likely to be in the middorsal line where there is an adequate flow of haemolymph very close to the body wall. Oxygenation of the haemolymph could be increased in several ways, one of which is the projection of tracheoles into the pericardial cavity. This is essentially the condition found in *Scutigera* where the atrial cavity and tracheal lungs accomplish this very effectively. It could have been the condition in *Peripatus* if Kästner's (1968) comments are well founded. Thus the middorsal position in *Scutigera* is a primitive position. Migration away from the area could occur as the haemolymph plays a less prominent part in respiration and the tracheae conduct air directly to the body tissues, and the system shifts to take up positions giving the shortest diffusion paths for the organs they supply. Migration laterally and ventrally of the spiracles would lead to the positions they now occupy, which can be seen as part of an evolutionary pattern arising from a common process.

A further problem with the tracheal system in Uniramia is the different pat-. terns of segmental distribution found in the different groups. It is not hard to see in the very generalized distribution of tracheae in *Peripatus* a pattern from which all these others could be derived. Their distribution above the mandible, on the neck, and in different combinations of the thorax and abdomen cannot really be taken as evidence of independent origin or of deep divisions between the groups, since within a single genus, *Japyx*, several different patterns occur, and quite major variations occur in the apterygote insects.

8.6.6.3. Crustacea

In the Crustacea it seems inescapable that respiration through the general body surface was adequate in the early primitive members of the group. The presence of the miscellaneous structures that act as gills indicates their independent development within the group.

8.7. CIRCULATORY SYSTEM

8.7.1. Introduction

The circulatory system in arthropods has been examined, named, and interpreted under the conviction that it must be derived from the perfect closed circulatory system of annelids. In order to break free from this restriction I propose that we deal with the circulatory system as it appears in the arthropods, without preconceived notions and starting with the simplest condition, working toward the complex patterns that occur.

The basic structure of the circulatory system present in all arthropods is the haemocoele. In the adult this is a fluid-filled cavity. It functions as a body cavity containing the internal organs, and the fluid within transports materials between the organs and about the body. Embryologically, it is formed from two components, the schizocoele and the coelomic cavities. If it is assumed that a closed circulatory system was present in the arthropod ancestor, then this system of cavities too must be added to those mentioned above.

The haemocoelic space in its simplest form is found in the Tardigradia, where it is a cavity containing fluid, the haemolymph, in which occur a number of cells, the haemocytes and globules of other material. There is no heart or vessels of any sort. Movement of the fluid results from movement of the body wall, or of organs within the haemocoele. Among the myriapods, the order Pauropoda also shows a simple body cavity of this type. Haemocytes are rather fewer and it appears that here, and probably in the Tardigradia also, there is not an epithelial lining to the cavity, the organs being limited by their basement membranes to which a few haemocytes may adhere. In the Pauropoda also there is no trace of a heart or blood vessels.

Both these groups of animals are of small size, and the absence of a circulatory system is usually thought to result from its being redundant in small arthropods. In other arthropods, however, it is present, or when it is absent, the animals are derived from creatures in which its presence has been established. For example, in Acari a heart with one or two pairs of ostia and an anterior and posterior aorta is present in many of the Ixodidae and less well developed in the Argasidae and Mesostigmata: elsewhere in the group it is absent. Simplicity in the circulatory system, as in other organs, can result from its being primitive or being secondarily simplified. That both are correlated with size does not alter this, but only makes the separation of the two much more difficult.

The next step forward would seem to be illustrated by the condition of the circulatory system in the Pycnogonida and the Onychophora.

In the Pycnogonida the haemocoele is divided in two by a horizontal septum that lies above the gut and is attached laterally to the dorsolateral wall. The septum extends into the appendages, nearly reaching the tips of the limbs. The haemocoele is thus divided into a perivisceral sinus and a dorsal pericardial sinus. Communication between the two occurs at the limb tip, by the membrane being incomplete anteriorly and posteriorly, and by spaces at its edges between the points where it is fastened to the body wall.

In the pericardial sinus two other longitudinal septa are present. They originate from the body wall close to the middorsal line and come together where they are inserted into the pericardial septum in the midline. These two septa enclose a V-shaped space. They form the sides of the V, which is closed dorsally by the body wall. The septa are contractile and do not extend to the anterior end of the animal. They and their enclosed space form the heart of the animal. Haemolymph enters the heart through vertical slits in its walls; these form the primitive ostia. No blood vessels are present. The course of circulation is outward in the preivisceral sinus into the dorsal sinus through the spaces in the septum or around the tip of the limbs and inward towards the pericardial sinus. Haemolymph enters the heart through the ostia and is discharged anteriorly.

In the Onychophora the haemocoele is also divided into a dorsal pericardial sinus and a ventral perivisceral sinus by a horizontal membrane, the pericardium. The lateral edges of the pericardium are fused with the musculature of the body wall much nearer the dorsal midline than is the case in Pycogonida. Early in development it arises in a more ventral position. In the adult the pericardium is attached to the body wall in the intersegmental region, and the perivisceral and pericardial sinuses are confluent segmentally, except where the segment is limbless. A long tubular vessel lies in the pericardial sinus, being closely attached ventrally to the pericardial septum. Compared with the heart in Pycnogonida, it is an inverted V, the point of the V being toward the dorsal body wall, with which it has no definite connection. There is no aorta, and there are no arterial vessels. Haemolymph enters the heart through laterally placed segmentally arranged ostia, and through ostia penetrating the pericardium and heart wall, putting the perivisceral and cardiac cavities in direct communication. Haemolymph is pumped forward by the heart, then flows posteriorly and enters the pericardial sinus through the spaces in the pericardial septum.

Lateral muscles arising from near the midventral line and inserted on the dorsolateral region of the body wall occur in each segment in Onychophora and tend to form a pair of longitudinal septa that separate the perivisceral cavity from a pair of lateral cavities in which lie the coelomic organs, and the coxal glands, and which open to the haemocoele of the legs. A ventral septum is absent.

With increasing complexity of the circulatory system it is better to examine this system group by group from this stage onward.

8.7.2. Heart

8.7.2.1. Uniramia

In the Myriapoda other than the Pauropoda there is a well defined pericardial membrane separating a pericardical sinus from a perivisceral one. A tubular heart lies in the pericardial cavity attached ventrally to the pericardium. A pair of ostia opens into the heart for each segment in which the heart lies. The heart may form a tube in which chambers are produced by segmental dilatation of the heart lumen, or it may be of nearly uniform diameter throughout. Primitively there are no values across the lumen of the heart, but in some instances flaps developed in connection with the ostia may meet across the lumen and can act as a valve.

In the Diplopoda anteriorly the heart opens to a short cephalic aorta, which has a valve, the aortic valve, as its base, whose function is to prevent blood flow back to the heart. Anteriorly the aorta widens out to form a blood sinus, which surrounds the foregut and opens ventrally to a sinus separated dorsally from the perivisceral sinus by a ventral septum or membrane. Blood entering the ventral

sinus anteriorly flows posteriorly over the nerve cord, seeping back to the perivisceral and pericardial sinuses through gaps in the septa and spaces between the
organs. A pair of thin-walled vessels, the "lateral arteries," arises from the heart
in each segment, and extends a short distance ventrally to open to the haemocoele in a region approximately lateral to the gut. The direction of blood flow
in these arteries does not seem to have been determined; by analogy with insects it is assumed to be away from the heart.

In the Symphyla the cephalic aorta passes to the head, where it branches; two
branches go to the antennae. Near its origin from the heart a pair of lateral
arteries passes to the sacculus of the maxillary gland, and a ventral artery passes
around the gut, unites with its fellow, and passes posteriorly above the nerve
cord as the supraneural artery.

In Chilopoda the Scutigeromorpha have a well-developed arterial system. In
Scutigera an elongate tubular heart lies in the pericardial sinus. A cephalic aorta
arises from its anterior end and passes forward to the head, where it branches to
supply the antennae and the mandibles. Near the origin of these vessels a pair of
ventral vessels passes around the gut and unites to form a supraneural vessel,
which extends the whole length of the body. A pair of vessels branches off
from the supraneural artery in each segment to supply the limbs; branches also
arise that supply the ventral nerve cord. A pair of ostia opens into the heart
from the pericardial sinus in each segment, and a pair of lateral arteries arises
in each segment and extends to the region of the fatty tissue around the gut.

8.7.2.2. Insecta.   In the Protura a pericardial septum is present above the gut
and attached laterally high up on the body wall of the animal. According to
Berlese (1909) a tubular heart is absent, its place being taken by a troughlike
filament, the pericardial cord. Aubertot (1939) describes a heart having contractile chambers in the mesothorax and seven abdominal segments in *Acerentomon.* No other vessels appear to be present.

In Collembola a short tubular heart of six chambers is present in the Poduridae
and Entomobryidae, the anterior two chambers lying in the meso- and metathorax, the others in the first four abdominal segments. In the Symphypleona
the heart may shorten, and only two chambers are present. A pair of ostia
opens to each chamber. The pericardium is well developed and contains alary
muscles. Anteriorly a cephalic aorta passes forward and opens to a sinus that
surrounds the aorta just behind the brain. No other vessels and no ventral
membrane appear to be present.

In the Diplura a well-developed pericardial membrane and dorsal vessel are
present. The dorsal vessel in *Japyx* has two chambers in the thorax, one in the
meso- and one in the metathorax, and eight chambers in the first eight abdominal
segments. Alary muscles are said to be absent. A cephalic aorta passes forward.

In the Thysanura, the pericardial septum is well developed posteriorly, less so anteriorly. The heart in *Machilis* has 11 chambers, one in the mesothorax, one in the metathorax, and one in each of the first nine abdominal segments. Alary muscles in the pericardial septum are associated with each chamber. In addition to the paired lateral ostia two pairs of ventral ostia occur in the posterior region of the heart opening into the perivisceral cavity. Lateral ostia occur in the chambers bearing these ventral ones. Anteriorly a cephalic aorta passes forward and divides to form two or more cephalic arteries, which may further subdivide. Posteriorly, a posterior vessel continues backward from the heart to form a vessel supplying the medial caudal filaments. Lateral arteries appear to be absent.

In the Pterygota there is much variation in the details of the circulatory system. Its main features appear to have already evolved in the Apterygota. The heart is nearly always confined to the abdomen, and this shortening can be correlated with the development of the flight muscles and extrinsic muscles of the legs. Similarly the pericardial septum is also largely or entirely confined to the abdomen. Chambering of the heart can be very marked, and there is a rather rough correlation between good development of the chambered heart and aquatic habitat of the insect. Primitively, eight chambers are present, a pair of lateral ostia opening to the pericardial sinus in each segment. Ventral ostia are frequently present. In the Orthoptera these have been shown to have an exhalant function. The lateral arteries, one pair per segment, are present in some insects. Their arterial nature has been shown by observations of lateral movement of cells and inclusions away from the heart during systole. These arteries extend about halfway around the segment. Anteriorly, a cephalic aorta is developed and passes forward to the brain. It may end in a sinus just behind the brain, or just beneath it, or in front of it, in the well-know aortic funnel of orthopteroid insects. Posteriorly, the heart is closed except where caudal filaments or gills are present, where a posterior aorta passes back to these structures. In a few insects, e.g., dragonflies, there may be a pair of ostia halfway along the cephalic aorta in the thorax. This may indicated that the aorta in this region is a reduced part of the heart and not the separately developed vessel found elsewhere.

8.7.2.3. Crustacea

The primitive condition of the circulatory system in the Crustacea is best illustrated by reference to the condition in the Branchiopoda. The body cavity is divided by an incomplete transverse septum lying above the gut separating off a dorsal pericardial sinus from a more ventral preivisceral one. In *Artemia* the heart is an elongate tube lying in the pericardial sinus and extending from the first to the penultimate trunk segment. In *Branchipus* a terminal ostium opens the heart, but elsewhere the heart is closed posteriorly. From the anterior end a median cephalic aorta passes forward and opens to the body cavity just behind the brain. Laterally, 18 pairs of ostia open to the heart, one pair in each segment. No other blood vessels have been described. Blood circulation is anterior in the heart and posterior in the perivisceral sinus.

In many of the suborders of the Branchiopoda and the other orders of the Crustacea except the Malacostraca, the circulatory system is reduced. The pattern of reduction is shortening of the heart, reduction and occassionally the disappearance of the pericardial septum, and the disappearance of the blood vessels. The heart may shorten until it is a small capsulated sac whose original segmental design is only indicated by the presence of several pairs of ostia. At the tissue level the saccular heart has a very complex pattern of muscle fibres, quite different from the simple fibre pattern of short tubular primitive hearts. Its presence helps to indicate that these hearts are secondarily reduced.

In the Branchiura a short tubular heart is present. Haemolymph enters the heart through two pairs of lateral ostia and perhaps also from a ventral pair, and leaves the heart through anterior and posterior apertures very similar to the ostia.

In the Copepoda, a saccular heart is present in the Pontellidae and the Calanidae, but in general a heart is absent in most copepods, and haemolymph circulation depends upon regular oscillations of the gut. Similarly in the Ostracoda a saccular heart is present in the species of *Myodocopa*, but absent from most members of this order. In the Cirripedia the circulatory system is a very poorly developed system of sinuses; a contractile heart and definite vessels are absent.

The blood circulatory system reaches its highest degree of development in the Malacostraca.

In *Nebalia*, a pericardial septum is present, and a long tubular heart extending from the anterior thorax to the abdomen lies in the pericardial sinus. Posteriorly, a vessel passes back from the heart to the telson, and, anteriorly, a cephalic aorta passes forward toward the brain. From its anterior end arteries arise that supply the antennules and antennae before the aorta opens to the perivisceral body cavity. Seven pairs of ostia, three large and four small, open laterally into the heart. A pair of lateral arteries arise from the heart in each heart segment. Immediately after their origin from the heart each artery divides to form a visceral branch, which turns toward the midline and runs with its fellow toward the gut. The other branch passes out laterally and goes toward the limbs. This branching of these lateral arteries occurs in most of the Malacostraca that possess a well-developed blood system, and may be regarded as characteristic of the group.

In *Anaspides*, the Stomatopoda, and many of the higher Malacostraca, one pair of visceral arteries, usually a pair originating near the posterior end of the heart, passes ventrally around the gut and nervous system and unites to form a subneural artery, which passes anteriorly and posteriorly beneath the ventral nerve cord.   In the Decapoda, this ventral artery, the sternal artery, is formed by only one member of the pair, the other remaining unmodified from the general pattern of visceral arteries.

8.7.2.4. Chelicerata

Unlike the Crustacea, the most primitive chelicerates that we can examine for the structure and function of their circulatory system are large animals with very complex systems, namely the Xiphosura.   In the Arachnida, too, many of the primitive groups like the scorpions are large; and of the small groups that could be regarded as primitive, the Palpigradia are said to have a heart with one to two pairs of ostia opening to it, but this needs reexamination, and the Ricinulei appear to lack heart and blood vessels.   Elsewhere, the small hearts, usually of a saccular type, and the remaining aorta are likely to be secondarily simplified, although the evidence for this is not so clear as it is in the Crustacea.   On balance it seems probable that the Chelicerata at an early stage had a very well-developed circulatory system, and it is reflected more accurately in the existing members, such as the Xiphorsura, the Scorpiones, and the Araneae, than it is in the Acari, the Pseudoscorpiones, and the Opiliones.

In *Limulus*, as elsewhere in the Chelicerata, the haemocoele consists of a perivisceral sinus and a pair of ventral sinuses, their separation being due more to the position of muscles and organs than the existence of a definite septum between them.   A pericardial membrane, i.e., a membrane surrounding the heart so that this organ lies in a separate sinus, is present.   Descriptions of *Limulus* and figures elsewhere in the arachnids suggest that it forms a bag around the heart, which may not be complete on the dorsal side.   If this is to be homologized with the pericardial septum of other arthorpods, then it is considerably modified.   More definitive information on the pericardium of Chelicerates is needed.

The heart in *Limulus* is elongate, extending almost the whole length of the body.   Eight pairs of ostia open into it laterally; it ends blindly posteriorly. Four pairs of lateral arteries arise from the heart opposite the first four pairs of ostia.   They open to a co-lateral artery, which runs each side of the heart.   A cephalic aorta arises from the anterior end of the heart; a cardio-aortic valve is present.   Near the origin of the cephalic artery a pair of large arteries curves ventrally and posteriorly to meet just behind the oesophagus.   A branch from each arch supplies the crop, gizzard, and oesophagus.   A number of transverse arterial commissures join the lateral arteries in front of the oesophagus, where they approach each other.   These form part of a vascular ring that surrounds the oesophagus.   From this ring, arteries radiate out to supply the six pairs of prosomal appendages, the operculum, and the chilaria.   From the anterior edge of the ring a large artery goes to the eye and then joins the digestive gland (hepatic) artery. A large ventral artery passes back from the vascular ring, giving off branches to the five gill-bearing appendages of the mesosoma, the abdominal muscles, and the gut. Near the posterior end of the opisthosoma the ventral artery divides: a branch passes each side of the gut to join the superior abdominal artery.

Anteriorly, the co-lateral arteries end in a few branches close to the cephalic aorta; posteriorly they join behind the heart, where they form the superior abdominal artery. Branches from the co-lateral artery supply their adjacent muscles and tissues and send a series of branches to the gut. A large hepatic artery arises from the co-lateral arteries near their junction with the second lateral artery. It curves forward to supply the digestive gland, sending a branch to anastomose with the eye artery. A large branch arising from the posterior angle of the hepatic artery passes forward around the margin of the prosoma and is called the marginal artery.

The cephalic artery is not very well developed and can be absent in some specimens.

Haemolymph is discharged from the end of the fine branches of the arteries and flows and permeates through the tissues, eventually being conducted to the pair of ventral sinuses lying on each side of the nerve cord. From each ventral sinus channels conduct the haemolymph toward the book-gills. Further channels, the branchio-cardiac canals, guide the haemolymph back to the pericardial sinus surrounding the heart.

There is, then, in *Limulus* a very complex arterial system, part of which (heart–anterior aorta–ventral vessel–superior abdominal artery–co-lateral artery) appears to form a closed loop.

In the Arachnida the heart and circulatory system is well developed in the orders Scorpiones, Amblypygi, and Araneae, where it is a long tube of relatively large diameter. A single aorta conducts blood forward from the heart. The aorta dips down ventrally and branches just dorsal to the gut. These branches pass forward to supply the chelicera and the anterior prosoma. Laterally branches pass around the gut and unite to form a ventral supraneural artery, which passes posteriorly dorsal to the nerve cord. It branches and passes round the gut; just in front of the telson the branches reunite and form a superior abdominal artery, which passes back to the heart.

The heart has a number of pairs of ostia opening laterally; the members of a pair are not always symetrically placed. Near the opening of each ostia a small lateral artery ramifies to supply the surrounding tissues. The myocardium of the heart consists of two muscle layers, having an outer epithelial layer and an inner thin endothelial lining. A series of suspensory muscles and ligaments brings about diastole, while the myocardium brings about systole.

Haemolymph empties from the smaller arteries into the tissue spaces and from there passes to a single large ventral sinus where it bathes the book-lungs. A series of channels conduct blood from the ventral sinus and book-lungs to the pericardial cavity.

In other arachnids the heart is much reduced, often forming only a small sac into which one or two pairs of ostia open, and which has anteriorly a single short aorta.

8.7.2.5. Trilobita

In the trilobite, *Cryptolithus*, fossil evidence suggests that there was a dorsal axial heart from which branches or vessels spread out onto the surface of the pleura between the insertions of the muscles and apodemes. A marginal canal, which perhaps contained a blood vessel, runs around the lateral margin in all trilobites.

8.7.3. Haemolymph

The haemolymph in arthropods is the tissue by which the metabolic intermediates of the chemical processes of the animal are passed around the body. It is composed of fluid in which these substances are in solution and a number of cells, the haemocytes. As might be expected, a very large number of chemicals can be, and have been, isolated from the fluid part of the haemolymph. Their varying concentrations reflect the physiological condition of the animal and the environmental stresses that are placed upon it. Detailed analysis has been performed in relatively very few cases, and statements that assume the presence of a chemical in one species, or that differences between two species of different groups mean differences are reflected at higher taxonomic levels, need to be treated with more suspicion in this field than in any other. The following difference appears to have some validity at higher levels.

8.7.3.1. Crustacea

In Crustacea the osmotic pressure of the haemolymph is largely due to the presence of inorganic ions, principally sodium and chloride. Their total concentration is somewhat less than in seawater, and there is a shift in cation patterns, calcium being present in higher concentrations and magnesium in lower ones than in seawater.

Glucose is the only blood sugar present in all the species that have been analyzed, though as many as eight sugars have been isolated from the haemolymph of a single species. Traces of lactic acid are to be found in the haemolymph. Amino acids are present in relatively low concentrations.

Haemocyanin is the respiratory pigment found in the blood of principally the Decapoda and the Malacostraca. It is in solution in the haemolymph and functions as the normal oxygen carrier. It shows a Bohr effect, that is, the oxygen-carrying capacity of the blood decreases with an increase in acidity, or lower pH, and increases with increase in pH, or increasing alkalinity.

Haemoglobin is found in solution in the blood in a number of lower Crustacea. It is normally saturated at very low oxygen concentrations, but even so acts as a normal oxygen carrier in *Artemia*. In *Daphnia* it is of high molecular weight (360,000) which is normal for respiratory pigments found in solution in the blood. Its appearance and concentration are greatly affected by oxygen pressure of the environment, temperature, and activity of the animal.

8.7.3.2. Chelicerata

As in Crustacea, the osmotic pressure of the haemolymph in Chelicerata is largely dependent upon the presence of inorganic ions, but the cation pattern is different, the calcium being in lower and the magnesium in higher concentration than is found in seawater.

Glucose appears to be the main reducing sugar in the blood, but determinations are few. Trehalose has been reported in one instance, but its presence has not been confirmed. Amino acids are present in low concentrations.

The respiratory pigment haemocyanin has been reported in *Limulus*, scorpions, and some spiders. It apparently is not universally present or even commonly present in the group. It is in solution in the haemolymph. Where it has been studied, it shows some differences from the haemocyanin of Crustacea. There are small differences in its amino acid composition, and it shows a reverse Bohr effect under normal physiological conditions. It is impossible to say whether these differences are true at the higher phyletic levels, or whether they are simply an expression of adaptations at the lower level. Haemoglobin does not seem to have been reported in any chelicerate.

8.7.3.3. Uniramia

There is a great paucity of chemical information on the haemolymph in the Onychophora and Myriapoda, and most of these remarks will be confined to the Insecta.

In the primitive aptergote insects such as *Machilis*, inorganic ions are responsible for almost all the osomtic pressure of the haemolymph. This also seems to be true for the Onychophora. In the pterygote insects this is also true of the primitive orders, Ephemeroptera and Odonata, and often primitive families of more advanced orders. In further evolution small organic molecules play an increasing role in determing haemolymph osmotic pressure, and are responsible for about half its value in Orthoptera and Hemiptera and almost its entire value in the Lepidoptera, Coleoptera, and Hymenoptera. Correlated with this is a high amino acid content of the haemolymph. The ionic pattern is also distinctive: relatively high phosphate concentrations are present, and sodium is replaced in many instances by magnesium. In insects the cation pattern is strongly correlated with the phylogeny of the group, but details indicate that it clearly arose in the insects and does not help at the levels being discussed here.

The blood sugar of insects is trehalose, but glucose is the dominant sugar in scattered species.

Haemoglobins occur in isolated cases in certain insects and are clearly grade 3 adaptations. In the fly *Gastrophilis*, haemoglobin occurs in the tracheal cells and, when isolated, has a molecular weight of only 34,000; such low molecular weights are characteristic of intracellular respiratory haemoglobins. In the fly *Chironomus*, haemoglobins are found in the haemolymph in the larval forms. This haemoglobin too has a molecular weight of 31,000, uncommonly low for a respiratory pigment in solution and perhaps indicative of the secondary nature and origin of insect haemolymph haemoglobins from intracellular pigments.

The haemocytes have been extensively studies in the different groups of arthropods, but no common ground between the major groupings appears to be agreed on (see Gupta, Chapter 13.)

Coagulation of the haemolymph has also been extensively studied, and four systems are recognized as occurring in the arthropods (Grégoire, 1971). All four occur in each group, and there appears to be no phyletic correlation with their distribution.

8.7.4. General Discussion on the Circulatory System

This discussion will be divided into five sections: a) origin of the haemocoele, b) origin of the heart, c) relationships between the arterial vessel patterns, d) "arthropod loop," and e) annelid and arthropod systems.

8.7.4.1. Origin of the Haemocoele

The simplest expression of the haemocoele in the arthropods is as a fluid-filled space occurring between the organs. In the Pauropoda, the haemocoele is the remains of the schizocoele together with the coelomic cavities, which fuse with the schizocoele when their walls break down. The contribution of the coelomic cavities to the haemocoele space is small. There is no epithelial lining to the haemocoele in these animals. This appears also to be true of the Tardigradia. In neither case is there any indication of a contribution of a closed circulatory system to this body cavity.

In the Pentastomida the haemocoelic cavity is well developed. There is no heart, and circulation within the haemocoele is accomplished by movement of the body organs. Such movements are adequate in many small arthropods that lack a heart, but the Pentastomida are not small in this context.

Except in the Tardigradia, all arthropods show, or have evidence of, the haemocoele being divided into two sinuses by a longitudinal horizontal septum. This septum is attached laterally to the body wall in most Pycnogonida, but in a more dorsolateral position in nearly all other arthropods. In the Pentastomida and the Pycnogonida and the development of many others, it is attached in its midline to the dorsal wall of the gut. The haemocoelic space ventral to this septum and surrounding the gut is the perivisceral sinus; that dorsal to the membrane is the pericardial sinus. This name is kept, since it is in this sinus that the heart lies, but it should be noted that in the absence of the heart it is the gonads that lie in this sinus in the pentastomids, and the sinus is judged to be dorsal because it lies opposite the nerve cord and the rudimentary appendages. Strictly speaking, it should be called the gonadial sinus and the septum the perigonadial septum, but to avoid a multiplicity of names, and since the proposal here is that this horizontal septum is homologous in all arthropods, the most commonly used name will be adhered to.

It should be noted that never at any stage is there present a dorsoventral mesentery in which lies the gut, a feature that might be expected to have left some imprint if ever it had existed.

The division of the haemocoele by the pericardial septum makes possible a definite circulatory path in the haemocoele. Peristaltic movements in the gut, together with whole gut movements caused by muscles within the pericardial septum, will cause the fluid to flow posteriorly in the perivisceral sinus and give a forward flow in the pericardial sinus, a flow path kept to this day.

Other septa may well have been present. Certainly other longitudinal septa are found in the different groups of arthropods. In the Pycnogonida a pair of septa occurs in the pericardial sinus; in the Onchophora a pair occurs in the perivisceral sinus; in many Crustacea and Insecta a ventral septum separates a perineural sinus from the perivisceral one. Full descriptions of these septa are seldom given, but they rarely form complete membranes. Gaps occur in them, putting the contents of the sinuses in communication with each other. Primitively, it would appear that the pericardial membrane is attached to the body wall at each intersegmental membrane, and that the gaps are formed in relationship to the haemocoele of the appendages.

The haemocoele then may be regarded as a fluid-filled body cavity formed from the schizocoele and the coelom, crossed by at least one horizontal septum, and with its own circulatory system. There are no transverse membranes and no evidence of a contribution from a blood-filled closed vascular system.

8.7.4.2. Origin of the Heart

In the Pentastomida circulation of the haemolymph is by movement of the organs and the pericardial (perigonadial) septum, and no special organs are developed to aid haemolymph movement. The structure of the heart in the Pycnogonida suggests how this organ could have evolved within the arthropods. The lateral walls of the heart are two longitudinal septa arising from the dorsal wall somewhat lateral to the middorsal line and meeting in the midline at the pericardial septum where it joins the gut. Contraction of these septa would aid in the general movement of the gut and pericardial septum. Development of "peristaltic" muscular contractions down their length rather after the manner of the ventral membrane in lepidopterous caterpillars would aid in the fluid flow forward in the pericardium. It would in fact create a relatively fast forward flow, within the channel lumen. Slits in the septa would allow fluid to enter this forward stream from the pericardial sinus. Such slits would be likely to develop opposite the limbs, where haemolymph would be returning into the pericardial sinus from the dorsal limb sinus. These slits form the primitive ostia. The heart in Pycnogonida differs from that of other arthropods in that the dorsal wall of the heart is also the dorsal wall of the body. The final step to form the heart would be for the two septa to meet in the middorsal line and so form a closed muscular tube.

Such a heart is found in the Onychophora, where it consists of a muscular tube lying in a pericardial sinus. Haemolymph enters the heart through ostia opening between the pericardial sinus and the heart lumen at each segment, and posteriorly through ostia opening between the perivisceral sinus and the heart lumen. Blood is discharged anteriorly from the end of the heart as in Pycnogonida; there is no aorta. There are no other blood vessels, and circulation in the rest of the body consists of posterior flow and lateral flow through the septa.

The heart of arthropods is visualized as a new structure, developed within the group. It is formed from a pair of the longitudinal septa which primitively divided the haemocoele into sinuses, and whose movements aid haemolymph circulation. The ostia arose from slits in the septa through which haemolymph could pass from one sinus to another. The heart is not basically a segmental structure, since it lacks transverse valves as a primitive feature and its chambered appearance is largely superficial. The segmental arrangement of the ostia is secondarily impressed on the heart owing to the mechanics of haemolymph flow associated with the limbs. Perivisceral nonsegmental ostia occur.

8.7.4.3. Haemolymph Vessels

The vessels in the arthropods are all distributive vessels carrying haemolymph away from the heart to the organs. There are no portal systems, and, with the exception of the arthropod loop to be discussed later, all end in sinuses or lacunae between tissues. These distributive vessels, which correspond to arteries in other vascular systems, are tubes lined with a membrane or thin cuticle, with a low columnar or cubical epithelium, resting on a basement membrane. They are elastic but lack muscles.

The pattern of distributive vessels or arteries is quite variable, and comparative studies within a group show that it is readily adapted to the requirements of the animal. In the Crustacea, for example, the system is greatly influenced by the position of the respiratory structure, as shown by the Stomatopoda and the Isopoda compared with *Anaspides* or *Nebalia* or the Amphipoda. This makes a detailed comparison of arteries to try to establish a basic pattern throughout the arthropods very difficult. I doubt that such a pattern exists, but some advance can be made if the distributive pattern is broken up into functional groups.

An arterial blood supply to the limbs occurs in the malacostracan Crustacea, in the Scutigeromorpha in the Uniramia, and in the primitive Chelicerata. In the Crustacea such as *Nebalia* a pair of arteries arises from the heart in each trunk segment. Each artery divides: one branch goes to the limbs and one to the gut. In *Anaspides* the maxillipeds and thoracic limbs receive their supply from a subneural vessel that is supplied from an enlarged podial artery in the last thoracic segment, and a similar condition occurs in the Lophogastridea. In the decapod Crustacea the subneural artery also supplies the abdominal limbs, as well as the thoracic ones and the maxillipedes. In the Stomatopoda the subneural artery is supplied by branches arising from the limb arteries, which continue to the appendages.

In *Scutigera* the limbs are supplied from a ventral supraneural vessel, a pair of arteries going to each limb. The supraneural vessel receives its haemolymph from the heart via the anterior aorta and a pair of ventral vessels that pass around the gut just behind the brain. In the Symphyla there is a supraneural vessel supplied from a pair of ventral vessels arising from the anterior aorta just in front of the heart. No arteries supply the limbs in this animal. In the Diplopoda a pair of sinuses just behind the head puts the anterior aorta into communication with the ventral sinus in a rather more definite way than the usual pattern of haemocoele channels allows. Elsewhere in the Uniramia there are no arteries supplying the limbs. *Scutigera* is a somewhat specialized and very active animal with unusually long limbs. It is likely that the arterial supply here is an adaptation developed within the Myriapoda to meet this requirement. While the pattern is similar to the pattern in some Decapoda, the limbs in Myriapoda were never supplied by arteries arising from the heart, and the ventral vessel is supraneural, not subneural.

In the Chelicerata the limbs are supplied with arteries in the Xiphosura, the Scorpionoidea, the Amblypygi, and some of the Araneae. In all cases the vessels arise from a ventral vascular ring formed by two cephalic arteries that curve around the gut and join to form a ventral supraneural artery. This does not correspond with the pattern in the Crustacea and must be independent of the pattern in *Scutigera*. Elsewhere the limbs that have a circulation of haemolymph have accessory hearts and small vessels, or transverse membranes, which pump haemolymph through their cavities. As the need to circulate haemolymph through them is clearly there, it seems unlikely that if they once had an arterial connection, it would have been lost. In addition, their diversity suggests independent evolution of accessory pumping organs, which is compatible with a primitive absence of a closed vascular system.

Many of the segmental structures, coelomic organs, muscles, and nerve ganglia have an arterial supply, as might be expected; the arteries too are segmentally distributed. This is equally expected whether the arteries are derived from a primitively closed, segmentally evolved vascular system or from an adaptive pattern developed to supply these structures.

The visceral vessels present more of a problem. These are found in all three major groups and basically consist of thin-walled vessels, arising from the heart, segmentally arranged and passing to tissues in the lateral region of the gut, where they open to the haemocoele. As the gut and its tissues appear to be largely uninfluenced by the segmental design of the animal, there seems no reason why these should be segmentally arranged. Their function is uncertain. Anatomically they are suitably arranged to pass haemolymph to the "fat body" cells that in one form or another lie in this and other regions. The haemolymph they discharge would presumably be carried back to the heart in the general flow of fluid from the perivisceral to the pericardial sinus. Haemolymph that passes right to the posterior end especially through the perineural sinus may not pass through these tissues or the nephrocytes before it is recirculated. This system could serve as a means of ensuring that most of the haemolymph had been subjected to these tissues and to their many metabolic functions before being recirculated. These vessels are absent in Onychophora which do not have a fat body, and developed in insects and myriapods that do. Nevertheless, this does not account for their segmental distribution. Perhaps at one stage they were associated with segmentally arranged organs like the coelomic organs, now much reduced in arthropods.

8.7.4.4. Arthropod Loop

An unusual arterial pattern occurs in the Chelicerata and some of the decapod Crustacea, but not in the Uniramia. In the arachnids that have a well developed arterial system, arteries arising from the anterior end of the heart pass forward and ventrally and join to form a ventral vessel, which passes back to the telson, divides there and passes round the gut, and is joined by a supra-abdominal artery arising from the posterior end of the heart. In Xiphosura this loop is more complex, since the supra-abdominal artery divides to form the co-lateral arteries, which receive lateral branches from the heart. In *Limulus* there is no question of haemolymph returning to the heart via the supra-abdominal artery; the co-lateral arteries clearly carry haemolymph to the visceral region. In the Arachnida there seems to be no definite determination of the direction of the flow of blood in the supra-abdominal artery. If it is to, instead of away from, the heart, then we have an arterial loop in which blood is circulated without passing through any tissue whatever—hence the designation "arthropod loop." Two such loops in fact occur in *Limulus*—the one described, together with a marginal loop formed from the common hepatic and marginal artery, which, after division, is continued as the marginal artery around the edge of the prosoma and joins the frontal artery, meeting its fellow from the other side. Here the flow of the blood is out from the heart through the common hepatic artery, and must be out from the heart through the frontal artery. The aortic valve stops return flow to the heart anteriorly.

A loop also occurs in the decapod Crustacea. Haemolymph leaving the heart through the sternal artery at its posterior end passes down the sternal artery to the subneural vessel. The posterior branch of the subneural vessel goes posteriorly and then dorsally to join the supra-abdominal artery arising from the heart.

The arthropod loop can be likened to a ring main where haemolymph can be pumped to the tissues and organs on the loop from either direction. This will ensure a constant supply to all tissues, even if some on the route take more than their normal supply of the haemolymph. This is quite unlike the annelid circulatory system loops where blood flow is unidirectional. Much of the similarity between the arthropod and annelid circulatory systems has depended upon the belief that the loop systems functioned in the same manner. The above interpretation of the arthropod loop depends almost entirely upon anatomical evidence backed with a few observations of haemolymph flow in the vessels. Experimental work needs to be done, both in arthropods and annelids, establish this concept properly.

8.7.4.5. Annelid-Arthropod Systems

Most theories about the nature of arthropod circulatory systems are governed by thinking of the Annelida as being the ancestors of the Arthropoda; that, at some stage, the arthropod ancestor had a fairly complete closed blood vascular system, relics of which form the systems found in present-day arthropods.

The haemocoele was formed principally from the venous side of this blood vascular system expanding, its walls breaking down, and its cavity fusing with that of the coelom and the schizocoele. In some annelids the veins swell to form aneurisms just before they enter the heart. A further extension of this process would lead to fusion of veins in successive segments and eventually to the complete fusion and breakdown of the venous system to form the haemocoele. This process was called phelbodesis by Lankester (1881). Something of the sort happens to give the body cavity in some annelids, but there is no evolutionary evidence of it happening in the arthropods.

The account of the circulatory system in arthropods given earlier shows the possibility that the system evolved within the group to bring about circulation and distribution of fluid in an already existing haemocoele. Steps by which this could have happened have been outlined above.

COELOMIC ORGANS

Introduction

Fundamentally the coelom is a system of secondary body cavities, one pair developing in the mesoderm of each segment. Each cavity is lined by epithelia and opens, via an intercellular coelomoduct and coelomopore situated on the ventral surface close to the base of an appendage, to the exterior. In arthropods the coelomic cavities are best developed and most obvious in the early embryos, and the manner of origin and the details of their fate are matters for embryology rather than postembryonic comparative anatomy. However, one or two points need to be borne in mind to help orientate us to the complexes that arise in association with this system.

At their first appearance in animals the coelomic cavities and genital cells have a separate origin. The genital cells separate very early from the somatic tissues before any sign of a coelom has appeared. Later, the genital cells and the coelom become closely associated and this association persists in all animals that have a coelom at any stage in their development.

In arthropods, the coelomic cavity does not remain intact very long in the developing embryo. After forming, it usually becomes divided by membranes into three cavities, one lying in the dorsal, one in the ventral, and one in the limb regions of the original coelomic sac. The mesodermal tissues surrounding the dorsal and limb cavities disperses, but the ventral cavity tends to remain intact and forms together with the coelomoduct the coelomic or segmental organs. In a few cases coelomic organs are present in the majority of trunk segments, but more often they are found in only one of a few regions. They are best now considered group by group.

8.8.2. Chelicerata

No chelicerate is known that has coelomic organs in all its segments. The maximum number of segments to have coelomic organs is four, found in the Xiphosura, Scorpiones, and some Araneae. In the majority of cases where more than one pair is present, while the coelomosacs remain as separate entities, the coelomoducts fuse to form a composite duct with only one pair of openings to the exterior. While this can help in water conservation in the terrestrial members of the group, it is also found in *Limulus*, which is aquatic and for which there is no suggestion of a terrestrial ancestor.

The coelomic organs in the chelicerates are seen in their simplest form in those members of the group in which only a single pair occurs. While somewhat specialized, the coelomic organ of the ticks, especially *Ornithodorus*, will be considered first.

The coelomic organs in *Ornithodorus* consist of a thin-walled sac opening to a coelomoduct that is tightly coiled proximally, but which widens to form a reservoir at its distal end, then finally opens to the exterior through a short duct. An accessory gland of unknown function opens near the distal end of the reservoir. The coelomosac has a number of muscle fibers originating from the surrounding tissues inserted onto its basement membrane. On the coelomic side of the basement membrane are specialized cells, the podocytes. Contraction of the muscles expands the lumen of the coelomosac, haemolymph filtering through the fine walls of the sac into the expanding sac lumen. When the muscles relax, the sac decreases in size, the fluid passing to the lumen of the coelomoduct. Here fluid may either a) progress down the length of the tubule, chloride and other salts being absorbed during this passage; or b) pass directly from the proximal dilation to the distal dilation across the membrane where the two vessels touch. This description applies to an animal that is adapted to deal with a large fluid intake. The two alternative passages for the fluid are a modification for this purpose. This example serves to illustrate the fundamental properties of coelomic organs. The end sac acts as a filtration chamber, preventing the passage into its lumen of all large protein molecules, and the fluid subsequently passes down the coelomoduct, where substances required by the animal are absorbed. The final excretory product is a fluid containing waste materials. Primitively waste nitrogenous products are not excreted from animal by the coelomic organs. Their primitive function would appear to be to deal with the salt balance and osmoregulatory needs of the animal.

In Xiphosura (*Limulus*) the coelomosacs are brick red in color and contain a mass of fine, partly intracellular ducts. There are four pairs, those of the most anterior segment possessing coelom-organs that are smaller than than the others. All receive an arterial vessel and are adequately supplied with haemolymph. The coelomosacs open to a labyrinth of ducts, which form a stolon linking the sacs together. Posteriorly the stolon opens to the duct of the fourth coelomic organ, which forms a much convoluted tube widening at its distal end, before opening by a short narrow duct to the exterior through a coelomopore situated on the articular membrane of the fifth prosomal appendage.

In the Solifugae, Ricinulei, Schizopeltidae, and Palpigradia only one pair of coelomic organs is present, and these open by a coelomopore situated on the first pair of legs. In the Amblypygi, Opiliones, and some Araneae there are two pairs of coelomic organs, but their coelomoducts fuse and there is only a single coelomopore, whose opening may be on either the second or third pair of legs.

8.8.3. Crustacea

The coelomic organs occur on only two segments in the Crustacea, the antennal segment and the second maxillary segment. Both pairs are present in the adults of the Leptostraca, the lower Myscidacea, and the marine Ostracoda. The antennal coelomic are present in the adult Malacostraca and the larval stages of the Copepoda. The maxillary coelomic organs occur in the adult Entomostraca, Copepoda, and the Cumacea and Isopoda. In Cirripedia the antennal coelomic organs form the cement gland.

The coelomic organs consist of an end sac opening to a tube, the coelomoduct, which expands distally to form a reservoir that opens by a short duct to the exterior. In some of the higher Crustacea part of this canal appears to be of ectodermal origin, and in the Ostracoda the whole duct from coelomosac to external pore is derived from the ectoderm.

The coelomosac has been studied in *Artemia* (Tyson, 1968), *Uca* (Schmidt-Nielson et al., 1968), and *Cambarus* (Kummel, 1964). It appears similar in all three animals. The sac is thin-walled and lined with podocytes on its inner surface. Haemolymph passes through the sac wall into the sac-lumen. Large molecules are filtered off and retained in the haemolymph. In the lower Crustacea there is no special blood supply, and in *Artemia* the maxillary gland floats in the haemolymph. However, Tyson (1968) points out that it is suspended by ligaments which might hold the sac open. If these were muscular or contractile, then dilatation of the sac might well occur in the same manner as has been described for *Ornithodorus*; there is, however, no evidence that this is so. In higher Crustacea there is a good arterial supply of haemolymph, not only to the coelomosac, but also to the coelomoduct.

The coelomosac opens to a tube that expands distally to form a reservoir and finally opens by a short duct to the exterior. Some absorption of sodium, potassium, and chloride occurs in the tube. Excretion of formed bodies into the lumen has been described. Analysis shows that the fluid within the tube is iso-osmotic with the haemolymph, but nevertheless has a very different ionic makeup.

In *Astacus* the gland is further complicated by adaptations for life in a fresh-water environment, but these are clearly secondary and need not concern us here. Although excretion of nitrogenous waste through the coelomic organs occurs in some Crustacea, it is clearly not a primitive function of the organs, and in no case does it account for a high percentage of the total nitrogen excreted by the animal.

8.8.4. Uniramia

8.8.4.1. Onychophora

The Onychophora have an almost complete set of coelomic organs, a pair developing in almost every segment of the body. Each organ consists of a thin-walled coelomosac opening to a coelomoduct. The duct is narrower proximally and expands to form a reservoir distally. It is folded on itself. The reservoir opens by a coelomopore situated on the ventral base of each leg. A unique feature in the coelomic organs of these animals is the presence of a funnel of flagellated cells at the junction of the coelomosac and coelomoduct. The flagella are sometimes figured directed down the lumen of the duct and sometimes projecting into the lumen of the coelomosac.

Fluid is probably filtered through the thin walls of the end sac either under the normal hydrostatic pressure of the body, which is about 10 cm of water

(Picken, 1936), or under increased pressure brought about by muscular contractions of the body wall. Material is absorbed back into the haemolymph from the fluid in the coelomoduct. Relatively little fluid is produced by each coelomic organ. The animals discharge a small amount from each organ simultaneously, often at intervals of two to three weeks. A second discharge rarely occurs within 24 hr.

The main function of these organs appears to be regulation of the salt content of the body. The amount of water produced is too small to be important in fluid regulation. Ammonia can be detected in small amounts but is of little or no importance in the total nitrogen excretion of the animal. It is possible that the organs have the ability to change urea into ammonia, this being an important step in the excretion of phosphate and sulfate ions.

The slime glands are modified coelomic organs with a small end-sac and relatively enormous coelomoduct. The coelomoduct of the posterior pair of coelomic organs forms the gonadial duct.

8.8.4.2. Myriapoda

Two pairs of coelomic organs are present in the Pauropoda, one pair in the premandibular and one pair in the maxillary segments. The maxillary organs are said to retain their functions in water and ion balance, while the premandibular organs now have a salivary function.

In the Symphyla, *Hanseniella* also has two pairs of coelomic organs, one pair on the premandibular and one on the maxillary segment. However, in *Scutigerella* the organs are on the first and second maxillary segments.

In the Diplopoda there is a single pair of coelomic organs on the maxillary segment. In the Julidae, the ducts of these organs are extremely well developed, forming long loops that extend almost the entire length of the body. In other Diplopoda they are confined to the head tagmata. There the organs function as salivary glands, and also supply the animal with fluid that it uses for grooming purposes.

In the anamorph Chilopoda coelomic organs are present in the head. In *Lithobius* and *Scutigera* they consist of one pair of sacs from which arise two pairs of ducts. The coelomoducts open ventrally near the base of maxillae 1 and dorsolaterally on the proximal margin of maxilla 2. The fluid they secrete is used by the animals for grooming (Bennet and Manton, 1963).

The coelomic organs of the Myriapoda appear to conform to the standard pattern and consist of an end- sac, coelomoduct, and distal reservoir.

8.8.4.3. Insecta

Typical coelomic organs are found in the apterygote insects and have been described in the Protura, Collembola, Diplura, and Thysanura, where they occur in the labial segment. They have not been identified so far in the pterygote insects, although they have been described in the embryos of the sawfly, *Pontania* (Hymenoptera), and in the Psocoidea.

Coelomic organs are absent in the Tardigradia, Pentastoma, and Pycnogonida. In the Pycnogonida it is doubtful if any coelomic cavities are formed at all in the transient mesoderm.

8.8.5. General Discussion on Coelomic Organs

The coelomic organs perform a number of functions—ionic regulation, fluid excretion, nitrogen excretion, salivary secretion, and production of grooming fluids in the different groups and species of arthropods. The common function performed by them all is ionic regulation of the haemolymph, and this is thought to be their primitive function. This ionic regulation is not necessarily accompanied by excretion of unwanted ions or of excess fluid. Where ionic regulation is their only function, their fluid output in marine and terrestrial forms appears low. The need to have special organs for ionic regulation of the haemolymph may well be associated with the massive bathing of the nerve cord and muscles that this condition brings about, and the dependence of these tissues on the correct sodium and potassium ratios for their proper functioning. During evolution the permeability barriers between organ and haemolymph will develop, and then there will be less dependence on correct haemolymph ionic balance for proper organ functioning, and the coelomic organs can be reduced in number without injury to the animal. Complete disappearance is unlikely until some other tissue has fully taken over ionic regulation, such as the malpighian tubules in insects.

The absence of coelomic organs from the Tardigradia, Pentastoma, and Pycnogonida, if a primary feature, may indicate that these animals are less closely related to the Crustacea, Chelicerata, and Uniramia than their similarity in other systems suggests; or that coelomic organs arose within the Arthropoda. In the Uniramia, Crustacea, and Chelicerata the organs are remarkably similar in structure, and on present evidence it is hard to believe that they were separately evolved. But how does the mass of intracellular tubules in the end-sac of the organs in *Limulus* correspond to the structure of the end-sac elsewhere? A comparison of these organs with these questions in mind would certainly repay study.

Only in the Onychophora is a full set, or nearly a full set, of coelomic organs present. Elsewhere they are borne only on a few or one segment. Does this imply that the ancestor to the Crustacea, Chelicerata, and Uniramia had a complete set retained only in Onychophora and the arrangement elsewhere is achieved by reduction? On the whole I think one must accept this basis. The coelomic organs, possibly at the level of development found in the Onychophora, were segmentally arranged in their ancestors, and these regressed either as their function was rendered unnecessary in the manner suggested above, or as one or a few pairs developed and had the capacity to serve the whole body. On which segments the organ survived would depend upon the general organization of the internal structure. Their survival at the anterior end in the Crustacea and the Uniramia (except *Peripatus*) can be correlated with the early design of the circulatory system, which tended to concentrate haemolymph flow in that region. In the Chelicerata their survival on the prosomal limb segments can also be correlated with the distribution of arterial blood, and the development of the vascular complexes in that region. The survival of the complete pattern in the Onychophora may be correlated with the relatively poor circulation and the impediments that its type of body form distortion may impose on fluid transport, and perhaps the nature of the membranes separating organs from haemocoele.

In general the coelomic ducts in the Chelicerata tend to fuse and open by a common pore in successive segments that bear them. This is an uncommon occurrence elsewhere.

8.9. REPRODUCTIVE SYSTEM

8.9.1. Introduction

The function of the reproductive system is to provide a place where ova and sperm can be produced, stored, and expelled from the body.

The production of ova follows a well-defined sequence of changes. The germ cells are separated from the somatic cells very early on in the development of most embryos and eventually come to lie in the gonads. Their further development may be delayed until the adult stage is approached, or until the animal actually is an adult. The germ cells in the female are called oogonia, and are rather small cells with large nuclei. After a phase of rapid multiplication, their daughters are called secondary oogonia, and these transform to form the oocytes. The early oocytes show little growth, but the nucleus undergoes a series of premeiotic changes. Then the nucleus swells to form the germinal vesicle, the oocyte meanwhile growing slowly by an increase in its cytoplasmic component. This phase may last a long time. Finally, the oocyte enters a period of rapid growth when most of the yolk is deposited, called vitellogenesis. In arthropods, as in the majority of other animals, a layer of cells surrounds the developing egg to form a follicle. In many arthropods several secondary oogonia may associate together; only one then forms the oocyte, the others helping in its nutrition and being eventually absorbed into it. This may be a very primitive feature in arthropods, since it is found in the Onychophora, as well as in many Myriapoda, Insecta, and Crustacea.

The production of sperm from the germ cells also follows a well-developed sequence of changes. The early spermatogonia are also small cells with large nuclei. These undergo meiotic division to form the primary spermatocytes, which rapidly multiply by mitotic division to form the secondary spermatocytes. These secondary spermatocytes transform to the spermatozoa.

In the primitive arthropods the sperm appears to be of the typical flagellate type found in most other animals. Flagellate sperm are found in the Pycnogonida, Uniramia, Crustacea, and Chelicerata. In the Uniramia, Crustacea, and Chelicerata the tail of the sperm has the typical central pair of microtubules surrounded by a ring of nine tubules. In the Insecta there is a secondary ring of nine tubules outside the inner ring. In the Diplopoda many genera produce a nonmotile disc-shaped sperm. In the Merostomata the sperm tail is typical, but in the Arachnida the sperm type is more variable; there are three central tubules in the Araneae, while in the Acarina the sperm is cylindrical and nonflagellate. In the Crustacea the variety of sperm is greater than elsewhere. In the Anostraca it is spherical with many arms and nonmotile. In the Decapoda the sperms consist only of the "head," which has many cytoplasmic extensions that enable it to move slowly. The head is further differentiated into a perforating organ, which, by what can only be described as an explosion, penetrates the wall of the egg (for more details see Baccetti, Chapter 11).

8.9.2. Tardigradia, Pentastomida, Pycnogonida

A single gonad is present in the Tardigradia, lying in the posterior dorsal region of the trunk, and attached to the body wall anteriorly by a terminal filament. In the female there is a single oviduct, which passes posteriorly and ventrally on one side of the rectum, into which it opens ventrally. In the male there is a pair of vasa deferentia, which passes round the rectum and again opens into it ventrally. A dorsal accessory gland opens into the rectum on the dorsal side opposite the opening of the gonoducts. Many of the ova go to nourishing other ova in the ovary. Males are comparatively rare. It is possible that the paired gonoducts in the male point to the fact that the gonads were once also paired but have fused together as they have elsewhere in the arthropods. This possibility needs further examination.

In the Pentastomida the gonads lie in the dorsal region of the body, partly in the "pericardial" or gonadial sinus. In the female there is a single median ovary in the form of an elongate sac; anteriorly it gives rise to a pair of oviducts, which encircle the gut and join ventrally and form a long convoluted uterus, which opens by a narrow gonopore in the median ventral line a little in front of the anus. At the junction of the oviducts and uterus a pair of receptacula seminalis opens into them. In the male the testis may be single or paired lying in the gonadial (pericardial) sinus. A pair of vasa deferentia arises from the testis anteriorly, encircles the gut, and passes posteriorly in the perivisceral sinus. Toward their posterior end the vasa deferentia are modified to form a complex copulatory apparatus; the two ducts then unite and open by a median ventral pore just behind the mouth. Once again there is evidence of pairing of the gonads and gonoducts, but further investigation is needed to see if the single gonads are primitive or secondary structures.

In the Pycnogonida the gonads lie on either side of the heart dorsal to the gut and hence in the pericardial sinus. They fuse in the adult to form a U-shaped structure. Diventricula pass from the gonads to the ambulatory legs, the ovaries reaching to the fourth podomere and the testis to the third in each limb. In *Pycnogonium* the ova develop in the trunk region of the gonads, but in other genera in the limbs, often in the region of the femur. A gonoduct arises from each diventriculum and opens through a pore situated on the second coxae. A closing mechanism may develop at the gonopore. There is a single pair of gonopores on the last pair of limbs in *Pycnogonium* and *Rhynchothorax*, in most genera on the last two pairs of legs, the last three pairs in *Nymphon* and *Phoxichilus*, and on all five pairs in *Decolopoda* and *Phoxchilidium*. It is reasonable to consider that increase in gonopores is an adaptation taking place within the group, and there is no evidence here of a primary segmental origin of five pairs of gonads.

8.9.3. Crustacea

The primary germ cells differentiate in the ventral walls of each pair of coelomic cavities, where they form small groups and thus initially take up a segmental distribution in *Nebalia* and *Anaspides* (Manton, 1934; Hickman, 1937). During further development the groups link up to form a pair of strands suspended from the pericardial floor. Each strand becomes surrounded by mesoderm cells originating from the pericardial septum. These strands split to form the gonadial cavity, which is clearly formed independently of the coelomic cavity. In *Astacus* the rudiments of the gonads form dense accumulations of cells lying in the 12th, 15th, and 16th trunk segments, appearing at a stage when the coelomic pouches no longer exist. Cavities form in these masses, and they eventually fuse to form a single unpaired rudiment (Reichenbach, 1886). In the copepod, *Learnea*, the gonadial cavity is formed from the schizocoelic space between two coelomic pouches. The gonadial cavity in Crustacea is not formed from the coelomic cavity, but the gonoducts are formed from the coelomoducts, which originate independently from their coelomic sac. There is much variation within the Crustacea as to which segment contributes the coelomoduct to the gonads; it is often different in the two sexes of the same species.

In the adult Crustacea the gonads in the primitive orders form a pair of tubular ovaries or testis, opening to the exterior by a pair of gonoducts. In the female the gonoduct often arises from the subterminal position of the tube, not a surprising feature in view of the manner of formation of the gonads and their ducts. Within the gonad the germinal zone is associated with the dorsal wall of the gonadial cavity. In *Mysis* the dorsal germinal zones appear fused across the middorsal line feeding into the paired ovaries about one-third from their anterior end. This perhaps represents a stage in the development of the "fused" ovary so common in the Decapoda.

8.9.4. Chelicerata

The early development of the gonads in *Limulus* is unclear, and most of the studies come from the Arachnida.

In scorpions the germ cells segregate from the somatic cells at an early stage and form a large compact mass beneath the germ disc. During further development they come to migrate to the second opisthosomal segment, where in all these animals, as in all arachnids, the gonads develop. In the Araneae the germ cells segregate much later and become recognizable as large cells in the coelomic sacs of the third to sixth opisthosomal segments.

The origin of the gonads in arachnids is said to be from the coelomic sac of the second and adjacent opisthosomal segments, the gonoducts growing out from the somatic walls of the coelomic sac and meeting to open in the ventral midline. In the pedipalp, *Thelyphonus*, Schimkewitsch (1906) concluded that the coelom in later development disappeared and that parts of its walls containing the germ cells became associated with folds of mesentery which grew together enclosing a schizocoelic space which formed the gonadial cavity in these animals. The cavity still opened to the coelomoducts of the second opisthosomal segment. If both these views are accepted, it would appear that there is a major modification in the nature of the gonadial cavity in the arachnids. Similar changes in other evolutionary lines also occur.

In the adult Chelicerata, the gonads in *Limulus* consist of an intricate reticulum of fine tubes, in the walls of which develop many small follicles containing ova in the female and sperm in the male. The ova tend to clump into groups, which greatly distend the thin transparent walls of the ovary.

In the adult arachnids three major designs of gonads can be distinguished:

a) *The Ladder Type.* In this design each gonad consists of a pair of longitudinal tubes opening into each other posteriorly and joining to form a gonoduct anteriorly. A number of transverse tubes join these tubes together at intervals throughout their length. In the female ova develop in sacs arising anywhere on the longitudinal or transverse tubes. Similarly sperm develop in sacs in the males. The gonoducts open separately into a pouch on the second opisthosomal segment. This type of gonad occurs in the scorpions and the Uropygi. In some cases the median longitudinal tubes of the gonads fuse to form a single tube in the midline, giving a triplex of tubes; in some cases the transverse tubes are not complete and do not join the longitudinal tubes together. It is difficult to see what this structure can accomplish. In many cases it is associated with viviparity, the embryo developing in the sac wherever it occurs on the tube lattice structure. As these animals have internal fertilization, perhaps it is an adaptation to enable the sperm to reach all parts of the ovary regardless of the possible blockage of any channel by a nearly fully developed embryo.

b) *The Saccular Type.*   Here the gonads consist of a single pair of sacs, a coelomoduct arising from each and opening in the midline of the gonadial segment.   Ova and embryos develop in pouches that project from the surface of the ovary into the haemocoele.   This type occurs in the Araneae, Solifugae, and Acari and perhaps represents the primitive type for the Chelicerata in spite of the more complex gonads in *Limulus* and the scorpions.

c) *The single saccular type.*   In the pseudoscorpions the gonad is a single median tube, again with a follicle projecting into the haemocoele in the female.

8.9.5. Uniramia

8.9.5.1. Onychophora

The primary germ cells segregate early from the somatic cells and come to lie with the posterior midgut cells beneath the developing mesoderm. From here they migrate to come to lie in the dorsal edge of the seventh and following trunk segment. The gonads are formed from the splanchnic wall of the dorsolateral region of the tripartite coelom, which merges with the pericardial floor and spreads over the surface to form the splanchnic muscle. The region containing the germ cells separates as a closed vesicle beneath the pericardial floor. The space the vesicle encloses is part of the original coelomic cavity. The vesicles of successive segments fuse to form a pair of longtudinal tubes, which are the gonads. The germinal cells migrate into their lumen. In the female the ovaries fuse just before they open to the paired gonoducts; in the male the testis remains separate. The gonoducts are the coelomoducts of the preanal segment, which grow forward to fuse with the posterior end of the gonads.

In the adult Onychophora the gonads are a pair of long tubes lying above the gut in the perivisceral cavity. In the female they are bound together in a single connective tissue sheath; in the male they remain separate. In the male the gonoducts form a swollen vesicula seminalis immediately following the testis; a long coiled vas deferens passes backward and unites with its fellow to form a short ejaculatory duct opening on the preanal segment. In the female a receptaculum seminalis occurs at the point where the gonoduct opens to the ovary in some species, and a long oviduct, forming a uterus in viviparous species, passes posteriorly and joins its fellow just before opening on the preanal segment.

8.9.5.2. Myriapoda

In the Pauropoda the genital rudiment develops from the unsegmented median mesoderm, which before the somites have been formed contains a single primordial germ cell. In the first instar the rudiment is unpaired and lies below the gut in the median groove. This rudiment is not derived from the segmented mesoderm, and no coelomoduct develops in connection with it. In the adult the ovary is a thin-walled sac lying between the gut and the nerve cord; the ova develop from an unpaired germinal epithelium lying on the ventral floor about halfway along its length. Tiegs (1947) records that in an unidentified species the germarium is terminal. The ova do not lie loose in the ovary but are enclosed in a follicle attached to the ovarian wall. At the anterior end the ovary opens to a thick-walled gonoduct, which passes to a thin-walled duct opening by a small pore on the third trunk segment. In the male there are four large testes forming a series, one behind the other, lying above the alimentary canal. Each testis is a hollow sac leading to a vas deferens; the vasa deferentia pass on each side of the gut and unite to form two exit channels in the sixth segment. These channels are formed from the epidermis, and are differentiated into three regions, but eventually pass to the penes which lie just median to the second coxae.

In the Symphyla the primordial germ cells come to lie in the gonads, which are formed from the median-ventral walls of the tripartite coelom. These retain their coelomic cavities in the fifth to seventh trunk segments; the cavities fuse to form two longitudinal tubes, the gonads, which lie dorsal to the gut. The gonoducts are formed by ectodermal ingrowths from the third trunk segment. The posterior coelomoducts do not form.

In the Diplopoda the gonads develop in a similar manner to those of the Symphyla. No posterior coelomoducts are formed, and the gonoducts are ectodermal ingrowths of the third trunk segment. In the adult Diplopod the ovary is usually a single tube, dorsal to the gut; the gonoducts open from it anteriorly and pass around the gut to the median genital pore. The paired nature of the ovary can still be distinguished in the adult, since if the ovary is opened, two longitudinal bands of germinal epithelium, one on each side of the midline bearing the developing ova, can be seen.

In the Chilopoda the gonad is formed from the somatic walls of the dorsolateral lobes of the tripartite coelom. These form a sac enclosing a part of the coelomic cavity; the sacs of successive segments fuse to form a pair of longitudinal gonads whose lumen is part of the original coelomic cavity. The mediolateral lobes of the tripartite coelom of the last two segments develop ventrolateral outgrowths, which are true coelomoducts. These pass to each side of the gut and meet in the midventral line to open to the gonopore on the last trunk segment. In the adult the ovary is a single tubular organ situated in the perivisceral sinus above the gut. In the male, one to 24 pairs of testes may be present, each testis opening to a single pair of sperm ducts, which opens to a median gonopore. In the Lithomorpha the sperm ducts are paired on each side, and each is joined to the other at intervals by transverse ducts giving a ladder-type structure. Accessory glands may be present in both sexes.

8.9.5.3. Insecta

In the Insecta the germ cells are segregated from the somatic cells early in development. Eventually they form a pair of strands lying beneath the splanchnic mesoderm in the middle abdominal segments. These strands become covered with a thin layer of splanchnic mesoderm, which splits to form the cavities of the gonads. These cavities are formed independently of the coelomic cavities and thus the gonadial cavity in insects is a new formation not comparable with that formed in other Uniramia. The gonoducts are formed from coelomoducts.

8.9.5.3.1. The Apterygota.   In the Protura, the gonads are ventral to the gut, and in the adult are a pair of tubes with the germinal epithelium located at the anterior end and the gonoducts arising from the posterior end. In the male the gonads are confluent anteriorly. In the female the lateral oviducts fuse to form a common oviduct in the region of the eighth abdominal segment. The common oviduct opens by a gonopore on the membrane between the 11th and 12th sterna. In the male the vasa deferentia are modified to form a pair of seminal vesicles before joining to open to a common ejaculatory duct.

In the Collembola the gonads lie in the ventrolateral region of the perivisceral sinus. In the males of the arthropleonus Collembola a pair of testes is present. Each testis tapers anteriorly to a long terminal filament attached to tissue lying in the mesothorax. The germarium lies in the dorsolateral region of each testis. In the symphypleonus Collembola a single testis is present lying above the gut, and is thought to be formed by fusion of a pair of testes present in the more primitive forms. The germarium is restricted to the anterior region of the organ, as it is in all other insects. The ovaries are simple tapering tubes with an anterior terminal filament, which helps to support the ovary, attached to the fat body in the middorsal line of the mesothoracic segment. The germinal area of the ovary is mediodorsal to the site of vitellogenesis, not terminal as in other insects. Posterior gonoducts derived from coelomoducts are present.

In the Diplura there appears to be an evolutionary series in which the gonads are modified from the simple tubular form of other Uniramia to give the complex pattern of ovarioles characteristic of all higher insects.

In *Campodea* the gonads consist of a pair of elongate sacs lying in the abdomen. Posteriorly the gonads open to a short gonoduct which unites with its fellow to open in the midventral line. The gonopore in both sexes is on the posterior margin of the eighth abdominal sternum.

In *Japyx* the testes consist of a pair of large sacs bent back on themselves lying in the anterior region of the abdomen.   Posteriorly a long coiled vas deferens passes posteriorly, expands distally, and joins its fellow to open by a gonopore on the ninth abdominal segment.

In the female *Japyx* the ovaries consist of seven pairs of segmentally arranged tubes.   Each tube consists of a terminal filament that passes to the midline dorsally where it joins its fellow and runs anteriorly, the paired terminal filaments from each segment joining the others as they pass forward.   The germarium lies in the tube close to the terminal filament.   Laterally the tubes consist of a vitellarium and then open to an oviduct, which enters the lateral oviduct. The lateral oviduct is formed from the oviduct on the most anterior pair of tubes.   The lateral oviducts open to a pouch in the midventral line situated on the posterior margin of the eighth segment.

It is sometimes held that each of these tubes represents an ovary and here is evidence that the gonads were segmentally arranged in the primitive arthropods. As has been indicated in the other groups, the germ cells often come quite early into contact with the developing coelom, but usually any segmental arrangement that this may give is lost as the genital cells or their coelomic associates fuse, to give a single pair of gonads.   It seems here that this fusion is lost or delayed, and segmental distribution of germ cells persists into the adult.   To facilitate comparsion with other arthropods it is proposed to refer to each lateral group of tubes as an ovary and the individual tubes as ovarioles, as is the custom in entomology.

In *Anajapyx* there are two pairs of testes present as two elongate tubes, one pair much longer than the other.   Two pairs of ovarioles are present lying in the middle abdomen; the lateral oviducts continue for a short distance past the anterior ovariole.   Posteriorly the oviducts meet in the midventral line.

The Thysanura.   In *Lepisma* the primordial germ cells come to lie in the somites 2–6 in the abdomen in females and 4–6 in males.   In this position they are carried dorsally and become individually sheathed by splanchnic mesoderm, which splits to form the gonadial cavity.   Mesodermal gonoducts are formed from the splanchnic mesoderm on the abdominal segments and are considered modified coelomoducts.

In the early instars of *Lepisma* the ovarioles lie lateral to the midline with their germinal ends directed medially.   From this end a terminal filament passes to the midline in its appropriate segment and joins its fellow turning anteriorly, where it is joined by terminal filaments from the other ovarioles as it passes forward.   Laterally each ovariole opens to a lateral oviduct, which passes posteriorly and turns to the midline in the seventh segment, where it joins its fellow and opens to the gonopore.   As development proceeds, the ovarioles enlarge and the terminal filaments remain short.   The segmental arrangement of the ovarioles is thus lost, and the ovary comes to resemble that typical of the majority of insects. The system in the male is very similar in general anatomy to that in the female.

In *Petrobius* the reproductive system is similar to that of *Lepisma*, but the male gonoduct has a ladder-like design for most of its length.

8.9.5.3.2. Pterygota. The reproductive system in the pterygote insects is basically similar to that described for *Lepisma*, though there is considerably better development of accessory structures. These aid one in considering evolutionary pathways in the Insecta, but not in considering the broader relationships being dealt with here. They are too diverse and of too great an adaptive significance to allow a short summary to be given. Perhaps two features might be mentioned on pterygote reproductive systems. In some orders the gonoducts do not fuse to form a common duct but open separately on the appropriate segment. This condition is considered primitive, but as can be seen it is not commonly encountered in present-day apterygote insects. Secondly, most students are familiar with the numerous ovarioles that occur in locusts and cockroaches or in termites and bees. These are adaptations to the huge reproductive potential of these animals, and the majority of insect groups have far fewer ovarioles, five to eight per ovary being a commonly found number.

8.9.6. General Discussion on the Reproductive System

The proper evaluation of the different patterns of gonad formation that occur within the arthropods really belongs to the field of comparative embryology, not comparative morphology. However, it was thought necessary to make some reference to it; otherwise comparisons between structures that are not really homologous might seem appropriate but would in fact cloud the main issues. These issues are considered to be: a) What is the design of the most primitive reproductive system from which arthropods and their immediate ancestors can be derived? b) Did segmentally arranged gonads ever exist in arthropods, or their immediate ancestors? c) How similar are the systems in the Uniramia, Crustacea, and Chelicerata?

In the adult a single median gonad is found in the Tardigradia, Pauropoda, Diplopoda, Chilopoda, Pseudoscorpionida, and sporadically in other arthropods. In the majority of cases, either on dissection or from studies of its development, it is clearly formed by the fusion of two gonads. In the case of the Tardigradia and the Pauropoda this is less certain, and there is a real possibility that the single gonad represents a primitive condition. The situation in the Pauropoda is particularly interesting. A single primordial germ cell can be found in the median unpaired and unsegmented mesoderm. A single gonad is formed from this mesoderm in the midventral line beneath the gut. In the female a single oviduct, not formed from a coelomoduct, is present. In the male four testes are formed, but these appear in sequence and their bilaterally symmetrical ducts are again ectodermal.

In the Tardigradia the ovary is again single and has a single duct. In the male there is a single testis, but the duct divides and passes each side of the gut before reuniting ventrally. The origin of the gonads and ducts in these animals would well repay further investigation. A unique feature in the Tardigradia among arthropods is the gonoducts opening into the rectum.

Taking the two groups together and combining the unique features of their reproductive system suggests that the gonad may originally have been unpaired. A single noncoelomic gonoduct, or a pair of such ducts, was present, and opened to the rectum. Elsewhere in the animal kingdom we find such a set of structures in the Aschelminthes such as the Rotifera and Nematoda. It is possible from this starting point to make a series in which the primordial germ cells of the median mesoderm become associated with eventual segmentation and formation of the coelomic cavities, to be carried bilaterally to form the gonads typical of the majority of arthropods. To make a detailed case now would exceed the length of this article, and in any case much original work has to be done before it can be substantiated or denied. The series indicates that the reproductive system may have been very conservative in evolution and a very primitive situation came to coexist with advanced body design in the early arthropods, and may perhaps form a pointer to the ancestors of the group.

In the majority of arthropods the germ cells become associated with the coelom at some stage in their development. In nearly every case in which the gonadial cavity is derived from the coelomic cavity, the gonadial cavity is formed by the fusion of a number of successive coelomic cavities to give a pair of bilaterally arranged gonads. Only one pair of coelomoducts develops to form the gonoducts, either from the middle segments of the trunk or from the posterior ones. Although the segments that develop the ducts vary from group to group and sometimes between the sexes of the same species, except in some Chilopoda there is no evidence in any case that more than one pair of ducts ever evolved to act as gonoducts.

Although the germ cells became associated with the coelom in most arthropods and their contributions from several segments and coelomic cavities went to make up the gonads, there is no evidence from comparative anatomy that the early arthropods or their immediate ancestors ever had more than a single pair of gonads.

The major differences between the gonads of the arthropods lie in the possibility that the gonads of the Tardigradia and the Pauropoda do not have a coelomic component. Their gonads are mesodermal sacs surviving from noncoelomic ancestors. Elsewhere the association of germ cells with the coelom occurs, and it is the primitive condition in the present-day members of the Chelicerata, Crustacea, and Uniramia (except Pauropoda).

In the female the ovary, whether formed from coelomic cavities or not, has the germinal epithelium limited to only a portion of its surface. It often forms a longitudinal dorsal strip, probably formed by longitudinal fusion of the germ cells carried into position by the separate segmental mesodermal migrations. I know of no account indicating that the longitudinal strip remains as a band of isolated germal cell groups, though such may well occur. Where a single ovary is formed from fusion of two ovaries as in the Diplopoda, its dual nature is indicated by a pair of longitudinal germ bands on its dorsolateral surfaces.

In general the ovary remains in a rather primitive state. Ova are retained until mature at their site of origin on the germ band, and are then released into the ovary cavity. A relatively simple duct leads to the outside. This condition is frequently found in the Crustacea, where the duct often seems to be derived from one of the middle segments of the sequence that form the ovary, a feature often indicated in the adult by its subterminal position.

In the Uniramia the primitive ovary is not very different from that described above, the germ band being a median dorsal strip. In several lines it becomes confined to one end of the ovary, which takes on a tubelike form, the end away from the germinal epithelium leading to the oviduct, which is anterior in the progoneate Myriapoda but posterior in all other Uniramia. In the Insecta the tubelike ovarioles have the germinal cells at their proximal ends, and as the ovum moves toward the oviducts, it undergoes a well-defined sequence of changes. This separation and orderly sequence of maturation and vitellogenesis is very well developed in insects and occurs elsewhere in the Uniramia and sporadically in some Crustacea.

In the Chelicerata the germinal epithelium appears not be be confined to a single band in the ovary epithelium, but further work is needed to establish whether this is generally true of the group or not. The ova develop anywhere on the surface of the ovary; vitellogenesis occurs where the ova happen to be situated, the developing eggs projecting from the ovary surface into the haemocoele giving the mature ovary a very pimply aspect.

Viviparity develops on many occasions in the arthropods. In the Chelicerata the embryo develops wherever the egg happens to be situated on the ovary surface. This leads to huge sacs projecting out into the haemocoele. The peculiar ladder-like shape of the ovary, in addition to allowing sperm passage to the ovary regardless of the random pattern of embryo development, may also serve to distribute these sacs throughout the haemocoele instead of clustering them in a single place. This would probably lead to more embryos being able to develop in the animal at any one time. In the Uniramia viviparity also occurs. In the Onychophora the eggs discharged from the ovum are fertilized and develop into embryos in the long oviducts, which are called uteri. In the insects the ova are also discharged from the ovarioles and usually pass to a median common oviduct, usually of ectodermal origin, also called the uterus, in which the embryo develops.

Viviparity where it occurs in Crustacea appears largely to be due to the retention of the eggs in brood pouches until they hatch and not to their remaining within the body of the parent.

These differences indicate that the ovaries do have different basic properties in the three groups.

The testes appear to form in a manner not unlike that of the ovary. A pair of simple large saclike testes does occur in all three groups, but often the testis consists of a number, sometimes a large number, of small sacs, each containing spermatocytes, which within each sac mature into spermatozoa. Ducts, the vasa efferentia, conduct the sperm to the main vas deferens, the definite gonoducts of the animal. This breakup of the single testis to a number of smaller units may be regarded as an adaptation for the more efficient processing of the large numbers of sperm that are required. It occurs in nearly all the groups of arthropods including the Pauropoda. Where, as in this instance, only a few are formed, care must be taken not to suppose that this situation necessarily reflects the segmental organization of the animal.

The structure of the reproductive system in arthropods points to very simple primitive origins and may assist in taking the origin of the group back to an almost precoelomate stage. There is no evidence that segmental gonads ever existed in these animals or their ancestors, or that the role of the coelom in their formation in most arthropods succeeded in altering their primitive saclike design.

8.10. GENERAL DISCUSSION

The general tenet of the accounts already given is that the internal anatomy of the organs and systems serving the visceral functions of the primitive arthropod was very simple. It consisted of a gut, which, in addition to functioning in the digestion, absorption, resynthesis, and storage of food, also excreted the nitrogenous waste and governed the ion and water balance of the animal; and there was a haemocoelic body cavity containing haemolymph, which was agitated by movements of the gut and acted as the distributive agent of the body. Nephrocytes present in the haemolymph acted as phagocytic cells and for the storage of accumulated waste products. A single one or a pair of saclike gonads was present; the early gonoducts were not coelomoducts. There were no special organs for respiration, which was through the body surface, for haemolymph circulation, or for excretion. There were no nephridia or coelomic organs.

It is perhaps unexpected that such a simple visceral anatomy should coexist with a relatively complex segmental body plan in which the segments bear well-defined appendages and have impressed their design firmly upon the neuro-muscular systems of the animal.

At this early stage the visceral systems are relatively untouched by the basic segmental plan of the rest of the body. The gut shows no evidence of it at all. The body cavity, although fairly early on receiving a contribution from the segmentally arranged coelomic cavities, is not much influenced in design by them. There are no transverse membranes or septa of any sort, and the penetration of the haemocoelic cavity into the lumen of the appendages is the only segmental imprint on the final body cavity. The saclike gonads show no sign of segmental imprint in the postembryonic animals. The germinal epithelium forms a longitudinal strip within the sac unaffected by segmental boundaries. When coelomic cavities and their ducts develop, a pair of coelomoducts takes over the function of gonoducts; but while in the different groups coelomoducts from different segments do this, the gonad is untouched by the segmental design of the animal.

Only those structures that developed in close relationship with the segmental pattern of the animal really bear the firm imprint of segmentation. These were principally the coelomic organs, to a lesser extent the respiratory organs, and the heart and haemolymph vessels, which serve the segmentally arranged muscles and ganglia. These all evolved after the segmental pattern had become firmly established, in the skeletal, muscular, and neural design of the group. Some effect on the gut, especially in the Chelicerata, occurs in the more advanced members of the group. The characters of the visceral system point most firmly to the Aschelminthes such as the Rotifera and the Nematoda as the probable nonsegmental ancestors of the arthropods. In any case, the early arthropods had a very simple visceral architecture from which more elaborate patterns evolved within the group.

A most important key to the internal design of the arthropods lies in the origin and development of the haemocoele. Its possession by the Tardigradia, Pentastomida, Pycnogonida, Onychophora, Myriapoda, Insecta, Crustacea, Merostomata, and Arachnida is one of the major reasons for linking them together.

The haemocoele is the body cavity in which lie the internal organs, and which is bounded externally by the body wall and integument. Traditionally, it was thought to be formed by the fusion of three primary cavities that appeared during development: the schizocoele or primary body cavity generating from the blastocoele, which was simply a space left between the organs; the coelomic system of cavities, primarily paired sacs appearing in the mesoderm in each segment; and the blood vascular system, a closed system of tubes and vessels through which blood was circulated. There is no doubt that the schizocoele and the coelomic system of cavities unite to form the haemocoele. However, evidence from the Pauropoda and the Pycnogonida suggests that the schizocoele is the important primary cavity. Later, the coelom is added to it, but was never a large component and was never as dominant a feature of adult arthropod anatomy as it was of adult Annelida. There is no evidence for a closed vascular blood system in the arthropods or for any such system of cavities being included in the haemocoele.

In the Tardigradia, the body cavity is well developed and lacks septa, heart, and circulatory system, movement of the haemolymph being solely due to gut movements. The Pauropoda also lack septa, heart, and blood vessels; the body cavity is perhaps less well developed, being readily visible in starved animals but nearly obliterated in well-fed ones. In the Pentastomida there is a well-developed body cavity divided by a horizontal transverse septum into a dorsal sinus and a ventral perivisceral one. There is no heart or blood vessels, but the gut is attached to the transverse septum in the midline and perhaps can be oscillated laterally by contractions of the horizontal septum after the manner of the nerve cord/ventral membrane complex of lepidopterous larvae. This longitudinal horizontal septum makes possible the flow of haemolymph anteriorly in the dorsal sinus and posteriorly in the perivisceral one. In the Pycnogonida a primitive heart occurs in the dorsal sinus above the horizontal membrane, which here, besides being attached to the gut, penetrates into the limbs. The heart consists of a V-shaped lumen between two longitudinal septa, the open end of the V being closed by the dorsal wall of the animal and the ventral point being attached to the horizontal septa. Slits in all these septa put the sinuses they bound in communication with one another. The horizontal septum now becomes pericardial membrane; the sinus it bounds, the pericardium; the slits penetrating the heart septa, the ostia. The ostia develop in relationship to the returning circulation of haemolymph from the limbs, and thus the segmental imprint of the general body design first manifests itself on the circulatory system. No other circulatory vessels exist in the Pycnogonida. It would be interesting to know the circulatory pattern of such large members of this group as *Colossendis*.

The pericardial septum is to be found in all the other major groups of arthropods, the Crustacea, the Chelicerata, and the Uniramia, and again is one of the features that link these three together.

The further development of the circulatory system has already been commented upon, but it is useful to make the following points here: a) The arterial system delivers blood to the different organs of the body as a simple distributive function. It does not ensure that blood goes first to one organ, then to another. After delivery the blood goes to a set of ventral sinuses and thence to the pericardial sinus and to the heart. There are no portal systems in arthropods. b) The arteries are simple elastic tubes that lack muscles and thereby differ very greatly from the vessels of closed circulatory systems, such as those of the Annelida. c) Arterial loops are formed such that the blood is fed into a ring artery from both ends of the vessel where it arises from the heart. In its course the ring artery gives rise to numerous blood vessels draining through lacunae in the tissues to the sinuses. Such a system provides for unusual blood flow from anywhere along the ring to the areas it supplies. Examples are the marginal arteries of *Limulus* and the posterior loops of Arachnida and Crustacea. I know of nothing like this in other animals, and have called this arrangement of vessels the "arthropod loop."

The evidence from comparative anatomy supports the view that the circulatory system in arthropods evolved in connection with more efficient circulation of haemolymph in the haemocoele. The septa, the heart, the blood vessels, and their patterns evolved within the group and are not remnants of a closed circulatory system possessed by some arthropod ancestor. The circulatory system has no features in common, either in the design of the heart, the nature of the blood vessels, or the pattern of the system, with that of the Annelida, and has features that are unique.

The main characteristics of the simple tubular gut of the primitive arthropod are: the first appearance of the tripartite structure of the gut, ectodermal inturned foregut, endodermal midgut, and ectodermal inturned hindgut; and the development of longitudinal muscle layers inside the circular muscles next to the epithelial basement membrane in the foregut and the hindgut. In subsequent evolution the foregut and the hindgut often become highly differentiated and make up most of the length of the gut. The midgut lacks support from extrinsic muscles or dorsoventral membranes and lies free in the haemolymph, where it is tensioned between the ends of the foregut and hindgut. This is especially true in the Tardigradia; elsewhere it may derive support from connective tissue, or the haemocoele may be packed with fat body. It is possible to correlate the unusual position of the longitudinal muscle in the fore- and hindguts with the rather weak nature of the midgut tissue design and its need for support and tensioning in the haemocoele. The overall layout of this gut is a characteristic that binds the different arthropod groups together.

The simple tubular midgut has evolved into four major designs, each characteristic of a major arthropod group, as follows: a) *The Uniramia:* Here the midgut remains basically of simple tubular design. Caeca are present in a few cases as a secondary adaptation. A peritrophic membrane is almost universally present and is characteristic of the group. Midgut shortening may be correlated with the development of a counter-current mechanism correlated with the presence of a peritrophic membrane. b) *The Crustacea:* The midgut is basically of simple tubular design but with a pair of unbranched caeca at its anterior end. A peritrophic membrane occurs in widely different members of the group and is perhaps characteristic of its design. Shortening of the midgut may be very considerable and is usually associated with the development of a hepatopancreas by the branching and ramification of the anterior midgut caeca. c) *The Chelicerata:* In the Chelicerata the midgut has a number of paired caeca arising at intervals throughout its length. In the Palpigradia (*Koenenia*) the midgut in the opisthosoma is large, and pouches protrude laterally between the segmental muscles, which is perhaps an indication of how these caeca arose. Their pattern is quite different from that of Crustacea, and they have probably evolved within the Chelicerata. In a number of groups one or more pairs of caeca have branched and ramified to form a complex digestive gland quite reminiscent of the hepatopancreas in Crustacea but of independent origin; hence the different name. In

Limulus a large digestive gland is present, opening from the midgut by two pairs of caeca. Shortening of the midgut does not seem to occur in any chelicerate. d) In the Pycnogonida the midgut has five pairs of caeca, which enter the five pairs of limbs. It is likely that this is governed, as is much else in these animals, by the very short trunk of small diameter characteristic of the group. This development is thought to have occurred independently of any of the modifications mentioned above and represents a separate line of evolution originating from the simple tubular gut of the primitive arthropods.

In all three groups—the Uniramia, Chelicerata, and Crustacea—the midgut tends to show particular specializations, or to shed some of its many functions. In the Uniramia much of the chemical processing of the body and food storage that occurs in the Onychophora is transferred to the fat body cells, which form a well-defined sheath around the gut in the Myriapoda and Insecta. Nitrogenous excretion is taken over by the malpighian tubules, which are said to be of ectodermal origin. Part of the ion and water regulation is transferred to the hindgut in insects and myriapods, and part of the ion balance to the coelomic organs in Onychophora.

In the Crustacea much of the chemical processing of the food remains with the midgut and not associated with a well-defined fat body. Nitrogenous excretion mostly lies with the midgut, but nitrogen is lost across the body surfaces, such as the gills. Water regulation is primary with the gut, but coelomic organs take over some of the ion and water balance of the body, and in some forms nitrogenous excretion is partly transferred to them.

In the Chelicerata chemical processing is still mainly with the gut, but connective tissue cells take over some of it and the food storage and form a loosely organized fat body largely in association with the digestive gland. Nitrogenous excretion is taken over by the malpighian tubules, which originate from the midgut and may be considered caeca specialized for this purpose. Ion and water balance are in part taken over by the coelomic organs. In one instance water balance may be transferred to the hindgut, but this is not characteristic of the group.

The types of gut and the pattern of their further development can each be derived from the simple primitive gut; but one is not easily transposed into the others, and they are considered independent lines of evolution.

Respiration was primitively through the body surface, and further development involved the formation of three quite different basic systems.

In the Uniramia tracheae and tracheoles developed very early on and are characteristic of the group. Their precursors are uncertain but could have been the integumental glands that are commonly found, the inter- and intracellular ducts of which are of a type that could be easily transformed into tracheae and tracheoles. Their organization into segmentally arranged tracheae so characteristic of the Myriapoda and Insecta can be followed within the group.

In the Crustacea the body surface, including the large surface area of the limbs, could make surface respiration adequate for quite active moderate-sized animals. The typical respiratory organs, where they develop, are epipodites, large platelike extensions from the dorsal surfaces of the limb-bases. In the Malacostraca an elaborate system of gills develops in this area.

In the Chelicerata the basic respiratory organ is the book-gill. This develops on the ventral side of the mesosoma appendages. Its transformation to book-lung and book-trachea can be readily followed in the group. Its origin is uncertain. In the Eurypterida vascularized areas on the ventral mesosoma plates appear to form specialized respiratory areas. It is difficult to believe that these can be really adequate for such large animals, and furthermore that they would be secondarily simplified from the bookgills of the Xiphosura under these circumstances. No other respiratory structures have been described for the Eurypterida, and again it is difficult to believe that surface respiration could have been sufficient for such large and apparently active animals. Functionally, it is more reasonable to accept these areas as the beginning of book-lungs that were formed by their developing the complex leaflife folds of these organs. Tracheae do appear to have originated in some arachnids independently of the book-gill, book-lung, trachea sequence. They do not seem to be segmentally organized structures, and perhaps represent separate adaptations to meet local specialized requirements within the group and are off the main stem of respiratory organ phylogeny. It should be noted that if tracheae and tracheoles originate from integumental glands, the degree of trans-formation required is very small, and caution must be exercised in using tracheae to establish phylogenetic relationships.

Tracheae transport air directly to the tissues they supply in many cases, but in others appear simply to form organs that aid in saturating the haemolymph with oxygen. In all three groups respiratory pigments may be found in the haemo-lymph. In the Uniramia haemoglobin is found in a few cases, its low molecular weight being indicative of an originally intracellular pigment secondarily released into the haemolymph. In the Crustacea haemoglobin also occurs sporadically in the haemolymph, its high molecular weight being characteristic of respiratory pigments occurring in solution in the blood; haemocyanin occurs in the Malacostraca. In the Chelicerata, haemocyanin is also present as a blood-borne respiratory pigment in a number of forms. The majority of Crustacea and most of the Chelicerata that lack tracheae depend upon oxygen in physical solution in the haemolymph and not in combination with respiratory pigments to convey oxygen from the exterior to their tissues (see also Gupta, Chapter 13).

The coelomic organs are absent in the Tardigradia, Pentastomida, and Pycnogonida, but are present in the Chelicerata, Crustacea, and Uniramia and are one of the features that serve to link these groups together. Their develop-ment in the embryo is fairly clear and their homology well established, but, in other contexts, they raise a number of problems. They consist of an end-sac, which is part of the medioventral coelomic cavity, and a duct leading to the exterior, which is the coelomoduct. The presence of flagella at the junction of the end-sac and coelomoduct in the Onychophora is taken to be evidence of a nephridial funnel so that the organ is really a combination of coelom, nephridium, and coelomoduct similar to the type of association between these three that is found in the Annelida. There seems to be no other evidence to support this possibility.

In the Onychophora transient coelomoducts occur on the first and second segments and paired segmental organs on all the others. The third pair is modified to form salivary glands and the last pair to form gonoducts. In the Crustacea vestigial segmental organs occur on a number of segments in *Artemia* (Benesch, 1969) and on the prosomal segments in the Chelicerata. The primitive pattern is probably one of a pair of coelomic organs associated with each segment of the body. In the majority of arthropods only a few segments develop coelomic organs, frequently only one or two pairs occurring in the adult. The pattern of this reduction is different in the different groups: considerable differences may occur in fairly closely related animals.

The function of the coelomic organs, except where they have become specialized for salivary purposes, is one of ion and water balance, a function they share with the gut. In a few cases, such as the fresh-water Decapoda, they become the main organs for salt and water balance and may also play a major role in excreting nitrogenous waste from the body. In the majority of cases, however, they play a relatively small role in the total salt and water balance mechanisms of the animal. Their origin is hard to account for. In a number of cases they are best developed in the late embryo and degenerate when it has hatched. The gut is usually late in development in arthropods, and perhaps the initial function of the coelomic organs was to deal with the salt and water regulation at this stage when the gut was nonfunctional.

The reproductive system consists of the gonads, the gonoducts, and their associated glands; there is some diversity in their origins in the different groups of arthropods. In the Pauropoda the gonads arise out of the nonsegmented median mesoderm before any somites have been formed (Teigs, 1947). In the Onychophora, Myriapoda, Chelicerata, and Crustacea the gonadial cavity is formed from the coelomic cavities; but in the Insecta, and perhaps in the Arachnida, it is formed as a new cavity appearing in the mesoderm after the coelomic cavities have disappeared, or is formed by folds of mesoderm growing together and enclosing part of the schizocoele to become the gonadial cavity. Where it is formed from the coelom, several coelomic sacs on each side fuse together to form the gonads. The form of the adult gonad, whether tubular or saclike, remains, regardless of the manner of its formation, and, except in the transient role played by the segmented coelom in the mechanics of gonadial formation, there is no evidence of segmental imprint on the gonads, or that more than one pair of gonads ever existed in the arthropods or their ancestors.

In a number of arthropods only a single gonad occurs in the adult animal. In the Diplopoda and the Chilopoda the presence of paired germinal tracts within the gonad indicates that this is a secondary state formed by the fusion of paired gonads in the midline. In the Tardigradia, Pentastomida, and Pauropoda there is no evidence that this is so, but further work needs to be done to determine whether primitively single gonads do exist within the group. In the Pycnogonida the five pairs of gonads are derived from a single medial genital plate of cells. Five pairs occur only in some members of the group and are considered a secondary adaption to the body form of these animals. In the insects the gonad is thought to become secondarily divided to form a number of ovarioles that in some apterygotes take up a segmental distribution, and this is not indicative of primarily segmentally arranged gonads occurring within the arthropods.

The gonoducts also are of diverse origin in the arthropods. In the Tardigradia, Pentastomida, and Pycnogonida they are not coelomoducts. This is also true of the Pauropoda and the other progoneate Myriapoda where they are of ectodermal origin. Elsewhere the gonoducts are coelomoducts and are formed from different segments in the different groups. In the Chelicerata they arise from the second opisthosomal segment; in the Crustacea the gonoduct segment, while usually the middle trunk region, is variable. In the Uniramia it is usually the last trunk segment, and in insects it is usually the seventh and eighth abdominal segments in females and the eighth and ninth in males that contribute to the terminal gonoducts.

In the Tardigradia, Pentastomida, and elsewhere where there is only one gonad, there may be only a single gonoduct passing either to the left or right side of the gut, or the single duct may divide to encircle the gut and fuse again on the ventral side to form a single duct leading to the gonopore. This division of the gonoduct may or may not indicate the original paired nature of the gonads. In the Tardigradia the gonopore opens on the ventral side of the rectum. This condition is found elsewhere only among the Aschelminthes and may be an important clue to the nearest nonarthropod relations of these animals and perhaps also to the origin of the arthropods.

Four major patterns of visceral design can be detected among the arthropods: a primitive pattern, a chelicerate pattern, a crustacean pattern, and a uniramian pattern.

In the primitive pattern the gut is a simple tube. There are no special respiratory organs; gaseous exchange is through the body surface. The haemocoele is either a simple body cavity lacking a horizontal septum and special organs of circulation, or it may have a horizontal septum, or a horizontal septum and special organs of circulation in the form of longitudinal septa that form with the body wall a V-shaped heart with ostia. There are no coelomic organs. The gonads are either single or paired with single or paired gonoducts that are not coelomoducts. In the Tardigradia the gonoducts open to the rectum. This primitive pattern is found not only in the Tardigradia, Pentastomida, and Pycnogonida, but also in some of the smaller and more primitive members of the Chelicerata, Crustacea, and Uniramia, such as the Pauropoda.

In the chelicerate pattern, the midgut has paired caeca at intervals throughout its length.   In the larger members of the group the caeca ramify to form a complex digestive gland. Intracellular digestion occurs, and a peritrophic membrane is absent. Two functions of the primitive midgut have been transferred to specialized organs. Nitrogenous excretion in the Arachnida is at least in part taken over by malpighian tubules, which may be simple specialized caeca of the posterior midgut region. In both the Merostomata and the Arachnida storage of food and probably much of the general metabolic chemistry are performed by connective tissue cells forming a loosely organized fat body associated with the midgut. Special respiratory organs occur. The main series of these is the book-gill, book-lung, and book-trachea organs, but other tracheae not part of a segmental series are found in a few arachnids. The haemocoele contains a well-developed tubular heart with ostia and blood vessels. A pericardial septum is present but appears to surround the heart as a sac rather than form a horizontal membrane.   An elaborate series of arteries is present, forming a series of "arthropod loops." The ventral vessel lies above the nerve cord and sends arteries to the prosomal limbs. A series of coelomic organs confined to the opisthosoma occurs, but usually only one to four pairs remain in the postembryonic stage. The gonads may assume one of three patterns: saclike, ladder-like, or ramifying tubules. The germinal epithelium forms all or nearly all the epithelium lining the gonadial cavity.

In the crustacean pattern, the midgut is primitively a long tube with a single pair of caeca opening anteriorly. Other caeca may develop in the Malacostraca, but never along the length of the midgut as in the Chelicerata. A peritrophic membrane is present. The midgut (including the hepatopancreas) retains most of the functions of the primitive midgut. Nitrogenous excretion is occasionally taken over by the coelomic organs, as are water balance and ionic regulation in some fresh-water forms.

Respiration in these mainly aquatic animals is still primarily through the body surface, which, owing to the presence of bi- (or tri-) ramous appendages on each segment, represents a quite considerable area. It was probably sufficient to maintain a reasonable increase in size and activity before special respiratory organs needed to be developed. These were primary epipodites developing at the base of the thoracic appendages to form flattened leaflike gills. In the Malacostraca, especially the Decapoda, an elaborate series of gills occurs. Gills may also develop elsewhere, as on the abdominal appendages in isopods and stomatopods. Their origin is quite different from that of the Chelicerata. In primitive Crustacea the haemocoele is divided by two membranes to form dorsal pericardial, middle perivisceral, and ventral perineural sinuses. A well developed tubular ostiate heart is present in the pericardial cavity, leading to a short dorsal aorta opening to the sinus in the head. In more advanced Crustacea a more elaborate system of arteries is present. Paired segmental arteries branch immediately on leaving the heart and send a dorsal branch to the limbs and a ventral one to the gut. In further development a subneural ventral artery is formed, which takes over the supply to the limbs. A single posterior "arthropod loop" is present. Two pairs of coelomic organs occur in the postembryonic stages of some Crustacea, but usually only one of these pairs is present in the adult. There is evidence of a more extensive series in some embryos. The organs are concerned with the salt and water balance of the animal and may play a major role in this function. Very rarely do they become involved in nitrogenous excretion. The gonads are a pair of sacs whose cavity is formed by the fusion of a number of coelomic cavities. There is a single pair of gonoducts opening in the midtrunk region.

In the Uniramia the gut is a simple tube. It lacks caeca except in a few specialized cases, and has a peritrophic membrane. The gut retains all the functions of the primitive midgut in the primitive members of the Uniramia, but becomes considerably specialized in the Myriapoda and Insecta. Nitrogenous excretion is taken over by malpighian tubules, here developed from the hindgut. (In some insects their midgut origin seems beyond dispute.) Water regulation and salt balance are also passed, at least in part, to the hindgut. Food storage and much of the general chemical metabolism of the animal is also passed from the midgut, where it occurs in Onychophora, to fat body cells that form a well-organized sheath around the gut in the Myriapoda and Insecta.

In some small and primitive forms respiration is through the general body surface, but undoubtedly the typical respiratory organ of the Uniramia is the tracheae and associated tracheoles. In primitive tracheate Uniramia it is likely that tracheae were widely scattered over the general body surface and came to occupy their segmental associations through further evolution within the group. The haemocoele is divided by horizontal septa to give a dorsal pericardial, a middle perivisceral, and a ventral perineural sinus. A tubular ostiate heart lacking aorta or blood vessels occurs in the Onychophora. The ostia can open from the perivisceral as well as the pericardial sinus into the heart lumen. An anterior aorta develops in the Myriapoda and the Insecta. A very elaborate arterial system occurs in the Symphyla and the Scutigeromorpha. In the latter a ventral supraneural vessel is developed, which sends arteries to the limbs. Elsewhere segmental arteries arise from the heart to supply the gut. These vessels are often ill-defined and may be little more than spaces between the tissues. An almost complete series of coelomic organs occurs in the Onychophora, where they are associated with ion balance. They are reduced in number in the Myriapoda, where their main function is either salivary or to supply grooming fluid. They have disappeared completely in the pterygote insects. The gonads are tubular, paired, and formed from coelomic cavities except in the insects, where the cavity is a new formation. The gonoducts are coelomoducts except in the Progoneata. In the Insecta the ovarioles are highly organized for the sequential processing of the eggs as they pass down to the lateral oviducts.

The chelicerate, crustacean, and uniramian patterns can each be derived from the primitive pattern; but, as each one began to emerge from this level, so much change would have been needed to convert into the other that it is judged unlikely that this happened. Emphasis must be placed on each pattern rather than on single features, since convergence and divergence make their similarities and differences poor indicators of true phylogenetic relationships. For example, the trachea type of respiration is clearly fundamental to the Uniramia; it is secondarily evolved from book-gills and book-lungs in the Chelicerata, and is an individual adaptation (grade 3) in very, very few Crustacea that have pseudotracheae. The evidence is that the patterns of internal visceral anatomy of the Chelicerata, Crustacea, and Uniramia have evolved independently of each other.

A major difficulty in accepting the primitive pattern as the basic simple design from which the others have evolved lies in the fact that it is always associated with small size, and there is a strong correlation between small size and simplicity. This correlation does not mean that small animals of simple structure have to be derived from large, complex ones. Undoubtedly the small, simple arthropods are a mixture of those that are primitively simple and those that are secondarily simple. It is not always easy to distinguish between the two. Here, I am reasonably satisfied that those animals I have called primarily simple are primarily simple; but there is much work to be done to produce satisfactory criteria for this distinction.

There is no intention that the "primitive pattern" should receive any systematic recognition, or be identified as equal to the other three. Its importance is in helping us to recognize the simplicity shown in the visceral anatomy of some groups of arthropods that do not fit into the chelicerate, crustacean, and uniramian patterns, as well as the simple anatomy shown by the primitive members of those groups.

This primitive pattern forms a link between the Chelicerata, Crustacea, and Uniramia, possibly at a level before their distinctive locomotory characteristics had evolved. On the other hand, it bears indications of having links with the Aschelminthes, and thus gives a firm pointer to the possible origins of the arthropods. Many of the adaptive features shown by members of the Aschelminthes bear much resemblance to those shown by the Arthropoda. Though far beyond the scope of this chapter, it is interesting to speculate about whether the setting up of an Aschelminthes "matrix" and comparing it with an arthropod matrix (Clarke, 1973) would show very good agreement and thus enable the entire characteristics of whole phyla to be used in determining relationships. I think it is quite likely that this would show close links between the two designs and their potentials.

8.11. SUMMARY

Within the groups of animals that have a segmented body, segmental appendages, and muscular and neural systems, the latter consisting of a dorsal cerebral ganglion, circumoesophageal connectives, and ventral ganglia, there may be found an extremely simple internal anatomy, which represents the primitive condition of the visceral design.

A haemocoele is a very primitive characteristic of the group. It is formed mainly from the schizocoele, with a relatively small coelomic component. At an early stage the haemocoele became divided by a horizontal longitudinal septum.

All further evolution of the viscera above the primitive pattern occurred within the arthropods on the following lines: 1) A heart and circulatory system developed within the haemocoele. The tubular ostiate heart links the Chelicerata, Crustacea, and Uniramia together, but the development of the arterial system occurred independently in the three groups. The system called the "arthropod loop" is peculiar to these groups. 2) The simple tubular midgut evolved in three independent lines. In the Uniramia the simple tubular nature was further developed; in the Crustacea paired caeca confined to the anterior end characterize the line; and in the Chelicerata paired caeca occurring at intervals along the midgut are peculiar. 3) The coelomic organs appeared within the arthropods, probably in connection with salt and water balance in the late embryo. They appeared very early in development and serve to forge further links between the Chelicerata, Crustacea, and Uniramia. 4) The respiratory organs evolved independently in each of the three major groups. Tracheae are characteristic of the Uniramia, epipodite gills of the Crustacea, and book-gills and their derivatives of the Chelicerata. Grade 3 adaptations acquired independently confuse this picture.

The gonads retain their simple paired sac or tubular structure, and were never segmentally arranged organs, even though in the embryo the segmental mesoderm plays a part in their formation. The evolution of coelomoducts developed within the group.

The primitive visceral pattern found in some arthropod groups and the primitive members of the Chelicerata, Crustacea, and Uniramia have their nearest similar pattern in the Aschelminthes, and it is among these animals that lines leading to full arthropod design originated.

The visceral anatomy shows features that unite the Chelicerata, Crustacea, and Uniramia, namely: a) the haemocoele with horizontal pericardial septum and tubular ostiate heart, b) the coelomic organs, c) the paired saclike gonads, and d) the ectodermal foregut and hindgut with their unusual muscular coats contrasting with those of the midgut. Certain negative characters also link them together: the absence of flame cells or nephridia, and the lack of specialized organs for nitrogenous excretion or salt and water balance.

Visceral anatomy also links them to the Tardigradia, Pentastomida, and Pycnogonida, giving a picture of a shared primitively simple visceral design from which three main evolutionary lines arose and developed independently of each other, a haemolymph circulatory system, a respiratory system, and detailed midgut modifications.

This picture supports the view that the Tardigradia, Pentastomida, Pycnogonida, Chelicerata, Crustacea, and Uniramia are all members of an evolutionary line of which the Tardigradia, Pycnogonida, and Pentastomida are groups that have remained near its base, and that the Chelicerata, Crustacea, and Uniramia are three major separate lines of evolution radiating out from this base. This evolutionary complex is properly designated as a single phylum, the Arthropoda, comparable in its distinctness and the range of diversity it encompasses with other major phyla such as the Mollusca and the Chordata. In other words, I conclude that the Arthropoda appear to be a monophyletic group.

This view is based on the visceral anatomy of the animals, which has been examined independently of other anatomical features, since it was thought that this was the most useful way in which a study of comparative anatomy of these organs and systems could contribute to the question of the homogeneity or diversity of this vast and complex group. The final view must, of course, result from the integration of all aspects of knowledge of these animals, in which the visceral anatomy, relatively lightly imprinted with the segmental design, has a major but as yet largely unrecognized role to play. There is a great need for modern work on arthropod internal anatomy to be undertaken specifically with questions of phylogeny in mind and not as a handmaiden to other studies. Only then can the truth or falsity of the proposition made above be finally determined.

REFERENCES

Aubertot, M. 1939. Présence d'un vaisseau dorsal contractile chez les Protoures du genre *Acerentomon*. *C. R. Acad. Sci. Paris* 208: 120–22.

Balss, H. 1926. Crustacea, pp. 277–1078. *In* W. Kükenthal and T. Krumbach (eds.) *Handbuch der Zoologie*. Bd. 3, Halfte 1 : de Gruyter, Berlin.

Benesch, R. 1969. Zur Ontogonie und Morphologie von *Artemia salina* L. *Zool. Jahrb. Anat. Ontog.* 86: 307–458.

Bennett, D. S. and S. M. Manton. 1963. Arthropod segmental organs and Malpighian tubules, with special references to their function in the Chilopoda. *Ann. Mag. Nat. Hist.* 13(5): 545–56.

Berlese, A. 1909. Monagrafia dei Myrientomata. *Redia* 6: 1–182.

Bullock, T. H. and G. A. Horridge. 1965. *The Structure and Function of the Nervous Systems of Invertebrates. I and II.* W. H. Freeman, Reading, Berkshire.

Cisne, J. L. 1973. Anatomy of *Triarthrus* and the relationships of the Trilobita, pp. 45–64. *In* A. Martinsson (ed.) *Evolution and Morphology of the Trilobita, Trilobitoidea and Merostomata*. Proc. N.A.T.O. Adv. Study Inst., Oslo.

Clarke, K. U. 1973. *The Biology of the Arthropoda*. Edward Arnold, London.

Cuénot, L. 1949. Les Tardigrades, pp. 39–59. *In* P. P. Grassé (ed.) *Traité de Zoologie*. VI. Masson et Cie, Paris.

Curry, A. 1974. The spiracle structure and resistance to dessication of centipedes. *Symp. Zool. Soc. Lond.* 32: 365–82.

Grégoire, C. 1971. Hemolymph coagulation in arthropods, pp. 145–90. *In* M. Florkin and B. T. Scheer (eds.) *Chemical Zoology*. VI. Academic Press, New York, London.

Henson, H. 1932. The development of the alimentary canal in *Pieris brassicae* and the endodermal origin of the Malpighian tubules in insects. *Q. J. Microsc. Sci.* **75**: 283-309.

Heymons, R. 1897. Über die Bildung und den Bau des Darmkanals bei niederen Insekten. *Sitzb. Ges. Naturf. Freunde Berlin* **1897**: 111-19.

Hickman, V. V. 1937. The embryology of the syncarid crustacean *Anaspides tasmaniae*. *Pap. Proc. R. Soc. Tasm.* **1936**: 1-36.

Hughes, C. P. 1973. Redescription of *Burgessia bella* from the Middle Cambrian Burgess Shale, British Columbia, pp. 415-36. *In* A. Martinsson (ed.) *Evolution and Morphology of the Trilobita, Trilobitoidea and Merostomata.* Proc. N.A.T.O. Adv. Study Inst., Oslo.

Kästner, A. 1968. *Invertebrate Zoology II*, transl. by H. W. and L. R. Levi. Interscience Publishers (John Wiley and Sons), New York, London, Sydney.

Kummel, G. 1964. Das Coelomosacken der Antennendruse von *Cambarus affinis* Say (Decapoda) Crustacea. *Zool. Beitr.* **10**: 227-52.

Lankester, E. R. 1881. *Limulus*, an arachnid. *Q. J. Microsc. Sci.* **21**: 504-48, 609-49.

Lawrence, R. F. 1953. *The Biology of the Cryptic Fauna of Forests.* A. Balkenna, Capetown.

Manton, S. M. 1934. On the embryology of *Nebalia bipes*. *Phil. Trans. R. Soc.* **B 223**: 168-238.

Manton, S. M. 1937. Studies on the Onychophora II. The feeding, digestion, excretion and food storage of *Peripatopsis*. *Phil. Trans. R. Soc.* **B 227**: 411-64.

Manton, S. M. 1965. Functional requirements and body design in Chilopoda. *J. Linn. Soc. Zool.* **46**: 251-501.

Manton, S. M. 1973. Arthropod phylogeny—A modern synthesis. *J. Zool. Lond.* **171**: 111-30.

Moore, P. F. 1941. On gill-like structures in the Eurypterida. *Geol. Mag. Lond.* **78**: 62-70.

Mukerji, D. 1932. Description of a new species of Collembola and its anatomy. *Rec. Indian Mus.* **1932**: 34.

Oudemans, J. T. 1888. Beitrage zur Kenntnis der Thysanura und Collembola. *Bijdr. Dierk.* **16**: 147-226.

Picken, L. E. R. 1936. The mechanism of urine formation in invertebrates, I. The excretion mechanisms in certain arthropods. *J. Exp. Biol.* **13**: 309-28.

Reichenbach, H. 1886. Studien zur Entwicklungsgeschichte de Flusskrehses. *Abh. Senckenb. Ges.* **14**: 1-137.

Richards, A. G. 1951. *The Integument of Arthropods.* University of Minnesota Press, Minneapolis.

Ripper, W. 1931. Versuch einer Kritik der Homologiefrage der Arthropodentracheen. *Z. Wiss. Zool.* **138**: 303-69.

Schimkewitsch, W. 1906. Über die Entwicklung von *Thelyphonus caudatus* (L) vergleichen mit derjenigen einiger anderer Arachniden. *Z. Wiss. Zool.* **81**: 1-95.

Schmidt-Nielsen, B., K. H. Gertz and L. E. Davis. 1968. Excretion and ultrastructure of the antennal gland of the fiddler crab, *Uca mordax*. *J. Morphol.* **125**: 473-96.

Shimkevich, V. M. 1908. Meterozis kak embriologich eskoi printsip. *Izv. Akad. Nauk.* **1908**: 997-1008.

Siewing, R. 1960. Nevere Ergebnisse der Verwandtschaftsforschung bei den Crustaceen. *Wiss. Z. Univ. Rostock. Math.-Nat. Reihe* **9**: 343-58.

Størmer, L. 1944. On the relationships and phylogeny of fossil and recent Arachnomorpha. *Skrift. Vid.-Akad. Oslo, I. Math.-Nat. Kl.* **5**: 1-158.

Størmer, L. 1963. *Gigantoscorpio willsi* a new scorpion from the Lower Carboniferous of Scotland and its associated preying microorganisms. *Skrift. Vid.-Akad. Oslo, I. Math.-Nat. Kl.* **8**: 1-171.

Tiegs, O. W. 1947. The development and affinities of the Pauropoda, based on a study of *Pauropus silvaticus*. Part I. *Q. J. Microsc. Sci*. **88(2):** 165–268. Part II. *Q. J. Microsc. Sci*. **88(3):** 275–336.

Tillyard, R. J. 1917. *The Biology of Dragonflies*. Cambridge University Press, Cambridge.

Tyson, G. E. 1968. The fine structure of the maxillary gland of the Brine Shrimp, *Artemia salina:* The end-sac. *Z. Zellforsch. Mikrosk. Anat*. **86:** 129–38.

Waterston, C. D. 1973. Gill structures in the Lower Devonian eurypterid *Tarsopterella scotica*, pp. 241–54. *In* A. Martinsson (ed.), *Evolution and Morphology of the Trilobita, Trilobitoidea and Merostomata*. Proc. N.A.T.O. Adv. Study Inst., Oslo.

9

Significance of Intersegmental Tendon System in Arthropod Phylogeny and a Monophyletic Classification of Arthropoda

H. B. BOUDREAUX

9.1. INTRODUCTION

Debate on the evolution of arthropods has been going on for nearly a century, and there is still disagreement about whether arthropods are monophyletic. A monophyletic taxon can be defined as one that includes all and only the descendants of a presumed common ancestor. The common ancestor can only be defined hypothetically, as a species exhibiting characters of a primitive nature. These characters must be such that many of the specializations of the descendants can be derived directly or by intermediate stages from the primitive states of the characters of the ancestor. A polyphyletic taxon is one whose members appear to have descended from more than one hypothetical ancestor. These ancestors would have given rise to two or more evolutionary lines whose survivors share similar but independently derived convergent characters. Polyphyletic taxa are undesirable in phylogenetic studies, and should be abandoned when the convergent characters on which they are based have been recognized (Hennig, 1953).

S. M. Manton (1950 to 1973) has argued most vigorously in support of the idea of a polyphyletic origin of arthropods from independent wormlike, nonarthropod stocks. In an extended study she has described for the first time and in detail the locomotor and gnathal mechanisms of selected examples of all types of modern arthropods and of Onychophora. The emphasis on specializations has led Manton to ignore any attempts to find a ground plan among the locomotor and jaw mechanisms of arthropods. She incorrectly assumed that phylogenetic study involves being able to show that the ancestor of any arthropod class can be found as a member of another class. Since she does not accept the transformation of the specializations of one class into those of another class, her conclusion is that no member of any known class of arthropod could have been ancestral to any other class. Although her basis for such a conclusion is in error, the conclusion is easily reached when one assumes that evolutionary development is inescapable other than by extinction. Most taxa of living animals include extinct fossil species, but the candidates for ancestral species are extinct and unknown. Even the extinct trilobites are too specialized to be ancestral to anything but other trilobites. Every species that Manton studied functionally is a modern one, and thus a present product of evolution. The ancestral state can only be hypothesized through comparison of homologous character states, and homologous characters can be recognized only if it is possible to recognize a ground plan from which each set of homologies evolved.

The ground plan characters of a hypothetical ancestral arthropod from a monophyletic viewpoint are listed in Table 9.1. These features in common can be found in almost any arthropod, in almost any state from primitive to highly derived, including suppression. These are the features that Manton assumed must have resulted from convergent evolution when she espoused a polyphyletic origin of arthropods. It is possible, for most of the 17 characters, to discuss how in each arthropod class the ground plan characters have evolved. In this chapter I will discuss mostly the intersegmental tendon system, together with trunk and limb musculature, and illustrate how it is possible to homologize these systems to support a monophyletic concept of arthropod evolution. The evidence that Manton used to support a polyphyletic origin of arthropods will be reinterpreted in terms of homology and monophyly.

9.2. INTERSEGMENTAL TENDON SYSTEM

9.2.1. Basic Intersegmental Tendon System

The intersegmental tendon system consists of strands of true connective tissue, which can be identified in the bodies of some members of all the arthropod classes except pycnogonids (Manton, 1964). The tendon system can be reduced to a common ground plan that would serve as a workable muscle attachment system in an ancestral arthropod (Fig. 9.1). The tendons, indicated by wavy dots

Table 9.1. Ground Plan Characters of the Hypothetical Ancestral Arthropod.

1. Epidermal cuticle consisting of chitin and protein.
2. Localized cuticular sclerotization through protein tanning.
3. Typical periodic moulting of cuticle in a characteristic sequence controlled by hormones.
4. Ecdysial glands in anterior end producing similar ecdysones.
5. Typical segmented metameric locomotor appendages, with a coxal base plus a distal segmented telopodite.
6. Coxal endites (gnathobases) related to feeding.
7. Coxal exite lobes (suppressed in terrestrial forms).
8. Metamerically segmented dorsal and ventral longitudinal muscles.
9. All muscles striated.
10. Intersegmental tendon system.
11. Tonofibrillae penetrating epidermis.
12. Suppression of all motile cilia except in spermatozoa.
13. Suppression of all circular somatic muscles.
14. Compound eyes.
15. Development of protocerebrum as ocular center and deutocerebrum as antennal center.
16. Tergal scutes extended laterally into paratergal folds.
17. Cephalic tagmosis.

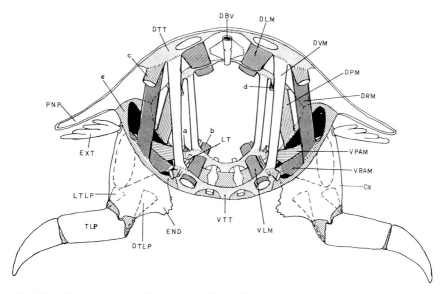

Fig. 9.1. Reconstruction of hypothetical primitive arthropod body segment, showing minimum operational somatic and extrinsic leg muscles, viewed from posterior aspect. a, Stump of VRAM at anterior intersegment; b, stump of VLM at anterior intersegment; c, stump of DPM at posterior intersegment; d, stump of DRM at anterior intersegment; e, unsclerotized pleuro-sternum. (see Table 9.2 for abbreviations used in figures.)

(DTT, VTT), typically are found at our near the intersegment dorsally and ventrally. Dorsally, the transverse tendons serve as points of origin for the ends of the segmental dorsal longitudinal muscles (DLM), indicated by the broader parallel lines, the dorsoventral muscles (DVM), indicated by lack of shading, and the dorsal extrinsic limb muscles. Promotor muscles (DPM) are indicated by wavy lines, and remotor muscles (DRM) are indicated by narrow parallel lines in the figure. In all subsequent figures, individual muscles are not separately indicated.

Table 9.2. Abbreviations Used in Figures.

AB AP	Abductor apodeme of MND	LT	Longitudinal ventral tendon
AD AP	Adductor apodeme of MND	LTLP	Levator muscle of telopodite
ANT	Antenna	MND	Mandible
AP	Sternal apodeme	MS	Mandibular suspensory rod
AT	Anterior tentorial arm	MX	Maxilla
ATP	Anterior tentorial pit	MXL	Maxillule
CA	Cardo of MND	MXP	Maxilliped
CR	Cuticular rod of transverse	NC	Nerve cord
	tendon	ODVM	Oblique dorsoventral muscle
CX	Coxa	OVLM	Oblique ventral longitudinal
CX AP	Coxal apodeme		muscle
CXC	Coxal cavity	P	Pleurite or pleuron
DA	Dorsal arm of transverse tendon	PM	Promotor muscles
DBV	Dorsal blood vessel	PNP	Paranotal process
DDM	Dorsal diaphragm muscle	PT	Posterior tentorial arm
DLM	Dorsal longitudinal muscle	RM	Remotor muscles
DPM	Dorsal promotor muscle	SCL	Adductor sclerite of MND
DRM	Dorsal remotor muscle	SKS	Skeletal strut
DTLP	Depressor muscle of telopodite	SPM	Sternopleural segment of DVM
DTT	Dorsal transverse tendon	ST	Stipes of MND
DVM	Dorsoventral muscle	STT	Sternal fragment of VTT
EA	Endosternal apodeme	T	Trunk terga
EM	Adductor muscle of endite	TENT	Anterior tentorium
	(segment of DPM)	TLP	Telopodite
END	Endite lobe of coxa	TRM	Trochanteral muscles
ES	Eye stalk	VDVM	Vertical dorsoventral muscle
EXT	Exite lobe of coxa	VLM	Ventral longitudinal muscle
GP	Gnathal pouch	VPAM	Ventral promotor-adductor
H	Heart		muscle
HPHY	Hypopharynx	VPM	Ventral promotor muscle
I AB AP	Insertion of abductor apodeme	VRAM	Ventral remotor-adductor
	of MND		muscle
IM	Intrinsic muscle of MND	VRM	Ventral remotor muscle
L	Leg base	VTT	Ventral transverse tendon
LB	Labium	VVM	Vestigial ventral muscles of
LM	Labrum		former appendages

The labels refer to muscle groups that are presumed to be homologous, but whose functions or positions may have become altered. Thus, dorsal promotor muscles (DPM) of the mandible may be presently employed as adductor muscles, as in pterygote insects (see below, Fig. 9.25).

Ventrally, the tendons serve as points of origin of the ends of the segmental ventral longitudinal muscles and the ventral extrinsic limb muscles, and the insertion of the dorsoventral muscles. In the primitive arthropod, the limb bases were probably articulated with the lower sides of the body by way of arthrodial membrane rather than through fixed condyles, as seems to be the case in trilobites. An acondylic limb joint with the body is a workable beginning, since this is the case in modern collembolans, xiphosurans, and some crustaceans (Manton, 1972). The dorsum of each segment was probably sclerotized into tergal plates, but originally the venter may have been unsclerotized. Use was probably made of local hemocoelic pressure to oppose the pull of appendicular muscles. Hemocoelic pressure is employed in this fashion in collembolans and arachnids (Manton, 1973a,b), and this was probably important in trilobites, in which there was no sternal sclerotization.

In subsequent figures, homologous muscles are indicated by the same labels, although the functions and origins of individual muscles may have been altered, particularly in the mandibular system. Most of the figures have been adapted from published work and relabeled to indicate homologies. (Abbreviations used in the figures are listed in Table 9.2.)

From the ground plan, there evolved the various ways in which the limb articulates with the body by way of articular condyles, after sclerotization of the sternal area. The tendon system became variously consolidated, fragmented, shifted, or eliminated in the various lines of evolution. Mouthparts have evolved from limb bases accompanied by changes in the musculature.

9.2.2. Chelicerata

In *Limulus* (Fig. 9.2), the coxae are surrounded by arthrodial membrane, and can be moved in any direction including protrusion and retraction, apparently in the case of protrusion as a result of local relaxation of the extrinsic limb muscles of the protruded coxa and simultaneous contraction of coxal muscles elsewhere, causing protrusion by hemocoelic pressure. Thus, in addition to a locomotor function for its legs, *Limulus* is capable of biting with its coxal gnathobases, and of shoving food into the mouth by alternately protruding and retracting its coxae. The chelicerates have specialized in consolidating the ventral tendons of the prosoma into a massive tendinous plate, known as the "endosternite," on which the ventral limb muscles originate. *Limulus* and nearly all arachnids possess a tendinous endosternite suspended by the dorsoventral muscles (VTT).

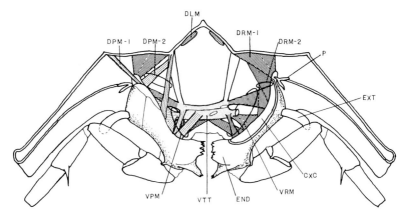

Fig. 9.2. Cross section of *Limulus polyphemus* (Xiphosurida), viewed from anterior aspect. Front edge of left coxa is cut away to show remotor muscles inserting on posterior border of coxa. DPM-1 is segment of DPM converted to abduction of coxa; DRM-1 is segment of DRM converted to abduction of coxa. Clear oval areas are insertion points of left VPM. Adapted from Manton (1964).

There are no known dorsal prosomal tendons in any chelicerate. The muscles of the prosoma became attached to the cuticle as these tendons were lost, but dorsal and ventral tendons are found in the opisthosoma of primitive spiders (Fig. 9.3A), and ventral tendons occur in the opisthosoma of most spiders (Millot, 1949) (Fig. 9.3B).

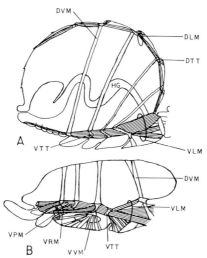

Fig. 9.3. Longitudinal sections of abdomens of spiders (Arachnida). A, *Liphistius batuensis*, with persistent dorsal and ventral intersegmental tendons; B, *Uroctea durandi*, with only ventral tendons persisting. VPM and VRM operate spinnerets. Adapted from Millot (1949).

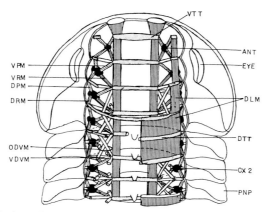

Fig. 9.4. Dorsal view of head and first three thoracic segments of Ordovician trilobite, *Triarthrus eatoni*. About 1 cm of animal is shown. Adapted from Cisne (1974).

9.2.3. Trilobita

Recent studies of trilobites (Fig. 9.4) employing X-rays (Cisne, 1974) suggested that trilobites had a segmental muscle and transverse tendon system very similar to my hypothetical ground plan. If Cisne's reconstruction is accurate, the dorso-ventral muscles became elaborated with the addition of oblique muscles (ODVM), similar to the condition in the mandibulates.

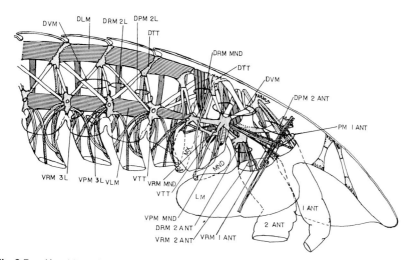

Fig. 9.5. *Hutchinsoniella macracantha* (Crustacea, Cephalocarida). Section showing longitudinal view of head and first three trunk segments. First leg is sometimes considered to be second maxilla, and is articulated on the head. Adapted from Hessler (1964).

9.2.4. Crustacea

Among modern arthropods the primitive crustacean *Hutchinsoniella macracantha* agrees most closely with the ground plan (Fig. 9.5). The crossed, oblique, dorso-ventral muscles appear to be a novelty in the mandibulate line, since they occur in crustaceans, myriapods, and some insects. A small amount of consolidation of the mandibular with maxillary ventral tendons occurs in *Hutchinsoniella*. The thoracic limbs bear dorsal and ventral promotor and remotor muscles originating on their respective tendons. The maxillule (MX) bears a stronger gnathobase than do the legs, but the coxae bear muscles similar to those of the legs. The mandible of the adult (MND) lacks a telopodite, and articulates with the lower surface of the head at its upper end, but the legs join the body by arthrodial

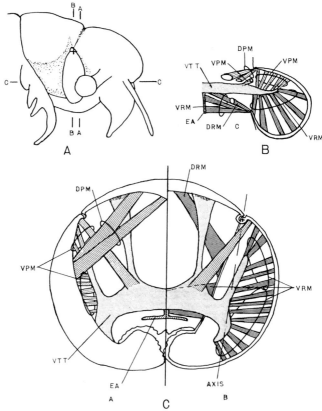

Fig. 9.6. *Chirocephalus diaphanus* (Crustacea, Anostraca). A, Lateral view of head showing levels of sections *a*, *b*, and *c*; B, right mandible as seen in Section *A*, *c–c*, viewed dorsally; C, cross section through mandible, anterior view. At left, section *a–a* is near anterior border, while at right section *b–b* is through axis. Symbol + indicates primary dorsal articulation of rolling mandible. Adapted from Manton (1964).

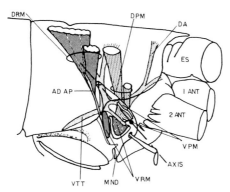

Fig. 9.7. *Anaspides tasmaniae* (Crustacea, Anaspidacea). Lateral view of head shown as if transparent. Note inclined axis of mandible, with a secondary articulation ventrally indicated by solid black circle. Adapted from Manton (1964).

membrane only. The dorsal condyles of primitive crustacean mandibles permit only a promotor-remotor roll around a vertical axis. The gnathobases rub together as the mandible rolls (Fig. 9.6), but there is no biting action. The opening and closing of higher crustacean jaws evolved with a shift of the vertical axis of roll to a more inclined position (Fig. 9.7), as shown in the malacostracan genus *Anaspides*. The lower front border became secondarily articulated with the head, so that a former promotor roll has become an abduction movement, opening the jaw. The remotor roll has become the biting action as the gnathobase shifted backwards. The transverse tendon of the mandible of *Anaspides* (VTT) is a broad expanse of tissue bearing the origins of some of the mandibular ad-

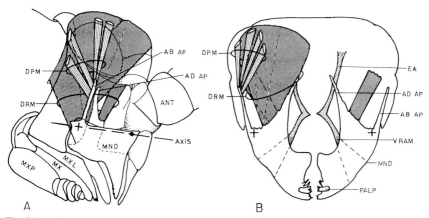

Fig. 9.8. *Ligia oceanea* (Crustacea, Isopoda). A, Head shown laterally as if transparent. Mandibular axis is almost horizontal, and is dicondylic. B, Posterior view of head in cross section posterior to mandibles. Ventral muscles of MND are small, and insert on a sternal apodeme (EA) rather than on the ventral tendon, which is suppressed. Adapted from Manton (1964).

ductors, but an additional muscle crosses directly from one mandible to the other without joining the tendon (this muscle is shown as a cross-lined oval in Fig. 9.7). The transverse tendon of the mandible of higher crustaceans has been suppressed, and in some (Fig. 9.8) it is replaced by cuticular apodemes. The axis of swing has further shifted, so that the axis is nearly horizontal in isopods. The large oceanic isopod, *Bathynomus giganteus*, bears a well-developed mandibular telopodite or palp, but it is reduced in most or absent in the land isopods. In *Ligia* (Fig. 9.8), the palp is reduced to a "lacinia mobilis."

9.2.5. Myriapoda

In the myriapods, the Pauropoda have eliminated the dorsal body tendons (Fig. 9.9), and remnants of ventral tendons remain ventrally in the trunk and between the solid, one-piece mandibles. The diplopods have further reduced the ventral intersegmental tendons, except for the mandibular tendon, and most trunk tendons and dorsal limb muscles are suppressed, except in the primitive pselaphognathid millipede genus *Polyxenus* (Fig. 9.10). In this genus the ventral tendons are strong and have been invaded by cuticualr rods, and dorsal coxal muscles are retained.

The jaw structure of myriapods and insects was presumed to be a whole-limb jaw, with tip biting rather than gnathobasic biting. This idea is a fundamental concept for Manton's claim that arthropods are at least diphyletic, and possibly tetraphyletic. There are two joints and three gnathomeres in the mandibles of diplopods (Fig. 9.11). Each of these gnathomeres is considered to be a full-fledged podomere by Manton (1964). But the only podomere of any arthropod limb that opens directly into the hemocoel is the coxa. There is never any extrinisic muscle inserting on the terminal podomere in an arthropod leg, as would have to be the case if the mandible were a whole-limb. Fragmentation of the

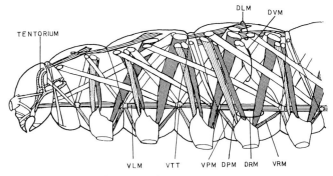

Fig. 9.9.   *Pauropus sylvaticus* (Pauropoda).   Lateral view of trunk and limb musculature. Only ventral transverse tendons remain. Adapted from Manton (1966).

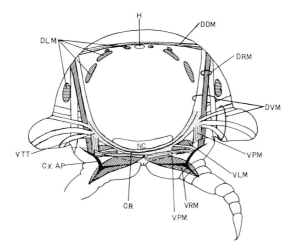

Fig. 9.10. *Polyxenus lagurus* (Diplopoda, Pselaphognatha). Cross section of trunk showing musculature. Note that dorsal tendons are suppressed, and that a cuticular rod (CR)
is embedded in ventral transverse tendon. Adapted from Manton (1956).

coxa into coxomeres is not unknown in arthropods. The coxae of collembolan
legs are fragmented into two coxomeres (see below, Fig. 9.18a,b). The coxae of
insect maxillae and labia also consist of two coxomeres. The musculature of
myriapod mandibles is consistent with the idea of a fragmented coxa bearing a
movable endite lobe operated by cranial extrinsic muscles. The muscles originating on the ventral transverse tendon all insert on the coxal walls and the movable
gnathal lobe. In diplopods, the cranial and ventral muscles cause strong transverse biting action. Myriapod jaws are opened or abducted by a new mechanism.
A swinging anterior tentorial apodeme (AT, Figs. 9.11, 9.13, 9.16) is operated
by segments of the dorsoventral mandibular segmental muscles so that the anterior tentorium is swung forward and outward, bearing on the inside of the jaw,
causing abduction.

The symphylan body muscles are also modified from the ground plan (Fig.
9.12). The body transverse tendons are highly suppressed, and most muscles
originate on the body wall. The dorsal and ventral longitudinal muscles are further segmented, especially dorsally, where there are median and lateral bundles
extending to the edges of the hemitergites. Some muscles extend across two
body segments. The symphylid jaw (Fig. 9.13) is solid except for the movable
endite. But the biting muscles, as in millipedes, are all variations of the ground
plan extrinsic coxal muscles. Jaw opening is performed by the swinging forward
and outward of the anterior tentorial apodeme, whose muscles are converted
dorsoventral muscles.

The absence of a palpus or telopodite in myriapod and insect mandibles is
here regarded as the result of evolutionary suppression. Partial suppression of

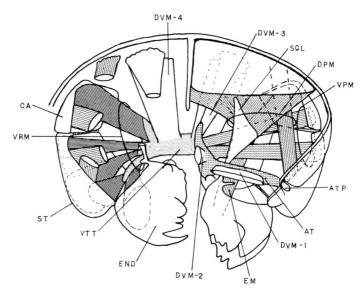

Fig. 9.11. *Poratophilus punctatus* (Diplopoda, Juliformia). Cross section of head, anterior view. Promotor muscles of right side are not drawn in, to show remotor muscles. Most DPM and DRM operate as adductors. DVM-1 and DVM-2 insert on movable tentorium, and cause indirect abduction of mandible. DVM-3 and DVM-4 aid in adduction. DRM suppressed. Adapted from Manton (1964).

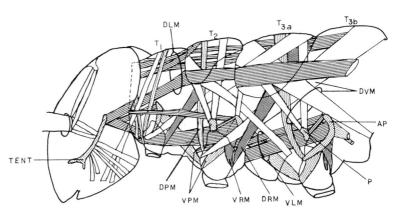

Fig. 9.12. *Hanseniella agilis* (Symphyla). Lateral view of head and first three trunk segments. Transverse tendons are suppressed, and muscles attach ventrally on pleuron (P) or sternal apodemes (AP). Third tergum consists of two hemitergites (T3a and T3b). Adapted from Manton (1966).

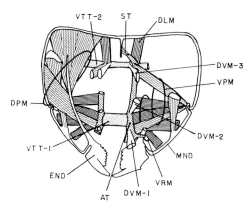

Fig. 9.13. *Scutigerella immaculata* (Symphyla). Section through head showing mandibular musculature. DRM suppressed. Left DPM excluded to show other muscles. DVM-1 and DVM-2 operate swinging tentorium in abduction of MND. VTT-1 is mandibular tendon and VTT-2 is maxillary tendon. Upper sector of VPM may act as an abductor muscle of MND. Adapted from Manton (1964).

the telopodite is the rule in crustaceans, and many crustacean jaws lack a palpus in the adult stage. In myriapods and insects, the mandibular palpus is suppressed even in the embryonic stage. Manton argued that the absence of an embryonic palp was proof that the myriapod-insect mandible was a whole limb. I suggest that this is not proof of anything. If one insisted on embryonic recapitulation of all ancestral characters, then one would have to argue that the wingless conditions of biting and sucking lice and of grylloblattids is not the result of suppression of wings, since there is no ontogenetic indication of any wing rudiments in these insects, although they are all pterygote-related insects.

The body muscles of centipedes are generally more elaborate and numerous (Fig. 9.14). But even so, they can be derived from the arthropod ground plan. The dorsal tendons (DDT) have become small areas of fascia-like substance between the ends of the dorsal longitudinal muscles, but ventrally the tendons (VTT) are more elaborate, serving as anchor points for the ventral extrinsic coxal muscles and for the dorsoventral muscles (Fig. 9.15). In centipedes, the mandibles consist of sclerotic plates jointed to one another (Fig. 9.16), suggesting coxomeres. A new feature was acquired by centipedes, a suspensory bar (MS), which permits the jaw to be protruded and retracted, while it closes by means of the usual dorsal cranial and ventral muscles, most of which cause adduction. Such a bar has evolved convergently in Collembola (see Fig. 9.19, MS). The swinging tentorial apodeme of centipedes is involved in the opening of the jaw, as in other myriapods. The dorsoventral mandibular muscles have shifted from

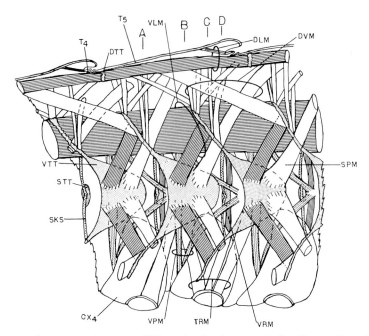

Fig. 9.14. *Scutigera coleoptrata* (Chilopoda, Scutigeromorpha).   Dorsal view of body musculature.   Fourth, fifth, and sixth body segments are shown as if opened dorsally and spread out.   Somatic muscles were omitted in lower (left) side to reveal some of leg muscles. Note reduced terga of fourth (T4) and sixth segments.   *A, B, C,* and *D* indicate level of sections shown in Fig. 9.15. Adapted from Manton (1965).

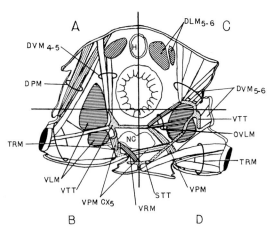

Fig. 9.15. *Scutigera coleoptrata* (Chilopoda).   Cross sections through body segment 5 at approximate levels *A, B, C,* and *D* of Fig. 9.14.   Each quadrant represents one-fourth of respective section. Adapted from Manton (1965).

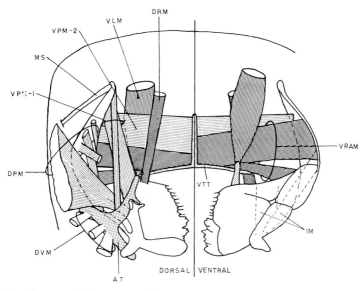

Fig. 9.16. *Cormocephalus nitidosus* (Chilopoda, Scolopendromorpha). Right mandible and associated muscles from ventral and dorsal views. Muscles DVM insert on tentorium instead of tendon, and aid in abduction. VPM-1 represents sectors that transferred insertions to tentorium, acting as retractors of tentorium, aided by VLM. Adapted from Manton (1964).

the ventral tendon to become the operators of the swinging anterior tentorium, pushing the jaw open.

9.2.6. Insecta

The dipluran body muscles are also derivable from the ground plan (Fig. 9.17). Dorsal tendons remain between the ends of the dorsal longitudinal muscles, but the dorsoventral and dorsal extrinsic limb muscle origins have shifted to the tergal cuticle. Ventrally, there are intersegmental tendons somewhat similar to those of thysanurans (Carpentier and Barlet, 1951) between the ends of the longitudinal muscles. Some of the ventral leg muscles and the ventral ends of the dorsoventral muscles originate on variously developed sternal cuticular apodemes. These apodemes are quite large, and are surrounded by ventral tendon material in the Japygidae (Barlet, 1965).

The body muscles of collembolans (Fig. 9.18) consist of powerful longitudinal muscles (DLM, VLM) adapted for jumping by increasing the hemocoelic pressure, which causes extension of the leaping organ. Their segmental ends are attached to each other through the dorsal or ventral tendons. As in diplurans, the dorsal extrinsic leg muslces originate on the terga. Ventrally, the tendon system (VTT)

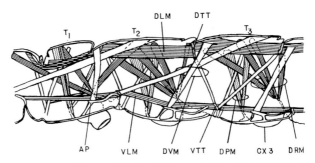

Fig. 9.17. *Campodea* sp. (Insecta, Diplura). Internal view of right half of thorax. Only small intersegmental tendons persist dorsally, and ventrally tendons are more extensive than shown. Origins of DVM, DRM, and DPM shifted to cuticle dorsally. Insertion of vertical DVM shifted to endosternal apodeme (AP). Adapted from Manton (1972).

is elaborated into a complicated system of struts (Carpentier, 1949), which resist the hemocoelic pressure, which might cause ballooning of the soft integument, especially at the acondylic coxa-body joints (Manton, 1964). The middle and hind coxae of collembolans are frequently subsegmented, with extrinsic muslces on the upper edges of both coxomeres (Fig. 9.18a,b). The Collembola are regarded as neotenous insects, and the upper coxomere (a) is here interpreted as

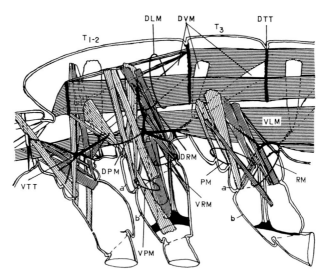

Fig. 9.18. *Tomocerus longicornis* (Insecta, Collembola). Thoracic musculature and tendons. Origins of DRM, DPM, and vertical DVM shifted from dorsal tendon to tergum. Ventral transverse tendons elaborated. T 1-2, combined prothoracic and mesothoracic nota. a, Upper coxomere (= catapleurite); b, lower coxomere (= definitive coxa of other insects). Adapted from Manton (1972).

the homolog of the catapleurite of the other apterygote insects, not yet incorporated into the pleuron. The anapleurite is usually very small.

Collembolan, proturan, and dipluran mandibles are simple, unjointed structures, as in other insects, and their operation is similar to the mandibular movements of machilids, with some specializations of their own, such as entognathy. Thus in the Collembola the mandibles have a rotating action, but also can be protruded and retracted, and weakly abducted, by judicious use of various sectors of the extrinsic coxal muscles (Fig. 9.19). The upper (inner) end of the mandible is suspended by a cuticular fold, which permits protrusion and retraction. Distally near the teeth, the anterior tentorium forms a support for the mandibles, emulating the secondary condyle of the pterygote mandible.

Dipluran mandibles operate as the collembolan mandibles do, except that protrusion and retraction are caused by muscles arising on a large transverse tendon. The anterior tentorium is suppressed in diplurans (Manton, 1964). In the Protura, the entognathous mandibles are slender, solid picks whose motions are mainly protrusion and retraction. Their muscles are simple, but consist of cranial muscles originating on the tentorium and transverse tendon.

The body muscles of machilids are highly specialized for leaping (Fig. 9.20), in that the dorsal and ventral longitudinal muscles form heavy bundles of twisted,

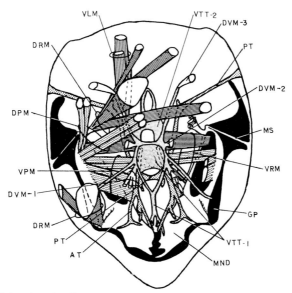

Fig. 9.19. *T. longicornis.* Mandibular musculature from above. Some muscles on left of head omitted to reveal underlying mandibular muscles. Mandibular VPM insertions shifted from mandibular VTT to anterior tentorium. DVM serve as suspensors of the VTT. VLM insert on mandibular (VTT-1) and maxillary (VTT-2) transverse tendons. Adapted from Manton (1964).

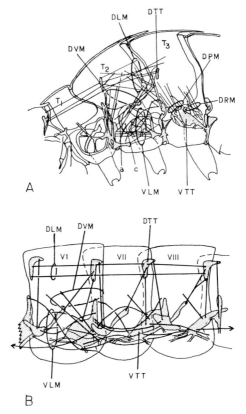

Fig. 9.20.   A, *Lepismachilis y-signata* (Insecta, Microcoryphia). Thoracic musculature and tendons. No muscles are shown in prothorax except for some dorsal muscles. Somatic muscles are shown in mesothorax, and extrinsic coxal muscles in metathorax. Anapleurite (a) and catapleurite (c) are shown in mesothorax. Muscles are so numerous that each is represented by a line. Adapted from Barlet (1967). B. *Trigoniophthalmus alternatus* (Insecta, Microcoryphia). Semischematic representation of abdominal musculature and tendons in sixth, seventh, and eighth abdominal segments. Note twisted ventral muscles (VLM). Dorsal and ventral tendons are separated medially and appear paired in each segment. Adapted from Bitsch (1973).

ropelike strands (Manton, 1972), which cause leaping through the strong pushing action of the abdomen on the ground together with the use of the legs. The dorsal and ventral transverse tendons are separated into laterally paired structures providing attachment points for most muscles, except a few dorsoventral muscles and dorsal extrinsic leg muscles. The ventral thoracic tendons form a complicated endosternite in each segment (Carpentier, 1949).

The machilid mandible is a primitive monocondylic rolling type with an anterior-posterior roll about a roughly vertical axis (Fig. 9.21). The upper end

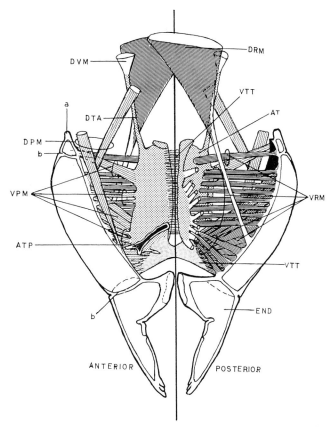

Fig. 9.21. *Machilis strenua* (Insecta, Microcoryphia). Right mandible and associated structures in anterior and posterior views. There is only one mandibular condyle (a). Internal sclerotic ridges (b) similate cardino-stipital and stipito-lacinial unions. Remnants of VTT persist between and below anterior tentorial arms (AT). Unshaded stumps on rear of tentorium at right are maxillary muscles.

of the mandible forms a ball lying in a socket of the head. Machilids cannot open or abduct the mandibles. In feeding, the picklike terminal endites scrape the food from surfaces. The particles are suspended in saliva, and the food particles may be treated by the milling action of the gnathal lobes before being swallowed (Manton, 1964). The anterior tentorial arms have invaded the region of the ventral mandibular tendon, so that many of the ventral muscles of the coxa originate on the anterior tentorium, except for a lower segment of muscles that originate on what is left of the tendon.

In thysanurans (Fig. 9.22), the dorsal tendons of the abdomen are entirely suppressed, but small patches remain dorsally in the thorax at the intersegments,

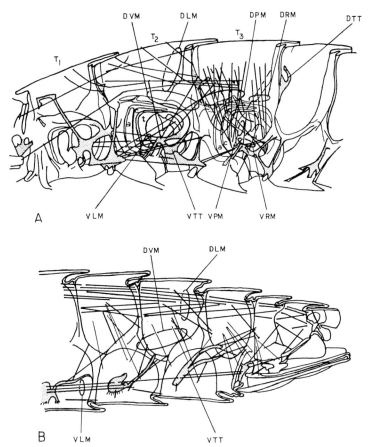

Fig. 9.22. *Lepisma saccharina* (Insecta, Thysanura).   A, Schematic representation of thoracic musculature and skeleton. Prothoracic muscles not shown. Longitudinal musculature (DLM, VLM) and dorsoventral muscles (DVM) only shown in mesothorax. Extrinsic leg muscles (DPM, DRM, VPM, VRM) only shown in metathorax. Ventral thoracic tendons are well developed, but dorsal tendons have been reduced and separated into lateral fragments. Adapted from Barlet (1951, 1953, 1954). B, *Thermobia domestica* (Insecta, Thysanura). Schematic representation of musculature and associated structures of sixth through eleventh abdominal segments. Dorsal tendons have been suppressed, and ventral tendons are reduced and laterally separated. Adapted from Rousset (1972). In both (A) and (B), muscles are represented by lines.

as in machilids (DDT).   Tendons remain ventrally, on which some ventral leg muscles and ventral longitudinal muscles originate.   These tendons, as in machilids, are separated medially in the abdomen, forming paired structures.   Other muscles attach to the cuticle at various places.   The thysanuran jaw (Fig. 9.23) is transitional between the monocondylic jaw of machilids, diplurans, and collem-

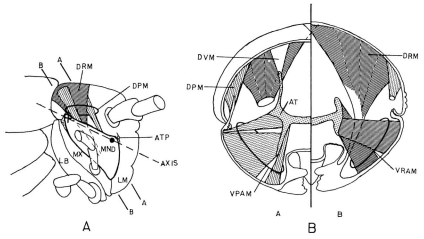

Fig. 9.23. *Ctenolepisma ciliata* (Insecta, Thysanura). A, Head, side view, represented as a transparent object. Sections *a-a* and *b-b* are shown in (B). B, Semivertical cross sections through mandible, with left side sectioned near front (a) and right side more posteriorly (b). There are no more head transverse tendons. Ventral muscles (VPAM, VRAM) originate on anterior tentorium (AT), and serve as adductor muscles, while dorsal muscles (DPM, DRM) serve as strong adductors and abductors. Mandibular axis is tilted nearly horizontally, and there is a secondary mandibular articulation shown by the solid black circle in (A). Adapted from Manton (1964).

bolans, and the typical insectan dicondylic jaw. The situation is apparently in parallel to the shift from a rolling type of coxa in the lower Crustacea to a strong biting dicondylic jaw in the Malacostraca. The axis of the thysanuran jaw is inclined, with a new ventral anterior articulation, convergent with the condition in *Anaspides*. The forward roll has become an opening action. The gnathal endite is shifted backward, and has become a biting surface rather than a grinding surface as in machilids. The posterior dorsal remotor (**DRM**) is huge, and is a powerful closing muscle, since the remotor roll has become a biting motion. The transverse mandibular tendon has disappeared except in *Tricholepidion*, leaving the anterior tentorium as the chief point of origin of the ventral muscles.

The pterygote insects have eliminated all their transverse tendons except for small thoracic ventral fragments at the intersternites. The muscles are attached to the cuticle by way of microtubules through the epidermis to muscle attachment fibers in the procuticle (Neville, 1975). In neopterous insects and ephemerids, the dorsal thoracic intersegments are deeply invaginated as phragmata providing attachment for the powerful dorsal longitudinal muscles and others that provide for the downstroke of flight action. However, fragments of the ventral tendons persist in the thorax of *Corydalus cornutus* (Fig. 9.24), situated over the intersternites bearing the spinae (Barlet, 1977).

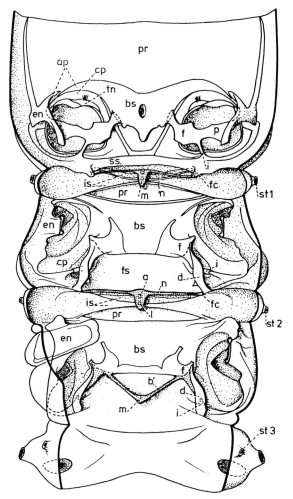

Fig. 9.24. *Corydalus cornutus* (Insecta, Megaloptera). Dorsal view into ventral side of larval thorax, showing persistent fragments of ventral transverse tendons (*m, n, b'*) at the intersegments associated with the spina (*a, l*). The labels are: *a* = first spinal attachment; *ap* = anapleural arc; *b'* = lateral arm of intersegmental transverse tendon of metathorax; *bs* = basisternite; *cp* = catapleural arc; *d* = postanapleural attachment to furca; *en* = endopleurite; *f* = cuticular furca; *fc* = furcilla; *fs* = furcisternite; *i* = pleural attachment; *is* = intersternite; *l* = second spinal attachment; *n* = lateral intersegmental tendon of pro- and mesothorax; *p* = prothoracic pleural process; *pr* = presternite; *ss* = spinisternite; *st* = stigmata; *tr* = trochantin. Figure courtesy of Barlet (1977).

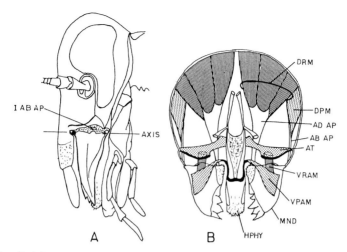

Fig. 9.25. *Periplaneta americana* (Insecta, Blattaria). A, Side view of head showing dicondylic-horizontal mandibular axis and insertion of abductor muscle (I AB AP), which is converted from mandibular DPM; B, cross section through mandible. As in *Ctenolepisma*, ventral muscles (VRAM, VPAM) have shifted their origins to tentorium and hypopharynx respectively, in absence of ventral transverse tendon, and are now adductors only. Huge DRM serves as main adductor. Adapted from Snodgrass (1952).

The mandibles (Fig. 9.25) are operated mostly by the dorsal extrinsic coxal muscles in adduction and abduction, similar to the condition in thysanurans. A few pterygotes such as cockroaches have retained some ventral muscles that have shifted their origins to the tentorium and hypopharynx in the absence of a mandibular tendon.

9.3. CONCLUSIONS

I have demonstrated that it is possible to interpret all the specializations of arthropod limb and jaw musculature as having a common origin from a simple ground plan (Fig. 9.1). A satisfactory beginning from a limb base articulated with the body through arthrodial membrane would account for the various types of coxa-body joints which include arthrodial membrane only (xiphosurans, trilobites), dorsal coxal condyles (some insects), ventral coxal condyles (myriapods), both dorsal and ventral condyles (proturans and higher insects), or fusion of the coxae to the body (arachnids). The muscles have become more elaborate, shifted in position, or eliminated in various ways. The transverse tendon system has survived in a modified fashion in primitive members of all classes of arthropods except in pycnogonids. Arthropod mandibles are all derived from coxae, retaining the basic leg extrinsic musculature with modifications. There seems to be no real basis for maintaining a theory of polyphyletic origin of arthropods based on functional morphology (for a different viewpoint see Manton, Chapter 7).

A new classification of the arthropods is proposed, illustrating a monophyletic system derived by the phylogenetic method (Hennig, 1953). The system can be illustrated with a typical cladogram, but it is here presented in the form of an outline. Monophyletic taxa are characterized by their unique, shared, derived characters (synapomorphies), which serve as ground plan characters. These synapomorphies form the basis for later transformation series of homologous character states, and in that sense become the ancestral characters (plesiomorphies) of descendant groups. Within taxa, the plesiomorphies often become autapomorphic, in that they are specializations restricted to a portion of the members of the taxon. Old names have been used whenever possible, but sometimes at a different taxonomic rank than originally proposed. The Tardigrada and Pentastomida are not treated in this paper. A discussion elaborating on the 17 ground plan characters of arthropods (Table 9.1) on a monophyletic basis is in preparation.

9.4. A PHYLOGENETIC CLASSIFICATION OF THE ARTHROPODS

Superphylum LOBOPODIA Snodgrass, 1938. Hypothetical Prearthropod

Loss of coelom except in embryos or in gonads and coelomoducts; acquisition of hemocoel, dorsal tubular heart with metameric ostia; meroblastic yolky eggs, with ventral germ band developing from blastoderm; loboform paired metameric appendages moved by extrinsic muscles, and with intrinsic muscles operating a terminal sclerotized piece; cuticle with chitin, primitively unsclerotized; specialized epidermal cells forming typical sensory setae; body muscles consisting of paired longitudinal dorsal lateral and ventral muscles, plus primitively a circular muscle layer; septal muscles reduced to paired dorsoventral intersegmental muscles; acronal palps enlarged into antenna-like structures.

I. Phylum ONYCHOPHORA Grube, 1853.
Early Cambrian to Recent.
Development of hemocoelic hydrostatic skeleton with: additional oblique muscle layer of body wall; extensive collagenous tissue under epidermis; conversion of dorsoventral intersegmental muscles into continuous ostiate diaphragms; development of circular vascular channels ("rings") under epidermis of body and legs; use of claw of leg I as jaw; leg II converted to slime papilla; originally eversible tubelike organs on limbs serving aquatic respiration (*Ayscheaia pedunculata*), presently restricted to one eversible vesicle per leg in some, used as water absorbing organ. Recent species with aerial respiration through numerous fine tracheal openings in grooves be-

tween body "rings"; reduction of gonads to one pair, the gonoducts uniting to a median posterior gonopore; head of acron plus two somites. Relatively primitive: unjointed lobopod (= coxa plus simple telopodite or claws); unsclerotized cuticle; ciliated gonoducts and nephric tubes; all muscles smooth except those of jaw; eyes simple.

II. Phylum ARTHROPODA Siebold and Stannius, 1848.
Sclerotized chitinous cuticle with periodic moults; ecdysial glands originally in cephalic region; limbs basically a coxopodite (limb base) formed from the main lobopod plus a telopodite consisting of several podomeres derived by segmentation of the terminal segment; eversible vesicle of limb base elaborated into respiratory gill (epipodite); endite lobes formed on medial part of coxopodite (= gnathobases); all muscles striated; longitudinal muscles became segmented, attaching to intersegmental tendons; somatic and extrinsic leg muscles primitively attached to intersegmental tendons; acronal palp formed into a short segmented antenna (= antennule); development of compound eyes; initial tagmosis of acron with somite 1; lateral tergal folds above limb bases (= paranota); subepidermal collagen elaborated into an intersegmental tendinous "endoskeleton" system, variously modified; loss of somatic circular muscle.

A. Subphylum CHELICEROMORPHA new subphylum.
Tagmosis of body into a trophic-locomotor prosoma of 6 or 7 somites, plus an abdominal opisthosoma of 12 or 11 somites; epipodites suppressed on prosomal limbs; opisthosomal limbs modified as gill-bearing structures or reduced; acronal antennules suppressed; limb of first metamere converted to a three-segmented pincer (chelicera); intracellular digestion in vacuoles of cells of midintestinal diverticula; meroblastic cleavage of eggs retained in the first several mitoses, secondarily becoming total in various ways, but always forming a blastoderm and embryonic ventral germ band plus extraembryonic dorsal blastoderm.

1. Infraphylum PYCNOGONIDA Latreille, 1810, new status. (Devonian to Recent.) Sea spiders.
Opisthosoma reduced or suppressed, without limbs; second legs palplike; third legs modified as ovigers; prosoma typically in two parts, the acron and first four somites fused into a cephalon, the next three somites typically not fused together; cephalon extended forward as a stiff proboscis with terminal mouth; compound eyes, paranota, coxal endites, intersegmental tendons all suppressed; precocious protonymphon larva with only three or four somites; tarsi subsegmented. A persistent relatively primitive condition is in the multiple paired gonads with gonopores on trochanters of locomotor legs.

2. Infraphylum CHELICERATA Heymons, 1901, new status.
Prosomal terga and paranota fused into a solid structure; opisthosomal

appendages of 9th through 13th somite modified into respiratory organs consisting of epipodites constructed in the form of flat lamellae (book-gills–book-lungs); appendages of 14th through 18th somites suppressed; gonopores fixed on genital segment (8th somite); ventral intersegmental prosomatic tendons united into a tendinous "endosternum"; ganglia of prosoma concentrated forward near pharynx; sulfur-bonding acquired as an additional cuticular tanning process; compound eyes without externally separate lens facets.

a) Superclass XIPHOSURIDA Latreille, 1802, new status. (Cambrian to Recent.) Horseshoe crabs.

Opisthosomal appendages shortened, flattened, serving as gill covers, their motion as paddles ventilating the book-gills, used in swimming, or burrowing; seventh metamere (pregenital) united with prosoma; stomodaeum with a masticatory gastric mill. Evolution within the group (autapomorphy) resulted in: restriction of opisthosoma to six metameres; fusion of opisthosoma into a solid tagma; elongation of telson spine; development of chelae on prosomatic legs; reduction of trochanters to one podomere; enlargement of coxal endites of five prosomal legs into gnathobases; reduction of seventh leg to a lobelike structure (chilarium).

b) Superclass CRYPTOPNEUSTIDA new superclass.

Opisthosomal book-gills or book-lungs in covered chambers; seventh somite free of prosoma, becoming first opisthosomatic; coxal mobility restricted; paranota reduced or suppressed.

(1) Class EURYPTERIDA Burmeister, 1843. (Ordovician to Permian, extinct.) Sea scorpions.

Changed to fresh-water life; posterior prosomal limbs usually elongated into swimming paddles; body scorpion-like, with prominent telson spine.

(2) Class ARACHNIDA Latreille, 1810. (Siluran to Recent.) Scorpions, spiders, mites, etc.

Changed to terrestrial, air-breathing animals; book-gills converted to sunken book-lungs, and/or tracheae developed; second legs modified into sensory-feeding palps (pedipalps); typically four pairs of ambulatory legs (on third through sixth metameres); first opisthosomal metamere (seventh, pregenital) highly reduced or suppressed; compound eyes suppressed; coxal mobility lost, with main leg-body joint between coxa and trochanter; tarsi subsegmented; food ingested suspended in intestinally produced liquid or ingested in liquid form; semen enclosed in spermatophores in most; ability to excrete nitrogen as insoluble guanine acquired.

B. Subphylum GNATHOMORPHA new subphylum.

Functional tagmosis into an anterior sensory-gnathal region including the acron and at least three (secondarily in various lines four, five, or seven) metameres, and a posterior trunk section with locomotor, respiratory, digestive, etc., functions; acronal antennae developed into the primary tactile-chemosensory organs; coxal gnathobases always well-developed on appendage of second somite, but primitively present on most coxae.

1. Infraphylum TRILOBITOMORPHA Størmer, 1944, new status.

Sensory-gnathal region fused into a solid head capsule including first three metameres; acron expanded laterally and posteriorly, partly surrounding the head somites; gnathobases of head limbs larger than other coxal endites. This group apparently retained respiratory epipodites on all the limbs.

a) Class TRILOBITA Walch, 1771. (Lower Cambrian to Devonian, extinct.) Trilobites.

Terga heavily sclerotized, with prominent paratergal lobes ("pleura"); terga firmly articulated permitting only dorsoventral bending. Early trilobites without tagmosis in trunk, later developing a pygidium consisting of unseparated variable numbers of posterior metameres. The three head limbs with telopodite fully developed, but with gnathal endites on coxae.

b) Class TRILOBITODEA Størmer, 1959. Name emended from Trilobitoidea. (Mid-Cambrian Burgess Shales, extinct.)

Paraterga not markedly set off from body, joining more or less smoothly; head appendages variously enlarged and adapted to seizing food.

2. Infraphylum MANDIBULATA Snodgrass, 1938.

First appendage modified by suppression of epipodite and gnathobase, and may be changed to function as accessory sensory structure (second antennae); telopodite of second appendage reduced to three podomeres (mandibular palp); epipodite of mandible lost; coxal endites suppressed on legs of trunk, but well developed at least on mandible; food digestion in lumen of midintestine through action of enzymes released by rupture of epithelial cells.

a) Subterphylum CRUSTACIFORMIA new subterphylum.

Tagmosis initially included union of part of the first metamere with the acron, forming a protocephalon, and union of second and third metameres (mandibular and maxillulary) into a single part. The first appendages became primarily tactile-chemosensory second antennae.

(1) Class CHELONIELLIDA Broili, 1933. (Devonian, extinct.)

Strong gnathobases on coxae of second, third, fourth, and fifth gnathal legs, with reduction of one podomere (trochanter?); paranota widely extended laterally resulting in forming a broad, disc-shaped body form.   A patella (?) appears to have been retained in the trunk appendages.

(2) Class CRUSTACEA Pennant, 1777. (Ordovician to Recent.)
Gnathal tagmosis extended to include the fourth metamere (second maxillary); epipodites were further suppressed on appendages behind the twelfth metamere (abdomen); the first trochanter (basipodite) of all appendages developed a prominent exite lobe, the exopodite; the patella became suppressed in all forms; antennules became reduced to eight or fewer musculated antennomeres; the mandible became the main biting organ; the first maxilla (maxillule) was highly reduced; posterior growth of the maxillary tergum produced a carapace in some; typical locomotor legs became reduced posteriorly on an abdominal tagma distinct from a postcephalic locomotor tagma (thorax); hatching became precocious with establishment of the typical nauplius larva with only three metameres.   Modifications typical of the various subclasses and orders.

b) Subterphylum MYRIAPODOMORPHA new subterphylum.
Derived characters related to emergence from aquatic to terrestrial life: air-breathing developed by invaginated metameric tracheae; return to uniramous limbs by suppression of epipodites (? = coxal eversible vesicles?); limbs adapted to terrestrial locomotion.

(1) Superclass ARTHROPLEURIDA Waterlot, 1934, new status. (Carboniferous, extinct.)
Very large (up to 180 cm long) centipede-like plant eaters; paraterga prominent, demarked from body; legs with all podomeres bearing spiny endites; heavy pleural plates around legs.   Head poorly preserved, telson unknown. Legs still retained patella. Relationship indicated provisional.

(2) Superclass ATELOCERATA Heymons, 1901.
First metamere suppressed except for its persistent ganglia (tritocerebrum) and embryonic limb buds; head capsule of acron plus three or four somites; levator muscle of pretarsus eliminated; patella suppressed; malpighian tubules from proctodaeum; diverticula of midintestine, when present (caeca), devoid of digestive function; telopodite of mandible eliminated; anterior tentorium established; coxal eversible

vesicles redeveloped as water-absorbing organs, variously modified or secondarily repressed in each class; distinct storage organ, the fat body, established; metameric tracheae established (also in Arthropleurida?); spermatophore established, with indirect transfer to female (convergent with Onychophora, Arachnida, some Crustacea). Primitive features retained but variously modified or later eliminated in each class include: individually musculated antennomeres; transverse intersegmental tendons; precocious hatching into a larva of nine or more metameres with anamorphic growth; continued moulting beyond sexually adult stage.

(*a*) Class MYRIAPODA Leach, 1814.

Mandibles bearing movable, articulated endite (gnathal lobe, gnathobase), abducted through outward thrust of swinging anterior apodemes; paratergal lobes reduced or eliminated; coxae with single ventral articulation with sternum; digestive diverticula entirely suppressed; ecdysis through transverse split behind head followed by exit from rear of head capsule and from front of trunk exuviae. Primitive atelocerate features; body behind head not formed into locomotor and nonlocomotor tagmata; limbs retained on most postcephalic somites.

i) Subclass COLLIFERA, new subclass.

Collum segment without legs, formed from fourth metamere, not joined with head as in other atelocerates; first (and only) maxillae consisting of a pair of united coxae whose telopodites (palps) became highly reduced; gonopore ventral between second legs (sixth metamere); antennules of eight or fewer musculated antennomeres; larvae at hatching with fully developed head and usually four postcephalic somites with three pairs of legs, plus two or more incomplete somites and the growth zone; dorsal intersegmental tendons suppressed.

(i) Infraclass PAUROPODA Lubbock, 1866, new status. (Fossils unknown.)

Trunk segment number reduced to 11 or 12; leg number reduced to 9, 10, or 11 pairs; antennae with four or six antennomeres, bearing two branches at tip on which a ringed flagellum forms; tracheal openings when present on coxae of legs, but tracheae usually entirely suppressed; tergal plates usually suppressed on collum and fourth, sixth, eighth, and tenth trunk segments (*Millitauropus* bears 12 trunk somites, each with a tergal plate); mandibles entognathous, consisting of a solid unsegmented unit; eyes suppressed; eversible coxal vesicles restricted to one pair ventrally on legless collum segment; intersegmental tendons lost except between mandibles; trochanter single; tarsus subsegmented with two tarsomeres.

(ii) Infraclass DIPLOPODA Blainville-Gervais, 1844, new status. (Carboniferous to Recent.) Millipedes.

Trunk somites beyond the fourth fused in pairs, forming diplosomites each bearing two pairs of ventral ganglia and two pairs of heart ostia; paraterga reduced or absent; mandible with coxa divided into two, with flexible gnathal lobe; tracheae in tufts on inner ends of hollow pleural apodemes; intersegmental ventral trunk tendons (Pselaphognatha) lie below the nerve cord (above, in all other arthropods), or are suppressed except between the mandibles; anamorphic growth from larva bearing head, four trunk segments, three pairs of legs, two diplosomites with limb buds and telson with zone of growth in most, but secondarily a trend toward epimorphosis in some (*Pachyiulus* hatches with 17 pairs of legs); compound eyes highly reduced to a few facets; lateral repugnatorial glands on side of body. Burrowing established by pushing into substrate, correlated with loss of tergo-coxal muscles and segmental tendons, sclerotization of cuticle into solid segmental rings, and extension of anamorphic growth to provide up to 100 or more diplosomites. The Pselaphognatha retained ancestral characters such as the persistent ventral intersegmental tendons, tergo-coxal muscles, paranota, fewer trunk segments (collum, three single and five double somites), soft exoskeleton, and palps on the gnathochilarium, as well as inability to burrow.

ii) Subclass ATELOPODA new subclass.

Fourth metamere united with head to form a head capsule of four somites plus acron; typical legs suppressed on last two trunk somites; antennae elongated, consisting of many musculated antennomeres; anamorphic larvae hatch with head and five to seven pairs of trunk appendages plus two or more partly formed somites.

(i) Infraclass SYMPHYLA Ryder, 1880, new status. (Oligocene Baltic Amber to Recent.)

Gonopore simple, anterior, on fourth trunk segment (eighth metamere); attain sexual maturity before completing anamorphosis; first maxillae palpless, with two endite lobes similar to galea and lacinia of insects, but without division into flexible cardo and stipes; coxae of second maxillae also palpless, and flexibly united into a labium-like structure; eyes suppressed; first trunk legs generally reduced, sometimes totally suppressed; tergites doubled and jointed flexibly on three or more trunk segments; intersegmental trunk tendons suppressed, replaced by sternal and coxal apodemes (mandibular and maxillary transverse tendons retained); tracheae restricted to head, with openings on side of head; 13th trunk segment legless, bearing tergal spinning organs; 14th trunk segment small, legless and undetached from telson; eversible vesicles and stylus-like pegs borne on venter near coxae; legs with single trochanter; tarsus undivided.

(ii) Infraclass CHILOPODA Leach, 1814, new status. (Cretaceous to Recent.) Centipedes.

Mandible retracted in head pouch, loosely articulated basally through sclerotic rod, protrusible and retractile, with several coxal sclerites; coxae of both maxillae generally united; first maxilla with short flat two-segmented telopodite serving as a labium; second maxillae with elongate palp-like telopodite; first leg with robust coxae bearing strong endites ("can openers" used in opening arthropod prey); telopodite of first leg modified into poison fang, with poison gland; compound eyes generally reduced or suppressed, but well developed in scutigeromorph centipedes; pleural sclerites acquired around leg bases; minimum number of metameres in trunk 22, up to 170 in some; last two metameres without legs (may be gonopods); gonopore on last metamere; anameric forms complete anamorphosis before sexual maturity; fat body from "mesoderm" cells rather than from "endoderm" as in other myriapods; gonads dorsal to intestine rather than ventral as in other myriapods; silk glands in genital chamber, except in scutigeromorphs; tarsus subsegmented into two or more tarsomeres.

(b) Class INSECTA Linnaeus, 1758, s.s. (= HEXAPODA Latreille, 1825). (Devonian to Recent.) Insects.

Head capsule consisting of four metameres fused with acron (convergent with pycnogonids, crustaceans, and atelopods); initial mandibular specialization a dorsal articulation with promotor-remotor rolling motion; coxa of first maxilla (insect maxilla) subsegmented into cardo and stipes, gnathobases articulated (galea and lacinia), telopodite (palp) shortened; coxae of second maxilla subsegmented, joined mesally into a flexible platelike labium, palp shortened; posterior tentorium acquired; first three trunk somites (fifth, sixth, and seventh metameres) specialized into a locomotor thorax; thoracic legs with a single trochanter; abdomen of 11 somites with limbs initially reduced to coxal plates bearing short segmented telopodite (abdominal styli) and eversible coxal vesicles; total number of metameres limited to 18 or less; 11th abdominal styli modified into sensory cerci; 10th abdominal styli and coxae suppressed; pleural areas of thorax initially sclerotized forming "anapleure"; thoracic coxae subsegmented, the upper portion joining the anapleure as a "catapleure" (anapleure and catapleure became pleuron in pterygotes); insectan median and lateral ocelli established in addition to compound eyes; secondary, simple gonopore posterior on abdomen formed by a median invagination meeting the two gonoducts; ecdysis through median dorsal split from head through thorax; midintestine diverticula lost the digestive function; spermatozoa modified in acquiring an extra ring of axonemes forming a 9 + 9 + 2 pattern, with two mitochondrial derivatives (see Baccetti, Chapter 11), *Ancestral atelocerate characters* in primitive insect, variously modified or suppressed in more advanced forms: individually musculated antennomeres; transverse intersegmental tendons in entire body; paranotal expansions; unsegmented tarsus; simple pretarsus; embryo not originally developing amnion; moulting continued beyond sexual adulthood; no external genitalia; simple rolling mandible; compound eyes; spermatophore with indirect sperm transfer. *New features* originating in various evolving lines: paranota becoming wings; return to aquatic life; social organization; external genitalia; internal fertilization; parasitism; pupation; extraembryonic membranes ("amnion" and "serosa"); dicondylic biting mandibles; sucking mouthparts; segmented tarsi; suppression of pretarsus; tarsal claws; secondary winglessness; filter chamber of intestine; and so on.

9.5. SUMMARY

A comparison of the various intersegmental tendons of members of major arthropod groups was made. A characteristic feature of the intersegmental tendons is that they serve as points of origin for somatic and extrinsic appendicular muscles. The primitive position is at the intersegments dorsally and ventrally. The tendons may be found in many members of all the arthropod groups, in variously derived states. In the primitive state, the dorsal intersegmental tendons are anchor points for the dorsal longitudinal muscles and the dorsoventral muscles, and serve as the origins of the extrinsic dorsal coxal promotor and remotor muscles. The dorsal tendons are attached to the cuticle through the epidermis at or near the intersegmental lines. The dorsal tendons have been lost in xiphosurans and most arachnids, but persist in the opisthosoma of primitive spiders. They persist also in primitive crustaceans and insects, and in some centipedes, but have become dorsally suppressed in most arthropod heads and in the trunks of higher crustaceans, symphylans, proturans, diplopods, lepismatids and pterygote insects. The ventral tendons, which are the most persistent, are generally suspended over the nerve cord by way of short muscles or tendon strands, serve as ventral anchor points for the ventral longitudinal muscles and the ventral ends of the dorsoventral muscles, and are the points of origin of the extrinsic ventral coxal promotor and remotor muscles. In chelicerates the ventral tendons have become elaborated and united into a fused prosomal plate known as the "endosternum." The opisthosoma may retain ventral remnants of the tendons, such as in primitive spiders, but generally the tendons are suppressed in the opisthosoma. Ventral tendons may be found in many crustaceans, myriapods except symphylans, the entognathous insects, machilids, and thysanurans. When the tendons are absent, the somatic and extrinsic coxal muscles are attached to the cuticle or to apodemes by way of transepidermal tonofibrillae. A discussion of the various ways the coxal and body muscles and the tendon system have become modified in various arthropods is given, supporting a concept of a monophyletic origin of arthropods. Analysis of the gnathal appendages and their muscles and tendons based on a monophyletic concept refutes the theory of the biting whole-limb ascribed to the myriapods and insects. All arthropod mandibles consist of modified coxae and their gnathobasic endites. A phylogenetic classification based on a monophyletic concept is proposed for the Arthropoda.

REFERENCES

Barlet, J. 1951. Morphologie du thorax de *Lepisma saccharina* L. (Apterygote Thysanoure). *Bull. Ann. Soc. Entomol. Belg.* **87**: 253–71.

Barlet, J. 1953. Morphologie du thorax de *Lepisma saccharina* L. (Apterygote Thysanoure). La musculature (1e partie). *Bull. Ann. Soc. Entomol. Belg.* **89**: 214–36.

Barlet, J. 1954. Morphologie du thorax de *Lepisma saccharina* L. (Apterygote Thysanoure). La musculature (2me partie). *Bull. Ann. Soc. Entomol. Belg.* **90**: 299–321.

Barlet, J. 1965. L'endosquelitte thoracique d'un japygide. *Proc. 12th Int. Congr. Entomol. Lond. Sec. 2. Morphol.* **1964(1965)**: 145–46.

Barlet, J. 1967. Squelette et musculature thoraciques de *Lepismachilis y-signata* Kratochvil (Thysanoures). *Bull. Ann. Soc. R. Entomol. Belg.* **103**: 110–57.

Barlet, J. 1977. Thorax d'apterygotes et de pterygotes holometaboles. *Bull. Ann. Soc. Entomol. Belg.* **113**:229–39. Paper presented at 15th Int. Congr. Entomol., Wash., D.C., 1976.

Bitsch, J. 1973. Morphologie abdominale des machilides (Insecta Thysanura). I. Squelette et musculature des segments pregenitaux. *Ann. Sci. Nat. Zool. 12th Ser.* **15(2)**: 12–199.

Blainville-Gervais. 1844. (Origin of the name Diplopoda) *Vide:* Laurentiaux, D. 1953. Classe des Myriapodes, pp. 385–96. *In* J. Piveteau (ed.) *Traité de Paleontologie.* Vol. III. Masson et Cie, Paris.

Broili, F. 1933. Ein zweites Examplar von Cheloniellon. *Sitz. Math-Nat. Abteil. Bayer Akad. Wiss. München* **1933**: 11–32.

Burmeister, H. C. C. 1843. (Origin of the name Eurypterida) *Vide:* Waterlot, G. 1953. Classe des Merostomes, pp. 529–54. *In* J. Piveteau (ed.) *Traité de Paleontologie.* Vol. III. Masson et Cie, Paris.

Carpentier, F. 1949. A propos des endosternites du thorax des collemboles. *Bull. Ann. Soc. Entomol. Belg.* **85**: 41–52.

Carpentier, F. and J. Barlet. 1951. Les sclerites pleuraux du thorax de Campodea (Insectes, Apterygotes). *Inst. R. Sci. Nat. Belg.* **27(4)**: 1–7.

Cisne, J. L. 1974. Trilobites and the origin of arthropods. *Science (Wash., D.C.)* **186(4158)**: 1–7.

Grube, A. E.   1853.   (Origin of the name Onychophora) *Vide:* Cuénot, L. 1949. Les Onychophores, pp. 32–37. *In* P. P. Grassé (ed.) *Traité de Zoologie.* Vol. VI. Masson et Cie, Paris.

Hennig, W.   1953.   Kritische Bemerkungen zum phylogenetische System der *Insekten. Beitr. Entomol.* 3 *(Sonderh.)*: 1–85.

Hessler, R. R.   1964.   The Cephalocarida. Comparative skeleto-musculature. *Mem. Conn. Acad. Sci.* **16**: 1–97.

Heymons, R.   1901.   Die Entwicklungsgeschichte der Scolopender. *Zoologica (Stuttg.)* **13**(33): 1–244.

Latreille, P. A.   1802.   Familles naturelles et generes. *In: Histoire naturelle, generale et particuliere des Crustaces et des Insectes.* Vol. III. Dufart, Paris.

Latreille, P. A.   1810.   *Considerations generales sur l'ordre naturel des animaux composant les classes des Crustaces, des Arachnides et des Insectes avec un tableau methodique de leurs genres disposes en familles.* Schoell, Paris.

Latreille, P. A.   1825.   *Familles naturelles du regne animale; succinctment et dans un ordre analytique, avec l'indication de leurs genres.* Bailliere, Paris.

Leach.   1814.   (Origin of the names Myriapoda and Chilopda) *Vide:* Laurentiaux, D. 1953. Classe des Myriapodes, pp. 385–96. *In* J. Piveteau (ed.) *Traité de Paleontologie.* Vol. III. Masson et Cie, Paris.

Linnaeus, C.   1758.   *Systema Naturae. I. Regnum Animale.* Edition 10. Holmiae, Upsala.

Lubbock, J.   1866.   On *Pauropus*, a new type of centipede. *Trans. Linn. Soc. Lond.* **26**: 181–90.

Manton, S. M.   1950.   The evolution of arthropodan locomotory mechanisms. Part 1. The locomotion of *Peripatus. J. Linn. Soc. (Zool.)* **41**: 529–70.

Manton, S. M.   1952a.   The evolution of arthropodan locomotory mechanisms. Part 2. General introduction to the locomotory mechanisms of the Arthropoda. *J. Linn. Soc. (Zool.)* **42**: 93–117.

Manton, S. M.   1952b.   The evolution of arthropodan locomotory mechanisms. Part 3. The locomotion of Chilopoda and Pauropoda. *J. Linn. Soc. (Zool.)* **42**: 118–66.

Manton, S. M.   1953.   The evolution of arthropodan locomotory mechanisms. Part 4. The structure, habits and evolution of the Diplopoda. *J. Linn. Soc. (Zool.)* **42**: 299–368.

Manton, S. M.   1956.   The evolution of arthropodan locomotory mechanisms. Part 5. The structure, habits and evolution of the Pselaphognatha (Diplopoda). *J. Linn. Soc. (Zool.)* **43**: 153–87.

Manton, S. M.   1958.   The evolution of arthropodan locomotory mechanisms. Part 6. Habits and evolution of the Lysiopetaloidea (Diplopoda), some principles of design in Diplopoda and Chilopoda, and limb structure in Diplopoda. *J. Linn. Soc. (Zool.)* **43**: 487–556.

Manton, S. M.   1961.   The evolution of arthropodan locomotory mechanisms. Part 7. Functional requirements and body design in Cobolognatha (Diplopoda), together with a comparative account of diplopod burrowing techniques, trunk musculature and segmentation. *J. Linn. Soc. (Zool.)* **44**: 383–461.

Manton, S. M.   1964.   Mandibular mechanisms and the evolution of Arthropods. *Phil. Trans. R. Soc. Lond. B. Biol. Sci.* **24**: 1–183.

Manton, S. M.   1965.   The evolution of arthropodan locomotory mechanisms. Part 8. Functional requirements and body design in Chilopoda, together with a comparative account of their skeleto-muscular systems and an appendix on a comparison between burrowing forces of Annelida and Chilopoda and its bearing upon the evolution of the arthropodan haemocoel. *J. Linn. Soc. (Zool.)* **45**: 251–484.

Manton, S. M. 1966. The evolution of arthropodan locomotory mechanisms. Part 9. Functional requirements and body design in Symphyla and Pauropoda and the relationships between Myriapoda and Pterygota. *J. Linn. Soc. (Zool.)* **46**: 103–41.

Manton, S. M. 1972. The evolution of arthropodan locomotory mechanisms. Part 10. Locomotory habits, morphology and evolution of the hexapod classes. *Zool. J. Linn. Soc.* **51**: 203–400.

Manton, S. M. 1973a. Arthropod phylogeny—A modern synthesis. *J. Zool. Proc. Zool. Soc. Lond.* **17**: 111–30.

Manton, S. M. 1973b. The evolution of arthropodan locomotory mechanisms. Part 11. Habits, morphology and evolution of the Uniramia (Onychophora, Myriapoda, Hexapoda) and comparisons with the Arachnida, together with a functional review of uniramian musculature. *Zool. J. Linn. Soc.* **54**: 257–375.

Millot, J. 1949. Ordre des Araneides (Araneae), pp. 589–743. *In* P. P. Grassé (ed.) *Traité de Zoologie*. Vol. VI. Masson et Cie, Paris.

Neville, A. C. 1975. *Biology of the Arthropod Cuticle*. Springer-Verlag. Heidelberg, Berlin, New York.

Pennant. 1777. (Origin of the name Crustacea) *Vide:* Dechaseaux, C. 1953. Classe des Crustaces, 255–56. *In* J. Piveteau (ed.) *Traité de Paleontologie*. Vol. III. Masson et Cie, Paris.

Rousset, A. 1973. Squelette et musculature des regions genitales et post-genitales de la femelle de *Thermobia domestica* (Packard). Comparaison avec la region genitale de *Nicoletia* sp. (Insecta: Apterygota. Lepismatida). *Int. J. Insect Morphol. Embryol.* **2**: 55–80.

Ryder, J. A. 1880. Scolopendrella as the type of a new order of Articulata. *Amer. Nat.* **14**: 375–76.

Siebold, C. T. E. von and H. Stannius. 1848. *Lehrbuch der vergleichenden Anatomie der wirbellosen Tiere*. Veit, Berlin.

Snodgrass, R. E. 1938. Evolution of the Annelida, Onychophora and Arthropoda. *Smithson. Misc. Collect.* **97**(6): 1–59.

Snodgrass, R. E. 1952. *A Textbook of Arthropod Anatomy*. Cornell University Press, Ithaca, N.Y.

Størmer, L. 1944. On the relationships and phylogeny of fossil and recent Arachnomorpha. *Skrift. Vid.-Akad. Oslo, I Math.-Nat. Kl. No.* **5**: 1–158.

Størmer, L. 1959. Trilobitomorpha, pp. O22–O540. *In* R. C. Moore (ed.), *Treatise on Invertebrate Paleontology*. Part O. Arthropoda. Geol. Soc. Amer. and University Press of Kansas, Lawrence, Kansas.

Walch, J. E. E. 1771. (Origin of the name Trilobita) *Vide:* Størmer, L. 1949. Classe des Trilobites, pp. 160–97. *In* P. P. Grassé, (ed.) *Traité de Zoologie*. Vol. VI. Masson et Cie, Paris.

Waterlot, G. 1934. Étude de la faune continentale du terrain houillier sarro-lorrain. *Etud. Gites Miner. Fr.* **2**: 1–320.

10

Significance of Sperm Transfer and Formation of Spermatophores in Arthropod Phylogeny

F. SCHALLER

10.1. INTRODUCTION

Annelid ancestors of arthropods were aquatic. In aquatic habitats three possibilities for sperm transfer can be found: 1) emission of sperm cells into the water and external insemination of eggs (a, without pair formation; b, with pair formation), 2) formation of spermatophores (a, with indirect transfer to the females—without or with pair formation; b, with direct transfer to the female—with pairing or even with mating), and 3) internal insemination of eggs (a, through gonopodial appendages—improper copulatory organs; b, through true copulatory organs). In recent annelids, types 1a and 1b, 2b, and 3b are found.

During the transition from aquatic to terrestrial life, type 1 proved to be ineffective. Therefore, the ancestors of terrestrial arthropods had to be annelids that either formed spermatophores or had copulatory organs. Formation of spermatophores and/or copulatory mechanisms are therefore considered as preadaptive requirements for the transition from aquatic to terrestrial life (Ghilarov, 1956, 1959; Schaller, 1965; Sharov, 1965, 1966). Actually, polychaets forming spermatophores can be found. Also, some terrestrial oligochaets and hirudineans transfer their sperm in packets (Plate 10.I).

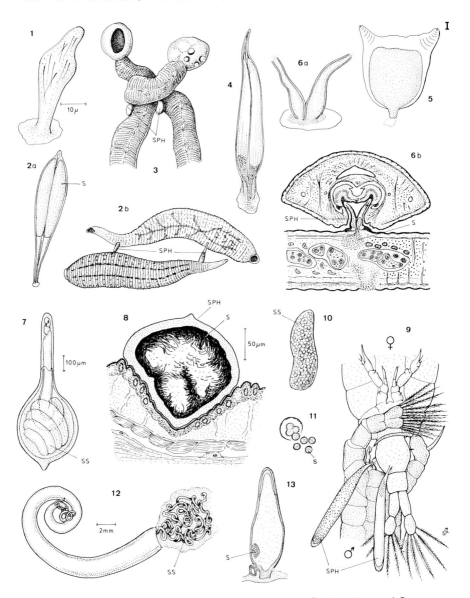

Plate 10.I   Spermatophores and sperm transfer in Annelida, Onychophora, and Crustacea. Annelida: Figs. 1–6b.   1. Spermatophore of *Microphthalmus aberrans.* From Westheide (1967).   2a.   Spermatophore of *Glossiphonia complanata.*   After Brumpt (1901), from Wesenberg-Lund (1939). 2b. Pairing of *G. complanata.* After Brumpt (1901), from Wesenberg-Lund (1939). 3. Pairing of *Piscicola geometra.*   (After Scriban-Autrum) (1934), from Grassé (1959). 4. Spermatophore of *Haementeria parasitica.* After Whitman (1891), from Grassé (1959). 5. Spermatophore of *Trachelobdella punctata.* After Whitman (1891), from

10.2. SPERM TRANSFER IN VARIOUS ARTHROPOD GROUPS

10.2.1. Primitive Arthropods

Most primitive arthropods most probably lived in aquatic habitats. This is certain for Gigantostraca, Trilobita, Xiphosura, and Crustacea. In these groups, external insemination was and still is generally possible and actually found in the Xiphosura. In *Limulus*, the male mounts the female and inseminates the freshly laid eggs from her back.

10.2.2. Crustacea

Commensurate with their richness in forms, crustaceans have developed many modes of sperm transfer. Many of them copulate using their extremities as copulatory organs, or as a means of sperm transfer. Also, spermatophores are often formed and transferred more or less directly to or into the female genital opening. Figure 10.1 shows the various inseminating behaviors found in crustaceans.

Spermatophores of crustaceans, however, are of primitive nature, without any specific covering or supporting elements (Plate 10.I). This is especially true for the gelatinous sperm balls of the Decapoda. Especially fresh-water crustaceans have developed organs and methods to guarantee to a great extent a direct means of sperm transfer. From a sexual-biological viewpoint, they have in many ways reached the stage where a direct transition to terrestrial life seems possible. Crabs and wood lice are especially successful in the occupation of terrestrial habitats. Their gonopods are so perfectly specialized that the sperm packets, being hardly exposed, can be transferred into the genital opening of the female. The great variability of structures and methods used for sperm transfer in crustaceans suggests the many analogies and convergences found in the development of the external sexual mechanisms. Sessile Cirripedia and parasitic Isopoda clearly demonstrate the predominant influence of ecological factors in the development of sexual appendages and pairing methods. Males of these two—

Grassé (1959). 6a. Spermatophore of *Erpobdella octoculata,* After Whitman (1891), from Grassé (1959). 6b. Pairing of *E. octoculata,* section. After Brumpt (1901), from Wesenberg-Lund (1939). Onychophora: Figs. 7, 8. 7. Spermatophore of *Peripatus corradoi*. After Bouvier (1907), from Grassé (1949). 8. Spermatophore of *Peripatopsis sedgwicki* on integument of a female. From Manton (1938). Crustacea: Figs. 9–13. 9. Pairing of *Diaptomus gracilis*. After Wolf (1905), from Meisenheimer (1921). 10. Spermatophore of *Cyclopsine castor.* From Bronn (1879). 11. Sperm capsule opened. From Claus (1889). 12. Spermatophore of *Scyllarides martensii.* From Matthews (1954). 13. Spermatophore of *Eupagurus meticulosus.* From Bronn (1895). *Abbreviations used on Plates 10.I–10.IV:* S = sperm; SD = sperm drop; SM = sperm mass; SPH = spermatophore (sperm package); SS = sperm sac; ST = spermatophore stalk; LD = liquid drop (with attractive pheromone?).

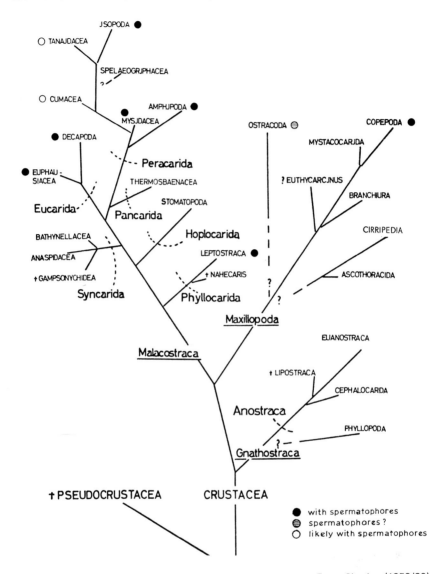

Fig. 10.1 Occurrence of spermatophores in recent crustaceans. From Siewing (1959/60).

not at all closely related—groups have evolved proper copulatory organs (penes). This demonstrates that reproductive biology apparently cannot contribute much to the clarification of the course of phylogeny and systematic relationships of Crustacea. This statement unfortunately is also valid when we consider the other groups of arthropods, which turned completely to a terrestrial life, namely, the Chelicerata, Myriapoda, and Hexapoda.

10.2.3. Chelicerata

There is no known transitional form from aquatic to terrestrial life for Chelicerata (Xiphosura, Eurypterida, Pycnogonida, Arachnida, Acarina) (cf. Fig. 10.2). *Limulus* cannot be considered as such. However, many spiders are confined to humid biotopes for ecological and physiological reasons; this means that they obviously have not achieved full emancipation from humid biotopes. This also accounts for their sexual biology. It is characterized in general by a) strongly indirect forms of sperm transfer, b) frequent usage of spermatophores, and c) formation of many structural and functional analogies.

Pairing is common in most Chelicerata (Abalos and Baez, 1963). Exceptions are found in pseudoscorpions and mites. The latter groups, commonly placed at the beginning of the chelicerate system, have strikingly similar types of spermatophores (Plate 10.II). These are found in scorpions, Pedipalpida (Uropygida and Amblypygida), and pseudoscorpions (Plate 10.III) (Kew, 1930). These groups form stalked spermatophores that are attached to the ground and are generally opened and/or emptied actively by the females. This leads to long and complicated pairing behaviors, with close (in scorpions, Uropygida, and many pseudoscorpions), less close (in Amblypygida and some pseudoscorpions)

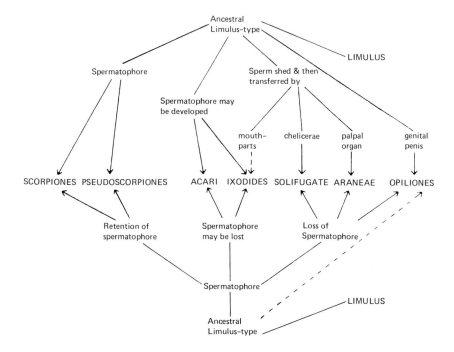

Fig. 10.2 Origin of different mating behavior types in Chelicerata. From Alexander and Ewer (1957).

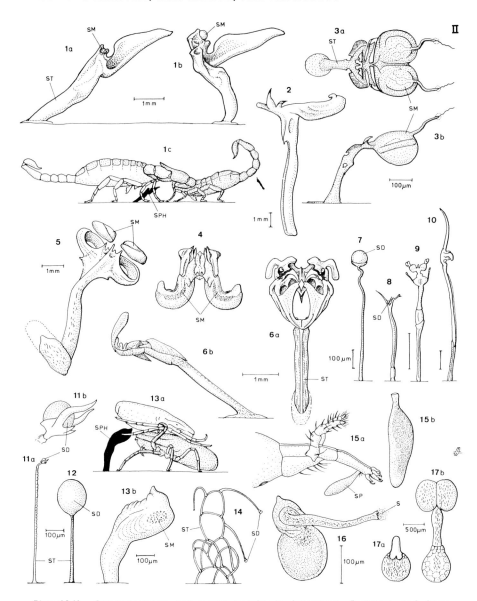

Plate 10.II. Spermatophores and sperm transfer in Scorpiones, Pedipalpi, and Acari. Scorpiones: Figs. 1, 2. 1. *Euscorpius italicus*; 1a, spermatophore (lateral view); 1b, spermatophore (opened); 1c, pairing. After Angermann (1957), from Schaller (1962). 2. Spermatophore of *Heterometrus petersii* (lateral view). From Nemenz and Gruber (1967). Pedipalpi: Figs. 3–6. 3. Spermatophore of *Trithyreus sturmi*; 3a, from above; 3b, lateral view. From Sturm (1958). 4. Spermatophore of *Mastigoproctus brasilianus,* from above. From Weygoldt (1972). 5. Spermatophore of *Damon variegatus,* with detached sperm

(Alexander, 1962a), and no bodily contact at all (some pseudoscorpions). Forms of the spermatophores and the modes of behavior lead to the assumption that these phenomena may be considered homologous. These modes of indirect spermatophore transfer can be considered as a requirement for the occupation of terrestrial environments in the above-mentioned groups of Chelicerata, as assumed by Alexander and Ewer (1957). This applies especially to the ancestors of recent scorpions, which probably had already developed spermatophores during their aquatic stage.

The sexual behavior in other chelicerate groups may derive from spermatophore transfer in scorpions, pedipalps, and pseudoscorpions, as presumed by Weygoldt (1966a,b). At any rate, they use their extremities as tools in sperm transfer, whereby it apparently becomes unnecessary to form complicated spermatophores (Plate 10.II). Beginnings of this direct form of sperm transfer are already apparent in amblypygids and pseudoscorpions, where, for example, the males of Cheliferidae use their forelegs to insert the spermatophore into the female genital opening. Without the formation of spermatophores, the males of Solifugida use their chelae (Muma, 1966), the male Araneae their pedipalps, and the males of Ricinulei their third pair of legs to transfer their sperm. Yet there are authors who also use the term spermatophore in the Solifugida (Cloudsley-Thompson, 1961), Ricinulei (Cooke, 1967), and some mites, even if only a most simple sperm mass of a probable mucilaginous composition is transferred. Remarkably, there are cases among pseudoscorpions where pair formation does not occur. The males deposit spermatophores without the females' being present. This means that the females are left on their own to find the sperm packets. This is mainly the case among the Chthoniidae, Neobisiidae, Garypidae, and Cheridiidae, where males therefore produce a great number of spermatophores.

Along with Schaller (1954b, 1971) and Weygoldt (1966a,b, 1970a,b), one can assume that sperm transfer without pair formation represents the more primitive mode. Increasing complexity of sexual structures and behavior patterns does fit

---

masses. From Alexander (1962b). 6. Spermatophore of *Tarantula marginemaculata;* 6a, frontal view; 6b, lateral view. From Weygoldt (1969). Acari: Figs. 7–17. 7. Spermatophore of *Belba gemiculosa*. After Pauly (1952), from Schaller (1962). 8. Spermatophore of *Cyta latirostris,* lateral view. From Alberti (1974). 9. Spermatophore of *Biscirus silvaticus,* frontal view. From Alberti (1974). 10. Spermatophore of *Bdella longicornis,* lateral view. From Alberti (1974). 11. Spermatophore of *Anystis baccarum.* From Schuster and Schuster (1966). 12. Spermatophore of *Calyptostoma velutinus.* After Theis and Schuster (1974). 13a, Pairing; 13b, spermatophore of *Saxidromus delamarei,* From Coineau (1976). 14. Combined spermatophores of *Nanorchestes amphibius.* From Schuster and Schuster (1977). 15. *Haemogamasus hirsutus*; 15a, spermatophore grasped by chelicera of a male; 15b, spermatophore. 16. Spermatophore of *Uroobovella marginata.* From Faasch (1967). 17. Spermatophore of *Ornithodorus savignyi*; 17a, at beginning; 17b, at end of evagination. From Feldman-Muhsam et al. (1973). Abbreviations same as on Plate 10.I.

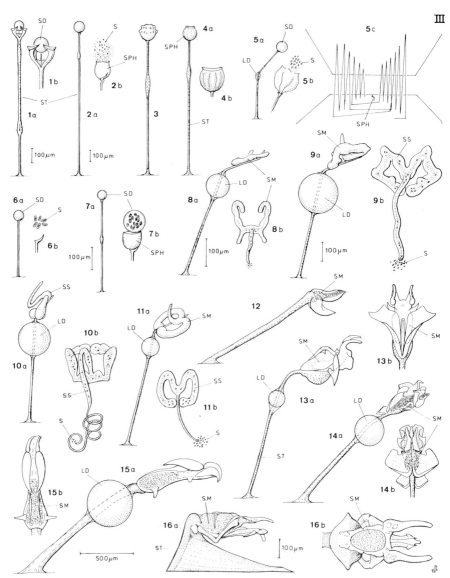

Plate 10.III.   Spermatophores of Pseudoscorpiones.   Figs. 1–16.   From Weygoldt (1966a,b, 1969, 1970a,b, 1975).   1. *Chthonius tetrachelatus.*   2. *Neobisium muscorum.*   3. *Pseudogarypinus marianae.*   4. *Pseudogarypus banksi.*   5. *Serianus* sp., spermatophore between hedge of web.   6. *Apocheiridium ferum.*   7. *Cheridium museorum.*   8. *Dinocheirus tumidus.*   9. *Chernes cimicoides.*   10. *Dendrochernes morosus.*   11. *Lasiochernes pilosus.*   12. *Paratemnus braunsi.*   13. *Chelifer cancroides.*   14. *Dactylochelifer latreillei.*   15. *Rhacochelifer disjunctus.*   16. *Withius subruber;* 16a, lateral view; 16b, from above.   Abbreviations same as on Plate 10.I.

very well with the conventional, morphologically consolidated system of pseudo-scorpionid families.   Both morphologically and ecologically primitive families, Chthoniidae and Neobisiidae, also demonstrate the more primitive form of sexual behavior.   Of course, one cannot ignore the consideration that pairing may be the primary type of sexual behavior (cf. *Limulus*), and that production of spermatophores without pairing is a secondary form, repeatedly derived there-from (cf. Fig. 10.3).   In any case, the morphological and functional specializa-tions of Solifugida, Araneida, and Ricinulei must be considered as parallel; this means that they positively have developed independently of each other.   This applies even more for the Opilionidae and families of mites, where males have developed true penes.   Also, from the sexual-biological viewpoint it is a com-plicated task to judge clearly the systematic placing of mites.   Just about all modes that can be considered for arthropods are found among them.   The sys-tematic occurrence of these phenomena does not lead to any conclusions con-cerning their evolutionary importance.   It rather seems that ecological factors may have played a great part in the differentiation of certain modes of reproduc-tion.   This means that multiple analogous mechanisms of sperm transfer have been developed.   This is especially true of the different cases of spermatophore production and indirect transfer of spermatophores among Acari (Plate 10.II) (Putmann, 1966; Feldman-Muhsam, 1967), especially in soil-dwelling oribatids, in different families of soil-inhabiting Trombidiformes (Schuster and Schuster,

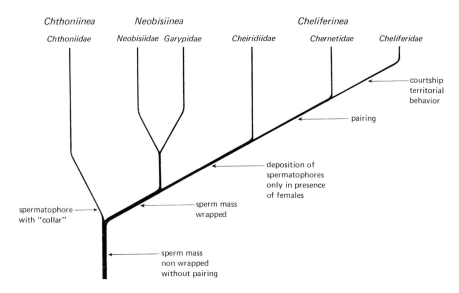

Fig. 10.3   Phylogeny of different forms of spermatophores and modes of mating in pseudo-scorpions. From Weygoldt (1966a).

1966, 1969, 1970, 1977), and in fresh-water Hydrachnellidae (Mitchell, 1958). The stalked spermatophores of these inhabitants of humid or aquatic biotopes are of the most simple structure. Sperm transfer with a difficult pairing ceremony very seldom occurs. It is hardly possible to come to conclusive results regarding the phylogenetic importance and evaluation of these differentiations. This applies even more for those mites that developed different copulatory mechanisms, such as many Parasitiformes, which use their chelicerae as gonopods, or even have developed a true penis, as in the Tarsonemi. Some Sarcoptiformes also copulate with a penis (e.g., Acaridae). In the oribatids, the most primitive method of indirect sperm transfer by means of simple stalked sperm drops appears in the systematically lower as well as higher groups. Only in a few exceptional cases is pair formation found (in *Collohmannia*, Schuster, 1962).

10.2.4. Onychophora

The derivation of the Tracheata (Onychophora, Myriapoda, Insecta) from their annelid ancestors is a well-known problem with nearly infinite possibilities for speculation. At any rate, there is a group of so-called Protracheata, which can be taken as an example of possible transitional characters. These are the Onychophora, which possess both annelid and arthropod characteristics to various extents. From a sexual-biological viewpoint, the Onychophora offer another good example, showing that the formation of spermatophores must have been advantageous for the transition from aquatic to terrestrial life. Males of *Peripatopsis* produce small, covered sperm packets, which are attached to the skin of the females. This behavior calls to mind similar observations in Polychaeta, where spermatophores also are attached to some part of the female skin (Westheide, 1967). The sperm have to penetrate through the body wall into the body cavity and female genital ducts (Manton, 1938). Bottle-like spermatophores of the genus *Peripatus* are supposed to be implanted by their neck into the female genital opening. Finally, males of *Paraperipatus* are reported to produce no spermatophores at all. Their sperm are transferred directly into the female spermatheca by an "intermittent organ" (Cérénot, from Grassé, 1949). Thus, in the Onychophora, three steps in the mode of sperm transfer can be found (Davey, 1960). However, the viviparity in Onychophora does not appear to be a primitive character. Therefore, the sexual conditions in different species of *Onychophora* cannot be used to come to any conclusions concerning the phylogenetic derivation of the Tracheata (Ruhberg and Storch, 1976).

10.2.5. Myriapoda

It is interesting to compare the sexual biology of the members of the Myriapoda. They are commonly known to appear in two groups, which differ mainly in the number of segments and legs. Their poly- or monophyletic origin is still debated (Snodgrass, 1938; Tiegs, 1940, 1947; Siewing, 1960; Dohle, 1964). The millipedes are obviously not a uniform group phylogenetically. This is also true for the Apterygota. The pterygote hexapods, however, are certainly monophyletic. The sexual-biological differentiation of the millipedes also has both evolutionary and adaptive causes. This applies especially for the small euedaphic forms of Pselaphognatha, Pauropoda, and Symphyla, which also exhibit the most primitive form of sperm transfer (Juberthie-Jupeau, 1963). They deposit bare sperm drops without direct bodily contact between the sexual partners. The usually larger and more hemiedaphic Chilopoda and Diplopoda generally form pairs or use their gonopods as copulatory tools (Klingel, 1960, 1962). However, no millipede has ever developed true copulatory organs. The different positions of the genital openings in pro- and opisthogoneate millipedes definitely influence the course of their sexual behavior, but seemingly have no meaning concerning the principal mode of sperm transfer.   Most millipedes produce spermatophores that are deposited and taken up more or less actively by the females. In a few cases, they are transferred directly (Plate 10.IV).

Within the different major groups of millipedes, phylogenetic series of increasingly complex and safe methods of sperm transfer can be demonstrated. These, however, have no value for deductions concerning the descent and relationship of the groups among themselves. This holds true especially for the Diplopoda, where increasingly complex gonopods are found.

Because of their eye and tracheal structures, the Notostigmophora are generally considered to be the original form of the Chilopoda. It is noteworthy that they also possess the most primitive form of sperm transfer.

Symphyla deserve special attention, since they are generally considered as a possible basic group from which the Hexapoda originated. Their males produce most primitive stalked sperm drops, which have to be taken up actively by the females. The mode of production and the structure of these spermatophores are strikingly similar to those of the Diplura and Collembola. Even so, definite conclusions of a close relationship cannot be drawn, since exactly the same simple type of spermatophore can be found also among mites, as well as in pseudoscorpions. Again, in this case this instead refers to the common ecological situations of these systematic groups, which have led to principally similar reproductive methods. At any rate, the deposition of concentrated sperm drops seemed to be one of the most preferred preadaptive conditions for the transition from aquatic to terrestrial life for Arachnida, as well as for the primitive forms of Antennata.

For phylogenetic reasons, it is significant to note that the fertilization of eggs in Symphyla is performed externally, in that the sperm-containing drops from the female gnathal sperm pockets are smeared onto the eggs during the process of laying.

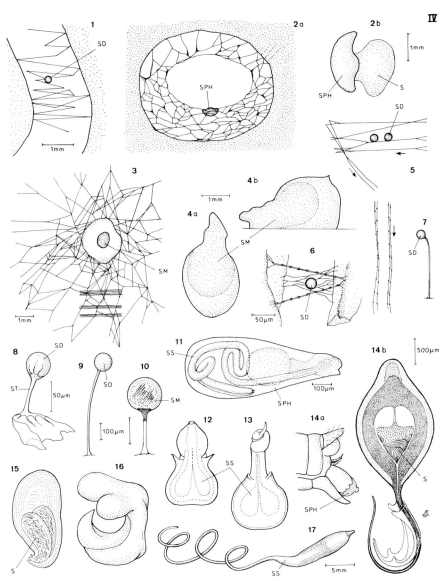

Plate 10.IV. Spermatophores and sperm drops in Myriapoda and Insecta. Myriapoda: Figs. 1–7. 1. *Geophilus longicornis,* sperm drop on web in soil slot. After Klingel (1959), from Schaller (1971). 2. *Scolopendra cingulata*; 2a, spermatophore in web in soil gallery; 2b. spermatophore. From Klingel (1960). 3. *Lithobius forficatus,* web with spermatophore and slime strip. From Klingel (1959). 4. *Scutigera coleoptrata,* spermatophore; 4a, from above; 4b, lateral view. From Klingel (1960). 5. *Polyxenus lagurus,* web with two sperm drops and signal thread. After Schömann (1956), from Schaller (1971). 6. *Stylopauropus*

10.2.6. Hexapoda

10.2.6.1. Apterygota

Sexual behavior of apterygotes is fully characterized by sperm drops or sperm packets, which may be stalked or deposited on webs. The uptake of these sperm containers by the female may occur with or without contact between sexes. Unfortunately, information about the Protura is totally lacking. Male Collembola produce stalked sperm drops in great numbers, because receptive females have to find them without any help or guidance from the males. A smaller number of spermatophores also are produced in case the males partici-pate more actively during spermatophore transfer. Thus, frequent and intensive contacts (true pairing) in more specialized Collembola (Sminthuridae: Bretfeld, 1970; Betsch-Pinot, 1976; Massoud, 1976) also occur. Direct sperm transfer by copulation, however, does not occur within the Collembola. Diplura (*Campodea* and *Japyx*) usually produce stalked sperm drops similar to those in Collembola (Schaller, 1954a; Bareth, 1964, 1965, 1966). The pairing behavior of the ectognathous Thysanura (Machilidae and Lepismatidae) is much more complicated. Their males spin threads and "help" during the sperm transfer. The sperm is exposed for a certain time during this procedure, making this a process that can only be conducted in a humid environment such as on or within soil (Tuxen, 1970). It is quite remarkable that both sexes of Thysanura (Archaeognatha and Zygentoma) already have paired external genital appendages with structures coinciding with corresponding appendages of pterygote insects. Yet unlike those in the latter, they are not used during pairing. As long as one does not accept coincidental convergences (as some authors do), this seems to represent an interesting case of complex preadaptive differentiation. It is very important to note that the Zygentoma (*Lepisma*) produce a bottle-like sper-matophore, which can be compared to the sperm containers in Orthoptera (Plate 10.IV).

---

pedunculatus, web in soil slot with sperm drop. After Laviale (1964), from Schaller (1971). 7. Spermatophore of *Scutigera immaculata*. After Juberthie-Jupeau (1959), from Schaller (1971). Insecta: Figs. 8–17. 8. Spermatophore of *Podura aquatica*. From Schliwa and Schaller (1963). 9. Spermatophore of *Orchesella flavescens*. From Schaller (1952). 10. Spermatophore of *Campodea remyi*. From Bareth (1964). 11. Spermatophore of *Lepisma saccharina,* lateral view. From Sturm (1956). 12. Spermatophore of *Leptophyes albovittata.* After Boldyrev, from Grassé (1977). 13. Spermatophore of *Poecilimon bosphoricus*. After Boldyrev from Grassé (1977). 14. *Liogryllus campestris;* 14a, end of abdomen with sper-matophore; 14b, spermatophore, partially longitudinal section. From Regen (1924). 15. Spermatophore of *Blattella germanica*. From Beier (1974). 16. Spermatophore of *Leuco-phaea maderae*. From Beier (1974). 17. Spermatophore of *Pimelia sardea* (Tenebrionidae). From Dzimirski (1963). Abbreviations same as on Plate 10.I.

10.2.6.2. Pterygota

All pterygotes copulate, whereby many primitive forms still transfer spermatophores (Davey, 1960; Schlee, 1969). The pairing behavior of palaeopteran Odonata is quite peculiar. The males have developed special gonopodial appendages on the frontal part of the abdomen, which are used for sperm transfer. These appendages are situated on the second and third abdominal segments—thus roughly at the same place where the genital opening is found in the progoneate millipedes.

10.3. PHYLOGENETIC CONSIDERATIONS

A comparative review of the varieties of sexual behavior in arthropods clearly demonstrates that, especially at the beginning of their terrestrial differentiation, formation and transfer of spermatophores are predominant. During the transition from purely aquatic to terrestrial life, many arthropod forms remained in humid terrestrial environments, especially soil, as their adaptations proved to be sufficient for such biotopes. Such ecological confinements of certain arthropod groups exceedingly complicate the phylogenetic explanation and evaluation of their structural and functional characteristics. This is especially true for their sexual-biological differentiation and adaptations. The ability to produce spermetaphores seems to be a form of preadaptive differentiation, which was reached in all annelid-like arthropod ancestors long before their occupancy of land. This also seems to be the case in lower Chelicerata (Xiphosura, Eurypterida, Pycnogonida, Arachnida, Acarina), as well as lower Antennata (= Tracheata) (Myriapoda, Hexapoda), whose recent forms have been derived from that transitional stage. The type and formation of spermatophores and the methods of their transfer show no signs of a mono-, di-, or polyphyletic origin. During the periods after the transitional phase, all arthropod groups have developed a variety of adaptations and differentiations to such an extent that their possible prototypes can only be thought of as purely abstract constructions.

More can be said about the systematic position of the other classes of arthropods. The terrestrial Chelicerata primarily exhibit indirect transfer of spermatophores. This is demonstrated through striking similarities in the structure of their stalked spermatophores. Many other aspects also lead to a common ancestry of scorpions, pedipalps, and pseudoscorpions. From the sexual-biological viewpoint no clear statements can be made on the ancestry of the other Chelicerata. At least from that point of view, a monophyletic origin of terrestrial Chelicerata also seems possible.

Possible relationships between Chelicerata and Crustacea can be neither denied not confirmed on that basis. This also applies to the controversial relationship between Antennata (= Tracheata) and Diantennata (Crustacea). However, some interesting aspects of sexual biology of the Antennata, having some bearing on its systematic position, can be found. Primitive spermatophores (stalked sperm drops) of Symphyla, Diplura, and Collembola on one side and similar primitive sperm webs of Pauropoda and Pselaphognatha on the other may be assumed to be purely ecologically conditioned analogies or morphological homologies. After all, there are still other characteristics that are common in Symphyla and entognathous hexapods. Nevertheless, it would be incorrect to regard the living Symphyla as ancestors of the entognathous Apterygota of today. Similarly, the ectognathous Zygentoma (Thysanura: Lepismatidae) cannot be considered ancestors of Pterygota on the basis of the similarity in their genitalia. It is indisputable that the hexapods must have had myriapod-like polypodous ancestors. This is specially proved by their embryology. The primitive method of indirect sperm transfer, as found in living Myriapoda and lower Apterygota, also suggests a common ancestry. Since the tracheal system of Myriapoda, apterygote, and pterygote insects cannot apparently be considered homologous, it would be better to avoid the term Tracheata and use the term Antennata instead. From the sexual-biological viewpoint also, no indications are found for a close relationship between the Diantennata (Crustacea) and Antennata that would justify a class such as Mandibulata (Tillyard, 1930; Manton, 1938, 1965, 1969a,b, 1972; Tiegs and Manton, 1958; Matsuda, 1958, 1976). Pre-Cambrian ancestors of euarthropods, however, were definitely preadapted for the transition to a terrestrial life, even in their ability to produce spermatophores (Lauterbach, 1973, 1975; Kristensen, 1975; Kraus, 1976).

It is quite uncertain whether there was a primitive, completely unspecialized, ancestral form from which the prototype of the Crustacea, Trilobita, Xiphosura, Chelicerata, and Antennata descended, since parallel, synchronously or metasynchronously appearing ancestors also could be possible. According to Hennig (1953, 1969) the phenomena of spermatophore production must have been a symplesiomorphic basic characteristic in archeo-arthropods.

The method of indirect spermatophore transfer apparently appeared repeatedly as an apomorphic characteristic being convergently specialized. Without any doubt, one of the most striking convergences is the formation of spermatophores in Collembola and oribatids. Since crustaceans, generally speaking, never left aquatic environments, they do have the ability to produce spermatophores; yet they have never reached the stage of indirect transfer of spermatophores. This also demonstrates that crustaceans must have descended separately from all other euarthropods. In accordance with that, they have developed more direct forms of sperm transfer; frequently they even produce sperm without flagella. Most sexual differentiations in arthropods have to be considered as phylogenetically recent, and as varied and graded acquisitions. This also applies to the remarkable shift in their genital apertures. Pro- and opisthogoneate forms are found not only in Myriapoda, but also in all the three main groups, such as the Diantennata, Chelicerata, and Antennata. Yet, the most striking differentiations are found in Myriapoda. It is questionable whether the position of the genital opening has such a significant systematic value as is assumed by several authors. The position of the genital openings varies in annelids, probably owing to the formation of the epitoke segments, which primarily may have served in vegetative reproduction. It is quite possible that such epitoke regions were dropped after the arthropodization at various developmental stages. In addition to that, it seems likely that genital apertures were shifted frequently even at the arthropod stage. Naturally, the shifting cannot be taken literally. It seems possible that from the original plurisegmental genital openings one finally remained in a more progoneate or opisthogoneate position.

Thus, again it has to be stated that the question of the mono- or polyphyletic origin of arthropods cannot be answered by means of their reproductive biology. It is certain that sperm transfer through spermatophores was used in all groups during the transition from aquatic to terrestrial life. However, it remains debatable whether spermatophores of crustaceans, arachnids, millipedes, and insects are homologous or not. However, in the discussed groups, types of spermatophores and the modes of sperm transfer provide strong support for clarifying systematic problems at the familial and generic levels. This has been clearly demonstrated by the extensive comparative studies on pseudoscorpions by Weygoldt (1966a,b, 1969, 1970a, 1975).

The following statements may contribute to the discussion on whether or not the insects have mono- or polyphyletic origin. The so-called Ur-Insekten (Apterygota) are of a possible polyphyletic origin. Spermatophores of Collembola and Diplura are of similar structure, yet compared with the Thysanura (Machilidae and Lepismatidae) they show fundamental differences. Only in Zygentoma (Lepismatidae) can sperm containers be found, which seem to be comparable to simple spermatophores in lower pterygotes. Thus the Zygentoma also can be considered, by means of their reproductive biology, as a possible ancestral group of Pterygota with orthopteran "Legeapparat" and dicondylic mandibles. More primitive forms of Hemimetabola produce spermatophores that consist of a gelatinous matrix, which covers one to several sperm sacs. If the spermatophores are attached only externally at the genital opening of the female, they can be furnished with additional special covers, as in the Gryllidae (Khalifa, 1949b). More commonly, however, they are secondarily simplified, and consist of only a gelatinous or mucous capsule, such as in bugs (*Rhodnius*; Davey, 1959).

All orders of Orthoptera produce spermatophores. Those of the Phasmida are shaped in the form of a gelatinous club, which is attached at the female genital opening with a short stalk (Chopard, 1934). Similar spermatophores of Mantidae are shown under the subgenital plates of the females and seem to be better protected that way. Spermatophores of cockroaches are unstalked and also covered by the female subgenital plates. In some species, they disappear under the plates (Gupta, 1947; Khalifa, 1950). Tettigoniidae and Gryllidae have spermatophores similar to those of Phasmida (Khalifa, 1949a,b). The gelatinous matrix is reduced in the Acridiidae. The sperm mass is formed into a pseudo-spermatophore ("sperm-sac") through the penis tube, which is expanded during copulation.

In some members of Hemiptera, spermatophores are produced that are of a normally quite simple structure: gelatinous masses forming sperm plugs, which are inserted into the female genital opening (e.g., *Rhodnius*). Other bugs do not produce spermatophores any more. Yet, they still are provided with homologous appendical glands, which produce sperm secretions (*Oncopeltus*; Bonhag and Wick, 1953).

Spermatophores are described for many Lepidoptera. They always consist of one or several sperm sacs that are embedded within a gelatinous matrix and inserted into the bursa copulatrix. This is also true for some trichopteran families. Most Neuroptera, however, produce typical bottle-like spermatophores, which are inserted into the female genital opening only by their neck portion. Yet, it is certainly unjustifiable to consider them as direct relics of the Onychophora, as suggested by Davey (1960).

True spermatophores are extremely rare in Hymenoptera. Sporadically, they are found in Coleoptera, such as Tenebrionidae and Scarabaeidae. The same is true for Diptera, where spermatophores are especially found in lower forms (Nematocera) (Davies, 1965; Downes, 1968; Pollock, 1972).

Generally, it can be said that pterygote insects demonstrate a tendency toward reduction of spermatophores and toward the development of copulatory organs, so that the sperm could be transmitted in a liquid form (Imms, 1937, 1945; Alexander, 1964). Yet, in all apterygote insects spermatophores are transmitted directly to or into the female genital opening (Gerber, 1970). In apterygote insects, only the simple bottle-like spermatophores of *Lepisma* can be compared to the spermatophores of lower pterygote insects (Orthoptera and cockroaches). This is an indication that the Zygentoma (Lepismatidae) come closest to the pterygote insects.

Finally, it should be stated that the term spermatophore is used uncritically by all authors. Complicated structures such as the spermatophores in scorpions, simple bare sperm drops on stalks in Collembola, or those on webs in Machilidae, primitive bottle-like sperm containers in *Lepisma*, complicated sperm capsules with special attaching or opening mechanisms in Gryllidae, and finally the primitive gelatinous sperm ball or sperm tubes in higher pterygote insects, are neither morphologically nor functionally truly comparable. Questions of homologies may be answered only after the modes of formation, and microscopic and ultrastructural details, as well as the chemical nature of the components of the various types of spermatophores, have been sufficiently analyzed.

10.4. SUMMARY

Indirect sperm transfer is commonly found in recent terrestrial arthropods, especially in soil-inhabiting forms. Their aquatic ancestors released their eggs and sperm into the water and therefore exhibited "external insemination." During the transition to terrestrial life, this primitive form of fertilization became impossible. The delicate sperm cells had to be covered with protective secretion and coverings. Thus, sperm balls, sperm packets, and sperm containers on or without stalks were developed, their various forms being referred to as spermatophores. Such spermatophores are found in annelids, crustaceans, Arachnomorpha, Onychophora, Myriapoda, and Hexapoda.

The modes of formation, structure, and method of transfer of spermatophores are so variable that important phylogenetic conclusions cannot be drawn on their bases. Adaptation to life in soil, however, has led to striking similarities, which can be looked at only as analogies, such as the spermatophores of Collembola and oribatid mites. However, among arachnids, millipedes, and lower insects, it is possible to postulate homologies in the formation and transfer of spermatophores. This applies especially to the basically uniformly stalked spermatophores in scorpions, Uropygi, Amblygi, and pseudoscorpions, which are firmly considered to be of monophyletic origin. The question of monophyletic or polyphyletic origin of Apterygota and Pterygota seems rather difficult to answer on the bases of their sexual biology. It can be stated only that spermatophores were produced by all their ancestors, their primitive recent forms still producing them today. In most cases these spermatophores are transmitted directly. Higher insects generally possess morphologically reduced spermatophores, transferring sperm balls or fluid sperm directly.

ACKNOWLEDGMENTS

I thank Prof. Maria Mizzaro-Wimmer for illustrations, and Dr. Walter Hödl for translation into English.

REFERENCES

Abalos, J. W. and E. C. Baez. 1963. On spermatic transmission in spiders. *Psyche* **70**: 197–207.

Alberti, G. 1974. Fortpflanzungsverhalten u. Fortpflanzungsorgane der Schnabelmilben (Acarina: Bdellidae, Trombidiformes). *Z. Morphol. Tiere* **78**: 111–57.

Alexander, A. J. 1962a. Courtship and mating in amblypygids (Pedipalpi, Arachnida). *Proc. Zool. Soc. Lond.* **138**: 379–83.

Alexander, A. J. 1962b. Biology and behavior of *Damon variegatus* Perty of South Africa and *Admetus barbadensis* Pocock of Trinidad, W. I. (Arachnida, Pedipalpi). *Zoologica (Stuttg.)* **47**: 25–37.

Alexander, A. J. and D. W. Ewer. 1957. On the origin of mating behavior in spiders. *Amer. Nat.* **91**: 311–17.

Alexander, R. D. 1964. The evolution of mating behavior in arthropods, pp. 78–94. *In* K. Highnam (ed.), *Insect reproduction*. R. Entomol. Soc., London.

Bareth, C. 1964. Structure et dépôt des spermatophores chez *Campodea remyi*. *C.R. Acad. Sci. Paris* **259**: 1572–75.

Bareth, C. 1965. Le spermatophore de *Lepidocampa* (Diploures Campodéidés). *C.R. Acad. Sci. Paris* **260**: 3755-57.

Bareth, C. 1966. Études comparatives des spermatophores chez les Campodéidés. *C.R. Acad. Sci. Paris* **262**: 2055-58.

Beier, M. 1974. Blattariae (Schaben). *Handb. Zool.* **4(2)**: 1-127.

Betsch-Pinot, M.-C. 1976. Le comportement reproducteur de *Sminthurus viridis* (L.) (Collembola, Symphypleona). *Z. Tierpsychol.* **40**: 427-39.

Bonhag, P. F. and J. R. Wick, 1953. The functional anatomy of the male and female reproductive organs of the milkweed bug. *J. Morphol.* **93**: 177-284.

Bretfeld, G. 1970. Grundzüge des Paarungsverhaltens europäischer Bourletiellini (Collembola, Sminthuridae) und daraus abgeleitete taxonomisch-nomenklatorische Folgerungen. *Z. Zool. Syst. Evolutionsforsch.* **8**: 259-73.

Bronn, H. G. 1879. *Klassen und Ordnungen des Tierreichs.* Bd. 5, 1.Abt. Winter'sche Verlagshandlung, Leipzig.

Bronn, H. G. 1895. *Klassen und Ordnungen des Tierreichs.* Bd. 5, 2.Abt. Winter'sche Verlagshandlung, Leipzig.

Chopard, L. 1934. Sur la présence d'un spermatophore chez certaines inséctés orthoptères de la famille des Phasmides. *C.R. Acad. Sci. Paris* **199**: 806-7.

Claus, C. 1889. Über den Organismus der Nebaliiden und die systematische Stellung der Leptostraken. *Arbeiten aus dem Zool. Inst. Wien* **8**.

Cloudsley-Thompson, J. L. 1961. Observations on the natural history of the "Camel-Spiders," *Galeodes arabs* C. L. Koch (Solifugae: Galeodidae) in the Sudan. *Entomol. Mon. Mag.* **97**: 145-52.

Coineau, Y. 1976. Les parades sexualles Saxidrominae (Acariens Prostigmates, Adamystidae). *Acarologia* **18**: 234-40.

Cooke, J. A. L. 1967. Observations on the biology of *Ricinulei* (Arachnida) with description of two new species of *Cryptocellus*. *J. Zool. Lond.* **151**: 31-42.

Davey, K. G. 1959. Spermatophore formation in *Rhodnius prolixus*. *Q. J. Microsc. Sci.* **100**: 221-30.

Davey, K. G. 1960. The evolution of spermatophores in insects. *Proc. R. Entomol. Soc. Lond.* **35**: 107-13.

Davies, L. 1965. On spermatophores in Simuliidae (Diptera). *Proc. R. Entomol. Soc. Lond.* **(A) 40**: 30-34.

Dohle, W. 1964. Über die Stellung der Diplopoden im System. *Verh. Dtsch. Zool. Ges. Kiel.* **1964**: 587-606.

Downes, J. A. 1968. Notes on the organs and processes of sperm-transfer in the lower Diptera *Can. Entomol.* **100**: 608-17.

Dzimirski, J. 1963. Kopulationsverhalten und Abgabe einer Spermatophore beim Feistkäfer (*Pimelia sardea* Solier, Tenebrionidae). *Z. Tierpsychol.* **20**: 10-15.

Faasch, H. 1967. Beitrag zur Biologie der einheimischen Uropodiden *Uroobovella marginata* und *Uropoda orbicularis* und experimentelle Analyse ihres Phoresieverhaltens. *Zool. Jahrb. Syst.* **94**: 521-608.

Feldman-Muhsam, B. 1967. The components of the spermatophore, their formation and their biological functions in Argasid ticks. *Proc. 2nd Int. Congr. Acarol.* **1967**: 357-62.

Feldman-Muhsam, B., S. Boruth, S. Saliternik-Givant and C. Eden. 1973. On the evacuation of sperm from the spermatophore of the tick *Ornithodoros savignyi*. *J. Insect Physiol.* **19**: 951-62.

Gerber, G. H. 1970. Evolution of the methods of spermatophore formation in pterygote insects. *Can. Entomol.* **102**: 358-62.

Ghilarov, M. S. 1956. L'importance du sol dans l'origine et l'évolution des insectes. *Proc. 10th Int. Congr. Entomol.* **1**: 443-51.

Ghilarov, M. S. 1959. Evolution of the insemination type in insects as the result of transition from aquatic to terrestrial life in the course of the phylogenesis, pp. 50–55. *In* I. Hrdy (ed.), *The Ontogeny of Insects.* Academic Press, New York, London.

Grassé, P. P. (ed.). 1949, *Traité de Zoologie,* Vol. VI, Masson et Cie, Paris.

Grassé, P. P. (ed.). 1959. *Traité de Zoologie,* Vol. V, Masson et Cie, Paris.

Grassé, P. P. (ed.). 1977. *Traité de Zoologie,* Vol. VIII, Masson et Cie, Paris.

Gupta, P. D. 1947. On the structure and formation of the spermatophore in the cockroach *Periplaneta americana. Indian J. Entomol.* **8:** 79–84.

Hennig, W. 1953. Kritische Bemerkungen zum phylogenetischen System der Insekten. *Beitr. Entomol.* **3:** 1–85.

Hennig, W. 1969. *Die Stammesgeschichte der Insekten.* Waldemar Kramer, Frankfurt.

Imms, A. D. 1937. The ancestry of insects. *Nature* (Lond). **1937:** 399–400.

Imms, A. D. 1945. The phylogeny of insects. *Tijdschr. Entomol.* **88:** 63–6.

Juberthie-Jupeau, L. 1963. Recherches sur la reproduction at la mue chez les Symphyles. *Arch. Zool. Exp. Gén.* **102:** 1–172.

Kew, H. W. 1930. On the spermatophores of the pseudoscorpions *Chthonius* and *Neobisium. Proc. Zool. Soc. Lond.* **1930:** 253–56.

Khalifa, A. 1949a. Spermatophore production in Trichoptera and some other insects. *Trans. R. Entomol. Soc. Lond.* **100:** 449–79.

Khalifa, A. 1949b. The mechanism of insemination and the mode of action of spermatophore in *Gryllus domesticus. Q. J. Microsc. Sci.* **90:** 281–91.

Khalifa, A. 1950. Spermatophore production in *Blattella germanica. Proc. R. Entomol. Soc. Lond.* **(A) 25:** 53–61.

Klingel, H. 1959. Die Paarung des *Lithobius forficatus* L. *Verh. Dtsch. Zool. Ges. Münster* **1959:** 326–32.

Klingel, H. 1960. Vergleichende Verhaltensbiologie der Chilopoden *Scutigera coleoptrata* L. ("Spinnenassel") und *Scolopendra cingulata* Latr. (Skolopender). *Z. Tierpsychol.* **17:** 11–33.

Klingel, H. 1962. Paarungsverhalten bei Pedipalpen (Telyphonus und Sarax). *Verh. Dtsch. Zool. Ges. Wien* **1962:** 452–59.

Kraus, O. 1976. Zur phylogenetischen Stellung und Evolution der Cheliceraten. *Entomol. German.* **3:** 1–12.

Kristensen, N. P. 1975. The phylogeny of hexapod "orders". A critical review of recent accounts. *Z. Zool. Syst. Evolutionsforsch.* **13:** 1–44.

Lauterbach, K.-E. 1973. Schlüsselereignisse in der Evolution der Stammesgruppe der Euarthropoda. *Zool. Beitr. (NF)* **19:** 251–99.

Lauterbach, K.-E. 1975. Über die Herkunft der Malacostraca (Crustacea). *Zool. Anz.* **194:** 165–79.

Manton, S. M. 1938. Studies on the Onychophora IV. The passage of spermatozoa into the ovary in *Peripatopsis. Phil. Trans. R. Soc. Lond.* **228:** 421–42.

Manton, S. M. 1965. The evolution of arthropodan locomotion. *J. Linn. Soc. (Zool.)* **45:** 251–484.

Manton, S. M. 1969a. Introduction to classification of Arthropoda, pp. R3–R15. *In* R. C. Moore (ed.), *Treatise on Invertebrate Palaeontology,* R. (Arthropoda 4). University Press of Kansas, Lawrence, Kansas.

Manton, S. M. 1969b. Evolution and affinities of Onychophora, Myriapoda, Hexapoda and Crustacea, pp. R15–R56. *In* R. C. Moore (ed.), *Treatise on Invertebrate Palaeontology,* R. (Arthropoda 4). University Press of Kansas, Lawrence, Kansas.

Manton, S. M. 1972. The evolution of arthropodan locomotory mechanisms. Part 10. Locomotory habits, morphology and evolution of the hexapod classes. *J. Linn. Soc.* **51:** 203–400.

Massond, Z. 1976. Essai de synthèse sur la phylogenie des Collemboles. *Rev. Ecol. Biol. Sol.* 13: 241-52.

Matsuda, R. 1958. On the origin of the external genitalia of insects. *Ann. Entomol. Soc. Amer.* 51: 84-94.

Matsuda, R. 1976. *Morphology and Evolution of Insect Abdomen.* Pergamon Press, Oxford.

Matthews, D. C. 1954. A comparative study of the spermatophores of three scyllarid lobsters. *Q. J. Microsc. Sci.* 95: 204-15.

Meisenheimer, J. 1921. *Geschlecht und Geschlechter.* Vol. I. Fischer, Jena.

Mitchell, R. 1958. Sperm transfer in the water-mite *Hydryphantes ruber* Geer. *Amer. Midl. Nat.* 60: 156-58.

Muma, M. H. 1966. Mating behavior in the solpugid genus *Eremobates* Banks. *Anim. Behav.* 14: 346-50.

Nemenz, H. and J. Gruber. 1967. Experimente und Beobachtungen an *Heterometrus longimanus* Petersii (Thorell) (Scorpiones). *Verh. Zool.-Bot. Ges. Wien* 107: 5-24.

Pauly, F. 1952. Die "Copula" der Oribatiden (Moosmilben). *Naturwissenschaften* 39: 572-73.

Pollock, J. N. 1972. The evolution of sperm transfer mechanisms in the Diptera. *J. Entomol.* (A) 47: 29-35.

Putmann, W. L. 1966. Insemination in *Balaustium* sp. (Erythraeidae). *Acarologia* 8: 424-26.

Regen, J. 1924. Anatomisch-physiologische Untersuchungen über die Spermatophore von *Liogryllus campestris* L. Sitzungsber. *Akad. Wiss. Wien Math.-Nat. Kl., Abt. I.* 133: 347-60.

Ruhberg, H. and V. Storch. 1976. Zur Ultrastruktur von männlichem Genitaltrakt, Spermiocytogenese und Spermien von *Peripatopsis moseleyi* (Onychophora). *Zoomorphologia* 85: 1-15.

Schaller, F. 1952. Das Fortpflanzungsverhalten apterygoter Insekten (Collembolen und Machiliden). *Verh. Dtsch. Zool. Ges. Freiburg* (= *Zool. Anz. Suppl.*) 16: 184-89.

Schaller, F. 1954a. Indirekte Spermatophorenübertragung bei *Campodea.* *Naturwissenschaften* 41: 406-7.

Schaller, F. 1954b. Die indirekte Spermatophorenübertragung und ihre Probleme. *Forsch. Fortschr. Dtsch. Wiss.* 28: 321-26.

Schaller, F. 1962. *Die Unterwelt des Tierreichs.* Springer, Heidelberg, Berlin, New York.

Schaller, F. 1965. Mating behavior of lower terrestrial arthropods from the phylogenetical point of view. *Proc. 12th Int. Congr. Entomol. Lond.* 1965: 297-98.

Schaller, F. 1971. Indirect sperm transfer by soil arthropods. *Annu. Rev. Entomol.* 16: 407-46.

Schlee, D. 1969. Sperma-Übertragung in ihrer Bedeutung für das phylogenetische Studium an Hemiptera I. Psylliformes (Psyllina + Aleurodina) als monophyletische Gruppe. *Z. Morphol. Tiere* 64: 95-138.

Schliwa, W. and F. Schaller. 1963. Die Paarbildung des Springschwanzes *Podura aquatica* (Apterygota, Collembola). *Naturwissenschaften* 50: 698.

Schuster, R. 1962. Nachweis eines Paarungszeremoniells bei den Hornmilben. *Naturwissenschaften* 49: 502.

Schuster, R. and I. J. Schuster. 1966. Über das Fortpflanzungsverhalten von Anystiden-Männchen (Acari, Trombidiformes). *Naturwissenschaften* 53: 162.

Schuster, R. and I. J. Schuster. 1969. Gestielte Spermatophoren bei Labidostomiden (Acari, Trombidiformes). *Naturwissenschaften* 56: 145.

Schuster, R. and I. J. Schuster. 1970. Indirekte Spermaübertragung bei Tydeidae (Acari, Trombidiformes). *Naturwissenschaften* 57: 256.

Schuster, R. and I. J. Schuster. 1977. Ernährungs- und fortpflanzungsbiologische Studien an der Milbenfamilie Nanorchestidae (Acardi, Trombidiformes). *Zool. Anz.* (in press).

Sharov, A. G. 1965. Origin and main stages of the arthropod evolution. 1. From annelids to arthropods. 2. Origin and phylogenetic relationships of principal groups of Arthropods. *Zool. Zh.* **44**: 803-17, 963-79.

Sharov, A. G. 1966. *Basic Arthropod Stock with Special Reference to Insects.* Pergamon Press, Oxford.

Siewing, R. 1959/60. Neuere Ergebnisse der Verwandtschaftsforschung bei den Crustaceen. *Vortrag. Als Ms. gedruckt. Wiss. Z. Univ. Rostock 9 Math.-Nat. Reihe, H.* 3.

Siewing, R. 1960. Zum Problem der Polyphylie der Arthropoden. *Z. Wiss. Zool.* **164**: 238-70.

Snodgrass, R. E. 1938. Evolution of the Annelida, Onychophora, and Arthropoda. *Smithson. Misc. Collect.* **97(6)**: 1-159.

Sturm, H. 1956. Die Paarung beim Silberfischchen *Lepisma saccharina. Z. Tierpsychol.* **13**: 1-12.

Sturm, H. 1958. Indirekte Spermatophorenübertragung bei dem Geißelskorpion *Trithyreus sturmi* Kraus (Schizomidae, Pedipalpi). *Naturwissenschaften* **45**: 142-43.

Theis, G. and R. Schuster. 1974. Gestielte Tröpfchenspermatophoren bei Calyptostomiden (Acari, Trombidiformes). *Mitt. Naturwiss. Ver. Steiermark* **104**: 183-85.

Tiegs, O. W. 1940. The embryology and affinities of the Symphyla, based on a study of *Hanseniella aquilis. Q. J. Microsc. Sci.* **82**: 1-225.

Tiegs, O. W. 1947. The development and affinities of the Pauropoda, based in a study of *Pauropus silvaticus. Q. J. Microsc. Sci.* **88**: 165-336.

Tiegs, O. W. and S. M. Manton. 1958. The evolution of the Arthropod. *Biol. Rev. (Cambridge)* **33**: 255-337.

Tillyard, R. J. 1930. The evolution of the class Insecta. *Pap. Proc. R. Soc. Tasmania* **1930**: 1-89.

Tuxen, S. L. 1970. The systematic position of entognathous apterygotes. *Ann. Esc. Nac. Cienc. Biol. Mex.* **17**: 65-79.

Wesenberg-Lund, C. 1939. *Biologie der Süßwassertiere.* Springer, Berlin, New York.

Westheide, W. 1967. Monographie der Gattungen *Hesionides* und *Microphthalmus. Z. Morphol. Ökol. Tiere* **61**: 1-159.

Weygoldt, P. 1966a. Vergleichende Untersuchungen zur Fortpflanzungsbiologie der Pseudoskorpione. Beobachtungen über das Verhalten, die Samenübertragungsweisen und die Spermatophoren einiger einheimischer Arten. *Z. Morphol. Ökol. Tiere* **56**: 39-92.

Weygoldt, P. 1966b. Spermatophore web formation in a pseudoscorpion. *Science (Wash., D.C.)* **153**: 1647-49.

Weygoldt, P. 1969. Beobachtungen zur Fortpflanzungsbiologie und zum Verhalten der Geißelspinne *Tarantula marginemaculata* C. L. Koch (Chelicerata, Amblypygi). *Z. Morphol. Tiere* **64**: 338-60.

Weygoldt, P. 1970a. Vergleichende Untersuchungen zur Fortpflanzungsbiologie der Pseudoskorpione II. *Z. Zool. Syst. Evolutionsforsch.* **8**: 241-59.

Weygoldt, P. 1970b. Courtship behavior and sperm transfer in the Giant Whip Scorpion, *Mastigoproctus giganteus* (Lucas) (Uropygi, Thelyphonidae). *Behaviour* **36**: 1-8.

Weygoldt, P. 1972. Spermatophorenbau und Samenübertragung bei Uropygen (*Mastigoproctus brasilianus*) und Amblypygen (*Charinus brasilianus* und *Admetus pumilio*) (Chelicerata, Arachnida). *Z. Morphol. Tiere* **71**: 23-51.

Weygoldt, P. 1975. Die indirekte Spermatophorenübertragung bei Arachniden. *Verh. Dtsch. Zool. Ges.* **1974**: 308-13.

11

Ultrastructure of Sperm and Its Bearing on Arthropod Phylogeny

B. BACCETTI

11.1. INTRODUCTION

A glance at the general evolutionary tree of sperm structure in animals (Fig. 11.1) reveals a basic aquatic model, with minor differences in many primitive phyla, and a series of lines arising from it, culminating in filiform or aflagellate sperm models, typical of the more advanced phyla; the latter have developed internal fertilization, and colonized land. While in the simplest animal phyla, the phyletic spermatologic tree is relatively simple, consisting of one or few lines, in arthropods the situation is enormously complicated. No other phylum has a comparable variety of sperm models. By comparison Chordata appear as a homogeneous and simple group.

I will first describe the basic aquatic sperm (Fig. 11.2). It is round, with a spherical nucleus, has several vesicular mitochondria near its base, a bilayered (acrosome and perforatorium) acrosomal complex, located anterior to the nucleus, and a long flagellum with a simple 9 + 2 (made up of tubulin and dynein) axoneme. This model, although definitely preserved in the aquatic life, is not the most primitive one. The acrosome, for example, progressively evolved from Mesozoa, through Poriphera and Cnidaria (see Baccetti and Afzelius, 1976). However, in the arthropods the basic aquatic model had already evolved when the phylum started. Sipunculida, Echinodermata, Tunicata, Cephalocordata, some of Annelida, and Mollusca also possess this model.

What was the situation in the arthropodan ancestor? Tardigrads (Baccetti

Paper performed under C.N.R. project "Biology of Reproduction."

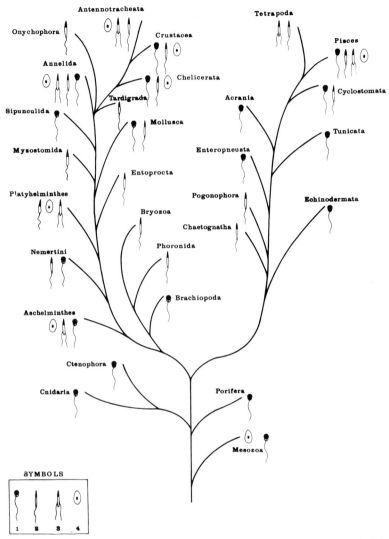

Fig. 11.1. Phylogenetic tree showing different types of sperm in Metazoa. 1, Primitive type of sperm; 2, modified sperm; 3, biflagellate sperm; 4, aflagellate sperm. Modified from Franzen (1970) and from Baccetti and Afzelius (1976).

et al., 1971a) maintain unchanged several primitive aquatic characteristics, including the 9 + 2 axoneme, scattered mitochondria and the bilayered acrosomal complex evolved, just as the nucleus does, in a helical shape. Onychophorans, on the contrary, have a highly evolved sperm, showing some characteristics typical of annelids (such as a condensed mitochondrial coil interposed between nucleus and axoneme—a synapomorphic character) and other character-

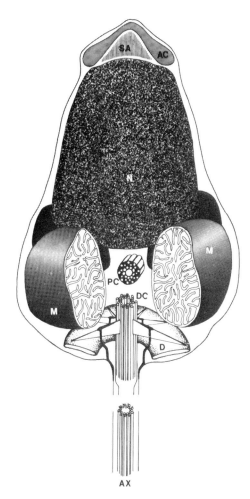

Fig. 11.2.   Classic marine sperm of *Golfingia gouldi.*   Ac = acrosome; Ax = axoneme; D = basal disk; DC = distal centriole; M = mitochondria; N = nucleus; PC = proximal centriole; SA = subacrosomal material or perforatorium.

istics that were apparently acquired independently and are typical of pterygote insects (axoneme with nine accessory tubules and a peripheral manchette). They represent, therefore a "sister-group" of Arthropoda, and should not be considered ancestral to them (Baccetti et al., 1976).

11.2. BASIC AQUATIC SPERM MODEL AND ITS PRESENCE IN VARIOUS ARTHROPOD GROUPS

The aquatic sperm model, or its immediate derivatives, are present in the lowest members of most of the arthropod classes, but only *Limulus* shows a typical picture (Fig. 11.3) of the basic aquatic sperm—plesiomorphic characters. It has a spherical nucleus, 9 + 2 axoneme, and long everting subacrosomal rod, whose principal component is actin (Tilney, 1975). The same situation obtains in *Euscorpius* among arachnids (André, 1959), mystachocarid, cirriped, and branchiuran crustaceans (Brown, 1970; Brown and Metz, 1967), and Symphyla among Myriapoda (Rosati et al., 1970); all of them have features in common with this category (all symplesiomorphic characters) but appear slightly evolved from it. This situation (Fig. 11.4) indicates a monophyletic (sensu lato) origin of arthropods, starting from a *Limulus*-like aquatic ancestor. It is impossible to relate any one of the above-mentioned groups to others, all of them being very different from one another, with many minor differences. They appear to have independently branched out from the ancestor, resulting in different lines. We can easily follow each of these radiations on the basis of comparative sper-

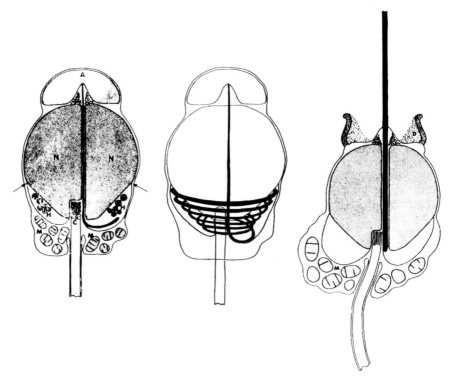

Fig. 11.3.   Acrosome reaction with a preformed acrosomal filament (*Limulus polyphemus*). Acrosome filament penetrates nucleus and is wound six turns around base of middle piece. After completion of acrosome reaction filament projects anteriorly to a length of about 50 μm.   A = acrosome; C = centriole; D = dense acrosome content; F = filament; H = helical filament; M = mitochondria; N = nucleus.   From André (1963).

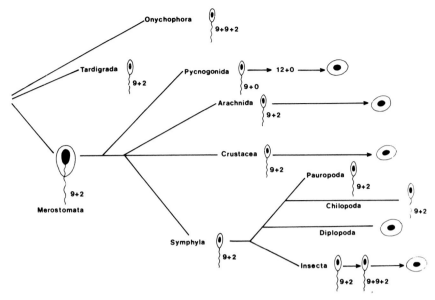

Fig. 11.4. Phylogeny of arthropodan sperm.

matology. We find in each of them—as a consequence of a very specialized internal fertilization—some common evolutionary pathways with similar steps.

11.3. EVOLUTION OF VARIOUS SPERM MODELS IN ARTHROPODS

11.3.1. Immotility and Eventual Secondary Motility

The first step is the acquisition of unusual axoneme patterns. This is due to the fact that selection is not so favorable for the 9 + 2 model (which provides the best motility) when the sperm is released so close to the egg, frequently coiled in "encysted" condition. During the next stage, one or both arms of the doublets are lost (owing to loss of dynein), resulting in the loss of axoneme movement. And finally the flagellum is eliminated, and the sperm cell becomes immotile. Only in Insecta do we observe increased motility due to the extreme elongation of the flagellum and the acquisition of an extra set of nine tubules as we have seen in the Onychophora; nevertheless, aflagellarity—as the final evolutionary step—is commonly found among the higher members of the most specialized insect groups. Finally, in rare cases the motility is secondarily acquired by some immotile models, but the axoneme never reappears. Substitutive devices for the lost axoneme have been invented in different cases. Several examples of this can be given.

11.3.1.1. Arachnida

Arachnida are a well-studied group, and furnish a good example of the above-mentioned history (Fig. 11.5). In the most primitive order, Scorpiones, the sperm is filiform, the acrosome bilayered, and the flagellum long. But the 9 + 2 pattern (a plesiomorphic character) is preserved only in *Euscorpius* (André, 1959), while other genera such as *Hadrurus* (Fig. 11.6) or *Centruroides* (Jespersen and Hartwick, 1973; Hood et al., 1972) have a 9 + 1 axoneme, and others such as *Tityus*, *Vejovis*, *Anuroctonus*, *Uroctonus* have a 9 + 0 (Hood et al., 1972; da Cruz-Landim and Ferreira, 1972; Jespersen and Hartwick, 1973) pattern. All the studied Pseudoscorpiones seem to belong to the basic 9 + 2 category, but the sperm acquires the encysted condition (Fig. 11.7), having a spheroidal shape, the axoneme being completely embedded in the cytoplasm and coiled with the nucleus. The sperm is therefore immotile. A good scheme is given by Legg (1973), who also found typical minor differences in the general sperm shapes of the various families. The bilayered acrosomal complex and encysted conditions are conserved by Uropygi (Phillips, 1976) and Araneida (Osaki, 1969; Baccetti et al., 1970b; Reger, 1970a), but a new characteristic appears: both groups have a 9 + 3 axoneme (synapomorphic characters).

In Opiliones a further step is reached: the sperm cell is disk-shaped, without acrosome and axoneme, with the nuclear envelope fused with the plasma membrane, and mitochondria embedded in the nuclear content (Reger, 1969; Baccetti, 1970). However, an example of the ancestral situation is found in the most primitive opilion, the cyphophtalm *Siro rubens* (Fig. 11.8). This species maintains the plesiomorphic characteristics of a bilayered acrosomal complex and 9 + 2 axoneme, even though the latter is nonfunctional, being four times coiled around the nucleus, and devoid of dynein arms (Juberthie et al., 1976). It seems certain that the encysted condition always culminates in a functionally aflagellate sperm.

Finally, the highest arachnid order, Acarina, possess only fusiform, aflagellate

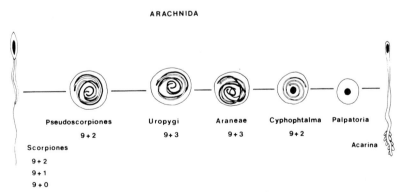

Fig. 11.5. Phylogeny of arachnid sperm.

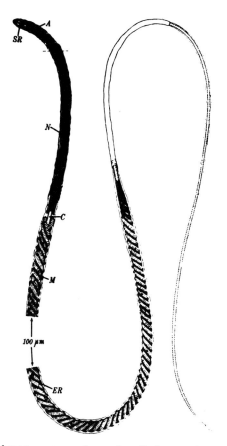

Fig. 11.6. Diagram of mature sperm of scorpion, *Hadrurus arizonensis.* A = acrosome; SR = subacrosomal rod; N = nucleus; C = centriole; M = mitochondrion; ER = endoplasmic reticulum. Note that outer membrane has been removed from anterior and end piece, and acrosome region is in longitudinal section. From Jespersen and Hartwick (1973).

sperm, where the acrosome is simplified to a monolayered condition (Fig. 11.9). A new situation nevertheless appears: motility (abandoned after the Scorpiones level by the encysted and disk-shaped sperm in all the examined arachnid orders) is secondarily acquired in mites and ticks by new organelles: a series of motile appendages devoid of microtubules but full of fibrils (Reger, 1963, 1974; Alberti and Storch, 1976; Feldham-Muhsam and Filshie, 1976).

11.3.1.2. Pycnogonida

Pycnogonida is a separate arthropodan class, related to Arachnida. The sperm was recently investigated by Van Deurs (1973, 1974a). *Nymphon* seems to possess the basic sperm type, and it is still motile; the axoneme reveals a 9 + 0

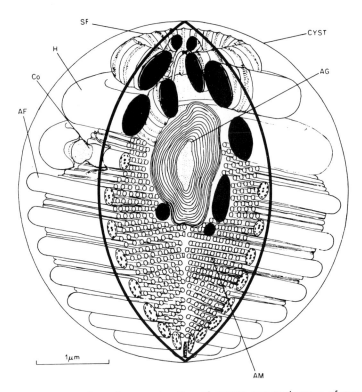

Fig. 11.7. Cut-away schematic reconstruction of mature encysted sperm of pseudoscorpion, *Chthonius ischnocheles.* AF = axial filament; AG = possible acrosome; AM = annular mitochondrion; Co = possible centriole; cyst = cyst; H = head and SF = spiral filament. From Legg (1973).

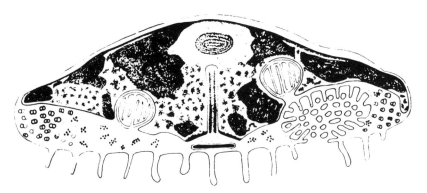

Fig. 11.8. Mature spermatid of opilionid cyphophtalm, *Siro rubens.* Free doublets can be seen sectioned on both sides of cell. From Juberthie et al. (1976).

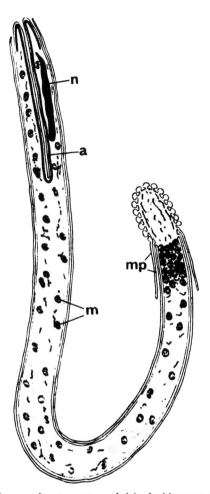

Fig. 11.9.   Schematic diagram of mature sperm of tick, *Amblyomma dissimilis.* Tick sperm
are devoid of flagellum and have an organization which is quite different from that of other
animals.   Length of sperm is 200-300 μm. There is a great deal of cytoplasm containing
scattered mitochondria (m).   One region contains an acrosome (a) which is invaginated into
a U-shaped structure.   Nucleus (n) is situated close to acrosome.   Near distal end of sperm
there are motile precesses (mp), which are able to pull sperm forward.   Distal end thus acts
as anterior end during locomotion. From Reger (1963).

or 12 + 0 pattern according to species, with some variations from 8 + 0 to
11 + 0. Only the inner arm seems to be present in the figures of Van Deurs. The
situation clearly demonstrates an evolution towards the loss of movement–a
situation we will find, with more steps, in the insect order Protura. In fact, in
Pycnogonum, Van Deurs (1974b) reveals an elongate immotile sperm full of
only longitudinal, isolated microtubules, and devoid of any other organelle.

11.3.1.3. Crustacea

The picture of the crustacean sperm evolution is more confused than that of arachnids, but one can show the same major steps. In Crustacea also some primitive groups preserve the plesiomorphic character of a motile 9 + 12 flagellum. They are Mystacocarida (Brown and Metz, 1967), Cirripedia (Fig. 11.10), Ascothoracida (Pochon-Masson et al., 1970; Munn and Barnes, 1970; Wingstrand, 1973), and Branchiura (Fig. 11.11), which could include Pentastomida (Wingstrand, 1972). All of them have the basic bilayered acrosome complex, except Branchiura, where the acrosome disappears during spermiogenesis, and is substi-

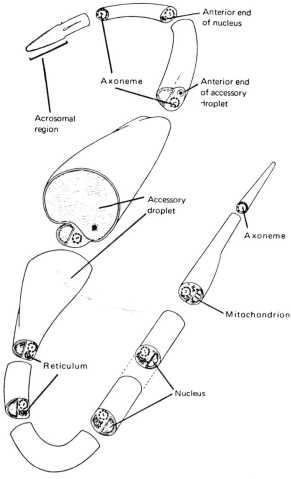

Fig. 11.10. Diagram showing structure of mature sperm of cirriped *Balanus*. From Munn and Barnes (1970).

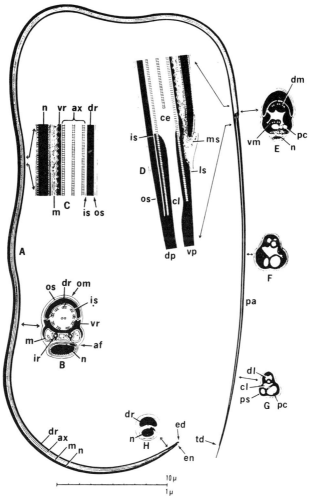

Fig. 11.11. Diagram of mature sperm of branchiuran, *Argulus foliaceus*. A, Sperm; B, cross section of body; C, median section of body; D, median section of transitional region between body and pseudoacrosome; E, cross section through anterior end of nucleus; F, cross section of pseudoacrosome, basal part; G, cross section of pseudoacrosome, terminal part; H, cross section through posterior end of body. af = remnants of acrosome filament in nuclear membranes; ax = axoneme; ce = centriole; dl = dorsal lumen of pseudoacrosome; dm = dorsal extension of pseudoacrosomal granular matter, containing axonemal filaments nos. 9, 1, 2; dp = dorsal rod of pseudoacrosome; dr = dorsal ribbon; en = posterior end of nucleus; is = inner membranous sac of dorsal ribbon; ir = intermitochondrial light rods; ls = limit of pseudoacrosomal membranous sac on ventral side; m = mitochondria; n = nucleus; o = oblique membrane between axonemal doublet 1 and dorsal sheath; os = outer membranous sac of dorsal ribbon; pa = pseudoacrosome; ps = pseudoacrosomal membranous sac; td = top of dorsal rod of pseudoacrosome; pc = light cores in pseudoacrosome; vm = ventral extension of pseudoacrosomal granular matter, including axonemal filaments 4 and 5; vp = ventral rod of pseudoacrosome; vr = ventral ribbon. From Wingstrand (1972).

tuted by a pseudoacrosome, apparently secondarily acquired. Mitochondria are vesicular and scattered in Mystacocarida, but only one in Cirripedia and Ascothoracida and three in Branchiura. It appears, therefore, that the above 9 + 2 crustacean groups, except perhaps Mystacocarida, are more evolved than *Limulus*, and that no crustacean has the basic aquatic sperm model. The other primitive crustacean groups have some primitive characteristics, and some

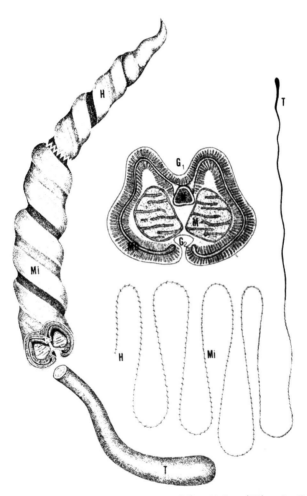

Fig. 11.12.   Schematic drawing of enlarged head (H), midpiece (Mi) and tail (T) segments of sperm of ostracod, *Cypridopsis*. Enlarged cross section taken from middle region of sperm is shown in center of figure. Two grooves (G_1, G_2) present on surface are staggered 180° with respect to each other and exhibit a pitch angle of approximately 45°. N = nucleus; M = mitochondrion; MO = membranous organelle. Length of sperm is approximately 1.0 mm. From Reger (1970b).

evolved ones. Cephalocarida (Brown and Metz, 1967) have a typical bilayered acrosomal complex, with a long rod, similar to that of *Limulus*, and a round nucleus, but lack the flagellum and mitochondria; Branchiopoda and Copepoda also have a spheroidal nucleus and lack not only the flagellum but also the acrosome, and have mitochondria fused in a nebenkern that in one species, *Tanymastix,* is full of crystalline material (Brown, 1970). Ostracoda is a more complicated group (Fig. 11.12): the flagellum is always absent, but many species have filiform motile sperm with helicoidal mitochondria and nucleus, and an enigmatic multilaminar, acrosome-like organelle (Zissler, 1969, 1970; Reger, 1970b). However, a small subfamily, the Asteropinae (Wingstrand, personal communication), has disk-shaped, aflagellate, immotile sperm, like Branchiopoda. It is difficult to say whether this model is more primitive than the filiform motile ostracodan one or not. All ostracodan models are really very specialized, and

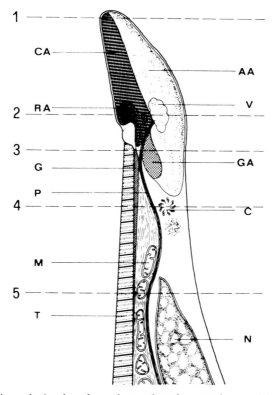

Fig. 11.13. Schematic drawing of anterior portion of sperm of terrestrial isopod, *Armadillidium vulgare*. CA = crystalline part of acrosome; AA = amorphous part of acrosome; RA = reniform part of acrosome; V = vesicle; GA = granular part of acrosome; G = groove; P = perforatorium; C = centrioles; M = mitochondria; N = nucleus; T = tail. From Cotelli et al. (1976).

the basic aquatic pattern is lost. In my opinion the disk-shaped model of Astero-pinae is the most primitive, and the motile, filiform ones of *Notodromas*, *Cypri-dopsis*, and *Candona* are derived from it, acquiring motility secondarily like Acarina among Arachnida.

Peracarida have peculiar sperm, which are whiplike, aflagellate, with a conventional bilayered acrosomal complex (Fig. 11.13) and scattered mitochondria, an elongate nucleus, and a tubular proteinaceous tail, consisting of a crystalline cylindrical wall (Fain-Maurel et al., 1975a, b; Cotelli et al., 1976). More peculiar is the eucarid model: it is starlike (Fig. 11.14), with the nuclear envelope fused with the plasma membrane, and the bacteroid nucleus devoid of protamines and histones (Chevaillier, 1970). The acrosome complex is bilayered, with the subacrosomal rod protruding in *Eupagurus*, or totally everted in *Carcinus* (Pochon-Masson, 1968a, b).

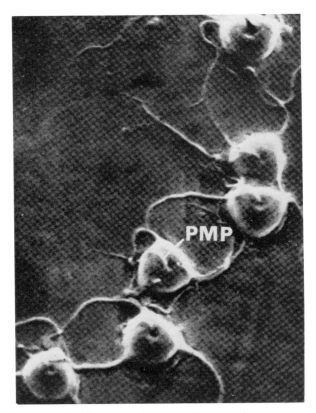

Fig. 11.14. Sperm of crab, *Libinia emarginata*, at early stages of sperm penetration. Radial arms of sperm are twisted and bent over adjacent sperm and their arms. Posterior median processes (PMP) of sperm appear to emerge from opening at base of sperm. From Hinsch (1970).

Evidently, Peracarida and Eucarida are the highest crustacean groups, since they do not show all the preliminary steps of sperm evolution. But all the other orders show traces of the ancestral models (symplesiomorphy), and different models gradually evolved from them. Crustacea, at any rate, are the most heterogeneous arthropodan class.

11.3.1.4. Myriapoda

Myriapoda show a more classic phyletic picture. Symphyla (Fig. 11.15) in their types of sperm conserve, with many transformations, most of the basic, aquatic

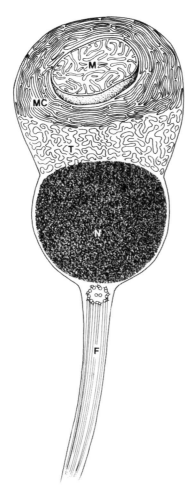

Fig. 11.15. Schematic drawing of smaller model of sperm of *Symphylella vulgaris.* F = flagellum; M = mitochondrion; MC = concentric acrosomal membranes; N = nucleus; T = tubular subacrosomal system.

sperm features (plesiomorphic characteristics): 9 + 12 motile axoneme, and spheroidal nucleus. The acrosome is monolayered, and only one vesicular mito-chondrion with normal cristae exists (Rosati et al., 1970). Pauropoda and Chilo-poda still have the somewhat basic sperm, but distinctive and more highly evolved than in Symphyla. The former lack the acrosome, show a fibrous sheath around the long 9 + 2 axoneme, and have scattered normal mitochondria behind the nucleus (Rosati et al., 1970); Chilopoda also have a fibrous sheath (an apomorphic character present only in these groups), which bears nine reinforcements, recall-ing accessory fibers around the 9 + 2 axoneme; moreover, the acrosome, even if monolayered, is present, and mitochondria are fused in a helicoidal derivative (Horstmann, 1968; Descamps, 1972; Camatini et al., 1973).

The only peculiar, highly evolved myriapodan sperm is that of Diplopoda, which was until recently a matter of controversy among spermatologists. It was one of the first immotile sperm ever described (Silvestri, 1898), and it has been claimed, both in *Polyxenus* (Fabre, 1855; Sokoloff, 1914; Tuzet and Manier, 1955) and in *Julus* (Oettinger, 1909), that the sperm produces a conventional flagellum when it penetrates the female genital apparatus. This claim has been disproved by Baccetti et al. (1974a, 1977b). In Pselaphognatha (*Polyxenus*), in fact, a strong metamorphosis of the sperm in the female body does occur. The ripe ejaculated sperm is barrel-shaped; has a filamentous nucleus, several scat-tered mitochondria, and a series of "spongy chambers" beneath the plasma

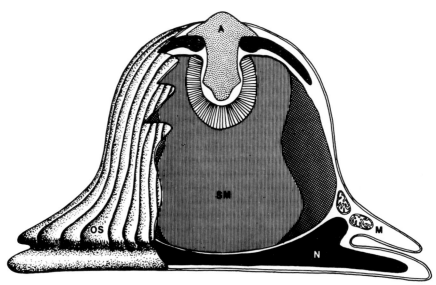

Fig. 11.16. Schematic drawing of unreacted sperm of diplopod, *Pachyjulus*.   A = acrosome; M = mitochondria; N = nucleus; OS = outer surface; SM = subacrosomal material or perforatorium.

membrane; and lacks the acrosome. In the female body, the sperm becomes ribbon-like, with a long nucleus flanked by a long acrosome, and a few mitochondria. The flagellum is, however, lacking. The sperm elongation convinced the early light microscopists of the late production of a tail. In Chilognatha no metamorphosis occurs, although a flagellum has been claimed (Oettinger, 1909). The sperm are barrel-shaped and arranged in couplets. The acrosome and mitochondria are lacking or very simple in Polydesmoidea (Reger and Cooper, 1968; Manier and Boissin, 1975) and present in Spirostreptoidea and Spiroboloidea (Horstmann and Breucker, 1969a, b; Reger, 1971; Manier et al., 1974; Manier and Boissin, 1976). In Glomeroidea, where sperm are isolated, the flagellum is certainly lacking, and only a caplike acrosome and a few small mitochondria have been described (Chevaillier and Prigent, 1969; Boissin et al., 1972). But in Juloidea, Oettinger (1909) observed a long tail protruding from the spheroidal sperm after fecundation. We have demonstrated (Baccetti et al., 1977b) that the flagellum and centriole are always lacking (Fig. 11.16), only a few mitochondria are present near the flattened, basal nucleus, and the main portion of the sperm is occupied by an acrosome and a large proteinaceous, subacrosomal crystal, both of these protruding as a long filament at the time of the acrosomal reaction (Fig. 11.17). No actin or tubulin is involved (as in *Limulus*) in this process, but several other proteins (with a molecular weight between 40,000 and 90,000) make up the filaments.

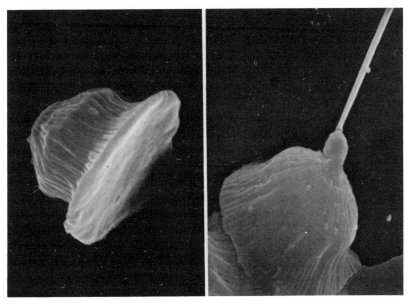

Fig. 11.17.   Unreacted (at left) and reacted (at right) sperm of *Pachyjulus* as seen by scanning electron microscopy. 10,000 diam.

In conclusion, the myriapod classification and phylogeny based on spermatology seem to be in good agreement with the commonly accepted one, but cannot provide an explanation of the origin of Diplopoda, which are an independent and homogeneous group, with characters indicating their highest specialization.

11.3.1.5. Protura, Collembola, Diplura (Japygidae)

Insects are a class quite rich in forms, and being also the best known, are suitable for many evolutionary considerations on a spermatologic basis. The most primi-

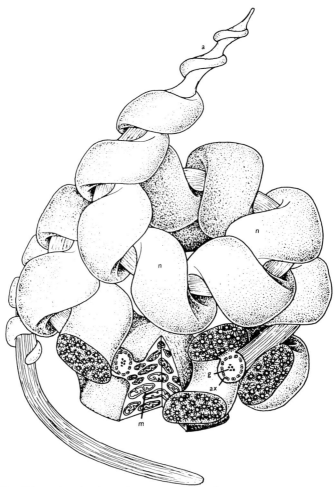

Fig. 11.18. Schematic drawing of mature sperm of proturan, *Acerentulus trägardhi*. a = acrosome; ax = axoneme; g = glycogen granules; m = mitochondria; n = nucleus. From Baccetti et al. (1973b).

tive stage is represented by Collembola and japygid Diplura. The situation in Protura, however, is quite peculiar, as stated by Baccetti et al. (1973b). In this group, the 9 + 2 model is lost, dynein lacking, and the sperm are immotile; some genera, such as *Acerentulus* (Fig. 11.18) and *Acerentomon*, develop unusual axoneme patterns (12 + 0, 14 + 0) devoid of arms; the final stage, found in *Eosentomon* (Fig. 11.19), is the development of a disk-shaped cell, devoid of axoneme and acrosome. The situation is similar to that in Pycnogonida, where also one could predict the occurrence of a disk-shaped, aflagellate model. In Collembola (Fig. 11.20), on the contrary (Dallai, 1967, 1970), the sperm maintains some less specialized characters (bilayered acrosomal complex with endonuclear rod, scattered normal mitochondria, 9 + 2 flagellum), but it is in an encysted condition, corresponding to the most common condition found in Arachnida (from Pseudoscorpions to Opilions). In Japygidae (Baccetti and Dallai, 1973), the sperm is motile with a 9 + 2 axoneme, and encysted (and therefore more primitive). But the compact and elongate nucleus, extranuclear subacrosomal rod, and the two mitochondrial derivatives suggest a slightly higher evolution than in Collembola. In conclusion (Fig. 11.21), the sperm structure of insects starts at a step somewhat higher than that of the lowest Arachnida, Crustacea, or Myriapoda. A form similar to that of *Limulus* is lacking. Protura are an independent group evolved towards aflagellarity. The Collembola, which avoid aflagellarity, preserve some aquatic, plesiomorphic characters by acquiring the encysted condition. On the contrary, the japygid model reveals, in its actual shape, the potential for the evolution of motility in dense fluids; it is filiform and maintains an extra set of microtubules, in addition to the axoneme, which is located in the mitochondrial compartment. From this model, it seems, the classical insect sperm has been derived.

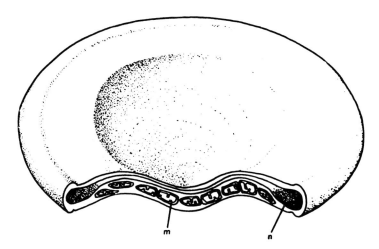

Fig. 11.19. Schematic drawing of mature sperm of proturan, *Eosentomon transitorium*. m = mitochondria; n = nucleus. From Baccetti et al. (1973b).

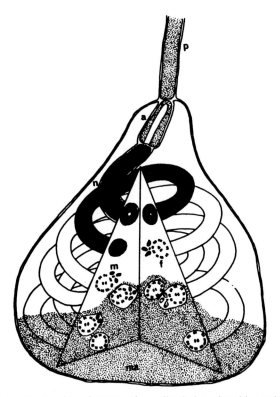

Fig. 11.20. Schematic drawing of sperm of a collembolan, *Anurida maritima.* a = acrosome; f = flagellum; m = mitochondria; ma = granular material; n = nucleus; p = peduncle. From Dallai (1970).

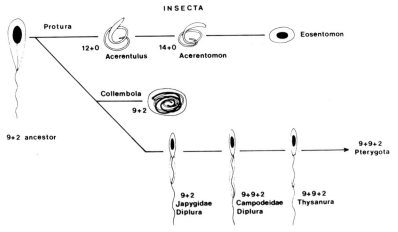

Fig. 11.21. Phylogeny of insect sperm.

11.3.2. Increased Motility, Sometimes Followed by Immotility and Secondary Motility in Higher Insects

The next stage is found in campodeids, where the japygid model acquires nine accessory tubules in the axoneme compartment. These tubules are formed in the spermatid against the nine doublets, but migrate, at maturity, on one side of the axoneme (Baccetti and Dallai, 1973). The same situation occurs in the Machilidae (Fig. 11.22) (Dallai, 1972; Wingstrand, 1973), which have also two crystalline accessory bodies, flanking the axoneme, that are found in several pterygotes (Fig. 11.23). Finally, Lepismatidae maintain the extra set of nine tubules regularly spaced around the axoneme, reaching the 9 + 9 + 2 condition (Bawa, 1964; Wingstrand, 1973)—a synapomorphic characteristic of the insect

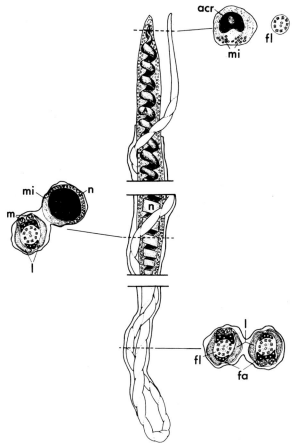

Fig. 11.22. Diagram of sperm of *Machilis distincta*, removed from spermatheca. acr = acrosome; fa = amorphous material enclosing accessory fibres; fl = axial filament; l = osmiophilic laminae; m = mitochondria; mi = microtubules; n = nucleus. From Dallai (1972).

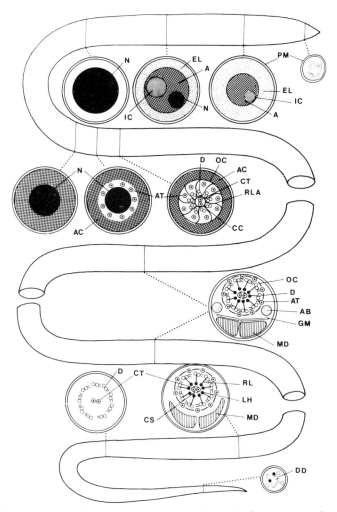

Fig. 11.23. Schematic drawing of a conventional model of a pterygote insect sperm; A = acrosome; AB = accessory bodies; AC = centriolar adjunct; AT = accessory tubules; CC = central cylinder; CF = coarse fibrous material; Cs = central sheath projections; CT = central tubules; D = doublets; DD = dissociated doublets; IC = inner cone or perforatorium; EL = extraacrosomal layer; GM = Golgi-derived membranes; LH = link-heads; MD = mitochondrial derivatives; N = nucleus; OC = outer cylinder; PM = plasma membrane; RL = radial links; RLA = radial laminae. Modified from Baccetti (1972).

sperm. Concomitantly, proteinaceous crystals appear in the lepismatid mitochondrial derivatives; they are made of crystallomitin, typical of the pterygote insects and described by Baccetti et al. (1977a). The role of the enormous, crystalline mitochondrial derivatives of the insect sperm is uncertain. Crystallomitin is a protein devoid of enzymatic activity (which is confined to the mitochondrial cortex); it has a molecular weight and amino acid composition similar to those of tubulin, but is stabilized by disulfide bridges. Probably it plays a role in sustaining and stabilizing the flagellar beat, which, according to the sliding filament model, can become disorganized when the axoneme becomes too long. It is a synapomorphic characteristic.

The transition to Pterygota is not so dramatic. In certain respects, the first orders appear more primitive than lepismatids. In fact, Ephemeroptera (Baccetti et al., 1969a) are strongly aberrant in the axoneme, which is of the 9 + 9 + 0 pattern, having lost the two central tubules. Moreover, two accessory bodies, like those of Machilidae, flank it, but the mitochondrial derivative lacks crystallomitin. The acrosome complex is regressed to a monolayered condition. Odonata have a conventional 9 + 9 + 2 axoneme (Kessel, 1966; Phillips, 1970) and two accessory bodies, but the two mitochondrial derivatives are still devoid of crystalline protein (Rosati et al., 1976). Ephemeroptera and Odonata seem to be separate from all the other pterygote orders.

Plecoptera (Baccetti et al., 1970a) also have conventional sperm, 9 + 9 + 2 with crystallized mitochondrial derivatives, but they have only one accessory body and a monolayered acrosome complex.

The first orders showing the complete set of the typical insect sperm organelles are the paurometabolic orders Blattaria, Orthoptera, and Dermaptera (Baccetti, 1970, 1972; Baccetti et al., 1971b, c; Eddleman et al., 1970; Rosati et al., 1976). They have a three-layered acrosome complex, 9 + 9 + 2 axoneme, two crystallized mitochondrial derivatives. Phasmatodea represent a different situation (Baccetti et al., 1973a). They have a 9 + 9 + 2 axoneme, but they lack perforatorium in the acrosomal complex, and also the mitochondria, having instead two enormous accessory bodies endowed with ATPase activity, which play an active role in the axoneme bending. The composition of these bodies is still under investigation in our laboratory. The question of the absence of the mitochondria is important. We have already mentioned sperm devoid of mitochondria; the size of such sperm did not permit metabolic studies. We (Baccetti et al., 1973d) have carefully studied the phasmid, *Bacillus*, however; in this insect the enzyme lactate dehydrogenase has been detected in the exiguous membrane system located in the axoneme matrix, and shows kinetics adapted to an anaerobic metabolism, i.e., high affinity to pyruvate, and low affinity to lactate, which inhibits the reaction.

Embioptera (Baccetti et al., 1974c) possess simpler sperm with a monolayered acrosomal complex and without accessory bodies. Isoptera are more interesting. We do not know the ultrastructure of termopsid (Isoptera) flagellate sperm, but on the basis of light microscopy they seem to possess an Embioptera-like simple model.

In a slightly higher family (Mastotermitidae), we find a unique, gigantic multi-

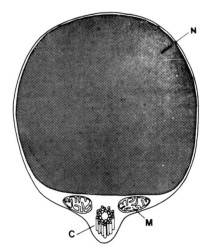

Fig. 11.24. Schematic drawing of sperm of *Reticulitermes lucifugus*.   C = centriole; M = mitochondria; N = nucleus.

flagellate model (Baccetti and Dallai, 1977b) with 9 + 0 axonemes devoid of inner arms. In higher groups, we find aflagellate models (Baccetti et al., 1974c). In *Reticulitermes*, in fact, there are a spheroidal nucleus, absence of acrosome and axoneme, and two normal, noncrystalline mitochondria (Fig. 11.24). In *Calotermes*, we find a general elongate shape, an elongate nucleus, with exiguous monolayered acrosome, scattered normal mitochondria, absence of axoneme, and a well-developed manchette of microtubules in many laminar posterior protrusions, giving the sperm the shape of an ear of corn with many leaves. Apparently, Isoptera appear to have evolved towards immotility, showing at least four models: 1) a primitive, motile (Termopsidae) model; 2) an aberrant, multiflagellate model with 9 + 0 axonemes (*Mastotermes*); 3) one devoid of axoneme, and involuted (*Reticulitermes*); 4) finally one secondarily evolved to a conical shape with rich systems of microtubules, and immotile (*Calotermes*).

A similar but more complicated history is shown by rhynchotoid insects (Fig. 11.25), the Paraneoptera. Psocoptera show a somewhat conventional insect sperm (Baccetti et al., 1969c), consisting of a 9 + 9 + 2 axoneme, crystalline mitochondrial derivatives, an acrosomal complex without perforatorium, and lack of accessory bodies. We know that one genus has a 9 + 9 + 1 axoneme pattern (Phillips, 1969), and biflagellarity frequently occurs (Baccetti et al., 1969c). In Mallophaga (Baccetti et al., 1969c) and Anoplura (Ito, 1966), the sperm consistently have two 9 + 9 + 2 axonemes (synapomorphy), and, in other respects, resemble Psocoptera. The same is true in Heteroptera, where—although some species are biflagellate with two axonemes of different length (Furieri, 1963)—the normal monoflagellate structure prevails. But evolution towards

RHYNCHOTOIDS

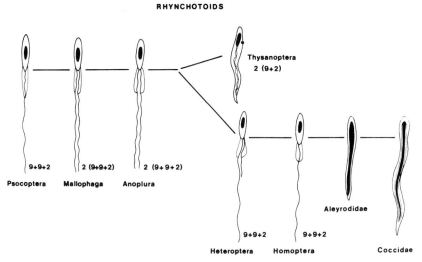

Fig. 11.25. Phylogeny of rhynchotoid sperm.

immotility is evident.   Thysanoptera (Baccetti et al., 1969c) have the same characteristics (only one mitochondrial derivative) but a strange flagellum, made up of two 9 + 2 fused and disorganized axonemes, devoid of arms.   All the lowest Homoptera have a highly motile and classical sperm with the rhynchotoid character of the acrosome (absence of perforatorium), large accessory bodies (lacking only in cercopids and cicadids), enormous crystalline mitochondria, and one 9 + 9 + 2 axoneme (see Folliot and Maillet, 1970; Phillipps, 1970); the same is true in aphidids (Mazzini, 1970), but aleyrodids have aflagellate, immotile sperm (Baccetti and Dallai, 1977a); coccids also have aflagellate sperm, but with a large helix of parallel longitudinal microtubules surrounding the nucleus (see Robison, 1970).   ATPase activity has been detected between the tubules (Moses, 1966), and the sperm is motile.

    Thus, rhynchotoids show an initial monoflagellarity, according to the classical insect sperm model, then biflagellarity, followed by direct evolution towards immotility (Thysanoptera) or return to monoflagellarity (Heteroptera and lower Homoptera) and loss of flagellum (aleyrodids).   The final step consists of the reacquisition of motility, with a new kind of microtubule system, different from the axoneme (coccids), which, when lost, is never regained.   We can consider the coccids as the final step of the rhynchotoid sperm evolution.

    No particular patterns can be observed in the four orders, Neuroptera, Coleoptera, Strepsiptera, and Hymenoptera.   All of them belong to the primitive neopteran sperm model, having a perforatorium that is endonuclear in Hymenoptera (Hoage and Kessel, 1968) and extranuclear in Coleoptera (Baccetti et al., 1973c) under the acrosome, and the synapomorphic characters, such as two fully crystallized mitochondrial derivatives, one or more accessory bodies, and one 9 + 9 + 2 axoneme. Only Neuroptera (Baccetti et al., 1969b) lack the acrosome. In this respect, the Strepsiptera seem to belong to coleopteroid insects, quite far from Diptera. A peculiar situation stressed by me earlier (Baccetti, 1970, 1972) is that of Mecoptera and Aphaniptera. Basically, the sperm of these two orders are quite similar, confirming the close relationship between them. There are, however, minor differences. For example, Mecoptera have only one mitochondrial derivative, and Aphaniptera two; and the perforatorium is fully extranuclear in Mecoptera (Baccetti et al., 1969b), but partially intranuclear in Aphaniptera (Baccetti, 1968). But a peculiar feature characterizes the two groups: the axoneme, even though motile, belongs to the 9 + 2

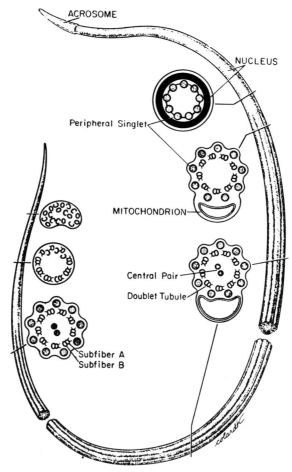

Fig. 11.26. Diagram of a sperm of caddis fly, *Neuronia*, illustrating some of features common to insect sperm. From Phillips (1970).

pattern. It would be interesting to inquire how this could have happened. One can assume the possibility of a back mutation, resulting in the lack of the outer set of tubules typical of insects (without any further evolution toward immotility), or that this is a symplesiomorphic condition, which would imply a direct origin of the mecopteran ancestor from 9 + 2 dipluran-like apterygote insects, without a passage through heterometabolic insects.

Trichoptera and Lepidoptera also are closely related. The former (Fig. 11.26) are either devoid of an acrosomal complex (Baccetti et al., 1970a), or have only a small acrosome (Phillips, 1970), while the latter are only devoid of the perforatorium (Phillips, 1971); both have only one, fully crystallized mitochondrial derivative and a peculiar appearance in the outer accessory tubules of the 9 + 9 + 2 axoneme; the tubules are full of glycogen and very thick (Baccetti et al., 1970a). The outer tubules of these orders are so peculiar that they could have originated separately from the other pterygotes. It would be tempting to assume an origin from a mecopteran-aphanipteran–like ancestor.

Diptera are by far the most peculiar insect order for the spermatologist. Lower Diptera start in a conventional pattern. Tipulidae and Limoniidae have in

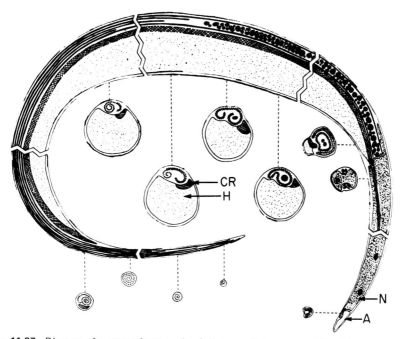

Fig. 11.27. Diagram of a sperm from testis of dipteran, *Sciara coprophila*. Discontinuities in diagram indicate that cell is much longer relative to its width than depicted. An axial filament of singlets and doublets is represented as two rows of dots and of doublets alone as one row of dots. A = acrosome; N = nucleus; CR = mitochondrial crystalloid; H = mitochondrial homogeneous material. From Phillips (1966).

fact a typical insect sperm, with the synapomorphic characters of a 9 + 9 + 2 axoneme and one or two crystalline mitochondrial derivatives. The only peculiarity is that the acrosome complex is monolayered (unpublished). This characteristic is constant in the whole order. The one or two mitochondrial derivatives are always fully crystallized, while the axoneme undergoes a drastic evolution

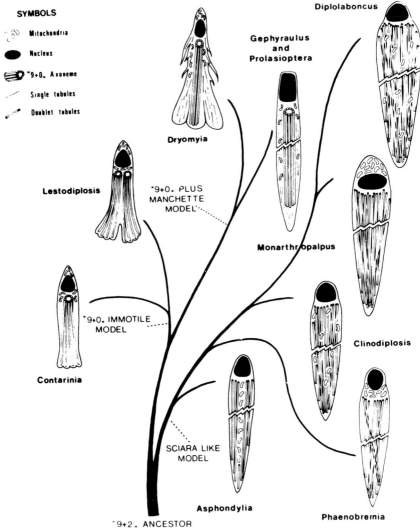

Fig. 11.28. Schematic drawing of sperm evolution in Cecidomyidae.   From Baccetti and Dallai (1976).

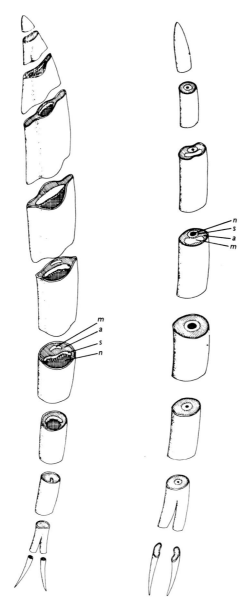

Fig. 11.29. Sperm of psychodid flies have a structure that is unknown from other animal groups. Flagellum, centriole and microtubules are missing, and remaining organelles are organized as shown in figure. Two sperm are from *Telmatoscopus albipunctatus* (left) and *Psychoda cinerea* (right). a = acrosome; m = mitochondrion; n = nucleus; s = subacrosomal material. From Baccetti et al. (1973a).

in Nematocera. Some families have an additional central unit, as the Simuliidae do (9 + 9 + 3: Baccetti et al., 1974b) and the Fungivoridae (9 + 3: Phillips, 1969); others have only one tubule in the center, e.g., Culicidae (9 + 9 + 1: Breland et al., 1966); still others have no tubules at all in the center, e.g., Bibionidae (9 + 9 + 0: Trimble and Thompson, 1974, and unpublished). Strangely, the Chironomidae, which have a relatively high systematic position, have the basic 9 + 9 + 2 pattern (unpublished). Sciaridae, Cecidomyidae, and Psychodidae show more aberrant situations. Sciaridae (Phillips, 1966, 1970) have 70 doublets or more, and 70 accessory tubules (Fig. 11.27). They have, therefore, a 70 + 70 + 0 pattern. Since arms are present, motility is possible. Cecidomyids (Fig. 11.28) are a peculiar family (Baccetti and Dallai, 1976, and unpublished): here we find in the somewhat primitive species, *Contarinia*, a 9 + 0 pattern; some other biflagellates (*Lestodiplosis*) also have a 9 + 0 pattern; others have a big manchette around the axoneme (9 + 0 + manchette: *Dryomya*, *Gephyraulus*, *Dasineura*, *Cystiphora*); finally, some (*Asphondilia*, *Clinodiplosis*, *Monarthropalpus*, *Diplolaboncus*, *Lestremia*) show a model with up to 100 doublets, similar to those of sciarids but with only one arm and without accessory tubules. So in Diptera we have two main groups (9 + 0 and *Sciara*-like) and several lines within them. Motility is always either very limited, or absent. Psychodid (Baccetti et al., 1973a) sperm are aflagellate, elongate, and immotile (Fig. 11.29). This family seems to be very homogeneous and the most evolved of nematocerans. No similar evolution is found in the Brachycera, in which all the species have the classical insect sperm 9 + 9 + 2, with one or two fully crystallized mitochondrial derivatives. The only characteristic is a monolayered acrosome (Perotti, 1969; Warner, 1970; Gassner, 1970)—a synapomorphic character typical of all Diptera.

11.4. CONCLUSIONS

In conclusion, it seems that comparative spermatology can offer a good phylogenetic approach. However, one must be careful to choose the best characteristics and reject apparent similarities. That means a good deal of submicroscopic analysis. Superficial research, or insufficient ultrastructural resolution, can lead to opposite conclusions, such as recently reached by Bargallo' (1975) and Lopez i Camps (1975), because of apparent outer general similarities in shape.

11.5. SUMMARY

Sperm ultrastructure has important significance in the discussion of the main evolutionary steps and lines in Arthropoda. Among the phyla related to Arthropoda, the basic aquatic sperm is present only in Tardigrada, while the Onychophora show a highly evolved spermatozoon, and appear as a sister-group to the arthropods. Among arthropods, only *Limulus* has the basic aquatic sperm, while the lowest Arachnida, Crustacea, and Myriapoda appear slightly evolved from it. This situation indicates a monophyletic origin of arthropods, starting from a *Limulus*-like aquatic ancestor. It is impossible to relate any of the above-mentioned groups to others; they appear to have independently branched out from the ancestor as different radiations. In each of them, as a consequence of a very specialized internal fertilization, we can follow some common evolutionary pathways with similar steps. The first step is the acquisition of unusual axoneme patterns, different from the basic 9 + 2 model, and frequent coiling in an encysted condition. The second step involves the loss of one or both the dynein arms, resulting in the lack of axoneme movement. Only in insects do we find an increased motility due to the extreme elongation of the flagellum and the acquisition of an extra set of nine tubules. Nevertheless, aflagellarity, as the final evolutionary step, is found among the higher members of the most specialized insect groups. Finally, in rare cases, the motility is secondarily acquired by some immotile models, but the axoneme never reappears, and substitutive devices for the lost axoneme have been invented in different cases.

For the various arthropodan classes, a classification and phylogeny based on spermatology are in good agreement with the commonly accepted ones.

REFERENCES

Alberti, G. and V. Storch. 1976. Ultrastruktur Untersuchungen am mänliche Genitaltrakt und an Spermien von *Tetranychus urticae* (Tetranychidae, Acari). *Zoomorphologie* **83**: 283-96.
André, J. 1959. Étude au microscope eléctronique de l'evolution du chondriome pendant la spermatogenèse du scorpion *Euscorpius* flavicaudis. *J. Ultrastruct. Res.* **2**: 288-308.
André, J. 1963. À propos d'une leçon sur la Limule. *Ann. Fac. Sci.* **26**: 27-38.
Baccetti, B. 1968. Lo spermatozoo degli artpodi. V. Aphaniptera. *Redia* **51**: 153-58.
Baccetti, B. 1970. The spermatozoon of Arthropoda. IX. The sperm cell as an index of arthropod Phylogenesis, pp. 169-82. *In* B. Baccetti (ed.) *Comparative Spermatology*. Academic Press, New York, London.
Baccetti, B. 1972. Insect sperm cells. *Adv. Insect Physiol.* **9**: 315-97.
Baccetti, B. and B. A. Afzelius. 1976. The biology of the sperm cell. *Monogr. Dev. Biol.* **10**: 1-254.
Baccetti, B. and R. Dallai. 1973. The spermatozoon of Arthropoda. XXV. New accessory tubule patterns in the sperm tail of Diplura. *J. Miscrosc. (Paris)* **16**: 341-44.
Baccetti, B. and R. Dallai. 1976. The spermatozoon of Arthropoda. XXVII. Uncommon axoneme patterns in the different species of cecidomyid dipterans. *J. Ultrastruct. Res.* **55**: 50-69.
Baccetti, B. and R. Dallai. 1977a. The spermatozoon of Arthropoda. XXIX. The degenerated axoneme and branched acrosome of Aleyrodids. *J. Ultrastruc. Res.* **60**: 260-70.
Baccetti, B. and R. Dallai. 1977b. The first multiflagellate animal spermatozoon in *Mastotermes darwiniensis*. (The spermatozoon of Arthropoda. XXX). *J. Cell Biol.* **76**: 569-76.
Baccetti, B., R. Dallai and A. G. Burrini. 1973a. The spermatozoon of Arthropoda. XVIII. The non-motile bifurcated sperm of Psychodidae flies. *J. Cell Sci.* **12**: 287-311.

Baccetti, B., R. Dallai and A. G. Burrini. 1976. The spermatozoon of Onychophorans: *Peripatopsis. Tissue Cell* **8**: 659–72.

Baccetti, B., R. Dallai and B. Fratello. 1973b. The spermatozoon of Arthropoda. XXII. The 12 + 0, 14 + 0 or aflagellate sperm of Protura. *J. Cell Sci.* **13**: 321–35.

Baccetti, B., R. Dallai and F. Giusti. 1969a. The spermatozoon of Arthropoda. VI. Ephemeroptera. *J. Ultrastruct. Res.* **29**: 343–49.

Baccetti, B., R. Dallai and F. Rosati. 1969b. The spermatozoon of Arthropoda. III. The lowest holometabolic insects. *J. Microsc. (Paris)* **8**: 233–48.

Baccetti, B., R. Dallai and F. Rosati. 1969c. The spermatozoon of Arthropoda. IV. Corrodentia, Mallophaga and Thysanoptera. *J. Microsc. (Paris)* **8**: 249–62.

Baccetti, B., R. Dallai and F. Rosati. 1970a. The spermatozoon of Arthropoda. VII. Plecoptera and Tricoptera. *J. Ultrastruct. Res.* **31**: 212–28.

Baccetti, B., R. Dallai and F. Rosati. 1970b. The spermatozoon of Arthropoda. VIII. The 9 + 3 flagellum of spider sperm cells. *J. Cell Biol.* **44**: 681–83.

Baccetti, B., F. Rosati and G. Selmi. 1971a. Electron microscopy of tardigrades. IV. The spermatozoon. *Monit. Zool. Ital.* **5**: 231–40.

Baccetti, B., F. Rosati and G. Selmi. 1971b. The spermatozoon of Arthropoda. XVI. The acrosome of Orthoptera. *J. Submicrosc. Cytol.* **3**: 319–37.

Baccetti, B., F. Rosati and G. Selmi. 1971c. The spermatozoon of Arthropoda. XV. An unmotile 9 + 2 pattern. *J. Microsc. (Paris)* **11**: 133–42.

Baccetti, B., R. Dallai, F. Bernini and M. Mazzini. 1974a. The spermatozoon of Arthropoda. XXIV. Sperm metamorphosis in the diplopod *Polyxenus. J. Morphol.* **143**: 187–246.

Baccetti, B., R. Dallai, F. Giusti and F. Bernini. 1974b. The spermatozoon of Arthropoda. XXIII. The 9 + 9 + 3 spermatozoon of simuliid Diptera. *J. Ultrastruct. Res.* **46**: 427–40.

Baccetti, B., R. Dallai, F. Rosati, V. Pallini and B. Afzelius. 1977a. Protein of insect sperm mitochondrial crystals. Crystallomitin. *J. Cell Biol.* **73**: 594–600.

Baccetti, B., R. Dallai, F. Rosati, F. Giusti, F. Bernini and G. Selmi. 1974c. The spermatozoon of Arthropoda. XXVI. The spermatozoon of Isoptera, Embioptera and Dermaptera. *J. Microsc. (Paris)* **21**: 159–72.

Baccetti, B., A. G. Burrini, R. Dallai, V. Pallini, M. Camatini, E. Franchi and L. Paoletti. 1977b. The delayed flagellum of milliped sperm is a reacted acrosome. (The spermatozoon of Arthropoda. XXVIII) *J. Submicrosc. Cytol* **9**: 187–219.

Baccetti, B., A. G. Burrini, R. Dallai, F. Giusti, M. Mazzini, T. Renieri, F. Rosati and G. Selmi. 1973c. Structure and function in the spermatozoon of *Tenebrio molitor*. The spermatozoon of Arthropoda. XX. *J. Mechanochem. Cell Motility* **2**: 149–61.

Baccetti, B., A. G. Burrini, R. Dallai, V. Pallini, P. Periti, F. Piantelli, F. Rosati and G. Selmi. 1973d. Structure and function in the spermatozoon of *Bacillus rossius*. The spermatozoon of Arthropoda. XIX. *J. Ultrastruct. Res.* **44** (*suppl. 12*): 1–73.

Bargallo', R. 1975. Polimorfisme dels gàmetes masculins. I. Convergència dels espermatozoides de Cirripedes i Quetognats. *Butlleti Soc. Catalana Biol.* **1**: 53–57.

Bawa, S. R. 1964. Electron microscopic study of spermiogenesis in a firebrat insect, *Thermobia domestica* Pack. I. Mature spermatozoon. *J. Cell Biol.* **23**: 431–46.

Boissin, L. J. F. Manier and O. Tuzet. 1972. Étude ultrastructurale de la spermatogenèse de *Glomeris marginata* Villers (Myriapode, Diplopode). *Ann. Sci. Nat. Zool. (12e sér.)* **14**: 221–39.

Breland, O. P., G. Gassner, R. W. Riess and J. J. Biesele. 1966. Certain aspects of the centriole adjunct, spermiogenesis and the mature sperm of insects. *Can. J. Genet. Cytol.* **8**: 759–73.

Brown, G. G. 1970. Some comparative aspects of selected crustacean spermatozoa and crustacean phylogeny, pp. 183–205. *In* B. Baccetti (ed.) *Comparative Spermatology*. Academic Press, New York, London.

Brown, G. G. and C. B. Metz. 1967. Ultrastructural studies on the spermatozoa of two primitive crustaceans, *Hutchinsoniella macracantha* and *Derocheilocaris typicus*. *Z. Zellforsch. Mikrosk. Anat.* 80: 78-92.

Camatini, M., A. Saita and F. Cotelli. 1973. Spermiogenesis of *Lithobius forficatus* L. at ultrastructural level. *Symp. Zool. Soc. Lond.* 32: 231-35.

Chevaillier, P. 1970. Le noyau du spermatozoide et son évolution au cours de la spermiogenèse, pp. 499-514. *In* B. Baccetti (ed.) *Comparative Spermatology*. Academic Press, New York, London.

Chevaillier, P. and J. L. Prigent. 1969. Analyse structurale de spermatozoide de *Glomeris marginata* Villers (Myriapode, Diplopode). *C. R. Hebd. Séances. Acad. Sci. Paris* 269: 1202-4.

Cotelli, F., M. Ferraguti, G. Lanzavecchia and C. Lora Lamia Donin. 1976. The spermatozoon of Peracarida. I. The spermatozoon of terrestrial isopods. *J. Ultrastruct. Res.* 55: 378-90.

Cruz-Landim, C. da and A. Ferreira. 1972. Sperm differentiation in the scorpion *Tityus bahiensis* (Perty). *Caryologia* 25: 125-35.

Dallai, R. 1967. Lo spermatozoo degli Artropodi. I. *Anurida maritima* (Guézin) ed *Orchesella villosa* (Geoffroy) (Insecta Collembola). *Atti Accad. Fisiocr. Siena (ser. 13)* 16: 468-76.

Dallai, R. 1970. The spermatozoon of Arthropoda. XI. Further observations on Collembola, pp. 275-79. *In* B. Baccetti (ed.), *Comparative Spermatology*. Academic Press, New York, London.

Dallai, R. 1972. The arthropod spermatozoon. XVII. *Machilis distincta* (Insecta Thysanura). *Monit. Zool. Ital.* 6: 37-61.

Descamps, M. 1972. Étude ultrastructurale du spermatozoide de *Lithobius forficatus* L. (Myriapoda, Chilopoda). *Z. Zellforsch. Mikrosk. Anat.* 126: 193-205.

Eddleman, C. D., O. P. Breland and J. J. Biesele. 1970. A three dimensional model of the acrosome in the American cockroach, *Periplaneta americana*, pp. 281-88. *In* B. Baccetti (ed.) *Comparative Spermatology*. Academic Press, New York, London.

Fabre, M. 1855. Recherches sur l'anatomie des organes reproducteurs et sur le développement des Myriapodes. *Ann. Sci. Nat. (4e sér.)* 3: 257-316.

Fain-Maurel, M. A., J. F. Reger and P. Cassier. 1975a. Le gamète mâle des Schizopodes et ses analogies avec celui des autres Péracarides. I. Le spermatozoide. *J. Ultrastruct. Res.* 51: 269-80.

Fain-Maurel, M. A., J. F. Reger and P. Cassier. 1975b. Le gamète mâle des Schizopodes et ses analogies avec celui des autres Péracarides. II. Particularité de la différenciation cellulaire au cours de la spermatogenèse. *J. Ultrastruct. Res.* 51: 281-92.

Feldman-Musham, B. and B. K. Filshie. 1976. Scanning and transmission electron microscopy of the spermiphores of *Ornithodorus* Ticks: an attempt to explain their motility. *Tissue Cell* 8: 411-19.

Folliot, R. and P. L. Maillet. 1970. Ultrastructure de la spermiogenèse et du spermatozoide de divers insectes homopteres, pp. 289-300. *In* B. Baccetti (ed.) *Comparative Spermatology*. Academic Press, New York, London.

Franzen, A. 1970. Phylogenetic aspects of the morphology of spermatozoa and spermiogenesis, pp. 29-46. *In* B. Baccetti (ed.) *Comparative Spermatology*. Academic Press, New York, London.

Furieri, P. 1963. Osservazioni preliminari sulla ultrastruttura dello spermio di *Pyrrhocoris apterus* Fall. (Hemiptera, Pyrrhocoridae). *Redia* 48: 179-87.

Gassner, G. 1970. Studies on the housefly centriole adjunct. *J. Cell Biol.* 47: 69A.

Hinsch, G. W. 1970. Penetration of the oocytes by spermatozoa in the spider crab *Lebinia emarginata* L. *J. Ultrastruct. Res.* 35: 86-97.

Hoage, T. R. and R. G. Kessel. 1968. An electron microscope study of the process of differentiation during spermatogenesis in the drone honey bee (*Apis mellifera* L.) with special reference to centriole replication and elimination. *J. Ultrastruct. Res.* **24**: 6-32.

Hood, R. D., O. F. Watson, T. R. Deason and C. L. B. Benton. 1972. Ultrastructure of scorpion spermatozoa with atypical axonemes. *Cytobios* **5**: 167-77.

Horstmann, E. 1968. Die Spermatozoen von *Geophilus linearis* Koch (Chilopoda). *Z. Zellforsch. Mikrosk. Anat.* **89**: 410-29.

Horstmann, E. and H. Breucker. 1969a. Spermatozoen und Spermiohistogenese von *Graphidostreptus* sp. (Myriapoda, Diplopoda). I. Die reifen Spermatozoen. *Z. Zellforsch. Mikrosk. Anat.* **96**: 505-20.

Horstmann, E. and H. Breucker. 1969b. Spermatozoen und Spermiohistogenese von *Spirostreptus* sp. (Myriapoda, Diplopoda). II. Die Spermiohistogenese. *Z. Zellforsch. Mikrosk. Anat.* **99**: 153-84.

Ito, S. 1966. Movement and structure of louse spermatozoa. *J. Cell Biol.* **31**: 128A.

Jespersen, Å and R. Hartwick. 1973. Fine structure of spermiogenesis in scorpions from the family Vejovidae. *J. Ultrastruct. Res.* **45**: 366-83.

Juberthie, C., J. F. Manier and L. Boissin. 1976. Étude ultrastructurale de la double spermiogenèse chez l'Opilion cyphophtalme, *Siro rubens* Latreille. *J. Microsc. Biol. Cell.* **25**: 137-48.

Kessel, R. G. 1966. The association between microtubules and nuclei during spermiogenesis in the dragonfly. *J. Ultrastruct. Res.* **16**: 293-304.

Legg, G. 1973. The structure of encysted sperm of some British Pseudoscorpiones (Arachnida). *J. Zool. Lond.* **170**: 429-40.

Lopez i Camps, J. 1975. Polimorphisme dels gametes masculins. II. Diversitat en la morfologia dels espermatozoides dins una mateixa linia evolutiva monofilética. Ostràcodes i Cirripedes. *Butlleti Soc. Catalana Biol.* **1**: 59-63.

Manier, J. F. and L. Boissin. 1975. Aspects ultrastructuraux de la spermiogenèse des Polydesmoidea (Myriapodes, Diplopodes). *Ann. Sci. Nat. Zool. Paris* **17**: 505-20.

Manier, J. F. and L. Boissin. 1976. Nouvelle contribution a l'étude ultrastructurale de la spermiogènese des Diplopodes Spirobolides. *Ann. Sci. Nat. Paris* **18**: 437-48.

Manier, J. F., L. Boissin and O. Tuzet. 1974. Etude ultrastructurale de la spermiogenèse de *Plethocrossus acutiformis* Demange et *Isoporostreptus bouixi* Demange (Myriapodes Diplopodes Spirostreptoidea). *Bull. Biol.* **108**: 169-83.

Mazzini, M. 1970. Lo spermatozoo di un afide: Megoura viciae Kalt. *Atti Accad. Fisiocr. Siena (ser. 14)* **2**: 1-6.

Moses, M. J. 1966. Cytoplasmic and intranuclear microtubules in relation to development, chromosome morphology and motility of an aflagellate spermatozoon. *Science (Wash., D.C.)* **154**: 424.

Munn, E. A. and H. Barnes. 1970. The fine structure of the spermatozoa of some cirripedes. *J. Exp. Mar. Biol. Ecol.* **4**: 261-86.

Oettinger, R. 1909. Zur Kenntnis der Spermatogenese bei den Myriapoden. Samenreifung und Samenbildung bei *Pachyjulus varius* Fabr. *Arch. Zellforsch.* **3**: 563-626.

Osaki, H. 1969. Electron microscope study on the spermatozoon of the liphistid spider, *Heptathela kimurai. Acta Arachnol. Tokyo* **22**: 1-12.

Perotti, M. E. 1969. Ultrastructure of the mature sperm of *Drosophila melanogaster* Meig. *J. Submicrosc. Cytol.* **1**: 171-96.

Phillips, D. M. 1966. Fine structure of *Sciara coprophila* sperm. *J. Cell Biol.* **30**: 499-517.

Phillips, D. M. 1969. Exceptions to the prevailing pattern of tubules (9 + 9 + 2) in the sperm flagella of certain insect species. *J. Cell Biol.* **40**: 28-43.

Phillips, D. M. 1970. Insect sperm: Their structure and morphogenesis. *J. Cell Biol.* **44**: 243-77.

Phillips, D. M. 1971. Morphogenesis of the lacinate appendages of lepidopteran spermatozoa. *J. Ultrastruct. Res.* **34**: 567–85.

Phillips, D. M. 1976. Nuclear shaping during spermiogenesis in the whip scorpion. *J. Ultrastruct. Res.* **54**: 397–405.

Pochon-Masson, J. 1968a. L'ultrastructure des spermatozoïdes vésiculaires chez les crustacés décapodes avant et au cours de leur dévagination experimentale. I. Brachyoures et anomoures. *Ann. Sci. Nat. (ser. 12)* **10**: 1–100.

Pochon-Masson, J. 1968b. L'ultrastructure des spermatozoïdes vésiculaires chez les crustacés décapodes avant et au cours de leur dévagination experimentale. II. Macroures. Discussion et conclusions. *Ann. Sci. Nat. (ser. 12)* **10**: 367–454.

Pochon-Masson, J., J. Bocquet-Vedrine and Y. Turquier. 1970. Contribution à l'étude du spermatozoïde des crustacés cirripèdes, pp. 205–19. *In* B. Baccetti (ed.) *Comparative Spermatology.* Academic Press, New York, London.

Reger, J. F. 1963. Spermiogenesis in the tick, *Amblyomma dissimili*, as revealed by electron microscopy. *J. Ultrastruct. Res.* **8**: 607–21.

Reger, J. F. 1969. A fine structure study on spermiogenesis in the arachnid *Leiobunum* sp. (Phalangida, Harvestmen). *J. Ultrastruct. Res.* **28**: 422–34.

Reger, J. F. 1970a. Spermiogenesis in the spider, *Pisuarina* sp.: A fine structure study. *J. Morphol.* **130**: 421–34.

Reger, J. F. 1970b. Some aspects of the fine structure of filiform spermatozoa (Ostracod, *Cypridopsis* sp.) lacking tubule substructure, pp. 237–46. *In* B. Baccetti (ed.) *Comparative Spermatology.* Academic Press, New York, London.

Reger, J. F. 1971. Studies on the fine structure of spermatids and spermatozoa from the millipede *Spirobolus* sp. *J. Submicrosc. Cytol.* **3**: 33–44.

Reger, J. F. 1974. The origin and fine structure of cellular processes in spermatozoa of the tick *Dermacentor andersoni. J. Ultrastruct. Res.* **48**: 420–34.

Reger, J. F. and D. P. Cooper. 1968. Studies on the fine structure of spermatids and spermatozoa from the milliped *Polydesmus* sp. *J. Ultrastruc. Res.* **23**: 60–70.

Robison, W. G., Jr. 1970. Unusual arrangement of microtubules in relation to mechanisms of sperm movement, pp. 311–320. *In* B. Baccetti (ed.) *Comparative Spermatology.* Academic Press, New York, London.

Rosati, F., B. Baccetti and R. Dallai. 1970. The spermatozoon of Arthropoda. X. Araneids and the lower myriapods, pp. 247–254. *In* B. Baccetti (ed.) *Comparative Spermatology.* Academic Press, New York, London.

Rosati, F., G. Selmi and M. Mazzini. 1976. Comparative observations on the mitochondrial derivatives of insect sperm. *J. Submicrosc. Cytol.* **8**: 51–67.

Silvestri, F. 1898. Ricerche sulla fecondazione di un animale a spermatozoi immobili. *Ricerche Lab. Anat. R. Univer. Roma* **6**: 255–65.

Sokoloff, J. 1914. Uber die Spermatogenese bei *Polyxenus* sp. *Zool. Anz.* **44**: 558–66.

Tilney, L. G. 1975. The role of actin in non muscle cell motility, pp. 339–88. *In* S. Inoué and R. E. Stephens (eds.) *Molecules and Cell Movement.* Raven Press, New York.

Trimble, J. J., III and S. A. Thompson. 1974. Fine structure of the sperm of the lovebug, *Plecia neartica* Hardy (Diptera: Bibionidae). *Int. J. Insect Morphol. Embryol.* **3**: 425–32.

Tuzet, O. and J. F. Manier. 1955. Achèvement de la spermiogenèse dans les voies génitales femelles: cas de *Polyxenus lucidus* Chalande (Myriapode Diplopode). *Ann. Sci. Nat. Zool.* **17**: 351–56.

Van Deurs, B. 1973. Axonemal 12 + 0 pattern in the flagellum of the motile spermatozoon of *Nymphon leptocheles. J. Ultrastruct. Res.* **42**: 594–98.

Van Deurs, B. 1974a. Pycnogonid sperm. An example of inter and intraspecific axonemal variation. *Cell Tissue Res.* **149**: 105–11.

Van Deurs, B.   1974b.   Spermatology of some Pycnogonida (Arthropoda), with special reference to a microtubule–nuclear-envelope complex. *Acta Zool. (Stockh.)* **55:** 151–62.

Warner, F. D.   1970.   New observations on flagellar fine structure. *J. Cell Biol.* **47:** 159–82.

Wingstrand, K. G.   1972.   Comparative spermatology of a pentastomid, *Raillietiella hemidactyli*, and a branchiuran crustacean, *Argulus foliaceus*, with a discussion of pentastomid relationship. *K. Danske Vidensk. Selsk. Skr.* **19:** 5–72.

Wingstrand, K. G.   1973.   The spermatozoa of the thysanuran insects *Petrobius brevistylis* Carp. and *Lepisma saccharina* L. *Acta Zool. (Stockh.)* **54:** 31–52.

Zissler, D.   1969.   Die Spermiohistogenese des Süsswasser-Ostracoden *Notodromas monacha* O. F. Müller. I. and II. *Z. Zellforsch. Mikrosk. Anat.* **96:** 87–133.

Zissler, D.   1970.   Zur Spermiohistogenese im Vas deferens von Süsswasser-Ostracoden. *Cytobiologie* **2:** 83–86.

12

Comparison of Arthropod Neuroendocrine Structures and Their Evolutionary Significance

A. S. TOMBES

12.1. INTRODUCTION

The value of studying an internal organ system and relating the observations to the general question of arthropod phylogeny depends upon the extent of our knowledge of that system in the representative taxonomic groups. Our understanding of arthropod neuroendocrine structures began around the turn of the century and is most complete in the insects, followed by crustaceans, myriapods, and chelicerates. Practically nothing is known about the onychophorans, and, of course, the fossil record of trilobites provides no clues to internal anatomy. Nevertheless, by assessing the available information and establishing possible structural relationships where they may exist, we not only identify the voids in the comparative neuroendocrine anatomy of the major arthropod taxa that may attract future researchers, but we can identify one or several features that appear to be indicative of the presence, or absence, of an evolutionary continuum in the endocrine system within the arthropods.

In those taxa where scant information is available, many of the original papers will of necessity be reviewed in order to determine the morphological patterns characteristic of the group. However, with others, such as the crustaceans and insects, effort must be expended to focus on the general and not the specific because the objective of this paper is to examine the phylogeny of classes within the phylum and not of orders within the classes.

A review of the chemical coordinating system in arthropods reveals a fair degree of uniformity. The generalized arthropod possesses: 1) clusters of paired, unipolar, neurosecretory cells within most ganglia of the central nervous system which stain positive to either paraldehyde fuchsin, Azan, chrom-hematoxylin phloxine, or alcian blue, and which contain electron-dense granules or vesicles between 80 and 300 nm in diameter; 2) groups of axons which extend centrifugally from these neurosecretory cells toward the ganglionic periphery and through which will flow the detectable neurosecretion; 3) nonspecific areas or specific neurohemal structures receiving the neurosecretory axon terminals from which will pass the active neurohormone; and 4) various true endocrine glands of ectodermal origin which are responsive to the presence of hormones. Thus, variations will be of interest in the comparative structure of these broad characteristics which may support the theory of a common, monophyletic origin of the arthropods, or support a polyphyletic origin with a rather sharp distinction between the marine (chelicerates and crustaceans) and the land representatives (onychophorans, myriapods, and insects).

As electron microscopic and biochemical techniques become more widely used in endocrine research, earlier designations of neurosecretory elements and endocrine glands based only on optical microscopy are frequently being subjected to critical reinterpretation. Likewise, new elements of the neuroendocrine network are continually being identified. Thus, whenever appropriate, especially concerning the designation of neurosecretory cells and neurohemal areas, preference will be placed on the more recent ultrastructural observations while remembering that these techniques are not always infallible.

12.2. ARTHROPOD GROUPS

12.2.1. Onychophora

12.2.1.1. Neurosecretory Cells

Such cells have been identified with chrome hematoxylin-phloxine and paraldehyde-fuchsin in the cerebral ganglion and along the two ventral cords (Gabe, 1954; Sanchez, 1958). They comprise five paired groups: first a single pair of cells in the dorsal cellular cortex, then two groups in the laterodorsal and two groups in the lateroventral region of the cerebral ganglion. The axonal paths from these cells as well as areas for neurosecretion release are unknown.

12.2.1.2. Neurohemal Areas

Some speculation has centered on the possible endocrine nature of the paired infracerebral organs closely aligned to the ventral aspect of the cerebral ganglion. Suggestions have been made that they are neurohemal in nature (Sanchez, 1958) or that they have an endocrine function (Cuénot, 1949). However, no axons bearing neurosecretion have been identified within these organs. Further optical and electron microscopy would be of considerable value in assessing the endocrine nature of these structures in this group of primitive invertebrates, but such techniques have not been employed.

12.2.2. Chelicerata (Table 12.1)

12.2.2.1. Neurosecretory Cells

The chelicerates have well-developed and apparently well-compartmentalized neurosecretory systems, and although most of our descriptive information is derived from optical microscopy conducted in the fifties, the few recent electron microscopic studies have supported generalized patterns with the exception of that in Xiphosura. Since the early seventies a number of excellent articles have appeared describing the neuroendocrine structure of our most notable ancestral anthropod, thus correcting earlier misconceptions derived solely from optical microscopy before the advent of specific stains. In an ultrastructural study of *Limulus polyphemus*, Fahrenbach (1973) identified four pairs of large neuro-secretory cells in close proximity with the visual association center of the protocerebrum. Electron microscopic observations were also made by Herman and Preus (1973) but referred largely to the abdominal ganglia. The arrangement of cells in *Limulus* is in line with the other chelicerate observations which show clusters of neurosecretory cells in the proto- and tritocerebrum.

The elementary neurosecretory granules in the protocerebral cells are electron-dense and membrane-bounded, but according to Fahrenbach (1973) vary greatly in diameter, averaging between 30 and 400 nm. Herman and Preus (1973) observed distinctly different granules in each of the two separate clusters of neurosecretory cells in the abdominal ganglia of *Limulus*. One cell type had dense, spherical granules approximately 230 nm in diameter, and the other had dense ovid, ellipsoid, or dumbbell granules measuring approximately 540 by 160 nm. These two types will be referred to later in the discussion of neuro-hemal organs.

All other observations on neurosecretory cells in chelicerates present basically the same observations, i.e., clusters of cells in the supraesophageal ganglia and each ganglionic region of the fused subesophageal nervous mass. However, many variations are apparent, which reflect differences between the classes Mero-stomata and Arachnida as well as between members of the arachnids.

12.2.2.2. Neurohemal Areas

In contrast to the similarities in the location of the neurosecretory cells, there is a considerable degree of variance in what appears to be the nature of neurohemal structures. Thus far axonal paths in the protocerebrum of *Limulus* have been

Table 12.1. Summary of Neuroendocrine Structures in Chelicerata.

	Neurosecretion			Endocrine Structures		
	Neurohemal Component					
	Cell Location	Area or Structure	Intrinsic Secreting Cells	Target Cells Directly Innervated	Glands	Secretion
Xiphosura	Protocerebrum: vision association area (18) Abdominal ganglia: anterior-lateral-ventral regions, and mid-ventral regions (26).	Diffuse, at ventral area of proto. (18) Unknown, but axons leave ganglia (26).		Ommatidial pigment cells and receptor cells of all eyes (18).		Ecdysone extracted from whole larvae (30) and blood (73). Ecdysone initiates molting cycle (31, 25).
Araneae	Protocerebrum Abdomial ganglia	Organs of Schneider I and II (20).	Present, and their axons may extend to "Tropfenkumplex"			Ecdysone initiates molting cycle (6).
Opilionida	Syncerebrum, composed of brain, supraesophageal or cerebral mass (39).	Paraganglionic plates (39) adjacent to syncerebrum and subesophageal nervous	They are not present (39)			

	Subesophageal nervous mass	mass. Not detached as separate structures Seven types of axon terminals.	
Pseudoscorpionida	Proto- and tritocerebrum (37) Subesophageal nervous mass	Paraganglionic plates not detached as separate structures. Four types of axon terminals (37).	
Scorpionida	Proto- and tritocerebrum (22) Subesophageal ganglion.	Sympathetic or stomatogastric ganglion. On anterior face of protocerebrum (22).	Present (22)
Acarina	Supraesophageal part (17) Subesophageal part (17)	Paraganglionic plates (21)	

impossible to map, and no structure or organ innervated by nerves from the supraesophageal ganglia has been found. Fahrenbach (1973) did observe axon endings among glial cells of the protocerebrum adjacent to sinusoids of the circulatory system. The endings contained spherical electron-dense granules, which measured 120 to 280 nm in diameter, a measurement more limiting in range than the granules observed in the protocerebral cells. Synaptoid specializations are also observed adjacent to the plasma membrane, an indication of neurosecretion release.

Fahrenbach (1973) also found axonal terminals (neurochromatomotor junctions) with synaptoid terminals on the proximal, intraommatidial, and distal pigment cells of the compound eye of *Limulus*. The point of interest here is that the granules are cylindrical, up to 300 nm long and between 100 and 500 nm in diameter, measurements which overlap with those reported by Herman and Preus (1973) for abdominal ganglia.

The opilionids, pseudoscorpions, and acarina possess paraganglionic plates that are specializations on the posterior face of the supraesophageal mass adjacent to the anterior dorsal edge of the subesophageal nervous mass. The plates are composed of glial cells and neurosecretory axon terminals, which are separated from the hemocoel by a thick neural lamella. Juberthie and Juberthie-Jupeau (1974b) identified seven types of axon terminals in the paraganglionic plates of an opilione *Trogulus nepaeformis*.

Based only on the optical microscopy by Habibulla (1970, 1975), the scorpionids have a neurohemal area, the single sympathetic ganglion (stomatogastric ganglion) along the anterior ventral region of the supraesophageal ganglion which receives axonal tracts from the proto- and tritocerebral cell cluster. This structure is more completely set apart from the brain than the paraganglionic plates, and because of its ganglionic nature it also possesses intrinsic neurosecretory cells, not observed in any of the plate structures.

This comparative view leads to the Araneae, where the neurohemal structures, also referred to as retrocerebral complexes, are very distinct and consist of the primary and secondary organs of Schneider. These organs are composed of neurosecretory axonal endings as well as intrinsic glandular cells. The endings are from the cells within the supraesophageal ganglia, but no ultrastructural data are available on the nature of their ending.

12.2.2.3. Endocrine Organs and Secretions

The search for a glandular source of a molting hormone in Xiphosura has been unproductive, and what appears to be an ecdysial gland in arachnids is based only on histological rather than biochemical studies. However, there have been several recent reports indicating that ecdysones are present in the bodies of chelicerates. These observations are based on positive bioassay analyses for the molting hormone in whole *Limulus* larvae (Jegla, 1974) and in the blood of juveniles (Winget and Herman, 1976). Also α- and β-ecdysones and several analogs are effective in stimulating molting in larvae of *Limulus* (Jegla and Costlow, 1970; Jegla *et al.*, 1972). Again, however, the source of the hormone or even the body region from which the chemical might originate is not known.

12.2.3. Crustacea (Table 12.2)

12.2.3.1. Neurosecretory Cells

Few nonmalacostracans have been seriously examined endocrinologically (Mc-Gregor, 1967; Lake 1969, 1970; van den Bosch de Aguilar, 1972, 1976; Davis and Costlow, 1974), leaving several major taxonomic groups, representing a variety of physiological strategies, yet to be studied. From the somewhat limited information available a pattern of neurosecretory cell distribution is suggested, which includes the cerebral ganglia, optic medulla, subesophageal ganglion, and ganglia of the ventral nerve cord. In the Anostraca (Lake, 1969, 1970) a pair of X-organs, or dorsal frontal organs, extends anteriorly from the surface of the mid-brain. They contain neurosecretory cells, which may release their contents locally into the hemolymph or at a more distant release area.

In contrast, the neurosecretory systems in the malacostracans are well known, and the vast majority of studies have focused on the decapods (Aiken, 1969; Tombes, 1970; Highnam and Hill, 1977). Ganglia within the brain, eyestalk, circumesophageal connectives, and ventral nerve cord all show several types of neurosecretory cells. The three organs of Hanström (ganglionic X-organs) associated with the optic ganglia have been established as the most important source of neurohormones because of the diversity of cell types and their hormonal effects. As many as six cell types can be distinguished histologically in the organ of Hanström on the medulla terminalis. Axons from these neurosecretory cells can be followed histologically, and they contribute exclusively to the neurohemal structure of the eyestalks, the sinus gland.

Neurosecretory cells are also present in several less distinct areas of the cerebral ganglia, whose locations and the specific tinctorial characteristics of the cells vary in different species. The neurohemal area for these cells is also the sinus gland.

The ganglia of the circumesophageal connectives and the ventral nerve cord all contain neurosecretory cells which illustrate similar morphology to that of cells found elsewhere in the malacostracan nervous system.

12.2.3.2. Neurohemal Areas

In Anostraca the sinus gland (Lake, 1969) and the optic neurohemal organs (Lake, 1970) are clusters of neurosecretory axon endings positioned among

Table 12.2. Summary of Neuroendocrine Structures in Crustacea.

	Neurosecretion				Endocrine Structures	
		Neurohemal Component		Target Cells Directly Innervated		
	Cell Location	Area or Structure	Intrinsic Secreting Cells		Glands	Secretion
Non-Malacostracan Branchiopoda	Proto-, deuto-, and tritocerebrum and with x-organ on eyestalks (47, 48) Subesophageal ganglia	Sinus gland, located between optic medulla and lamina ganglionaris, receives axons from cells of the brain and organs of Hanstrom. (47, 48)	Not present			α and β Ecdysone (2) (extracted from whole bodies)
Cirripedia	Cerebral ganglion and ventral nerve cord (15)					
Malacostracan	Supraesophageal ganglion and connectives. Optic ganglionic x-organs or Organs of Hanstrom ventral ganglia (3, 21)	Sinus Gland an aggregation of axonal endings from organs of Hanstrom and brain, postcommissure organs, pericardial organs, and along the ventral nerve cord (3, 21)			Molting gland (y-organ) (64, 65) Androgenic gland	Molting hormone (α- and possibly β-ecdysone) (72, 10) Possibly a steroid (4)

supporting cells between the terminus of the optic medulla and the lamina ganglionaris at the distal area of the eyestalk. These structures have morphological and histological features that are quite similar to those of the sinus gland of higher Crustacea. Release may also occur from other structures which are suspected of having a neurohemal function, i.e., organs of Bellonici (Lake, 1969).

The principal neurohemal organs of Malacostraca are the paired sinus glands located on the periphery of the optic ganglia. Each gland receives extensive neurosecretory axon endings from cells within the organs of Hanström, and these endings form a thickened disk or cup-shaped structure adjacent to the body cavity but just under the neurilemma or epineurium. All axons contain neurosecretory granules, which are of several types based on electron density and granule diameter. The paired postcommissure organs also serve as neurohemal areas for the cells located within the tritocerebrum and circumesophageal connective ganglia. Within these organs the axons divide into smaller branches and terminate as bulbous endings under the neurolemma adjacent to the blood sinus just as in the sinus gland. The third neurohemal structure, the pericardial organ, hangs freely within the pericardial cavity, so that it is bathed by blood moving from the gills to the heart. The axons extend from cell bodies that are either within the organ or are located in the central nervous system, such as the subesophageal ganglion.

12.2.3.3. Endocrine Organs

Two paired epithelial endocrine glands are present in Malacostraca: the molting glands in the anterior body region and the androgenic glands attached to the male reproductive structures. These glands have not been identified in the non-malacostracans, although α- and β-ecdysones have recently been reported in the Cirripedia (Bebbington and Morgan, 1977). Cells producing the molting hormones are assumed to be present throughout the lower crustaceans because of the obvious importance of a well-coordinated molting cycle, but they may be rather diffused. However, a similar assumption should not be made in reference to the presence of an androgenic gland, since other mechanisms are available for the coordination of male reproductive development (Charniaux-Cotton, 1975a,b).

The identification and homology of the molting glands within the decapod group is under current debate (Sochasky et al., 1972; Sochasky and Aiken, 1974; Willig and Keller, 1976), and the well-established "Y-organ" designation may prove to possess broad anatomical rather than specific functional significance. Thus, the glands that secrete the molting hormone may have been identified as glands concerned with ingestive or digestive functions, and likewise not all so designated "Y-organs" may be secreting a molting hormone.

The histology and ultrastructure of the gland cells producing the molting hormone are typical for steroid-synthesizing cells and show close similarity to the molting glands (prothoracic gland, thoracic glands, and so on) in insects (Herman, 1967).

The androgenic glands, identified in all orders of the subclass Malacostraca, appear as compact strands of epithelial cells which are attached, via a conjunctival sheath, to different areas of the male reproductive tissue. These holocrine glandular cells from several species have been studied with optical and electron microscopy, and in very broad terms illustrate features that are suggestive of a non-protein hormone synthesis. The accumulation of this secretory product is then followed by a holocrine release to the hemocoel and cellular degeneration (Payen, 1972; Rader and Cracium, 1976).

12.2.3.4. Cellular Secretions

In malacostracans when the molting glands are not under the influence of the molt-inhibiting neurosecretion, they release to the body fluid a molting hormone. Unlike the unequivocal data for the source and role of α-ecdysone in insects, the crustacean molting hormone has been suggested to be β-ecdysone, primarily because of its high hemolymph concentration. Chang and O'Connor (1977), using organ-cultured crab Y-organs with radioimmunoassay and chromatographic analyses, have now presented evidence that refutes this suggestion, and they suggest that α-ecdysone is secreted by the molting gland. They agree with others (King and Siddall, 1974) that β-ecdysone is the conversion product originating in the target organ via the enzyme ecdysone 20-hydroxylase. If their conclusions are supported in other crustaceans, then the secretion of the molting glands for both crustaceans and insects would be identical. The recent evidence for the presence of α- and β-ecdysone in whole-body extracts of Cirripedia (Bebbington and Morgan, 1977) extends to the nonmalacostracans as a working hypothesis the concept that α-ecdysone is the primary molting hormone in arthropods.

The chemistry of the androgenic gland hormone has not been studied. Indirect experimental and ultrastructural evidence suggests that the secretion may be a steroid.

12.2.4. Myriapoda (Table 12.3)

12.2.4.1. Neurosecretory Cells

There is considerable similarity among the Diplopoda, Chilopoda, and Symphyla in the general appearance and arrangement of the cerebral neurosecretory cells. These cells are found in all three primitive ganglia, proto-, deuto-, and tritocerebrum, and some investigators have divided the cells into three cell types—A, B, and C—with further subtyping in some species (Jamault-Navarro and Joly, 1977; Nair, 1974). Axons extend laterally and ventrally through the cerebral

Table 12.3. Summary of Neuroendocrine Structures in Myriapoda.

	Neurosecretion				Endocrine Structures	
		Neurohemal Component		Target Cells		
	Cell Location	Area or Structure	Intrinsic Secreting Cells	Directly Innervated	Glands	Secretion
Diplopoda	Cerebral ganglia: protocerebrum deutocerebrum tritocerebrum Subesophageal ganglia Ventral nerve cord	Cerebral neuro-hemal organ (38) Gabe organ (55, 41, 61) Paraesophageal body (55) Connective body (56, 51)	None are present (38) Present (41) Present (55) None are present (56, 51)			
Chilopoda	Cerebral ganglia: protocerebrum deutocerebrum tripocerebrum (29) Subesophageal ganglia Ventral nerve cord	Cerebral gland (34, 35)	Present (34, 35)		Ecdysial gland (62, 58)	Suggested to be the molting hormone
Symphyla	Cerebral ganglia (40)	Cephalic gland or neurohemal organ, containing eight types of granules in axons endings (40, 42, 43)	None are present (42)			

ganglia, exit at several locations, and form distinct neurohemal structures. Cells are also present in the subesophageal and ventral ganglia of the nerve cord, and their areas of release are probably along the margin of the cord.

12.2.4.2. Neurohemal Areas

In the Diplopoda there are four neurohemal structures in close association with the cerebral ganglia. Both the Gabe organ (cerebral gland) and the paraesophageal body (visceral ganglion) contain intrinsic neurosecretory cells along with cerebral neurosecretory axon endings (Juberthie-Jupeau, 1973; Petit and Sahli, 1975). On the contrary, the connective body, located on the lateral esophageal connective, is composed of only axon endings with their cell bodies in both the brain and the subesophageal ganglion (Nair, 1974). Likewise, the fourth structure, the cerebral neurohemal organ, has no intrinsic secretory cells and receives axons from only the brain (Juberthie and Juberthie-Jupeau, 1974a,b).

In the most extensively studied chilopod, *Lithobius forficatus*, there is only one pair of neurohemal structures, the cerebral glands. They are attached to the underside of each optic lobe and are in contact with the hypodermis (Joly, 1966). The histology and ultrastructure of the cerebral gland is very similar to the Gabe organ in Diplopoda. The gland contains numerous axon endings which are believed to originate in neurosecretory cells in the pars intercerebralis and frontal lobe of the protocerebrum.

In Symphyla the cephalic gland or cephalic neurohemal organ is located on the posterior lateral aspect of the brain. It has been studied optically and ultrastructurally and is homologous to the cerebral gland and the Gabe organ of the other myriapods. It is constructed of numerous neurosecretory axon terminals which are separated from the hemolymph by a thin neural lamella. Seven types of neurosecretory glanules have been identified, based on their size and electron density (Juberthie-Jupeau, 1973).

12.2.4.3. Endocrine Organs

On structural grounds alone "Glandula ecdysalis" have been identified in *Lithobius*. Although tentative, they constitute the first reputed molting glands in Myriapoda (Seifert and Rosenberg, 1974; Rosenberg and Seifert, 1975). There does appear to be a cycle in the changes of cellular fine structure, which may be related to the animal's molting cycle.

12.2.5. Insecta (Table 12.4)

12.2.5.1. Neurosecretory Cells

The first of two groups of protocerebral neurosecretory cells is located as paired cell clusters close to the midline of the pars intercerebralis. Axons from each group extend posteriorly and ventrally, decussate, and issue from the surface of the brain as two distinct nerves (internal nervi corporis cardiaci or NCC I). In most insects a second cluster of cells is located laterally to the first and near the corpora peduculata. The axons from these also proceed posteriorly, and form a second pair of nerves to the corpora cardiaca (external nervi corporis cardiaci or NCC II).

In Thysanura, of the primitive subclass Apterygota, the neurosecretory cell bodies are separated from the cerebral ganglion, thus forming paired frontal organs. The Collembola, which are considered more primitive than the Thysanura, do not show this segregation of protocerebral cells (Cassagnau *et al.*, 1968; Cassagnau et Juberthie, 1970).

Using several standard histological methods, neurosecretory cells have been classified by numerous investigators into cell types, and the four that seem to be most consistently found are A, B, C, and D. Because of the extensive literature that has accumulated over the years on the description and classification of insect material, the pterygote insect has served as a useful model to investigators of other arthropods.

As in all arthropods, neurosecretory cells are present in the subesophageal and ventral nerve cord ganglia.

12.2.5.2. Neurohemal Areas

The principal terminals for the cerebral nerves, NCC I and II, are the paired corpora cardiaca (CC). They are present in Collembola of the Apterygota, but only as swellings at the end of the cerebral nerves. In the Pterygota the cardiaca are situated behind the brain, lateral to the esophagus and anterior to the corpora allata (CA), with which they are connected by the nervi allati. The glands are connected medially with the hypocerebral ganglion and also with the subesophageal ganglion. Neurosecretion accumulates in the endings of axons and passes across the plasma membrane and stroma to the hemolymph. Intrinsic secretory cells, absent in the Apterygota, are present in the Pterygota, and they appear to have all characteristics of neurosecretory cells that release their neurohormones to the hemolymph.

Other areas, which are more diffused than the CC, exist for the accumulation and release of neurosecretion, and they are found along the ventral nerve cord stemming from the medial nerve. These structures appear as thin axonal projections extending into the hemolymph and are called either perisympathetic organs or perivisceral neurohemal organs.

As a consequence of numerous ultrastructural investigations into insect tissue, profiles of axons containing neurosecretion in close proximity with muscle, epithelial, and glandular cells have been identified. In the CA, for example, secretomotor junctions have been observed with an approximately 20-nm gap between the release area of the axon and allatal cell. In most cases the location of the cell bodies for these axons has not been determined. Two exceptions

Table 12.4. Summary of Neuroendocrine Structures in Insecta.

	Neurosecretion	Neurohemal Component		Target Cells Directly Innervated	Endocrine Structures	
	Cell Location	Area or Structure	Intrinsic Secreting Cells		Glands	Secretion
Apterygota	Cerebral ganglion: prontocerebrum and deutocerebrum (9). In Thysanura the medial protocerebral cells are set apart as lateral frontal organs (69) Subesophageal ganglion and ventral nerve cord ganglia with several cell types (69)	Corpora cardiaca, connected to brain by a composite of NCC I and NCC II, appear as bilateral lobes or swellings along the lateral and dorsal wall of the aorta (69, 7)	None in Collembola (9, 7), but present in Thysanura (69)		Corpora allata, are innervated by nerves from the cardiacum and subesophageal ganglion (7, 36, 53) Ventral gland in thoracic region (69)	Juvenile hormone (69, 53) Molting hormones (69, 53)

| Pterygota (71, 16, 52, 27) | Cerebral ganglion: protocerebrum deutocerebrum tritocerebrum Subesophageal ganglion Ventral nerve cord | Corpora cardiaca, paired or fused medially, located immediately behind cerebral ganglion and connected to protocerebrum by two pairs of nerves NCC I and NCC II Perisympathetic organs along the median ventral nerve | Neurosecretory cells are present throughout structure or in a specific area. | Muscles, glands and epithelial cells | Corpora allata, attached to distal edge of corpora cardiaca Prothoracic gland, in ventrolateral area of the prothorax | Juvenile hormone in immature insects which has a gonadotropic function in adults. Molting hormone, ecdysone |

are the axons within the CA which originate within the protocerebrum and axons that extend to the heart muscle from the CC.

12.2.5.3. Endocrine Organs

There are two epithelial endocrine glands, the CA and prothoracic (ventral, thoracic, or ecdysial) glands, which arise embryonically as ectodermal invaginations in the region of the first and second maxillary pouches. In the Apterygota the paired CA remain along the ventral epidermis and are not connected to the CC in all species. The allata are present in all Pterygota and are connected to the CC by the nervi allati, which contain axons extending from the protocerebrum.

In all insects, the prothoracic glands are paired, or only partially fused, glandular structures, often located in the ventrolateral areas of the prothorax, innervated by a few small nerves from the subesophageal and thoracic ganglia. Neurosecretory granules have been observed in some of the axons. In the pterygotes there are clear signs of glandular degeneration within a few days after adult ecdysis, which is not seen in apterygotes (Watson, 1963, 1964).

12.2.5.4. Cellular Secretions

The juvenile hormone is believed to be the single product of the CA in all insects during all developmental stages. It has been called the inhibitory hormone because it suppresses metamorphosis in the immature, and the gonadotropic hormone because it influences oocyte development. The structure of several juvenile hormones has been determined as terpenoid compounds.

The ecdysones are associated with the molting process and are synthesized within the molting glands under the control of the activation neurohormone. Two molting hormones can be extracted from most insects: α- and β-ecdysones; the second is also referred to as 20-hydroxyecdysone or ecdysterone. The most recent information supports the hypothesis that α-ecdysone is secreted by the molting gland and is converted to β-ecdysone by the target organs (King et al., 1974; Bollenbacher et al., 1976).

12.3. POSSIBLE PHYLOGENETIC SIGNIFICANCE

With the examination of arthropod neuroendocrine structures completed (see Table 12.5 for summary), we now turn our attention to the question of whether these observations support the polyphyletic theory of Manton (1964) or the monophyletic theory of Sharov (1966). The available data are insufficient to answer the question and no convincing arguments are presented for either theory. Nevertheless, some support for the monophyletic theory could be derived from

Table 12.5. Summary of Neuroendocrine Structures in the Five Arthropod Groups.

	Neurosecreting Cells	Neurohemal Areas or Organs	Neurohemal Organs with Intrinsic Neurosecretory Cells from Transformed Nerve Ganglia	Epithelial Endocrine Glands and Secretions
Onychophora	Cerebral ganglia and ventral nerve cord	Not known	Not known	Not known
Chelicerata	Cerebral ganglia and ventral nerve cord	Diffused area at base of brain Paraganglionic plates Organ of Schneider Stomatogastric ganglion	Organ of Schneider Stomatogastric ganglion	Gland not known but ecdysone is present
Crustacea	Cerebral ganglia and ventral nerve cord	Sinus gland Postcommissure organs Pericardial organs		Androgenic gland-androgenic hormone Y-organs—ecdysone
Myriapoda	Cerebral ganglia and ventral nerve cord	Cerebral neurohemal organ Connective body Cephalic gland Gabe organ Paraesophageal body Cerebral gland	Gabe organ Paraesophageal body Cerebral gland	Ecdysal gland—ecdysone
Insecta	Cerebral ganglia and ventral nerve cord	Perisympathetic organs Corpora cardiaca	Corpora cardiaca	Corpora allata—juvenile hormone Ecdysal gland-ecdysone

the lack of dissimilarity in the data concerning the neurosecretory cells, the ecdysial glands, and products of each. On the other hand, support for the polyphyletic theory could be derived from the arrangement of the neurohemal organs which show substantial variation between the two major groupings. There are groups of arthropods on which no observations have been made and groups in which more observations on additional species are needed in order to increase our confidence in the generalized picture.

It is apparent that the fundamental endocrine unit—the neurosecretory cell, its axon, and its ending—is similar among the arthropods and follows closely the pattern found throughout the invertebrates (Gabe, 1966; Bern and Hagadorn, 1965). An important integrative function must have been provided by these cells in the primitive multicellular animal, for they have been retained from the Cnidaria, through all the invertebrates, the vertebrates, and are, in a general sense, similar in their morphology, histology, and ultrastructure. Further, since they were almost certainly present in animals in the pre-Cambrian, these cells are of little value in providing support for theories concerning arthropod evolution since the Cambrian.

The presence of ecdysial glands and their product, ecdysone has been of considerable interest to investigators by providing observations of phylogenetic application. This chemical has now been found in three of the four major arthropod groups, is assumed to be in Chelicerata, and is probably also present in Onychophora. Moreover, not only has ecdysone been detected in nematodes (Horn et al., 1974), but phytoecdysones, chemicals with ecdysone activity, are also present in many vascular plants. Therefore, the strength of what was once thought to be a good indicator of the strategies of arthropod evolution (Krishnakumaran and Schneiderman, 1970; Schneiderman, 1972) has now been weakened by the widespread occurrence of the ecdysone molecule.

The area that offers some support for the polyphyletic theory concerns the arrangement of the neurohemal organs. The Myriapoda and Insecta groups show similar structures, which contain compact axon endings with or without intrinsic neurosecretory cells. The identity of the neurohemal structure in Onychophora will be interesting whether it proves to be the infracerebral organ or one yet to be discovered.

The Crustacea, representing one of the evolutionary arms in the polyphyletic scheme, are without intrinsic neurosecretory cells in any neurohemal organ. In addition, the sinus gland is not located posterior to the brain along the aorta as is the case in the Myriapoda-Insecta group, but rather is structured as a disc or plate on the surface of the eyestalk, showing some similarities to the paraganglionic plates in Chelicerata. Dahl (1965) used these differences in neurohemal organ locations in a discussion concerning the homologies of insect and crustacean neurosecretory organs and concluded that the two were not homologous.

In contrast to these observations in the Crustacea and in the Myriapoda–

Insecta group, the Chelicerata show, in the Merostomata, a very primitive neuro-endocrine system containing random axon endings on the base of the brain with-out any neurohemal organ, and, in some Arachnida, a neurohemal system con-taining intrinsic neurosecretory cells. It might then be argued that these observa-tions support a triphyletic, rather than a diphyletic, arrangement, with the Chelicerata serving as the least advanced taxon with the widest variety of types.

Providing additional support for the separate crustacean category, and thus the polyphyletic theory, is the inhibitory nature of the neurohormone, for the hormone that serves the same function in the Myriapoda–Insecta group is stimulating to the ecdysial glands.

In conclusion, it is judged that the neuroendocrine data available on the arthropods are not representative enough to offer the comparative data neces-sary to support one theory to the exclusion of the other. The lack of stronger evidence for the polyphyletic theory could be interpreted by some as support by default for the monophyletic theory. I do not share that view. Assistance in solving the phylogenetic problem can be obtained, but appropriate questions

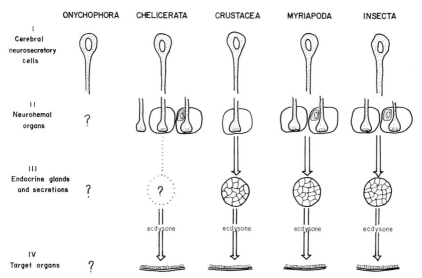

Fig. 12.1. Comparative diagrammatic schemes of neuroendocrine systems in arthropods. I. Neurosecretory cells in cerebral ganglia common in all taxa. II. No neurohemal structures identified in Onychophora; diffused axon endings and compact neurohemal areas in Chelic-erata; neurosecretory endings in distinct organs or structures in Crustacea, Myriapoda, and Insecta; no intrinsic neurosecretory cells in Crustacea. III. No nonneural, epithelial, endocrine organs identified in Onychophora or Chelicerata; probable structure in Myria-poda; structures identified in Crustacea and Insecta. IV. Ecdysone, the molting hormone, has been identified in whole-animal extracts, in hemolymph, or from *in vitro* culture of ec-dysal glands in all groups except in Onychophora, which has not been examined.

still need to be formulated. It is hoped that this review will direct interested investigators toward this area of study.

12.4. SUMMARY

Arthropod neuroendocrine structures (Fig. 12.1) consist of neurosecretory cells, which are within ganglia of the central nervous system; axons from cells within the cerebral ganglia, which extend to diffused or concentrated neurohemal areas; neurohemal structures, which may or may not possess intrinsic neurosecretory cells; and true endocrine glands, which are responsive to neurohormones. There appears to be considerable morphological similarity among the neurosecretory cells and the ecdysial glands, and molecular similarity among the molting hormones. These observations provide little opportunity for charting phylogenetic relationships among the arthropod classes. However, substantial variations exist in the morphology of the neurohemal organs. Three morphological types, based on the arrangements of the axonal endings, have been found in the chelicerates, two in the insects and myriapods, and only one in the crustaceans. These observations are interpreted as support for either the monophyletic or the polyphyletic theory of arthropod evolution with no clear distinction that one theory is favored over the other.

REFERENCES*

Aiken, D. E. 1969. Photoperiod, endocrinology and the crustacean molt cycle. *Science (Wash., D.C.)* **164**: 149–55. (1)

Bebbington, P. M. and E. D. Morgan. 1977. Detection and identification of moulting hormone (Ecdysones) in the barnacle *Balanus balanoides. Comp. Biochem. Physiol.* **56B**: 77–79. (2)

Bern, H. A. and I. R. Hagadorn. 1965. Neurosecretion, pp. 353–429. *In* T. H. Bullock and G. A. Horridge (eds.), *Structure and Function in the Nervous System of Invertebrates.* Vol. I. Freeman, San Francisco. (3)

Blanchet, M. F., R. Ozon and J. J. Meusy. 1972. Metabolism of steroids, *in vitro*, in the male crab *Carcinus maenas* Linné. *Comp. Biochem. Physiol.* **41B**: 251–61. (4)

Bollenbacher, W. E., W. Goodman, W. V. Vedeckis and L. I. Gilbert. 1976. The *in vitro* synthesis and secretion of α-ecdysone by the ring gland of the fly, *Sarcophaga bullata. Steroids* **27**: 309–24. (5)

Bonaric, J. C. 1976. Effects of ecdysterone on the moulting mechanisms and duration of the intermoult period in *Pisaura mirabilis* Cl. *Gen. Comp. Endocrinol.* **30**: 267–72. (6)

Cassagnau, P. and C. Juberthie. 1967. Structures nerveuses, neurosécrétion et organes endocrines chez les Collemboles. II. Le complex cérébral des Entomobryomorphes. *Gen. Comp. Endocrinol.* **8**: 489–502. (7)

Cassagnau, P. and C. Juberthie. 1970. Structures nerveuses, neurosécrétion et organes endocrines chez les Collemboles. Neurosécrétion dans la chaîne nerveuse d'un Entomobryomorphe, *Orchesella kervillei* Denis. *C. R. Acad. Sci. Paris* **270**: 3268–71. (8)

*Numbers in parentheses at end of each reference refer to numbers in parentheses in Tables 12.1–12.4.

Cassagnau, P., C. Juberthie and G. Raynal. 1968. Structures nerveuses, neurosécrétion et organes endocrines chez les Collemboles. III. Le complexe cérébral des Symphypléones. *Gen. Comp. Endocrinol.* **10**: 61–69. (9)

Chang, E. S. and John D. O'Connor. 1977. Secretion of α-ecdysone by crab Y-organs *in vitro. Proc. Natl. Acad. Sci. USA* **74**: 615–18. (10)

Charniaux-Cotton, H. 1975a. Contrôle hormonal de la différenciation sexuelle et de l'activité génitale chez les Crustacés malacostracés. *Publ. Staz. Zool. Napoli* **39** *(Suppl.)*: 480–509. (11)

Charniaux-Cotton, H. 1975b. Hermaphroditism and gynandromorphism in malacostracan Crustacea, pp. 91–105. *In* R. Reinboth (ed.) *Intersexuality in the Animal Kingdom.* Springer-Verlag, Berlin, New York. (12)

Cuénot, L. 1949. Onychophores, pp. 1–37. *In* P. P. Grassé (ed.) *Traité de Zoologie.* Masson, et Cie, Paris. (13)

Dahl, E. 1965. Frontal organs and protocerebral neurosecretory systems in Crustacea and Insecta. *Gen. Comp. Endocrinol.* **5**: 514–17. (14)

Davis, C. W. and J. D. Costlow. 1974. Evidence for a molt inhibiting hormone in the barnacle, *Balanus improvisus* (Crustacea, Cirripedia). *J. Comp. Physiol.* **93**: 85–91. (15)

Doane, W. W. 1973. Role of hormones in insect development, pp. 291–497. *In* S. J. Counce and C. H. Waddington (eds.) *Developmental Systems–Insects.* Vol.II, Academic Press, New York, London. (16)

Eisen, Y., M. R. Warburg and R. Galun. 1973. Neurosecretory activity as related to feeding and oogenesis in the fowl-tick, *Argus persicus* (Oken). *Gen. Comp. Endocrinol.* **21**: 331–40. (17)

Fahrenbach, W. H. 1973. The morphology of the *Limulus* visual system V. Protocerebral neurosecretion and ocular innervation. *Z. Zellforsch.* **144**: 153–66. (18)

Gabe, M. 1954. La neurosécrétion chez les Invertébres. *Ann. Biol.* **30**: 5–62. (19)

Gabe, M. 1955. Données histologiques sur la neurosécrétion chez les Arachnides. *Arch. Anat. Microsc. Morphol. Exp.* **44**: 531–83. (20)

Gabe, M. 1966. *Neurosecretion.* Pergamon Press, Oxford. (21)

Habibulla, M. 1970. Neurosecretion in the scorpion *Heterometrus swammerdami.* *J. Morphol.* **131**: 1–16. (22)

Habibulla, M. 1975. Neurosecretion in the scorpion *Heterometrus swammerdami.* *J. Morphol.* **140**: 53–62. (23)

Herman, W. S. 1967. The ecdysial glands of arthropods. *Int. Rev. Cytol.* **22**: 269–347. (24)

Herman, W. S. 1972. Molt initiation in response to phytoecdysones and low doses of animal ecdysones in the chelicerate arthropod, *Limulus polyphemus. Gen. Comp. Endocrinol.* **18**: 301–5. (25)

Herman, W. S. and D. M. Preus. 1973. Ultrastructural evidence for the existence of two types of neurosecretory cells in the abdominal ganglia of the chelicerate arthropod, *Limulus polyphemus. J. Morphol.* **140**: 53–62. (26)

Highnam, K. C. and L. Hill. 1977. *The Comparative Endocrinology of Invertebrates.* Edward Arnold, London. (27)

Horn, D. H. S., J. S. Wilkie and J. A. Thomson. 1974. Isolation of β-ecdysone (20-hydroxy ecdysone) from the parasitic nematode *Ascaris lumbricoides. Experientia* **15**: 1109–10. (28)

Jamault-Navarro, C. and R. Joly. 1977. Localisation et cytologie des cellules neuro-sécrétrices protocérébrales chez *Lithobius forficatus.* L. (Myriapode, Chilopode). *Gen. Comp. Endocrinol.* **31**: 106–20. (29)

Jegla, T. C. 1974. Ecdysone activity in *Limulus polyphemus. Amer. Zool.* **14**: 1288. (30)

Jegla, T. C. and J. D. Costlow, Jr. 1970. Induction of molting in horseshoe crab larvae by polyhydroxy steroids. *Gen. Comp. Endocrinol.* **14**: 295–302. (31)

Jegla, T. C., J. D. Costlow and J. Alspaugh. 1972. Effects of ecdysones and some synthetic analogs on horseshoe crab larvae. *Gen. Comp. Endocrinol.* **19**: 159–66. (32)

Joly, R. 1966. Sur l'ultrastructure de la glande cérébrale de *Lithobius forficatus* L. (Myriapode, Chilopode). *C.R. Acad. Sci. Paris* **263**: 374–77. (33)

Joly, R. 1970. Evolution cyclique des glandes cérébrales au cours de l'intermue chez *Lithobius forficatus* L. (Myriapode, Chilopode). *Z. Zellforsch.* **110**: 85–96. (34)

Joly, R. and G. Devauchelle. 1970. Étude cytochimique de la glande cérébrale de *Lithobius forficatus* L. (Myriapode, Chilopode); nature des sécrétions. *J. Microsc. (Paris)* **9**: 631–42. (35)

Juberthie, C. and P. Cassagnau. 1971. L'évolution du systéme neurosécréteur chez les insectes; l'importance des Collemboles et des autres Aptérygotes. *Rev. Ecol. Biol. Sol.* **8**: 59–80. (36)

Juberthie, C. and J. Heurtault. 1975. Ultrastructure des plaques paraganglionnaires d'un pseudoscorpion souterrain, *Neobisium cavernarum.* *Ann. Spéléol.* **30**: 433–39. (37)

Juberthie, C. and L. Juberthie-Jupeau. 1974a. Étude ultrastructurale de l'organe neurohémal cérébral de *Spelaeoglomeria doderoi* Silvestri, Myriapods Diplopode Cavernicole. *Symp. Zool. Soc. Lond.* **32**: 199–210. (38)

Juberthie, C. and L. Juberthie-Jupeau. 1974b. Ultrastructure of neurohemal organs (paraganglionic plates) of *Trogulous nepaeformis* (Scopoli) (Opiliones, Troguliadae) and release of neurosecretory material. *Cell Tissue Res.* **150**: 67–78. (39)

Juberthie-Jupeau, L. 1961. Données sur la neurosécrétion protocérébrale et mise en évidence de glandes céphaliques chez *Scutigerella pagesi* Jupeau (Myriapode, Symphyle). *C.R. Acad. Sci. Paris* **253**: 3081–83. (40)

Juberthie-Jupeau, L. 1973. Étude ultrastructurale des corps paraoesophagiens chez un Diplopode Oniscomorphe *Loboglomeris pyrenaica* Latzel. *C.R. Acad, Sci. Paris* **276**: 169–72. (41)

Juberthie-Jupeau, L. and C. Juberthie. 1973a. Étude ultrastructurale de l'organe neurohémal cephalique chez un Symphyle *Scutigerella silvatica* (Myriapode). *C.R. Acad. Sci. Paris* **276**: 1577–80. (42)

Juberthie-Jupeau, L. and C. Juberthie. 1973b. Decharge par exocytose d'une categorie de granules de neurosécrétion dans l'organe neurohemal d'un Symphyle. *C.R. Acad. Sci. Paris* **277**: 1357–60. (43)

King, D. S. and J. B. Siddall. 1974. Biosynthesis and inactivation of ecdysone, pp. 147–152. *In* W. J. Burdette (ed) *Invertebrate Endocrinology and Hormonal Heterophylly.* Springer-Verlag, Berlin, New York. (44)

King, D. S., W. E. Bollenbacher, D. W. Borst, W. V. Vedeckis, J. D. O'Connor, P. I. Ittycheriah and L. I. Gilbert. 1974. The secretion of α-ecdysonie by the prothoracic glands of *Manduca sexta in vitro. Proc. Natl. Acad. Sci. USA* **71**: 793–96. (45)

Krishnakumaran, A. and H. A. Schneiderman. 1970. Control of molting in mandibulate and chelicerate arthropods by ecdysones. *Biol. Bull. (Woods Hole)* **139**: 520–38. (46)

Lake, P. S. 1969. Neurosecretion in *Chirocephalus diaphanus* Prévost (Anostraca) I. Anatomy and cytology of the neurosecretory systems. *Crustaceana* **16**: 273–87. (47)

Lake, P. S. 1970. Histochemical studies of the neurosecretory system of *Chirocephalus diaphanus* Prévost (Crustacea: Anostraca). *Gen. Comp. Endocrinol.* **14**: 1–14. (48)

Manton, S. M. 1964. Mandibular mechanisms and the evolution of arthropods. *Phil. Trans. R. Soc. Lond. (Ser. B)* **247**: 1–183. (49)

McGregor, D. B. 1967. The neurosecretory cells of barnacles. *J. Exp. Mar. Biol. Ecol.* **1**: 154–67. (50)

Nair, V. S. K. 1974. Studies of the probable neurosecretory control of moulting and vitellogenesis in the millipede *Jonespeltis splendidus* Verhoeff. Ph.D. thesis, University Kerala, Trivandrun, India. (51)

Novak, V. J. A. 1975. *Insect Hormones.* Chapman and Hall, London. (52)

Palévody, C. and A. Grimal. 1976. Variations cytologiques des corps allates au cours du cycle reproducteur du collembole, *Folsomia candida. J. Insect Physiol.* **22:** 63–72. (53)

Payen, G. 1972. Étude ultrastructurale de la degenerescence cellulaire dans la glande androgene du crabe *Ocypode quadrata* (Fabricius). *Z. Zellforsch.* **129:** 370–85. (54)

Petit, J. and F. Shali. 1975. Cytochemical and electronmicroscopic study of the para-oesophageal bodies and related nerves in *Schizophyllum sabulosum* (L.) Diplopoda, Judidae. *Cell Tissue Res.* **162:** 367–75. (55)

Prabhu, V. K. K. 1962. Neurosecretory system of *Jonespeltis splendidus* Verhoeff (Myriapoda: Diplopoda), pp. 417–20. *In* H. Heller and R. B. Clark (eds.) *Neurosecretion.* Academic Press, New York, London. (56)

Rader, V. G. and C. Cracium. 1976. The ultrastructure of the androgenic gland in *Porcellio scaber* Latr. (Terrestrial Isopods). *Cell. Tissue Res.* **175:** 245–63. (57)

Rosenberg, J. and G. Seifert. 1975. Ist allein die glandula ecdysalis die häutungsdrüse von *Lithobius. Experientia* **31:** 1100. (58)

Sanchez, S. 1958. Cellules neurosécrétrices et organs infracérébraux de *Peripatopsis moseleyi* Wood-Mason (Onychophore) et neurosécrétion chez *Nymphon gracile* Leach (Pycnogonide). *Arch. Zool. Exp. Gén.* **96:** 57–62. (59)

Schneiderman, H. A. 1972. Insect hormones and insect control, pp. 3–27. *In* J. J. Menn and M. Beroza (eds.) *Insect Juvenile Hormones.* Academic Press, New York, London. (60)

Seifert G. and E. El-Hifnauri. 1972. Die ultrastruktur des neurohamalorgans am nervus protocerebralis von *Polyxenus lagurus* (L.) (Diplopoda, Penticillata). *Z. Morphol. Tiere* **71:** 116–27. (61)

Seifert, G. and J. Rosenberg. 1974. Elektronenmikroskopische untersuchungen der häutungsdrüsen ("Lymphstrange") von *Lithobius forficatus.* L. *Z. Morphol. Tiere* **78:** 263–79. (62)

Sharov, A. G. 1966. *Basic Arthropodan Stock.* Pergamon Press, Oxford. (63)

Sochasky, J. B. and D. E. Aiken. 1974. The Y organ–molting gland problem in decapod Crustacea: The "organ by Madhyastha" and "organ of Carlisle," two putative Y organs. *Can. J. Zool.* **52:** 1251–57. (64)

Sochasky, J. B., D. E. Aiken and N. H. F. Watson. 1972. Y organ molting gland and mandibular organ: A problem in decapod crustacea. *Can. J. Zool.* **50:** 993–97. (65)

Tombes, A. S. 1970. *An Introduction to Invertebrate Endocrinology.* Academic Press, New York, London. (66)

van den Bosch de Aguilar, P. 1972. Les caractéristiques tinctoriales des cellular neuro-sécrétrices chez *Daphnia pulex* (*Crustacea: Chadocera*). *Gen. Comp. Endocrinol.* **18:** 140–45. (67)

van den Bosch de Aguilar, P. 1976. Étude histochimique du systeme neurosécréteur de *Balanus perforatus* et *B. balanoides* (Crustacea: Cirripedia). *Gen Comp. Endrocrinol.* **30:** 228–30. (68)

Watson, J. A. L. 1963. The cephalic endocrine system in the Thysanura. *J. Morphol.* **113:** 359–73. (69)

Watson, J. A. L. 1964. Moulting and reproduction in the adult firebrat, *Thermobia domestica* (Packard) (Thysanura, Lepismatidae). I. The moulting cycle and its control. *J. Insect Physiol.* **10:** 305–17. (70)

Wigglesworth, V. B. 1970. *Insect Hormones.* Freeman, San Francisco. (71)

Willig, A. and R. Keller. 1976. Biosynthesis of α- and β-ecdysone by the crayfish *Orconectes limosus in vivo* and by its Y-organs *in vitro. Experientia* **32:** 936–37. (72)

Winget, R. R. and W. S. Herman. 1976. Occurrence of ecdysone in the blood of the chelicerate arthropod, *Limulus polyphemus. Experientia* **32:** 1345. (73)

13

Arthropod Hemocytes and Phylogeny

A. P. GUPTA

13.1 HEMOCYTE TERMINOLOGY

Hemocytes of arthropods and several other invertebrate groups have been studied. Among arthropods, they have been most extensively studied in insects (Gupta, 1978), followed by crustaceans, arachnids, and myriapods. Hemocytes of a few onychophorans also have been described. It is not surprising, therefore, that the need for a reliable, uniform classification of various hemocyte types has been felt more keenly by insect hematologists than those of other arthropod groups. Fortunately, a generally acceptable hemocyte classification in insects, based largely on morphological characteristics, now exists. Several workers (Kanungo and Naegele, 1964; Gupta, 1968; Vostal and Pirčová, 1968; Dolp, 1970; Sundara Rajulu, 1970, 1971a,b; Vostal, 1970; Brinton and Burgdorfer, 1971; Ravindranath, 1973, 1974a, b, 1975; Sherman, 1973; Fujisaki et al., 1975; and Lavallard and Campiglia, 1975) have recently adopted the insect hemocyte classification, based more or less on Jones's (1962) classification, for various arthropod and onychophoran groups.

Hemocyte classifications both in insects and arthropods have been variously based on morphology, functions, and staining or histochemical reactions of hemocytes. Thus, it is not unusual to find the same hemocyte type or its various forms being referred to by different names in various arthropods, by different authors—a situation that has inevitably resulted in a confusing mass of terminology. Consequently, it becomes very difficult to compare hemocytes of one species to those in others. This has particularly hindered any phylogenetic consideration of the evolution of hemocyte types in various arthropod groups and Onychophora. Clearly, there is a need for a uniform classification for all arthropod groups; and since the insect hemocyte classification has been successfully adopted by several authors for several arthropod groups, it is reasonable to suggest that, with some minor modifications, Jones's (1962) classification be used for all arthropod groups. This classification has evolved over a period of more than half a century. According to Millara (1947), Cuénot (1896) was the first to classify insect hemocytes into four categories, and was later followed in this attempt by Hollande (1909, 1911) and others. Wigglesworth (1939) summarized most of the earlier classifications, and presented a classification that was widely accepted.

In order to adopt a uniform classification for discussing hemocytes and their phylogenetic significance in various arthropod groups, it would be necessary to homologize terminologies used by different authors on the bases of descriptions, observed functions, line drawings, and micrographs of hemocytes studied by those authors. A summary of the main hemocyte types in various arthropod groups and Onychophora is presented in Table 13.1. Cells that could not be homologized have been included in the category of "others." In some cases cells included in this category may not necessarily be hemocytes. Terms in parentheses were not used by the original authors, but have been adopted by me after scrutinizing original descriptions and figures, and in many instances represent only approximations to the category they have been assigned to in the table. Hemocytes categorized as amoebocytes and/or phagocytes by the original authors have been assigned mostly to the category of plasmatocyte (PL), although they could be included in granuloctye (GR) or spherulocyte (SP), and/or adipohemocyte (AD), inasmuch as these latter three forms also are supposed to be phagocytic. Since all the hemocyte types have been described in insects, a general description of various types, based on both light and electron microscopic studies, is presented below, along with their synonymies and interrelationships.

13.2. Types of Hemocytes in Various Arthropod Groups

13.2.1. Insecta

As presented below, many hemocyte types have been described in insects. Ultrastructurally however, only seven types have so far been identified in various insects: prohemocyte (PR), plasmatocyte (PL), granulocyte (GR), spherulocyte (SP), adipohemocyte (AD), oenocytoid (OE), and coagulocyte (CO). Of these seven, AD has been reported only be Devauchelle (1971) and CO by Goffinet and Grégoire (1975) and Ratcliffe and Price (1974). Podocyte (PO) and vermicyte (VE) have not been recognized as distinct types in electron microscope studies so far, primarily because ultrastructurally they appear similar to PLs (Devauchelle, 1971).

Table 13.1. Summary of Hemocyte Types in Various Groups of Arthropods and Onychophora.

Groups	Prohemocyte PR	Plasmatocyte PL	Granulocyte GR	Spherulocyte SP	Adipohemocyte AD	Coagulocyte CO	Oenocytoid OE	Others
Aquatic Chelicerata (Xiphosura)	—	—	GR	—	—	—	—	Cyanoblast Cyanocyte
Crustacea	PR	PL	GR	SP	AD	CO	—	Cyanocyte
Terrestrial Chelicerata	PR	PL	GR	SP	AD	CO	OE	(Cyanocyte)
Myriapoda	PR	PL	GR	SP	AD	CO	OE	Crystal cell
Insecta	PR	PL	GR	SP	AD	CO	OE	For numerous other terms see section 13.2.1.
Onychophora	PR	PL	GR	SP	—	—	OE	—

Cells that could not be homologized have been included in the category of "others."

Prohemocyte (PR)

Structure. These are small, round, oval or elliptical cells with variable sizes (6–10 μm wide and 6–14 μm long). The plasma membrane is generally smooth (Fig. 13.1A) but may show vesiculation (Fig. 13.3A). The nucleus is large and centrally located, almost filling the cell; nuclear size is variable (3.6–12 μm) in various insects; more than one nucleus and nucleoli may be present. A thin or dense, homogeneous, and intensely basophilic layer of cytoplasm surrounds the nucleus, the nucleocytoplasmic ratio being 0.5–1.9 or more. The cytoplasm may contain granules, droplets, or vacuoles (Fig. 13.3A).

The laminar nature of the plasma and nuclear membranes may not be evident. The cytoplasm generally contains a low concentration of endoplasmic reticulum (ER), mitochondria, and Golgi bodies. However, free ribosomes, rough endoplasmic reticulum (RER), and even mitochondria may be numerous. Centrioles—indicating the mitotic nature of PRs—and microtubules have been observed in the cytoplasm.

PRs are generally found in groups, and appear indistinguishable from young or small PLs. They may be numerous, rare or absent, depending on the developmental and physiological state of the insect at the time of observation. PRs are seldom seen *in vivo*.

Synonymy. The term that has survived to date with little or no change since its adoption by Hollande (1911) is *proleucocyte.* Yeager (1945) used the term *proleucocytoid,* and Jones and Tauber (1954) *prohemocytoid.* I believe Arnold (1952) was the first to use the term *prohaemocyte.* Other synonyms for PRs are *"macronucleocyte"* (Paillot, 1919); *formative cells* (Müller, 1925), "jeune globule" (Bruntz, 1908); *smooth-contour chromophilic cells* (Yeager, 1945); *"jeune leucocyte"* (Millara, 1947); *plasmatocytelike* cells (Jones, 1959); *young plasmatocyte* (Gupta and Sutherland, 1966; Gupta, 1969); *young granulocyte* (François, 1974); and *proleucocyte,* used by many authors.

Interrelationship with Other Hemocyte Types. The controversial questions that are often raised regarding PRs are: 1) whether they are the stem cells that transform into other hemocyte types; 2) and if they are, are they the main postembryomic source of hemoctyes. Although there are substantiating reports that PRs do transform at least into a few other hemoctye types, evidence of their being the main postembryonic source is inconclusive. The term *prohemocyte* suggests that these cells give rise to other types, but it has not yet been demonstrated conclusively that all hemocyte types are derived from PRs. The most generally accepted view is that PRs transform into PLs (Yeager, 1945; Arnold, 1952, 1970, 1974; Jones, 1954, 1956, 1959; Shrivastava and Richards, 1965; Mitsuhashi, 1966; Wille and Vecchi, 1966; Beaulaton, 1968; Devauchelle, 1971; Lai-Fook, 1973; Beaulaton and Monpeyssin, 1976). Several authors have suggested that PRs transform into other types as well (Muttkowski,

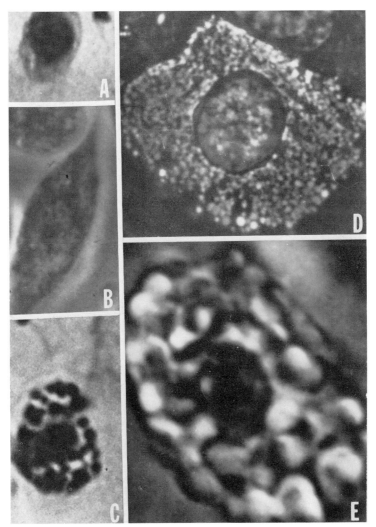

Fig. 13.1. *A.* Prohemocyte of *Periplaneta americana.* ca ×8,000. *B.* Plasmatocyte of *P. americana.* ca ×10,000. *C.* Spherulocyte of *Nauphoeta cinerea.* ca ×8,500. From Gupta and Sutherland (1967). *D.* Granulocyte of *Locusta migratoria.* From Costin (1975). *E.* Spherulocyte of *N. cinerea.* ca ×25,000. From Gupta and Sutherland (1967).

1924; Bogojavlensky, 1932; Yeager, 1945; Arvy and Lhoste, 1946; Ashhurst and Richards, 1964). Arnold (1952) stated that "haemocytes, with the possible exception of the Oenocytoids, apparently develop originally from a common source, the prohaemocytes." Wille and Vecchi (1966), however, suggested that PR can give rise to OE. Arnold (1970) stated that PRs are stem cells for PLs, GRs

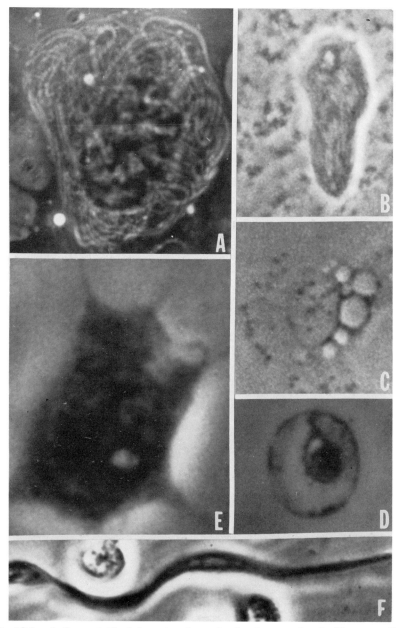

Fig. 13.2.   *A.* Oenocytoid of *Locusta migratoria*, showing cytoplasmic filaments.   From Costin (1975).   *B.* Oenocytoid of *P. americana.*   ca ×8,000.   *C.* Adipohemocyte of *P. americana* nymph.   ca ×8,500.   From Gupta and Sutherland (1966).   *D.* Coagulocyte of *Epicauta cinerea.*   ca. ×8,000.   *E.* Podocyte of *P. americana* nymph.   ca ×8,500.   *F.* Vermicyte of *P. americana.*   ca ×1,400.   D, E, F, from Gupta (1969).

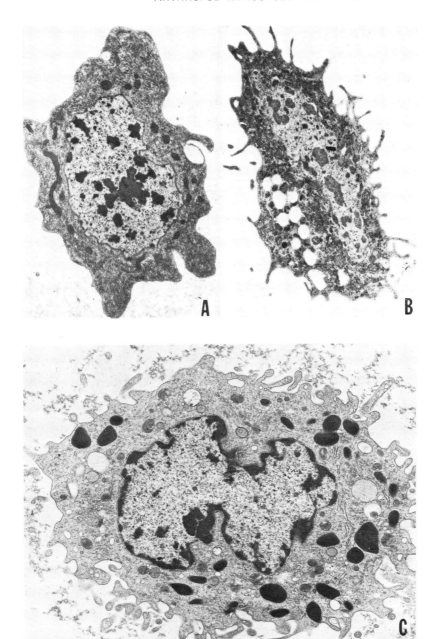

Fig. 13.3. *A.* Prohemocyte of *Pectinophora gossypiella*. ca X 12,000. From Raina (1976).
B. Plasmatocyte of *P. gossypiella*, showing micropapillae. ca X 15,000. From Raina (1976).
C. Plasmatocyte of *Carausius morosus*, showing micropapillae, and lobate nucleus. X 9,000.
From Goffient and Grégoire (1975).

and SPs. Yeager (1945) and Jones (1959) reported that PRs can give rise to SPs and ADs. Devauchelle (1971) reported that PRs, PLs, GRs, and ADs are derived from each other. François, (1974) found that PRs transform into GRs. Recently, Sohi (1971) indicated by subculturing that PRs are the germinal cells from which other categories develop.

As I stated earlier, the evidence on whether PRs constitute the main postembroynic source of hemocytes is inconclusive. There is growing evidence that PRs reside in the hemopoietic organs (Hoffmann et al., 1968; Arnold, 1970; Akai and Sato, 1973; Zachary and Hoffmann, 1973; François, 1974), and differentiate into other hemocyte types. Hoffmann (1967), Arnold (1974), and Beaulaton and Monpeyssin (1976) stated that PRs are germinal cells and often seen in mitotic division. Earlier (1970), Arnold stated that PRs appear in the hemolymph only intermittently, and often in groups, suggestive of their release from hemopoietic tissue. Wille and Vecchi (1966) reported that PRs are abundant in newly emerged bees, but rare in old ones. Gupta and Sutherland (1968) reported an increase in PLs, GRs, SPs, and COs (= CYs) in *Periplaneta americana* following treatment with sublethal doses of chlordane.

Plasmatocyte (PL)

Structure. These are small-to-large polymorphic cells with variable sizes (3.3-50 μm wide and 3.3-40 μm long). The plasma membrane may have micropapillae, filopodia, or other irregular processes, as well as pinocytotic or vesicular invaginations (Figs. 13.1B, 13.3B,C). The nucleus may be round or elongate, and is generally centrally located. It may be lobate (Fig. 13.3C), vary in size (3-9 μm wide and 4-10 μm long) in various insects and appear punctate. Scattered chromatin masses may be present along with the nucleolus (Fig. 13.3C). Occasionally, binucleate PL may be found.

The laminar nature of the plasma and nuclear membranes may not be visible. The cytoplasm is generally abundant, may be granular or agranular, and is basophilic and rich in organelles. Generally, there is a well-developed and extensive RER (Fig. 13.4B), which may form greatly distended cisternae or a vacuolar system. Golgi bodies (= dictyosomes = golgiosomes or internal reticular apparatus) (Fig. 13.4A) and lysosomes (membrane-bounded, electron-dense bodies, 0.1-1.30 μm in size) may be numerous; lysosomes can be identified by the presence in them of the reaction products of the hydrolytic marker enzymes, acid phosphatase and thiamine pyrophosphatase (Scharrer, 1972) and are often associated with the RER or the vacuolar system. The Golgi bodies produce the electron-dense granules (generally 0.5 μm in diameter) that one observes in the PLs. Microsomes and cisternae of ER (= "ergastoplasme" of French authors) may be present. Free ribosomes (polysomes or polyribosomes) or those at-

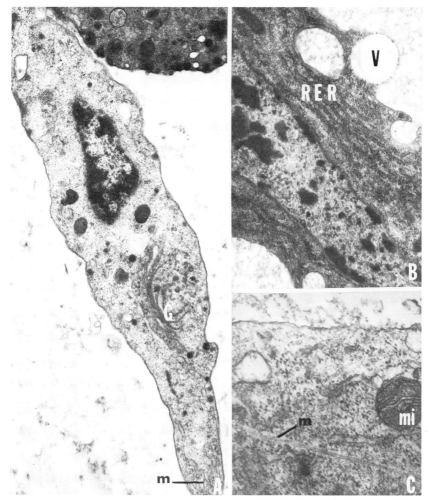

Fig. 13.4. *A.* Plasmatocyte of *Melolontha melolontha*, showing Golgi (G) and intracytoplasmic microtubules (m). ×18,000. Reinterpreted from Devauchelle (1971). *B.* Portion
of plasmatocyte of *P. gossypiella*, showing rough endoplasmic reticulum (RER) and vacuoles
(V). ca ×30,500. From Raina (1976). *C.* Portion of plasmatocyte of *Ephestia kühniella,*
showing mitochondrion (mi) and intracytoplasmic microtubule (m). ca ×63,000. Reinterpreted from Grimstone et al. (1967).

tached to microsomes or RER may be present; intracytoplasmic microtubules
are present, sometimes arranged in bundles (Fig. 13.4A,C).

PLs are generally abundant, and may be indistinguishable from PRs and GRs.
Several types (most often the transitional forms) of PLs have been described on
the bases of their sizes and shapes.

Synonymy.   Yeager and Munson (1941) first introduced the term *plasmatocyte.*   Some of the commonly used synonyms of PL are: *"leucocyte"* (Kollman, 1908; Metalnikov, 1908), *"micronucleocyte"* (Paillot, 1919), *phagocytes, amoebocytes,* and *lymphocytes* (of many authors),*podocyte* (Devauchelle, 1971), and *vermiform cells* (Lea and Gilbert, 1966; Devauchelle, 1971). PLs also include the *lamellocytes* of some authors, and the *nematocyte* of Rizki (1953).

Interrelationship with Other Hemocyte Types: The first real problem which one encounters with PLs is that of distinguishing them, particularly the so-called young PLs, from the PRs. This situation is further complicated by the presence of many transitional forms between these two types. The distinction between the PRs and PLs is generally based on the relative cell and nuclear sizes, intensity of cytoplasmic basophilia, and the extent and development of the intracellular organelles.

The question that is often raised regarding PLs is whether they are the primary type of cells that give rise to other forms by secondary transformation. Taylor (1935) claimed that amoebocytes (= mostly PLs), and not the chromophils (= PRs), are the basic types.   Gupta and Sutherland (1966) and Gupta (1969) supported Taylor's claim, and considered PRs as young PLs.   Direct transformation of PLs into GRs (Yeager, 1945; Jones, 1956; Gupta and Sutherland, 1966; Hoffmann, 1967; Devauchelle, 1971), SPs (Devauchelle, 1971), ADs (Yeager, 1945; Shirvastava and Richards, 1965; Gupta and Sutherland, 1966; Raina, 1976), COs (Gupta and Sutherland, 1966; Devauchelle, 1971; Breugnon and Le Berre, 1976), OE (Gupta and Sutherland, 1966; Hoffmann, 1967), VE (Tuzet and Manier, 1959; Gupta and Sutherland, 1966; Lea and Gilbert, 1966; Devauchelle, 1971; François, 1974, 1975), and PO (Gupta and Sutherland, 1966; Nappi, 1970; Devauchelle, 1971; François, 1974, 1975) has been reported.   Devauchelle (1971) considered the VEs and POs ultrastructurally simlilar to PLs.   That it is the PL which transforms into other types is indicated also by the corresponding decrease of PLs and increase of other types in differential hemocyte counts.   For example, in *Prodenia*, when PLs fall in number, spheroidocytes (= ADs) increase (Yeager, 1945); in *Drosophila melanogaster*, when POs increase, PLs decrease (Rizki, 1962); and in *Periplaneta americana*, within four hr of antennal hemorrhage, GRs increase while PLs decrease (personal observation).

It has also been suggested (Gupta and Sutherland, 1966; Moran, 1971; Scharrer, 1972; Price and Ratcliffe, 1974; Beaulaton and Monpeyssin, 1976) that insects have only one basic type of hemocyte, and that the commonly recognized types of hemocytes are merely different physiological manifestations of the same type, depending on the physiological needs of the insect at different times. Although the PL has been regarded as the primary type in insects, survey of hemocyte types in other arthropod groups reveals that the GR, not the PL, is the basic type (see Section 13.4).

Granulocyte (GR)

Structure. These are small-to-large, spherical or oval cells (Figs. 13.1D, 13.5A,B, 13.6A,B with variable sizes (10–45 μm long and 4–32 μm wide, rarely larger). The plasma membrane may or may not have micropapillae, filopodia, or other irregular processes. The nucleus may be relatively small (as compared with that in the PL), round or elongate, and is generally centrally located. Nuclear size is variable (2–8 μm long and 2–7 μm wide).

The laminar nature of the plasma and nuclear membranes may not be visible. The cytoplasm is characteristically granular (Fig. 13.1D, 13.5A,B, 13.6A). Several types of membrane-bounded granules have been described in the GRs of various insects (Figs. 13.6A,B, 13.7A,B,C, 13.8A,B). Recently, Goffinet and Grégoire (1975) summarized and synonymized various types of granules into three categories, based on their observations in *Carausius morosus.* The following summary and synonymy of granules are based on these and other works: 1) *structureless, electron-dense granules* [= unstructured inclusions (type 1) of Baerwald and Boush, 1970; melanosome-like granules of Hagopian, 1971; opaque body of Moran, 1971; type 2 bodies of Scharrer, 1972; and electron-dense granules of Raina, 1976, and others] ; 2) *structureless, thinly granular bodies* [= type 3 of Scharrer, 1972; electron-lucent granules of Raina, 1976] ; and 3) *structured granules* [= "globules" or "granules multibullaires" of Beaulaton, 1968; "grains denses structures" of AD of Devauchelle, 1971, and Landureau and Grillet, 1975; "corpus fibrillaires" of Hoffmann et al., 1968, 1970; cylinder inclusions (type 2), regular-paced inclusions (type 3), and inclusions with bandlike units (type 4) of Baerwald and Boush, 1970; "Granula mit tubularer Binnenstruktur" of Stang-Voss, 1970; premalanosome-like granules of Hagopian, 1971; tubule-containing bodies or TCB of Moran, 1971; type 1 of Scharrer, 1972; and granules with a microtubular structure of Ratcliffe and Price, 1974] . The length or the diameter of the structureless granules varies from 0.15 to 3 μm, or more, in various insects, while that of the structured granules varies from 0.5 to 2 μm. The shape of the granules may be spherical, ovoid, elongate, or irregularly polygonal (Figs. 13.5A,B, 13.6A,B, 13.7A,B,C, 13.8A). The diameter of the microtubules within the structured granules varies from 15 to 80 nm) in various insects. Internally, the microtubules may show micro-microtubules about 5 nm in diameter (Hagopian, 1971) (Fig. 13.8B). Akai and Sato (1973) also have described "subunits of fibrils" in their so-called secretory vesicles. The number of microtubules/granule may vary from 9 to 80.

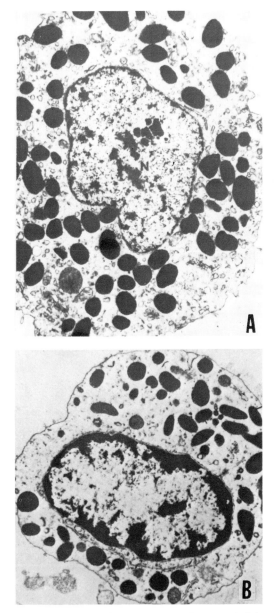

Fig. 13.5. *A.* Granulocyte. ×7,000 (Courtesy of Dr. G. Devauchelle). *B.* Granulocyte of *Thermobia domestica.* ×10,500. From François (1975).

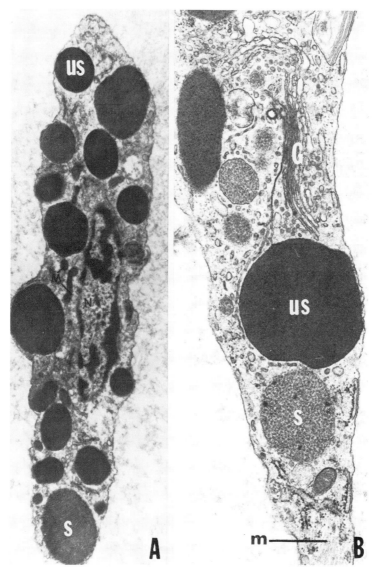

Fig. 13.6.   *A.* Granulocyte of *P. americana*, showing structured (s) and unstructured (us) granules. ca ×16,000. Reinterpreted from Baerwald and Boush (1970). *B.* Portion of granulocyte of *Leucophaea maderae*, showing derivation of structured (= premelanosome-like) granule (s) from Golgi (G), structureless or unstructured (= melanin-like) granule (us), and intracytoplasmic microtubules (m). ×3,000. Reinterpreted from Hagopian (1971).

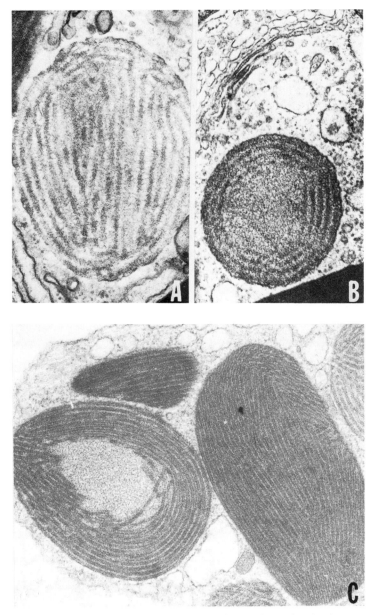

Fig. 13.7. *A.* Structured granule from granulocyte of *L. maderae*, showing internal microtubules. ×40,000. *B.* An earlier stage of development of internal microtubules. ×50,000.
C. Section of a structured granule showing concentric arrangement of internal microtubules.
×38,000. A, B, C, from Hagopian (1971).

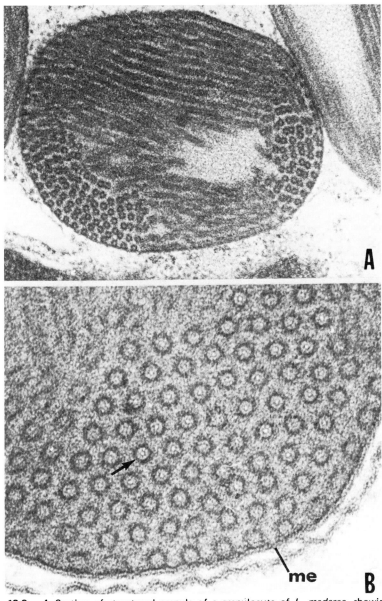

Fig. 13.8. *A.* Section of structured granule of a granulocyte of *L. maderae*, showing arrangement of microtubules about 25 nm in diameter. X63,000. *B.* Highly magnified view of microtubules of structured granules. Note microtubules of structured granules. Note micro-microtubules (5 nm in diameter, arrow) within microtubules and limiting membrane (me) of granule. X240,000. A, B, from Hagopian (1971).

From the accounts provided by Hagopian (1971), Scharrer (1972), Akai and Sato (1973), and François (1975), it appears that the granules are derived from the Golgi bodies (Fig. 13.6B), the microtubules developing during the later stages of morphogenesis. It is conceivable that the structureless, electron-dense granules represent the final stages of development of these granules, in which the structured nature becomes obliterated. Supposedly, the granules are eventually released into the hemolymph. Histochemically, most of the granules contain sulfated, periodate-reactive sialomucin and other glycoproteins or neutral mucopolysaccharides (François, 1974, 1975; Costin, 1975). Occasionally, lipid droplets may be present.

In addition to the structureless and structured granules, the cytoplasm is rich in free ribosomes (polysomes), Golgi bodies, both ER and RER, and lysosomes. Mitochondria are generally few in number. Marginal bundles of intracytoplasmic microtubules are also present (Fig. 13.6B).

Synonymy.   Jones (1846) first established the category of *granular cells*, and later Cuénot (1896) mentioned *"amoebocytes"* with finely granular cytoplasm. Lea and Gilbert (1966) treated GRs and ADs. The so-called *pycnoleucocytes* of Wille and Vecchi (1966) are probably GRs. Recently, Devauchelle (1971) synonymized GRs with cystocytes (= COs) and François (1975) did with SPs. GRs have been referred to as *phagocytes, amoebocytes,* and *hyaline cells.*

Interrelationship with Other Hemocyte Types.   GRs have been widely misinterpreted and confused with SPs, ADs, and COs (= cystocytes). François (1975) considered the SPs, described by Gupta and Sutherland (1966) and Price and Ratcliffe (1974), as GRs. Goffinet and Grégoire (1975) reported separate categories of GRs and COs in *Carausius morosus.* As a matter of fact, the separate existence of GR (not to be confused with PL, SP, AD, and CO) is now recognized by most authors, although Devauchelle (1971) has included both GR and CO in his type III.

How are GRs formed? Are they derived from PRs or PLs? Both sources of origin have been suggested (Arnold, 1974). Gupta and Sutherland (1966) indicated that the derivation of GR from PL is a short step. Takada and Kitano (1971) reported that GRs showed a trend to increase and PLs to decrease with time. Are GRs capable of transforming into other types of hemocytes? There are reports that indicate that GRs do indeed give rise to SPs, ADs and COs (Gupta and Sutherland, 1966). Arnold (1974) has suggested that GRs "might be considered basic units from which more precisely structured and functioning classes of cells have developed." This is supported by my present survey of hemocyte types in Arthropoda (see Section 13.4).

The presence of microtubules in the granules, and in the cytoplasm of the GRs, also has caused debate. According to Crossley (1975), the microtubules of the granules do not have the "dimensions of typical cytoplasmic microtubules (24-27 nm diameter), nor have been demonstrated to be sensitive to colchicine or vinblastine . . . , and therefore, they should not be called micro-tubules." According to him, only in *Leucophaea* are the dimensions of the inclusion tubules (25 nm) comparable to intracytoplasmic microtubules.

The intracytoplasmic microtubules have been described in several insects (Grimstone et al., 1967; Baerwald and Boush, 1970, 1971; Devauchelle, 1971; Hagopian, 1971; Scharrer, 1972; to mention a few), and are generally nar-rower in diameter than the microtubules of the granules. These intracytoplasmic microtubules may be arranged into marginal bundles (Hagopian, 1971), or may be randomly distributed in the cytoplasm (Devauchelle, 1971), and supposedly are found in all hemocyte types, except the OEs (Devauchelle, 1971), although Raina (1976) has described them in OEs.

Spherulocyte (SP)

Structure. These are ovoid or round cells (Figs. 13.1C,E, 13.9A,B) with variable sizes (9-25 μm long and 5-10 μm wide), and usually larger than GRs. The plasma membrane may or may not have micropapillae, filopodia, or other irregular processes. The mucleus is generally small (5-9 μm long and 2.5-6 μm wide), central or eccentric, rich in chromatin bodies and generally obscured by the membrane-bounded, electron-dense, intracytoplasmic spherules, which are characteristic of these cells.

The number of the spherules may vary from few to many, and the diameter from 1.5 to 5 μm. The spherules contain granular, fine-textured, filamentous or flocculent material (Raina, 1976). The granules within the spherules may vary from 15-17 nm in diameter (Akai and Sato, 1973). In addition to the spherules, the cytoplasm contains polyribosomes (Fig. 13.10C), Golgi bodies (moderately to well developed) (Fig. 13.10A), membrane-bounded vacuoles (= lysosomes) (Fig. 13.9C), numerous, randomly distributed microtubules, elongated mitochondria, and RER (Figs. 13.9C, 13.10A,C). Devauchelle (1971) has also described a more or less loose network of fibrils in the cytoplasm (Fig. 13.10B). SPs release the material in their spherules into the hemolymph by exocytosis.

Histochemically, the spherules have been reported to contain neutral or acid mucopolysaccharide and glyco-mucoproteins by several authors (Vercauteren and Aerts, 1958; Nittono, 1960; Ashhurst and Richards, 1964; Gupta and Sutherland, 1967; Costin, 1975). Much earlier, Hollande (1909) reported that the spherules contain "lipochrome" (a kind of carotenoid lipid). The presence of tyrosinase has been reported by Dennell (1947), Jones (1956), and Rizki and Rizki (1959). Most recently, Costin (1975) reported the presence of non-sulfated sialomucin, in addition to glycoproteins and neutral mucopolysaccharides.

Synonymy. Hollande (1909) was the first to use the term *spherule cells*,

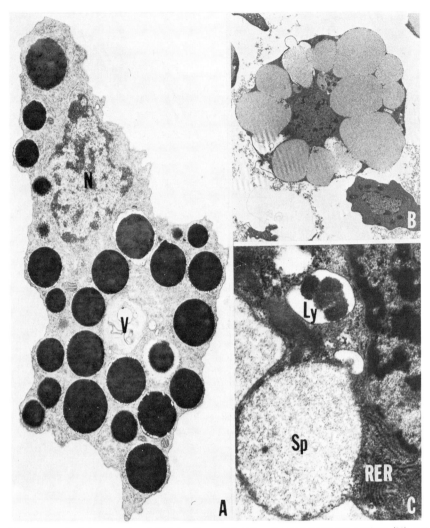

Fig. 13.9. *A.* Spherulocyte of *Melolontha melolontha*, showing eccentric nucleus (N) and numerous spherules and a vacuole (V). ×13,500. From Devauchelle (1971). *B.* Spherulocyte of *Bombyx mori*. ca 10,000. From Akai and Sato (1973). *C.* Portion of spherulocyte of *P. gossypiella*, showing rough endoplasmic reticulum (RER), spherule (Sp) with granular contents, and lysosome (Ly). ca ×30,000. From Raina (1976).

which is now generally used by most workers. Other terms that have been used by various authors are *"cellules spheruleuses"* or *"cellules à sphérules"* (Paillot, 1919; Paillot and Noel, 1928), *spherocytes* (Bogojavlensky, 1932), *eruptive cells* (Yeager, 1945), *oenocytoid* (Dennell, 1947), *rhegmatocyte* (?) (Hrdy, 1957), and *hyaline cells* (Whitten, 1964). Harpaz et al. (1969) classified SPs as ADs. I suggest that *spherule cells* be called *spherulocytes.*

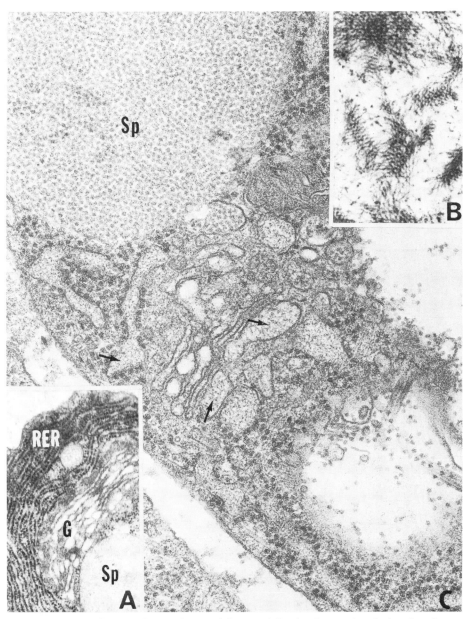

Fig. 13.10. *A.* Portion of spherulocyte of *P. gossypiella*, showing rough endoplasmic reticulum (RER) and Golgi (G) involved in formation of spherule (Sp). ca X 25,000. From Raina (1976). *B.* Portion of spherulocyte of *M. melolontha*, showing loose network of intracytoplasmic fibrils. From Devauchelle (1971). *C.* Portion of spherulocyte of *B. mori*, showing spherule (Sp) with fine granules, rough endoplasmic reticulum containing fibrous material in its cisternae (arrows), and ribosomes. X 80,000. From Akai and Sato (1973).

Interrelationships with Other Hemocyte Types. The main controversies about SPs concern the transformation of these cells into other types, formation of the spherules, and the functions of these cells. The transformation of SPs into other types is not well documented. Gupta and Sutherland (1966) suggested that SPs are capable of transforming into ADs and COs (= cystocytes) and that SPs are themselves derived from GRs. Millara (1947) and Arnold and Salkeld (1967) also considered SPs as a phase in the life of a GR. Later, Arnold (1974) stated that "they seem to be another specialized cell within the granular hemocyte complex." Beaulaton (1968) has suggested that SPs are degenerated OEs.

Little information is available on the formation of the spherules. According to Akai and Sato (1973), the material in the spherules is first observed in enlarged cisternae of the RER, then transferred into the Golgi complex, where it is packaged into the membrane-bounded spherules.

The role of SPs is highly controversial. Hollande (1909) considered these cells respiratory in function because of the presence of the so-called lipochrome. It has been demonstrated by Åkesson (1945), Ashhurst and Richards (1964), Gupta and Sutherland (1967), and Costin (1975) that the material contained in the spherules is neutral or acid mucopolysaccharide, and not a carotenoid lipid. Hollande (1909) stated also that these cells contained an oxidase. Dennell (1947), Jones (1959), and Rizki and Rizki (1959) reported tyrosinase in the spherules of various Diptera, but Gupta and Sutherland (1967) found no tyrosinase in the SPs of cockroaches. Gupta and Sutherland (1965) were supposedly the first to report SPs in cockroaches. Whitten (1964) suggested that SPs (= her hyaline cells) may play a role in the darkening of the puparium in some cyclorrhaphous Diptera. Perez (1910) reported that SPs took part in histolysis. Although this has been disputed by Åkesson (1945), the histolytic role of SPs should not be surprising, considering the fact that before and after molting several SPs are observed to congregate on histolysing tissue. Gupta (1970) has suggested the probable histolytic or phagocytic functions of SPs. It is probable that SPs both histolyse and phagocytize tissues in insects. The phagocytic function was reported by Kollman (1908), Cameron (1934), and Åkesson (1945), but further work will be needed to demonstrate clearly the histolytic role of SPs. Raina (1976) found no evidence of their role in phagocytosis. Metalnikov and Chorine (1929) and Metalnikov (1934) found that the SPs in *Galleria* are related to bacterial immunity. Nittono (1960) stated that strains of silkworm larvae that lacked SPs completely or incompletely tended to produce relatively smaller quantities of silk. Wigglesworth (1959) suggested that SPs are involved in the uptake and transport of other substances such as hormones. Akai and Sato (1973) suggested that SPs are sources of some blood proteins.

Adipohemocyte (AD)

Structure.   These are small-to-large, spherical or oval cells (Fig. 13.2C) with variable sizes (7–45 μm in diameter). The plasma membrane may or may not have micropapillae, filopodia, or other irregular processes. The nucleus is relatively small (as compared with that in PL or SP), round or slightly elongate, and is centrally or eccentrically located. Nuclear size is variable (4–10 μm in diameter). It may appear to be concave, biconvex, punctate, or lobate.

The laminar nature of the plasma and nuclear membranes may not be visible. The cytoplasm contains characteristic small–to–very large refringent fat droplets (0.5–15 μm in diameter) and other nonlipid granules (0.5–9 μm in diameter), and vacuoles, which, according to Arnold (1974), become filled with lipids under certain conditions. In addition, the cytoplasm contains well-developed Golgi bodies, mitochondria, and polyribosomes.

Histochemically, ADs are reported to contain PAS-positive substance in the granules (Ashhurst and Richards, 1964; Lea and Gilbert, 1966). Costin (1975) did not recognize ADs as a type in her study.

Synonymy.   Hollande (1911) first introduced the term *"adipoleucocyte,"* although Kollman (1908) had earlier used the term *adipo-spherule cell* for some hemocytes of invertebrates. Other terms used for ADs are: *spheroidocytes* (Yeager, 1945; Arnold, 1952; Rizki, 1953; Jones, 1959), *later stages of spherules* (Whitten, 1964), and *adipocytes* (Wigglesworth, 1965).

Interrelationship with Other Hemocyte Types.   The main controversy about ADs concerns their identity as a distinct category of hemocytes. Scrutiny of the literature leads one to believe that they are not a distinct type. Several authors have reported that it is difficult to distinguish them from the GRs (Jones, 1970; Arnold, 1974), and many others did not recognize the category of ADs in their studies (Wittig, 1968; Akai and Sato, 1973; Costin, 1975; François, 1975; Goffinet and Grégoire, 1975; Boiteau and Perron, 1976; Raina, 1976, to mention a few recent ones). Raina (1976) noted a progressive accumulation of lipid drops in GRs, and on that basis considered ADs as functional stages of GRs. Gupta and Sutherland (1966) also have reported the transformation of GRs into ADs. Almost all of the ultrastructural studies of hemocytes do not include ADs as separate category, the exception being that by Devauchelle (1971). However, his micrographs (his Figs. 20, 21, 24) are strikingly similar to the GRs (cf. Fig. 13.11 with Fig. 13.6B) described by Hagopian (1971) and other authors.

On the basis of the histochemical nature of these cells also it is difficult to justify the term, and hence the category of ADs. According to Crossley (1975), "ultrastructural studies of so-called adipohaemocytes include cells which contain no reported lipid (Devauchelle, 1971, Fig. 18-22), lipid of doubtful authenticity (Pipa and Woolever, 1965) or material believed to be mucoprotein or mucopoly-

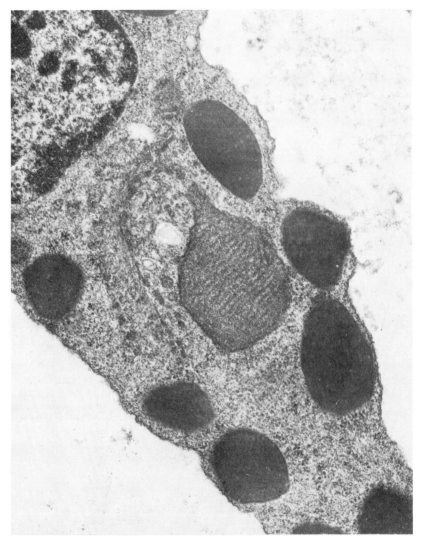

Fig. 13.11. Supposedly adipohemocyte of *M. melolontha*. But note that it looks like granulocyte. X50,000. From Devauchelle (1971).

saccharide (Beaulaton, 1968, Fig. 7)." Costin (1975) also did not recognize AD on the basis of the histochemistry of hemocytes in *Locusta migratoria*.

The resemblance of ADs to fat body cells is also confusing, and adds to the difficulty of identifying ADs in fresh hemolymph samples. For example, one may find all gradations between small ADs and fat body cells (Wigglesworth,

1955).   Jones (1965) suggested that hemocytes "with excentric nucleus and many brilliant fatlike droplets should be termed ADs only when they can be clearly distinguished from fat body cells." According to him (Jones, 1975), ADs are at least 10 times smaller than fat body cells.

Gupta and Sutherland (1966) suggested that under certain conditions, such as chilling, starvation, and diapause (periods of non-feeding and reduced metabolic rate), the PLs respond by changing into ADs with "lipid" droplets. This was based on their observation that mealworm larvae, when chilled for 20-24 hr at 5°C, showed numerous "lipid" droplets and ADs, other types of hemocytes being rare.   Such larvae subsequently recovered.   Ludwig and Wugmeister (1953) noted an increased amount of free fats in the hemolymph of the starving Japanese beetle, *Popillia japonica*, and Clark and Chadbourne (1960) reported a greater number of ADs (= their spheroidocytes) in diapausing larvae of the pink bollworm, *Pectinophora gossypiella*.   Thus, it seems very likely that the appearance (or transformation from PLs) of ADs in the hemolymph at certain times is governed by the physiological state of the insect.

Oenocytoid (OE)

Structure.   These are small-to-large, thick, oval, spherical or elongate cells (Figs. 13.2A,B, 13.12A,B) with widely variable sizes (16-54 μm or more) and shapes.   The plasma membrane is generally without micropapillae, filopodia, or other irregular processes.   The nucleus is generally small, round or elongate, and generally eccentrically located (Figs. 13.2B, 13.12A).   Nuclear size may vary (3-15) μm).   Occasionally, 2 nuclei may be present.

The laminar nature of the plasma and nuclear membranes may not be visible. The cytoplasm is generally thick and homogeneous, and has several kinds of plate-, rod-, or needle-like inclusions.   According to Costin (1975), OE is distinguished by an elaborate system of filaments that fills the cytoplasm (Fig. 13.2A, 13.12C,D), and is visible under a phase-contrast microscope.   Histochemically, the filaments resemble the cytoplasm, and hence are not visible in stained preparations.   Hoffmann (1966) and Hoffmann et al. (1968) also have reported such filaments.   In addition to the above intracytoplasmic inclusions and filaments, a few electron-dense spherules may be present in the cell periphery (Devauchelle, 1971).   With the exception of polyribosomes and the abundant, large mitochondria, which are conspicuous, other organelles, such as ER and Golgi, are poorly developed.   Supposedly, lysosomes are absent.

Histochemically, OEs are reported to contain tyrosinase (Dennell, 1947), protein (Akai and Sato, 1973), and PAS-positive-only granules, indicating the presence of glycoproteins or neutral mucopolysaccharides, and sulfated, periodate-reactive sialomucin (Costin, 1975).

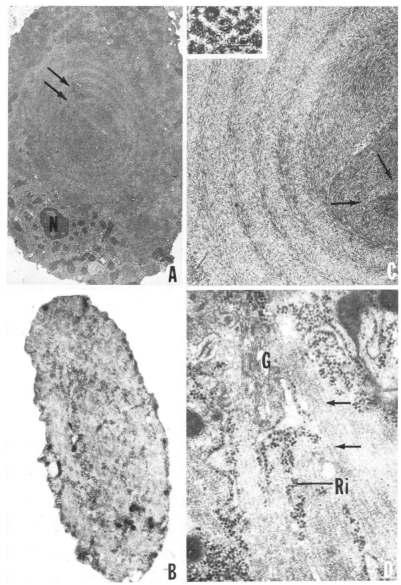

Fig. 13.12.   *A.* Oenocytoid of *B. mori*, showing concentric arrangement of intracytoplasmic fibrils (arrows) and eccentric nucleus (N). X 4,000. From Akai and Sato (1973). *B.* Oenocytoid of *P. gossypiella.* X 8,800. From Raina (1976). *C.* Portion of oenocytoid of *B. mori*, showing highly magnified view of concentric rings of intracytoplasmic fibrils. Note also unoriented fibrils (arrows) ca X 40,000. Inset (ca X 300,000) shows fibrils in cross section. From Akai and Sato (1973). *D.* Portion of oenocytoid of *P. gossypiella*, showing longitudinally arranged intracytoplasmic microtubules (arrows), Golgi (G), and ribosomes (Ri). X 47,300. From Raina (1976).

One peculiarity of OEs seems to be their highly labile nature. They are particularly fragile *in vitro*, and lyse quickly, ejecting material in the hemolymph. They are nonphagocytic.

Synonymy. No two terms have caused as much confusion as *oenocytes* and *oenocytoids*. Oenocytes differ from oenocytoids in that they are ectodermal in origin, usually segmentally arranged, yellow in color, and are not hemocytes. Poyarkoff (1910) first introduced the term OE, later followed by Hollande (1911). In order to avoid any confusion between oenocytes (proposed by Wielowiejsky, 1886) and oenocytoids, Hollande (1914) proposed to replace the term oenocyte by "Cerodecytes." It is not surprising that several earlier authors mistook oenocytes for OEs. Even after Hollande's (1920) detailed description of OEs, several authors (Metalnikov and Gaschen, 1922; Müller, 1925; Tateiwa, 1928; Metalnikov and Chorine, 1929; Bogojavlensky, 1932; and Cameron, 1934) used the term oenocyte instead of OE in their respective works.

Other synonyms used for OEs are *oenocyte-like cells* (Yeager, 1945), large *non-granular spindle cells* and *non-phagocytic giant hemocytes* (Wigglesworth, 1933, 1955; see discussion in Jones, 1965), *crystalloid* and *dark hyaline hemocytes* (Selman, 1962), *crystal cells* (Rizki, 1953, 1962; Nappi, 1970), and COs of Hoffmann and Stoekel (1968).

Interrelationship with Other Hemocyte Types. The main controversy about OEs concerns their identity as a separate category, particularaly their distinction from COs. The view that OEs are part of the CO complex has received support owing to the observation by some authors (Lea and Gilbert, 1966) that OEs are unstable *in vitro*, and that they undergo rapid and drastic transformation into hyaline cells (= COs). These authors report that in *Hyalophora cecropia,* OEs begin to transform within 15–30 sec, and fully transformed OEs are found within 15 min. Jones (1959) and Nittono (1960) also have reported such transformation of OEs in *Prodenia* and *Bombyx*, respectively. Coupled with these observations are the reports by many authors that OEs either are found in very small numbers or are absent. This may partly explain why several authors either have not reported OEs in their studies or do not recognize OEs as a distinct category. Crossley (1975), however, stated that ultrastructurally OEs and COs are different, and indeed some authors (Hoffmann et al., 1968) have described both OEs and COs in their ultrastructural studies. The ultrastructural identity of OEs is also supported by the fact that although these cells eject material into the hemolymph as COs do, this does not result in plasma gelation (Arnold, 1974).

The origin or derivation of these cells is also controversial. Gupta and Sutherland (1966) suggested that OEs are differentiated from PLs, although they stated that they did not actually observe the direct transformation of PL into OE. Devauchelle (1971) has indicated that OEs might be derived from PRs. Arnold (1974) stated, "The cells seem to be allied with the complex of granular cells, but their origins and relationships are not understood."

Coagulocyte (CO)

Structure. These are generally small-to-large (3–30 μm long), spherical, hyaline, fragile, and unstable cells, combining the features of GRs and OEs (Arnold, 1974). The plasma membrane is without any micropapillae, filopodia, or other irregular processes. The nucleus is relatively small (5–11 μm long), generally eccentric, oval, sharply outlined, and under phase-contrast may appear cartwheel-like owing to the arrangement of the chromatin in that fashion (Fig. 13.2D). According to Goffinet and Grégoire (1975), there is a pronounced perinuclear cisterna (Fig. 13.13A), which supposedly distinguishes these cells from other types.

The laminar nature of the plasma and nuclear membranes may not be visible. The plasma membrane may show microruptures. The cytoplasm is hyaline and rich in polyribosomes, but has few mitochondria, and moderately developed ER. In addition, the cytoplasm has some spherical or elongate granular inclusions, about 1 μm in diameter [Fig. 13.13A). François (1975) has described four types of such granules in *Thermobia domestica*: 1) electron-dense, homogeneous granules, generally resembling those in GRs; 2) moderately electron-dense, homogeneous granules; 3) heterogeneous granules with a central or lateral dense zone, the remaining portion being homogeneously granular; and 4) structured granules, with internal microtubules (15 nm in diameter), arranged in a parallel fashion and 40 nm apart. Goffinet and Grégoire (1975) also described structured granules in the COs of *Carausius morosus*. It is obvious that there is a very close resemblance between GRs and COs.

Histochemically, COs are clearly distinguishable from PLs and GRs according to the PAS test (Costin, 1975). According to her, "compared with the cytoplasm of the other types of blood cells, that of coagulocyte has much reduced basophilia." It is very weakly PAS-positive.

Synonymy. Grégoire and Florkin (1950) for the first time introduced the term *coagulocyte* or *unstable hyaline hemocytes* in *Gryllulus* and *Carausius*. Earlier, Yeager (1945) for the first time used the term *cystocyte* for cells with cystlike inclusions. Jones (1950) used that term for "coarsely granular haemocytes," and later (Jones, 1962) suggested that the "term coagulocyte for these cells may be preferred to cystocyte because these cells are only identified by their function."

Interrelationship with Other Hemocyte Types. The main controversies about COs concern their identity, function, and origin. It is still debatable whether the COs are ultrastructurally different from GRs. Devauchelle (1971) found them indistinguishable from GRs, and synonymized them with the latter. Moran (1971) found a type of cell in *Blaberus discoidalis,* frequently in newly molted, untanned adult, with membrane-bounded, tubule-containing bodies (TCB) filled with rows of 34-nm tubules, which are quite different from cyto-

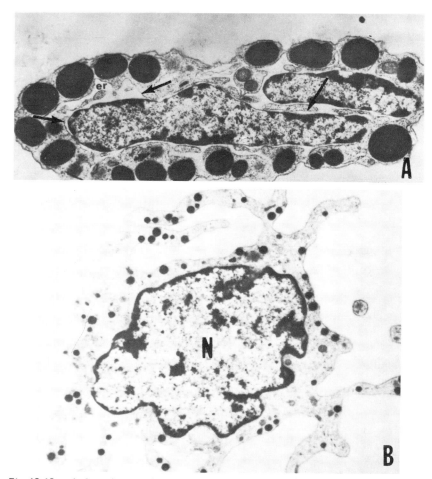

Fig. 13.13. *A.* Coagulocyte, showing perinuclear cisternae (arrows) and those of endo-plasmic reticulum (er).   Note presence of electron-dense (structureless) granules, as are found in granulocytes.   ×20,000.   Podocyte, showing pseudopodia.   Note resemblance to prohemocyte or young plasmatocyte.   ×9,500.   A, B, courtesy of Dr. G. Devauchelle.

plasmic microtubules.   He suggested that these cells are equivalent to COs (cystocytes).   Ratcliffe and Price (1974), however, have identified COs (their cystocyte) in their work.   Most recently, Goffinet and Grégoire (1975) and Grégoire and Goffinet (1978) claimed a separate identity for them.   According to them, the perinuclear cisternae of the COs are much more pronounced than those in PRs, PLs, and GRs, and their plasma membrane is ruptured during coagulation, those in the PRs, PLs, and GRs remaining intact.

The role of the COs in hemolymph coagulation is generally accepted, and has been recently reconfirmed by Grégoire (1974), François (1975), Goffinet and Grégoire (1975), and Grégoire and Goffinet (1978). Gupta and Sutherland (1966), however, have suggested that COs are the effect rather than the cause of coagulation on the basis of their observation that as soon as coagulation starts, several PLs transform into COs. This view, however, is not accepted by Grégoire. Supposedly, COs also contain phenol-oxidizing enzymes (Crossley, 1975). It must be mentioned here that in several arthropod groups coagulation of the hemolymph is caused by GR (see Section 13.4.4), and evidence is accumulating that this is also true in some insects (Rowley and Ratcliffe, 1976).

The origin of COs is still debatable. Grégoire and Goffinet (1978) have suggested that these cells originate in the hemopoietic organs. However, if we accept the premise that hemocytes respond to bodily injury in the insect, it is conceivable that some other type of hemocyte would produce COs by transformation. For more details on the structure, functions and origin of COs, the reader is referred to Grégoire and Goffinet (1978).

Podocyte (PO)

Structure. These hemocytes have not been recognized as a separate category in any ultrastructural study, and are not ordinarily observed in hemoctye samples under the light microscope. According to Arnold (1974), they have been correctly identified only in *Prodenia* (Yeager, 1945; Jones, 1959). However, I (Gupta, 1969) have observed them in *Periplaneta americana* nymph. More often than not, radiate PLs with pseudopodia are mistaken for POs.

These hemocytes are very large (Fig. 13.2E), extremely flattened, PL-like cells with several cytoplasmic extensions (Fig. 13.13B). The nucleus is generally large and centrally located, and may appear punctate.

Synonymy. Yeager (1945) introduced the term *podocyte*. Grabers's (1871) *star-shaped amoebocytes* and Lutz's (1895) *radiate cells* may have included POs.

Interrelationship with Other Hemocyte Types. Their origin and mode of differentiation are unknown. It is conceivable that they are derived from PLs. Gupta and Sutherland (1966) have suggested their transformation from PLs. This seems to be supported by Rizki's (1962) observation that in *Drosophila melanogaster* when POs increase in differential counts, PLs decrease. Rizki's (1953) POs appear to be PLs. Whitten (1964) questioned the concept of POs, and Devauchelle (1971) considered them variant forms of PLs.

Vermicyte (VE)

Structure. This form is generally called vermiform cell. As the name suggests, these are extremely elongated cells with slightly granular or agranular cytoplasm. The nucleus may be located centrally or eccentrically (Fig. 13.2F).

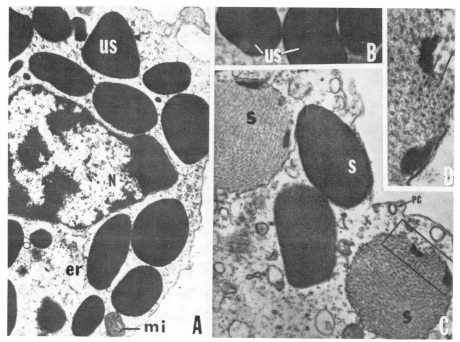

Fig. 13.14. *A.* Granulocyte of *Limulus polyphemus*, showing electron-dense, unstructured granules (us), nucleus (N), mitochondrion (mi), cisterna of endoplasmic reticulum (er). ca ×10,000. *B.* Magnified view of unstructured granules. ca ×20,000. *C.* With onset of coagulation granules in (B) become less dense and eventually reveal microtubules and appear structured (s). PC = pinocytotic caveola. ca ×22,000. *D.* High magnification of rectangular area in (C), showing tubular nature of microtubules in cross sections (arrows). ca ×60,000. A, B, C, D, from Dumont et al. (1966).

Synonymy.   Yeager (1945) introduced the term *vermiform cell,* but the term vermicyte is suggested here for the sake of consistency with all the preceding hemocyte types. Tuzet and Manier (1959) used the term *"giant fusiform cells"* for VEs.

Interrelationship with Other Hemocyte Types.   Their origin is unknown. However, it is conceivable that they are derived from PLs, as has been suggested by Gupta and Sutherland (1966). Lea and Gilbert (1966) considered them a variant form of PLs. According to Arnold (1974), "they seem to occur mainly just prior to pupation, but never in large numbers."

Other Hemocyte Types

In addition to the above nine hemocyte types, several authors have, from time to time, reported hemocytes, many or all of which have not been generally accepted—for example: *haemocytoblast* of Bogojavlensky (1932), *"leucoblast"* of Arvy and Gabe (1946) and Arvy (1954), *proleucocytoid* and *prohaemocytoid* of Yeager (1945) and Jones (1950), respectively. Yeager also introduced the term nematocyte. Rizki (1962) in his work used the term *lamellocyte* and *crystal cell*. The latter has also been adopted by Whitten (1964). According to Arnold (1974), the crystal cell and lamellocytes are considered to be the variants of OEs and PLs respectively. Gupta (1969) also suggested that the crystal cell is probably OE. Terms such as *seleniform cells* (Poyarkoff, 1910), *miocytes* (Tillyard, 1917), *splanchnocytes* (Muttkowski, 1924), *"teratocytes"* (Hollande, 1920, *"pycnonucleocyte"* (Morganthaler, 1953; Wille and Vecchi, 1966), *nucleocyte*, and *rhegmatocyte* (Hrdy, 1957) are rarely encountered in the literature; and most likely several of these cells are not even hemocytes. Jones (1965) introduced the term *granulocytophagous cell* in his work on *Rhodnius prolixus*, and Ritter (1965) and Scharrer (1965) *"anucleate crescent body"* and *cresent cell,* respectively, in the cockroach, *Gromphadorhina portentosa*. Zachary and Hoffmann (1973) described the hemocyte *"thrombocytoid"* that takes part in encapsulation in *Calliphora erythrocephala* (Zachary et al., 1975).

Several years ago, I (Gupta, 1969) reported that the hemocytes in several orders of insects have not been studied. For example, as of that year, among Apterygota, only Thysanura had been studied. Among the orthopteroid groups, Isoptera and Embioptera awaited study. In the hemipteroid complex, no account of hemocytes was available in Zoraptera, Phthiraptera, Corrodentia, and Thysanoptera; and finally in the neuropteroid group, hemocytes were yet to be studied in Raphidioidea, Mecoptera, and Siphonaptera. It seems the situation has changed very little since then, since hemocytes in most of the above groups are still awaiting studies.

In terms of the number of species studied in various orders, Orthoptera (sensu lato) appear to be the most extensively studied group. Next to them are perhaps Lepidoptera, Hymenoptera, Coleoptera, and Diptera, in that order. In addition to the Heteroptera, Homoptera, and Odonata, of which only a few species have so far been studied, Dermaptera, Plecoptera, Trichoptera, and Thysanoptera are the most poorly studied groups.

With the exception of PLs and GRs, all other types of hemocytes are not present in all insect orders. According to Arnold (1974), all nine types have been reported only in *Prodenia* (Yeager, 1945; Jones, 1959), and most insects seem to possess the PRs, PLs, and GRs.

13.2.2. Aquatic Chelicerata

Unfortunately, hemocytes of only one group (Xiphosura) have been described. Hemocytes of *Limulus* have been studied by many authors (Howell, 1885a,b; Copley, 1947; Dumont et al., 1966; Fahrenbach, 1968, 1970; Kenny et al.,

1972; Snyder and Stanley, 1973; Murer et al., 1975; Belamarich, 1976; White, 1976; Armstrong, 1977, to mention some). Only two types of hemocytes have been described in this animal (Table 13.1): the GR and the cyanocyte (CYN). The former is still referred to as amoebocyte by many authors. Structurally, the GR in *Limulus* is similar to the GR in insects (Fig. 13.14A, B, C, D). The description of the CYN, based on Fahrenbach's (1968, 1970) works, follows.

Cyanocyte (CYN)

Structure.   The immature stage of the CYN is called the cyanoblast (CYB). Young CYBs are 8-10 μm in diameter (Figs. 13.15A, 13.16A). Ultrastructurally, the CYB contains abundant polyribosomes; cisternae of ER are con-

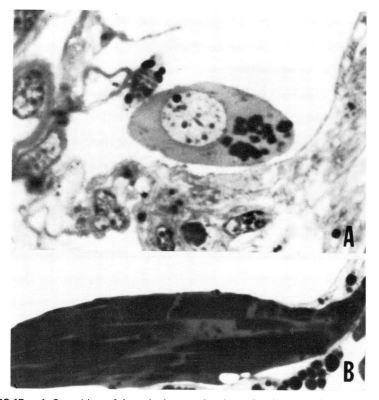

Fig. 13.15.   *A.* Cyanoblast of *L. polyphemus,* showing a few hemocyanin crystals. ca X 2,000. *B.* Cyanocyte of *L. polyphemus,* packed with hemocyanin crystals. At this stage cyanoctye breaks down and releases hemocyanin into hemolymph. X 1,500. A, B, from Fahrenbach (1970).

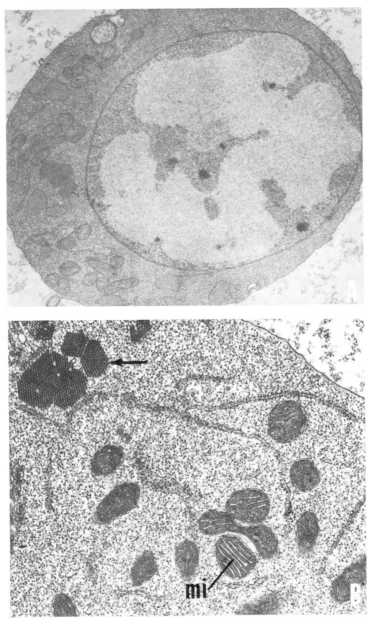

Fig. 13.16. *A.* Very young cyanoblast of *L. polyphemus.* ×12,000. *B.* Portion of an advanced cyanoblast with hemocyanin crystals (arrow), mitochondrion (mi), and endoplasmic reticulum. ×40,000. A, B, from Fahrenbach (1970).

tinuous with the perinuclear cisternae and both are often filled with dense granular material; a small Golgi apparatus and a number of mitochondria are present. The nucleus fills most of the cell as in PR. An occasional hemocyanin crystal indicates its unmistakable identity (Fig. 13.16A,B). According to Fahrenbach (1970), this ultrastructural morphology of the CYB is typical of an active protein-synthesizing and storing cell.

Synonymy. Fahrenbach (1968) suggested the term *cyanocyte* when he discovered this hemocyte for the first time in *Limulus.* He also likened it to the vertebrate erythroblast.

Interrelationship with Other Types. The site of the origin of the CYB is unknown, but Fahrenbach (1970) suspected that it is produced in the hepatopancreas. Its relationship with GR is unknown. He also made the interesting observation that CYNs are more abundant at molting time.

13.2.3. Crustacea

Among curstaceans, the hemocytes of decapods have been studied by many authors (Hardy, 1892; Tait, 1918; Tait and Gunn, 1920; George and Nichols, 1948; Toney, 1958; Hearing and Vernick, 1967; Wood and Visentin, 1967; Sawyer et al., 1970; Stang-Voss, 1971b; Bauchau and De Brouwer, 1972, 1974; Johnston et al., 1973; Bauchau et al., 1975; Chassard-Bouchaud and Hubert, 1975; Ravindranath, 1975; Williams and Lutz, 1975). Kollman (1908), Ravindranath (1974b), and Hoarau (1976) have studied a few species of Isopoda. The two most primitive branchiopods, *Artemia* and *Daphnia*, have been studied by Lockhead and Lockhead (1941) and Hardy (1892), respectively. On the basis of the above studies, it seems that within Crustacea, the more highly evolved subgroups (e.g., Isopoda and Decapoda) have developed all the hemocyte types (Table 13.1), compared with the primitive branchiopods, *Artemia* and *Daphnia*. As early as 1892, Hardy commented: "It seems a point of remarkable morphological interest that the archaic features of the anatomy of *Daphnia* should extend to interest that the archaic features of the anatomy of *Daphnia* should extend to its blood cells. . . . Each cell of *Daphnia* includes within itself the characteristics of the explosive and eosinophile blood corpuscles and of the basophile cells of the cell tissue of *Astacus.*" It would be interesting if future studies of hemocyte types in other subgroups of Crustacea supported the phylogenetic significance of the number of hemocyte types within Crustacea.

13.2.4. Terrestrial Chelicerata

As far as I know, the hemocytes of only one scorpion (*Palamneaus;* Ravindranath, 1974a, two tarantula spiders (*Phormictopus*, Deevey, 1941; and *Eurypelma*,

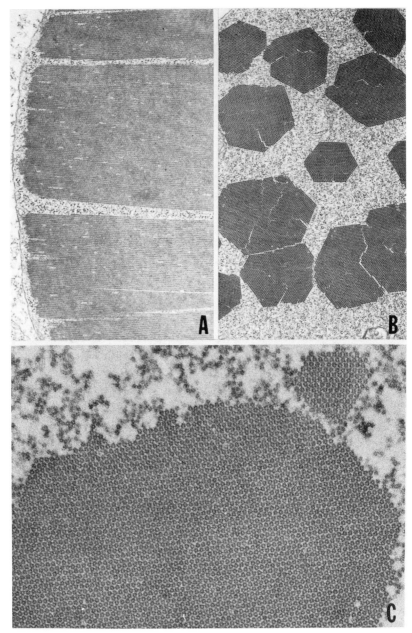

Fig. 13.17. *A.* Portion of mature cyanocyte, packed with hemocyanin crystals. ca ×40,000. *B.* Cross section of hemocyanin crystals in mature cyanoblast. ca ×22,000. *C.* Free-floating hemocyanin crystals in hemocoel. ×80,000. A, B, C, from Fahrenbach (1970).

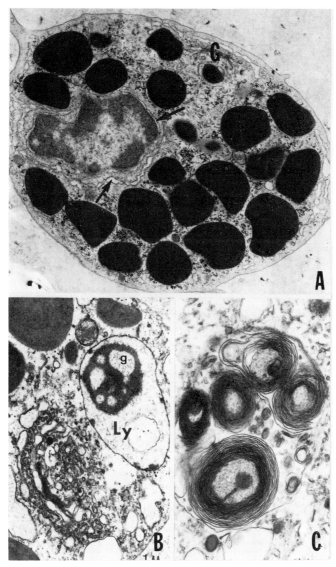

Fig. 13.18. *A.* Granulocyte of *Eriocheir sinensis*, showing electron-dense (structureless) granules, Golgi (G), and perinuclear and endoplasmic reticular cisternae (arrows). ×15,500. *B.* Portion of granulocyte of *E. sinensis*, showing lysosome (Ly) with a granule (g) in process of resorption, and Golgi (G). ×17,500. *C.* Granules with concentric filaments (myelin). ×23,400 A, B, C, from Bauchau and De Brouwer (1972).

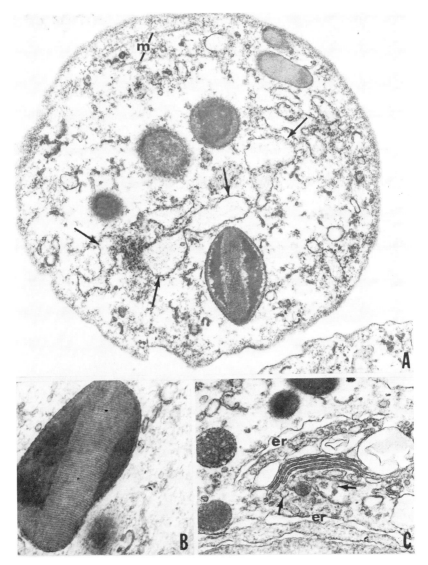

Fig. 13.19.   *A.* Granulocyte of *E. sinensis*, showing fusiform granule with a crystalloid, intracytoplasmic microtubules (m), and several cisternae of endoplasmic reticulum (arrows). ×29,000. *B.* Magnified view of granule with crystalloid. Note transverse banding pattern (micro-microtubules) in crystalloid. ×45,000. *C.* Portion of granulocyte of *E. sinensis*, showing details of Golgi apparatus: minute vesicles (arrows) with dense materials. Note newly formed granules of several sizes, and cisternae of endoplasmic reticulum (er). ×41,500. A, B, C, from Bauchau and De Brouwer (1972).

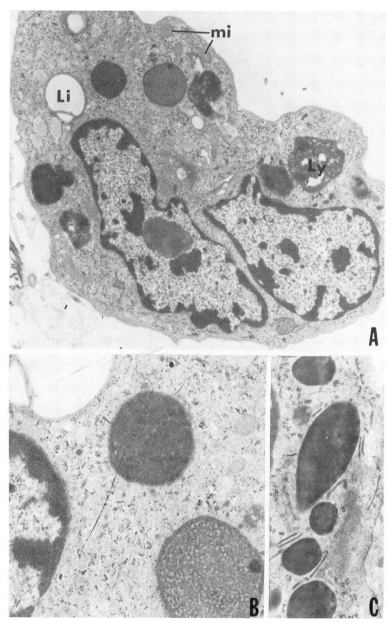

Fig. 13.20. *A.* Granulocyte of *Peripatus acacioi*, showing bilobed nucleus, lysosome (Ly), several granules, mitochondria (mi), and liposome (Li). × 11,000. *B.* Portion of granulocyte showing two granules and liposome. × 28,000. *C.* Portion of granulocyte, showing biclavate bodies in peripheral region. × 27,000. A, B, C, reinterpreted from Lavallard and Campiglia (1975).

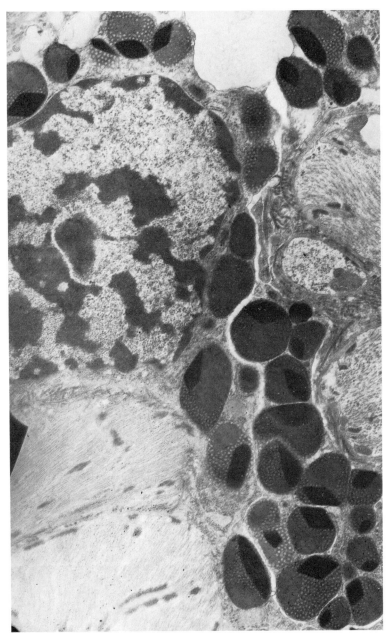

Fig. 13.21. Spherulocyte of *P. acacioi* (between muscle fibers), showing spherules. Note that most spherules contain lozenge-shaped bodies and networks of compacted material (microtubules?). X17,500. From Lavallard and Campiglia (1975).

Sherman, 1973), and a few ticks and mites are known (Nordenskiöld, 1905; Dolp, 1970; Douglas, 1943; Teravsky, 1957; Tsvilineva, 1959; Kanungo and Naegele, 1964; Brinton and Burgdorfer, 1971, Fujisaki et al., 1975). According to these studies, Scorpionida, Araneida, and Acarina have developed among them all the hemocyte types (Table 13.1), as well as supposedly the CYN, since hemocyanin has been reported in some scorpions (Lankester, 1871; Svedberg and Hedenius, 1933), spiders (Boyd, 1937; Loewe and Linzen, 1973; Angersbach, 1975; Markl et al., 1976), and one pedipalpid (Redmond, 1968). However, on the basis of their TEM works, Brinton and Burgdorfer (1971) have reported only PR, PL, SP (four categories), and OE in the tick (Figs. 13.22AB, 13.23AB, 13.24), and Sherman (1973) identified only PL, GR, and OE in the spider. OE has not been observed in the scorpion (Ravindranath, 1974a). According to Brinton and Burgdorfer (1971), Tsvilineva (1959) found it difficult to classify blood cells in several ixodid ticks on the bases of their appearance and inclusion bodies. Clearly, there is considerable need for hemocyte studies, including search for CYN, in various subgroups of terrestrial Chelicerata.

13.2.5. Myriapoda

In this group, hemocytes of Pauropoda are unknown, and only a few species of Diplopoda (Vostal and Pirčová, 1968; Vostal, 1970; Ravindranath, 1973), Chilopoda (Shukla, 1964; Sundara Rajulu, 1970, 1971b), and Symphyla (Gupta, 1968) have been studied.   All hemocyte types, (except CO in Chilopoda and OE in Symphyla) are present in Myriapoda (Table 13.1).

13.3. TYPES OF HEMOCYTES IN ONYCHOPHORA

Hemocytes of only three species of Onychophora are presently known. Tuzet and Manier (1958) briefly mentioned hemocytes of *Peripatopsis moseleyi*, and it is difficult to interpret from their figures the presently recognized categories of hemocytes, except that their Figs. 4a,b definitely look like SPs. Sundara Rajulu et al. (1970) reported PR, PL, GR, SP, and OE in *Eoperipatus weldoni*, and Lavallard and Campiglia (1975), on the basis of their TEM work, recognized PR, PL (their hyalocytes), GR (their macrophages), and SP in *Peripatus acacioi*. The ultrastructural details of the spherules of the SP (Fig. 13.21) seem to be different from those of arthropods.

13.4. HEMOCYTES AND ARTHROPOD PHYLOGENY

In the preceding account we noted that there are several types of hemocytes in arthropods, and that not all types are found in all arthropod groups. Indeed, the number of types varies within all major arthropod groups. We also observed

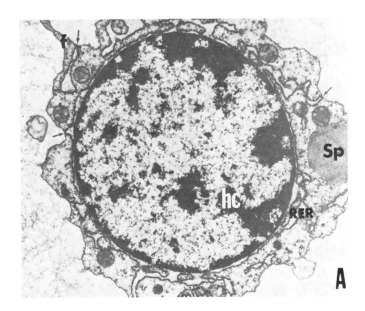

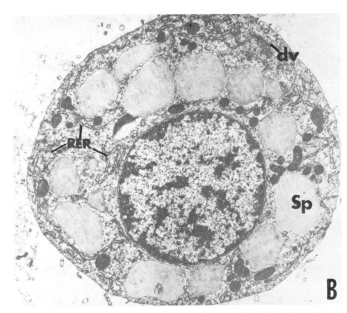

Fig. 13.22. *A.* Prohemocyte of *Dermacentor andersoni*, showing filopodia, cisternae of rough endoplasmic reticulum (RER) touching plasma membrane (arrows), heterochromatin (he), and some spherules (Sp). ×15,000. *B.* Spherulocyte of *D. andersoni*, showing rough endoplasmic reticulum (RER), spherules (Sp), free ribosomes, and Golgi (= dictyosomal vesicles, dv). ×7,000. A, B, from Brinton and Burgdorfer (1971).

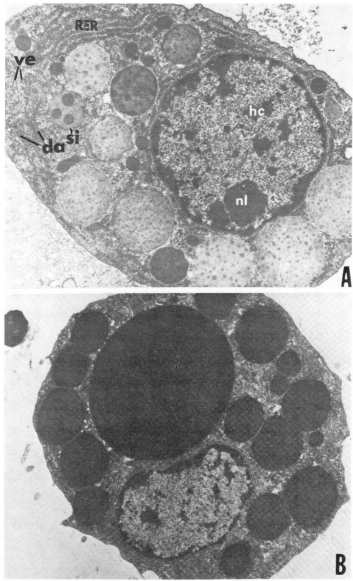

Fig. 13.23. *A.* Spherulocyte of *D. Andersoni*, showing several spherules with mottled appearance, immature spherules (si), rough endoplasmic reticulum (RER), Golgi (= dictyosomal apparatus, da), and vesicles of Golgi (ve), heterochromatin (he), and nucleolus (nl). Note that immature spherule is associated with Golgi apparatus, indicating that is is produced by it. ×11,000; *B.* spherulocyte of *D. andersoni*, showing electron-dense spherules. Note that this Spherulocyte can be interpreted as granulocyte. ca ×6,000. A, B, from Brinton and Burgdorfer (1971).

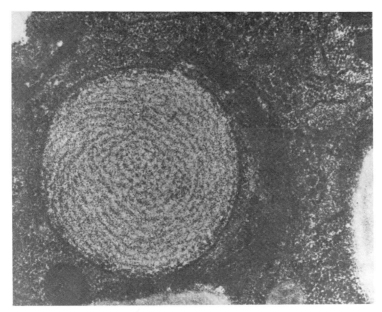

Fig. 13.24. Portion of spherulocyte of *D. andersoni*, showing spherule with fibrous material. ca ✕28,000. From Brinton and Burgdorfer (1971).

that certain hemocyte types are confined only to certain taxa. For example, the hemocyanin-producing hemocyte, CYN, is found only in Crustacea, aquatic Chelicerata, and a few terrestrial Chelicerata. Do these variations have any phylogenetic significance? Can hemocytology be used to elucidate phylogenetic relationship between taxa? I believe it does have promise, and it was with this belief that I (Gupta, 1968) suggested earlier the probable ancestry of Insecta from Symphyla on the basis of a comparison of the hemocyte types of the two groups. The prospect, however, is severely limited, primarily owing to: 1) lack of uniform terminology, and hence the difficulty of establishing comparisons, and 2) paucity of comprehensive studies of hemocyte types in large numbers of species within various groups to enable one to draw meaningful conclusions on the basis of important variations.

In any discussion of the significance of hemocyte types in the phylogeny of arthropods, it will be important first to identify the most primitive or plesiomorphic hemocyte type, and then to trace its evolutionary distribution, differentiation, and interrelationship with other hemocyte types in various arthropod groups. It is generally accepted that arthropods originated from some marine lobopod annelid ancestor. Is it possible to speculate as to what type of hemocyte that lobopod annelid ancestor had?

13.4.1. Plesiomorphic Hemocyte—Granulocyte (GR)

Cameron (1932) called the earthworm hemocytes "coelomic corpuscles" and recognized granular and agranular types among them. He also described the nonhemocytic chloragogen cells. Later, Liebmann (1942) described two types of "coelomocytes" in the coelomic fluid of the annelids, *Eisenia foetida* and *Lumbricus terrestris*: eleocytes and true leucocytes. According to him, the term eleocyte was first introduced by Rosa (1896) for those nonamoeboid cells that carried fat droplets. Earlier, Cuénot (1891) considered them as chloragogen cells, whose excretory and food storage roles are still controversial (Roots, 1957). Liebmann (1942) did not accept their excretory function (which is generally based on the presence of guanin granules in these cells), but considered them nutritive, since they carry lipids, proteins, and glycogen. Thus, the nonhemocytic eleocytes represent typical trephocytes, and as such play an important role in tissue regeneration (Liebmann, 1942, 1943). Roots (1957) also doubted the excretory function of the eleocytes, and suggested that they contain mainly phospholipids, and a yellow pigment which he did not characterize. Stang-Voss (1971a) reported that the eleocytes produce hemoglobin and ferretin, which are eventually released into the hemolymph. The second type of the coelomocytes, the leucocytes, according to Liebmann (1942), is composed of mainly eosinophilic amoebocytes. A critical look at his Figs. 28 and 29 leaves no doubt that his eosenophils are GRs in the present classification. Stang-Voss (1971a) and Valembois (1971) have described the blood cells of *Eisenia foetida* by electron microscope; and although they have not used the terminology proposed in this chapter, it can be interpreted that their amoebocytes contain GRs.

If these and other annelids have GRs, it is reasonable to assume that the lobopod ancestor also had this hemocyte in some form.

13.4.2. Occurrence of GR in Various Arthropod Groups

It is obvious from Fig. 13.25 that the GR is the only hemocyte which is found in all major arthropod groups as well as in the Onychophora. Sherman (1973) also commented that "perhaps the most widely found [hemocyte] in the arthropods is granulocyte. With their discovery in spiders, granulocytes now have been found in all major arthropod classes."

The primitive *Limulus* (Xiphosura) is believed to possess only GRs, although Fahrenbach (1968, 1970) demonstrated the presence of the CYN. All the works on Crustacea (Hardy, 1892; Kollman, 1908; Tait, 1918; Tait and Gunn, 1920; Lockhead and Lockhead, 1941; George and Nichols, 1948; Toney, 1958; Hearing and Vernick, 1967; Wood and Visentin 1967; Sawyer et al., 1970; Stang-Voss, 1971b; Bauchau and De Brouwer, 1972, 1974; Johnston et al., 1973; Ravindranath, 1974b, 1975; Bauchau et al., 1975; Chassard-Bouchaud and Hubert,

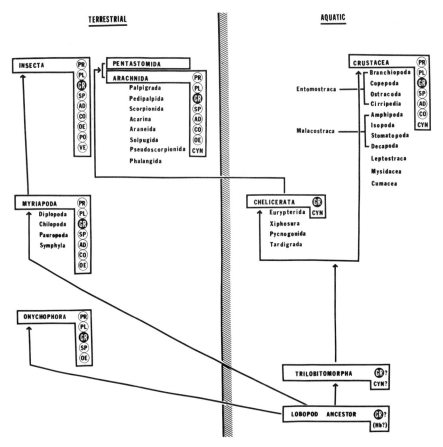

Fig. 13.25. Diagram showing distribution of plesiomorphic granulocyte and other hemocyte types and their phylogenetic significance (see text) in various groups of arthropods and Onychophora. Note presence of CYN only in aquatic groups and some terrestrial Chelicerata. AD = adipohemocyte; CO = coagulocyte; CYN = cyanocyte; GR = granulocyte; Hb = hemoglobin; OE = oenocytoid; PL = plasmatocyte; PO = podocyte; PR = prohemocyte; SP = spherulocyte; VE = vermicyte.

1975; Hoarau, 1976) either have reported GRs or can be interpreted to show these hemocytes (Figs. 13.18A,B,C, 13.19A,B,C). All workers (Deevey, 1941; Teravsky, 1957; Kanungo and Naegele, 1964; Sherman, 1973; Ravindranath, 1974a), except Nordenskiöld (1905), Douglas (1943), Dolp (1970), and Brinton and Burgdorfer (1971), have reported or can be interpreted to have reported GRs in the terrestrial Chelicerata. Of the four authors who did not, Nordenskiöld and Douglas recognized only one category of amoeboid corpuscles (which must include GRs), and Dolp, and Brinton and Burgdorfer have reported SPs; the latter authors recognized four categories of SPs, and it is most likely that these include GRs (see Fig. 13.23B). All studied myriapods, except those studied by Vostal (1970) and Vostal and Pirčová (1968), have been shown to possess GRs (Gupta, 1968; Shukla, 1965; Sundara Rajulu, 1970, 1971a,b; Ravindranath, 1973), and the same is true in Onychophora (Tuzet and Manier, 1958; Sundara Rajulu et al., 1970; Lavallard and Campiglia, 1975) (Fig. 13.20A,B,C).

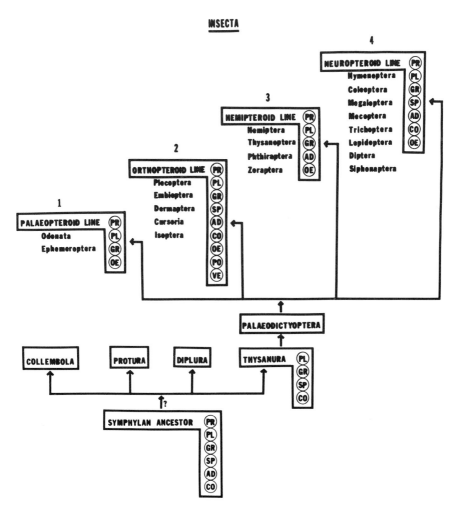

Fig. 13.26.   Diagram showing distribution of various hemocyte types in four evolutionary lines (groups) and in the symphylan ancestor of Insecta. Note that listed orders under each evolutionary line are only some examples of representative orders, not necessarily those in which hemocyte types designated under each evolutionary line are present. Abbreviations same as in Fig. 13.25.

The following account of the presence of GR in Insecta is based on Arnold's (1974) reinterpretation of various works, my own (Gupta, 1969) survey of the hemocyte literature in many insects, and the most recent transmission electron microscope (TEM) studies of hemocytes. As far as is known, only Bruntz (1908), Millara (1947), Barra (1969), Gupta (1969), and François (1974, 1975) have worked on the hemocytes of the Apterygota; and on the basis of these works, both Collembola (it is controversial whether they should be included in Apterygota) and Thysanura possess GR. All higher orders of insects possess GRs (Fig. 13.26). Some of the most recent TEM studies (Hoffmann et al., 1968, 1970; Baerwald and Boush, 1970; Hagopian, 1971; Moran, 1971; Scharrer, 1972; Ratcliffe and Price, 1974; Goffinet and Grégoire, 1976) have reported (or can be interpreted to show) GRs in various insects. The most highly specialized neuropteroid orders also possess GRs, with the possible exception of Trichoptera, but this could be due to the difficulty of interpreting LaCava's (1964; cited by Arnold, 1974) work, or lack of more studies.

13.4.3. Differentiation of the Plesiomorphic GR Into Other Hemocyte Types in Various Arthropod Groups

It is obvious from Fig. 13.25 that, with the exception of aquatic Chelicerata, all other arthropod groups have several hemocyte types. Since the fact that one hemocyte type can and does differentiate into another type has been reported by several authors it is reasonable to assume that during evolution the plesiomorphic GR differentiated into other hemocyte types. It can be postulated also that the GR originates from the so-called prohemocyte (PR) or stem cell and goes through the plasmatocyte (PL) stage before becoming a distinct GR type. In taxa in which only GRs have been observed, the PR and PL are merely evanescent stages, and have not achieved distinctness as types. In taxa that are reported to possess other types besides PR, PL, and GR, the last perhaps further differentiated into SP, AD, CO, and OE, not necessarily in that order. The post-GR differentiation is generally accompanied by distinct PRs and PLs. Furthermore, in the more highly evolved taxa any of the types may be suppressed. The main differentiation pathways as postulated above may be represented as follows:

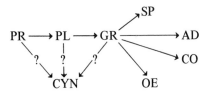

It is uncertain whether the GR differentiated into CYN in the aquatic Chelicerata, Crustacea, and some terrestrial Chelicerata.

13.4.4 Functional Adaptations of GRs and CYNs

In the preceding account we noted that 1) the plesiomorphic GR is ubiquitous; 2) the hemocyanin-producing CYN is present only in the aquatic Chelicerata, Crustacea, and some terrestrial Chelicerata. We shall now consider the functional adaptations of these two hemocyte types, and the significance of these adaptations in those arthropod groups which possess these hemocyte types.

13.4.4.1. Granulocyte (GR)

The two most commonly reported functions of the primitive blood cells are phagocytosis and their role in blood coagulation. The plesiomorphic GR (or another of its differentiated forms in some groups) performs these functions in all arthropod groups.

13.4.4.1.1. Phagocytosis. Metchnikoff (1892) was probably the first to recognize the existence of phagocytosis in invertebrates. It is generally difficult, although possible (Anderson, 1975), to observe phagocytosis *in vitro*, and consequently it has not been reported in many species. Among the extant arthropods, *Limulus* is unique in that it has only two types of hemocytes: GR and CYN. Several authors (Dumont et al., 1966; Fahrenbach, 1968, 1970; Kenny et al., 1972; Snyder and Stanley, 1973; Murer et al., 1975; Belamarich, 1976; White, 1976; Armstrong, 1977) have recently studied the GR in *Limulus*, but none has reported on the phagocytic nature of this hemocyte. However, it surely must perform that function, since no other hemocyte type (CYN has a different specific function) is present. Obviously, considerable work needs to be done to demonstrate the phagocytic function of GR in the aquatic Chelicerata.

Among Crustacea, phagocytosis by GRs has been reported by Metchnikoff (1884, as cited by Anderson, 1975), Bruntz (1905, 1907), Tait and Gunn (1920), McKay and Jenkins (1970a,b,c), Sawyer et al. (1970), Stang-Voss (1971b), Bauchau et al. (1975), Ravindranath (1974b), Bang (1975), Chassard-Bauchaud and Hubert (1975).

Recently, Lavallard and Campiglia (1975) reported in Onychophora a hemocyte type, the macrophage, which looks like an early GR (Fig. 13.20A,B).

Unfortunately, detailed accounts of the phagocytic activity of GRs in Crustacea, terrestrial Chelicerata, Myriapoda, and Onychophora are lacking.

It should be mentioned here that in animals in which PLs, SPs and ADs are present as distinct types, the phagocytic function may have shifted from GR to these hemocytes. Indeed, in insects, phagocytosis has been reported not only by PLs, but also by SPs and ADs. Phagocytosis plays an important role in cellular immunity in insects, and has been reported and reviewed by several authors (Wittig, 1962; Salt, 1970; Whitcomb et al., 1974; Anderson, 1975; Crossley, 1975). It should be noted here that until very recently the role of GR in phagocytosis in insects was not well documented. It was generally believed that mostly PLs take part in phagocytosis (Salt, 1970; Ratcliffe and Rowley, 1974, 1975). However, more and more evidence is accumulating that GR is the main phagocytic cell in insects as well (Takada and Kitano, 1971; Neuwirth, 1974; Bohn, 1977).

13.4.4.1.2. Coagulation. Unlike phagocytosis, which cannot generally be observed *in vitro*, hemolymph coagulation is easily observed, and has therefore been reported from all groups of arthropods. Coagulation plays an important role in wound healing. Hemolymph coagulation may be caused by 1) hemocyte agglutination, or 2) plasma gelation, or a combination of both.

The coagulation function of GRs in *Limulus* is well documented. Dumont et al. (1966) studied the clotting process in *Limulus* by electron microscopy. They described the ultrastructural changes in normal and clotted GRs (i.e., after clotting sets in). According to them, the plasma in *Limulus* is not capable of coagulation, and the granules of the GRs provide the material that participates actively in the formation of the clot. It should be noted that Howell (1885b), Loeb (1904, 1910), and Maluf (1939) suggested much earlier that the material which actually forms the clot is contained within the hemocyte and not in the plasma. This has been recently supported by Murer et al. (1975). They claim that the granules of the GR (their amoebocytes) contain all the factors, including the clottable protein, and that these factors are released when these granules rupture during cell aggregation. The nature of the clottable protein remains unknown. GRs in *Limulus* also aggregate in response to injuries (Kenny et al., 1972). White (1976) compared the microtubules within the granules of the GR in this animal to the "polymers" (= microtubules?) of sickled human erythrocytes.

The role of crustacean GR (or its differentiated form CO) in coagulation has been either directly reported or has been interpreted to be present by several authors (Halliburton, 1885; Hardy, 1892; Cuénot, 1895; Kollman, 1908; Tait and Gunn, 1920; George and Nichols, 1948; Toney, 1958; Hearing and Vernick, 1967; Wood and Visentin, 1967; Stutman and Dolliver, 1968; Wood et al., 1971; Stang-Voss, 1971b; Johnston et al., 1973; Bauchau and De Brouwer, 1974; Ravindranath, 1974b, 1975). Stang-Voss (1971b) recognized a separate category of clotting cells in *Astacus*, but these cells look like either PL or early stages of GR, since they show granules. She suggested that the granules in these hemocytes are the storage site of hemagluttinin. According to her, the clotting cells are very fragile and their cytolysis accelerates the clotting process of the hemolymph. Johnston et al. (1973) homologized their so-called β cells in *Carcinus* with the "explosive corpuscles" of Tait and Gunn (1920), which would indicate that these cells participate in coagulation. Stutman and Dolliver (1968) reported that the crustacean hemocyte seems to be more potent than vertebrate cells in clotting, and for a given number of cells, crustacean hemolymph coagulates

2 to 20 times faster than human blood. Durliat and Vranckx (1976) reported three soluble proteins in the hemocytes of *Astacus*, one of which they referred to as fibrinogen, which is the origin of the clottable plasmatic protein. If this is true, it indicates that in Crustacea also the clotting protein originates in the hemocytes as is the case in *Limulus* (Murer et al., 1975). Ghidalia et al. (1977) reported that in the decapod crustacean, *Macropipus*, the coagulation process depends on one or several plasmatic factors and on a thermo-labile cellular factor which does not seem to be specific.

The literature on coagulation in terrestrial Chelicerata is meager, but coagulation by COs has been reported by several authors (Deevey, 1941; Ravindranath, 1974a).

In Myriapoda, presence of COs (in addition to GRs) has been reported by Gupta (1968) and Ravindranath (1973). Apparently, coagulation in this group is brought about by COs and not GRs—an indication of a shift of function.

Grégoire (1955) described the coagulation patterns in Onychophora, and since COs are not reported in *Peripatus* and other onychophoran species, GRs are implicated in coagulation.

Among insects, coagulation has been extensively studied. For a review of COs and coagulation, see Grégoire and Goffinet (1978). It is generally accepted that COs in insects cause coagulation. However, Rowley and Ratcliffe (1976) have demonstrated that in the wax moth, *Galleria mellonella*, hemolymph coagulation is initiated by granules from the GRs. This finding revives the controversy whether GRs and COs are in reality two separate categories in insects.

13.4.4.2. Cyanocyte (CYN)

In most animals, the transport of oxygen is accomplished by the so-called chromoproteins that form loose combinations with oxygen and release the gas at lower tensions. The two most widely known chromoproteins are hemoglobin (Hb) and hemocyanin (Hn). Let us now examine how oxygen transport may have been accomplished in the arthropod ancestor; how the problem of oxygen transport was solved when the arthropods colonized land; and the situation as it presently obtains in some Crustacea. Both Hb and Hn occur in arthropods. According to Maluf (1939), Hb is found in several entomocostracan (Branchiopoda, Copepoda, Ostracoda, Cirripedia) crustaceans (*Apus, Branchipus, Artemia, Daphnia, Chirocephalus, Lernanthropus, Clavella*, and *Congericola*) and *Chironomus* larvae in insects. The presence of Hn in the blood of Crustacea was established by Krukenberg (1880a,b, 1882), and is now confirmed also in several crustaceans (Aphipoda, Stomatopoda, Decapoda) (Redfield, 1933; Florkin, 1934; Redmond, 1968; Ghiretti-Magaldi et al., 1973) and in Xiphosura (*Limulus*) (Fahrenbach, 1968, 1970), in the spiders, *Chaetopelma* (Maluf, 1939), *Latrodectus* (Boyd, 1937), *Dugesiella* (Loewe and Linzen, 1973), *Eurypelma*

(Angersbach, 1975), *Cupiennius* (Markl et al., 1976), in scorpions (Lankester, 1871; Svedberg and Hedenius, 1933), and at least one pedipalpid (*Mastigoproctus giganteus*) (Redmond, 1968). Muttkowski (1921a,b) suggested that insects contain Hn because grasshopper blood seemed to show copper in amounts equal to that of crayfish. Hemocyanin is found in some molluscs, but has not been reported in annelids. It is generally believed that in arthropods Hn and Hb are not located in the blood cells (Redfield, 1934; Maluf, 1939; Hoar, 1966). However, Fahrenbach (1968, 1970) demonstrated that in *Limulus* Hn originates in hemocytes called cyanoblasts (CYB), which when mature become cyanocytes (CYNs) (Figs. 13.15A,B, 13.16A,B, 13.17A,B,C) and release the Hn in the hemolymph. According to him, CYBs and CYNs are very rarely found in the general circulation because they are trapped in sinusoidal spaces pervading the neural plexus of the compound eyes. As a matter of fact, Fahrenbach discovered these cells while studying the visual system of *Limulus*. Recently, Stang-Voss (1971b) implicated the GR in Hn synthesis in *Astacus*, and Ghiretti-Magaldi et al. (1973) confirmed the presence of CYNs in *Carcinus maenas*.

According to Redmond (1968), Hn is the basic blood respiratory pigment in higher Crustacea and Arachnida. As compared with Hb, Hn is less efficient in terms of its oxygen-carrying capacity. This lesser efficiency is further reduced in crustaceans and *Limulus*, in which the blood copper is 4-9 mg/100 ml, as compared with 25 mg/100 ml in cephalopod molluscs (Prosser, 1973). It should be noted, however, that Hn with its reduced oxygen-carrying capacity is admirably suited for the less-active crustaceans inhabiting waters (e.g., saline water) that are low in oxygen. According to Weiland and Magnum (1975), the oxygen affinity of gastropod and xiphosuran Hn decreases with increasing salt concentration, while that of crustaceans increases. In other words, increasing salt concentration has a normal Bohr effect (the oxygen-carrying capacity of the blood decreases with increasing acidity or lower pH, and increases with increasing alkalinity or higher pH) in crustaceans, but induces a reverse Bohr effect in xiphosurans. Because the blood of both the crustaceans and xiphosurans is generally saturated at low oxygen tensions, they seem to be well adapted for life in their poorly aerated habitats. Poor aeration, however, of necessity limits their activity. Hemocyanin is not really suitable for active animals, although exception to this is known (Weiland and Magnum, 1975). Truchot (1975) suggested that in *Carcinus maenas* the oxygen affinity of the Hn is physiologically regulated to ensure adequate oxygen transport under varied temperature and salinity conditions. But what if the oxygen tension becomes dangerously low? There is evidence that many crustaceans have developed safeguard mechanisms to cope with such situations. According to Prosser (1973), a fresh-water cladoceran, *Moina*, synthesizes Hb during several days in low oxygen. Several crustaceans show Hb in blood when in water with low oxygen. In *Artemia*, the Hb is increased 20-fold in low oxygen. This animal is brilliant red with Hb when found in hypersaline pools. Similarly *Daphnia* collected or reared in water that is low in oxygen are red, while those from high oxygen are pale. It loses or gains Hb in about 10 days (Prosser, 1973). Miller (1966) has shown that even some insects (e.g., Notonectidae) have tracheal cells that are rich in Hb. They surface to breathe every 2–5 min, but this duration is reduced under CO_2.

Blood Hb concentration also increases in reduced oxygen in the dipteran larvae of *Chironomus* and *Anatopynia*.

If we postulate that the lobopod annelid ancestors of arthropods had Hb in their blood as the present-day annelids do (see Ghiretti-Magaldi et al., 1977), it is reasonable to suggest that the crustaceans evolved a hemocyanin-producing hemocyte, the *cyanocyte* (CYN), which equipped them for life in poorly oxygenated habitats and less-active life. If this suggestion is accepted, the presence of Hb in some crustaceans and its crucial role in the survival of these animals under poor oxygen conditions would have to be regarded as a secondary development.

If we accept the idea that both the aquatic (Pycnogonida, Xiphosura, Eurypterida, and Tardigrada) and the terrestrial (Pentastomida and Arachnida) Chelicerata evolved from Trilobitomorpha, as indicated by Hessler and Newman (1975) and Bergström (Chapter 1) (although Snodgrass, 1965, does not believe this), it would seem that CYN evolved during the trilobitomorph stage, since Xiphosura (future studies might reveal the presence of CYN in Pycnogonida and Tardigrada!) among the aquatic Chelicerata and the Scorpionida and Araneida, and Pedipalpida among the terrestrial Chelicerata, possess CYN. Interestingly, no CYN or Hn has been reported from Onychophora, Myriapoda, and Insecta. According to Maluf (1939), the blood of arthropods with a well-developed tracheal system contains no respiratory protein or other oxygen-combining substance. Three questions need answering here: 1) what role did the oxygen-deficient habitat of the progenitors of the terrestrial arthropods play in triggering the process of land colonization, 2) how did this process affect the evolution of the CYN in the terrestrial arthropods, 3) and why is it that of the two main terrestrial lines, one (terrestrial Arachnida) retained CYN and thus Hn, but the other (Onychophora-Myriapoda-Insecta) did not?

Part of the answer to the first question can be found in Carter's (1931) belief that "it is the area of stagnant water, especially the swampy shallow tropical bogs and pools where oxygen reaches a low level, which was the most forceful in the evolution of terrestrial groups." (see Hoar, 1966.) It is very likely that oxygen deficiency in the natural habitats of the progenitors of the terrestrial arthropods played an important role in transition from an aquatic to a land environment. Snodgrass (1965) has commented that trilobites "lived probably in the manner of modern *Limulus*, though some species are thought to have been pelagic or even deepwater inhabitants, ... and it has been thought that probably they were mud feeders." According to Størmer (1977), the arthropod invasion of land occurred during late Silurian and Devonian times, some 400 million years ago.

The answer to the second question is easy: Hn, being unsuitable for active terrestrial life, had no survival value for the terrestrial arthropods. The final blow in the disappearance of Hn was probably dealt by the evolution of the trachea, which attained its greatest development in insects. The insect tracheal system is most efficient in bringing oxygen directly to individual cells. It is not surprising, therefore, that the complete disappearance of Hn from the terrestrial arthropods is directly related to the degree of the evolution of an efficient tracheal system. This should explain why the terrestrial arachnids that do not possess a well-developed tracheal system still retain Hn, but the Onychophora, Myriapoda, and Insecta do not. That the early scorpions were probably aquatic is supported by the presence of gills in the fossil scorpion, *Waeringoscorpio hefteri*, as reported by Størmer (1977). According to him, the gill-books are regarded as homologous with the lung-books of scorpions and other arachnids. One could still find among the arachnids the probable evolutionary steps which may have occurred in the development of the trachea from the gill-books of the aquatic arachnids via the lung-books in terrestrial scorpions and some spiders. Among the latter, one would find any combination of lung-books and trachea (Kerkut, 1958, in Borradaile et al., 1958).

13.4.5. Taxonomic Significance of Hemocytes

Does the differentiation of GR into other hemocyte types in various taxa reflect any phylogenetic trends? Could it be that the number of hemocyte types increases in phylogenetically higher taxa? Is there any phylogenetic significance in the number of hemocyte types that an arthropod group possesses? Although a more reliable assessment of the phylogenetic significance of the hemocyte types will have to wait until more comprehensive and comparative studies, using uniform criteria for classification, are available in all the arthropod groups, the answers to the above questions on the basis of the limited information available could be in the affirmative. As far as I know, no attempts have been made to analyze the phylogenetic significance of hemocyte types in various arthropod groups. However, a few authors have suggested that there might be some significance. For example, Hearing and Vernick (1967) reported that in Crustacea the increase in number of hemocyte types is directly related to phylogenetic advancement. By adopting the insect hemocyte classification to hemocyte types in *Scutigerella*, I (Gupta 1968) suggested phylogenetic affinity of Insecta with Symphyla. Sundara Rajulu et al. (1970) and Sundara Rajulu (1971b) suggested on the basis of the presence of OE in Onychophora, Myriapoda, and Insecta, one line of evolution of these groups, although they erroneously stated that OE is not present in the Chelicerata. Arnold (1972a,b,) and Arnold and

Hinks (1975) have used hemocytes in insect taxonomy. Ravindranath (1973, 1974b) also indicated that comparison of hemocyte types in various groups would provide a phylogenetic relationship. The same view has been expressed by Bauchau et al. (1975), who stated that the branching off of arthropods corresponds with the differentiation of blood cells.

A review of the insect hemocyte literature indicates some phylogenetic trends in the diversity of the hemocyte types as we go up the evolutionary ladder in Insecta as a group (Fig. 13.26). This was observed also by Arnold (1974), although he attributed this diversity more to "shifts in the emphasis on certain functions or in the assignment of functions to different tissues," than to phylogeny. And at least in certain instances he may be right.

As far as known, only Bruntz (1908), Millara (1947), Barra (1969), Gupta (1969), and François (1974, 1975) have worked on the hemocytes of the Apterygota; and on the basis of these works, Collembola (it is controversial whether they should be included in Apterygota) possess only the plesiomorphic GR, whereas the Thysanura (Lepismatidae) seem to possess PL, GR, SP, and CO. Assuming that the Thysanura originated from Symphyla-like ancestors, and that the latter had hemocyte types comparable to *Scutigerella* (Gupta, 1968), we find that a reduction from six (PR, PL, GR, SP, AD, and CO) in the symphylan ancestor to four in the Thysanura has occurred. I have no evidence to suggest whether this is a secondary suppression and/or reduction, or, as Arnold (1974) suggested, due to shifts in function. It is also possible that future studies will reveal more types than presently known.

According to Carpenter (1976), the derivation of the Pterygota from the apterygote Thysanura is almost universally accepted. It is also generally believed that the pterygotes evolved along four evolutionary lines: palaeopteroid, orthopteroid, hemipteroid, and neuropteroid. It is interesting to note (Fig. 13.26) that in the Palaeoptera, although the number of hemocyte types has not increased from the ancestral thysanuran number, OE has already made its appearance in the very beginning of the evolution of winged insects. In addition, PR has achieved distinctness, and the CO is either suppressed or its function is taken over by GR, which is generally the case in some other arthropods (aquatic Chelicerata and some Crustacea).

It is also evident from Fig. 13.26 that beyond the palaeopteroid line the number of hemocyte types increased, and all the six to seven types were realized in the orthopteroid, hemipteroid, and neuropteroid lines. It seems that by the time the orthopteroid line evolved, the plesiomorphic GR had already differentiated into all the distinct types which are presently known in pterygote insects, and that no further evolution in the hemocyte types occurred beyond that point. It is doubtful whether within the orthopteroid group any phylogenetic significance of the hemocyte types exists. The number of hemocyte types reported in various orders of this group (see Gupta, 1969; Arnold, 1974) is so variable that it is very difficult to discern any phylogenetic trends. On the basis of the information presently available, the hemocyte types vary from three to eight or nine by light microscopy, and seven (PR, PL, GR, SP, CO, AD, OE) types have been demonstrated by TEM. Unfortunately, hemocytes of several hemipteroid orders have not been studied (Gupta, 1969; Arnold, 1974), most of the works being confined to a few species in the order Hemiptera (Poisson, 1924; Hamilton, 1931; Khanna, 1964; Wigglesworth, 1955, 1956; Jones, 1965; Lai-Fook, 1970; Zaidi and Khan, 1975). Five (PR, PL, GR, AD, OE) types by light microscopy and four (PR, PL, GR, OE) by TEM studies have been identified. The apparent absence of SP and CO in the hemipteroid group as a whole is probably due to lack of enough studied species. It is difficult to imagine that these two types have been suppressed in the hemipteroid orders.

In the neuropteroid group, we find the seven major types (PR, PL, GR, SP, AD, CO, OE), although all of these types have not been reported from each of the orders in this group (Gupta, 1969). As a matter of fact, the types of hemocytes reported vary widely even within an order in this group. For example, two to seven types have been reported in Lepidoptera, Coleoptera, and Diptera, and five types in Hymenoptera and Neuroptera, four in Megaloptera, and only two (PL, GR) in Trichoptera (see Gupta, 1969, for various listings, and Arnold, 1974). Since all the orders are highly evolved, it is quite conceivable that most or all major types would be found in all the orders as more studies become available.

13.4.6. Monophyly Versus Polyphyly

Whenever we evaluate morphological structures for phylogenetic considerations, we should indicate whether we are dealing with morphological similarities due to symplesiomorphy, convergence, or synapomorphy. In other words, whatever characters one uses to indicate phylogenetic relationship between groups, one should first establish the plesiomorphic and the apomorphic states of those characters. It should then be possible to show whether the groups in question possess in common any derivative (synapomorphic) character or a relatively plesiomorphic (symplesiomorphic) character, or whether the similarity is due to convergence. According to Hennig (1965), monophyletic groups share synapomorphic characters, while paraphyletic ones (e.g., Apterygota and Palaeoptera or Pisces and Reptilia) possess symplesiomorphic characters. And, finally, the groups in which the morphological resemblance is due to convergence are polyphyletic. Hennig (1965) believes that: 1) a phylogenetic system of classification should not consider characters whose resemblance rests on symplesiomorphy or convergence; 2) in a phylogenetic system there can be no solely primitive or solely derivative groups, and the possession of at least one derivative (relatively apomorphic) ground plan character is a precondition for a group to be recognized as monophyletic. In synapomorphy, it is always assumed that the compared characters belong to one and the same transformation series (Hennig, 1966). In other words, in synapomorphy, it makes no difference whether the derived character (a′) is present in the same derived (a′) state in all species, or whether it is present in different derived (a′ and a″) conditions.

Let us now test the above assumptions in considering the phylogeny of the various arthropod groups on the basis of hemocyte types. I established earlier that the granulocyte (GR) is the plesiomorphic hemocyte type. The homologous nature of this hemocyte on the basis of its ultrastructure and function in all extant arthropod groups is incontrovertible. And on the basis that it is present in present-day annelids, it is reasonable to assume that the so-called lobopod annelid ancestor of the arthropods also possessed this hemocyte, and by extrapolation so did the Trilobitomorpha from which supposedly the aquatic Chelicerata and Crustacea originated.

The cyanocyte (CYN) also may be considered a plesiomorphic hemocyte, since it is present, along with the GR, in both the aquatic Chelicerata and Crustacea. It is predictable that future studies would reveal the same combination (GR and CYN) in Pycnogonida and Tardigrada. In Crustacea, of course, in addition to the plesiomorphic GR and CYN, we find other hemocyte types (all apomorphic).

According to Bergström (Chapter 1), the transition to land among terrestrial Chelicerata may have occurred sometime during the Devonian or Carboniferous, and this perhaps may have occurred more than once. Bergström questions the inclusion of Scorpionida in Arachnida. Unfortunately, of all the subgroups included here in the terrestrial Chelicerata (Fig. 13.25) the hemocytes from only a few Scorpionida, Acarina, and Araneida are known. I have stated earlier that hemocyanin (Hn) (and hence CYN) has been reported from scorpions, spiders, and one pedipalpid. This clearly indicates that CYN in the terrestrial Chelicerata is a derived plesiomorphic character. The other hemocyte types (SP, AD, CO, OE) are synapomorphic hemocyte types shared in common by all terrestrial arthropod groups, and the Onychophora. Since Crustacea, terrestrial Chelicerata, Myriapoda, and Insecta share the synapomorphic hemocyte types (with the exception of OE by Crustacea), they are regarded as monophyletic groups. Onychophora are a sister-group and have not developed all the hemocyte types found in the Myriapoda, Insecta, terrestrial Chelicerata, and Crustacea.

13.5. SUMMARY

Hemocytes of arthropods and several other invertebrate groups are known. Among arthropods, they have been most extensively studied in insects. An acceptable classification of insect hemocytes exists, and has been successfully adopted by several authors for classifying hemocytes of other arthropod groups and Onychophora. However, a uniform hemocyte classification for all arthropod groups is lacking, and it is proposed in this review to adopt, with few modifications, the insect hemocyte classification for all arthropod groups. This classification recognizes seven hemocyte types: prohemocyte (PR), plasmatocyte (PL), granulocyte (GR), spherulocyte (SP), adipohemocyte (AD), oenocytoid (OE), and coagulocyte (CO). Podocyte (PO) and vermicyte (VE)—two other types reported in insects—are ultrastructurally similar to PLs.

All the seven types, and several others, have been described in various insects. Aquatic Chelicerata (Xiphosura) possess only GR and a hemocyanin-producing hemocyte, the cyanocyte (CYN). Among Crustacea, the more highly evolved groups such as Decapoda and Isopoda have all the seven types, but the primitive branchiopods (*Artemia* and *Daphnia*) possess only PL and GR. Among terrestrial Chelicerata, Scorpionida, Araneida, and Acarina have developed among them all the seven types as well as the CYN. Myriapoda also have the seven types. Onychophora, however, have only five types: PR, PL, GR, SP, and OE.

Of the seven types found in the arthropods, the ubiquitous GR has been designated as the plesiomorphic hemocyte type. This is the only type that is found in all arthropods as well as in Onychophora. It has been postulated that during evolution the plesiomorphic GR has differentiated into other hemocyte types. Supposedly, the GR originates from the PR and goes through the PL stage before becoming a distinct GR type. In taxa in which only GRs have been found, the PR and PL are merely evanescent stages, and have not achieved distinctness as types. In taxa that are reported to possess other types besides PR, PL, and Gr, the last further differentiates into SP, AD, and OE, not necessarily in that order. This post-GR differentiation is generally accompanied by distinct PRs and PLs. Furthermore, in more highly evolved taxa any of the types may be suppressed. The role of GR in phagocytosis and blood coagulation has been documented in all arthropod groups. Hemocyanin (Hn), with its reduced oxygen-carrying capacity, is admirably suited for the less-active crustaceans, and some arachnids. It is most likely, therefore, that these groups consequently evolved the CYN. Interestingly, no CYN or Hn has been reported from Onychophora, Myriapoda, and Insecta. Apparently, Hn, being unsuitable for active terrestrial life, had no survival value for terrestrial arthropods, and the final blow in the disappearance of Hn was probably dealt by the evolution of the trachea, which attained its greatest development in insects.

The variation of hemocyte types in various arthropod groups appears to have some taxonomic significance. An attempt has been made for the first time to evaluate the significance of hemocyte types in arthropod phylogeny. It has been established that the ubiquitous GR is the plesiomorphic type. Its homologous nature, on the basis of its ultrastructure and function, in all extant arthropod groups has been shown to be incontrovertible. The CYN also has been regarded as a plesiomorphic hemocyte, since it is present, along with the GR, in both the aquatic Chelicerata and Crustacea. Its presence in some terrestrial Chelicerata seems to be a derived plesiomorphic character. The other hemocyte types (SP, AD, CO, OE) are regarded as synapomorphic hemocyte types shared in common by all terrestrial arthropod groups, and the Onychophora. Since Crustacea, terrestrial Chelicerata, Myriapoda, and Insecta share the synapomorphic hemocyte types (with the exception of OE by Crustacea), they are regarded as monophyletic groups. Onychophora have been considered a sister-group that has not yet developed all the hemocyte types found in the Myriapoda, Insecta, terrestrial Chelicerata, and Crustacea.

ACKNOWLEDGMENTS

Without the published illustrations that I have included, this review would not have been possible. For providing me with prints or negatives and allowing me to reproduce their published illustrations, I am most indebted to Drs. H. Akai, R. J. Baerwald, A. Bauchau, L. P. Brinton, S. Campiglia, N. M. Costin, G. Devauchelle, J. M. Dumont, W. H. Fahrenbach, J. François, G. Goffinet, Ch. Grégoire, R. Lavallard, A. V. Loud (for the late Dr. M. Hagopian), A. K. Raina, G. Salt, and S. Sato. The entire credit and my deep appreciation for preparing all the illustrations of this review go to Dr. Y. T. Das and Mr. S. B. Ramaswamy. I sincerely appreciate the secretarial assistance of Mrs. Joan Gross.

REFERENCES

Akai, H. and S. Sato. 1973. Ultrastructure of the larval hemocytes of the silkworm, *Bombyx mori* L. (Lepidoptera: Bombycidae). *Int. J. Insect Morphol. Embryol.* **2(3):** 207-31.

Åkesson, B. 1945. Observations on the haemocytes during the metamorphosis of *Calliphora erythrocephala* (Meig.). *Ark. Zool.* **6(12):** 203-11.

Anderson, R. S. 1975. Phagocytosis by invertebrate cells *in vitro:* biochemical events and other characteristics compared with vertebrate phagocytic system, pp. 153-80. *In* K. Maramorosch and R. E. Shore (eds.) *Invertebrate Immunity.* Academic Press, New York, London.

Angersbach, D. 1975. Oxygen pressures in hemolymph and various tissues of the tarantula *Eurypelma helluo. J. Comp. Physiol. B. Metab. Transp. Funct.* **98(2):** 133-46.

Armstrong, P. B. 1977. Interaction of the mobile blood cells of the horseshoe crab, *Limulus.* Studies on contact paralysis of pseudopodial activity and cellular overlapping *in vitro. Exp. Cell Res.* **107(1):** 127-38.

Arnold, J. W. 1952. The haemocytes of the Mediterranean flour moth, *Ephestia kühniella* Zell. (Lepidoptera: Pyralididae). *Can. J. Zool.* **30:** 352-64.

Arnold, J. W. 1970. Haemocytes of the Pacific beetle cockroach, *Diploptera punctata. Can. Entomol.* **102(7):** 830-35.

Arnold, J. W. 1972a. Haemocytology in insect biosystematics: The prospect. *Can. Entomol.* **104:** 655-59.

Arnold, J. W. 1972b. A comparative study of the haemocytes (blood cells) of cockroaches (Insecta: Dictyoptera: Blattaria), with a view of their significance in taxonomy. *Can. Entomol.* **104:** 309-48.

Arnold, J. W. 1974. The hemocytes of insects, pp. 201–54. *In* M. Rockstein (ed.) *The Physiology of Insecta*. Vol. V, 2nd edition. Academic Press, New York, London.

Arnold, J. W. and C. F. Hinks. 1975. Biosystematics of the genus *Euxoa* (Lepidoptera: Noctuidae). III. Hemocytological distinctions between two closely related species, *E. campestris* and *E. declarata*. *Can. Entomol.* **107**: 1095–1100.

Arnold, J. W. and E. H. Salkeld. 1967. Morphology of the haemocytes of the giant cockroach, *Blaberus giganteus*, with histochemical test. *Can. Entomol.* **99**: 1138–45.

Arvy, L. 1954. Presentation de documents sur la leucopoïèse chez *Peripatopsis capensis* Grube. *Bull. Soc. Zool. Fr.* **79**: 13.

Arvy, L. and M. Gabe. 1946. Identification des diastases sanguines chez quelques insectes. *C.R. Soc. Biol. (Paris)* **140**: 757–58.

Arvy, L. and J. Lhoste, 1946. Les variations du leucogramme au cours de la mètamorphose chez *Forficula auricularia* L. *Bull. Soc. Zool. Fr.* **70**: 114–48.

Ashhurst, D. E. and A. G. Richards. 1964. Some histochemical observations on the blood cells of the wax moth, *Galleria mellonella* L. *J. Morphol.* **114**: 247–53.

Baerwald, R. J. and G. M. Boush. 1970. Fine structure of the hemocytes of *Periplaneta americana* (Orthoptera: Blattidae) with particular reference to marginal bundles. *J. Ultrastruct. Res.* **31**: 151–61.

Baerwald, R. J. and G. M. Boush. 1971. Vinblastine-induced disruption of microtubules in cockroach hemocytes. *Tissue Cell* **3(2)**: 251–60.

Bang, F. B. 1975. Phagocytosis in invertebrates, pp. 137–51. *In* K. Maramorosch and R. E. Shore (eds.) *Invertebrate Immunity*. Academic Press, New York, London.

Barra, J. A. 1969. Tégument des Collemboles. Presence d'hémocytes à granules dans le liquide exuvial au cours de la mue (Insectes, Collemboles). *C.R. Acad. Sci. Paris* **269**: 902–3.

Bauchau, A. G. and M. B. De Brouwer. 1972. Ultrastructure des hémocytes d'*Eriocheir sinensis*, Crustacé Décapode Brachyoures. *J. Microsc. (Paris)* **15**: 171–80.

Bauchau, A. G. and M. B. De Brouwer. 1974. Étude ultrastructurale de la coagulation de l'hémolymph chez les crustacés. *J. Microsc. (Paris)* **19(1)**: 37–46.

Bauchau, A. G., M. B. De Brouwer, E. Passelecq-Gerin and J. C. Mengeot. 1975. Étude cytochimique des hémocytes des crustacés décapodes brachyoures. *Histochemistry* **45**: 101–13.

Beaulaton, J. 1968. Étude ultrastructurale et cytochimique des glandes prothoraciques de vers à soie auz quatrième et cinquième âges larvaires. I. La Tunica propria et ses relations avec les fibres conjonctives et les hémocytes. *J. Ultrastruct. Res.* **23**: 474–98.

Beaulaton, J. and M. Monpeyssin. 1976. Ultrastructure and cytochemistry of the hemocytes of *Antheraea pernyi* Guer. (Lepidoptera, Attacidae) during the 5th larval stage. I. Prohemocytes, plasmatocytes, and granulocytes. *J. Ultrastruct. Res.* **55(2)**: 143–56.

Belamarich, F. A. 1976. Nature of the contents of the large granules of *Limulus* amebocytes, pp. 101–2. *In: Animal Models of Thrombosis and Hemorrhagic Diseases*. N.I.H., Bethesda, Maryland.

Bogojavlensky, K. S. 1932. The formed elements of the blood of insects (in Russian). *Arch. Russ. Anat. Hist. Embryol.* **11**: 361–86.

Bohn, H. 1977. Differential adhesion of the hemocytes of *Leucophaea maderae* (Blattaria) to a glass surface. *J. Insect Physiol.* **23(2)**: 185–94.

Boiteau, G. and J. M. Perron. 1976. Étude des hémocytes de *Macrosiphum euphorbiae* (Thomas) (Homoptera: Aphididae). *Can. J. Zool.* **54(2)**: 228–34.

Borradaile, L. A., F. A. Potts, L. Eastham and J. T. Saunders. 1958. *The Invertebrate* (revised by G. A. Kerkut). Cambridge University Press, Cambridge.

Boyd, W. C. 1937. Cross-reactivity of various hemocyanins with special reference to the blood proteins of the black widow spider. *Biol. Bull. (Woods Hole)* **73**: 181.

Breugnon, M. and J. R. Le Berre. 1976. Fluctuation of the hemocyte formula and hemolymph volume in the caterpillar *Pieris brassicae* (in French). *Ann. Zool. Ecol. Anim.* **8(1):** 1-12.

Brinton, L. P. and W. Burgdorfer. 1971. Fine structure of normal hemocytes in *Dermacentor andersoni* Stiles (Acari: Ixodidae). *J. Parasitol.* **57:** 1110-27.

Bruntz, L. 1905. Études physiologique sur les phyllopodes branchiopodes. Phagocytose et excretion. *Arch. Zool. Exp. Gén.* **4:** 37-48.

Bruntz, L. 1907. Remarques sur les organes globuligenes et phagocytaires des crustacés. *Arch. Zool. Exp. Gén.* **76:** 1-67.

Bruntz, L. 1908. Nouvelles recherches sur l'excrétion et la phagocytose chez les Thysanoures. *Arch. Zool. Exp. Gén.* **38:** 471-88.

Cameron, G. R. 1932. Inflammation in earthworms. *J. Pathol. Bacteriol.* **35:** 933-72.

Cameron, G. R. 1934. Inflammation in the caterpillars of Lepidoptera. *J. Pathol. Bacteriol.* **38:** 441-66.

Carpenter, F. M. 1976. Geological history and evolution of the insect. *Proc. 15th Int. Congr. Entomol.* **1976:** 63-70.

Carter, G. S. 1931. Aquatic and aerial respiration in animals. *Biol. Rev. (Cambridge)* **6:** 1-35.

Chassard-Bouchard, C. and M. Hubert. 1975. Étude ultrastructurale des hémocytes présents dans l'organe Y de *Carcinus maenas* L. (Crustacé: Décapode). *C.R. Acad. Sci Paris* D **281(12):** 807-10.

Clark, E. W. and D. S. Chadbourne. 1960. The hemocytes of non-diapause and diapause larvae and pupae of the pink bollworm. *Ann. Entomol. Soc. Amer.* **53:** 682-85.

Copley, A. F. L. 1947. The clotting of *Limulus* blood. *Fed. Proc.* **6:** 90-91.

Costin, N. M. 1975. Histochemical observations of the haemocytes of *Locusta migratoria*. *Histochem. J.* **7:** 21-43.

Crossley, A. C. 1975. The cytophysiology of insect blood. *Adv. Insect Physiol.* **11:** 117-221.

Cuénot, L. 1891. Études sur le sang et les glandes lymphatiques dans la série animale. *Arch. Zool. Exp. Gén.* (*ser. 2*) **9:** 13-90, 364-475, 591-670.

Cuénot, L. 1895. Études physiologiques sur les crustacés decapodes. *Arch. Biol. Liège* **13:** 245-303.

Cuénot, L. 1896. Études physiologiques sur les Orthoptères. *Arch. Biol.* **14:** 293-341.

Deevey, G. B. 1941. The blood cells of the Haitian tarantula and their relation to the molting cycle. *J. Morphol.* **68:** 457-91.

Dennell, R. 1947. A study of an insect cuticle. *Proc. R. Soc. Lond.* (B) **134:** 79-110.

Devauchelle, G. 1971. Étude ultrastructurale des hémocytes du Coléoptère *Melolontha melolontha* (L). *J. Ultrastruct. Res.* **34:** 492-516.

Dolp, R. M. 1970. Biochemical and physiological studies of certain ticks (Ixodoidea). Qualitative and quantitative studies of hemocytes. *J. Med. Entomol.* **7:** 277-88.

Douglas, J. R. 1943. The internal anatomy of *Dermacentor andersoni* Stiles. *Univ. Calif. Publ. Entomol.* **7:** 207-72.

Dumont, J. N., E. Anderson and G. Winner. 1966. Some cytologic characteristics of the hemocytes of *Limulus* during clotting. *J. Morphol.* **119:** 181-208.

Durliat, M. and R. Vranckx. 1976. Analysis of the hemocyte proteins from *Astacus leptodactylus*. *C.R. Acad. Sci. Paris* D. **282 (24):** 2215-18.

Fahrenbach, W. H. 1968. The cyanoblast: hemocyanin formation in *Limulus polyphemus*. *J. Cell. Biol.* **39:** 43a.

Fahrenbach, W. H. 1970. The cyanoblast. Hemocyanin formation in *Limulus polyphemus*. *J. Cell. Biol.* **44:** 445-53.

Florkin, M. 1934. Sur la teneur en oxygène et en CO_2 du sang des insectes a système trachéen ouvert. *C.R. Soc. Biol.* **115**: 1224.

François, J. 1974. Etude ultrastructurale des hémocytes du Thysanoure *Thermobia domestica* (Insecte, Aptérygote). *Pedobiologia* **14**: 157-62.

François, J. 1975. Hémocyte et organe hématopoietique de *Thermobia domestica* (Packard) (Thysanura: Lepismatidae). *Int. J. Insect Morphol. Embryol.* **4(6)**: 477-94.

Fujisaki, K., S. Kitoaka and T. Morii. 1975. Hemocyte types and their primary cultures in the argasid tick *Ornithodoros moubata* Ixodoidea. *Appl. Entomol. Zool.* **10(1)**: 30-39.

George, W. C. and T. Nichols. 1948. A study of the blood of some Crustacea. *J. Morphol.* **83**: 425-40.

Ghidalia, W., R. Vendrely, Y. Coirault, C. Montmory and O. Prou-wartelle. 1977. Hemolymph coagulation in *Macropipus puber* (L.) Crustacea Decapoda. Respective roles of hemolymph and plasma. *C.R. Acad. Sci. Paris* D. **284(1)**: 69-72.

Ghiretti-Magaldi, A., C. Milanesi and B. Salvato. 1973. Identification of hemocyanin in the cyanocytes of *Carcinus maenas*. *Experientia* **29**: 1265-67.

Ghiretti-Magaldi, A., F. Ghiretti and B. Salvato. 1977. The evolution of hemocyanin. *Sym. Zool. Soc. Lond. No.* **38**: 513-23.

Goffinet, G. and Ch. Grégoire. 1975. Coagulocyte alterations in clotting hemolymph of *Carausius morosus* L. *Arch. Int. Physiol. Biochim.* **83(4)**: 707-22.

Graber, V. 1871. Über die Blutkorperchen der Insekten. *Sitzb. Akad. Math.-Nat. Wiss. Wien* **64**: 9-44.

Grégoire, Ch. 1955. Blood coagulation in arthropods VI. A study of phase contrast microscopv of blood reactions *in vitro* in Onychophora and in various groups of arthropods. *Arch. Biol.* **66**: 489-508.

Grégoire, Ch. 1974. Hemolymph coagulation, pp. 309-60. *In* M. Rockstein (ed.) *The Physiology of Insecta*. Vol. V, 2nd edition. Academic Press, New York, London.

Grégoire, Ch. and M. Florkin. 1950. Blood coagulation in arthropods. I. The coagulation of insect blood, as studied with the phase contrast microscope. *Physiol. Comp. Oecol.* **2(2)**: 126-39.

Grégoire, Ch. and G. Goffinet. 1978. Controversies about the coagulocyte, pp. 000-000. *In* A. P. Gupta (ed.) *Insect Hemocytes: Development, Forms, Functions, and Techniques*. Cambridge University Press, Cambridge. (in Press)

Grimstone, A. V., S. Rotheram and G. Salt. 1967. An electron-microscope study of capsule formation by insect blood cells. *J. Cell Sci.* **2**: 281-92.

Gupta, A. P. 1968. Hemocytes of *Scutigerella immaculata* and the ancestry of Insecta. *Ann. Entomol. Soc. Amer.* **61(4)**: 1028-29.

Gupta, A. P. 1969. Studies of the blood of Meloidae (Coleoptera) I. The haemocytes of *Epicauta cinerea* (Forster), and a synonymy of haemocyte terminologies. *Cytologia* **34(2)**: 300-44.

Gupta, A. P. 1970. Midgut lesions in *Epicauta cinerea* (Coleoptera; Meloidae). *Ann. Entomol. Soc. Amer.* **63**: 1786-88.

Gupta, A. P. (ed.) 1978. *Insect Hemocytes: Development, Forms, Functions, and Techniques*. Cambridge University Press. Cambridge. (in Press)

Gupta, A. P. and D. J. Sutherland. 1965. Observations on the spherule cells in some Blattaria (Orthoptera). *Bull. Entomol. Soc. Amer.* **11**: 161.

Gupta, A. P. and D. J. Sutherland. 1966. *In vitro* transformations of the insect plasmatocyte in certain insects. *J. Insect Physiol.* **12**: 1369-75.

Gupta, A. P. and D. J. Sutherland. 1967. Phase contrast and histochemical studies of spherule cells in cockroaches. *Ann. Entomol. Soc. Amer.* **60(3)**: 557-65.

Gupta, A. P. and D. J. Sutherland. 1968. Effects of sublethal doses of chlordane on the hemocytes and midgut epithelium of *Periplaneta americana. Ann. Entomol. Soc. Amer.* **61(4):** 910-18.

Hagopian, M. 1971. Unique structures in the insect granular hemocytes. *J. Ultrastruct. Res.* **36:** 646-58.

Halliburton, W. E. 1885. On the blood of the decapod Crustacea. *J. Physiol.* **6:** 300-35.

Hamilton, M. A. 1931. The morphology of the water scorpion, *Nepa cinerea* Linn. *Proc. Zool. Soc. Lond.* **193:** 1067-1136.

Hardy, W. B. 1892. The blood corpuscles of the Crustacea together with a suggestion as to the origin of the crustacean fibrin ferment. *J. Physiol.* **13:** 165-90.

Harpaz, F., N. Kislev and A. Zelcer. 1969. Electron-microscopic studies on hemocytes of the Egyptian cottonworm, *Spodoptera littoralis* (Boisduval) infected with a nuclear-polyhedrosis virus, as compared to noninfected hemocytes. *J. Invert. Pathol.* **14:** 175-85.

Hearing, V. and S. H. Vernick. 1967. Fine structure of the blood cells of the lobster, *Homarus americanus. Chesapeake Sci.* **8:** 170-86.

Hennig, W. 1965. Phylogenetic systematics. *Annu. Rev. Entomol.* **10:** 97-116.

Hennig, W. 1966. *Phylogenetic Systematics.* University of Illinois Press, Urbana, Chicago, London.

Hessler, R. R. and W. A. Newman. 1975. A trilobitomorph origin of Crustacea. *Fossils Strata* **4:** 437-59.

Hoar, W. S. 1966. *General and Comparative Physiology.* Prentice-Hall, Inc. Englewood Cliffs, New Jersey.

Hoarau, F. 1976. Ultrastructure des hémocytes de l'oniscoide *Helleria brevicornis* Ebner (Crustacé Isopode). *J. Microsc. Biol. Cell* **27(1):** 47-52.

Hoffmann, J. A. 1966. Étude des oenocytoides chez *Locusta migratoria* (Orthoptère). *J. Microsc. (Paris)* **5:** 269-72.

Hoffmann, J. A. 1967. Étude des hémocytes de *Locusta migratoria* L. (Orthoptère). *Arch. Zool. Exp. Gén.* **108:** 251-91.

Hoffmann, J. A. and M. E. Stoekel. 1968. Sur les modifications ultrastructurales des coagulo-cytes au cours de la coagulation de l'hémolymphe chez un insecte Orthopteroide: *Locusta migratoria. C.R. Seances Soc. Biol. Strasbourg* **162:** 2257-59.

Hoffmann, J. A., A. Porte and P. Joly. 1970. On the localization of phenoloxidase activity in coagulation of *Locusta migratoria* (L.) (Orthoptera). *C.R. Hebd. Seances Acad. Sci. Paris* **270D:** 629-31.

Hoffmann, J. A., M. E. Stoekel, A. Porte and P. Joly. 1968. Ultrastructure des hémocytes de *Locusta migratoria* (Orthoptère). *C.R. Hebd. Seances Acad. Sci. Paris* **266:** 503-5.

Hollande, A. C. 1909. Contribution a l'étude du sang des Coléoptères. *Arch. Zool. Exp. Gén.* **(5)21:** 271-94.

Hollande, A. C. 1911. Étude histologiques comparée du sang des insectes a hémorrhée et des insectes sans hémorrhée. *Arch. Zool. Exp. Gén. (ser. 5)* **6:** 283-323.

Hollande, A. C. 1914. Le cerodecytes ou "oenocytes" des insectes consideres au point de vue biochimique. *Arch. Anat. Microsc.* **16:** 1-66.

Hollande, A. C. 1920. Oenocytoides et teratocytes du sang des chenilles. *C.R. Acad. Sci.* **170:** 1341-44.

Howell, W. H. 1885a. Observations upon the blood of *Limulus polyphemus, Callinectes hastatus,* and a species of Holothurian. *Stud. Biol. Lab. Johns Hopkins Univ.* **3:** 267-87.

Howell, W. H. 1885b. Observations upon the chemical composition and coagulation of the blood of *Limulus polyphemus, Callinectes hastatus,* and *Cucumaria* sp. *Johns Hopkins Univ. Cir.* **5:** 4-5.

Hrdy, I. 1957. Comparison of preparation and staining techniques of insect blood cells. *Acta Soc. Entomol. Cechoslav.* **54:** 305-11.

Johnston, M. A., H. Y. Elder and P. S. Davies. 1973. Cytology of *Carcinus* haemocytes and their function in carbohydrate metabolism. *Comp. Biochem. Physiol.* **46A**: 469–81.

Jones, J. C. 1950. The normal hemocyte picture of the yellow mealworm *Tenebrio molitor* Linnaeus. *Iowa State Coll. J. Sci.* **24**: 356–61.

Jones, J. C. 1954. A study of mealworm hemocytes with phase contrast microscopy. *Ann. Entomol. Soc. Amer.* **47**: 308–15.

Jones, J. C. 1956. The hemocytes of *Sarcophaga bullata* Parker. *J. Morphol.* **99(2)**: 233–57.

Jones, J. C. 1959. A phase contrast study of the blood cells in *Prodenia* larvae (Order Lepidoptera). *Q. J. Microsc. Sci.* **100(1)**: 17–23.

Jones, J. C. 1962. Current concepts concerning insect hemocytes. *Amer. Zool.* **2**: 209–46.

Jones, J. C. 1965. The hemocytes of *Rhodnius prolixus* Stål. *Biol. Bull.* (*Woods Hole*) **129**: 282–94.

Jones, J. C. 1970. Hematopoiesis in insects, pp. 7–65. *In* A. S. Gordon (ed.) *Regulation of Hematopoiesis*. Vol. 1. Appleton-Century-Crofts, New York.

Jones, J. C. 1975. Forms and functions of insect hemocytes, pp. 119–28. *In* K. Maramorosch and R. E. Shope (eds.) *Invertebrate Immunity*. Academic Press, New York, London.

Jones, J. C. and O. E. Tauber. 1954. Abnormal hemocytes in mealworms (*Tenebrio molitor* L.). *Ann. Entomol. Soc. Amer.* **47(3)**: 428–44.

Jones, T. W. 1846. The blood corpuscle considered in its different phases of development in the animal series. *Phil. Trans. R. Soc.* **136**: 1–106.

Kanungo, K. and J. A. Naegele. 1964. The haemocytes of the acarid mite *Caloglyphus berlesei* (*Mich.* 1903). *J. Insect Physiol.* **10**: 651–55.

Kenny, D. M., F. A. Belamarich and D. Shepro. 1972. Aggregation of horseshoe crab (*Limulus polyphemus*) amoebocytes and reversible inhibition of aggregation by EDTA. *Biol. Bull.* (*Woods Hole*) **143**: 548–67.

Khanna, S. 1964. The circulatory system of *Dysdercus koenigi* F. (Hemiptera: pyrrhocoridae). *Indian J. Entomol.* **26(4)**: 404–10.

Kollman, M. 1908. Recherches sur les leucocytes et le tissu lymphoïde des invertébres. *Ann. Sci. Nat. Zool.* **9**: 1–238.

Krukenberg, C. F. W. 1880a. Zur Kenntnis des Hämocyanins und seiner Verbreitung im Tiereiche. *Cbl. Med. Wissenseh.*

Krukenberg, C. F. W. 1880b. Beiträge zur Kenntnis der Respirationsorgane bei wirbellosen Tieren. *Vgl. Physiol. Studien an den Kusten Adria*, 1st ser., pt. 3.

Krukenberg, C. F. W. 1882. Zur vergleichenden Physiologie der Lymphe, der Hydro- und Hämolymph. *Vgl. Physiol. Studien an der Kusten Adria*, 1st ser., pt. 3.

Lai-Fook, J. 1970. Haemocytes in the repair of wounds in an insect (*Rhodnius prolixus*). *J. Morphol.* **130**: 297–314.

Lai-Fook, J. 1973. The structure of the haemocytes of *Calpodes ethlius* (*Lepidoptera*). *J. Morphol.* **139**: 79–104.

Landureau, J. C. and P. Grellet. 1975. Obtentions de lignées, permanentes d'hémocytes de Blatte. Caractéristiques physiologiques et ultrastructurales. *J. Insect Physiol.* **21**: 137–51.

Lankester, E. R. 1871. Ueber das Vorkommen von Hämoglobin in den Muskeln und die Verbreitung desselben in den lebendigen Organismen. *Arch. Ges. Physiol.* **4**: 315.

Lavallard, R. and S. Campiglia. 1975. Contributions a l'hématologie de *Peripatus acacioi* Marcus et Marcus (Onychophore). I. Structure et ultrastructure des hémocytes. *Ann. Sci. Nat. Zool.* (*12 ser.*) **17**: 67–92.

Lea, M. S. and L. I. Gilbert. 1966. The hemocytes of *Hyalophora cecropia* (Lepidoptera). *J. Morphol.* **118(2)**: 197–216.

Liebmann, E. 1942. The coelomocytes of Lumbricidae. *J. Morphol.* **71**: 221–45.

Liebmann, E. 1943. New light on regeneration of *Eisenia foetida* (Sav.). *J. Morphol.* **73**: 583-610.

Lockhead, J. H. and M. S. Lockhead. 1941. Studies on the blood and related tissues in *Artemia* (Crustacea Anostraca). *J. Morphol.* **68**: 593-632.

Loeb, L. 1904. On the spontaneous agglutination of blood cells of arthropods. *Univ. Penn. Med. Bull.* **16**: 441-43.

Loeb, L. 1910. Über die Blutgerinnung bei Wirbellssen. *Biochem. Z.* **24**: 478-95.

Loéwe, R. and B. Linzen. 1973. Subunits and stability region of *Dugesiella californica* hemocyanin. *Z. Physiol. Chem.* **354**: 182-88.

Ludwig, D. and M. Wugmeister. 1953. Effects of starvation on the blood of Japanese beetle (*Popillia japonica* Newman) larvae. *Physiol. Zool.* **26**: 254-59.

Lutz, K. G. 1895. Das Bluten der Coccinelliden. *Zool. Anz.* **18**: 244-45.

Maluf, N. S. R. 1939. The blood of arthropods. *Q. Rev. Biol.* **14**: 149-91.

Markl, J., R. Schmid, S. Czichos-Tiedt and B. Linzen. 1976. Hemocyanins in spiders. III. Chemical and physical properties of the proteins in *Dugesiella* and *Cupiennius* blood. *Z. Physiol. Chem.* **357**(12): 1713-25.

McKay, D. and C. R. Jenkins. 1970a. Immunity in the invertebrates. The role of serum factors in phagocytosis of erythrocytes by haemocytes of the freshwater crayfish (*Parachaeraps bicarinatus*). *Aust. J. Exp. Biol. Med. Sci.* **48**: 139-50.

McKay, D. and C. R. Jenkins. 1970b. Immunity in the invertebrates. The fate and distribution of bacteria in normal and immunised crayfish (*Parachaeraps bicarinatus*). *Aust. J. Exp. Biol. Med. Sci.* **48**: 599-607.

McKay, D. and C. R. Jenkins. 1970c. Immunity in the invertebrates. Correlation of the phagocytic activity of haemocytes with resistance to infection in the crayfish (*Parachaeraps bicarinatus*). *Aust. J. Exp. Biol. Med. Sci.* **48**: 609-17.

Metalnikov, S. 1908. Recherches experimentales sur les Chenilles de *Galleria mellonella*. *Arch. Zool. Exp.* (4th ser) **8**: 489-588.

Metalnikov, S. 1934. Rôle du système nerveux et des facteurs biologiques et psychiques dans l'immunité. *Monogr. Pasteur Inst.* Masson et Cie, Paris.

Metalnikov, S. and V. Chorine. 1929. On the natural and acquired immunity of *Pyrausta nubilalis* Hb. *Int. Corn Borer Investig.* (*Chicago*) **2**: 22-28.

Metalnikov, S. and H. Gaschen. 1922. Immunité cellulaire et humoral chez la chenille. *Ann. Inst. Pasteur* **36**: 233-52.

Metchnikoff, E. 1892. Leçons sur la pathologie de l'inflammations. Paris. (Reprinted by Dover Publications Inc., New York, 1968, as lectures on the comparative pathology of inflammation.)

Millara, P. 1947. Contributions à l'étude cytologique et physiologique des leucocytes d'Insectes. *Bull. Biol. Fr. Belg.* **81**: 129-53.

Miller, P. L. 1966. Function of hemoglobin in waterbug *Anisops*. *J. Exp. Biol.* **44**: 529-43.

Mitsuhashi, J. 1966. Tissue culture of the rice stem borer, *Chilo suppressalis* Walker (Lepidoptera: Pyralidae). II. Morphology and *in vitro* cultivation of hemocytes. *Appl. Entomol. Zool.* **1**: 5-20.

Moran, D. T. 1971. The fine structure of cockroach blood cells. *Tissue Cell* **3**: 413-22.

Morganthaler, P. W. 1953. Blutuntersuchungen bei Bienen. *Mitt. Schweiz. Entomol. Ges.* **26**: 247-57.

Müller, K. 1925. Über die korpuskularen Elemente der Blutflüssigkeit bei der erwaschsenen Honigbiene (*Apis mellifica*). *Erlanger Jahr. Blenenkunde* **3**: 5-27.

Murer, E. H., J. Levin and R. Holme. 1975. Isolation and studies of the granules of the amebocytes of *Limulus polyphemus*, the horseshoe crab. *J. Cell Physiol.* **86**: 533-42.

Muttkowski, R. A. 1921a. Studies on the respiration of insects. I. Gases and respiratory proteins of insect blood. *Ann. Entomol. Soc. Amer.* **14**: 150.

Muttkowski, R. A. 1921b. Copper: Its occurrence and role in insects and other animals. *Trans. Amer. Microsc. Soc.* **40:** 144.

Muttkowski, R. A. 1924. Studies on the blood of insects. II. The structural elements of the blood. *Bull. Brooklyn Entomol. Soc.* **19:** 4-19.

Nappi, A. J. 1970. Hemocytes of larvae of *Drosophila euronotus* (Diptera: Drosophilidae). *Ann. Entomol. Soc. Amer.* **63(5):** 1217-24.

Neuwirth, M. 1974. Granular hemocytes, the main phagocytic blood cells in *Calpodes ethlius* (Lepidoptera: Hesperiidae). *Can. J. Zool.* **52:** 783-84.

Nittono, Y. 1960. Studies on the blood cells in the silkworm, *Bombyx mori* L. (in Japanese, English summary). *Bull. Scricult. Exp. Sta. Tokyo* **16:** 171-266.

Nordenskiöld, E. 1905. Zur Anatomie und Histologie von *Ixodes reduvius*. *Zool. Anz.* **28:** 478-85.

Paillot, A. 1919. Le karyocinetose, nouvelle réaction d'immunité naturelle observé chez les chenillés de macrolépidoptères. *C.R. Acad. Sci. Paris* **169:** 396-98.

Paillot, A. and R. Noel. 1928. Recherches histophysiologiques sur les cellules pericardiales et les elements du sang des larves d'insectes (*Bombyx mori* et *Pieris brassicae*). *Bull. Hist. Appl. Physiol. Pathol.* **5:** 105-28.

Pérez, C. 1910. Recherches histologiques sur la métamorphose des muscides (*Calliphora erythrocephala* Mg.). *Arch. Zool. Exp. Gén.* **4:** 1-274.

Pipa, R. L. and P. S. Woolever. 1965. Insect neurometamorphosis. II. The fine structure of perineural connective tissue, adipohemocytes, and the shortening ventral nerve cord of a moth *Galleria mellonella*. *Z. Zellforsch. Mikrosk. Anat.* **68:** 80-101.

Poisson, R. 1924. Contribution a l'étude des Hémiptères aquatiques. *Bull. Biol. Fr. Belg.* **58:** 49-305.

Poyarkoff, E. 1910. Recherches histologiques sur la métamorphose d'un Coléoptère (La Galéruque de l'Orme). *Arch. Anat. Microsc.* **12:** 333-474.

Price, C. D. and N. A. Ratcliffe. 1974. A reappraisal of insect haemocyte classification by the examination of blood from fifteen insect orders. *Z. Zellforsch. Mikrosk. Anat.* **147:** 537-49.

Prosser, C. L. 1973. Respiratory functions of blood, pp. 317-61. *In* C. L. Prosser (ed.) *Comparative Animal Physiology*. W. B. Saunders Co., Philadelphia.

Raina, A. K. 1976. Ultrastructure of the larval hemocytes of the pink bollworm, *Pectinophora gossypiella* (Saunders) (Lepidoptera: Gelechiidae). *Int. J. Insect Morphol. Embryol.* **5(3):** 187-95.

Ratcliffe, N. A. and C. D. Price. 1974. Correlation of light and electron microscope hemocyte structure in the Dictyoptera. *J. Morphol.* **144:** 485-97.

Ratcliffe, N. A. and A. F. Rowley. 1974. *In vitro* phagocytosis of bacteria by insect blood cells. *Nature (Lond.)* **252:** 391-92.

Ratcliffe, N. A. and A. F. Rowley. 1975. Cellular defense reactions of insect hemocytes *in vitro:* Phagocytosis in a new suspension culture system. *J. Invert. Pathol.* **26:** 225-33.

Ravindranath, M. H. 1973. The hemocytes of a millipede, *Thyropygus poseidon*. *J. Morphol.* **141:** 257-68.

Ravindranath, M. H. 1974a. The hemocytes of a scorpion *Palamnaeus swammerdami*. *J. Morphol.* **144:** 1-10.

Ravindranath, M. H. 1974b. The hemocytes of an isopod *Ligia exotica* Roux. *J. Morphol.* **144:** 11-22.

Ravindranath, M. H. 1975. Effects of temperature on the morphology of hemoctyes and coagulation process in the mole-crab *Emerita* (= *Hippa*) *asiatica*. *Biol. Bull.* (*Woods Hole*) **148:** 286-302.

Redfield, A. C. 1933. The evolution of the respiratory function of the blood. *Q. Rev. Biol.* **8:** 31-57.

Redfield, A. C. 1934. The hemocyanins. *Biol. Rev. (Cambridge)* 9: 1-175.

Redmond, J. R. 1968. The respiratory function of hemocyanin, pp. 5-23. *In* F. Ghiretti (ed.) *Physiology and Biochemistry of Haemocyanins.* Academic Press, New York, London.

Ritter, H., Jr. 1965. Blood of a cockroach: Unusual cellular behavior. *Science (Wash., D.C.)* 147: 518-19.

Rizki, M. T. M. 1953. The larval blood cells of *Drosophila willistoni. J. Exp. Zool.* 123: 397-411.

Rizki, M. T. M. 1962. Experimental analysis of hemocyte morphology in insects. *Amer. Zool.* 2: 247-56.

Rizki, M. T. M. and R. M. Rizki. 1959. Functional significance of the crystal cells in the larva of *Drosophila melanogaster. J. Biophys. Biochem. Cytol.* 5: 235-40.

Roots, B. J. 1957. Nature of chloragogen granules. *Nature (Lond.)* 179: 679-80.

Rosa, D. 1896. I. Linfociti degli Oligocheti. *Mem. R. Acc. Tor.* 46(2): 149-72.

Rowley, A. F. and N. A. Ratcliffe. 1976. The granular cells of *Galleria mellonella* during clotting and phagocytic reactions *in vitro. Tissue Cell* 8: 437-46.

Salt, G. 1970. The cellular defense reactions of insects. *Cambridge Monogr. Exp. Biol.* 14: 1-118.

Sawyer, T. K., R. Cox and M. Higginbottom. 1970. Hemocyte values in healthy blue crabs, *Callinectes sapidus,* and crabs infected with the amoeba, *Paramoeba perniceosa. J. Invert. Pathol.* 15: 440-46.

Scharrer, B. 1965. The fine structure of an unusual hemocyte in the insect *Gromphadorhina portentosa. Life Sci.* 4: 1741-44.

Scharrer, B. 1972. Cytophysiological features of hemocytes in cockroaches. *Z. Zellforsch. Mikrosk. Anat.* 129: 301-13.

Selman, J. 1962. The fate of the blood cells during the life history of *Sialis lutaria* L. *J. Insect Physiol.* 8: 209-14.

Sherman, R. G. 1973. Ultrastructurally different hemocytes in a spider. *Can. J. Zool.* 51: 1155-59.

Shrivastava, S. C. and A. G. Richards. 1965. An autoradiographic study of the relation between haemocytes and connective tissue in the wax moth, *Galleria mellonella. Biol. Bull. (Woods Hole)* 128: 337-45.

Shukla, G. S. 1964. Studies on *Scolopendra morsitans Linn.* Part IV. Blood vascular system and associated structures. *Agra. Univ. J. Res. (Science)* 13: 227-32.

Snodgrass, R. E. 1965. *A Textbook of Arthropod Anatomy.* Hafner Publishing Co., New York, London.

Snyder, M. and M. Stanley. 1973. Are all hemocytes of the blue crab of 1 basic type? *Va. J. Sci.* 24(3): 129.

Sohi, S. S. 1971. *In vitro* cultivation of hemocytes of *Malacosoma disstria* Hübner (Lepidoptera: Lasiocampidae). *Can. J. Zool.* 49: 1355-58.

Stang-Voss, C. 1970. Zur Ultrastruktur der Blutzellen wirbelloser Tiere. I. Über die Haemocyten der Larve des Mehlkäfers *Tenebrio molitor. Z. Zellforsch. Mikrosk. Anat.* 103: 589-605.

Stang-Voss, C. 1971a. Zur Ultrastruktur der Blutzellen wirbelloser Tiere. IV. Die Hämocyten von *Eisenia foetida* L. (Sav.) (Annelidae). *Z. Zellforsch. Mikrosk. Anat.* 117: 451-62.

Stang-Voss, C. 1971b. Zur Ultrastruktur der Blutzellen wirbelloser Tiere. V. Über die Hämocyten von *Astacus astacus* (Crustacea). *Z. Zellforsch. Mikrosk. Anat.* 122: 68-75.

Størmer, L. 1977. Arthropod invasion of land during Silurian and Devonian times. *Science (Wash., D.C.)* 197: 1362-64.

Stutman, L. J. and M. Dolliver. 1968. Mechanism of coagulation in *Gecarcinus lateralis. Amer. Zool.* 8: 481-89.

Sundara Rajulu, G. 1970. A study of haemocytes of a centipede *Ethmostigmus spinosus* (Chilopoda: Myriapoda). *Current Sci.* **20**: 324-25.

Sundara Rajulu, G. 1971a. A study of the haemocytes of a millipede *Cingalobolus bugnioni* Carl (Diplopoda: Myriapoda). *Indian J. Zool.* **2**: 73-80.

Sundara Rajulu, G. 1971b. A study of haemocytes in a centipede *Scolopendra morsitans* (Chilopoda: Myriapoda). *Cytologia* **36**: 515-21.

Sundara Rajulu, G. N. Krishnan and M. Singh. 1970. The haemocytes of *Eoperipatus weldoni* (Onychophora: Arthropoda). *Zool. Anz.* **184**: 220-25.

Svedlberg, T. and A. Hedenius. 1933. Molecular weights of the blood pigments of invertebrates. *Nature (Lond.)* **131**: 325.

Tait, W. 1918. The blood of *Astacus*. *Q. J. Exp. Physiol.* **8**: 1-40.

Tait, J. and J. D. Gunn. 1920. The blood of *Astacus fluviatilus:* A study in crustacean blood, with special reference to coagulation and phagocytosis. *Q. J. Exp. Physiol.* **12**: 35-80.

Takada, M. and H. Kitano. 1971. Studies on the larval hemocytes of *Pieris rapae crucivora* Boisduval, with special reference to haemocyte classification, phagocytic activity and encapsulation capacity (Lep. Pieridae). *Kontyû* **39**: 385-94.

Tateiwa, J. 1928. Le formule leucocytaire du sang des chenilles normales et immunisées de *Galleria mellonella*. *Ann. Inst. Pasteur* **42**: 791-806.

Taylor, A. 1935. Experimentally induced changes in cell complex of the blood of *Periplaneta americana* (Blattidae: Orthoptera). *Ann. Entomol. Soc. Amer.* **28**: 135-45.

Teravsky, L. K. 1957. On the formed elements of the hemolymph of ticks of the family Argasidae. *Zool. Zh.* **36**: 1448-54.

Tillyard, R. J. 1917. *The Biology of Dragonflies (Odonata or Paraneuroptera)*. Cambridge University Press, Cambridge.

Toney, M. 1958. Morphology of the blood cells of some Crustacea. *Growth* **22**: 35-50.

Truchot, J. P. 1975. Factors controlling the *in vitro* and *in vivo* oxygen affinity of the hemocyanin in the crab *Carcinus maenas*. (L.). *Resp. Physiol.* **24(2)**: 173-89.

Tsvilineva, V. A. 1959. Formed elements of the hemolymph in ixodid ticks. *Dokl. Acad. Nauk Tadzhik. USSR* **2**: 45-51.

Tuzet, O. and J. F. Manier. 1958. Recherches sure *Peripatopsis moseleyi* Wood-Mason, Péripate du Natal. I. Étude sur le sang. II. La Spermatogenèse. *Bull. Biol. Fr. Belg.* **91**: 7-23.

Tuzet, O. and J. F. Manier. 1959. Recherches sur les hémocytes fusiformes géants (= "vermiform cells" de Yeager) de hémolymphe d'insectes holometaboles et heterometaboles. *Ann. Sci. Nat. Zool. Biol. Anim.* **12(1)**: 81-89.

Valembois, P. 1971. Étude ultrastructurale des coelomocytes du Lombricien *Eisenia foetida* Sav. *Bull. Soc. Zool. Fr.* **96**: 59-72.

Vercauteren, R. E. and F. Aerts. 1958. On the cytochemistry of the hemocytes of *Galleria mellonella* with special reference to polyphenoloxidase. *Enzymologia* **20**: 167-72.

Vostal, Z. 1970. Contribution to the classification of the hemocytes of Tracheata (in Czech.). *Biologia (Bratislava)* **25**: 811-18.

Vostal, A. and E. Pirčová. 1968. Zur Kenntnis der Hämocyten der Vielfüsser (Diplopoda). *Biologia (Bratislava)* **23**: 161-65.

Weiland, A. L. and C. P. Magnum. 1975. The influence of environmental salinity on hemocyanin function in the blue crab, *Callinectes sapidus*. *J. Exp. Zool.* **193(3)**: 265-73.

Whitcomb, R. F., M. Shapiro and R. R. Granados. 1974. Insect defense mechanisms against microorganisms and parasitoids, pp. 447-536. *In* M. Rockstein (ed.) *The Physiology of Insecta*. Vol. V, 2nd edition. Academic Press, New York, London.

White, J. G. 1976. A comment on the ultrastructure of amebocytes from the horseshoe crab (*Limulus polyphemus*), pp. 97-101. *In: Animal Models of Thrombosis and Hemorrhagic Diseases*. N.I.H., Bethesda, Maryland.

Whitten, J. M. 1964. Haemocytes and the metamorphosing tissues in *Sarcophaga bullata*, *Drosophila melanogaster*, and other cyclorrhaphous Diptera. *J. Insect Physiol.* **10**: 447-69.

Wielowiejsky, H. 1886. Über das Blutgewebe der Insekten. *Z. Wiss. Zool.* **43**: 512-36.

Wigglesworth, V. B. 1933. The physiology of the cuticle and of ecdysis in *Rhodnius prolixus* (Triatomidae, Hemiptera), with special reference to the function of the oenocytes and of the dermal glands. *Q. J. Microsc. Sci.* **76**: 269-319.

Wigglesworth, V. B. 1939. *The Principles of Insect Physiology.* 1st edition. Methuen and Co., London.

Wigglesworth, V. B. 1955. The role of haemocytes in the growth and moulting of an insect, *Rhodnius prolixus* (Hemiptera). *J. Exp. Biol.* **32**: 649-63.

Wigglesworth, V. B. 1956. The function of the amoebocytes during moulting in *Rhodnius*. *Ann. Sci. Nat. Zool. (ser. 11)* **18**: 139-44.

Wigglesworth, V. B. 1959. Insect blood cells. *Annu. Rev. Entomol.* **4**: 1-16.

Wigglesworth, V. B. 1965. *The Principles of Insect Physiology*, 6th edition. Methuen and Co., London.

Wille, H. and M. A. Vecchi. 1966. Étude sur l'hémolymphe de l'abeille (*Apis mellifica* L.). I. Les frottis de sang de l'abeille adulte d'été. *Mitt. Schweiz. Entomol. Ges.* **34**: 69-97.

Williams, A. J. and P. L. Lutz. 1975. Blood cell types in *Carcinus maenas* and their physiological role. *J. Mar. Biol. Assn. UK* **55**(3): 671-74.

Wittig, G. 1962. The pathology of insect blood cells: A review. *Amer. Zool.* **2**: 257-73.

Wittig, G. 1968. Electron microscopic characterization of insect hemocytes. 1968: 68-9. *26th Annu. Meeting of EMSA.* Claitor Publishing Co., Baton Rouge, Louisiana.

Wood, P. J. and L. P. Visentin. 1967. Histological and histochemical observations of the hemolymph cells in the crayfish, *Orconectes virilis. J. Morphol.* **123**: 559-67.

Wood, P. J., J. Podlewski and T. E. Shenk. 1971. Cytochemical observations of the hemolymph cells during coagulation in the crayfish. *Orconectes virilis. J. Morphol.* **134**: 479.

Yeager, J. F. 1945. The blood picture of the southern armyworm (*Prodenia eridania*). *J. Agric. Res.* **71**: 1-40.

Yeager, J. F. and S. C. Munson, 1941. Histochemical detection of glycogen in blood cells of the southern armyworm (*Prodenia eridania*) and in other tissues, especially midgut epithelium. *J. Agric. Res.* **63**(5): 257-94.

Zachary, D. and J. A. Hoffmann. 1973. The haemocytes of *Calliphora erythrocephala* Meig. (Diptera). *Z. Zellforsch. Mikrosk. Anat.* **141**: 55-57.

Zachary, D., M. Brehélin and J. A. Hoffmann. 1975. Role of the "Thrombocytoids" in the capsule formation in the dipteran *Calliphora erythrocephala. Cell Tissue Res.* **162**(3): 343-48.

Zaidi, Z. S. and M. A. Khan. 1975. Changes in the total and differential hemocyte counts of *Dysdercus cingulatus.* (Hemiptera Pyrrhocoridae) related to metamorphosis and reproduction. *J. Anim. Morphol. Physiol.* **22**(2): 110-19.

Taxonomic Index

Subject Index